Personal Armour Systems Symposium (PASS 2020)

Personal Armour

Postponed from 2020

Copenhagen, Denmark
11-15 October 2021

Print ISBN: 979-8-3313-1203-9
eISBN: 979-8-3313-1202-2

Printed from e-media with permission by:

Curran Associates, Inc.
57 Morehouse Lane
Red Hook, NY 12571

Some format issues inherent in the e-media version may also appear in this print version.

Royal Military Academy (Belgium)
Department of Weapon Systems
Renaissance Avenue 30
1000 Brussels
Belgium

Additional copies of this publication are available from:

Curran Associates, Inc.
57 Morehouse Lane
Red Hook, NY 12571 USA
Phone: 845-758-0400
Fax: 845-758-2633
Email: curran@proceedings.com
Web: www.proceedings.com

TABLE OF CONTENTS

ARMOUR MATERIALS

BEHIND ARMOUR BLUNT TRAUMA

BLAST INJURY AND MITIGATION

CASUALTY REDUCTION AND OPERATIONAL ANALYSIS

HUMAN FACTORS

HUMAN VULNERABILITY AND INJURY CRITERIA

TEST STANDARDS & METHODS OF ASSESSMENT

THREATS

An Investigation of Additive Manufactured Silicon Carbide Ceramics for Body Armor Applications

Tyrone L. Jones[1], Jerry LaSalvia[1], Nicholas Ku[1], Kristopher Behler[1], Douglas Harris[1], Bill Goodman[2], Shawn Kelso[2], Mehrdad N. Ghasemi Nejhad[3], Brenden M. Minei[3]

[1]*US Army Research Laboratory, Aberdeen Proving Ground, MD, 21005-5066, USA, tyrone.l.jones20.civ@mail.mil*
[2]*Goodman Technologies, Albuquerque, New Mexico 87109*
[3]*Hawai'i Nanotechnology Labs, Department of Mechanical Engineering, University of Hawai'i at Manoa, 2540 Dole St. Holmes Hall 302, Honolulu, Hawaii 96822*

Abstract. The US Army and a Goodman Technologies Small Business and University team evaluated the mechanical properties and ballistic integrity of 3-D printed ceramic armor plates, fabricated using a novel additive manufacturing method. The method involved a proprietary z-process of nanopastes and subsequent ceramization in a vacuum furnace using 5 cycles of polymer infiltration pyrolysis (PIP) to compare the results with traditionally manufactured armor. The processed armor formulations concluded from a design of experiments (DOE) had a composition of: SiC-Si-B6O-BN-SiC-whiskers- SMP-10 nanopaste. Next, plates (90 mm × 90 mm × 8 mm) were manufactured, employing the z-process method, for ballistic analysis in accordance with established ballistic characterization procedures, using a 50.8 mm thick Aluminum 6061 plate as backing and witness plates in the case of penetration or deformation. Six alumina plates were examined ballistically (one shot per plate) against the 12.7 mm APM2 projectile (45.9 g) at an impact velocity of 840 m/s.

1. Introduction

1.1 Filling a Critical Need of the Army Advanced Manufacturing Initiative

October 4, 2019 the Secretary of the Army Ryan D. McCarthy has approved a new policy on advanced manufacturing that will help the Army secure a competitive edge against near-peer adversaries [1]. It is Army Directive 2019-29 "Enabling Readiness and Modernization through Advanced Manufacturing, also called the Army Advanced Manufacturing Initiative (AAMI) [2]. The AAMI is in accordance with the President's Executive Order 13329 (69 FR 9181) entitled "Encouraging Innovation in Manufacturing", the Related (Research and Research and Development) [3].

1.2 The AAMI policy has three key principles:

1. Strategic Investment: The Army must develop a holistic, threat-based strategy for the investment in and use of advanced methods and materials. Executing the strategy will require partnership from the private sector. As such, the policy allows for incentives designed to promote industry investment in advanced technologies.

2. Systemic Adoption: The Army will incorporate advanced manufacturing upfront and throughout a system's lifecycle.

3. Deliberate and Thoughtful Use: When using advanced manufacturing, be mindful of things like: intellectual property implications and return on investment.

1.3 Advanced manufacturing will promote the Army's ability to:

- Increase system performance through lighter and stronger materials.
- Decrease design limitations imposed by traditional methods -- design for performance, not manufacturability.

https://doi.org/10.52202/078352-0001

• Produce complex components as one piece, reducing failure points and increasing reliability.
• Reduce development time by rapidly producing prototypes and quickly transitioning them to production.
• Reduce risk of obsolete parts and diminishing sources of supply.
• Fabricate closer to the point-of-need, when needed.

The work conducted during the Phase I and II projects by the US Army Research Laboratory and Goodman Technologies (GT), has retroactively addressed the AAMI policy. The armor technology that was developed, GT RoboArmor™, fills an Army Critical Gap need by the additive manufacturing of lightweight silicon carbide ceramic matrix nanocomposites armor in form fitting shapes, and at the Point-of-Need. This is a capability that does not presently exist in industry. "Advanced manufacturing enhances the supply chain and sustainment efforts, both forward in the field and in our maintenance depots, enabling Soldiers to quickly manufacture critical parts and supplies at the point of need," said Gen. Gus Perna, commanding general of U.S. Army Materiel Command [1].

2. Experimental Methodologies

2.1 Material Methodologies

All characterization of the material was conducted on the manufactured plates employing the proprietary z-process method developed with Goodman Technologies and the University of Hawaii. Sample densities were determined using the Archimedes' principle. Sample composition was determined using a Bruker D8 Discover (Billerica, MA) X-ray diffractometer. The plate was also machined and polished to expose cross-sections that were perpendicular as well as parallel to the z-process axis. The microstructure of the plates was then determined by optical and scanning electron microscopy imaging of the cross-sections.

2.2 Dynamic Impact Failure Mechanism Model

Depth of Penetration (DOP) or residual penetration experiments were designed to determine the relative ballistic performance of different ceramic materials. For DOP testing, a projectile is fired into a ceramic tile attached to a thick metal backer plate such that the projectile will not deform the back surface of the metal plate. These experiments avoid the fundamental problem of V_{50} ballistic dependence on armor design (e.g. front-to-back plate ratio and material), require fewer shots than V50 tests, and have a sensitivity equivalent to that of other ballistic test methods. The change in penetration into the metal plates provides a comparison by which to rank the performance of the ceramic materials.

The target configuration used for these experiments is illustrated in Figure 1. The target consisted of a 90mm x 90mm ceramic tile at a thickness of 8mm, backed by two backup plates of aluminum alloy 6061 (AA6061, MIL-DTL-32262) plates of 50.8mm (2-inch) thickness. An epoxy resin, Dureflex® Optical Aliphatic Polyether Polyurethane Grade A4700, was used to attach each tile to the first (50.8mm) 2-inch plate. AA6061 was chosen as a well-characterized and readily available backer material. The aluminum backer plates were also expected to provide better resolution than steel plates. No cover plate was employed.

https://doi.org/10.52202/078352-0001

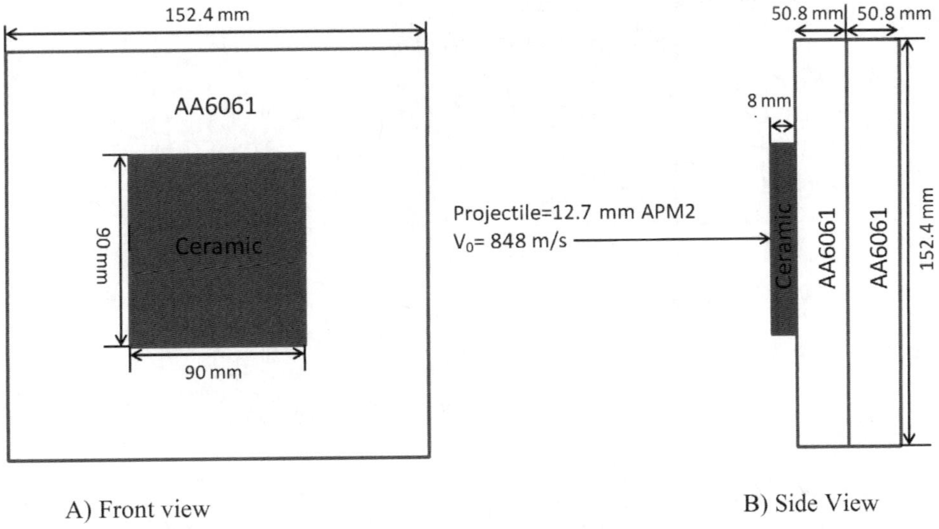

Figure 1. Ceramic DOP target assembly.

All ballistic impact experiments were conducted at the ARL, sample size n=6 per ceramic composite. The test projectile includes a hardened steel core penetrator with length of 47.6 mm (1.875 in), a diameter of 10.87 mm (0.428 in) and an aspect ratio of 4, and is also known as the 12.7 mm APM2, shown in Figure 2. The nominal projectile weight was 46 grams, and core density was 7.85 g/cc.

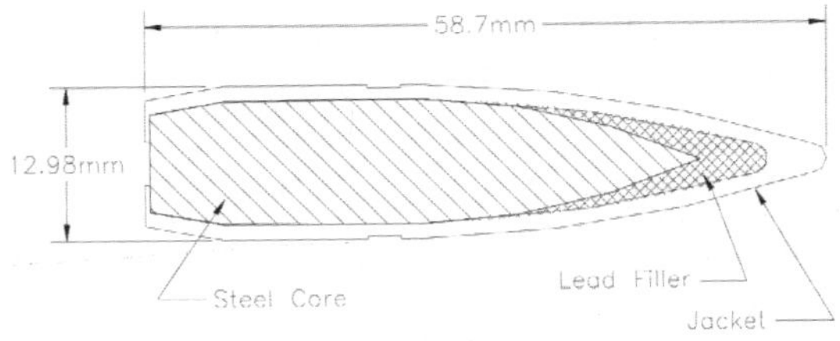

12.7-mm APM2

Projectile = 45.9 g

Hardened Steel Core = 25.9 g

Rc = 60-65

Figure 2. Cross Section of a 12.7-mm APM2.

The impact velocity used for all experiments was nominally 848 m/s (2782 ft/s), although some shots were varied from 824 m/s (2704 ft/s) up to 872 m/s (2861 ft/s) into the aluminum back plates alone to provide for DOP corrections for velocity variations. The velocity was chosen in order to produce a range of practical residual penetrations while being consistent with normal operating conditions.

Projectiles with 3 degrees or greater of total yaw were excluded from analysis; as previous studies had indicated this as an appropriate cutoff point for ballistic tests at zero obliquity. Measurement of the projectile yaw and velocity was accomplished using a Hewlett-Packard 150 kV Flash X-Ray System in 2 orthogonal planes

All residual penetration measurements were obtained by sectioning the AA6061 plates. A bandsaw was used to section all penetration cavities, and measurements were made using Vernier calipers to the deepest portion at the cavity, as indicated in Figure 3. Measurement of the "a" value avoids errors which could be caused by deformation of the aluminum block around the entrance cavity.

https://doi.org/10.52202/078352-0001

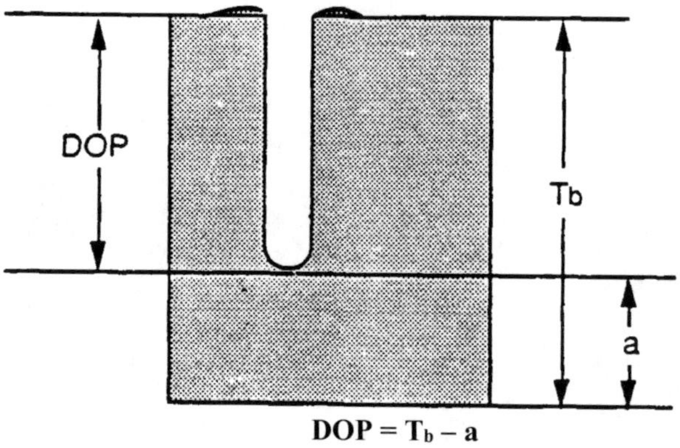

Figure 3. Measurement of Residual Penetration.

The contribution of the dynamic failure mechanisms (stiffness, deceleration) attributed to the aluminum tiles, which contributed to the performance of the aluminum plates, were normalized through deconstructing their interference employing a proprietary algorithm.

3.0 Results and Discussion

3.1 Material Characterization

Across nine plates, the density of the material was measured to be 2.370 g/mL with a standard deviation of 0.010 g/mL. This show the process results in material that is relatively homogeneous between plates, as there is little variation in density.

https://doi.org/10.52202/078352-0001

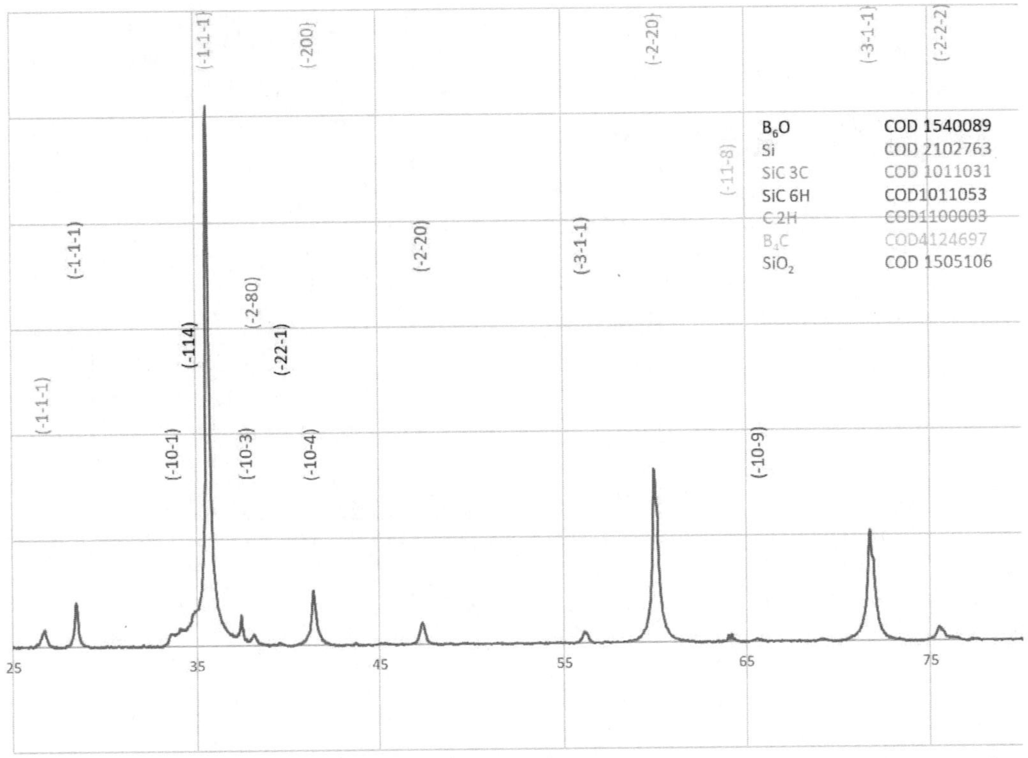

Figure 4. X-ray diffraction pattern of the sample plate with Miller indices of the identified phases for each peak listed.

The composition of the sample was determined by analyzing the X-ray diffraction pattern shown in Figure 4. Each peak is marked with Miller indices in a color corresponding the identified phase. Identification was made using the Crystallography Open Database (COD), with the pattern number used for identification listed in the legend. SiC-3C, SiC-6H, Si, B_6O, and graphite are present. The most common form of SiC in the samples is SiC-3C or β-SiC. This is the form of SiC that results from pyrolysis and crystallization of a polymer precursor. While BN was added to the constituent paste, no BN is present in the sample plate. While Si was added to react with residual carbon in the sample and aid densification, Both Si and graphite are present in the manufactured plate. These phases are generally considered detrimental to ballistic performance.

Figure 5. Optical photograph of polished cross-section that was cut parallel to the z-direction.

Figure 5 shows a cut and polished cross-section of a manufactured plate. The material in the plate was quite inhomogeneous, which was evident in the varying shades through the sample. This homogeneity seemed to correspond to the concentration of SiC whiskers. Significant cracking was visible throughout the microstructure. This would be highly detrimental to any mechanical or armor applications.

https://doi.org/10.52202/078352-0001

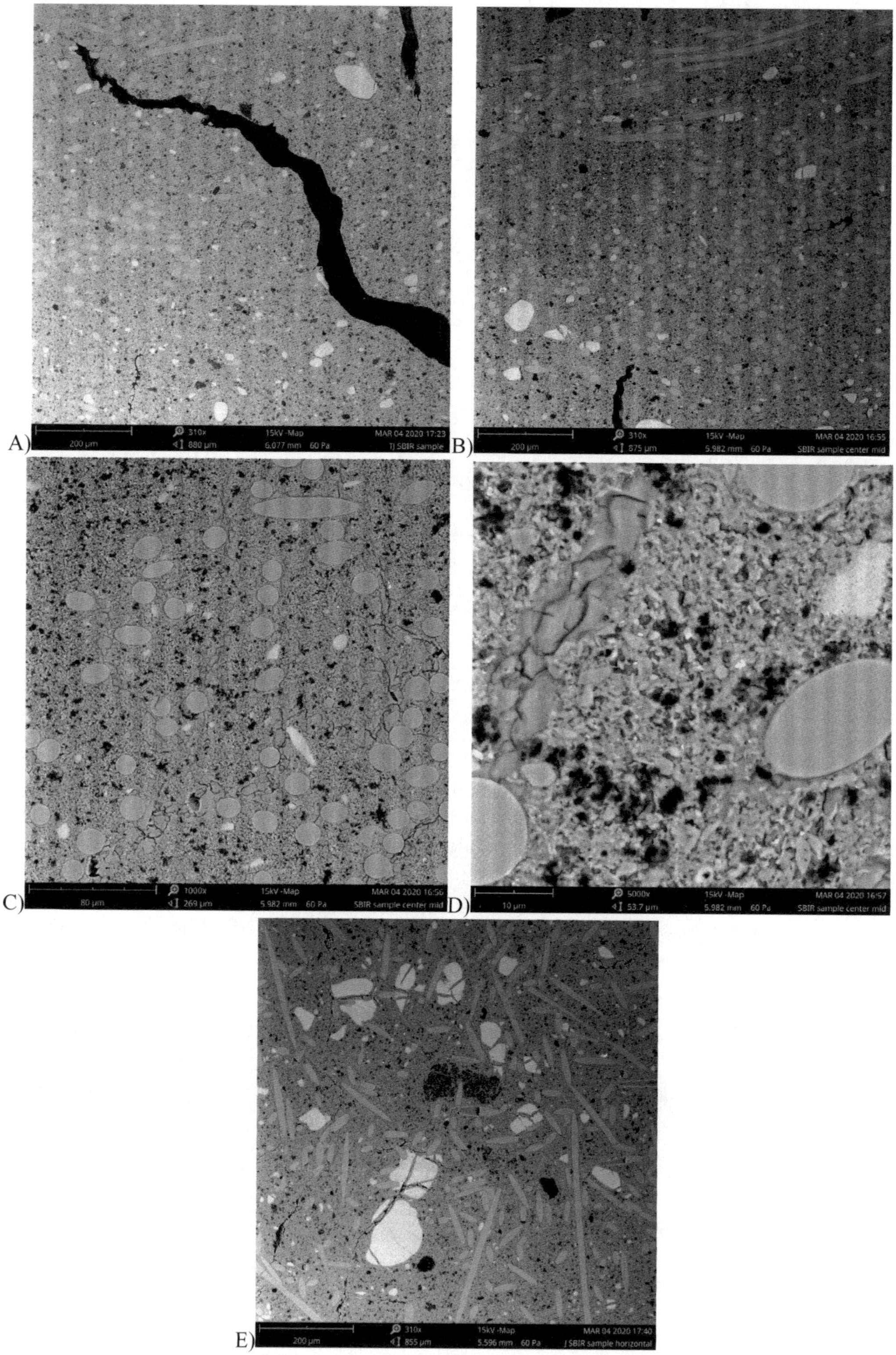

Figure 6. Microstructural images of the plate material. Image A was taken from a cross-section parallel to the z-direction. Taken from the same cross-section, Images B-D show another area of the sample at successively higher magnifications. Image E was taken from a cross-section perpendicular to the z-direction

https://doi.org/10.52202/078352-0001

Beyond optical imaging, SEM imaging was conducted to view the microstructure. Figure 6A shows the tip of a crack that was visible optically in Figure 6. The size of the crack opening makes the authors believe it is process related, likely from the polymer infiltration pyrolysis (PIP) process. Polymer derived ceramic precursors undergo significant shrinkage during conversion which can cause such fissures in the final ceramic. Figure 6B shows a separate area of the same cross-section, with many circular cross-sections of SiC whiskers of about 20 microns in diameter present. There is an increase in concentration of whiskers from area to area. Also visible are many dark black patches and large (50-60 micron) bright grains, which are identified as graphite and Si, respectively. Both materials are generally believed to be detrimental to armor performance, as they act as crack initiators. At the higher magnification shown in Figure 6C, fine cracks can be seen running around the interfaces between the SiC whiskers and fine-grained matrix material. These cracks are likely due to the coefficient of thermal expansion mismatch between the materials occurring during the heating cycles. At an even high magnification shown in Figure 6D, the individual micron-sized grains can be viewed.

Figure 6E shows a cross-section of the plate taken perpendicular to the z-direction. The cross-section of the SiC whiskers have a much higher aspect ratio when compared to the images in Figure 6A and 6B. This shows the orientation of the whiskers along this plane. Also visible in this area are large Si grains and agglomerates of graphite.

Hardness of the surface and cross section of the material were measured from a 6 x 6 grid of indents with 600 μm spacing in both X and Y directions, shown in Figure 7. The hardness of the cross sectioned material was 5.0 ± 1.57 GPa, with a large variability of 2.6 GPa to 7.6 GPa, in hardness from using 24 data points. The sample surface had a hardness of 6.4 ± 0.78 GPa with a variability from 520.5 to 8.2 GPa. The increase in hardness of the surface may be due to an increase of whiskers aligned parallel to the surface (long aspect of the whisker) leading to an increase in indent interaction with the whiskers. There was no observed excessive spall around the indents.

Data points were removed due to the indent being to asymmetric, >12%, and/or having a second phase at either the crack tips or center of the indent.

3.2 Dynamic Impact Characterization

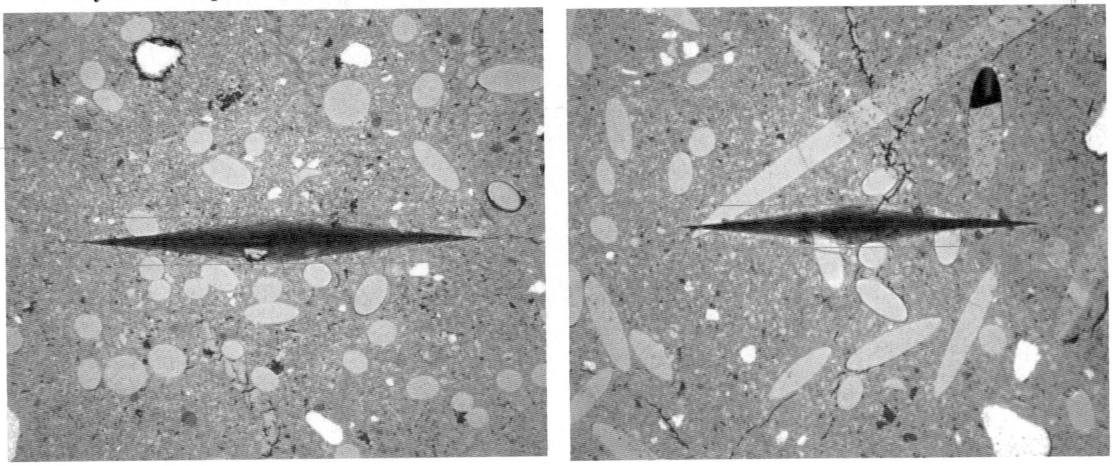

Figure 7. Optical images of 2 kgf Knoop indents on the cross section (left) and the surface (right) of the samples provided.

https://doi.org/10.52202/078352-0001

Six targets comprised of SiC tiles from Goodman Technologies were ballistically evaluated in accordance with experimental procedures reported in ARL-TR-7768 [4]. Each target was fully supported in the test rig by two C-clamps fixed on opposite sides of the target, to two vertical frames, as shown in Figure 8.

A) Front View B) Side View

Figure 8. Test setup for SiC targets in ballistic investigation.

The reference velocity for each target-penetrator interaction was 848 m/s. Post-test, the six targets were cut down the center of impact, labeled by shot number, and measured for depth of penetration, shown in Figure 9. The depth of penetration of the impacted targets ranged from 51 mm to 64 mm.

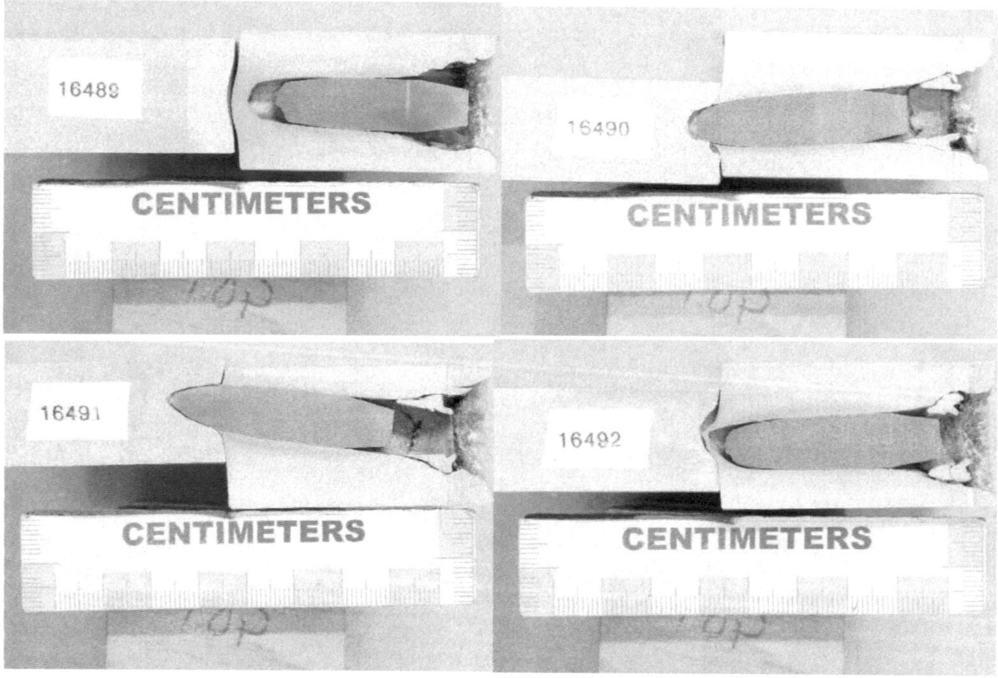

https://doi.org/10.52202/078352-0001

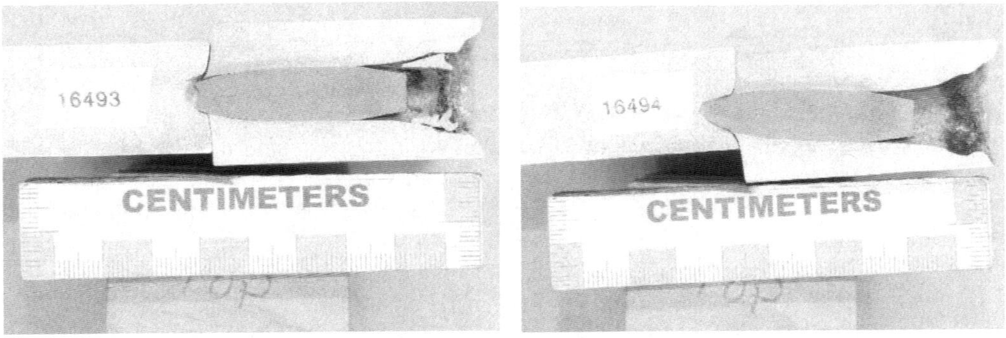

Figure 9. Cross section of 12-mm APM2 penetrator in target.

The depth of penetrations were normalized to the performance of the witness AA6061 witness plate. Finally, the ceramic efficiency, Cp, of each ceramic was calculated against the AA6061 witness plate, Figure 10. The density of versus the normalized depth of penetration into the witness plate was plotted for each SiC tile. The minima of each level of ceramic efficiency is shown as slash-dot lines. The ceramic efficiency of traditionally manufactured CoorsTek SiC, measured in an earlier experiment, is added for parametric analysis [4]. While the density of the z-processed SiC was lighter than the density of CoorsTek SiC (2.37 g/mL versus 3.21 g/mL), the ceramic efficiency of z-processed SiC was 42% of the CoorsTek SiC (6.94).

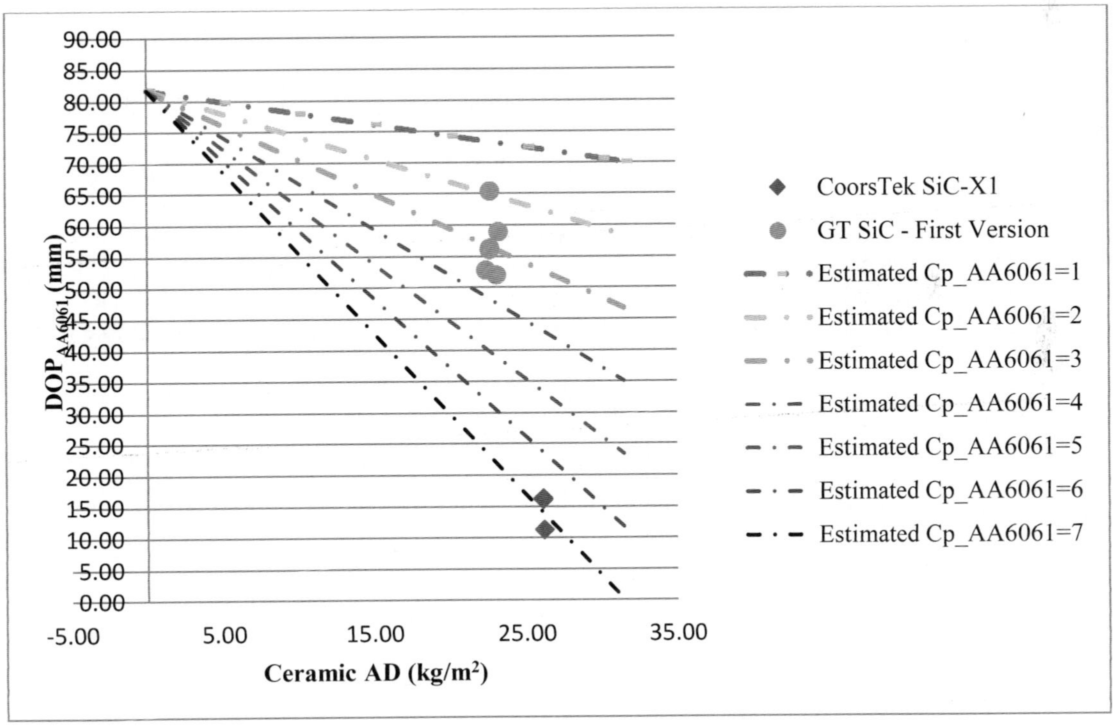

Figure 10. Ceramic efficiency of SiC tiles.

https://doi.org/10.52202/078352-0001

4.0 Conclusion

The manufactured plates consisted of a heterogeneous blend of β-SiC, α-SiC, Si, B_6O, and graphite. BN, one of the constituent material of the printed paste, was not found in the plate and may have reacted during processing. The microstructure of the material consisted of multiscale features, with large SiC whiskers and Si grains as well as fine, micron-sized SiC grains. Orientation of the whiskers was found to be align to the plane perpendicular to the z-process direction. Extensive multiscale cracking was also found through the sample. Larger fissure-like cracks are inferred to be caused by the volumetric shrinkage expected during the PIP process, with the fine crack around whisker interfaces due to thermal mismatch of the blended material. From a material perspective, the residual Si and graphite phase as well as multiscale cracking can be detrimental to ballistic performance of the plate. In ballistic studies, the 12.7 mm APM2 projectile penetrated the z-processed SiC tiles at 70% more depth than the CoorsTek SiC tiles. One primary reason for the degradation in performance is the lack of work on the nose of the projectile during the target-projectile interaction. Future research will examine the relationship between the multiscale cracks of the z-processed SiC and the effect on ceramic armor mechanisms during ballistic impact.

5.0 Acknowledgements

The authors would like to acknowledge Clifford Hubbard and William Gamble for their technical expertise and assistance. The authors would like to acknowledge the technicians Donnie Little and Perry Peregino for their ballistic assistance.

6.0 References

[1] Secretary of the Army approves new advanced manufacturing policy, U.S. Army Public Affairs, October 4, 2019, https://www.army.mil/article/228151/secretary_of_the_army_approves_new_advanced_manufacturing_policy.

[2] Enabling Readiness and Modernization Through Advanced Manufacturing, also called the Army Advanced Manufacturing Initiative (AAMI), September 18, 2019, https://armypubs.army.mil/epubs/DR_pubs/DR_a/pdf/web/ARN19451_AD2019-29_Web_Final.pdf.

[3] Encouraging Innovation in Manufacturing, A Presidential Document by the Executive Office of the President, February 26, 2004, https://www.federalregister.gov/documents/2004/02/26/04-4436/encouraging-innovation-in-manufacturing.

[4] Tyrone L Jones, Jeffrey Swab, Christopher S Meredith, and Benjamin Becker, The First Static and Dynamic Analysis of 3-D Printed Sintered Ceramics for Body Armor Applications, ARL-TR-7768, Aberdeen Proving Ground, MD, September 2016.

https://doi.org/10.52202/078352-0001

Effect of Moulding Pressure on High Strain Rate Performance of UHMWPE Composite

Hemant Chouhan[1], Neelanchali Asija[1,2], Kartikeya, Aisha Ahmed[1], Anil Yadav[1], Sanjay Prasad[1,3] and Naresh Bhatnagar[1]

[1]*Department of Mechanical Engineering, Indian Institute of Technology Delhi, New Delhi, 110016, India*
hemant.chouhan78@gmail.com
[2]*Department of Mechanical and Aerospace Engineering, Bennett University, Greater Noida, 201310, India*
[3]*Department of Mechanical Engineering, Indian Institute of Technology Dhanbad, Dhanbad, 826004, India*

Abstract. Resin starved fibre reinforced plastic (FRP) composites are very much in demand for defence applications. Based on the type of matrix, processing temperature for a given fibre-matrix system can be optimized. However, moulding pressure responsible for the consolidation of an FRP laminate may vary, which has to undergo a rate-dependent enactment in armour applications. The present study is focused on the high strain rate compressive stress-strain performance of Ultra-High-Molecular-Weight-Polyethylene (UHMWPE) fibre-based composite. UHMWPE laminates were moulded at two different pressures of 34.5 bar and 138 bar, respectively. Cylindrical specimens having ~0.5 aspect ratio and areal density of 4.01 kg/m^2 were cut-out and tested under high strain rate compressive loading. Test results of high strain rate loading revealed that increasing moulding pressure by four times enhanced the peak stress by 33% and the strain by 22%. Higher pressure moulded composites attained an 11% higher strain rate as well. Damage studies revealed relatively less damage to composite specimens as a function of higher compression moulding pressure. The study confirms that the mechanical performance of UHMWPE composites can be enhanced by increasing the moulding pressure, but to what limit, remains a mystery.

1. INTRODUCTION

Nowadays, ultra-high molecular weight polyethylene (UHMWPE) composite materials are extensively used as a raw material for the personal body armour applications [1]. UHMWPE fibres have very high specific tensile strength and modulus, thus superior ballistic performance than any other fibre. The UHMWPE fibres are arranged in a polymeric matrix material to form a lamina which is then stacked as $(0/90)_2$ cross-ply fabric. This cross-ply fabric is cut according to the required dimensions and pressed in a compression moulding machine to form a laminated armour panel.

The mechanical response of UHMWPE fibre-reinforced composite materials is rate sensitive. Koh et al. [2] tested UHMWPE fibre-reinforced composites under tensile loading at high strain-rate. It was determined that as the strain-rate increases, failure strain decreases whereas strength increases. A similar trend was also observed when UHMWPE fibre reinforced composites were tested in a Split-Hopkinson Pressure Bar (SHPB) at strain-rates up to 1200 /s [3]. Shaker et al. [4] studied out-of-plane compression of UHMWPE fibre reinforced polymer composites at strain-rates up to 8000 /s and found that out-of-plane compressive strength was ~800 MPa. Liu et al. [5] studied the in-plane compressive response of UHMWPE fibre reinforced polymer composites at a high strain rate using SHPB and observed that peak strength increased at high strain rates.

It is a well-established fact that fabrication temperature and pressure play a vital role in determining the ballistic performance of UHMWPE fibre reinforced composite materials. Zeng et al. [6] fabricated composite laminates at different temperatures ranging from 80 - 120 ºC and noticed that laminates fabricated at higher temperatures had better ballistic resistance. A ballistic impact study was conducted on laminated armour panels fabricated from Dyneema® HB 26 at two different pressures of 165 bar and 300 bar [7]. It was observed that laminated armour panels fabricated at higher pressure provided better ballistic resistance. Lassig et al. [8] stated that consolidation pressure affects void density, cracking in the matrix, fibre volume fraction of composite and fibre-fibre bonded joints during compression moulding. They reported that increasing moulding pressure from 20 bar to 165 bar resulted in ballistic resistance enhancement by 5.31% and further higher pressure of 300 bar could only improve ballistic resistance by 0.27%. This indicates that there could be some critical limit of moulding pressure.

It is evident from the known literature that the moulding pressure does play a vital role in establishing the high strain rate performance of UHMWPE fibre-reinforced composite materials. This study is conducted to evaluate the out-of-plane compressive high strain rate performance of UHMWPE

https://doi.org/10.52202/078352-0002

fibre-reinforced composite materials. The composite laminates were fabricated using compression moulding technique at two different pressures. Out-of-plane dynamic compression tests were performed on an in-house developed split-Hokinson pressure bar (SHPB) apparatus. The contrast between out-of-plane compressive strength and stress-strain response of UHMWPE fibre-reinforced composites fabricated at different pressure was studied.

2. EXPERIMENTS

2.1 Materials and Specimen

UHMWPE composite laminates were fabricated by compression moulding of 16 layers of a commercial grade of UHMWPE lamina. The commercial UHMWPE grade used for the laminate fabrication was Spectra Shield SR-3124, by Honeywell Inc., USA. Each layer of SR-3124 comprises of four unidirectional plies of UHMWPE laid orthogonal to each other. Compression moulding was performed at a temperature of 115 ^{0}C (240 F). The two different pressures selected for the study were 34.5 bar (500 psi) and 138 bar (2000 psi), respectively. The higher pressure for compression moulding is four-fold the lower pressure for a clear difference in properties as a function of moulding pressure. The moulded composites had an areal density of 4.01 kg/m^2. Rota broach machining was used to cut out 10mm diameter cylindrical specimens from the composite laminate. The cut-out specimen had the desired aspect ratio of in the range of 0.45 - 0.5 [9].

2.2 Split-Hopkinson pressure bar test

High strain rate testing in the thickness direction of the compression moulded composite specimens was performed using SHPB apparatus developed in-house at the Indian Institute of Technology Delhi, India. The SHPB set-up typically comprises of three Titanium (Ti6Al4V) bars of 16 mm diameter. The bars used in set-up were, a striker bar of length 240 mm, an incident bar of length 1200 mm and a transmission bar of length 1200 mm, respectively. The Titanium grade used for bars had a density of 4430 kg/m^3 and Young's modulus of 113.8 GPa and elastic wave speed in the bar material is 5068 m/s. It may be noted that the modulus of bar material had to be significantly higher than that of material being tested so that the bars remain within the elastic limit through-out the test. Also, to avoid severe impedance mismatch between the bars and specimen materials, high-density metals like steel should be avoided while testing low-density composites. To obtain the acceptable dynamic equilibrium in the SHPB test a pulse shaper having a thickness of 1.4 mm and 3.1 mm diameter was used between the striker bar and the incident bar. A commercial-grade of natural rubber (Linatex) was used to serve the purpose of the pulse shaper. The schematic of the SHPB set-up is shown in Figure 1.

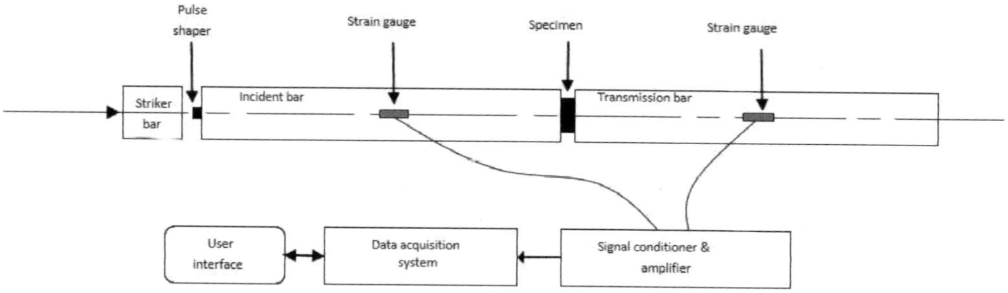

Figure 1. Schematic set-up of compressive split Hopkinson pressure bar

The composite specimen is placed in between the incident and transmission bar for the compressive high strain rate testing. The striker bar is accelerated by Nitrogen gas pressure acting on the striker bar. After travelling for a distance of 2000mm in a steel barrel, the striker bar hits the pulse shaper placed on the incident bar using a suitable lubricant for minimum friction condition. The impact of striker produces a compressive elastic wave in the incident bar, which travels through the incident bar as a function of density and velocity of sound of the bar material. When this incident wave crosses the mid-section of the incident bar, strain induced in the bar is recorded in the form of voltage change as a function of change in the resistance of strain gauge. On reaching the end of the incident bar, a part of the incident wave is reflected as a tensile wave and the rest of the elastic stress wave passes through the specimen and enters the transmission bar as a compressive wave. The resulting reflected and transmitted waves

https://doi.org/10.52202/078352-0002

in-turn produce reflection and transmission strain in the respective strain gauges. The voltage signals received from the incident bar in the form of reflection are responsible for the determination of specimen strain and strain rate. The voltage change recorded from the strain gauge mounted on the transmission bar is used to estimate specimen stress. A suitable signal conditioner amplifier is used to power up 350 Ω strain gauge, using 2 volts input. The data acquisition system records voltage variation at a rate of 2 Mega samples per second (Msps). Compressive high strain rate properties of composite specimens including stress (σ), strain (ε) and strain rate ($\dot{\varepsilon}$) were calculated based on one-dimensional propagation of elastic wave as discussed in the known literature using the following equations [10]:

$$\text{The strain rate, } \dot{\varepsilon}_s(t) = \left(\frac{2C_0}{Ls}\right) \varepsilon_r(t) \tag{1}$$

$$\text{The average strain, } \varepsilon_s(t) = \pm\left(\frac{2C_e}{Ls}\right) \int_0^t \varepsilon_r(t).\,dt \tag{2}$$

$$\text{The average stress, } \sigma(t) = \pm E \frac{A_B}{A_S} \varepsilon_t(t) \tag{3}$$

where A_B is the cross-sectional area of the bar, A_S is the cross-sectional area of the specimen, E is the modulus of elasticity of the bar material, C_e is the wave velocity in the bar, L_S is the length of the specimen, $\varepsilon_r(t)$ is the reflected strain gauge signal and $\varepsilon_t(t)$ is the transmitted strain gauge signal.

For the estimation of strain rate induced in the specimen slope of strain rate-time curve serves the purpose well for homogenous isotropic materials. However, in the case of fibre reinforced composites stress wave attenuation is an inevitable phenomenon; as a result dynamic equilibrium attainment is generally not possible [11]. For such cases, the scheme suggested in the known literature for materials showing stress wave attenuation is useful. The strain rate in this work is calculated by "dividing the area under the strain rate-strain curve, up to maximum strain under loading, by the maximum strain" [12].

3. RESULTS AND DISCUSSION

Through-the-thickness dynamic compression tests were performed on 16 layered UHMWPE composite compression moulded at low (34.5 bar) and high (138 bar) pressure. Nitrogen gas pressure used for propelling the striker bar was in the range of 0.5 – 3.0 bar. The resulting strain rates attained by the UHMWPE specimens were in the range of 2100 – 4770 /s, respectively. The maximum stress value of the stress-strain response before softening was considered as the peak stress; similarly, the strain was considered until it reaches maxima. The toughness was obtained by measuring the area under the stress-strain response curve.

3.1 High strain rate stress-strain response

Dynamic compression responses of low and high pressure moulded UHMWPE composites are shown in Fig. 2. The difference in rate-dependent behavior of low and high pressure moulded UHMWPE composites is evident from Fig. 2. For both the composites small initial section of the linear elastic region was followed by a relatively larger non-linear elastic stress growth. The stress-strain curve revealed a near plateau region after acquiring a stress value close to the peak for a given rate of loading. After acquiring the peak stress, the behavior of low and high pressure moulded composite differed significantly. However, after attaining the peak stress, the variation in the form of dropping stress in case of low pressure moulded specimens and rising stress in case of high pressure moulded specimens is noticeable, for intermediate loading rates. Further higher loading rates resulted in peak performance of both the composites, which was followed by macroscopic damage in the form of delamination. Nevertheless, it may be noted that the increasing rate of loading had barely any effect on the total strain induced in the low pressure moulded specimens. The low pressure moulded specimens attained the peak strain quite fast when the strain rate exceeded 3000/s and the total strain was in the range of ~0.27. On the contrary, high pressure moulded composites exposed a continuously rising total strain trend as a function of rising rates of loading, till the specimen macroscopically fractured at a strain of ~0.33. Thus, revealing the fact that increasing moulding pressure by four times has enhanced the total strain by 22%.

Figure 2 (a) depicts the identical rising stress-strain curve for low pressure moulded specimens, when impacted due to Nitrogen gas pressure in the range of 0.5 – 1.5 bar, respectively. At low Nitrogen gas pressure of 0.5 bar, a low strain rate of 2105 /s was induced in UHMWPE specimen. The resulting

https://doi.org/10.52202/078352-0002

specimen was physically intact with minimal damage. As the rate of loading increased, the composite exposed enhanced rate-dependent properties, until the gas pressure reached 1.2 bar pressure, resulting in a strain rate of 4050 /s. Associated with this growth in performance was slightly greater macroscopic damage on the cylindrical surface, but all the specimens till 4050/s rate of loading were physically intact. The further higher loading rate of 4220 /s, induced due to Nitrogen gas pressure of 1.5 bar resulted in a macroscopic failure of low pressure moulded specimen. The UHMWPE composite failure was noted in the form of delamination into two pieces. Also, the composite experienced the highest stress, strain and strain rate for these delaminated specimens.

Figure 2 (b) shows the rate-dependent behavior of high pressure moulded UHMWPE composites. Similar to low pressure moulded composite specimens, high pressure moulded composite specimens also revealed rising stress-strain behavior as a function of rising rates of loading. However, the slope of the stress-strain curve improved significantly. The strain rates attained by the high pressure moulded composite specimens were significantly lower for a given Nitrogen gas pressure when compared to low pressure moulded specimens. At a gas pressure of 1.5 bar, whereas low pressure moulded specimen delaminated at a strain rate of 4220 /s, the high pressure moulded specimens attained strain rate in the range of 3400 – 3500 /s, without significant damage.

Interestingly, increasing moulding pressure by four times enhanced the high strain rate performance of UHMWPE composite. At a strain rate of 4650 /s, composite exposed peak performance without delamination at a gas pressure of 2.5 bar. The further higher gas pressure of 3.0 bar resulted in the delamination of composite specimens identical to low pressure moulded specimen. However, the peak stress attained by the delaminated specimen was 595 MPa at a peak strain of 0.33, which is much higher than the intact specimen peak stress of 521 MPa at a peak strain of 0.32.

Though both the composites were derived from a single material system, the variation of a single parameter in the manufacturing cycle has significantly enhanced the properties of the resulting composite. Increasing compression moulding pressure by four times has resulted in just doubled gas pressure requirement for the successful damage of composites, depicted in the form of delamination in this case for high pressure moulded specimens. Whereas low pressure moulded composite resulted in constant peak strain, the high pressure moulded composite depicted constantly rising peak strain which in-turn will help in enhancing the area under stress-strain curve responsible for higher energy handling capacity of the composite material. An increment in peak stress by 33% and peak strain by 22% is a clear indicator of properties enhancement as a function of increasing moulding pressure.

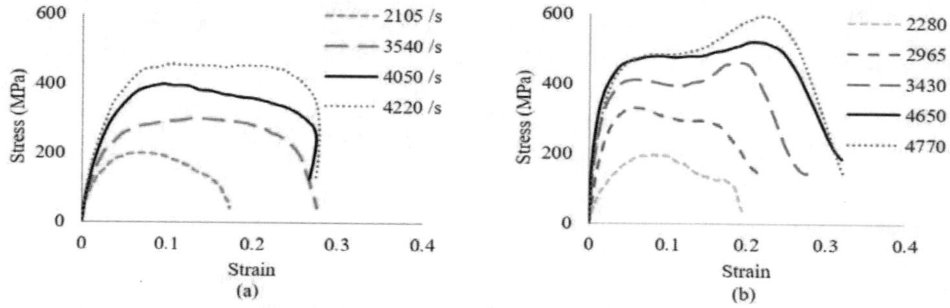

Figure 2. Dynamic stress-strain performance of composites moulded at (a) 34.5 bar and (b) 138 bar

3.2 Comparison of high strain rate response of low and high pressure moulded composites

The dynamic stress-strain responses of low pressure (LP) and high pressure (HP) compression moulded composites for identical incident energy and identical damage is shown in Figure 3. For identical Nitrogen gas pressure of 1.5 bar, a significant difference in the non-linear elastic stress growth was depicted, as shown in Figure 3 (a). The LP moulded composites attained peak stress, which was followed by a plateau until the decay of stress at constant strain. It may be noted that the LP moulded specimen delaminated at the loading rate of 4220/s. In the case of HP moulded specimen, the higher non-linear elastic modulus was followed by a secondary rise of the stress curve after a strain of 0.1. This peculiar phenomenon of minor stress reduction followed by a rise in the stress may be attributed to higher moulding pressure. High moulding pressure aids in better consolidation of plies due to the embedding of fibres in the adjacent plies in the resin starved laminate. HP moulding resulted in an intact specimen with a higher slope of the non-linear elastic stress-strain curve, lower strain rate, lower strain and identical peak stress as that of LP specimen.

https://doi.org/10.52202/078352-0002

Figure 3(b) depicts the dynamic stress-strain plots of LP and HP moulded specimen at the strain rate resulting in the physical breakdown of the composite specimen. Nitrogen gas pressure required to propel striker bar for the delamination of HP moulded specimen just doubled from 1.5 bar to 3.0 bar. The rate of stress growth improved significantly and the plateau region followed by a secondary rise in stress was much higher for HP moulded specimen. Associated with the higher impact energy is the enhanced performance of HP moulded composite, depicted in the form of higher dynamic modulus, higher peak stress and higher total strain to failure, resulting in higher material energy handling capacity.

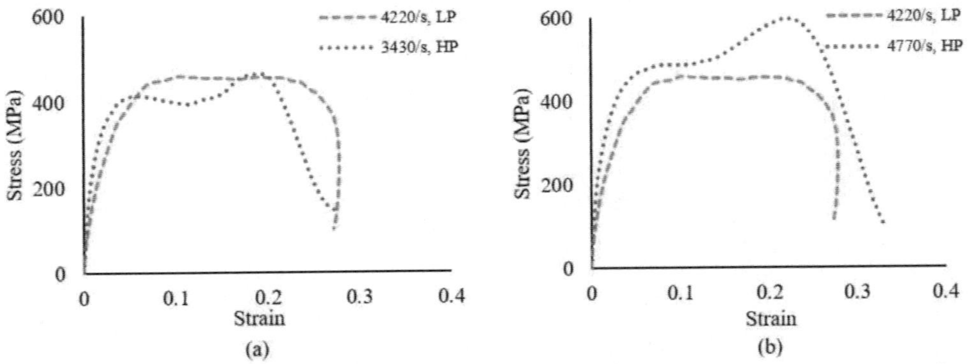

Figure 3. Dynamic stress-strain response comparison, (a) at identical impacting load, (b) at loading rate resulting in delamination

3.3. Influence of strain rate on dynamic properties

From the above dynamic stress-strain response of LP and HP moulded UHMWPE composites it is evident that there is a dependency of material properties on the moulding pressure and rate of loading. Therefore, the sensitivity of dynamic properties including peak stress, peak strain and toughness with respect to strain rate is discussed in this section for LP and HP moulded composites. Figure 4 depicts the influence of strain rate on the dynamic compressive properties of LP and HP moulded UHMWPE composites. The peak stress enhanced with the increment of strain rate in a linear fashion, as shown in Figure 4 (a). The peak stress of HP moulded composite always remained higher as compared to LP moulded composite. The relationship between the stress and strain rate for LP and HP moulded composites can be linearly established using Equation 4 and 5. An increment of 22.5% in the slope of the stress-strain rate curve for HP moulded composite is a clear indicator of property enhancement as a function of moulding pressure. Within the strain rate regime of 2100 – 4220/s for LP moulding and 2280 – 4770/s for HP moulding, the percentage increment in the stress with respect to strain rate was 128.2% and 201.3%, respectively. Thus, confirming the rate-dependent performance of UHMWPE composites.

LP moulded: $\text{Stress} = 0.1166\dot{\varepsilon} - 69.142 \ (R^2 = 0.9656)$ (4)

HP moulded: $\text{Stress} = 0.1422\dot{\varepsilon} - 99.736 \ (R^2 = 0.9552)$ (5)

Figure 4 (b) depicts the linear increment in the peak strain as a function of strain rate. Although, both the LP and HP moulded composites comprises an identical number of UHMWPE layers, a significant difference in terms of total maximum strain attained and slope of increment of strain is observed. Healthy rise in peak strain is a clear indicator of the deformation of composite, which in turn will also govern the area under the stress-strain curve. The percent increment in the peak strain of LP and HP moulded composites within the experimental range of strain rate is 60.8% and 90.2%, respectively. The relationship between the strain and strain rate for LP and HP moulded composite can be linearly estimated using Equation 6 and 7.

LP moulded: $\text{Peak strain} = 5E\text{-}05\dot{\varepsilon} + 0.0755 \ (R^2 = 0.9651)$ (6)

HP moulded: $\text{Peak strain} = 7E\text{-}05\dot{\varepsilon} + 0.0329 \ (R^2 = 0.9840)$ (7)

The toughness of HP moulded composites was slightly lower as compared to LP moulded composites at lower rates of loading. With an increase in strain rate above ~2900/s, the toughness of HP moulded composite took over the LP moulded composites. The increase in toughness within the strain

https://doi.org/10.52202/078352-0002

rate regime was three and six times the initial toughness value for LP and HP moulded composites. An increment of 36% in toughness, due to enhancement of moulding pressure by four times is an indicator of better ballistic performance of an identical number of UHMWPE layers. Linear growth of toughness with respect to strain rate can be given using Equation 8 and 9, respectively.

LP moulded: $$\text{Toughness} = 0.0363\dot{\varepsilon} - 50.188 \; (R^2 = 0.9664) \quad (8)$$

HP moulded: $$\text{Toughness} = 0.0519\dot{\varepsilon} - 92.429 \; (R^2 = 0.9849) \quad (9)$$

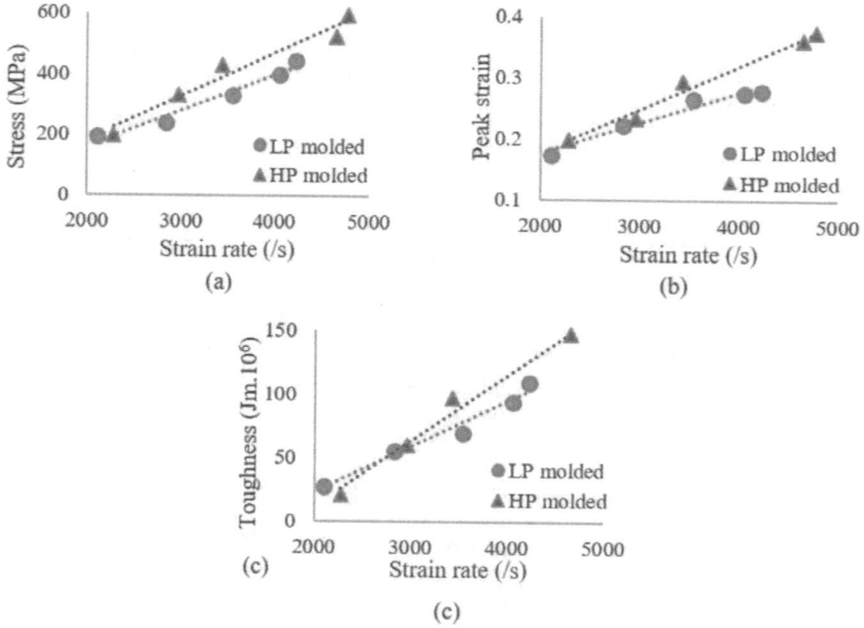

Figure 4. Effect of moulding pressure as a function of strain rate with respect to (a) peak stress, (b) peak strain, and (c) toughness

Figure 5 depicts the peak stress attained by the low and high pressure moulded UHMWPE composites at the first occurrence of macroscopic damage in the form of delamination. Peak stress improvement of 33% is recorded as a function of increment in the moulding pressure from 34.5 bar to 138 bar. Enhancement in peak stress, improved strain to failure and a higher rate of loading attained by high pressure moulded UHMWPE composites indicate the possibility of developing better ballistic resistant, lightweight UHMWPE composites without increasing the areal density of UHMWPE composites. Focused studies are required to establish the highest moulding pressure beyond which UHMWPE composite properties show no improvement or deterioration.

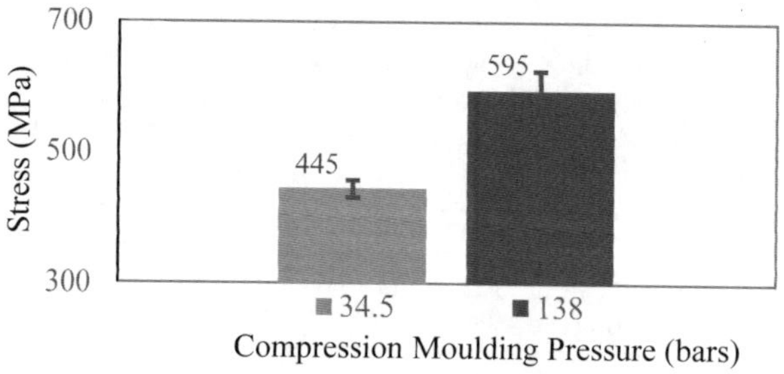

Figure 5. Effect of moulding pressure on peak stress of UHMWPE composite under dynamic loading

3.4. Macroscopic analysis

https://doi.org/10.52202/078352-0002

Macroscopic images of LP and HP moulded UHMWPE composites after dynamic testing are shown in Figure 6. Low pressure moulded specimens when loaded at 2105/s strain rate resulted in relatively higher damage on the specimen surface in contact with the incident/transmission bar (Figure 6(a)), as compared to high pressure moulded specimen under identical dynamic load condition as depicted from Figure 6(d). Damage in the form of fibre pull-out, individual ply bending due to matrix dislodging may be noted from these intact specimens loaded at lower dynamic load conditions. Lower damage due to high moulding pressure under identical load conditions indicates that higher consolidation pressures can further enhance the dynamic properties of UHMWPE composites. At a strain rate of 4220/s, LP moulded specimen experienced delamination along with higher ply damage resulting in larger fibre bundles getting dislocated as shown in Figure 6(b). The delaminated surface of the LP moulded specimen revealed ply level damage in the form of fibre bundles bending and twisting. However, fibre failure sites are negligible as the low interlaminar shear strength of laminate allows delamination before the dynamic loading could break the high tenacity fibre as depicted in Figure 6(c). Figure 6(c) shows the delaminated surface of a specimen. Delamination of HP moulded specimen revealed higher damage in the form of twisted, bent and entangled fibres at a strain rate of 4770/s, as shown in Figure 6(e). The delaminated surface further confirmed higher damage, as more number of individual twisted and entangled fibres can be seen in Figure 6(f), as compared to Figure 6(c) of LP moulded specimen. Thus, confirming the fact that higher consolidation pressure helps in enhancing the interlaminar shear strength of UHMWPE composites.

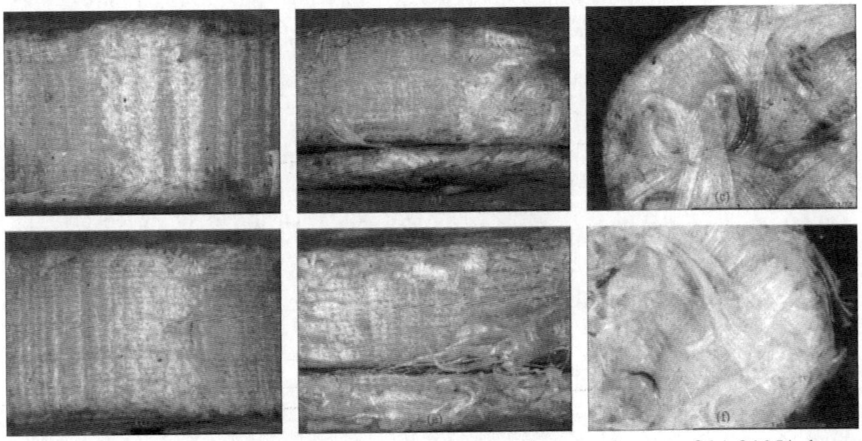

Figure 6. Macroscopic failure of UHMWPE composite tested at a strain rate of (a) 2105/s-low pressure moulded, (b) 4220/s-low pressure moulded, (c) delaminated surface at 4220/s of low pressure moulded, (d) 2280/s-high pressure moulded, (e) 4770/s-high pressure moulded, and (f) delaminated surface at 4770/s of high pressure moulded

4. CONCLUSIONS

Studies were carried out on low (34.5 bar) and high (138 bar) pressure compression moulded UHMWPE composites under high strain rate compressive loading. The high strain rate tests were conducted on SHPB apparatus in the thickness direction of composite laminates having an areal density of 4.01 kg/m2. The low and high pressure moulded composites were tested in the strain rate range of 2100 – 4220/s and 2280 – 4770/s, respectively. Based on high strain rate loading test results, it can be concluded that higher compression moulding pressures enhance the performance of UHMWPE composites. Following inferences can be made on the basis of presented experimental evaluations:

Irrespective of compression moulding pressure, the performance of UHMWPE composites enhances as a function of rising rates of loading.

For identical load conditions, composites moulded at higher pressure displays higher peak stress and the slope of the stress-strain curve improves as a function of rising moulding pressure.

The strain constantly grows with rising rates of loading in case of high pressure moulded composites. Low pressure moulded composites were easy to deform and a constant peak strain value of 0.27 was recorded for all the specimens loaded at strain rates above 3500/s.

Toughness was the only parameter that was recorded higher for low pressure moulded composites, at low rates of loading. With an increase in the strain rate above 2900/s, the toughness of high pressure moulded composites exceeded the toughness of low pressure moulded composites.

https://doi.org/10.52202/078352-0002

The peak stress value increased by 33% for the successful macroscopic failure of the UHMWPE composite moulded at 138 bar pressure. Incrementing compression moulding pressure by four times enhanced all the major composite properties, indicating the need for further studies on higher pressure moulded UHMWPE composites.

Acknowledgements

The authors are thankful to Honeywell Inc., USA, for contributing the UHMWPE composite specimens.

References

[1] O'Masta MR. Compton BG. Gamble EA. Zok FW, Deshpande VS and Wadley HNG., Ballistic impact response of an UHMWPE fibre reinforced laminate encasing of an aluminum-alumina hybrid panel. Int J Impact Eng, 2015; 86; 131–44.

[2] Koh CP. Shim VPW. Tan VBC. and Tan BL., Response of a high-strength flexible laminate to dynamic tension. Int J Impact Eng, 2008; 35; 559–68.

[3] Shi B Bin. Sun Y. Chen L. and Li JL., Energy Absorption of Ultra-High Molecular Weight Polyethylene Fibre-Reinforced Laminates at High Strain Rates. Appl Mech Mater, 2010; 34–35; 1532–35.

[4] Shaker K. Jabbar A. Karahan M. Karahan N. and Nawab Y., Study of dynamic compressive behaviour of aramid and ultrahigh molecular weight polyethylene composites using Split Hopkinson Pressure Bar. J Compos Mater, 2017; 51; 81–94.

[5] Liu BG. Kandan K. Wadley HNG. and Deshpande VS., High strain rate compressive response of ultra-high molecular weight polyethylene fibre composites. Int J Plast, 2019; 122; 115–34.

[6] Zeng Z. Guo B. Li M. Li J. and Zhou XD., Experimental and simulated investigation of temperature distribution of UHMWPE laminated composites during hot pressing process. J Appl Polym Sci, 2018; 135; 1–11.

[7] Greenhalgh ES. Bloodworth VM. Iannucci L. and Pope D., Fractographic observations on Dyneema® composites under ballistic impact. Compos Part A, 2013; 44; 51–62.

[8] Lässig TR. May M. Heisserer U. Riedel W. and Bagusat F. et al. Effect of consolidation pressure on the impact behavior of UHMWPE composites. Compos Part B, 2018;147; 47–55.

[9] Gray III GT (Rusty)., Classic Split-Hopkinson Pressure Bar Vol 8, Mechanical testing and Evaluation. ASM Handbook, ASM International; 2003.

[10] Gama BA. Lopatnikov SL. and Gillespie JWJ., Hopkinson bar experimental technique : A critical review. Appl Mech Rev, 2004; 57; 223–50.

[11] Naik NK. Ch V. and Kavala VR., Hybrid composites under high strain rate compressive loading. Mater Sci Eng A, 2008; 498; 87–99.

[12] Nakai K. and Yokoyama T., Strain Rate Dependence of Compressive Stress-Strain Loops of Several Polymers. J Solid Mech Mater Eng, 2008; 2; 557–66.

https://doi.org/10.52202/078352-0002

UHMWPE Composites Dynamic Property Variation Due to Moisture Ingress and Egress

Hemant Chouhan[1], Neelanchali Asija, Kartikeya[1], Aisha Ahmed[1], Sanjay Prasad[1] and Naresh Bhatnagar[1]

[1]*Department of Mechanical Engineering, Indian Institute of Technology Delhi, New Delhi, 110016, India*
hemant.chouhan78@gmail.com
[2]*Department of Mechanical and Aerospace Engineering, Bennett University, Greater Noida, 201310, India*
[3]*Department of Mechanical Engineering, Indian Institute of Technology Dhanbad, Dhanbad, 826004, India*

Abstract. Ultra-High-Molecular-Weight-Polyethylene (UHMWPE) fibre based composites are extensively used for armour applications. UHMWPE composites are usually processed under optimized conditions. However, the effect of ageing and in particular exposure to environmental moisture may deteriorate the performance of an armour. This study is conducted to understand the dynamic compressive behaviour of three different specimens of UHMWPE composite. Further, the objective of the study is to understand the relative change in the rate-dependent properties of UHMWPE composite as a function of moisture ingress and egress as compared to a fresh dry composite. UHMWPE composite laminate was compression moulded at 138 bar and cylindrical specimens were cut out from the laminate. Specimen types used for dynamic loading in a split Hopkinson pressure bar test were fresh dry, moist and dried specimens. The composite samples attained different strain rates under identical loading conditions as a function of moisture ingress and egress. Fresh dry composite achieved highest strain rates, stress and strain when compared to wet and dried composite. Dry specimens attained 42% and 29% higher stresses as compared to the wet and force dried specimen, respectively, indicating the importance of keeping UHMWPE composites away from moisture. Moisture egress enhanced dynamic performance by 18%. Although peak stress improved as a function of moisture egress, the loss in stress as compared to dry composite indicates an irrecoverable loss. The damage behaviour was identical in the form of delamination. However, the delamination initiates at a lower strain rate as a function of moisture ingress. Removal of moisture enhanced the composite properties but could not bring the performance of the composite equivalent to an initially dry composite, which indicates the need for the development of proper guidelines for the storage and usage of personal body armour.

1. INTRODUCTION

Defence and law enforcement personnel work in challenging environments. Particularly, extreme weather conditions add to the difficulties. Their fibre reinforced composite armours, vital to protecting against ballistic threats, also get subjected to those working environmental conditions. Generally, atmospheric temperature variation is within the acceptable range of polymer matrix properties, as far as their functionality for body armour applications is concerned. However, moisture concentration may vary the properties of the composite material system. Both polymers and fibres may interact with moisture at both macroscopic or microscopic level depending on their hydrophobic or hydrophilic nature. Moisture ingress at damage sites may lead to swelling or delamination, even in the case of hydrophobic composites [1][2].

UHMWPE fibre-reinforced polymer composites are the most widely used in personal body armour applications [3]. Thus, these composites are susceptible to temperature change, moisture and other liquid attacks. Water ingress is a major structural integrity degrading phenomenon for composite material. Allred and Roylance [4], studied Kevlar composite and reported a loss of 14%, 35% and 27% in stiffness, strength and elongation as a function of moisture ingress. The decline in properties was attributed to the degraded matrix and weakening of interface and tendency of fibre to fail as a function of moisture absorption. A similar loss in properties as a function of moisture is reported in the known literature for glass/epoxy, carbon/epoxy and kevlar/epoxy composites [5].

Mechanical response of UHMWPE fibre-reinforced composite materials is rate sensitive. Split-Hopkinson pressure bar (SHPB) is used to test materials under high strain rate load conditions. Thus, SHPB test becomes vital in determining the performance of UHMWPE composites [6]. Shaker et al. [7] tested UHMWPE fibre-reinforced polymer composites at strain-rates up to 8000 /s using the SHPB test.

https://doi.org/10.52202/078352-0003

Authors determined that out-of-plane compressive strength was around 800 MPa. Zhu et al. [6] studied the effects of moisture absorption on out-of-plane compressive strength of UHMWPE composites at various strain rates. The moisture absorption decreased strength at both low and high strain rates. In another study on the effect of moisture ingress in UHMWPE composites, it was revealed that high strain rate performance of compression moulded UHMWPE composites diminishes along with lower rates of loading for identical incident impact energy [8].

Composite panel serving as a personal protection gear may encounter minor damage while in operations. These minor damage sites may allow moisture ingress and hence can be the cause of property deterioration. Though moisture ingress is bound to affect composite performance, egress of moisture may further complicate the scenario. Therefore, a systematic study is required to study the effect of water ingress and egress on high strain rate properties of UHMWPE fibre-reinforced composites. The UHMWPE composite specimens were kept in a potable water container, for 24 hours. The composite specimens were then tested in out-of-plane compression direction at high strain rates using a split Hopkinson pressure bar (SHPB) apparatus. To study the effect of moisture egress, composite specimens were oven-dried followed by high strain rate loading on SHPB. High strain rate compression tests were performed on three different sets of composite specimens derived from a single composite laminate. The three types of specimens tested under high strain rate loading were, fresh dry compression moulded UHMWPE composites, moisture ingressed and moisture egressed specimens. Dynamic properties including peak stress, peak strain and toughness were evaluated for all type of composite specimens.

2. EXPERIMENTS

2.1 Materials and Specimen

UHMWPE composite laminate was fabricated using a high-pressure compression moulding machine. 16 layers of a commercial grade of Spectra Shield (Grade: SR-3124, by Honeywell Inc., USA) were compression moulded at a pressure of 138 bar and 115 °C, respectively. The areal density of the resulting composite was 4.01 ksm. Cylindrical specimens having 10 mm diameter were cut from the flat composite laminate using rota broach machining.

Three different types of composite specimens were used for experimental works. After machining, first set of specimens was weighted on a digital weighing scale and then kept in an oven at 65 °C and 700 mm of Hg for 24 hours, to measure if there was any moisture in the specimen. The temperature could not be increased beyond 65 °C, as the UHMWPE fibre/matrix interface may degrade at further higher temperatures. There was no change in the specimen weight before and after oven drying, specifying initial moisture-free specimens, being moulded at 115 °C. This set of specimens was treated as dry specimen and the same is supposed to unveil the exact UHMWPE composite response. Another set of specimens was placed in potable water for 24 hours in a clean room with the top edge of the specimen at least 100 mm below the surface of the water and with at least 50 mm clearance around the specimen, at 21 °C. Wetting time was kept much higher than NIJ 0101.06 standard to expose the property loss as a function of moisture ingestion, whereas the rest of the protocol was kept in conjunction with NIJ standards [9]. The high strain rate testing was performed on the wet composite specimens within one hour of removal of specimens from the water container. Before testing, the specimens were wiped and weighed on a digital balance. This set of the specimen was used to evaluate the performance of moisture ingressed UHMWPE composite. Another set of wet samples was kept for drying in an oven at 65 °C and 700 mm of Hg for 24 hours. Drying in oven brought down the weight of composite specimens to their original weight. These dried specimens served as the third set of composite specimens. The motive of this study is to reveal the high strain rate behaviour of the composite under dynamic high strain rate loading, to mimic the real-life threat of ballistic impact. Study of dried specimens was also conducted to reveal the composite behaviour as a function of moisture removal.

2.2 Test setup

A split Hopkinson pressure bar (SHPB) setup was designed and developed in-house for the high strain compressive loading of composite specimens. The SHPB setup typically comprises of a gas propelled striker bar, an incident bar, transmission bar, a suitable momentum trap and a signal conditioner amplifier in association with a data acquisition system as shown in Figure 1. Key specifications of bar material used in SHPB apparatus are given in Table 1. The specimen is kept in between the flat ground surface of Titanium incident and transmission bar. The high strain rate loading is induced in the specimen as a function of the impact of striker bar onto the incident bar. The striker bar is propelled by Nitrogen gas pressure. A suitable pulse shaper, natural rubber (Linatex) in this case is placed in between the striker

https://doi.org/10.52202/078352-0003

bar and incident bar to minimize high-frequency oscillations and to control the rise time of the incident pulse. The elastic stress wave travelling thorough the incident bar is picked up in the form of strain induced in the bar and is recorded by the strain gauge mounted at the mid-section of the bar. On reaching the incident bar/specimen interface, partially the stress wave is reflected and rest of the wave travels through the specimen into the transmission bar. The reflected and transmitted waves were recorded to give reflected and transmitted strains, respectively.

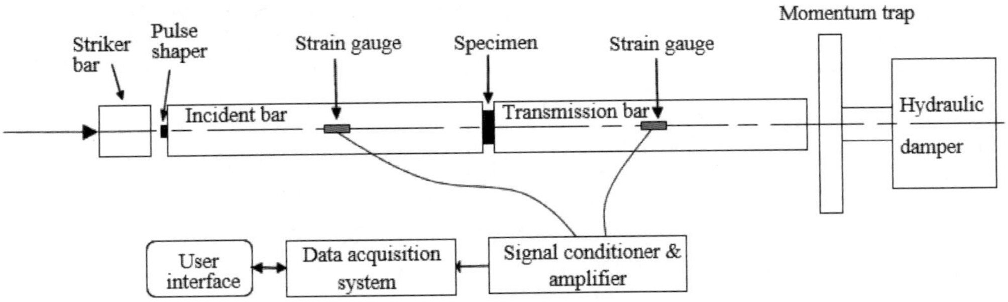

Figure 1. Basic arrangement of compressive split Hopkinson pressure bar test apparatus

One-dimensional wave propagation in elastic bars with particle motion in the longitudinal direction was used to establish the mathematical formulations of SHPB. Complete details of the one-dimensional wave propagation and necessary derivations are available in the known literature [10]. High strain rate properties of composite specimens including stress (σ), strain (ε) and strain rate ($\dot{\varepsilon}$) were calculated based on established one-dimensional propagation theory using the following equations,

$$\sigma = \frac{E_b A_b}{A_s} \varepsilon_t(t) \quad (1)$$

$$\varepsilon = \frac{-2C_e}{L_s} \int_0^t \varepsilon_r(t)dt \quad (2)$$

$$\dot{\varepsilon} = \frac{-2C_e}{L_s} \varepsilon_r(t) \quad (3)$$

where L_s is the length of the specimen, A_b and A_s are the cross-sectional area of the bar and specimen, E_b is the modulus of the bars, C_e is the wave velocity in the bar, $\varepsilon_r(t)$ is the reflected strain gauge signal and $\varepsilon_t(t)$ is the transmitted strain gauge signal.

Table 1. Specifications of SHPB Set-up

Bar properties	
Material	Titanium (Ti6Al4V)
Bar Diameter	16 mm
Striker length	240 mm
Incident bar length	1200 mm
Transmission bar length	1200 mm
Density	4430 kg/m3
Modulus of Elasticity	113.8 GPa
Elastic wave speed	5068 m/s

https://doi.org/10.52202/078352-0003

3. RESULTS AND DISCUSSION

High strain rate testing was performed in out-of-plane compression direction on UHMWPE composites. Three different sets of composite specimens were machined out from a 300 x 300 mm laminate moulded at 138 bar. The three different types of composite specimens tested were, (i) dry, (ii) moisture ingressed, and (iii) moisture egressed, respectively. The freshly moulded composite specimens were kept in potable water for 24 hours. An average increase in the weight of wet specimens was 5.5%. Table 2 depicts the increment in weight of a composite sample as a function of residence in potable water for 24 hours. It may be noted that heating wet composite in an oven at 65 °C and 700 mm of Hg for 24 hours expelled all the moisture content and the resulting dried specimens attained the original weight.

Table 2. Weight of UHMWPE specimens as a function of moisture ingress and egress

S. No.	Dry (gms)	Wet (gms)	Dried (gms)
1.	0.3262	0.3441	0.3262
2.	0.3254	0.3422	0.3254
3.	0.3265	0.3446	0.3265
4.	0.3260	0.3444	0.3260
5.	0.3270	0.3452	0.3270
6.	0.3220	0.3400	0.3220

The Nitrogen gas pressure used for propelling the striker bar was varied from 0.5 – 3.0 bar, the resulting striker bar velocity were in the range of 19.7 – 42.5 m/s for dry composite specimens. The gas pressure were varied from 0.5 – 2.0 bar for moisture ingressed and egressed specimens, resulting in maximum striker bar velocity of 37 m/s. Owing to severe macroscopic damage of wet and dried specimens, the higher striker bar velocities were sidestepped.

3.1 High strain rate stress-strain response

Through-the-thickness dynamic compression tests were performed on dry UHMWPE composites within the strain rate regime of 2280 – 4770 /s. Figure 2 (a) unveils the rate-dependent performance of UHMWPE composites under consideration. The composite underwent an insignificant linear elastic region followed by non-linear elastic region till a strain of ~0.035 is attained. The non-linear elastic region was followed by inelastic plateau region, followed by gradual stress decay. This inelastic plateau is the beauty of UHMWPE fibre with a rubber-based matrix which gives the composite a better ballistic resistance, owing to its tendency to retain constant stress with growing strain. Increasing gas pressure from 0.5 – 2.5 bar only resulted in enhanced strain rates with higher peak stress and strain. For the strain rates from 2280 – 4650 /s, all the composite specimens were recovered macroscopically intact. The further higher gas pressure of 3.0 bar resulted in macroscopic damage of the UHMWPE specimen in the form of delamination. Associated with this high rate loading was the peak performance of the composite. Since the composite specimen delaminated at this rate of loading, therefore, there was no need to enhance the rates of loading further. Also, the strain rate of 4770 /s can be treated at limiting strain rate of loading, at which the first sign of macroscopic damage is noted.

https://doi.org/10.52202/078352-0003

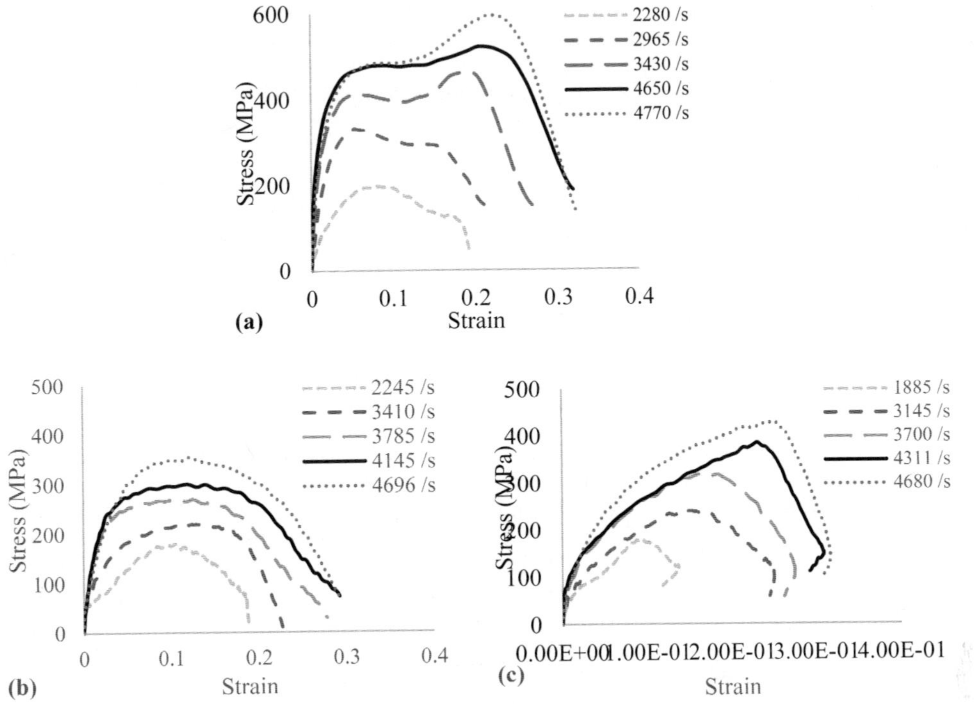

Figure 2. Dynamic stress-strain plots of UHMWPE composites (a) dry, (b) wet, and (c) dried

Figure 2 (b) depicts the high strain rate performance of UHMWPE composites after they were subjected to moisture exposure for 24 hours, resulting in a 5.5% weight gain. At the lowest rate of loading, the total specimen strain attained the values identical to dry specimens; however, the peak stress dropped by 20%, for these wet specimens. For very next rate of loading associated with a gas pressure of 0.8 bar resulting into strain rate of 3410/s, the composite specimen delaminated into two pieces. This early delamination of the specimen is attributed to the moisture residing in the voids and micro-cracks created during moulding and machining of the composite to get the test coupons. It is speculated that the moisture travelled to the interiors of the composite specimen by capillary action within micro-cracks and fibre-matrix interfaces. Reduction in the non-linear elastic stress growth was followed by lower peak stress for all the wet specimens under higher rates of loading. The plateau region shrunk on both ends, thereby decreasing the area under the stress-strain curve, responsible for governing the energy absorption capacity of the composite. This phenomenon of stress reduction is attributed to the plasticization of matrix material resulting in smooth ply movement due to reduced inter-laminar shear strength. However, insignificant loss of total strain is a unique phenomenon, as all other properties like peak stress, the slope of stress growth curve and the area under the stress-strain curve diminished. This unique phenomenon of attaining identical strain value is attributed to moisture expulsion induced deformation of the specimen. The peak stress attained by moist UHMWPE specimen was 353 MPa at a strain rate of 4696 /s. For identical loading conditions due to the gas pressure of 2 bar, a loss of 23% in stress is recorded for moisture ingressed UHMWPE composites.

The moist specimens were oven-dried and tested under identical dynamic loading conditions. Attainment of original specimen weight was a clear indicator of moisture removal from the composite specimens. However, loss in dynamic properties was expected as the space evacuated by moisture in the composite specimens is bound to serve as a site of void/local delamination. Figure 2 (c) shows the dynamic stress growth pattern of dried UHMWPE composite specimens as a function of increasing loading rates associated with a gas pressure of 0.5 – 2.0 bar. An entirely new pattern of the stress-strain curve under dynamic loading was revealed for moisture egressed specimens. All the specimens, irrespective of rates of loading revealed an inelastic stress growth until the stress maxima are attained. Interestingly, for the first time strain rate below 2000 /s was recorded for a gas pressure of 0.5 bar. The increasing rate of loading enhanced not only stress but also the strain to acquire peak stress kept rising. The strain required to attain peak stress increased from 0.1 to 0.25, as a function of rising rates of loading,

https://doi.org/10.52202/078352-0003

which was limited to 0.035 in previous cases. The peak stress increased from 181 MPa to 428 MPa as strain rates enhanced from 1885 – 4680 /s, in case of the dried composite.

3.2 Comparison of high strain rate response of dry, moisture ingressed and moisture egressed composites

Dynamic stress-strain response of dry, moisture ingressed and moisture egressed UHMWPE composites specimens at the loading rates resulting in the first occurrence of physical damage is shown in Figure 3. Freshly moulded specimen attained highest rates of loading at 3 bar of Nitrogen gas pressure resulting in striker bar velocity of 42.5 m/s. Associated with the highest rate of loading were highest peak stress, strain and area under the stress-strain curve. Moisture ingressed specimens could only take striker bar energy associated with 0.5 bar of gas pressure (striker velocity 19.7 m/s) without any damage. Firing at a gas pressure of 0.8 bar (striker velocity 24.7 m/s) onwards resulted in physical damage of the wet specimen. Moisture ingress diminished not only the stress-strain behaviour but also the strain rate dropped due to early physical damage of the composite specimen. Drying by placing the moisture ingressed specimens inside the hot oven assisted in enhancing the performance of composite specimens. Total strain and strain rates for dried composite approached the values close to that of dry composite. However, removal of moisture could not restore all the properties of a composite.

When compared to moisture ingressed composite properties at the first instance of delamination, moisture removal enhanced peak stress, strain and strain rate by 90%, 30% and 37 %, respectively. However, as compared to a fresh dry compression moulded composite, the dynamic composite properties were found significantly lacking, as the peak stress and strain were 28% and 15% lesser for the dried composite specimen. Interestingly, the strain rate could attain the values in the close proximity as that of dry composite. This unique phenomenon of partial strength regain is attributed to moisture removal, but the larger voids created due to moisture removal served as sites of larger local voids. These larger voids are responsible for diminished dynamic composite properties. In the case of wet specimens, dynamic compression of moisture sites increases the area of local delamination. When one moisture site meets another due to dynamic compression, a larger local delamination plane is created and this phenomenon continuous till complete delamination occurs. Hence, the resulting properties of moist specimens are lowest.

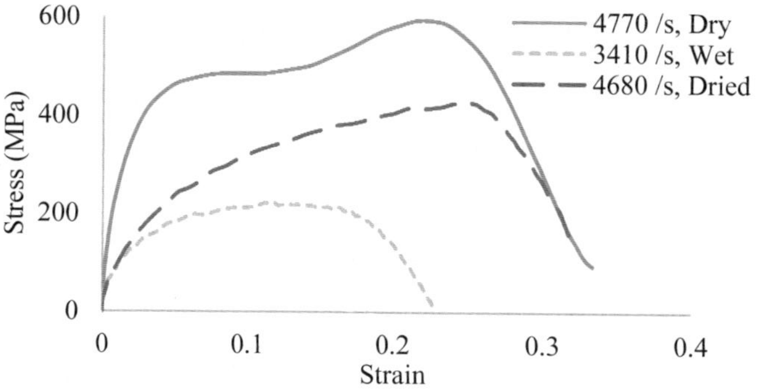

Figure 3. Comparison of dry, wet and dried UHMWPE composite specimens at the loading rates resulting in physical damage of the specimen

3.3 Influence of strain rate on dynamic properties of dry, moisture ingressed and moisture egressed composite

Rate sensitive UHMWPE composites exposed dependence to moisture ingress and egress. Figure 4 depicts the sensitivity of dynamic properties, including peak stress, peak strain and toughness with respect to strain rate for all the three cases. Table 3 presents the variation of stress, strain and toughness as a function of rising strain rates of loading of dry, wet and force dried composite. Irrespective of post-fabrication processing, all the UHMWPE composite specimens exposed increment in peak stress as a function of an increase in the rate of loading. Peak stress enhancement within the experimental regime was noted highest for freshly moulded composites. Increasing rates of loading from 2280 – 4770 /s, resulted in peak stress enhancement by 201% for dry composites. Similarly, for moisture ingressed and egressed composite, the peak stress growth with an increase in loading rates from 2245 – 4696 /s and 1885 – 4680 /s resulted in stress enhancement by 97% and 136%, respectively. However, the slope of stress growth is maximum for dry specimens and worst for moisture ingressed specimens. Removal of

https://doi.org/10.52202/078352-0003

moisture reduced strain rates at lower loading rates with minor increment in peak stress. Average growth of 20% stress was recorded due to the drying of specimen for the loading rates above 3500 /s. Figure 4 (a) depicts that moisture removal can only help marginally, as the slope of the stress-strain rate curve is negatively impacted when compared to the dry specimen's stress-strain rate growth. The relationship between the strain rate and stress for dry, moisture ingressed and moisture egressed composites can be linearly estimated using Equation 4, 5 and 6.

$$\text{Dry:} \quad \text{Stress} = 0.1424\dot{\varepsilon} - 98.557 \quad (R^2 = 0.9509) \tag{4}$$

$$\text{Wet:} \quad \text{Stress} = 0.0782\dot{\varepsilon} - 23.701 \quad (R^2 = 0.9677) \tag{5}$$

$$\text{Dried:} \quad \text{Stress} = 0.09\dot{\varepsilon} - 8.1577 \quad (R^2 = 0.9584) \tag{6}$$

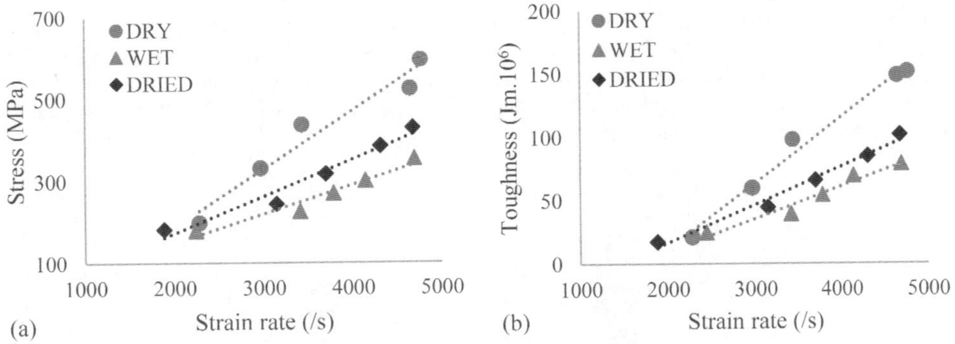

Figure 4. Effect of strain rate on (a) stress and (b) toughness

Table 3. Dynamic compressive properties of UHMWPE composites under different conditions

Specimen condition	strain rate	Max. strain	Stress (MPa)	Toughness $(Jm^3)*10^6$
Dry	2280	0.1976	197	20.78
	2965	0.2335	330	59.68
	3430	0.2940	437	97.54
	4650	0.3625	524	148.1
	4770	0.3758	594	150.84
Wet	2245	0.1906	179	24.2
	3410	0.2450	224	39.09
	3785	0.2900	269	53.97
	4145	0.3406	299	68.85
	4696	0.4000	353	78.22
Force Dried	1885	0.1375	180	17.13
	3145	0.2498	243	44.39
	3700	0.2754	317	65.54
	4311	0.3101	384	84.62
	4680	0.3181	428	101.22

The rate of toughness growth governing the energy absorption ability of UHMWPE composites can be established with respect to strain rate for dry, wet and dried composite using Equation 7, 8 and 9. At lowest rates of loading toughness values were identical for all the composites. As the rate of loading enhanced, dry specimens exposed much higher toughness growth rate. The rate of toughness growth was quite close for wet and dried composites. However, the toughness of dried composite always remained above the wet composites. Increasing rates of loading from 2280 – 4770 /s, resulted in toughness enhancement by 6.25 times for dry composites. Similarly, for moisture ingressed and egressed composite specimens, the toughness growth with an increase in loading rates from 2245 – 4696 /s and 1885 – 4680

https://doi.org/10.52202/078352-0003

/s resulted in enhancement by 2.2 and 4.9 times, respectively. Just like stress, toughness values also remained approximately 20% higher for dried composites as compared to wet composite at higher rates of loading within the experimental regime.

$$\text{Dry:} \quad \text{Toughness} = 0.0519\dot{\varepsilon} - 92.429 \quad (R^2 = 0.9849) \tag{7}$$

$$\text{Wet:} \quad \text{Toughness} = 0.0254\dot{\varepsilon} - 40.862 \quad (R^2 = 0.9611) \tag{8}$$

$$\text{Dried:} \quad \text{Toughness} = 0.0299\dot{\varepsilon} - 43.365 \quad (R^2 = 0.9813) \tag{9}$$

For all practical purposes, macroscopic damage of ballistic composite can be treated as the limiting condition of loading. As any further loading is bound to result in damage responsible for the loss of personal/material being protected. Peak stresses for dry, wet and dried UHMWPE composites at the dynamic load condition resulting into the first occurrence of macroscopic damage in the form of delamination are shown in Figure 5. The peak stress value falls by 60% due to moisture ingestion. Removal of moisture enhanced peak stress by 90%, as compared to wet composite, but the properties are far lower than initially dry composite. Yet, the results indicate that high-performance composites must be dried at the earliest to enhance the level of protection.

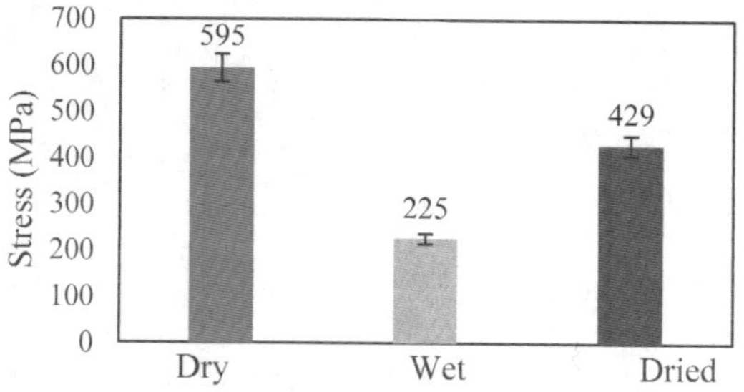

Figure 5. Stress attained by dry, wet and dried UHMWPE composites at dynamic load condition resulting in delamination

3.4 Macroscopic Damage Analysis

Macroscopic images of dry, moisture ingressed and moisture egressed UHMWPE composite specimens following dynamic tests at lowest strain rate and at limiting value of strain rate at which delamination occurs are shown in Figure 5. The dry specimen loaded at 2280/s strain rate experienced minimal macroscopic damage, primarily on both the faces in contact with the incident and transmission bar. All the in-between layers of the specimen revealed negligible damage, as noted in Figure 5(a). At a strain rate of 4770/s, dry specimen delaminated into two pieces. The two delaminated pieces were held together for imaging in Figure 5 (b). Apart from delamination, fibre pull-out and fibre bursting can be noted on the cylindrical surface of the impacted specimen. The delaminated surface of the specimen reveals failure at the lamina level, shown in Figure 5(c). Numerous sites of delamination in between the individual plies were noted with splitted fibre ends. Dynamic loading of moisture ingressed specimen at 2245/s strain rate resulted in more significant damage on both the surfaces in contact with the incident/transmission bar and cylindrical surface, as shown in Figure 5(d). For the very next gas pressure of loading resulting in strain rate of 3410/s, wet specimens showed multiple sites of delamination, as shown in Figure 5(e). Severe surface damage in the form of dislocation of plies, fibre bundle deformation, twist and plies of UHMWPE stuck on the bar surface were noted due to dynamic loading of moist specimens. Interestingly, the delaminated surface of the wet specimen revealed a clear opening in between the UHMWPE plies, as depicted in Figure 5(f). There was minimal damage in the form of UHMWPE ply pulling at all the delamination sites. The phenomenon of easier delamination of wet specimens is attributed to fibre/matrix interface degradation due to plasticization of matrix resulting in lower adhesive strength of the matrix.

https://doi.org/10.52202/078352-0003

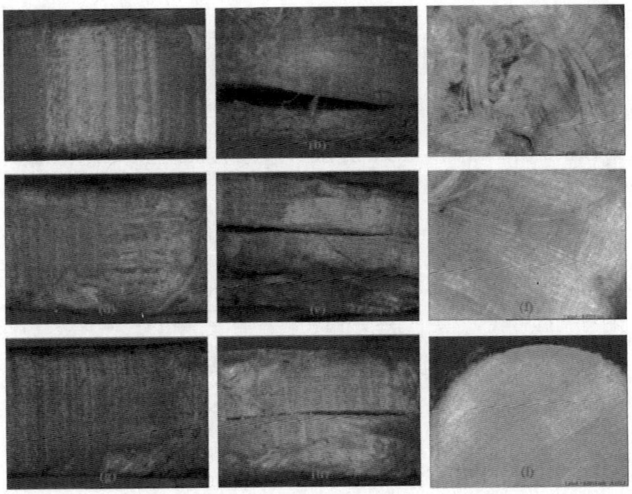

Figure 5. Macroscopic failure of UHMWPE composite (a) dry specimen at 2280/s, (b) dry specimen at 4770/s, (c) delaminated surface of dry specimen at 4770/s, (d) moisture ingressed specimen at 2245/s, (e) moisture ingressed specimen at 3410/s, (f) delaminated surface of moisture ingressed specimen at 3410/s, (g) moisture egressed specimen at 1885/s, (h) moisture egressed specimen at 4680/s, and (i) delaminated surface of moisture ingressed specimen at 4680/s

Moisture removal by oven drying non only enhanced the dynamic strength of composites but also improved the damage resistance as depicted in Figure 5 (g). For identical loading conditions, the strain rate acquired by dried specimens was minimum and the resulting damage was in between that of the dry and wet composite. Dynamic loading at 4680/s strain rate resulted in delamination of the dried composite. However, when compared to wet specimens damage, the delamination was limited to a single plane at the mid-section of the specimen. Numerous sites of fibre bundle pressing and deformation were noted on the surface in contact with the incident/transmission bar, as shown in Figure 5(h). The best outcome of moisture removal was revealed in the form of a clear UHMWPE ply delamination with minimal fibre pull-out on the delaminated surface, as shown in Figure 5(i). There were no sites of local ply damage on this surface, confirming the fact that moisture removal created large voids which served the purpose of delamination plane propagation resulting in ply opening without fibre damage. This type of delamination also confirms the fact that resin starved composites can easily degrade due to residence of foreign fluids and removal of the same may not necessarily bring the properties back to the original state.

5. CONCLUSIONS

Compressive high strain rate studies were carried out on dry, wet and dried UHMWPE composites in the thickness direction. For identical dynamic load conditions, different strain rates were attained by dry, wet and dried composite specimens. Based on compressive high strain rate loading experiments, the following inferences can be drawn:

a. Though UHMWPE composites were conditioned by three different schemes, the dynamic response in terms of peak stress, peak strain and toughness increased with an increase in the rate of loading.
b. For identical dynamic load conditions, a significant reduction in peak stress and toughness for identical strain as a function of moisture ingress confirms that 5.5% moisture retained by the voids and microcavities served as sites of the defect and helped in easy propagation of delamination plane across the UHMWPE plies.
c. Strain values are identical for dry and wet composites under high strain rate load conditions, but the removal of moisture resulted in lowering of strain at high strain rates of loading by 16%. Thus, confirming the fact that removal of moisture will further complicate the issue.
d. Damage studies revealed that matrix in dry composite specimens holds the fibre under dynamic loading whereas, moisture ingress and egress both aids in easy delamination due to reduced interlaminar shear strength.

https://doi.org/10.52202/078352-0003

e. Peak stress and rate of stress growth were highest for dry composites under dynamic loading conditions, indicating the importance of keeping the UHMWPE composites protected from the influence of moisture.

Acknowledgements

The authors are indebted to Honeywell Inc., USA, for providing the UHMWPE samples.

References

[1] Tual N. Carrere N. Davies P. Bonnemains T. and Lolive E., Characterization of sea water ageing effects on mechanical properties of carbon/epoxy composites for tidal turbine blades. Compos Part A, 2015; 78; 380–9.

[2] Haque A. and Hossain MK., Effects of Moisture and Temperature on High Strain Rate Behavior of S2-Glass-Vinyl Ester Woven Composites. Jr Compos Mater, 2003; 37; 627–47.

[3] Russell BP. Karthikeyan K. Deshpande VS. and Fleck NA., The high strain rate response of Ultra High Molecular-weight Polyethylene: From fibre to laminate. Int J Impact Eng, 2013; 60; 1–9.

[4] Allred RE. and Roylance DK., Transverse moisture sensitivity of aramid/epoxy composites. Jr Mater Sci, 1983; 18; 652–6.

[5] Sun B. Gu B. and Ding X., Compressive behavior of 3-D angle-interlock woven fabric composites at various strain rates 2005; 24; 447–54.

[6] Zhu L. Li Y. Zhu X. and Zhu Z., Compressive mechanics and failure mechanism for UHMWPE fibre reinforced composite laminates under hygrothermal environment., ICCM Int. Conf. Compos. Mater., China, 20-25 August, 2017.

[7] Shaker K. Jabbar A. Karahan M. Karahan N. and Nawab Y., Study of dynamic compressive behaviour of aramid and ultrahigh molecular weight polyethylene composites using Split Hopkinson Pressure Bar. J Compos Mater, 2017; 51; 81–94.

[8] Chouhan H. Asija N. Ahmed A. and Bhatnagar N. Effect of Moisture on High Strain Rate Performance of UHMWPE Effect of Moisture on High Strain Rate Performance of UHMWPE Fibre Based Composite Fibre Based Composite. Procedia Struct Integr, 2019; 14; 830–38.

[9] Mukasey MB. Sedgwick JL. and Hagy DW., Ballistic Resistance of Body Armour. NIJ Stand. - 0101-16; 2008; 23-45.

[10] Gama BA. Lopatnikov SL. and Gillespie JWJ., Hopkinson bar experimental technique: A critical review. Appl Mech Rev, 2004; 57; 223–50.

https://doi.org/10.52202/078352-0003

Anatomy and Anthropometry Analysis for Canine Injury and Body Armour Considerations

K. Loftis[1], A. Kulaga[1], K. Rafaels[2], L. Liberto[3], S. Sherman[4], C. Bir[5], A. Armstrong[6], J. Parker[6]

[1]*Army Futures Command, CCDC, DAC, Aberdeen Proving Ground, MD, USA,kathryn.l.loftis2.civ@mail.mil*
[2]*Army Futures Command, Army Research Laboratory, WMRD, Aberdeen Proving Ground, MD USA*
[3]*Joint Trauma Analysis and Prevention of Injury in Combat, Frederick, MD USA*
[4]*Exponent, Farmington Hills, MI, USA*
[5]*Wayne State University, Detroit, MI, USA*
[6]*US Army Medical Command, Joint Base San Antonio – Fort Sam Houston, TX*

Abstract. Military and police working dogs work alongside our uniformed service members as integral parts of the team. Much research has been completed on body armour for police and warfighters, but important information is lacking on coverage areas, injury risks, and body armour performance for canines. This study investigates vital thoracic organ anatomy and anthropometry between the human and canine for application with body armour development and coverage areas. Previous studies noted that coverage areas varied widely between different types of canine body armour currently on the market, but no standardised injury assessments were made regarding reduced injury risks with larger coverage areas. For this study, computed tomography scans of the human and canine thorax were obtained and measurements of the heart, lungs, and rib cage were assessed. Similarities and differences between the human and canine were discussed, especially related to body armour coverage areas. While humans need front and back protection to cover the majority of the thoracic vital organs, canines need protection on both sides as well as the front of the thorax. Additionally, the location of the heart within the thorax differs for the canine compared to the human, which is important when considering where to place the most armour protection. Future work includes the development of a standard canine injury scoring system and visualization tool to further enhance body armour coverage and protection for the working dog.

1. INTRODUCTION

For years, body armour has been developed to meet a standard of protection against various threats for our uniformed service members. As our understanding of materials and injuries evolve, so too does the armour which has become lighter and better at protecting against emerging threats. Also evolving are the roles and specialised tasks of military and police working dogs, increasing their risk of exposure to the same threats as their handlers. As working dogs are integral members of uniformed services teams, their protection and survivability is intrinsically tied to the survivability of their human team members. It is important to ensure the working dogs have the same level of protection for completing their jobs as their human counterparts. Currently there is a lack of standardised methods specific for canines for the certification and coverage of armour worn by military and police working dogs.

While there are some similarites when comparing the thoracic cavities of a human and canine, the differences are such that the coverage area for armour, injury severity, and injury risk need to be evaluated specifially for canines. The anatomical standing position for the canine is four paws on the ground with the thorax positioned parallel to the ground, where for a human the two feet are on the ground with the thorax perpendicular to the ground. The bony structures of the canine thorax are similar to that of a human and are comprised of rib pairs that form the bilateral limits of the thoracic cavity, sternum comprised of a segmented row of unpaired bones forming the thoracic cavity base, and the vertebral column. Both the canine and human thoracic cavity contain vital organs, including the heart and lungs. The canine's heart is very similar to the human heart and is located between the lungs. The lungs of a canine are not only important for respiration but also aid in regulating body temperature; to decrease body temperature, heavy breathing or panting is necessary. [1,2]

https://doi.org/10.52202/078352-0004

Developments in personal protective equipment (PPE) have primarily been related to advancements in the study and construction of materials and an increased understanding of the injury risk. With regards to material solutions, innovations designed for service members can also be implemented in PPE for working dogs. However, when it comes to understanding injury risk, the unique anatomy and physiology of canines does not directly translate to the human, necessitating specific research for canines and a comparison of internal anatomy, in particular the vital thoracic organs. One of the tools used to describe and study traumatic injuries in humans is the Abbreviated Injury Scale [3], but an equivalent system does not currently exist for canines Ideally this coding system would link back to human injury coding systems to allow for comparisons and injury assessments between handlers and working dogs. There are currently some veterinary injury coding schema available, such as the Animal Trauma Triage Scale (ATT) which assigns a severity score for trauma injuries and the modified Glasglow Coma Scale (mGSC) used just for neurologic injury, however these schemas use different injury binning techniques and offer no cross reference to human injury scoring schemas [4]. They also have no link to overall severities for comparisons between injuries. A new injury scoring system is needed for the establishment of a canine trauma registry to track and research traumatic injuries in working dogs [5].

The goal of this study is to investigate the thoracic anatomy and anthropometry between the human and canine to provide information about vital organ placement in relation to armour coverage areas and the potential for serious thoracic injury. This work will be used within a new project that aims to develop an injury coding schema and visualization tool for canine injury as well as body armour coverage analyses. This is an important tool for development to better understand traumatic injury to working dogs and their human counterparts.

2. METHODS

This study involved anlaysis of the canine anatomy and anthropometry using a representative computed tompography (CT) scan of a working dog. This scan of a 5 year-old, male Labrador Retriever was obtained from the U.S. Department of Defense Military Working Veterinary Hospital not for purposes of this study. The CT was acquired at 0.625 mm slice thickness using a 396 mm reconstruction diameter within 512 x 512 images, resulting in 0.773 mm in-slice resolution. The selection of the human CT scan used in this study was data-driven to identify an individual whose measurements are close to anthropometrics that represent the 50th percentile of a male U.S. Army soldier [6,7,8]. The CT was acquired at 1.25 mm slice thickness using a 500 mm reconstruction diameter within 512 x 512 images, resulting in 0.976 mm in-slice resolution. The U.S. Army representative male CT scan was acquired from U.S. Army Combat Capabilities Development Center Data & Analysis Center (CCDC DAC) via a collaborative research effort with Wake Forest University.

Tabel 1. Description of the measurements taken, the corresponding CT plane they were taken in and the human and canine nomenclature denoting anatomic orientation within that plane, along which the measurements were taken [7].

Measurement	CT Plane	Anatomic View Within Plane	
		Human	**Canine**
Depth	Sagittal	Anterior to Posterior	Dorsal to Ventral
Width	Axial	Left to Right	Left to Right
Length	Coronal	Superior to Inferior	Cranial to Caudal

Both scans were acquired separately from this research, therefore the scans were constucted using slightly different protocols. Given the difference in the CT-scan orientation metadata, the scans were compared using anatomical planes. Table 1 details which human and canine anatomical plane corresponds to this study's measurement label. CT analysis was completed using Mimics software (Materialise, 20.0); each scan was analysed independently. This study initially focused on describing the heart, lungs, and span of the rib cage since these areas had been previously identified as important when considering ballistic plate design [9]. Three-dimensional (3D) geometry was created to represent the skin, skeleton, heart, and lungs of the male human and canine independently using segmentation

https://doi.org/10.52202/078352-0004

tools (thresholding, multi-slice edit, etc.). The measurement tool was used to take in-plane CT measurements. The objective was to investigate the location and relative size between the thoracic cavity and essential organs (heart and lungs). The plane with the largest area of the heart was used for the measurements to develop an estimate of coverage area for the vital organs. For each plane a linear measurement was taken of the heart, the lungs, the rib cage, and the chest (skin) at their widest points. The 3D geometry was a visual means of ensuring the measurements aligned with the most extreme points of each organ since those points did not always exist in the same plane. To quantify the space that the heart and lungs occupy in the human and canine, the depth, width, and length of the heart and lungs were measured, as they have the highest severity associated with organ injury within the thorax [3].

3. RESULTS

The canine chest is largest dorsal-ventral (Figure 1) whereas the human chest (Figure 2) is widest left-right. As shown in an axial view, the canine heart encompasses a large percent of the thoracic width behind the sternum and encompasses the full width of presented area of the chest, whereas the human heart is found within the center of the chest and is only a fraction of the anterior presented area of the chest.

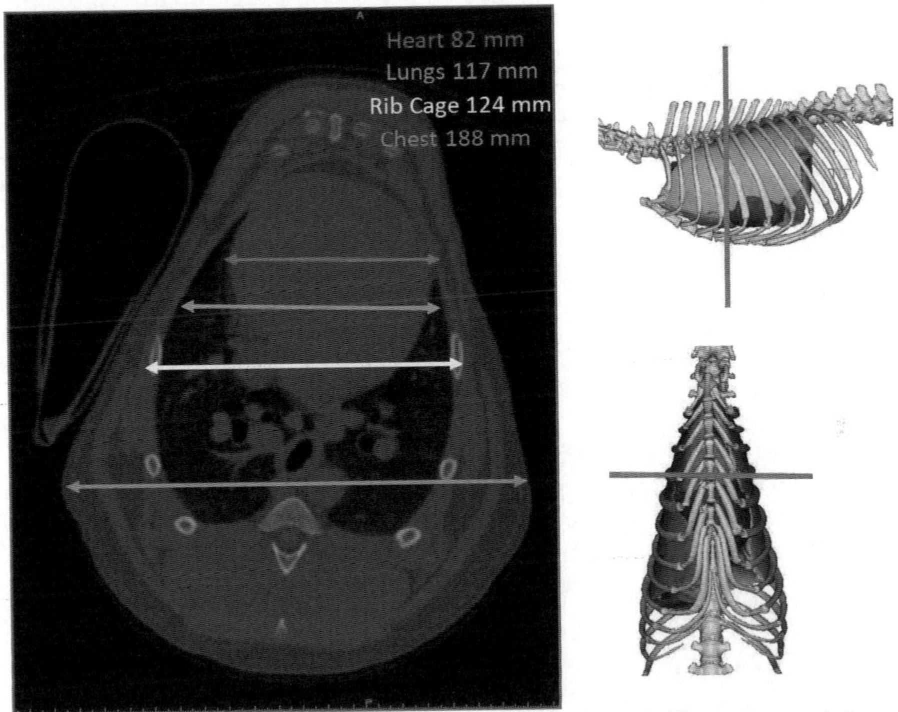

Figure 1. Axial cross section of the canine chest showing color coded lines where each tissue was measured within the slice. The arrows denote the width for each tissue: heart (red), lungs (pink), rib cage (pale yellow), and the chest (blue), measured at their widest point. Note that the canine in this view is oriented with the sternum at the top of the image, for comparison ease with the human CT shown in Figure 2.

https://doi.org/10.52202/078352-0004

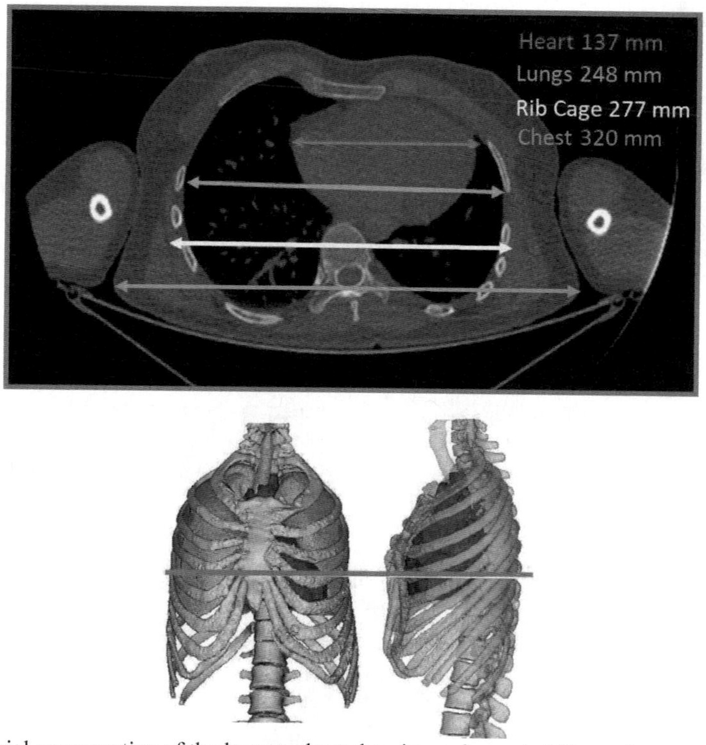

Figure 2. Axial cross section of the human chest showing color coded lines where each tissue was measured within the slice at each widest point. The arrows denote the width for each tissue: heart (red), lungs (pink), rib cage (pale yellow), and the chest (blue).

Figure 3 shows the thoracic anatomy and anthropometry from the ventral view (canine) and Figure 4 shows the anterior view of the human. The canine has 6 lung lobes – 4 on the right (cranial, middle, caudal, and accessory) and 2 on the left (caudal and cranial). In comparison, the human has 5 lung lobes – 3 on the right (superior, middle, inferior) and 2 on the left (superior and inferior).

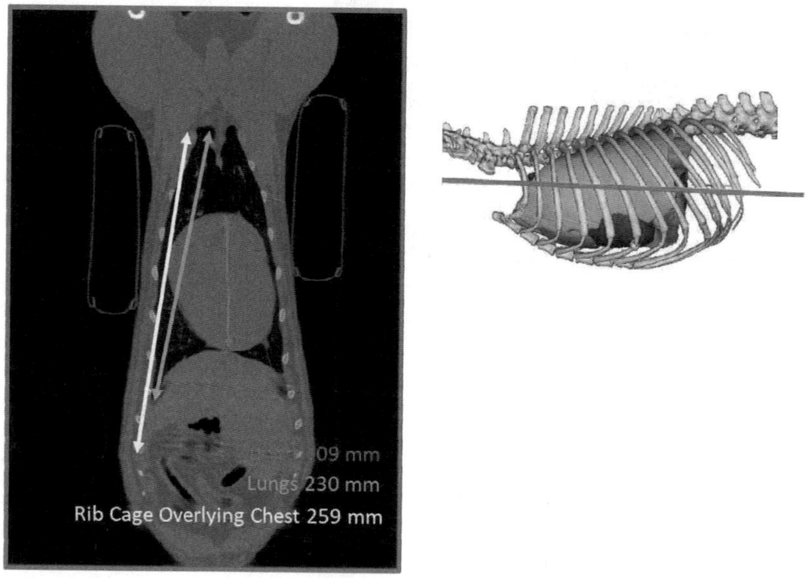

Figure 3. Coronal cross section of the canine showing color coded lines for where each tissue was measured within the slice. The arrows denote the length for each tissue: heart (red), lungs (pink), rib cage overlying the chest (pale yellow).

https://doi.org/10.52202/078352-0004

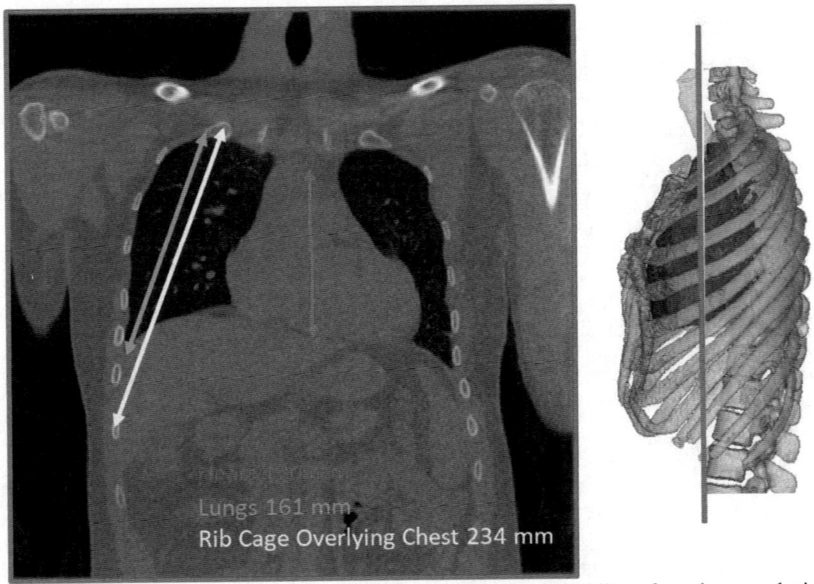

Figure 4. Coronal cross section of the human showing color coded lines for where each tissue was measured within the slice. The arrows denote the width for each tissue: heart (red), lungs (pink), and rib cage overlying chest (pale yellow)

Dorsal/posteriorly, the canine lungs encompass the areas underlying ribs 1-9, whereas in the human lungs encompass the areas underlying ribs 1-10. Ventral/anteriorly, this reduces to ribs 1-7 in the canine and ribs 1-6 in the human. This difference can be seen in Figure 5, where the rib cage, lungs, and heart are displayed for the canine and human.

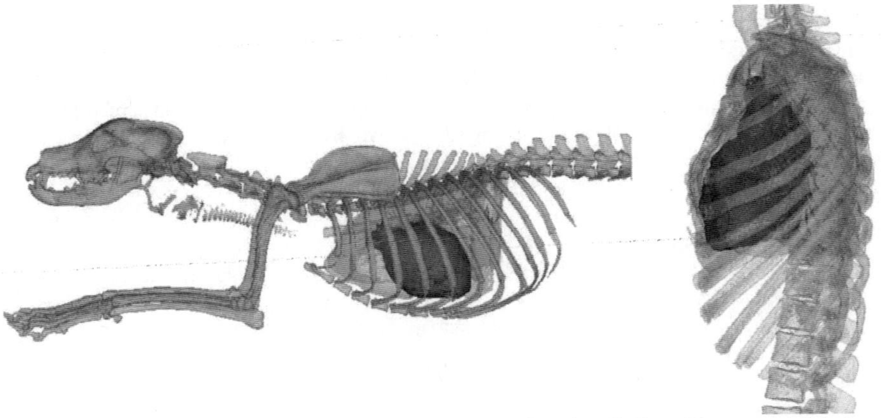

Figure 5. Anatomic view of heart, lungs, and skeleton of canine (left) and human (right) showing positioning of heart behind ribs, within chest cavity.

The apex of canine lungs is more narrow than that of the human, with the overall shape of the thorax in this view being more cone-shaped for the canine and more barrel-shaped for the human. Given this difference in shape, the canine heart is 70% of the width of the lungs, while in the human the heart is only 55% of the width of the lungs. The human heart is also more offset to the left compared to the canine. Table 2 details the final percentage of rib cage each organ encompassed for that measurement.

https://doi.org/10.52202/078352-0004

Table 1. Table of final measurements comparing percent coverage of the heart and lungs.

	Percentage of Rib Cage Length/Width	
	Human	**Canine**
Heart Length	51%	42%
Lungs Length	69%	89%
Heart Width	49%	66%
Lungs Width	90%	94%

4. DISCUSSION

With working dogs playing a significant role in our protection and safety alongside uniformed service members, it is important to understand their body armour requirements and the link between PPE and injury risk. Determining optimum coverage using anthropometrics and internal organ measurements can be difficult given the variablity within a species, let alone comparing anthropometrics across species. The last common classification in the phylogenic tree between humans and canines is the class Mammalia with humans then categorized as Primates and dogs as Carnivora. Both species' skeletal systems can be divided into axial and appendicular. Both axial skeletons have a rib cage to protect thoracic organs but with the human commonly having 12 ribs on each side and the canine having 13. Since both the human and canine have similar skeletal structures, the rib cage was used as a skeletal feature to compare organ location and anthropometry [7,10]. Furthermore, past studies have supported a higher correlation of skeletal landmarks to internal anatomy placement, as opposed to external soft-tissue landmarks (e.g., nipples and umbilica) [6,11].

After reviewing the anatomy and anthropometry between the male canine and human, the canine thoracic stature in a standing position more closely reflects the human in the prone position. When prone, the human head is presented first, with the neck and top of the chest the most vulnerable area of the thorax. For canines, both sides are exposed as well as the front of the chest, leading with their heads. For canines, body armour designs to fully protect against small arms and fragmenting threats will need side coverage as well as frontal coverage – potentially requiring 3 panels of protection versus the 2 that are currently offered with human body armour. This is especially true for situations that involve the canine advancing towards a threat, where their chest is the largest presented area and the majority of that surface area overlies the heart. As shown by the CT scan in Figure 1, the entire presented area of the chest for the canine is encompassed by their heart – leaving a vital area susceptable to lethal injury. While this initial study is developed from the CT scans of one male canine and one male human, they are representative of the standard population size that is relevant for the US male soldier and working dog. Future work would benefit from the inclusion of additional CT scans from representative subjects. Additionally, for the canine, this work is specific to dogs of a similar body size and shape to that of a working dog. There are many shapes and sizes of dogs where this data would not be relevant, such as brachycephalic dogs or much smaller/larger breeds.

There is not currently a standard test and body armour rating system that is specific to canines. Given the differences in anatomy and anthropometry, it is important to consider whether human test standards are appropriate for the canine, especially when investigating coverage areas with injury severity. Most canine body armours found for sale stated they were tested to the same NIJ standards as human body armour, but there is currently not a transfer function between canine and human thoracic injury [12]. To aid in this transition and comparison of injuries, it is necessary to apply a traumatic coding system to canine injury in the same way that AIS is used for humans. This task is currently underway and will result in a traumatic injury scoring system for the canine that is analagous to human AIS codes. There are important orientation and anatomical differences between the canine and human thoraces that need to be taken into account when developing a parallel injury coding schema. The lungs in the canine are structured with 4 on the right and 2 on the left, so that an injury to one or two lobes may not be as catatrosphic compared to human lung injury. Galer noted that pneumothorax in a canine can be treated with a chest tube very similarly to that of a human and canines have a fenestrated mediastinum so that a chest tube on one side may help with a pneumothorax on the other [13]. When analyzing shotlines and body armour coverage areas, the sides for the canine are a large presented area that leave the heart and lungs vulnernable to injury. While side impacts to the human are similarly damaging shotlines

https://doi.org/10.52202/078352-0004

encompassing both lungs and the heart, the coverage area for the side on a human is much smaller than that of the dog.

This work is contributing to the development of a canine injury coding system that will be used to standardise injury research and body armour protection for working dogs. This coding system is being developed from the human Abbreviated Injury Scale [3]. In human AIS codes, there are 191 codes for the chest. This includes the ribs, heart, lungs, bronchus, major arteries and veins, and trachea. Codes and descriptions must be altered for the canine. For example, the human AIS chest injury description of "breast avulsion" is not applicable to the canine. Once complete, it will be possible to compare canine injury and human injury from the same events in both the law enforcement and military scenarios. This is important for learning about emerging threats, body armour development, and injury comparisons to laboratory test data. Standard police and military working dogs stand at about knee-high for the human, making their injuries important for forming comparisons to human leg and foot injuries in blasts. The long-term goal of this effort is to develop a comprehensive canine traumatic injury scoring system for recording injuries and developing protective measures, increasing survivability for the working dog as well as the uniformed service members.

5. CONCLUSION

By investigating the comparative anatomy and anthropometry between the working canine and human, it is possible to develop better protection for our working dogs, making them more effective partners for their human counterparts. While much research has gone into human body armour protection areas, injury risk, and lighter-weight armour, the same is greatly lacking for the working canine. This research showed that the chest of the canine should be covered with body armour to best protect the dog when facing a threat, especially since the heart spans the entire width of the chest. Additionally, there is a need to develop a traumatic injury scoring system for the canine that is analagous to the system used in human injury research so that comparisons and crosswalks between injuries can be completed. Working dogs work alongside their human counterparts and the protection of the canine is vital for the team to function as a whole.

Acknowledgements

The views and conclusions contained in this document are those of the authors and should not be interpreted as presenting the official policies or position, either expressed or implied, of the U.S. Army Research Laboratory or the U.S. Government unless so designated by other authorized documents. Citation of manufacturer's or trade names does not constitute an official endorsement or approval of the use thereof. The U.S. Government is authorized to reproduce and distribute reprints for Government purposes notwithstanding any copyright notation hereon.

References

[1] Hermanson J.W., de Lahunta A., Evans H.E. Millers and Evans' Anatomy of the Dog, Fifth Edition, Elsevier 2020.

[2] Klein, B.G. Cunningham's Textbook of Veterinary Physiology, Sixth edition, Elsevier/Saunders, 2019

[3] Association for the Advancement of Automotive Medicine. The Abbreviated Injury Scale (AIS) 2015 Revision. AAAM. Chicago, IL, 2016.

[4] Ash K, Hayes G, Goggs R, Sumner J. Performance Evaluation and Validation of the Animal Trauma Triage Score and Modified Glasgow Coma Scale with Suggested Category Adjustment in Dogs: A VetCOT Registry Study. Journal of Veterinary Emergency and Critical Care, 23 (2018), pp.192-200 doi.10.1111/vec.12717

[5] Hall K, Boller M, Hoffberg J, McMichael M, Raffe N, Sharp C. ACVECC's Veterinary Committee on Trauma (VetCOT) Registry Subcommittee. ACVECC-Veterinary Committee on Trauma Registry Report 2013-2017 Journal of Veterinary Emergency and Critical Care, 28 (2018), pp. 497-502

 https://doi.org/10.52202/078352-0004

[6] Gayzik, E. Moreno, D., Geer P., Wuertzer R., Martin S. and Stitzel J. Development of a Full Body CAD Dataset for Computational Modeling: a Multi-modality Approach. Annals of Biomedical Engineering, 39 (10):2568-83, 2011.

[7] Paquette, S. G. (2009). Anthropometric Survey (ANSUR) II Pilot Study: Methods and Summary Statistics. *TR-09/014* .

[8] Reed M., Manary M., Scheider L. Methods for Measuring and Representing Automobile Occupant Posture. SAE Technical Paper 1999-01-0959, March 1999.

[9] Breeze J, Lewis EA, Fryer R, et al. Determining the Dimensions of Essential Medical Coverage Required by Military Body Armor Plates Utilizing. Injury Int. J Care Injured 2016;47:1932-38. https://dx.doi.org/10.1016/j.injury.2016.06.010

[10] White T., Folkens P. The Human Bone Manual. Elsevier Academic Press, 2005 (pp 181,185,256).

[11] Lockett, J.F., et al., Proposed Integrated Human Figure Modeling Analysis Approach for the Army's Future Combat Systems. 2005, SAE Technical Paper.

[12] Sherman, SS. A Biomechanical Assessment of Canine Body Armor. Dissertation at Wayne State University, Detroit, Michigan, 2015.

[13] Galer M, Magid D, Folio L. Gun Shot Wound to the Chest of a Military Working Dog. Military Medicine Radiology Corner, Volume 174, June 2009

https://doi.org/10.52202/078352-0004

Knitted Fabrics with Dyneema® Fibres for Ballistic Protection: Modelling and Experimental Validation of a Single Jersey Knit

M. Hazzard[1], U. Heisserer[1], M. van der Kamp[1], K. Freier[1]

[1]*DSM Protective Materials, Urmonderbaan 22, 6167 Geleen, The Netherlands*
mark.hazzard@dsm.com

Abstract. High strength UHMWPE fibres and their composites have been widely used in personal protection due to their excellent tensile strength and low weight. Typically, this protection is in the form of a hard or a soft ballistic panel of cross-ply unidirectional sheets which cover critical body parts, with protection level often requiring a balance between mobility and resistance. Knits based on UHMWPE provide the opportunity for increased personal protection by increasing the protection level on difficult to cover areas requiring high mobility and drapability. Knits intrinsically allow for more extensibility compared with weaves or cross-ply unidirectional sheets by making use of yarn curvature, something that is intrinsic to the yarn path geometry of a knit. Here, a simplistic finite element modelling approach has been used with tensile only beam elements to capture the behaviour of UHMWPE knits, branded as Dyneema®, in a ballistic setting with a single jersey knit architecture. The behaviour of the single jersey knit has been compared to performance of other form factors with experimental data and readily available data from literature, which highlights the potential of UHMWPE frag protective knits.

1. INTRODUCTION

UHMWPE fibre based personal protection, such as Dyneema® UD, has been offered in technology platforms such as soft ballistic sheets and hard ballistic panels for excellent lightweight protection due to the fibres intrinsic low weight and high tenacity [1]. The UD form factor is extremely well suited for ballistic performance as fibres are straight with no yarn cross-overs, providing minimal back face deformation and excellent ballistic limit. However, in order to increase the level of personal protection, form factors should be considered that cover areas of the body that require higher mobility/drapability. Knits are typically the go to form factor for comfort and are used for clothing and apparel in commodity textiles. Knits allow for extensibility through incorporation of yarn curvature, and without a binder system, are highly extensible and drapable. For high performance UHMWPE yarns that are intrinsically stiff in tension, comfort and extensibility can therefore be provided by exploiting hidden length within the knit geometry. This combines to allow for greater stretch, flexibility, and shear, providing increased comfort for those hard to cover areas of the body.

The simplest and most common type of knit is the single jersey knit comprising courses (loops side by side) and wales (interlocking of those loops vertically). Dwivedi *et al* [2], [3] compared multiple types of fabrics including felts, weaves, and knits through mechanical testing followed by ballistic trials. They showed that continuous filament knits with a high performance yarn provide a unique combination of extensibility and perforation levels. Aramid weaves, for example, provided an initial elastic modulus 4 orders of magnitude higher than a comparable aramid knit, yet the ballistic performance for a similar areal density knit was only reduced by 10-15% [3]. Architecture can therefore play a role in both ballistic performance and comfort levels, however for a given architecture, the intrinsic yarn properties are still expected to be a governing factor. Using the Cunniff parameter, a combination of specific tenacity and tensile wave speed of the fiber [4], Werff *et al* [5] highlights the potential of UHMWPE fibres in terms of ballistic performance due to their high tenacity, low density, and high modulus. This paper aims to show the performance of a continuous filament single jersey knit structure made with Dyneema® through an experimental approach. Using this data, a simplistic numerical modelling approach is outlined that is validated with this experimental data. The finite element modelling approach allows further investigation into the drivers of performance for knits for ballistic protection, with an example comparison to an aramid yarn provided.

 https://doi.org/10.52202/078352-0005

2. METHODOLOGY

2.1 Experimental

Single jersey knits were produced using a flatbed Shima Seiki with a Dyneema® 880 dtex yarn. Following knitting, 10 samples were cut to approximately 200×200 mm square patch, and to give some rigidity for handling, the sample was edged with duct tape. For ballistic testing, the sample was held to the frame with a small piece of duct tape to maintain its position, allowing numerical models to be considered free of boundary conditions. A 1.1 g fragment simulating projectile (FSP) was used, impacting approximately central to the target with a single shot. Velocity light gates were positioned in front and behind the target to capture an initial velocity (V_I) and a residual velocity (V_R) for the samples that were perforated. A ballistic limit was then estimated using a Lambert-Jonas approximation [6]. Additionally, a Photron fastcam SA5 was used to record the impact event with two illuminating Dedolight 400D lights. The camera was positioned at an angle of approximately 30° from the side on view of the knit in order to estimate the back face deformation whilst also have capability to measure wave propagation rates within the knit. Reference images were taken with reference geometry to extract absolute geometry and the precise angle of the camera relative to the sample. 3D position cannot be extracted with a single camera, hence the assumption was that wave propagation happens only in X and Y directions (in plane), whilst maximum back face deformation only in the Z direction (out of plane). The full setup is given in Figure 1. ImageJ was used to extract both the back face deformation with time as well as the shock wave velocity within the knit.

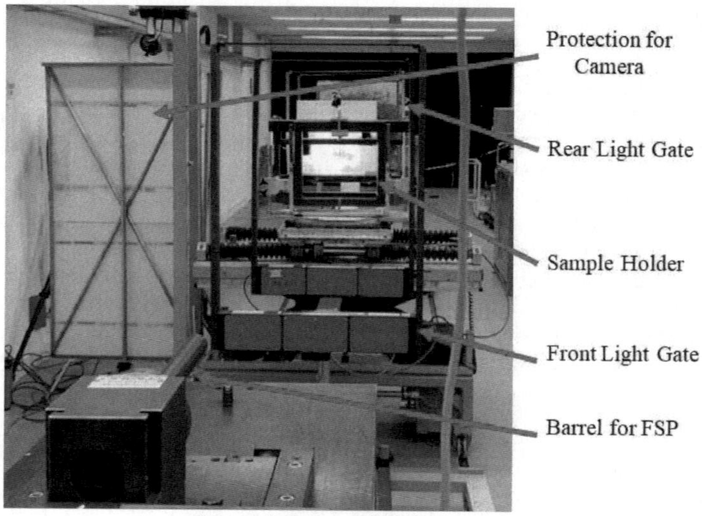

Figure 1: Image of ballistic setup.

2.2 Numerical

Recently, McKee *et al* [7] outlined a modelling method to first capture the geometry of a single jersey knit and meshing with a solid elements (Figure 2) using a transversely isotropic constitutive model for material behaviour to simulate ballistic impact. This provided reasonable correlation to experimental results in terms of wave propagation and back face deformation. One difficulty, particularly when capturing failure of yarns through a continuum approach, is the difficulty in homogenizing discontinuous filaments in a yarn, and the effects this has upon failure.

https://doi.org/10.52202/078352-0005

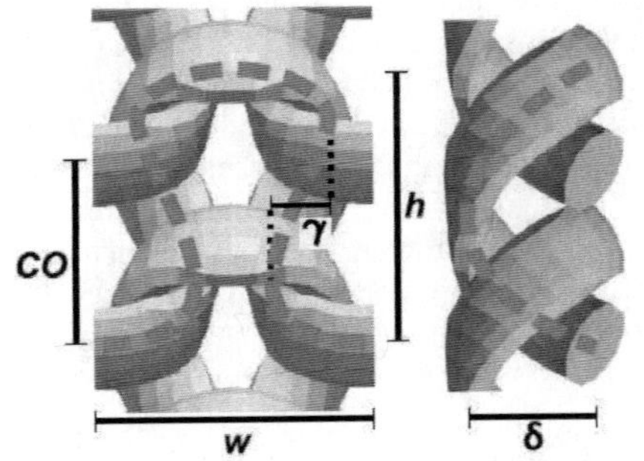

$$x(s) = \frac{s}{\lambda} w + \gamma \sin\left(4\pi \frac{s}{\lambda}\right) \quad (1)$$

$$y(s) = CO + \frac{h}{2} \cos\left(2\pi \frac{s}{\lambda}\right) \quad (2)$$

$$z(s) = \frac{\delta}{2} \cos\left(4\pi \frac{s}{\lambda}\right) \quad (3)$$

Figure 2: McKee *et al* [7] 3D solid representation, with parameters *CO* course offset, *w* loop width, γ loop curvature length, *h* loop height, δ thickness, and λ loop path length.

Here, we utilized the geometrical parameters outlined by McKee *et al* [7] but implement it using tensile only beam elements with an equivalent circular cross section within LS-DYNA (MAT CABLE DISCRETE BEAM). This model assumes perfect Hookean elasticity, with an UHMWPE modulus of 155 GPa and a density of 0.98 g/cm^3, and as yet, does not incorporate a failure model. This has reduced complexity and does not capture variations in yarn geometry along the loop path or variation in stress across a yarn, however this does allow for minimal run time whilst avoiding difficulties associated with continuum modelling of yarns. Measurements of knit geometry were performed via optical microscopy with the exception of δ (thickness) and λ (loop path length). These parameters can be solved analytically from a known areal density of the system (Table 1) and the yarn dtex. For this 1D approximation, the path length (and areal density) of the knit geometry can be dependent on the number of elements. For example, a coarse mesh has reduced length compared to the natural curvature of the knit that has not be discretised into elements. For the current construction, 20 elements were used per loop. The geometry of the model is provided in Figure 3 with the circular cross section visible, and the geometrical parameters provided in Table 1.

Table 1: Geometrical parameters used for the UHMWPE knit.

Geometrical Parameter	Value	Unit
w	1.9	mm
γ	0.4	mm
CO	0.71	mm
h	1.263	mm
δ	0.54	mm
Yarn	880	dtex
Areal Density	0.34748	(kg/m^2)
No. Elements per loop	20	-

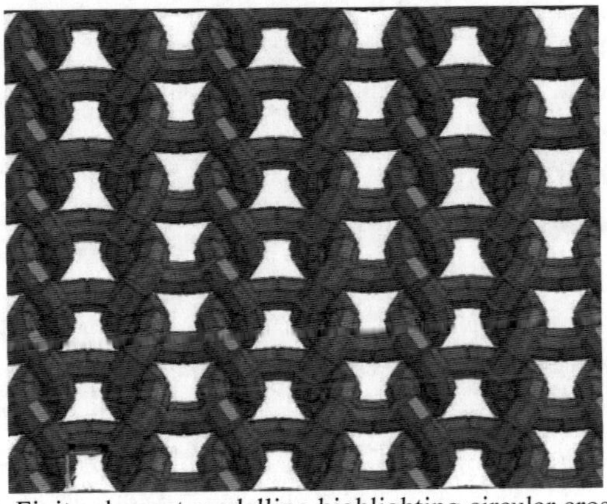

Figure 3: Finite element modelling highlighting circular cross section.

https://doi.org/10.52202/078352-0005

The projectile was modelled using geometry from the standard assuming no end flange [8]. The FSP was meshed with 0.5 mm solid quad-elements with a single integration point (formulation type 1 in LS-DYNA). A simplified Johnson Cook material model was used for the projectile with parameters taken directly from literature [9]. Automatic general contact was applied with SOFT=1 and a coefficient of static and dynamic friction of 0.1 within the model. An additional damping frequency range was utilised in order to damp out non-physical modes of deformation, particularly after initial contact, driven by natural frequencies of beam elements and the repeating unit of the knit structure. Care was taken to select damping frequencies that do not influence the result and induce additional energy loss. A frequency range of 500 to 10000 Hz was applied only to the knitted part with a damping coefficient of 0.1. Lower frequencies of damping, particularly below 500 Hz, were observed to induce faster projectile deceleration, indicating a non-physical energy loss (Figure 4).

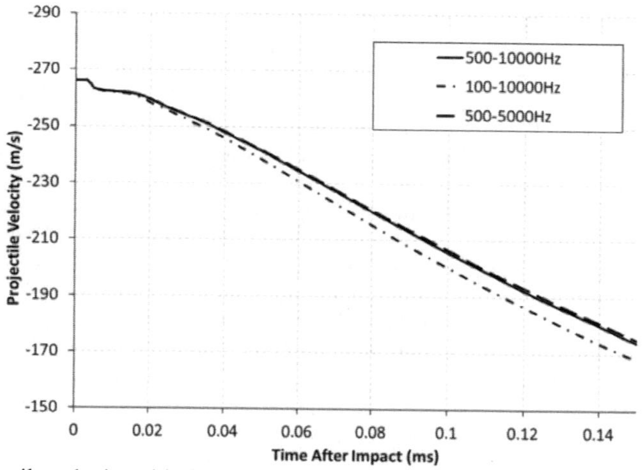

Figure 4: Projectile velocity with time from a 1.1 g FSP impact simulation at 266 m/s whilst varying the frequency damping range.

3. RESULTS

3.1 Experimental Results

Initial and residual velocities from 10 shots with a Lambert-Jonas fit are given in Figure 5, with the Lambert Jonas equation used for fitting also given (eq. 4). In total there were 5 stops and 5 penetrations, which indicate that the ballistic limit (V_{BL}) is approximately 290 m/s with a sharp stop transition for the non-commercial knit. This can be observed through high p values compared to other form factors and velocity ranges which typically have values in the range of 2-4 [10].

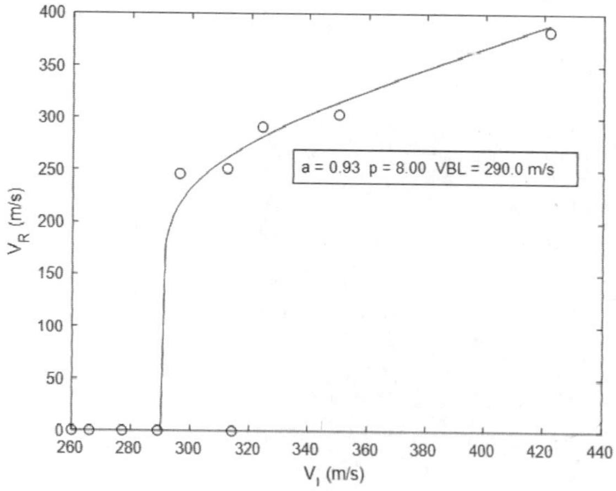

$$V_R = a\left(V_I^p - V_{BL}^p\right)^{1/p} \tag{4}$$

Figure 5: Initial velocity vs residual velocity of each shot with a Lambert-Jonas fit for ballistic limit. Note that the single jersey knit is non-commercial and does not consider additional requirements for application.

https://doi.org/10.52202/078352-0005

A high speed image sequence of a stop event with an impact velocity of 266 m/s is given in Figure 6. The sample can be observed at t = 0 prior to impact with surrounding duct tape around the sample. The centre of the sample is visualized by marking it with a pen. There was a small variation in impact location, however the location was always within 20 mm of the centre location, whilst the FSP had negligible pitch or yaw. Some waviness in the sample is observed and can be attributed to the low flexural stiffness of knits compared with other form factors whilst the knitting process itself can induce some residual stress following manufacture. After impact, the longitudinal waves (in-plane of the knit) can be observed propagating along the yarns ahead of the shear hinge (out-of-plane of the knit). For comparison to numerical results, the position of the waves in the vertical and horizontal directions have been tracked in each frame, with the location and orientation also highlighted in Figure 5 (P_{LV}, P_{LH}, P_{SV} P_{SH}). Once the longitudinal wave has reached the boundary, pull in of the fabrics can be observed from the edge. Locally around the projectile the knit exhibits additional stretch and opening up of the course and wale structure, where visually the gaps between yarns can become larger. This is shown in Figure 7a with an image taken after the impact event.

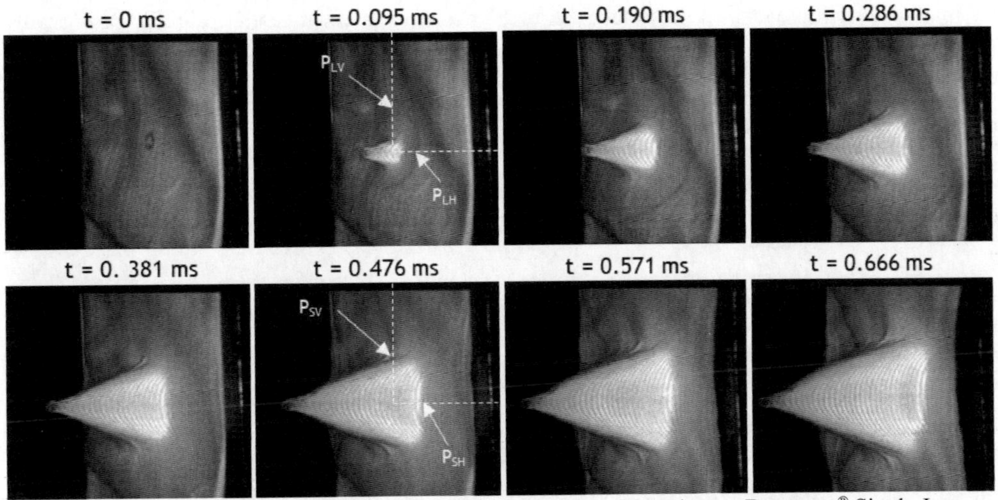

Figure 6: Highspeed image sequence of 1.1 g FSP impact at 266 m/s on a Dyneema® Single Jersey Knit with a stop. Position of measured wave position indicated on image for the Position Longitudinal Vertical wave (P_{LV}), the Position Longitudinal Horizontal wave (P_{LH}), the Position of the Shear-hinge Vertical (P_{SV}), and the Position of the Shear-hinge Horizontal (P_{SH}).

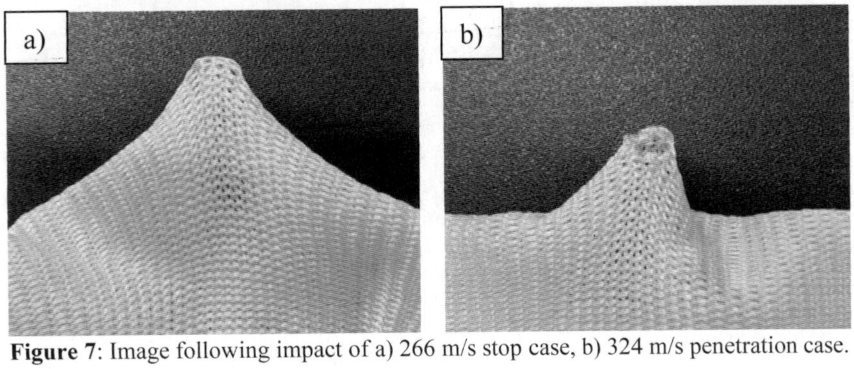

Figure 7: Image following impact of a) 266 m/s stop case, b) 324 m/s penetration case.

A highspeed image sequence is now also given for a penetration case in Figure 8. Its shown that the failure event happens early on in the impact process. A zoom view focusing on the impact location is also shown, and following inspection it appears as though a single yarn has failed which allows opening of the knit. The damage zone of the impact location is also provided in Figure 7b.

https://doi.org/10.52202/078352-0005

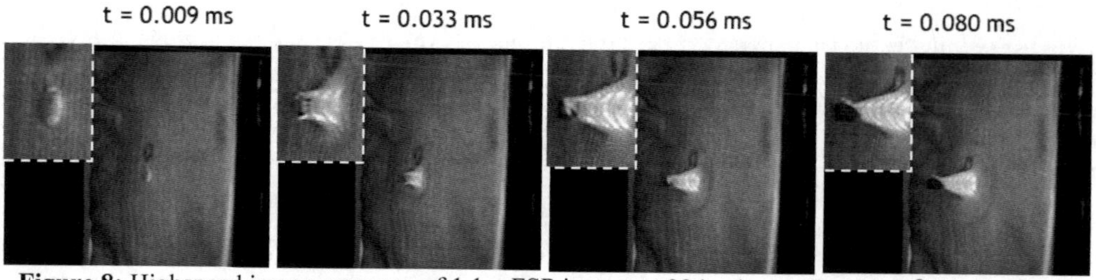

Figure 8: Highspeed image sequence of 1.1 g FSP impact at 324 m/s on a Dyneema® Single Jersey Knit with a penetration. Top left of each images isolates a close up of the impact location.

3.2 Numerical Validation

A stop case of 266 m/s was simulated with an image sequence following impact provided in Figure 9. Within the model, the longitudinal wave can be seen progressing ahead of the shear hinge as back face deformation develops. Maximum back face deformation is compared with that measured experimentally from highspeed images in Figure 10. The simple tensile only beam approach is surprisingly accurate. This seems to suggest that back face deformation is largely driven by conservation of momentum, and if the areal density is matched along with wave speeds, then a reasonable back face deformation can be expected. This also suggests that the low flexural rigidity of the knit provides little resistance to back face deformation, and that changes in back face deformation would be largely momentum driven as well as boundary dependent.

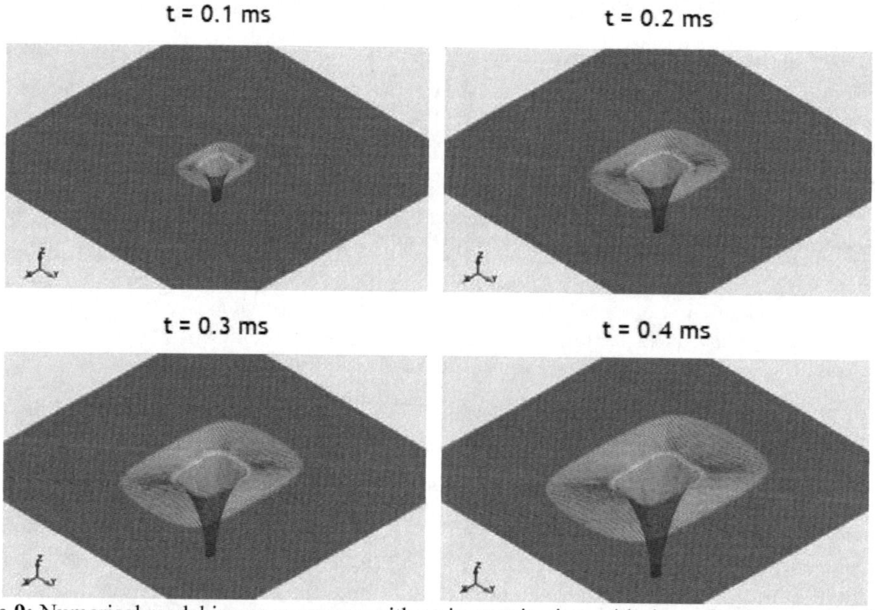

Figure 9: Numerical model image sequence with an isometric view with time after impact, t, indicated above each image. Red indicates more than 1 mm back face deformation, with the transition zone the approximate location of the shear hinge. In this image, the beam elements are visualised in only 1D.

https://doi.org/10.52202/078352-0005

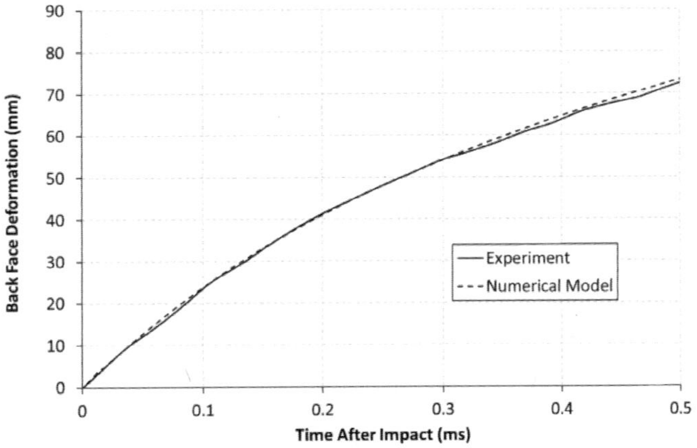

Figure 10: Back face deformation of Dyneema® Single Jersey Knit with a stop at 266 m/s.

Figure 11 compares the wave position for the horizontal and vertical longitudinal and shear waves measured experimentally and numerically. It was observed that longitudinal wave propagation was slower in the model than observed experimentally. This was observed in both the horizontal and vertical direction (Figure 11a and Figure 11b respectively). It is thought that the longitudinal wave speed difference is due to the difference in the simplified geometrical model used compared with that in reality. That being said, the model does capture the higher wave speed in the horizontal compared with the vertical direction. Note that the horizontal direction has a continuous filament as this is the course direction, whilst the vertical direction the wave needs to be transferred via interlocking loops (wales). This indicates that the modelled knit has a similar degree of anisotropy compared with the experimental result.

When considering the shear hinge speed, it was observed that the numerical model was reasonably accurate in both the horizontal and vertical direction until approximately 0.2 ms (Figure 11c-d). After this time, the results start to diverge, with the numerical model again progressing slower. One possible explanation for this is that boundary interaction within the experiment can start to occur due to faster longitudinal wave propagation observed in the experiment. The shot is not perfectly central, and duct tape may start to have an influence after approximately 60 mm travel of the longitudinal wave speed. There may also be some variability in true length to the boundary due to waviness in the knitted sample (observable in Figure 6). Again though, the difference in horizontal and vertical shear wave speed is captured in the simplistic model.

https://doi.org/10.52202/078352-0005

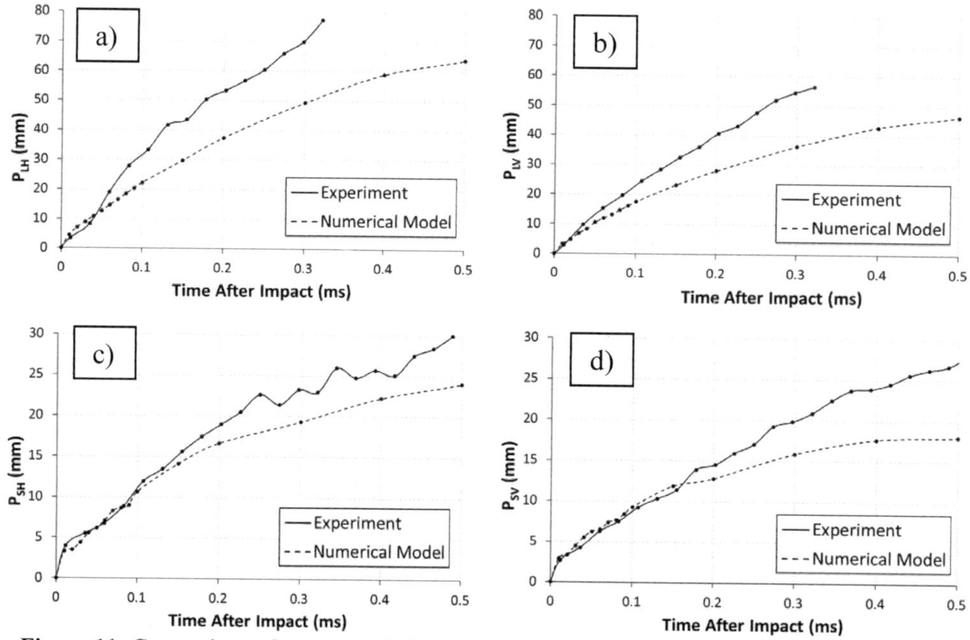

Figure 11: Comparison of wave speeds between experimental results and the numerical model. a) Position Longitudinal Horizontal wave (P_{LH}), b) Position of the Longitudinal Vertical wave (P_{LV}), c) the Position of the Shear-hinge Horizontal (P_{SH}), and d) the Position of the Shear-hinge Vertical (P_{SV}).

3.3 Comparison to Aramid

Numerical modelling allows direct comparison of individual parameters that are often difficult to perform experimentally. As such, a simulation was performed with an aramid yarn at the same dtex. This required a reduction in cross sectional area of the yarn, but the areal density and geometry of the knit are identical. This dictates that the aramid knit intrinsically has a reduction in coverage area, the frontal area of the knit that contains gaps, compared with UHMWPE yarns for the same construction and areal density.

Table 2: Table of yarn properties used for comparison

Parameter	Aramid	UHMWPE	Unit
Density	1.44 [7]	0.98	g/cm^3
Modulus	82.6 [7]	155	GPa
Tensile Strength*	3.88 [11]	4.2	GPa

* *Tensile strength not incorporated in the model, only for reference, aramid assumed Kevlar KM2.*

No failure/damage model was incorporated into the numerical modelling approach as tensile only beam elements assume constant stress over the cross sectional area. McKee *et al* [7] using solid elements has shown this is not the case, with stresses varying throughout the yarn from filament to filament. This would suggest that for our model, an equivalent force at failure would need to be defined for each yarn and each geometrical construction. However, we can compare maximum axial force observed in the yarn at a certain moment in time for a certain impact speed. The maximum force of the yarn was observed to vary dependent on the location in the loop and its location relative to the projectile. For this simple comparison, we compare the same element at an approximately central location under the impactor at the top of the loop (Figure 13). It was observed that there is some initial noise which has been assumed to be numerically driven and for the purpose of this comparison was assumed negligible. This is also supported by highspeed imagery, where failure did not occur until a later moment in time (Figure 8, 324 m/s impact example). The forces observed reach a similar maximum magnitude following some initiation noise (location highlighted), and assuming a similar stress distribution across a yarn at that location, then force at failure would largely be strength and cross sectional area dependent.

https://doi.org/10.52202/078352-0005

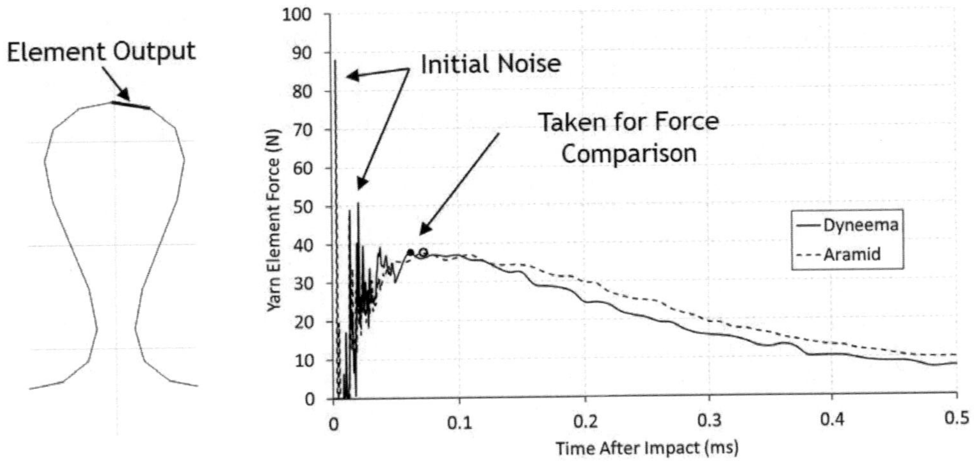

Figure 12: Force comparison within a single element

Whilst this comparison is not perfect as this does not consider several phenomena such as precise location of max force in the knit and variation of force over the yarn cross-section, it does suggest that as the tensile strength and cross-section of the yarn is higher for UHMWPE, on a weight basis, UHMWPE should outperform an aramid. An experimental comparison of V_{50} performance in Figure 13 highlights the potential a Dyneema® Knits. This is done through making use of non-dimensional areal density [4], which is a method to compare ballistic performance where differing projectiles have been used. This method also has its limitations as it is limited to blunt projectiles which themselves have variability in performance, whilst it makes no adjustment for the backing that was utilised (aramid comparisons have gelatin backing whilst here results are free hanging). However, the Dyneema® Single Jersey Knit shown experimentally here has similar ballistic limit to that of an aramid single jersey knit, yet with a drastic reduction in non-dimensional areal density. This also demonstrates that the application of knits compared with other form factors, such as hard ballistic panels (HB26 Dyneema® reference provided [10]), are best at providing protection in a lower areal density range.

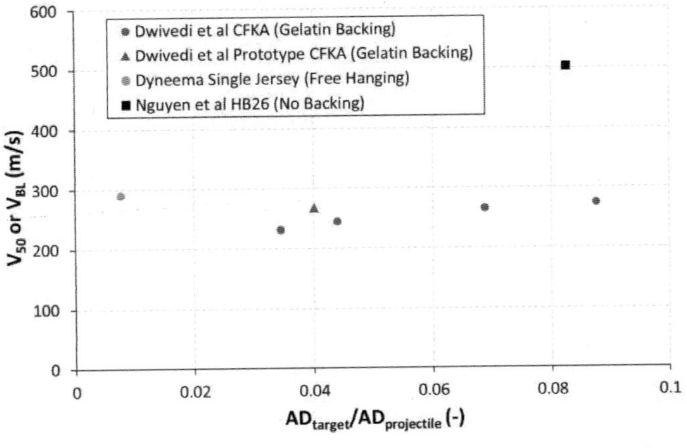

Figure 13: Ballistic performance of non-commercial Dyneema® single jersey knit compared with literature. Note that the single jersey knit is non-commercial and does not consider additional requirements for application.

4. CONCLUSION

This contribution has highlighted the potential of Dyneema® knits as a way to increase personal protection by providing protection for those hard to cover areas requiring high mobility and drapability.

https://doi.org/10.52202/078352-0005

We demonstrate:

- Excellent ballistic performance of a Dyneema® single jersey knit with a low areal density.
- Numerical modelling alongside ballistic experimental results suggest that Dyneema® single jersey knits outperform other knits where data was available in literature.
- Summarising, the form factor of knits for ballistic protection is well suited to those applications at lower areal densities requiring high mobility.

The numerical modelling method provided for investigating the behaviour of knits matched experimental back face deformation well, whilst trends in shock wave propagation were also captured. This type of 1D modelling has fast run times and avoids difficulties in approximating through thickness properties of yarns. As this method does not capture through thickness behaviour, element instabilities of solid elements under large deformations in a dynamic impact environment are also avoided. However, drawbacks include the inability to capture stress variation across the yarn thickness and its contribution to a desirable failure model. As such, a failure model is currently not incorporated.

Acknowledgments

The authors would like to acknowledge the support of Bart Schoonenberg, Mildred Janssens, and Rene Raets for the help in production and testing of single jersey knit fabrics. We would also like to thank Ilse Maes and Pierre Noyrez for their support in imaging and measuring the knitted constructions for details required for numerical modelling input.

References

[1] H. van der Werff and U. Heisserer, "High-performance ballistic fibers: ultra-high molecular weight polyethylene (UHMWPE)," in *Advanced Fibrous Composite Materials for Ballistic Protection*, Woodhead Publishing, 2016, pp. 71–107.
[2] a. Dwivedi *et al.*, "Continuous filament knit aramids for extremity ballistic protection," *28th Annu. Tech. Conf. Am. Soc. Compos. 2013, ASC 2013*, vol. 1, pp. 767–777, 2013.
[3] A. K. Dwivedi, M. W. Dalzell, S. a. Fossey, K. a. Slusarski, L. R. Long, and E. D. Wetzel, "Low velocity ballistic behavior of continuous filament knit aramid," *Int. J. Impact Eng.*, vol. 96, pp. 23–34, 2016.
[4] P. M. Cunniff, "Dimensionless parameters for optimization of textile-based body armor systems," in *Proceedings of the 18th International Symposium on Ballistics*, 1999, pp. 1303–10.
[5] U. Heisserer and H. van der Werff, "The relation between Dyneema® fiber properties and ballistic protection performance of its fiber composites," in *15th International Conference on Deformation, Yield and Fracture of Polymers*, 2012, vol. 3, pp. 242–246.
[6] J. Lambert and G. Jonas, "Towards standardization in terminal ballistics testing: velocity representation," no. Technical report AD-A021 389. Ballistic Research Laboratories Aberdeen Proving Ground, MD, 1976.
[7] P. J. McKee, A. C. Sokolow, J. H. Yu, L. L. Long, and E. D. Wetzel, "Finite element simulation of ballistic impact on single jersey knit fabric," *Compos. Struct.*, vol. 162, pp. 98–107, 2017.
[8] "MIL-DTL-46593B," no. July. 2006.
[9] M. K. Hazzard, R. S. Trask, U. Heisserer, M. Van Der Kamp, and S. R. Hallett, "Finite Element Modelling of Dyneema ® Composites : From Quasi- Static Rates to Ballistic Impact," *Compos. Part A Appl. Sci. Manuf.*, vol. 115, pp. 31–45, 2018.
[10] L. Nguyen, T. Lässig, S. Ryan, and W. Riedel, "A methodology for hydrocode analysis of ultra-high molecular weight polyethylene composite under ballistic impact," *Compos. Part A Appl. Sci. Manuf.*, vol. 84, pp. 224–235, 2016.
[11] M. Cheng, W. Chen, and T. Weerasooriya, "Mechanical properties of Kevlar® KM2 single fiber," *J. Eng. Mater. Technol. Trans. ASME*, vol. 127, no. 2, pp. 197–203, 2005.

https://doi.org/10.52202/078352-0005

IED-threat protective knits based on UHMWPE-Dyneema® fibres: A systematic review

[1]K. S. Freier

[1]DSM Protective Materials (DPM), Urmonderbaan 22, 6167 Geleen, The Netherlands
katrin-sabina.freier@dsm.com

Abstract. Secondary fragment wounds of high-energy blasts are one of the major causes of injury in modern warfare. An increase in wounding based on explosively accelerated fragments is seen among police officers, Law Enforcement, as well as First Responders. If applied, current ballistic Personal Protective Equipment (PPE) protects the vital organs of the torso's core and the head against gun-shot impacts, while the destructive blast forces of Improvised Explosive Devices (IEDs) simultaneously affect the unprotected body regions with potentially life threatening end results. This study provides a systematic overview on the research and development for knitted textile solutions made of UHMWPE fibres, branded as Dyneema®, to address this identified gap in personal protection. The resulting fragment protective knits are developed to serve the need for armouring against secondary fragmentation wounding. For those, systematic applied knit forming elements the loop, tuck, and float enable the fabric to respond to various fragment types in the most optimum way. The balanced knit structures level the geometric deformation response, identified as crucial for an optimized penetration resistance behaviour. Protective performance is achieved through various configurations while also addressing the added considerations of environmental exposure as well as the micro-climate of the human body in activity and rest. Knitted textiles are therefore developed to protect, while also providing comfort, identified as a being mission critical. Burden reduction is key for the wearer. The typical stretch characteristics of the knits give the protective textiles flexibility and drapability supporting the overall comfort while enabling integration into PPE systems at current unprotected areas. The specific yarn selection, knit patterns, and finishing processes result in knits providing the lowest possible burden on the wearer. Low weight, flexibility, moisture-, heat stress- and hygiene-management, create next-to-skin comfort without compromising mobility of the wearer. Fabric durability and launderability enable easy product maintenance.

1. BACKGROUND

One of the main threats faced in today's conflicts are Improvised Explosive Devices (IEDs). Besides the affected international military and peacekeeping missions, small IEDs are utilized in violent demonstrations causing accelerated fragmentation wounds to the local forces such as police officers and first responders. Though most vital body parts are covered by current ballistic protection systems, still about 70% of the body remains ballistically unprotected providing an extensive surface for serious detonation wounding, Figure 1 [1]–[3]. Resulting secondary injuries are characterized by high-energy fragmentation wounds containing a wide range of retained foreign materials within the body as the studies of Centeno *et al.* [4] show. However, only limited clothing system exist that addresses potential soft tissue wounding caused by IED weaponized soil and environmental debris, despite the fact that these fragments can cause traumatic amputation and/ or excessive injury leading to physical disabilities, long-term health care, and fatalities [4]–[6]. Consequently, protection against IED detonation threats is one key element for active personnel on combat missions to operate safely in an IED Environment.

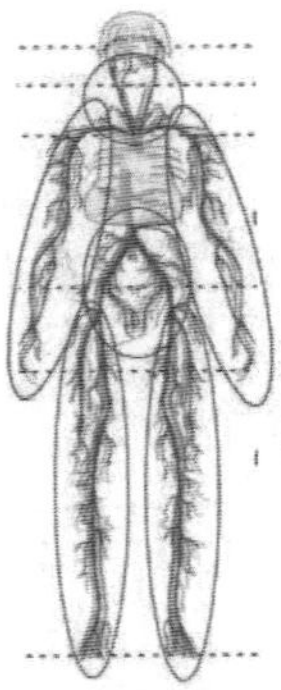

Figure 1: Fielded systems only cover vital organs (yellow) [3], while most of the human body remains unprotected (circled in blue)

In response to the identified protection gap, knitted textile structures are developed with the aim to broaden the protected area of the body for PPE clothing systems, complementary to the current ballistic body armour. The resulting knit constructions are made of high strength, low density, and high modulus Dyneema® yarn types. The knits respond to accelerated micro- to meso-mass fragments (referring to dirt, dust, debris up to 1.1 g fragment masses) holistically in the best possible way. At the same time, they provide high comfort for the wearer. Consistent performance properties under varied conditions have governed the development process as the textile solution will be integrated into PPE systems used in different environments under diverse mission activities. In order to effectively mitigate the ballistic fragmentation while enabling full mobility and comfort, DSM Protective Materials has studied the fragment impact into ductile knitted textile structures by the means of FEM modeling and supporting the results by empirical experiments leading to optimized textile fabric construction parameters, as e.g. introduced in the studies of Hazzard *et al.* [7].

 https://doi.org/10.52202/078352-0006

Most protective materials for soft armour are either UD type of materials, or woven constructions. Here, we demonstrate the performance of knitted structures made with Dyneema® fibre from the ballistics, but also in terms of comfort, and ability to integrate into PPE system solutions providing fragmentation protection to close the protection gap.

2. METHODICAL APPROACH ON FRAGMENT PROTECTIVE KNIT DEVELOPMENT

The stress strain behaviour of knitted textiles is one of the key-parameters in response to the forces of impacting fragments. Once the fragment contacts the target structure, it delivers kinetic energy which dissipates as momentum transfer as well as mechanical work to the textile. In response, the knit follows the characteristic stress strain behaviour of ductile materials, driven by the knit architecture. Generally, the load leads to change in length (Δl) of the fabric system allowing elastic and plastic deformation, absorbing the impact energy through mechanical work and momentum transfer. Single jersey knits are the most basic knit construction and have been applied previously for investigations of ballistic protective performances in comparison to other textile structures on a weight basis [7]–[9]. However, modifications of the knit architecture enable the fabric structure to respond to the impacting fragment in an optimized way. Considering the three main pattern forming elements loop, tuck, and float (missed stitch) **Figure 2**, the dynamics of the structural response can be tailored.

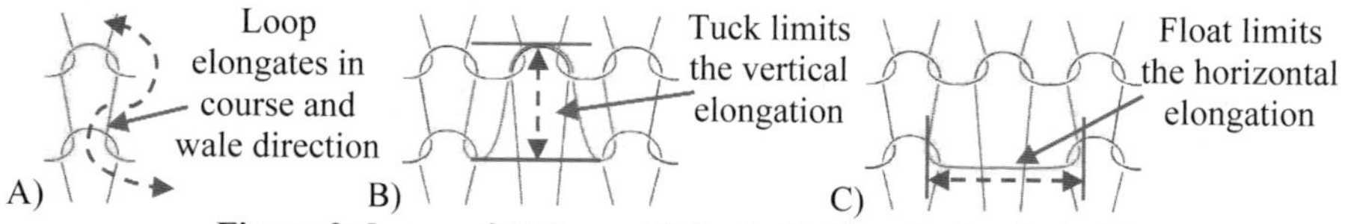

Figure 2: Image of A) Loop; B) Tuck; C) Float ('missed stitch')

In knits, each loop shapes along a 360-degree path (Figure 2 A) that enables high stretch of the structure in comparison to woven constructions where the warp and weft yarns interlaced at right angles limiting the natural extension. The loop course allow vertical and horizontal elongation (Figure 2A), while tucks set the boundary to the stretch in the wale (vertically) (Figure 2B), and floats in the course direction (horizontally) (Figure 2C). The fabric stretch is further steered with the loop height on the needle-bed and fabric pull-down within the manufacturing process that helps to determine the fabric density [10], [11]. By arranging the structural elements within the knit pattern, stiffness and elongation become controlled elements in the plane while the structural density and flexibility are balanced. The applied yarn type, thickness, and its surface structure further influence the characteristics. The present development process took advantage of these features to steer the physical properties in response to the impact forces of accelerated fragments.

Empirical ballistic studies:
Knit constructions have been developed varying in yarn type and knit pattern arrangement. To investigate the influence of the individual construction parameters, the V_{50} of the textiles have been determined according to STANAG 2920, with a 40 x 40 cm specimen size of which the edges were ducktaped for handling before framing, (Figure 3). 8- and 16-plot shooting patterns are shot with the RCC 0.13 g (2 grain) and 1.1g (17 grain) FSP creating empirical data sets. High-speed camera images show the structural behaviour in the process of collision allowing to draw further conclusions on the impact behaviour and knit structural response.

Figure 3: Image of 40 x 40 cm framed knit specimen in V_{50} testing with the 1.1 g FSP against air.

Test process:
Once the accelerated fragment is in contact with the target, the strain waves spread radially outward from the impact location along its surface. The elongation and tensile toughness of the knit fabric control the dynamics in response to the impacting fragment. During impact, compression initially occurs within the protective material at the impact point as momentum is transferred to the knit. The impacting forces spread longitudinal and perpendicular to the structure's axes. Stress-strain waves move through the

https://doi.org/10.52202/078352-0006

system producing shear and tensile forces. The energetic forces spread at the speed of sound engaging a larger area of the knit.
Figure 4 shows the fragment impact process into a knit structure with its physical reaction and dynamics to the penetration force of a 1.1g FSP tested against air, captured with the highspeed-camera.

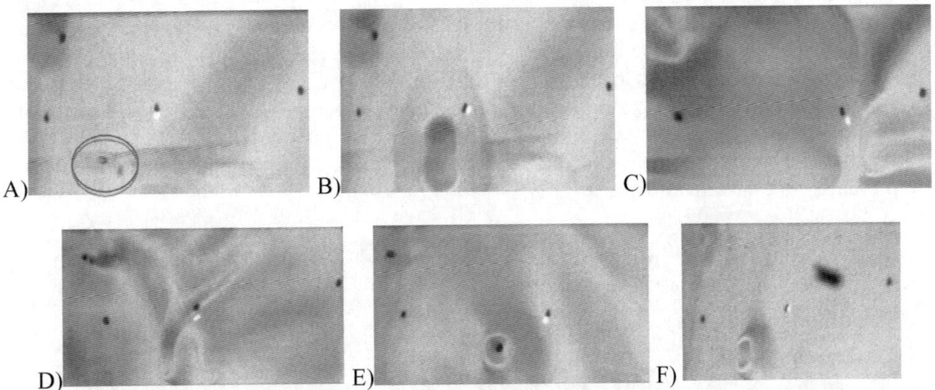

Figure 4: High speed camera sequence of 1.1g FSP impacting a knit structure placed against air. A) the fragment just before impact on the knit, B) once the fragment impacts, in-plane strain wave spreads outwards (light grey ellipse) followed by the (dark) transverse out-of-plane deflection, C) the knit structure elongates further D) at its maximum expansion the structural wave flows back to the centre of impact E, F) finally encapsulating the fragment within the structure.

Within the process of knit structure development general key findings have result:

A) The yarn properties and characteristics govern the fragment stopping process.
B) A loose construction leads to yarn pull out rather than stopping the fragment as well as constructions that offer too limited stretch. Figure 5 shows the yarn pull out in comparison to Figure 6 presenting a balanced stopping behaviour of the knit.
C) The failure criteria of the direction with least stretch in the plain define the resistance of the total construction.
D) The fabric areal density (AD) does not govern the stopping mechanisms alone, Figure 7 and Figure 8**Error! Reference source not found.** show the V_{50} levels of knit constructions on a weight bases, comparing the constructions A and B under impact of 2 grain (0.13 g) RCC and 17 grain (1.1 g) FSP, tested according to STANG 2920 (against air, 16-plot shooting pattern). The results show that the AD is not the leading factor for fragment resistance.

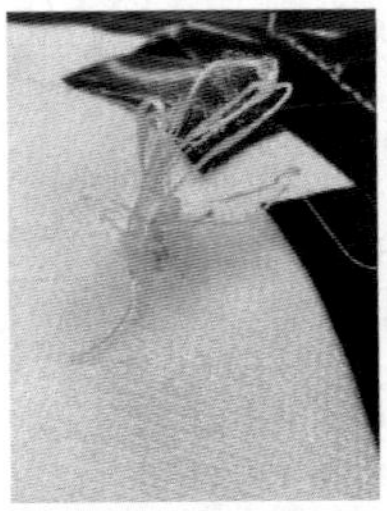

Figure 5: Yarn pull out after fragment impact.

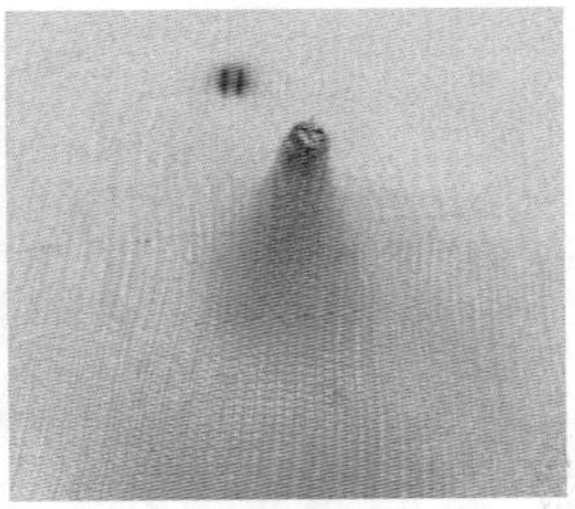

Figure 6: Balanced elongated knit has stopped the fragment.

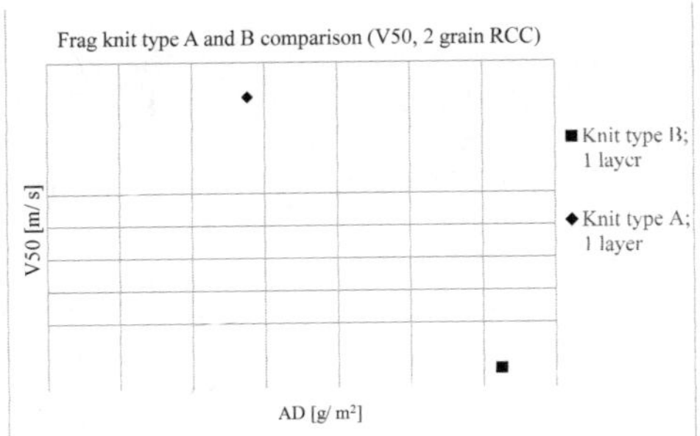

Figure 7: V_{50} (2 grain RCC) comparison of knit type A with B; (N = 5; 3-Sigma)

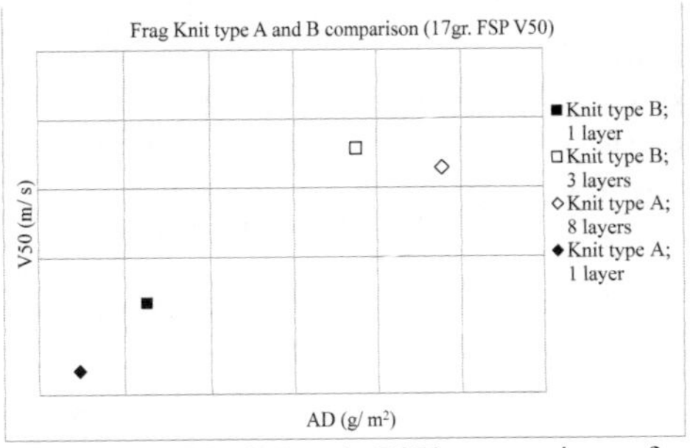

Figure 8: V_{50} (17 grain FSP) comparison of knit type A with B; single layer; multiple layer on a weight basis; (N = 5; 3-Sigma)

https://doi.org/10.52202/078352-0006

E) Higher elongation increases the contact duration with the impacting fragment which in return increases the energy transfer and therewith increases the V_{50} level.
F) The typical stretch characteristics of the knits give the protective textiles flexibility, and drapability supporting the overall garment comfort.

However, the elongation (Δl) of the structure is a critical aspect being considered in the developments, as pencilling of the fabric must be limited in response to tensile toughness of the body's soft tissue. Without limitation, what can occur is that although the fragment will not penetrate it pushes the fabric beyond its original plane of protection and into the protected tissue passing the tensile strength of the underlaying tissue creating penetrations/ injury as demonstrated by Breeze *et al.* [12], [13] within the development studies on related neck-protections. Pencilling and back face signature in clay blocks are shown in Figure 9 for the knit structures.

A) B) C)

Figure 9 shows a maximum of 20 mm pencilling back face signature after impact of 1.1 g FSP against A) air, and B) clay block. C) Analysis of the imprinted geometry in the clay witness.

Overall, the developments show that balanced elongation enables to increase the amount of kinetic energy being absorbed and dissipate into the knit structure. The structure's density (ρ), in alignment with its elastic modulus (E) dictate the energetic wave transmission within the knit. It is influenced by the yarn paths, and intersection points within the knit construction, while the yarn properties further govern the resistance. The tests show that immediately after impact, significant strain concentration develops under the centre of the impacting fragment, while the constructions allow longitudinal stress waves to spread into the plain. Transverse waves lead to deformation in fragment impact direction. The transverse wave speed (u) differs with changes dependent upon fragment size, mass, and geometry. It is identified that the knit structures respond differently under the change in dynamic loads. Consequently, the physical knit properties are adapted to the fragment types within the development process by differentiating the 0.13 g RCC threat and the 1.1 g FSP threat to provide optimized protection against those.

When the yarn is fully engaged by the projectile, the knit properties go hand in hand with the applied yarn characteristics. During elongation and initial shock, the applied yarn types in the construction take the lead to resist the impactor. The intrinsic yarn properties become the governing factor. The fibre must serve the resistance into the same manner as the textile construction does to achieve the maximum in protection. Knits made of high strength, low density and high modulus yarn absorb the kinetic energy of the intruding fragment in the best possible way, referring to Heisserer and Werff [14] highlighting the potential of UHMWPE fibres in terms of ballistic performance due to their specific properties. Within the knit structure, the yarn characteristics must support the fabric motion to allow optimized dynamics in response to the impacting threat. Low yarn on yarn friction of filament yarns enable a smooth structural movement and a quick structural response to the dynamics of the fragment motion.

Results:

Applying these findings resulted in two knit categories based on UHMWPE Dyneema® yarns and blends with such. Both categories offer a range of knit fabrics to also address individual secondary comfort properties. They are suitable to be applied as single as well as multiple layer systems or being hybridised subjected to the level of protection required and way of integration into the PPE systems.

1. **Knit category 1**: The base knits range between 230 – 300 g/ m² addressing the 0.13 g RCC fragment at a V_{50} of about > 250 - 320 m/ s suitable to be designed into light weight garment solutions as single layers, integrated as multiple layers, and used as lining systems.

2. **Knit category 2:** The advanced knits range between 450 – 650 g/ m² addressing the 1.1 g FSP at a V_{50} of about > 250 - 310 m/ s developed to provide advanced protection suitable to create flexible liners or inserts for PPE systems as single layers, multiple layer constructions, as well as hybrid solutions.

https://doi.org/10.52202/078352-0006

Differences in knit architecture, and underlying yarn mechanical properties result in different stopping values in response to the individual threat. Multiple layer constructions in Figure 10 as well as layer combinations of different knit-types Figure 11 can result in an optimized V_{50} subjected to their elongation behaviour.

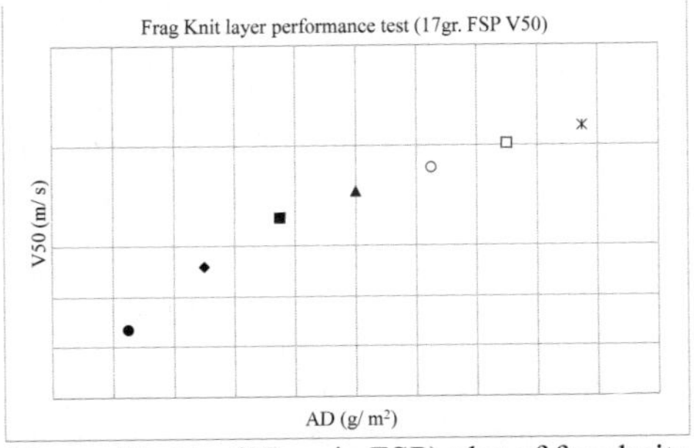

Figure 10: V_{50} (17 grain FSP) plot of frag knits subjected to step wise increase in number of layers (N = +1)

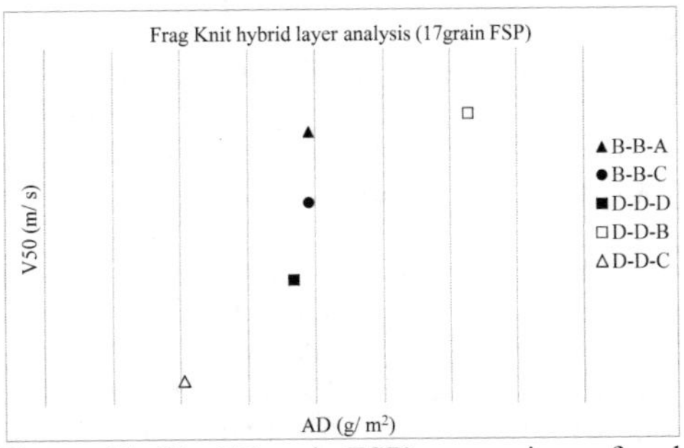

Figure 11: V_{50} (17 grain FSP) comparison of multiple layer combinations of different knit constructions on a weight basis.

3. AN OVERVIEW OF THE PROTECTIVE KNITS COMFORT PROPERTIES

For current PPE users such as military, police officers, and LE, action and rest alternate over prolonged duration, which also applies for first responders. They must carry heavy loads while being agile and mobile within urban terrain as well as confined spaces; therefore, additional burden must be avoided. Uncomfortable clothing can significantly decrease work performance and could become a critical issue for health and safety under these conditions. In worst cases, protective products are rejected to be worn. Thus, comfort has been identified as being mission critical in addition to the protective performance features required. However, comfort is determined by individual preferences in relation to personal wellbeing and individual needs, thereof it cannot be standardized in general. In that regard, three core influencers on comfort are identified that are addressed in the development process of the protective knit products:

1. The environmental/climate conditions
2. The person's active, rest, and stress conditions
3. The near environment (the clothing next to skin)

The human body's response and adaption to these influencers over short-term and prolonged wearing durations are considered for the development process. As a result, the overall knit performance requirements are defined in Figure 12 of which the comfort parameters are down selected to the attributes A – F, listed in Table 1.

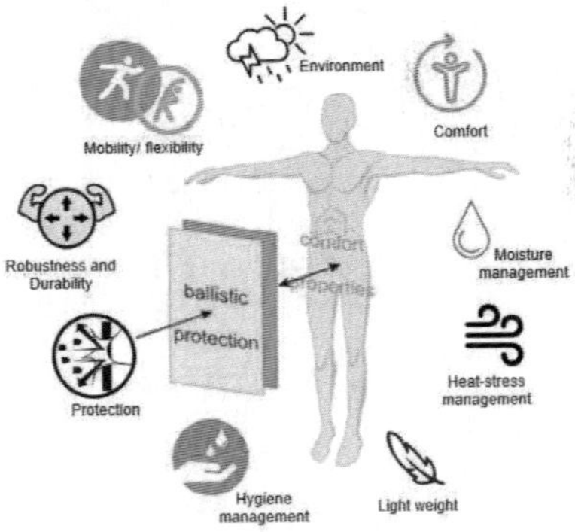

Figure 12: Protective knit performance properties

Comfort attributes for protective knit textiles	
A	Mobility
B	Durability
C	Heat- stress- management
D	Thermal management
E	Moisture- management
F	Sensory response to the textile constructions and hygiene management

Table 1: Comfort attributes for fragment protective knits

Empirical fabric comfort studies:

To investigate the comfort of the protective knits, different studies and testing processes have been carried out. Besides wear trials of cloth made of the fabric types, various testing methods have been down selected to empirically study and analyse the fabric behaviour simulating wear and use of the products.

 https://doi.org/10.52202/078352-0006

3.1 Comfort attributes for fragment protective knits

A) Mobility:
Active personnel might be on a mounted or dismounted mission. They must march, run, hide, crawl, lay in prone position, cross rivers, manoeuvre in confined spaces in urban terrain or within their armoured vehicles. It is of key importance for the present user-group being agile and mobile in their clothing system. In case the clothing retains movement, activities cannot be carried out to the level required, performance is limited. Moderate movement of the body requires stretch of the cloth with up to 45 % as Shishoo *et al.* [15] introduce, referring to active sports. These stresses might intensify once the product is worn with increasing dynamics of the wearer.

The textile flexibility and drapability response to the demand for mobility and is addressed with the typical stretch characteristics of the knits in the present case. For the knit pattern, the loop itself is key as it elongates in course and wale direction. The high system dynamics are supported by the low yarn on yarn friction of the UHMWPE fibre types that directly response under motion and follow the dynamic change. By this behaviour, the knitted fabrics drape and form alongside the body's individual ergonomic shape and adapts to the movement of the wearer giving full freedom of mobility at high durability. To study the mechanical response of knitted structures to mobility, the tensile testing method DIN ISO 139334-1, the soft armour flexibility test PED-IOP-008, and the Gelbo test ISO 7854-C (ASTM F392) are downs selected. The tests are described in the following section B) durability as mobility and durability of the knit structures go hand in hand.

Overall, in conclusion the high flexible UHMWPE Dyneema® fibres based knits adapt to the user's dynamics and therewith allow full mobility.

B) Durability:
Different testing methods have been down selected to empirically study and analyse the fabric behaviour by simulating the response to dynamic stresses as well as potential wear off due to motion and use.

Besides the investigations on general stress strain behaviour within the tensile tester in accordance with DIN ISO 139334-1, the flexibility on knit concepts are tested with the ball burst tester according to the PED-IOP-008 (Soft Armor Flexibility Test at U.S. ATC [16]) referring to TR-13-003L, ASTM D3787-07 (2001), ASTM D1777-96 (2011)e l, and ASTM E4 13 (Figure 13[16]) [16], [17]. By bending the fabrics under pressure load the force that is needed to deform the fabric gets determined within repetitive cycles, with the result that the less force that is applied to deform the fabric the higher the flexibility is ranked. The dynamic flex testing proves the extreme flexibility of the developed knit structures even for multiple layer constructions. As an example, a layered knit system bends under 1.045 N (3.76 lbf) compared to woven systems that bends only at about 3.89 N (14 lbf) on a weight to V_{50} basis.

Figure 13: Dynamic flex testing, PED-IOP-008 [16].

The Gelbo-test ISO 7854-C (ASTM F392) is generally applied to test the wrinkle resistance of foil materials and to define the fabric's permeability after dynamic stresses (Figure 14). In the present case the Gelbo-test ISO 7854-C (ASTM F392) has been chosen to investigate the knit behaviour under twisting dynamics. In specific, the test shell simulates the fabric dynamics within the bending regions of the body when groin, knees, arm pits, elbows, and joints are in motion. During the test, the fabrics get repeatably twisted and crushed (Figure 14). As of a result of the test the Dyneema® based knits show no fabric, nor yarn destruction after > 2700 cycles in the Gelbo tester.

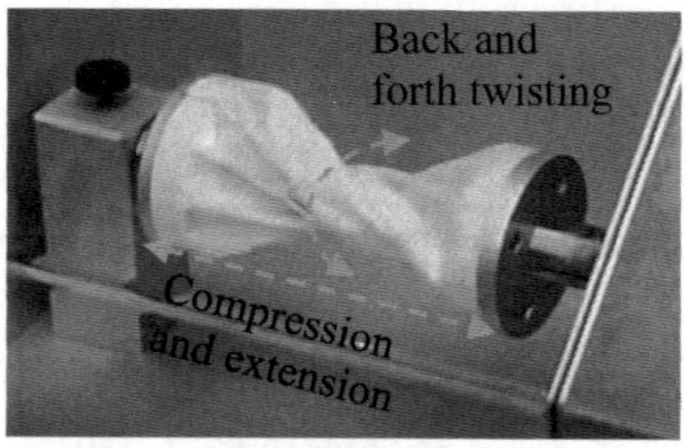

Figure 14: Gelbo test ISO 7854-C (ASTM F392) test twist durability of the knit sample.

https://doi.org/10.52202/078352-0006

The durability of the knits regarding fabric and or fiber destruction as well as pilling behaviour is key when they get exposed and must function under challenging wear conditions. The resistance to wear-off is further investigated with the inflated Diaphragm Test (ASTM D 3886) as well as with the Martindale test according to ISO 12947-2 (ASTM D 4966) Figure 15. Here, the use of different abradant materials have proven the high durability under friction showing no yarn nor loop destructions, and no pilling after > 150.000 cycles Figure 15 B), up to only slight fibral wear on the fabric surface was detected for some of the specimens Figure 15 C), that showed no change or reduction in strength and resistance within the tensile tester afterwards.

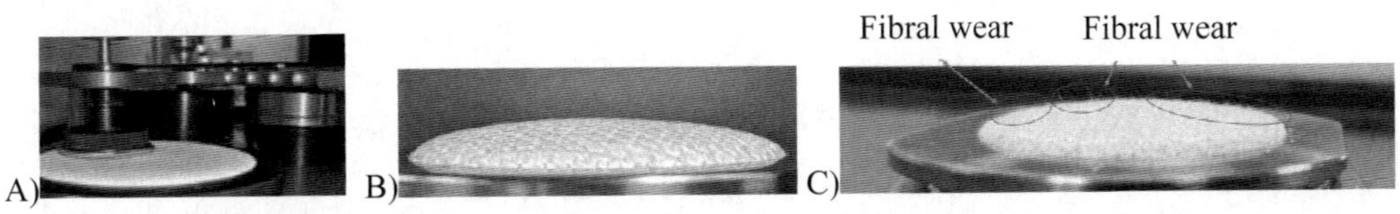

Figure 15: Martindale test ISO 12947-2 (ASTM D 4966); A) testing the abrasion resistance of the knit fabric sample; B), C) Knit specimen after abrasion > 150.000 test cycles, B) shows no wear, C) shows slight fibral wear.

The PPE user is commonly active in various environments in which abrasive, sharp objects or flora and fauna might work to compromise the clothing system. Besides the high abrasion resistance, the durable knit structures are constructed to resist mechanical impact forces such as cut, slash and tear based on the high tenacity of the UHMWPE Dyneema® fibres.

Besides the use, the manufacturing process of knitted textile goods is carefully studied regarding durability, as the loop forming processes as well as the high angled loop courses in the final structures create intense stresses on the applied yarn systems. High tensile strength yarn properties especially perpendicular to the yarn axis are key to resist the stress loads, avoid yarn and filament breakage and withstand the process stresses and abrasion.

Overall, in conclusion UHMWPE Dyneema® fibres based knits meet high durability standards as tested.

C) Heat-stress- management:

The user-groups wearing PPE systems encounter the full bandwidth of environmental and climatic conditions The human body constantly generates heat. Heat balance must be maintained to secure the body core temperature level and keep the work energy at moderate range. Heat exchange between the body and the near environment takes place at the skin surface by convection, radiation, conduction, and evaporation (
Figure 16) [18], [19]. Once the body becomes off balance the temperature rises or falls. The body counteracts that increase in the energy demand; a drop in performance and endurance with premature fatigue might result [20]. Consequently, it is key that the clothing system adapts to the climatic conditions of the user supporting to maintain the core temperature while balancing heat production and heat loss.

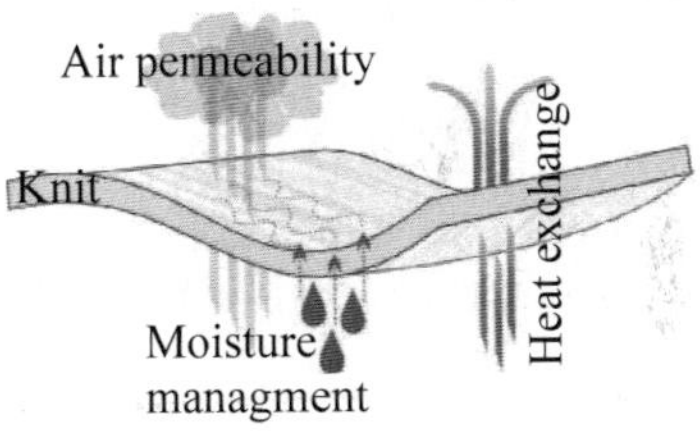

Figure 16: Climate management properties of the protective knit.

It was identified that moisture and thermal management are key parameters the knitted fabrics must manage in order not to create additional burden to the user. The impact on thermal comfort are the result of the textile parameters: fabric weight, thickness, porosity, moisture regain, air permeability and density -all being governed with the knit pattern design. Overall, subjected to the fabric application within the PPE system, the developed knits respond differently to the physical requirements. While the knit category 1 type of base knits is worn next to skin, climate exchange is key. As the knit category 2 type of knits are used in packages and inserts mainly of multiple layer solutions, climate exchange is defined as being secondary and might differ to the category 1 knits.

https://doi.org/10.52202/078352-0006

D) Thermal management:

Next-to-skin the thermal insulation and the barrier to moisture transport by the fabric layer affect comfort. The lab analysis chosen to assess the thermal physiological comfort index for the base knits of category 1 is the "Sweating guarded hot plate method" (Skin model) measuring the RCT (Thermal Resistance Coefficient) and the RET (Evaporative Resistance Coefficient), as well as the Imt (Water vapor permeability) in accordance with ISO 31092:1996 (ASTM F1868). The results have influenced the knit pattern design with the outcome that the lightweight base constructions provide a high level in air-permeability in parallel (Figure 17). Tested in accordance to DIN EN ISO 9237 (100 Pa) an average level of > 1000 l/ m²s is reached for the single layer fabrics, tested under non-stretched textile conditions. Here lower AD and lower fabric thickness increases the passage of air being transported through the textile construction in comparison to the category 2 knits of increased density reaching an average level of about > 250 l/ m²s. A general increase of the values are seen once the fabrics get stretched.

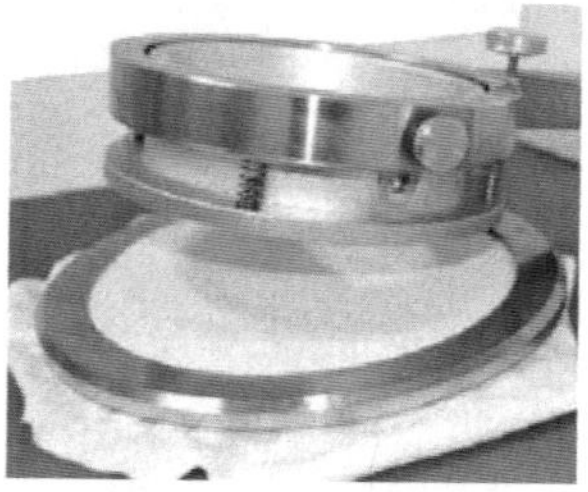

Figure 17: Air permeability test according to DIN EN ISO 9237.

Besides the physical properties of the knit construction, the thermal physiological comfort is addressed utilizing the applied UHMWPE Dyneema® yarns providing unique thermal properties due to their high thermal conductivity along the fibre. The orthorhombic PE enables the surplus heat to flow away from the body through the textile structure as the frag knits have intrinsically low thermal resistance. They provide a high thermal conductivity along the fibre axis of about 20 W/mk [21]. This translates into cool touch and feel while more heat is exchanged with the environment for body climate regulation Figure 18.

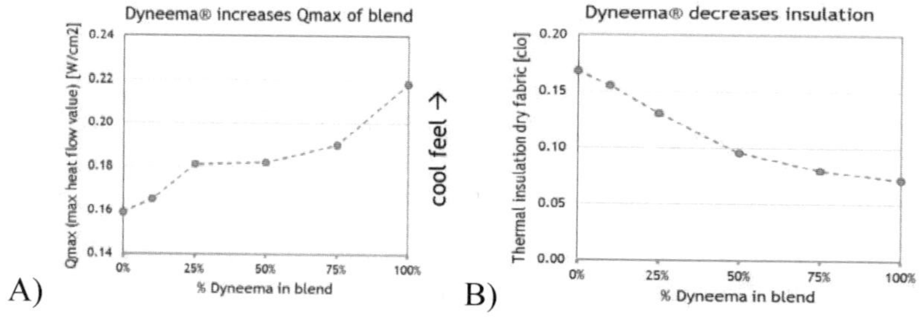

Figure 18: A) Heat Transfer Subjected to the Share of Dyneema® Fiber Influencing the Cool Feel of the Textile, B) Thermal Insulation in Relation to the Dyneema® Fiber Share in the Textile.

Overall, in conclusion the UHMWPE Dyneema® based knits show thermal management properties that are interesting / positive for protective garments.

E) Moisture management:

The protective knit constructions that are made to be worn next to skin provide a capillary effect due to the 3-dimensional pattern construction. Once sweat is produced it is absorbed into the porous open structure of the left fabric side Figure 19 and transport through the pores to the more close surface structure Figure 20 where the surplus in body moisture spreads and evaporates. A delay in moisture transport supports the body climate conditions, while configurations of the structure enable to steer the wetting process. Figure 21 shows the process of droplet absorption and wetting-effect of two different knit specimens made with Dyneema® fibre varying in its architecture regarding pore-size as well as permanent finishing showing a more hydrophilic behaviour in Figure 21: 1 A) – 1D) and rather hydrophobic properties for the knit in Figure 21: 2 A) – 2 D).

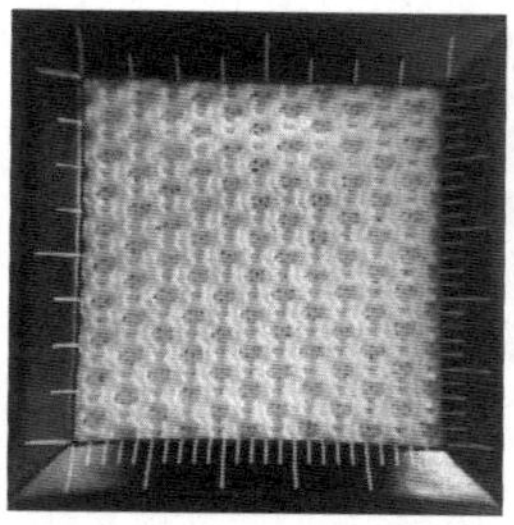

Figure 19: Example of left fabric side with porous structure

Figure 20: Example of right fabric side with closed surface structure

https://doi.org/10.52202/078352-0006

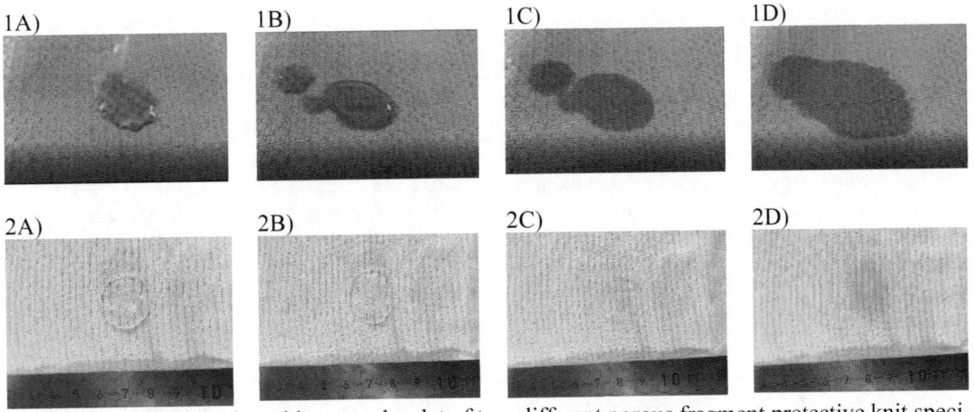

Figure 21: Wetting series with water-droplet of two different porous fragment protective knit specimens made with Dyneema® fibre, 1 A) droplet line breaks during application, sorption into the knit starts (0-1 s); 2 A) clear round droplet with sharp border lines (0-1 s), 1 B) droplet border line breaks up further, sorbs into the knit (2 s); 2 B) clear droplet border line, capillary sorption starts within the centre of the droplet into the structure pores (3 – 4 s), 1 C) wetting, the droplet spreads into the structure (2 – 3 s); 2 C) the droplet is transported into the pores (4 – 5 s); 1, 2 D) complete wetting (1 D) 5 – 8 s; 2 D) 3 – 4 s)

In conclusion, the UHMWPE Dyneema® based knits perform hydrophobic to hydrophillic properties based on the type of knit construction that is adapted to the requirements in the application areas.

F) Sensory response and hygiene management:
The moment the textile fabric gets in direct contact with the skin, comfort is affected by touch, friction, drape, and tactile characteristics, so called sensorial comfort perception [19]. The sensory feel of textile materials is related to mechanical stimulation of the sensory skin receptors by thermal affects, pressure, and friction forces [19]. Here, it is identified that the comfort next to skin is based on the sensory response of the individual and is therefore subjective to the wearer. Besides the described test methods, no additional test was identified to test, rate, level and empirically study those properties. It is concluded that the knit architecture in combination with the yarn selection applied for light weight, flexible structures subjectively result in a soft and smooth haptic with a cool touch as being comfortable to be worn next to skin. In addition, the Dyneema® fibre is ECO-TEX certified class II, and as such is certified as being compatible to be worn next to skin.
The chemical inert nature of the UHMWPE-fibre in general does not attract germs keeping the fabric hygienic. Hygiene management is further addressed with simple garment washing under regular household laundering conditions while keeping the full protective- and comfort properties. The moisture transport as well as quick drying characteristics of the knits supports the properties.

Overall, in conclusion the UHMWPE Dyneema® based knits show good sensory response and hygiene management properties that are interesting / positive for protective garment solutions.

4. CONCLUSION

This paper provides a systematic overview of fragment protective UHMWPE Dyneema® based knits. It highlights the key construction elements loop, tuck, and float applied in knitted textiles to reach fragment protective properties while addressing comfort aspects at the same time. It is identified that AD is not the dominating factor in ballistic resistance, but it is a combination of the fabric dynamic response to the impactor as well as the applied fibre properties. Although, the body's soft tissue stress-strain characteristics limit the knit elongation parameters and therewith the stretch that positively influences the energy absorption. The differences in knit architecture, and underlying yarn mechanical properties result in distinct stopping capabilities in response to the deviating threats: the 0.13 g (2grain) RCC fragment and 1.1 g (17 grain) FSP. Overall, this work has resulted in two knit categories based on UHMWPE Dyneema® yarns and its blends of which the category 1 type of knits, the base knits, address the 0.13 g RCC fragment at a V_{50} > 250 - 320 m/ s, and the category 2 type of knits, the advanced knits, address the 1.1 g FSP at a V_{50} of > 250 - 310 m/ s.

https://doi.org/10.52202/078352-0006

At the same time, these textile structures have the knit-typical features providing wear comfort to the individual. As such, the yarn selection, knit pattern, and finishing processes are chosen to provide the lowest possible burden on the wearer. Low weight, flexibility, moisture-, heat stress- and hygiene-management, and the sensory response to the textile constructions create comfort without compromising mobility. Fabric durability and launderability with quick drying capabilities enable easy product maintenance. The performance properties have been proven within in the broad experimental studies of which the present paper highlights the development and analyse on the key features levelling fragment protection and comfort as being relevant to the PPE user. As of a result, the UHMWPE Dyneema® based knit solutions are commercially available and are globally in evaluation and/ or applied in different PPE garment system solutions.

Acknowledgments
The authors would like to acknowledge the support of Mark Hazzard, Mirre van der Kamp, Dirk Louwers, Monte Nagy, Eva Osborne, James Rogers, William Roovers, Ross Rozansky, Sophie Wray, the Stanley TC Team, and the Heerlen TC Team, for the help in product research, analysis and support.

References

[1] L. Col, D. Tong, and R. Beirne, "Combat Body Armor and Injuries to the Head, Face, and Neck Region: A Systematic Review," *Mil. Med.*, vol. 178, no. 4, pp. 421–426, 2013.
[2] D. Carr and E. A. Lewis, "Ballistic-protective clothing and body armour," in *Protective clothing: managing thermal stress*, F. Wang and C. Gao, Eds. Cambridge, UK; Waltham, USA; Kidlington, UK: Woodhead Publishing, 2014, p. 500.
[3] K. Freier, "Prepared for the next mission abroad? - Blücher's fragment protective clothing (FPC) system closes the gap in current personal protective equipment (PPE)," *C-IED Report; Delta Bus. Media Ltd.*, no. Autumn 2015, pp. 91–96, 2015.
[4] J. A. Centeno *et al.*, "Embedded fragments from U.S. military personnel--chemical analysis and potential health implications.," *Int. J. Environ. Res. Public Health*, vol. 11, no. 2, pp. 1261–78, Jan. 2014.
[5] B. Kneubuehl, R. M. Coupland, M. A. Rothschild, and M. Thali, *Wundballistik: Grundlagen und Anwendungen*, 3rd ed. Heidelberg, Germany: Springer Medizin Verlag, 2008.
[6] P. F. Mahoney, J. M. Ryan, A. J. Brooks, and W. C. Schwab, *Ballistic Trauma. A Practical Guide*, 2nd ed. London, UK: Springer-Verlag, 2005.
[7] M. Hazzerd, U. Heisserer, M. van der Kamp, and K. Freier, "Knitted Fabrics with Dyneema® Fibers for Ballistic Protection: Modelling and Experimental Validation of a Single Jersey Knit," Geleen, The Netherlands, 2020.
[8] A. Dwivedi *et al.*, "Continuous filament knit aramids for extremity ballistic protection," in *28th Annu. Tech. Conf. ASC 2013*, 2013, pp. 767–777.
[9] A. K. Dwivedi, M. W. Dalzell, S. A. Fossey, K. A. Slusarski, L. R. Long, and E. D. Wetzel, "Low velocity ballistic behavior of continuous filament knit aramid," *Int. J. Impact Eng.*, vol. 96, pp. 23–34, Oct. 2016.
[10] K. F. Au, *Advances in knitting technology*. Cambridge, UK: Woodhead Publiching Limited, 2011.
[11] K.-P. Weber and M. O. Weber, *Wirkerei und Strickerei: technologische und bindungstechnische Grundlagen*. Dt. Fachverl, 2004.
[12] J. Breeze, L. C. Allanson-Bailey, N. C. Hunt, R. Delaney, A. E. Hepper, and E. A. Lewis, "Using computerised surface wound mapping to compare the potential medical effectiveness of Enhanced Protection Under Body Armour Combat Shirt collar designs.," *J. R. Army Med. Corps*, vol. 161, no. 1, pp. 22–26, Mar. 2015.
[13] J. Breeze, "Design validation of future ballistick neck protection through development of novel injury models," University of Birmingham, 2015.

https://doi.org/10.52202/078352-0006

[14] U. Heisserer and H. van der Werff, “The relation between Dyneema® fiber properties and ballistic protection performance of its fiber composites,” in *15th International Conference on Deformation, Yield and Fracture of Polymers*, 2012, vol. 3, pp. 242–246.

[15] R. Shishoo, *Textiles in sport*. Woodhead Pub. in association with the Textile Institute, 2005.

[16] “The United States Army Aberdeen Test Center.” [Online]. Available: https://www.atec.army.mil/atc/. [Accessed: 27-Aug-2020].

[17] “NTS Chesapeake (Belcamp, MD) | Ballistic & Materials Testing Lab.” [Online]. Available: https://www.nts.com/location/belcamp-md/. [Accessed: 28-Aug-2020].

[18] V. T. Bartels, “Physiological comfort of sportswear,” in *Textiles in sport*, R. Shishoo, Ed. Boca Raton FL,USA: Woodhead Publishing Ltd. and CRC Press LLC, 2005.

[19] A. Das and R. Alagirusamy, “Improving tactile comfort in fabrics and clothing,” in *Improving comfort in clothing*, G. Song, Ed. Cambridge, UK; Waltham, USA; Kidlington, UK: Woodhead Publishing Limited, 2011.

[20] F. Wang and C. Gao, *Protective Clothing - Managing Thermal Stress*. Cambridge, UK; Kidlington, UK; Waltham, USA: Woodhead Publishing Limited, 2014.

[21] X. Wang, V. Ho, R. A. Segalman, and D. G. Cahill, “Thermal Conductivity of High-Modulus Polymer Fibers,” *Macromolecules*, vol. 46, no. 12, pp. 4937–4943, Jun. 2013.

https://doi.org/10.52202/078352-0006

Determining the maximum acceptable length of a Hard Ballistic Plate

R. Molloy, **S. Laing**, M. Jaffrey and A. Furnell

Land Division, Defence Science and Technology Group, 506 Lorimer St, Fishermans Bend, Melbourne, VIC 3207, Australia, sheridan.laing@dst.defence.gov.au

Abstract. Optimising the dimensions of body armour requires consideration of the trade-off between coverage and mobility. However, the acceptable limits of plate length and width against a wearer's anthropometry remain unknown, limiting our ability to properly assess this trade-off. The aim of this work was to study encumbered soldier mobility to determine the maximum acceptable length of a hard plate. Four experimental plate conditions were worn by 45 male Australian infantry soldiers: no plate (Condition A); a reference body armour system (B), the maximum acceptable length plate (C1) and the minimum completely unacceptable length plate (C2). Participants completed four range of motion (ROM) activities and four functional movement tasks comprising rifle handling tasks and wall, window and crawling obstacles. Outcome measures included the ROM measures, obstacle completion time and an interference rating scale. Conditions C1 and C2 were determined by participant interference ratings when assessed with a subset of 27 variable length plates, differing in 10 mm increments. Forty participants met the inclusion criteria. The mean ± SD maximum acceptable plate length (C1) and minimum completely unacceptable plate length (C2) were determined as 29 ± 32 mm and 79 ± 38 mm longer than the wearer's front length respectively. The C2 plate condition resulted in significantly less ROM and longer times on all obstacles than the C1 condition ($p < 0.05$). Similarly, ROM and obstacle performance with C2 was worse than with B for all measures except the wall obstacle. Participants had significantly less ROM and took longer on the crawl obstacle with the C1 plate compared to cleanskin (A). Minimum detectable change values were provided to assess meaningful differences. This study shows how the maximum acceptable length of a plate is related to the wearer's front length and that exceeding the acceptable length limits will result in decrements to soldier mobility and task performance.

1. INTRODUCTION

Hard Ballistic Plates (HBPs) in body armour carriers are a key element of personal protective equipment for the modern soldier. When positioned and sized correctly, body armour provides coverage of important thoracoabdominal organs and structures of the torso. However, this protection is not without cost. HBPs are made from heavy and rigid materials and the use of HBPs may reduce wearers' mobility, ability to rapidly take cover, and their capacity to carry out essential lethality tasks, such as sighting and firing a weapon [1-3]. It is therefore a well-supported position that more protection (coverage) equates to decrements in soldier performance (mobility). Optimising the dimensions of chest-borne HBPs requires consideration of the trade-off between coverage and mobility.

1.1 Coverage requirements

Body armour coverage requirements have been established for Australian soldiers based upon the position of vital thoracoabdominal organs relative to anthropometric landmarks, as identified from supine and standing MRI data [4]. Consequently, coverage requirements state that the positioning of the HBP should protect important thoracoabdominal organs (e.g. heart, liver, spleen, and great vessels) from a front-on, perpendicular threat [4]. The plate should provide as much coverage as possible while remaining acceptable to the user. The likelihood of (perpendicular) coverage provided by HBPs of varying lengths and widths has been defined for a number of important organs [4]. These findings are based on the assumption that the top edge of the HBP is positioned at the sternal notch of the wearer.

1.2 Defining mobility requirements

Body armour is worn by soldiers performing complex and varied physically demanding tasks, often for prolonged periods. The size and shape of body armour has the potential to interfere with the performance of these tasks, and subsequently impact the user tolerance of a body armour system. The user-accepted limits of plate length and width against an infantry soldier's anthropometry remain unknown. The aim of this study is to determine the relationship between user anthropometry and the maximum acceptable length of a HBP, located appropriately at the sternal notch for infantry soldiers. These data may be used to inform the ergonomic acceptability of a new size range of plates and also inform injury models with

https://doi.org/10.52202/078352-0007

real-life data about the maximum coverage that is likely to be tolerated by soldiers. This study represents a new approach to these issues, combining elements of a regular body armour comparison study with principles used during Fitmapping [5] for protective equipment and clothing.

2. METHODS

2.1 Participant and anthropometric measures

Forty-five Australian infantry soldiers took part in the study. All participants were male with an average age of 24.5 ± 2.9 years and average time in the Army of 3.6 ± 2.6 years. Ethical approval to conduct the study was granted (protocol LD 02-17) in accordance with the DST Group low-risk human research ethics review process. The torso anthropometric measurements of the cohort well reflected the variance within the Australian Army male population [6] (Table 1).

Table 1. Torso anthropometric measures of the male Australian Warfighter Anthropometry Survey (AWAS) and study cohorts. 'Rank' indicates the percentile ranking of the min/max study cohort value within the normalised AWAS cohort.

	AWAS		Study Cohort					
Measures (mm)	Mean	SD	Mean	SD	Min	Max	Rank of min	Rank of max
Front Length	362	21	367	20	325	419	6th	99th
Chest Circumference	1010	74	1024	82	889	1285	5th	> 99th
Waist Circumference	888	94	880	91	735	1099	5th	99th

The front length is the vertical distance from the sternal notch to the top of the iliac crest. The chest and waist circumference measures are measured at the nipple and navel level respectively.

2.2 Participant assessments

2.2.1 Range of motion tasks

Three formal static ROM assessments were completed: seated torso flexion, lateral torso flexion and seated rotation with flexion. Measures were taken with participants sitting on a 600 mm high box, atop a 400 mm high platform. The base of the box was considered 0 mm and reaches measured below this point described as positive values. A further functional ROM activity was completed in a Bushmaster vehicle seat. Participants performed a seated forward reach; the horizontal distance of the reach was measured from the front of the bushmaster seat base (0 mm). These ROM movements were selected as they were anticipated to be most influenced by a change in plate length. ROM measures were recorded up to three times (average 2.3 measures) and the mean ROM value calculated and used in analysis.

2.2.2 Functional movement tasks

Following ROM tasks, participants completed four dynamic activities; selected for their criticality of function and expected ability to discriminate different plate length. The first three are obstacles from the Load Effects Assessment Program (LEAP): Wall, Window and Low Crawl. These activities were timed. The fourth dynamic task was a simulated marksmanship exercise with the F88 rifle. Participants took prone-, kneeling- and standing aim postures. The rifle exercises were not timed or measured, and were used solely to inform participants' overall interference rating for each system.

2.2.3 Outcome measures

Objective participant outcome measures comprised the measured ROM values and timed obstacle activities. Participants also completed a subjective interference rating assessment (Figure 1) at the conclusion of all ROM and functional movement tasks.

https://doi.org/10.52202/078352-0007

The length of this plate caused….				
No interference or degradation	Slight interference: easily worked around	Moderate interference; difficult, but able to work around	Severe interference, very difficult to work around	Extreme interference. Unable to work around; unacceptable
1	2	3	4	5

Figure 1. Interference rating scale, adapted from the scale used by Mitchell et al. (2017) [7]

2.3 Experimental design

The trial was a repeated measures partially-counterbalanced study across four experimental conditions; Cleanskin i.e. no plate (Condition A), a reference body armour system (Condition B) comprising a carrier, training HBP and training soft armour, and a variable-length hard plate and carrier (Condition C). Condition C comprised a maximum acceptable length plate (C1) and completely unacceptable length plate (C2).

Condition B was a fixed-length condition. One size of the reference body armour system was used for all participants. The training soft armour inserts for this system are longer and wider than the HBP, and the plate was held within the bounds of the soft armour, such that a minimum of 20 mm of soft armour bordered the plate in all directions.

Condition C was an adaptive condition determined using surrogate hard ballistic plates of 27 different lengths. The plates were modelled on the 3D shape of the reference front HBP (Condition B). The plates varied in length in 10 mm increments. The width of all plates was held constant (Figure 2). Trial plates were an approximate match in areal density to the training HBP, resulting in the mass of each system scaling with size; a 10 mm length increment corresponded to a mass delta of approximately 65 g. The soft armour inserts were designed to match the dimensions of the trial plates in length and width, were approximately 8 mm thick and had similar stiffness to real soft armour. Carriers were custom-made to house the plates and surrogate soft armour inserts, such that the external carrier length was 10 mm longer than the plate. Carriers were designed to enable maximum adjustability. Comparisons between Conditions B and C are presented in carrier length rather than plate length, in order to account for differences in system design.

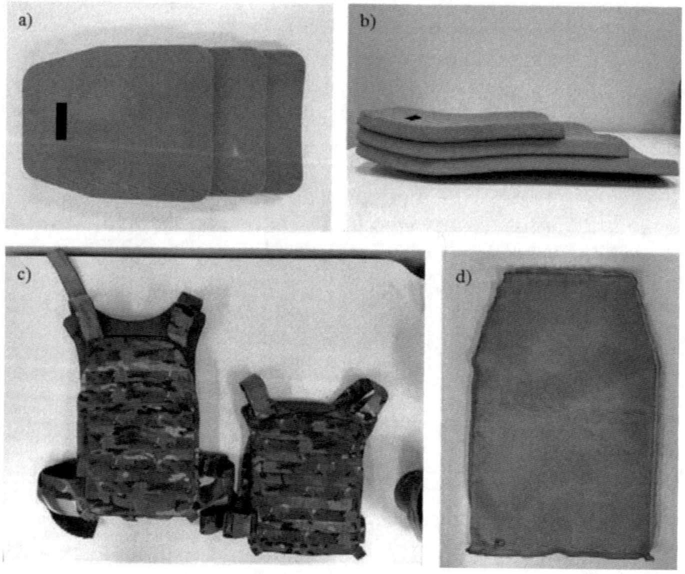

Figure 2. a) Top view of plates 1, 10 and 17, **b)** side profile of plates 1, 10 and 17, **c)** carriers for plates 10 and 1, and **d)** soft armour insert for plate 10. N.b. Plates 10 and 17 were 90 mm and 160 mm longer than Plate 1 respectively.

https://doi.org/10.52202/078352-0007

It was anticipated that the crossover point between ratings 2 and 3 on the interference scale (Figure 1) would represent the maximum acceptable length a wearer would be willing to accept (C1) and that the crossover point between ratings 3 and 4 would represent a completely unacceptable length for the wearer (C2). This assumption was verified by an initial questionnaire asking the participants how much interference they would be prepared to accept (with any body armour). This was completed prior to any participant assessments in the test conditions.

The plate lengths corresponding to interference ratings of 2, 3 and 4 were desired from every participant so that soldiers' subjective experience of wearing the plates could be captured and subsequently compared with the objective measures. Determining the thresholds of interference was the main aim of the study. However, because it was not clear where these points would occur for any individual, an adaptive approach was devised based on the principles of Fitmapping (Figure 3, [5].

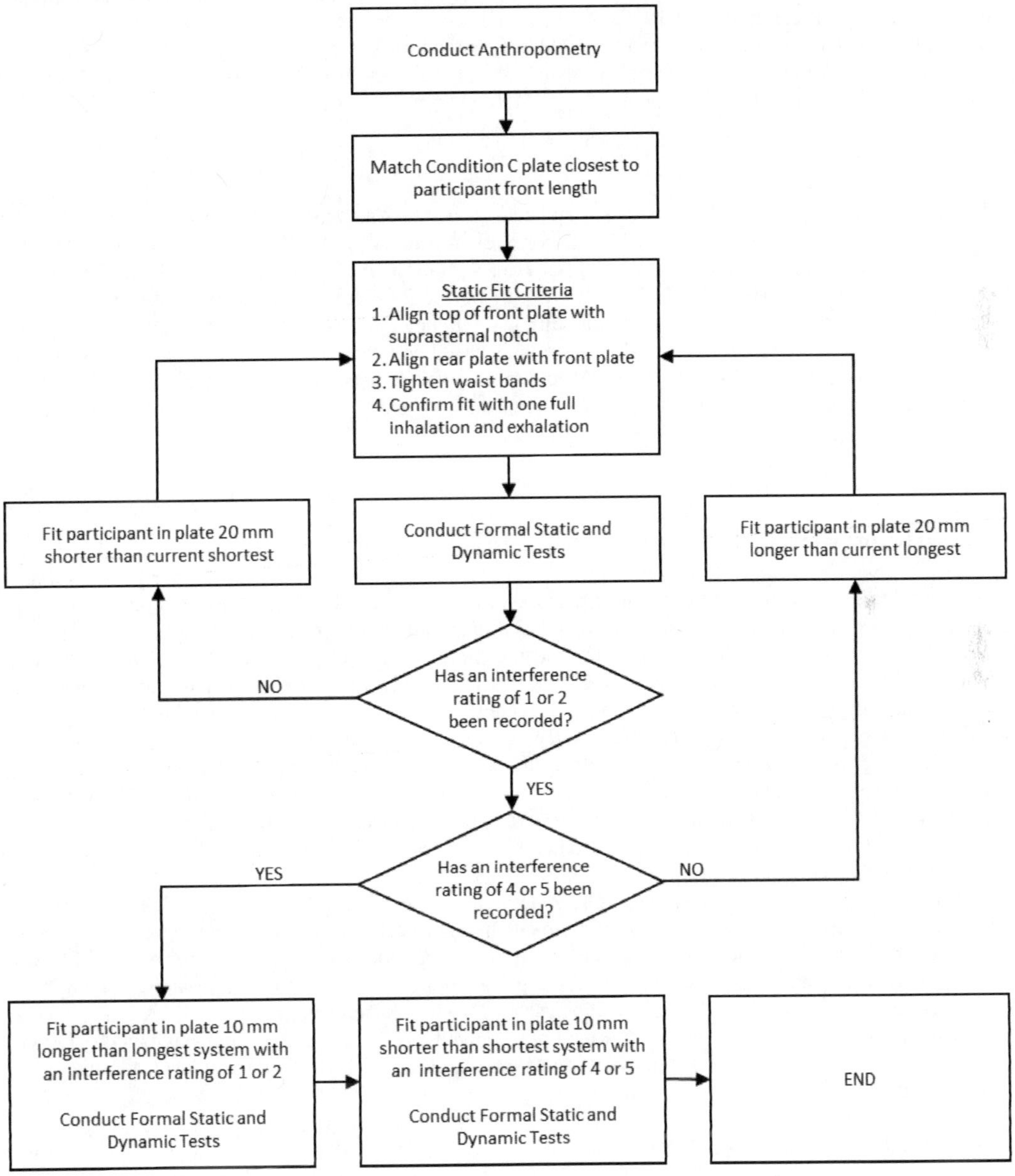

Figure 3. Condition C protocol and decision flow chart

https://doi.org/10.52202/078352-0007

The first plate length trialled in Condition C was chosen as the length closest to each participant's front length. Subsequently issued plates were either 20 mm longer or shorter, dependent on the subjective interference ratings of each trialled length. Once ratings of 2, 3 and 4 had been established at 20 mm increments, plate lengths at the crossover points from the ratings of 2 to 3, and then 3 to 4, were trialled in 10 mm length increments to increase the resolution at these key junctures. Once this data was collected, the protocol ceased. The decision to initially vary plate lengths in 20 mm rather than 10 mm increments was made to make more effective use of limited participant time and lessen the effects of boredom, fatigue and disengagement that may have affected results during a lengthier trial progressing in 10 mm increments only. The decision to run Condition C as an adaptive condition was deliberate. Although a regular counterbalanced approach is more regularly applied, this may not have yielded the results required to make decisions on maximum acceptable plate length since the conditions must be determined upfront and had to be relatively small in number. As such, resolution may be lost and the actual crossover points not established. It was not expected that trial activities would cause undue fatigue due to their short duration and discrete application, typically a cause of order effects in physical performance trials. Eighteen participants completed a randomised repeat run with a single Condition C plate which they had previously worn. This data was used to test for order effects caused by trial design, and to calculate the minimal detectable change (MDC) for each measure.

2.4 Data analysis

The data were analysed with a one-way repeated-measures Analysis of Variances (ANOVA). The Shapiro-Wilk and Fmax test statistics were used to test the assumptions of normality and homogeneity of variance. Mauchly's test was used to test the assumption of sphericity; Huynh-Feldt Epsilon is reported where violations occurred. Holm-Bonferroni corrections were applied to pairwise comparisons.

The MDC represents the minimum magnitude of change that exceeds measurement error and was calculated from the reliability data of 18 participants based on a 90% confidence interval (i.e. MDC_{90}). This value was calculated for the four ROM and three obstacle measures. The MDC values were used to provide a threshold over which the performance can be considered to have meaningfully changed.

3. RESULTS

3.1 Identifying C1 and C2

Prior to all tasks and test conditions, participants were asked to record their judgement about the amount of interference to their mobility that they would be prepared to accept in order to ensure appropriate coverage and protection. The results supported the initial assumption that the maximum acceptable length occurred at the last rating of 2; the majority of participants (69%) reported that this was the greatest impediment they would be prepared to accept. A further 18% of participants reported a rating of 3 was the greatest acceptable impediment. Therefore, the Condition C plate with the last rating of 2 was identified for all participants as C1. Where there was one deviation in the responses (i.e. a longer plate was rated as lower interference), a conservative approach was adopted whereby the shortest plate rated 2 was taken as the maximum acceptable length (C1). No participants responded in the questionnaire that a rating of 4 would be acceptable; consequently, the completely unacceptable plate length (C2) was classified as the first or only ratings of 4 or 5.

The mean number of variable lengths to achieve ratings of 2, 3 and 4 was 6.96 ± 1.3. This relatively high number supports the assertion that participants had not been cognizant of the progression criteria, or were not minded to engineer their responses to complete the trial with the least number of runs. Checks for consistency of rating response to increases in plate length were perfect for 30 participants (66.7%), deviated on one occasion only for 12 participants (26.7%) and deviated on more than one occasion for 3 participants who were hence removed from any further analysis. A further 2 participants failed to record a rating of 4 or 5, and were also removed from further analysis. Thus, the data of n = 40 participants were included in the subsequent analysis.

The variable-length dataset (Condition C) comprised a total of 276 runs completed by the 40 participants. The Condition C ratings for each plate length were standardised by each individual's front length, i.e. plate length minus front length. As the plate became longer against the participant's front length, the likelihood of obtaining an interference rating of 2 decreased (Figure 4).

https://doi.org/10.52202/078352-0007

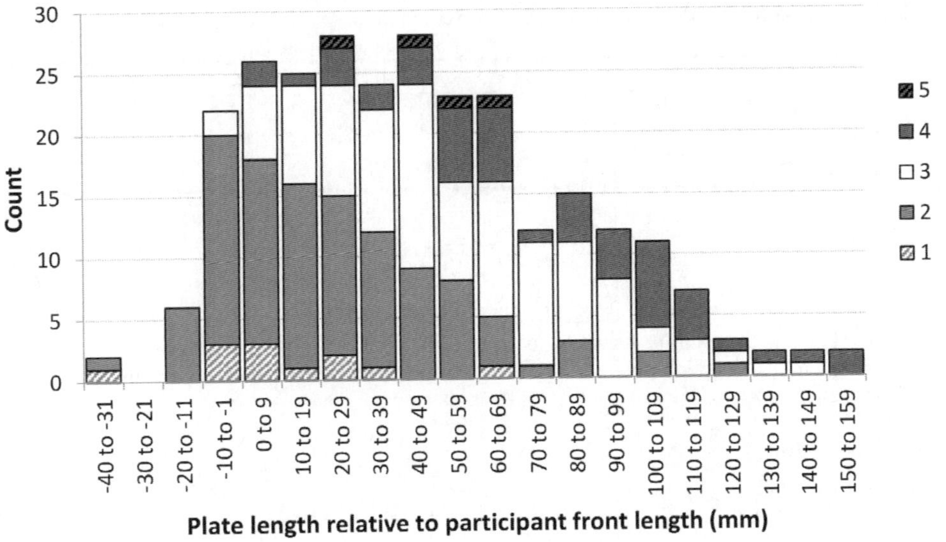

Figure 4. Interference ratings for each standardised plate length for Condition C (plate length minus front length) (total 276 runs)

The maximum acceptable plate length (C1) and minimum completely unacceptable plate length (C2) were determined as 29 ± 32 mm and 79 ± 38 mm longer than the wearer's front length respectively. There was a similar range for C1 and C2 values (142 and 155 mm respectively), indicating large variability and cross-over of the values across the participant sample (Figure 5). Both C1 and C2 datasets satisfied the criteria for normality (Shapiro-Wilk p-values of 0.272 and 0.567 respectively).

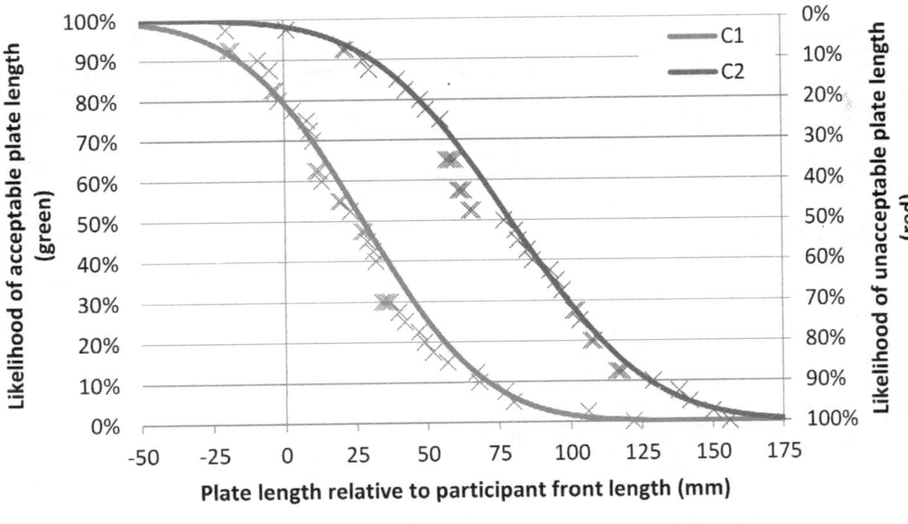

Figure 5. Normalised (line) and raw (cross markers) distribution of C1 and C2 standardised plate length thresholds across trial population (n = 40)

The trial carriers were 10 mm longer than the plates, therefore the maximum acceptable carrier length and minimum completely unacceptable carrier length were determined as 39 ± 32 mm and 89 ± 38 mm longer than the wearer's front length respectively (**Error! Reference source not found.**). By comparison, the fixed-length reference body armour system (B) was 6 ± 21 mm shorter than the wearer's front length.

https://doi.org/10.52202/078352-0007

Table 2. Relative lengths of reference system B, C1 and C2 carriers for n = 40 participants (carrier length less front length). All values in mm. Negative values indicate front length is longer than the carrier.

	Mean	SD	Min.	Max.	Range
B: Reference body armour system	-6	21	-59	35	94
C1: Acceptable carrier	39	32	-10	132	142
C2: Unacceptable carrier	89	38	11	166	155

3.1.2 Overall interference ratings

No interference ratings were completed for the cleanskin condition (A). For Condition B, the reference body armour system, the overall interference was most commonly rated a 2 (n = 27), with n = 11 participants rating it as 1 (no interference) and n = 2 participants rating it as 3 (moderate interference). By definition, all C1 plates were rated a 2 and all C2 plates were rated a 4.

3.2 Objective measures

3.2.1 Range of motion tasks

The MDC_{90} was calculated for each ROM task (Table 3). The four one-way repeated measures ANOVAs demonstrated significant main effects of body armour length on trunk flexion, lateral flexion, flexion/rotation and seated reach (all $p < 0.001$). Thus, pairwise comparisons were completed for all ROM tasks. Plate condition C2 caused statistically significant and meaningful restriction compared to conditions A, B and C1 for all ROM measures except lateral flexion (which was statistically, but not meaningfully, different). Compared to cleanskin (A), the C1 plate significantly decreased all ROM measures, however only flexion and flexion/rotation measures were meaningfully different. Plate condition C1 resulted in significantly less flexion, flexion/rotation and seated reach compared to B, however none of the measures were meaningfully different.

Table 3. Results of trunk ROM tasks (mean ± SD), all task and MDC_{90} values in mm (n = 40)

	Flexion	Lateral flexion	Flexion/Rotation	Seated reach
MDC_{90}	28.7	63.6	62.0	37.5
A: Cleanskin	69.6 ± 57.1	-105.5 ± 66.6	-110.2 ± 94.7	918.0 ± 83.0
B: Reference	41.4 ± 55.5	-112.0 ± 61.6	-152.4 ± 95.8	907.3 ± 83.9
C1: Acceptable	27.3 ± 60.1	-123.4 ± 67.7	-180.6 ± 100.2	888.7 ± 88.7
C2: Unacceptable	-20.2 ± 73.4	-144.4 ± 55.9	-262.0 ± 97.8	849.1 ± 99.5
Pairwise comparisons (p-values)				
A vs. B	< 0.001*	0.305	< 0.001*	0.078
A vs. C1	< 0.001*†	0.009 *	< 0.001*†	< 0.001*
A vs. C2	< 0.001*†	< 0.001*	< 0.001*†	< 0.001*†
B vs. C1	0.011*	0.069	0.010*	< 0.001*
B vs. C2	< 0.001*†	< 0.001*	< 0.001*†	< 0.001*†
C1 vs. C2	< 0.001*†	< 0.001*	< 0.001*†	< 0.001*†

*statistically significant at the Holm-Bonferroni-corrected alpha level
†difference greater than the MDC_{90}

3.2.2 Functional movement tasks

The MDC_{90} was calculated for each obstacle task (Table 4). The three one-way repeated measures ANOVAs demonstrated significant main effects of body armour length on the time taken to complete the wall, window and crawl obstacles (all $p < 0.001$). Thus, pairwise comparisons were completed for all functional movement tasks. All plate conditions were significantly different for the crawl obstacle. Plate condition C2 resulted in significantly longer time for all obstacles than Conditions A and C1, and for the window and crawl obstacles against Condition B. However, no differences between any conditions were deemed meaningful against the MDC_{90}.

https://doi.org/10.52202/078352-0007

Table 4. Results of functional movement tasks, task (mean ± SD) and MDC_{90} values in seconds (n = 40)

	Wall	Window	Crawl
MDC_{90}	2.07	1.57	2.77
A: Cleanskin	9.57 ± 1.66	8.52 ± 1.59	10.30 ± 2.56
B: Reference	9.96 ± 2.20	8.84 ± 1.93	11.30 ± 2.37
C1: Acceptable	9.87 ± 2.10	9.07 ± 1.96	12.05 ± 2.55
C2: Unacceptable	10.56 ± 2.39	9.97 ± 2.26	12.99 ± 2.89
Pairwise comparisons (p-values)			
A vs. B	0.107	0.119	0.007*
A vs. C1	0.129	0.026	< 0.001*
A vs. C2	0.001*	< 0.001*	< 0.001*
B vs. C1	0.637	0.245	0.029*
B vs. C2	0.016	< 0.001*	< 0.001*
C1 vs. C2	0.002*	< 0.001*	0.010*

*significant at the Holm-Bonferroni corrected alpha level
†difference greater than the MDC_{90}

3.3 Reliability testing

Eighteen participants conducted a repeat of one of their previously experienced conditions chosen at random. The test-retest correlation was assessed using Pearson's *r*. Most measures had good- to excellent reliability ($r > 0.8$ and $r > 0.9$ respectively) with acceptable reliability ($r > 0.7$) noted for the crawl time. All correlation coefficients were statistically significant with $p \leq 0.001$ for all measures. This data indicates that there were no observable order effects induced by the experimental protocol.

Table 5. Test-retest (mean ± SD) correlations for n = 18 participants

Measure	Test	Retest	Pearson's r	Significance
ROM Flexion (mm)	40.91 ± 41.03	46.54 ± 41.20	0.910	0.000
ROM Lateral flexion (mm)	-108.53 ± 78.38	-94.96 ± 65.25	0.870	0.000
ROM Flexion/Rotation (mm)	-198.35 ± 83.81	-190.51 ± 75.87	0.893	0.000
ROM Seated Reach (mm)	903.33 ± 64.85	908.36 ± 61.02	0.936	0.000
Wall Time (secs)	10.22 ± 2.33	9.93 ± 2.37	0.856	0.000
Window Time (secs)	9.15 ± 1.73	8.96 ± 2.00	0.878	0.000
Crawl Time (secs)	11.99 ± 2.13	12.30 ± 2.40	0.728	0.001

4. DISCUSSION

Measuring body armour length against user front length is a practical method of estimating the likelihood of user acceptability. Condition C1 results suggest that, as a plate extends below the user's front length, the likelihood the plate will be deemed accepted by the user decreases. A plate positioned at a user's sternal notch that is longer than front length will result in the bottom edge of the body armour extending below the top of the pelvis. Military body armour featuring a HBP is heavy and rigid, often worn tightly to the torso to prevent the armour system bouncing or shifting during walking. When extending below the top of the pelvis, the system might act akin to a splint, physically restricting or blocking movement of the torso over the pelvis (i.e. trunk flexion).

Choi et al. [8, 9] evaluated the effects of wearing Improved Outer Tactical Vests (IOTV) in the size above and below that identified by subject matter expert (SME) fit. The IOTV assessed was configured with hard plates at the front, back and sides and soft armour at the front, back yoke and collar. SME fit was based on visual inspection of ballistic plate and soft armour coverage in seating and standing. Standing carrier length was assessed relative to the navel. The average carrier length for SME fit was 76 mm below the navel. It was found increasing armour size beyond SME fit resulted in a movement penalty. User acceptance was not investigated. It is difficult to compare the results of the IOTV study with the current study due to the different anthropometric points of comparison. Additionally, IOTV systems increase in both length and width with each size, scaling primarily with chest circumference.

The current study used short, discrete measures of mobility to provide participants with exposure to movements and actions that would challenge any restrictions to trunk range of movement from body armour length, primarily to inform the subjective rating of interference. The measures themselves were

https://doi.org/10.52202/078352-0007

not intended to represent the full extent of movements performed by an infantry soldier wearing body armour. As such, although significant differences, including differences exceeding the MDC_{90}, were found between conditions, inferences on the degree to which this may functionally impact a soldier's performance in the conduct of their role are not made. However, the results do indicate that trunk ROM and performance in physical tasks reduced as body armour length increased, and that users may be prepared to accept longer HBP coverage than is currently provided by the reference armour used in the study.

The Condition C plate mass was not held constant. Instead, the mass was scaled to the plate dimensions by approximating the areal density of the Condition B training plate. This resulted in the acceptability of larger plates being negatively influenced by increased mass as well as length. As the ballistic performance of a HBP is a function its material composition, it was assumed that areal density would provide an applicable method of controlling the mass, i.e. any future recommendations for larger plate sizes will also result in heavier plates. Thus, the participant assessments herein, i.e. the interference rating, ROM and functional movement tasks, all consider both the change in geometric dimensions and change in mass associated with varying plate lengths.

The data herein can be used in conjunction with anatomical positioning data [4] to assess the protection and mobility afforded by body armour of various lengths. For example, the 50th percentile Australian Army male front length is 362 mm [6]; if the 50th male was assigned a HBP with length 400 mm, the normal distribution of the data herein would suggest that there is ~40% likelihood they would find the plate length acceptable. Organ mapping data suggests there is ~90% likelihood their inferior liver would be covered and ~80% likelihood their abdominal aorta down to the bifurcation would be covered by a plate of this length. Such assessments are limited to the consideration of perpendicular threats, the assumption that users are wearing their HBPs at the sternal notch and the assumption that the two small samples employed in the respective studies are representative of the wider Australian Army male population. However, these simple comparisons permit the rapid consideration of body armour length against the anthropometry of the target wearer.

This study was designed to investigate a relationship between user anthropometry and the maximum acceptable length of a HBP. Limitations to the applicability of these findings on body armour design include:

- Participants were from a single user group of male-only infantry soldiers. Thus, the results do not consider sex differences in anthropometry, or role-related mobility vs. protection trade-off requirements and preferences.
- The trial carrier was designed for quick and ready adjustment, and did not represent a deployable configuration. Integration with key soldier equipment, specifically load carriage (packs, webbing and pouches) was not attempted.
- Limited use of each system in discrete tasks was not representative of the requirements of an infantry soldier, with extended use in diverse environments, including both mounted and dismounted environments.
- HBP width variation was not included in this trial.
- Bespoke HBP and carrier dimensions will impose significant financial costs. Consolidation of protection and mobility requirements in a realistic range of sizes to suit a given population is not described.

Future research is planned to address these limitations. This study was funded by the Australian Department of Defence, as part of an ongoing effort to improve the fit, form, and function of equipment, to accommodate the diverse body shapes of Australian soldiers.

5. CONCLUSIONS

This study has shown how the maximum acceptable length of a plate is related to the wearer's front length and that exceeding the acceptable length limits will result in observable and objective decrements to soldier mobility and task performance. The subjective interference that users are willing to tolerate has been determined and the distribution of the plate lengths corresponding to these levels of interference provided. The data herein provides researchers with reference values for the evaluation of existing or proposed HBP lengths against male infantry populations of known anthropometry.

ACKNOWLEDGEMENTS

The authors would like to acknowledge the contributions of Dr Nathan Daniell in the conduct of the trial, and Dr Scott Michael and Dr Shahd Al-Janabi for assistance with the statistical analysis.

https://doi.org/10.52202/078352-0007

REFERENCES

1. Peoples G, Silk A, Notley S, Holland L, Collier B, Lee D. The effect of a tiered body armour system on soldier physical mobility. Centre for Human and Applied Physiology, Faculty of Health and Behavioural Sciences, University of Wollongong, UOW-HPL-Report-041, 2010.
2. Dempsey PC, Handcock PJ, Rehrer NJ. Impact of police body armour and equipment on mobility. Applied Ergonomics; 44(6):957-61, 2013.
3. Watson CH, Horsfall, I., Fenne, P. Ergonomics of body armour. Personal Armour Systems Symposium Quebec City, Canada, 13-17 September 2010.
4. Laing S, and Jaffrey, M. Thoraco-abdominal organ locations: Variations due to breathing and posture and implications for body armour coverage assessments, Land Division, Defence Science and Technology Group, Australia, DST-Group-TR-3636, 2019.
5. Choi HJ, Zehner G, Hudson J. A manual for the performance of protective equipment fit-mapping. Biosciences and Protection Division, Airforce Research Laboratory, AFRL-RH-WP-SR-2010-0005, 2009.
6. Edwards M, Furnell A, Coleman J, Davis S. A preliminary anthropometry standard for Australian Army equipment evaluation. Land Division, Defence Science and Technology Group DSTO-TR-3006, 2014.
7. Mitchell K.B, Choi HJ, Garlie TN. Anthropometry and range of motion of the encumbered soldier. Development and Engineering Center, U.S. Army Natick Soldier Research, NATICK/TR-17/010, 2017.
8. Choi HJ, Garlie T, Mitchell KB, Desimone L. Effects of body armor fit on warfighter mobility as measured by range of motion (ROM). In: Goonetilleke RS, Karwowski W (eds). Advances in physical ergonomics and human factors. AHFE 2018: Advances in intelligent systems and computing, vol. 789, Springer, 2018.
9. Choi HJ, Garlie T, Mitchell KB. Effects of body armor fit on encumbered anthropometry relative to bulk and coverage. In: Goonetilleke RS, Karwowski W (eds). Advances in physical ergonomics and human factors. AHFE 2018: Advances in intelligent systems and computing, vol. 789, Springer, 2018.

https://doi.org/10.52202/078352-0007

Oblique impact of a 7.62x39 mm projectile on ceramic-coated aramid plates

M.Seidl[1], H.Liao[2] and M.Kern[1,2]

[1]*French–German Research institute Saint-Louis (ISL), 5 Rue du Général Cassagnou, 68300 Saint-Louis, France*
[2]*University of Technology Belfort-Montbéliard (UTBM), Rue de Leupe, 90400 Sevenans, France*

Abstract. Projectile ricochet is an important protection mechanism. It refers to the deflection of a surface and the avoidance of perforation. It is predominantly used on armoured vehicles, where specific designs enhance the likelihood of ricocheting incoming threats. This study applies this knowledge to investigate specific structures and designs to enhance the likelihood of projectile ricochet on combat helmets. It investigates the potential of applying thin ceramic coatings on aramid plate. In cooperation with the University of Technology Belfort-Montbéliard (UTBM), using their different spray techniques, the ceramic layer was atmospheric plasma sprayed in different thicknesses on the aramid plates. Ballistic experiments were conducted at the French-German Research institute Saint-Louis (ISL), where a 7.62 x 39-mm projectile was launched at two different angles of obliquity. The impact velocity was approximately 600 m/s, simulating a shot from a 100m distance. A high-speed camera on the target plate strike face captured the ceramic layer spalling, while a digital image correlation measurement captured the dynamic back face deformation (BFD). This paper presents and discusses the differences in the maximum BFD of the ceramic-coated plates compared to non-coated ones.

1. INTRODUCTION

Helmets are an integral part of standard soldier protection equipment. The head, which is a vital zone, requires the best possible protection. However, the helmet mass must be kept to a minimum to ensure a high level of comfort. Heavy equipment during long missions could tire a soldier and affect their ability to move and detect threats. Recent international conflicts in Iraq and Afghanistan have demonstrated the need to increase the protection of soldiers' heads and necks [1]. The proportion of injuries in these parts of the body has increased by 30–50% within the evolution of conflicts, as shown in Figure 1. The increase in head and neck injuries was not further discussed in terms of the type of threats accrued.

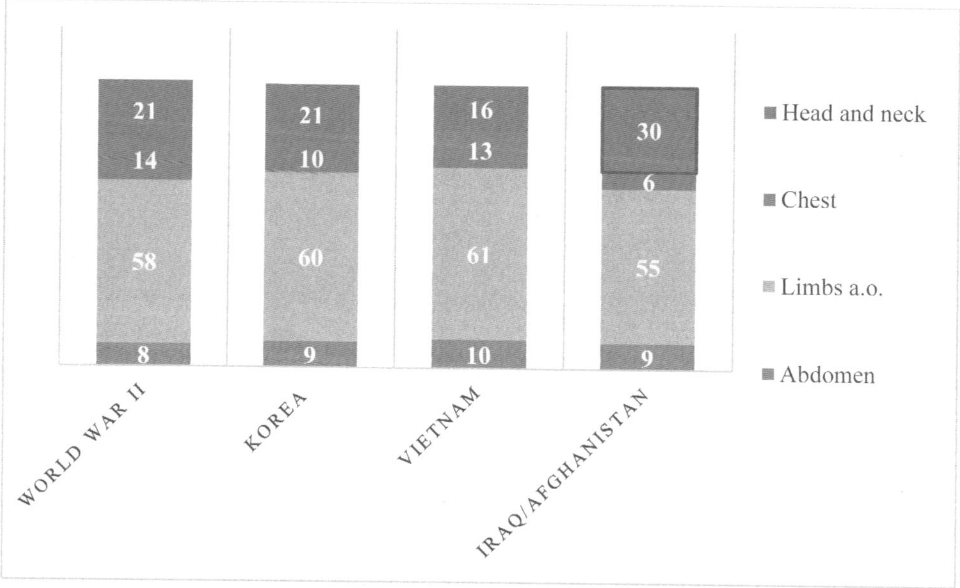

Figure 1. Percentage of various injuries of soldiers in different wars [1]

https://doi.org/10.52202/078352-0008

This work focuses specifically on the 7.62x39 mm projectile as there is a high probability of soldiers and armed forces facing such a threat during missions [2]. In order to design a helmet that can withstand a direct hit of such a threat, its mass would have to increase significantly, making it not feasible for daily use. One possibility to improve protection against a 7.62x39 mm without adding weight to the helmet is increasing the likelihood of projectile ricochet, e.g., through a hardened surface. In this study, ceramic spray coating was applied on aramid plates and helmets, and the protection capabilities were tested.

This work was conducted in cooperation with the University of Technology at Belfort and Montbéliard (UBTM) and the French-German Research Institute of Saint-Louis (ISL). The UBTM prepared and spray-coated the samples, while the ballistic tests were conducted at the ISL. This was a primarily experimental study to investigate the adhesion of a hard and brittle ceramic layer on a woven fibre composite plate – an aramid material. To ensure the adhesion of the ceramic coating on the composite plate, a thin layer of zinc was applied. The oblique impact and ricochet of 7.62.x39 mm was chosen as an applicable test scenario to test the performance of the ceramic coating. The hypothesis was that the harder the target surface, the lesser the penetration and the sooner the projectile would ricochet (Figure 2).

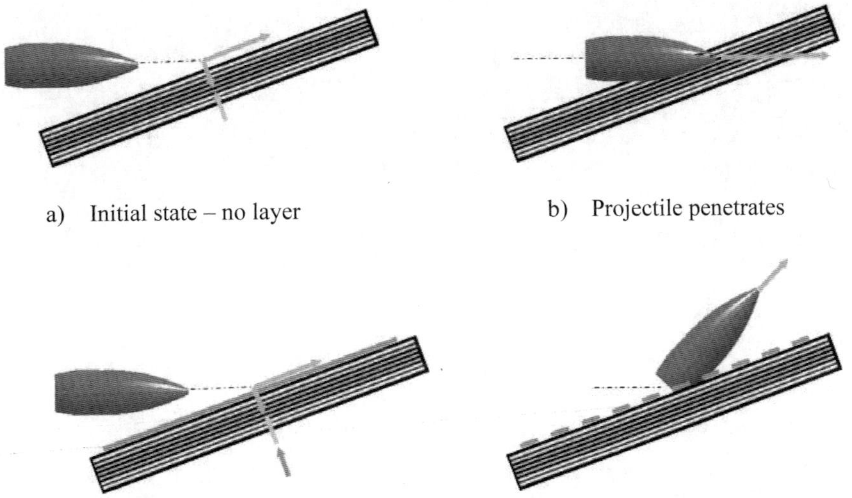

a) Initial state – no layer

b) Projectile penetrates

c) Initial state with layer (blue)

d) More resistance projectile ricochet upon layer fracture

Figure 2. Impact of the additional layer on ricochet

The influence of the ceramic coating on the projectile trajectory was measured using X-ray cinematography and back face deformation (BFD) of the aramid target plate using digital image correlation (DIC). Since the influence of experimental parameters, such as projectile nutation or target properties, on the projectile trajectory and, therefore, the ricochet event, might be higher than the influence measured due to the surface coating, the BFD was also measured. If the projectile ricochets earlier, it does not penetrate the target plate so extensively. As a result, there is less deformation on the back face. This experiment is a feasibility study to determine if the surface coating has an influence and, if so, if the measurement might be able to quantify it.

2. CERAMIC SPRAY-COATING

2.1 Sample preparation

For the ceramic spraying and ballistic testing, aramid plates were used, which are made from the same material and average thickness as combat helmets. The surface was roughened using a pneumatic sandblaster with pressure set at 3.5 bar to avoid damaging the fibres. The aramid plates were cleaned

 https://doi.org/10.52202/078352-0008

with ethanol to remove grease and other contamination to assure better adhesion of the ceramic coating during the projectile high-speed impact.

2.2 Binding layer

A layer of a zinc (Zn) binder was sprayed on the prepared and cleaned aramid plate surface for better adhesion between the aramid plate and its final ceramic layer. Zn was used due to its low melting temperature, which causes less surface damage on the aramid target strike face. The applied spray technique is called arc wire spray, wherein an electrical arc discharge is the source of energy melting the Zn (Figure 3, a). A 100 μm-thick layer of Zn was applied (Figure 3, b). This was the thinnest possible layer as the parameter was adhesion between the aramid plate and ceramic layer, which can be expected. The Zn layer was applied with a width of 100–150 mm over the full length. On average, an increase of mass of about 25–30 g was considered. A thorough investigation of the influence of the binder material, its thickness and the probable influence of different spraying techniques and settings was not part of this study. As a starting point, materials were chosen due to their availability.

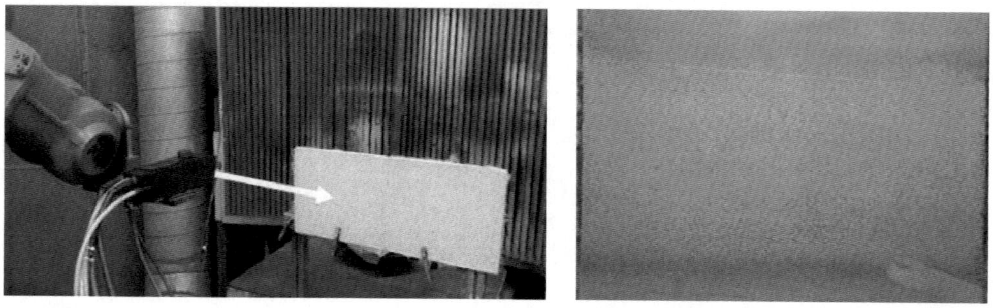

a) Spray coating at the UBTM b) Zn-layered sample

Figure 3. Arc-wired Zn coating

2.3 Ceramic coating

Two available zirconium oxide (zirconia, ZrO_2) powders were tested between 10–40 μm grain sizes. They were tested to determine if there was an influence of grain size on adhesion. A thickness from 300–1100 μm was applied (**Table 1**), where the larger grain size of ZrO_2 was deemed better for higher thicknesses. Other influences of the two different powders were not observed, which might be due to the limited sample size and primary test set-up.

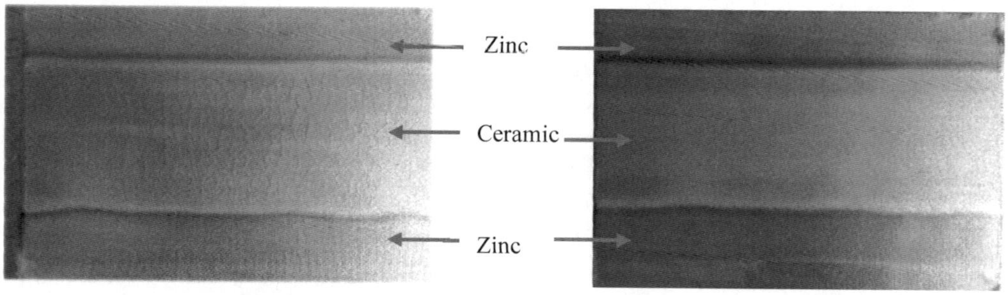

Figure 4. Plates coated with ZrO_2: (left) cracks and (right) after correction

To apply the ceramic coating on top of the Zn layer, atmospheric plasma spraying was chosen, as it is suitable to reach the high melting temperature of ceramics. In order to decrease the high temperature interacting with the aramid plate surface, the spray nozzle was placed further away, from the previous 110 mm distance to 165 mm. Additional fan cooling was applied. However, if the ceramics were cooled too fast, cracks appeared upon coming in contact with the Zn-layered composite plate (Figure 4).

https://doi.org/10.52202/078352-0008

2.4 Additional mass on combat helmet

The ceramic material had a density of 5.7 g/cm^3, which added a mass of 50–180 g to the roughly 750 g mass of those plates (Figure 4). To estimate the increase of helmet mass due to the added ceramic layers, the outer shell surface of a combat helmet was measured. The helmet was 3D-scanned, and its computer model was analysed. A complete spray coating of the outer shell surface was assumed. The 3D model was taken from a medium-sized helmet, keeping in mind that different sizes and models yield different results in terms of the mass increase. An outer shell of this type has an area of 1160 cm^2 (Figure 5). For ceramic coating of 300–1100 μm, the increase in mass was 200–700 g, where the radial increase of a rounded surface was neglected. In relation to a medium combat helmet with interior, the increase was 15–50%, depending on the thickness of the layer. This primary study did not consider the ergonomic aspects of additional helmet mass.

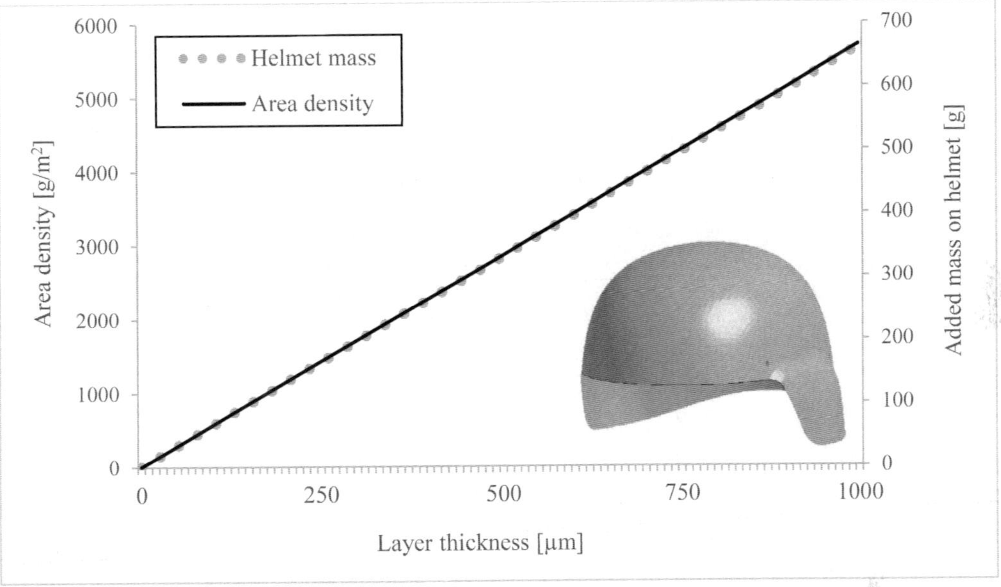

Figure 5. Increase in combat helmet mass based on the simulation model

3. BALLISTIC TESTS

3.1 Setup

The 7.62x39 mm projectile was launched at a constant initial velocity of v_i=610 m/s, simulating a shot from approximately a 100 m distance. The plates were inclined at 65° and 80° (NATO). These oblique angles were chosen based on previous studies, having 65°close to the critical ricochet angle and 80° being a grazing shot with as little deformation as possible. The DIC measurement technique was applied to quantify the BFD of the plates. Additionally, a high-speed camera was placed on the strike face to capture the fracture of the ceramic coating (Figure 6).

https://doi.org/10.52202/078352-0008

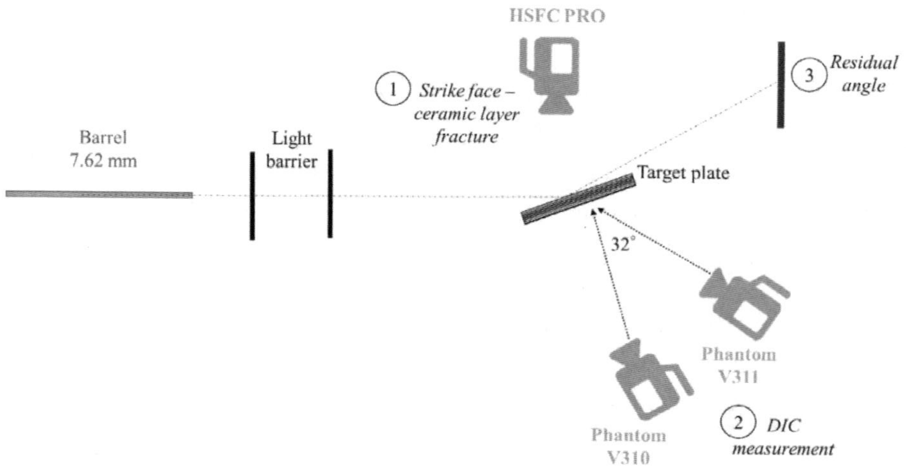

Figure 6. The experimental test setup

Table 1 summarises the test matrix of the investigated aramid plates. The plates were taken from one batch of 50 plates in order to decrease the potential influence of the production process as reference data were taken from plates with no layers. Plates at the upper 12 and 13 were chosen deliberately, and the middle ones were taken from the available samples. Four plates with only the Zn layer were tested in order to determine if the layer process had an influence on the ballistic plate performance. The twelve ceramic-layered plates and the type of powder utilised are presented in Table 1 with their different thicknesses.

Table 1. Test matrix of differently layered plates with initial conditions and residual properties

Layer	Plate number	Impact angle θ [°]	Ceramic thickness [μm]	Initial velocity [m/s]	Reflection angle ξ [°]	Max BFD [mm]	Note
No layer	22	80		606	77	7.0	
No layer	12	80		611	77	6.3	
No layer	13	65		615	58	11.0	
Zinc	8	65	100	613	-	12.0	D
Zinc	23	65	80	620	-	13.6	P
Zinc	24	65	80	621	66	12.1	
Zinc	25	65	160	615	-	-	P
ZrO_2 Y2O3	10	80	400	617	78	6.7	
ZrO_2 Y2O3	11	80	500	605	76	6.5	
ZrO_2 Y2O3	15	80	400	606	76	5.9	
ZrO_2 Y2O3	14	65	460	601	61	10.6	
ZrO_2 Y2O3	16	65	950	612	61	10.1	
ZrO_2 Y2O3	17	65	350	603	-	11.0	P
ZrO_2 Y2O3	18	65	290	606	61	-	
ZrO_2	9	80	400	606	76	5.9	
ZrO_2	19	65	350	606	57	-	
ZrO_2	20	65	620	605	-	12	B
ZrO_2	21	65	1100	611	-	10.7	P
ZrO_2	26	65	160	613	62		

Perforation P; Influence of boundary conditions B and Penetration D; else ricochet

https://doi.org/10.52202/078352-0008

3.2 Results

3.2.1 Ceramic layer fracture

Figure 7 shows plate impact at the two different angles with similar layer material, powder size, and layer thickness. Independent of the impact angle, the ceramic layer spalled after projectile penetration. Further investigation is required to determine if the additional ceramic layer reduces the impulse transition from the impacting projectile on the aramid plate.

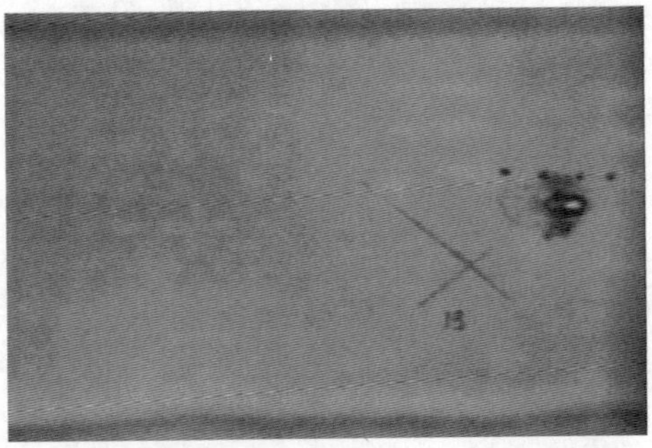

a) t=70 μs

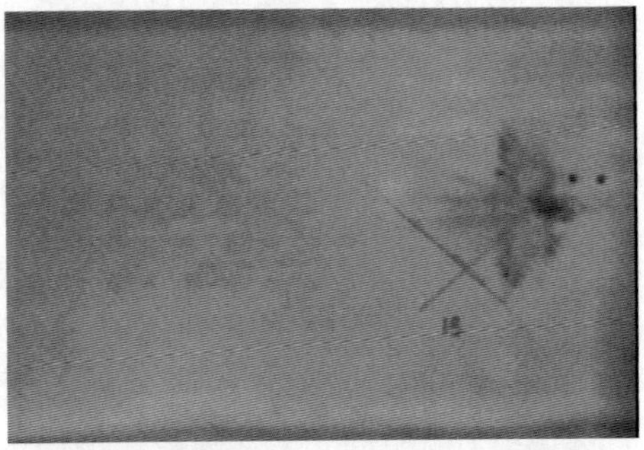

b) t=140 μs

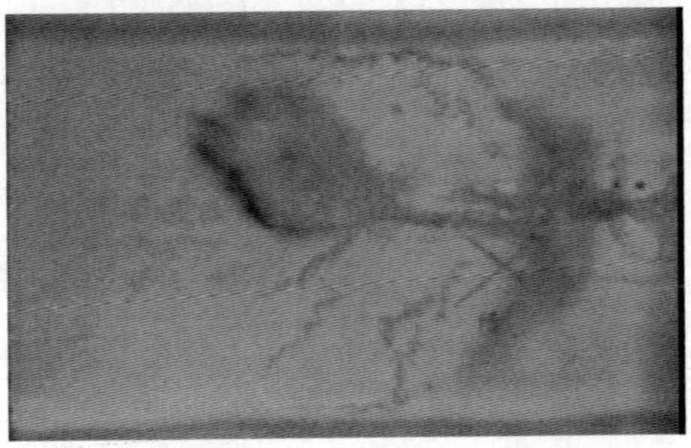

c) t=270 μs

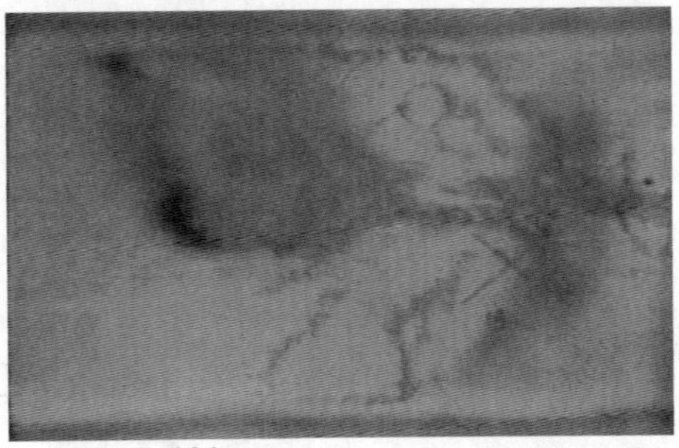

d) t=330 μs

Figure 7. Strike face of Plate 19 at 25°and 350 μm

Also, independent of the initial powder, at the true ricochet with θ=80°impact scenario, the pyrophoric material behaviour of the ceramic layer was observed (Figure 8). A burned surface on the remaining aramid plate was not observed.

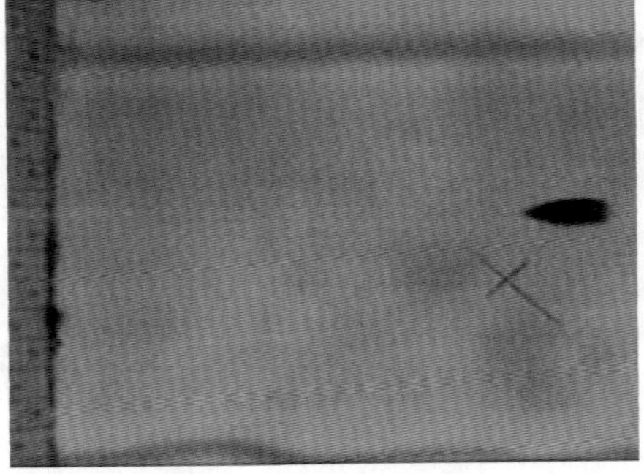

a) t=0 μs

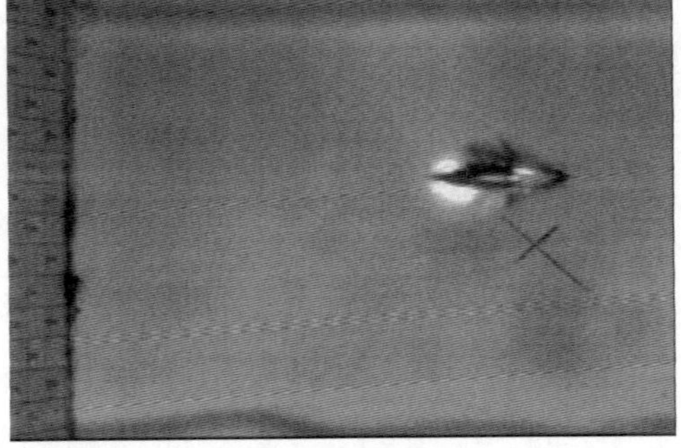

b) t=70 μs

https://doi.org/10.52202/078352-0008

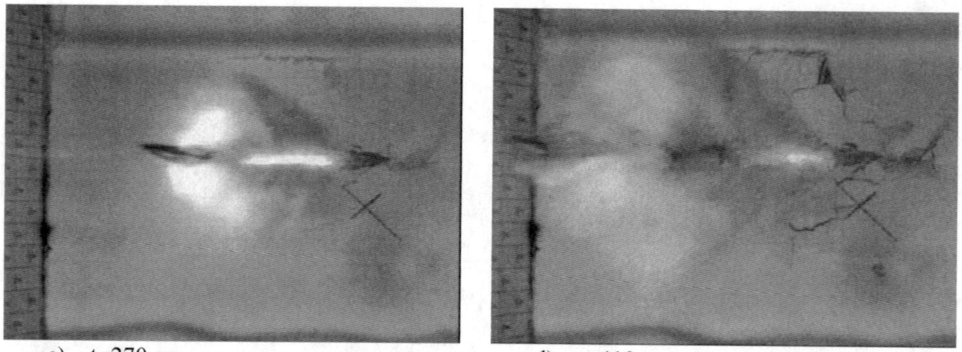

c) t=270 μs d) t=410 μs

Figure 8. Strike face of Plate 10 at 10° and 400 μm

3.2.2 Determination of the deflection angle

In order to determine the value of the refection angle β, a cardboard witness plate was positioned at the edge perpendicular to the aramid plate strike face (Figure 9a). Figure 9b shows the cardboard of Plate 21 with an impact angle θ of 65°.

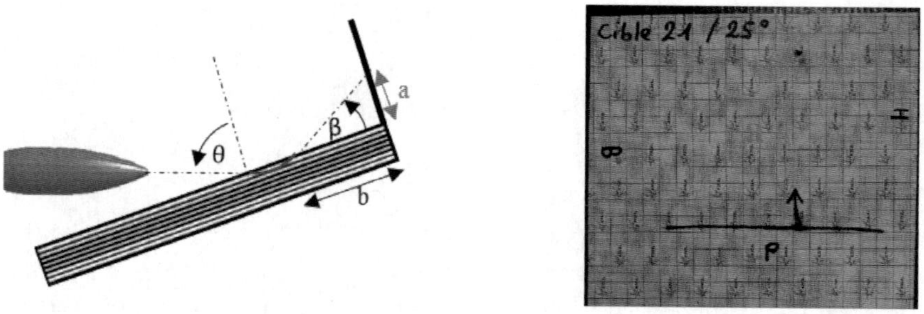

a) Sketch of impact b) Cardboard of Plate 29

Figure 9. Determination of residual angle β

Determining length b from the aramid target plate to the cardboard poses a challenge. The projectile penetrates or comes in contact with the plate for some distance before it ricochets [3]. Figure 10a shows the projectile penetration as it was observed for an impact angle θ=65°. The white arrow marks the direction. For an angle θ=80°, the projectile is in contact with the first aramid layer.

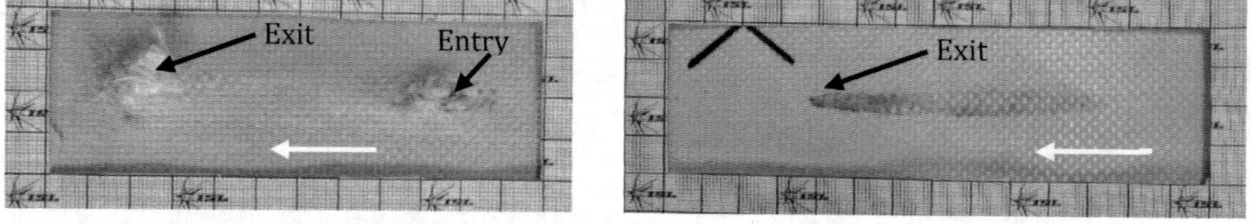

a) Oblique impact angle θ=65° b) Oblique impact angle θ=80°

Figure 10. Projectile trajectory on an aramid plate (arrow indicates impact direction)

The exit point is where the projectile leaves contact with the strike face and until the plate edge is defined at length b. Length a is perpendicular to the plate edge until the projectile perforates the witness cardboard. The angle β is shown in Equation 1.

https://doi.org/10.52202/078352-0008

$$\beta = \arctan \frac{a}{b} \tag{1}$$

Equation 2 shows the exit angles ξ. In cases without a value, the projectile did not ricochet – either they perforated or they were stuck in the plates.

$$\xi = 90° - \beta \tag{2}$$

An uncertainty of about $\xi \pm 2\text{-}4°$, depending on the impact case, was seen. The impact angle $\theta=65°$ is close to the critical ricochet angle θ_c [4]. In general, the impact θ and residual angle ξ correspond, as has been observed in previous works [5].

3.2.3 BFD

The dynamic BFD measurement was conducted with digital image correlation. As shown in Table 1, and as described in the previous section, the Zn layer seems to have an effect on the plate performance. However, more data is needed to confirm this statement. So far, several tests over a period of months have shown that all Zn-layered plates result in higher BFD. Perforation could also be observed under initial conditions in the case of untreated plates [3].

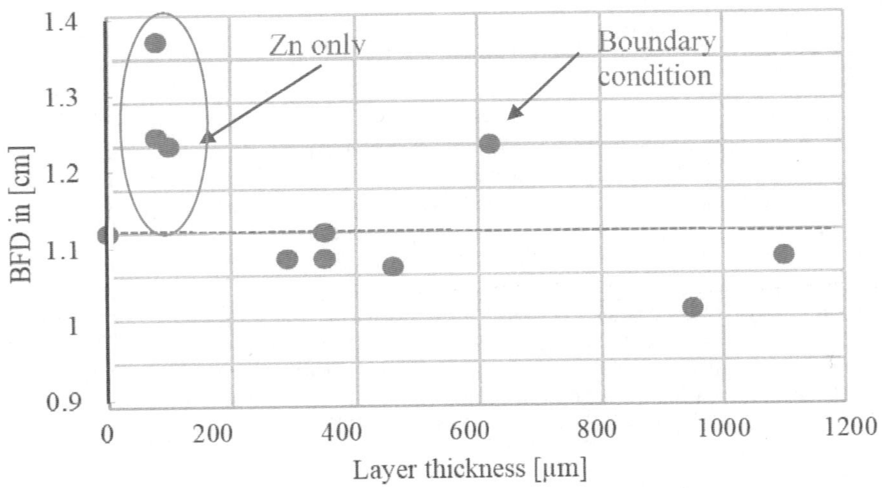

Figure 11. BFD depending on the ceramic thickness for θ=65°

When taking the average BFD from original plates to coated, a decrease of about 5% is achieved (Figure 11). For the impact angle of θ=65, at 0 µm layer thickness, the measurement of the original plate is shown as a reference state. This value aligns with previous experiments [3]. The 100 µm layer thickness comes from the Zn coating only, and the higher layer thicknesses came from Zn + ceramic coating. Although the Zn-coated plates showed less protective capability for the four tested cases than the untreated plates, if covered again with ceramic, all tests showed a reduction in BFD. The question that arises is that if the Zn layer showed less influence in the protective performance, would the ceramic layer BFD be even lower? Moreover, there seems to be an ideal coating thickness – at about 400 µm – for all tested initial conditions. The measurement error is too high to capture the influence of the thickness 700–1100 µm on the BFD. However, if the influence of the layer thickness was debatable, the additional mass chosen could be lower (Figure 5). The plate coated with 620µm in Figure 11 had a high BFD value due to the influence of boundary conditions. The projectile came in contact with the frame during the impact. Also, the influence of the different powder was not observed with the current testing capability. For the impact of θ=80° in Figure 12, fewer impact tests were conducted due to the pyrophoric behaviour of the ceramic layer (Figure 8). Two BFD values of the untreated plates were taken, and their middle value was used as a limit.

https://doi.org/10.52202/078352-0008

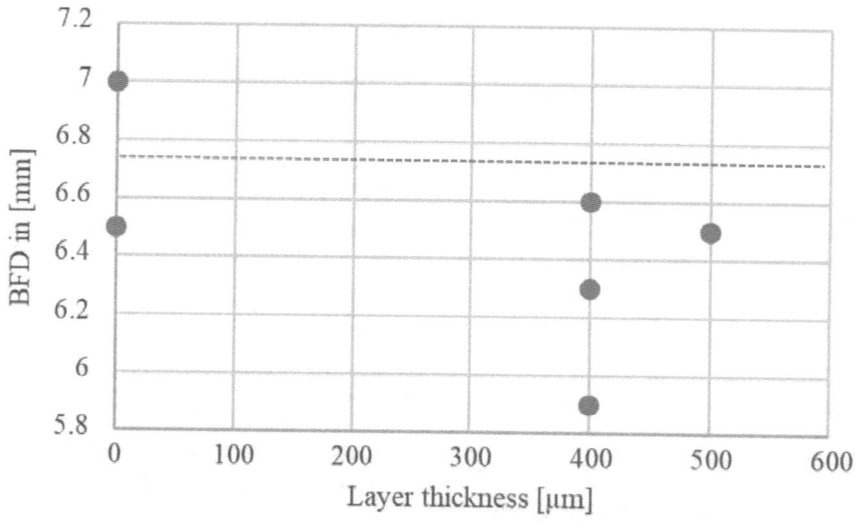

Figure 12. BFD depending on the ceramic thickness for θ=80°

4. CONCLUSION

This work shows a primary test on the ceramic spray-coating of aramid plates using an oblique impact and ricochet of a 7.62x39 mm projectile. Various parameters were measured: the BFD using DIC, the residual angle, and a high-speed video capturing the fracture of the ceramic coating. Two different impact angles – θ=65° and θ=80° (NATO) – were tested with a constant initial velocity 610 +/-10 m/s. The high-speed videos showed that the ceramic layer on the strike face for the higher θ showed pyrophoric behaviour. The influence of the surface coating on the residual angles could not be determined. The measured BFD showed that the Zn layer acted as an adhesive between the aramid plate and the ceramic layer and seemed to decrease their protective performance when compared to the original plates. The ceramic layer thickness showed repeatedly less measurable BFD. For a thorough investigation of the influence of spray technique, powder and thickness, a less complex target material is needed. However, the primary test showed better results than were initially expected in terms of the adhesion between the ceramic coating and the aramid plates.

References

[1] Jamroziak M.B. Ballistic head protection in the light of injury criteria in the case of the Wz.93 combat helmet, 2019; J. Appl. Sci. 9(13):2702.

[2] Laville S., Burke J. and Bogavac A., Why has the AK-47 become the jihadi terrorist weapon of choice? (The Guardian, London, UK) 2015.

[3] Seidl M., "Ricochet off helmets," in PASS, Washington D.C., USA, 2018.

[4] Rosenberg Z. and Dekel E., Terminal Ballistics (Springer-Verlag, Haifa, Israel) 2012.

[5] Becker M. et al., "Numerical ricochet model of a 7.62 mm projectile penetrating an armor steel plate," in 15th International LS-DYNA Conference, Detroit, USA, 2018.

[6] Seidl M. et al., "Modelling back face deformation of woven layered composite target under oblique impact," in 13th European LS-DYNA User Conference, Koblenz, DE, 2019.

[7] v. Hoof J., Modelling of impact-induced delamination in composite material, PhD thesis at Ottawa-Carlton Institute, Ontario, Canada, 1999.

https://doi.org/10.52202/078352-0008

Common Helmet Test System for Blast, Blunt, and Ballistic Testing

M. Bevan[1], J. Clark[1], J. Hrivnak[1], J. Herchek[1], V. Alphonse[1], Q. Luong[1] and K. Sedberry[2]

[1]*The Johns Hopkins University Applied Physics Laboratory, Laurel MD, 20723 USA, Matthew.Bevan@jhuapl.edu*
[2]*CFD Research Corporation, Huntsville AL, 35806 USA*

Abstract. A variety of disparate headforms and test devices have been utilized to test helmets against blast, blunt and ballistic insults. We have developed a common headform that can be used across all three insults and provides many unique features including on-board data acquisition that potentially eliminates the use of external cables, simplifying set-up and range operations. The common headform consists of a common base combined with modular crown portions customized for the insult of interest. For all test environments, the headform can be equipped with a neck that more closely matches the flexibility of a human neck – far less stiff than the Hybrid III neck. Two crowns were created for blast testing. The first blast crown configuration has an array of 52 potential surface pressure sensor locations, of which the on-board data acquisition system can support up to 18. This modularity allows for a variety of sensor configurations, enabling both general helmet performance studies as well as high-spatial resolution measurements in areas of interest. Along with a subset of these locations for external pressure sensors, the second blast crown configuration has a silicone brain surrogate, and instrumented external and intracranial pressure sensors. The ballistic crown configuration uses an array of load cells under concentric impact caps to provide both central and outer edge measurements of behind helmet blunt impact. This provides both spatial and temporal measurement of the impact forces. Five ballistic crowns were created to support this load cell assembly and enable the measurement of impact performance of a helmet in 5 shot orientations. The blunt crown configuration is designed to be used on a horizontal impactor to more closely simulate real-world impact events and is instrumented to measure kinematics in six degrees of freedom.

1. INTRODUCTION/BACKGROUND

A common headform and biofidelic neck surrogate were developed to evaluate the performance of helmets exposed to blast, blunt and ballistic insults. In addition to eliminating the need for disparate headforms, the common headform and biofidelic neck surrogate incorporate unique features and significant advancements over existing head surrogates. The system includes an internal data acquisition system which greatly simplifies field test setup, is highly modular allowing for many use cases, and uses extensive additive manufacturing fabrication techniques which allows for quick and efficient modifications to the headform designs. The modular system is comprised of the Neck, Common Headform System (CHS) Base, and CHS Crowns. There are separate Crowns for each insult modality:

- Blunt Crown
- Ballistic Crown (5 versions – one for each impact location)
- Blast Crown (2 versions – with and without brain simulant and skeletal features).

2. SYSTEM COMPONENT DESCRIPTIONS

2.1 Neck

Blunt impact and ballistic testing of headforms traditionally uses a rigid mounting to allow exact alignment of the headform to the insult. There is a growing body of evidence that angular acceleration/rotation is likely as important a component in evaluating risk of brain injury as linear acceleration. The Hybrid III Anthropomorphic Test Device (ATD) [1], developed for automotive testing, has a semi-flexible neck that allows the headform to move after impact and allows for measurement of neck forces and head kinematic response. The CHS system can interface with the Hybrid III neck, as well as the biofidelic neck that was developed to allow the headform to move more realistically during ballistic and blast events.

JHU/APL previously developed a human surrogate neck that is more biofidelic than the automotive industry standard Hybrid III neck [2]. JHU/APL's previous neck consists of a vertebral column enclosed in a silicone rubber, and it incorporates elastic cords to mimic the effect of muscles. While this neck

https://doi.org/10.52202/078352-0009

performs well in the anterior-posterior direction, it was not specifically designed for lateral or rotational motion. The previous neck required resetting the elastic cords between tests to ensure the proper tension. Furthermore, elastic cabling is susceptible to plastic deformation over time. In the current project, the previous surrogate neck was improved by 1) replacing the elastic cords with spring-loaded cables for a more repeatable and easily tunable response and 2) replacing the plastic vertebral column with a dense, flexible silicone core to improve the durability and reduce the cost and complexity of the part, as well as reducing the difficulty of fabrication.

Traditional drop tower blunt impact test protocols, such as those developed by the U.S. Department of Transportation (DOT) [3] and National Operating Committee on Standards for Athletic Equipment (NOCSAE) [4] do not utilize a neck surrogate, while a neck surrogate is used for a horizontal blunt impact testing [5]. Incorporating a neck into a monorail drop tower test system was examined, however there were several complications to consider. The additional length and moment arm created by the neck requires redesign of the anvil platform to ensure alignment with the headform. Maintaining appropriate carriage mass could result in significant changes to the carriage design. Additionally, a restraining device would likely be needed to limit potential damage to a biofidelic neck from excessive extension. The NOCSAE helmet horizontal blunt impact test protocol incorporates a head and compliant neck mounted to sliding platform. Upon impact from a pneumatically driven horizontal impactor, the neck is able to flex while platform slides, allowing angular acceleration to be induced. Due to the its ability to induce angular acceleration and better suitability of testing with a semi-flexible neck surrogate, the horizontal impactor was used to evaluate the CHS system.

Calculations show that springs need to be very stiff to absorb the energy from either a drop tower or a horizontal impactor. Sufficient stiffening of the neck to withstand this impact would significantly decrease biofidelity of the neck for blast, ballistic response, so a decision was made to use the Hybrid III neck for initial blunt impact testing and design the biofidelic neck surrogate primarily for blast and ballistic testing.

The resulting neck design is shown in Figure 1. The main element of the neck is a two-material silicone rubber column with a 63 mm diameter, moderate durometer, high elongation, center core (Silicones, Inc. XP-697) and a softer, more flexible 13-mm thick outer layer (Silicones, Inc. P-656) for a total diameter of 89 mm. It is equipped with five springs located on the periphery. The front spring is preloaded to 22 N and the others are preloaded to 44 N and can be individually pre-tensioned to the desired level. A steel cable is used to compress the spring and neck by transferring the load to the top plate. Flexible tubing encases the cable such that it does not cut into the neck material during bending. The top plate interfaces with the CHS headform using an interface similar to the Hybrid III ATD.

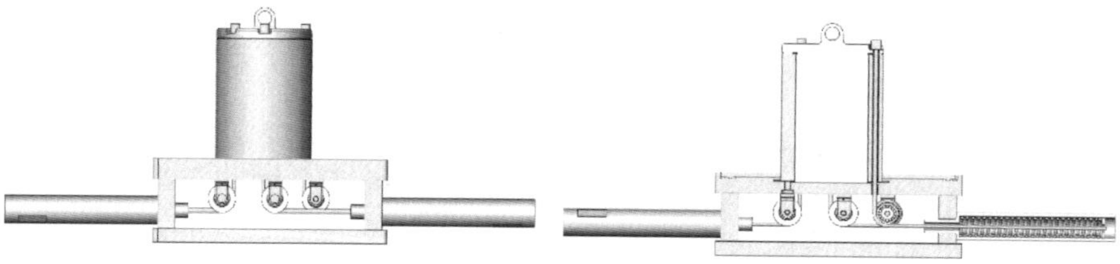

Figure 1. External and cross-section view of flexible neck and spring tension system.

2.2 Common Headform System Base

The three insult-specific crown types, blunt, ballistic, and blast, are attached to the common headform system (CHS) base, a platform that is shared for all of the test conditions. The ballistic and blast crowns have multiple versions to support different insult test locations (ballistic) and different sensing modalities (blast). The CHS base geometry was designed to be compatible with both the ANSUR II Army Anthropometric Survey (50th percentile male) [6] and Army Multi-sized Headform [7] (large) crown geometries. The blast headforms utilize the ANSUR II geometry as it incorporates ears, which allow for the donning of eyewear, and facial topography that is important to blast wave propagation. Since standoff is a major determinant of blunt and ballistic performance, the blunt and ballistic headform crowns reflect the Army Multi-sized Headform geometry, which provides a uniform 23-mm standoff for combat helmets of interest. The geometry of the lower portions of these crowns were modified slightly to blend with the ANSUR II geometry of the common headform system (CHS) base.

The test-specific crowns are attached to the CHS base, a platform that is shared by all of the test conditions (Figure 2). The CHS base shape is based on ANSUR II geometry which is shared with previously used blast headforms [8]. The CHS base is fabricated by selective laser sintering (SLS) of a glass-filled polyamide powder. Five bolts connect the headform crowns to the CHS base. It is equipped with 3 pressure sensor ports (mouth, left and right cheeks) and the 3-axis accelerometer and angular rate sensor package shared by all test configurations.

The CHS base is equipped with a DTS SLICE MICRO for on-board data collection (Figure 3). The SLICE MICRO is a ruggedized data acquisition system that can support up to 24 channels of synchronized data collection with a sampling rate of up to 500K samples per second. Six channels are dedicated to 3-axis acceleration (± 500 g max.) and angular velocity (± 140 rad/sec max.). The remainder are Integrated Electronics Piezo-Electric (IEPE) compatible for measuring additional channels of pressure and force. The system can be powered by an on-board battery, can be set to a buffered cycle data collection mode, and can be triggered by a sensor threshold, allowing for standalone data collection at the site of test. This can greatly simplify data collection for field tests where power and data transfer cables may be burdensome or at risk of damage or causing noise interference.

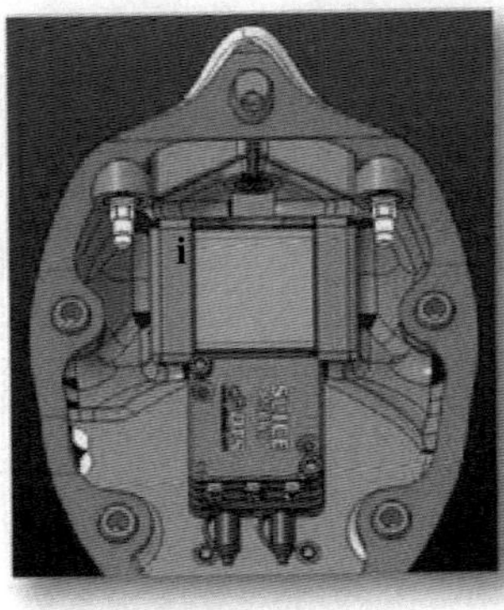

(a)

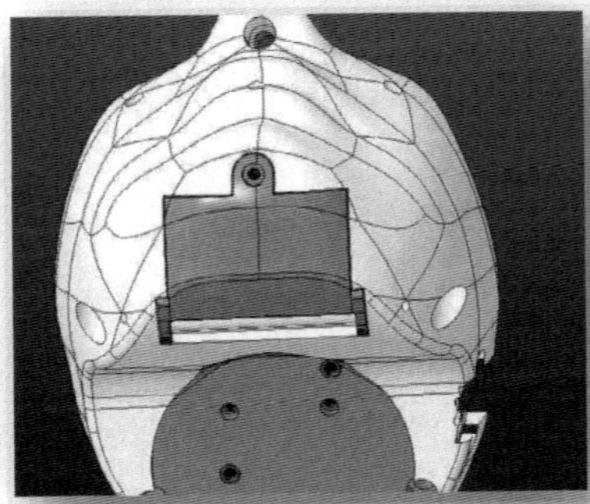

(b)

Figure 2. View of inside of CHS base (a) and battery compartment cover (b).

Figure 3. Reconfigurable data acquisition system.

2.3 Blunt Crown

Based on an assessment of peak accelerations from past drop tower test results, the blunt headform was designed to be able to withstand impact forces and resultant accelerations exceeding 500 g at 6 locations: front, rear, crown, right and left sides, and right and left nape. The blunt crown matches the geometry of the Army Multi-sized large headform.

The blunt crown was fabricated by selective laser sintering (SLS) of a glass-filled polyamide powder. It has an external shell supported with an array of cross-members to provide support and rigid response during impact (Figure 4). It is equipped with the accelerometer and angular rate SLICE sensors found in the SLICE MICRO data acquisition system located in the CHS base.

https://doi.org/10.52202/078352-0009

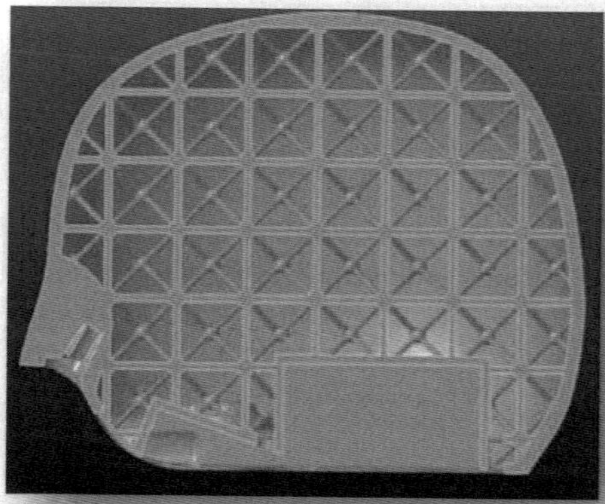

(a)

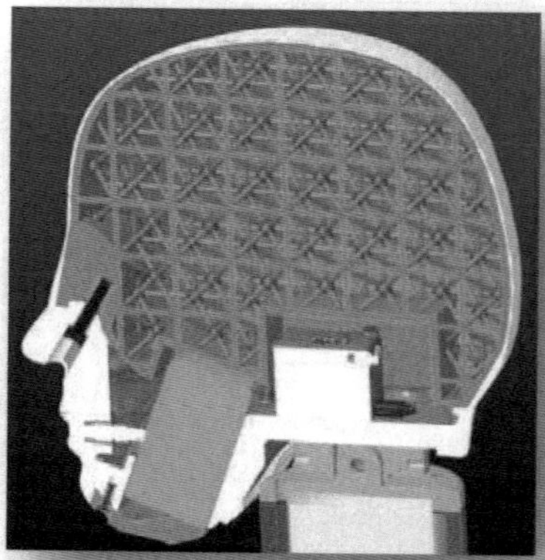

(b)

Figure 4. Section views of the blunt impact crown (a) and crown mounted on CHS base and neck (b).

2.4 Ballistic Crown

The ballistic crown was developed to measure ballistic impact forces behind helmets. The system was designed for a 9-mm NATO bullet at 427 m/s impacting a standard helmet equipped with pads. The crown geometry is based on the Army Multi-sized large headform to provide constant helmet standoff. The ballistic crown system has 5 variants, one for each impact location (front, rear, crown, right and left sides). It was designed to make two force measurements, one at the center, directly behind and in-line with the point of impact, and a second of the peripheral forces surrounding the center impact. The load cells used to make the force measurements have a very short response time to measure the ballistic impact force.

The ballistic crown design has five variants to allow ballistic testing of helmets from the five major directions, front, sides (2), rear and crown. The headform measures force at the center of impact (30-mm diameter cap) and peripheral forces (30 to 90 mm diameter concentric ring) from the axis of nominal impact (Figure 5). The headforms share a common impact module with a load cell array consisting of a center load cell (PCB 224C) and 5 peripheral load cells (PCB 201B05). This allows the center and peripheral caps to each measure forces up to 111kN.

The common impact module is fabricated from stainless steel to ensure a rigid response at the site of impact. The headform crown matches the Army Multi-size headform geometry, and is fabricated by selective laser sintering (SLS) of a glass-filled polyamide powder. Filling the gap between the impact caps and the nominal headform shape is a 12.7-mm thick soft neoprene impact pad molded specifically for the shape of each impact location.

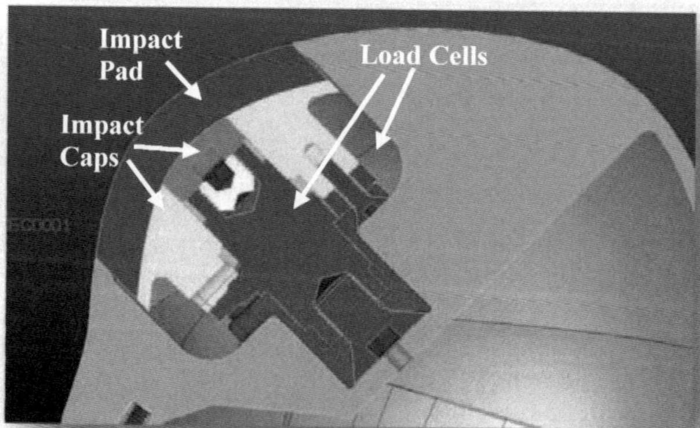

Figure 5. Section view of ballistic impact headform

https://doi.org/10.52202/078352-0009

2.5 Blast Crown - with External Pressure Sensors

Two variants of the blast crown were developed. The first was equipped with a large array of external pressure sensors distributed across the surface of the crown. A large number of possible sensor locations allow for highly tailorable sensor configurations for measuring head surface pressure near, under, and between helmet pads. The external geometry of the blast crown was based on the ANSUR II head geometry.

A blast crown was designed and fabricated that contains an array of 49 pressure ports across the headform surface (Figure 6). Combined with the 3 pressure ports in the CHS base, this provides 52 possible pressure sensor locations. Each pressure port is labelled, inside and out, with a unique identifier to simplify sensor configuration. While each pressure port can be populated with a PCB-113B26 pressure sensor capable of measuring transient pressures exceeding 3400 kPa, the internal data acquisition system limits it to 24 sensors while recording at 200 kilosamples per second. The blast crown was fabricated by selective laser sintering (SLS) of a glass-filled polyamide powder, and the external surface geometry is based on the 50th percentile male ANSUR II.

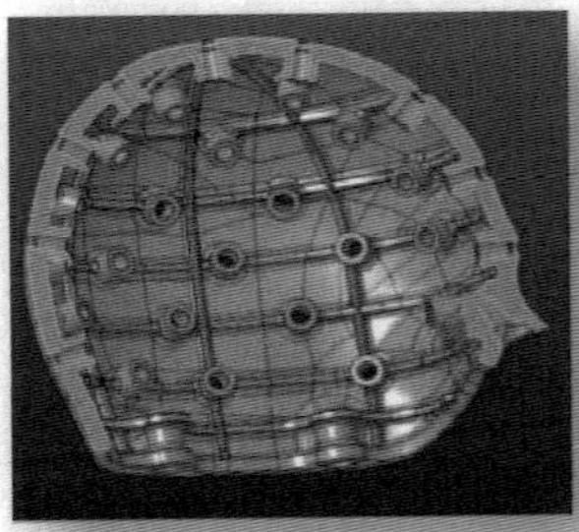

(a)

(b)

Figure 6. Blast crown with external pressure sensor ports showing (a) section view and (b) interior port labelling.

2.6 Blast Crown - with Brain Simulant and External Sensors

The second blast crown was equipped with a smaller array of external pressure sensors distributed across the surface of the crown as well as a silicone gel brain instrumented with pressure sensors (Figure 7). The external geometry of the blast crown was also based on the ANSUR II head geometry.

A blast crown was developed that measured not only the external surface pressure in 8 locations (not including 3 locations in the CHS base), but also has an anthropometrically representative skull surrogate and brain simulant equipped with 4 intracranial pressure sensors. This blast crown maintains the external geometry of the previous blast crown with external pressure sensors, but is equipped with thin, pressure sensors (Honeywell Model F) that are bonded to the external surface with internal wire routing. In the hollow cranial cavity, the headform is filled with silicone gel (Dow Corning Sylgard 184). Suspended in the brain simulant are 4 pressure sensors (TE Connectivity EPIH) which allow for the capture of intracranial pressure in both the time and spatial domains.

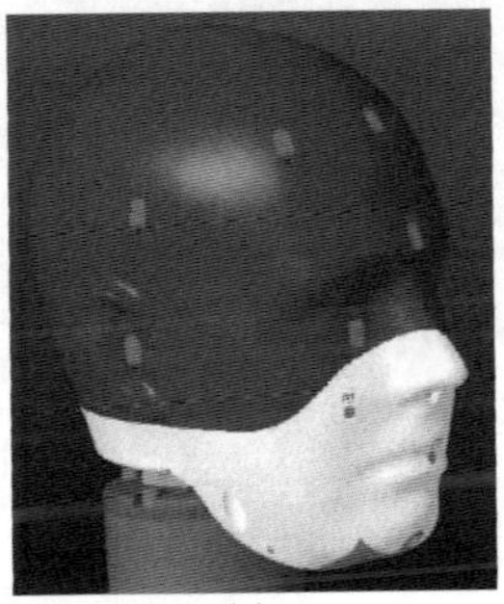

(a)

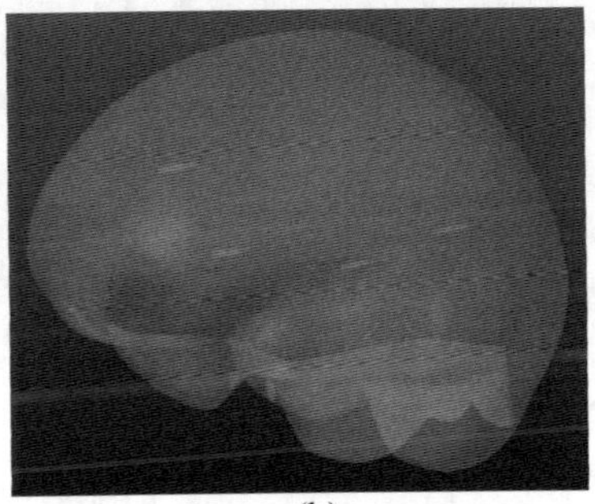

(b)

Figure 7. Blast Crown with brain simulant. Locations of external pressure sensors (a) and intracranial sensors (b) are shown.

https://doi.org/10.52202/078352-0009

3. TEST RESULTS

Prototype versions of the blunt, ballistic and blast (with external pressure ports) crowns have been tested to demonstrate their performance.

3.1 Blunt Crown Testing using Horizontal Impactor

In order to inform the final design, testing was conducted on an initial prototype of the blunt headform to evaluate the overall performance and durability of the blunt crown, as well as the response of the data acquisition system up to its rated shock resistance of 500 g. This level of impact testing greatly exceeds impact levels seen in typical drop tower tests [9] but was chosen to represent a worst-case testing scenario.

A total of 44 tests were conducted across 7 locations at 4 nominal velocities (2.5, 5.5, 7.44, and 9.35 m/s). A horizontal impactor (Cadex, SB202) was used with the HIII neck, the CHS base, and the blunt crown. The horizontal impactor uses a 96 mm diameter steel semi-hemispherical impactor weighing 14.6 kg which is accelerated to the desired impact velocity with a pneumatically driven piston. The HIII neck and blunt impact headform are mounted on a carriage that can freely slide away after impact. An Advanced Combat Helmet (ACH) with pads was mounted on the headform in accordance with the ACH operator's manual.

The data acquisition system performed well, and the CHS base was undamaged. At the highest two velocities, damage was observed on the blunt crown (Figure 8); the external shell showed slight signs of damage and analysis of computed tomography (CT) scans showed internal lattice struts were broken. The crown impact location was damaged at 7.1 m/s with 2 struts broken while the left nape and side locations were damaged at 8.9 m/s with 3 struts broken at each location. These tests suggest that the initial prototype blunt crown structure has an approximate 300 g limit in this horizontal test. It is important to note that these impact velocities and resulting accelerations are far greater than standard combat helmet drop tower test velocities of 3 m/s and 4.3 m/s [9]. There was also residual powder remaining from the rapid prototype process of the initial prototype.

The examples of the data collected are shown in Figure 9. Both angular rate and acceleration data were filtered at the standard CFC1000 filter specification. For a 5.5 m/s impact velocity, the peak accelerations ranged from 99 to 171 g depending on impact location. The peak angular velocity ranged from 18.8 to 38.5 rad/s. The accelerations measured are in a similar range as those measured by McEntire et al [9].

Based on lessons learned from testing the initial prototype blunt headform, the shell thickness and outer lattice thickness were increased to improve the impact resistance of the blunt impact shell. Additional and larger clearance holes were added to improve the removal of the residual processing powder.

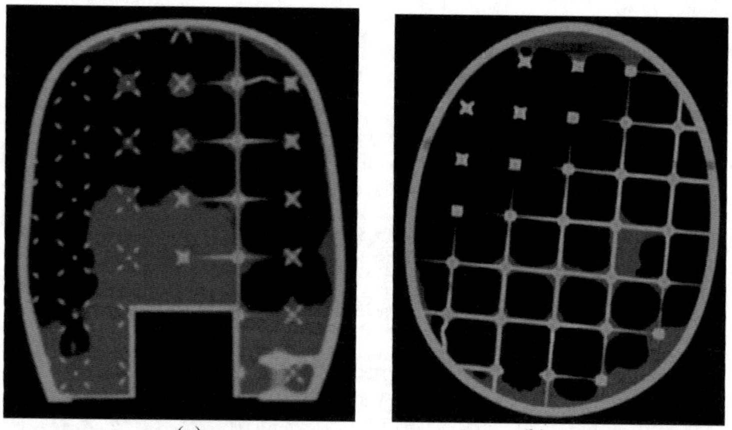

(a) (b)

Figure 8. Examples of damage observed on the left side (a) and nape (b) impact locations.

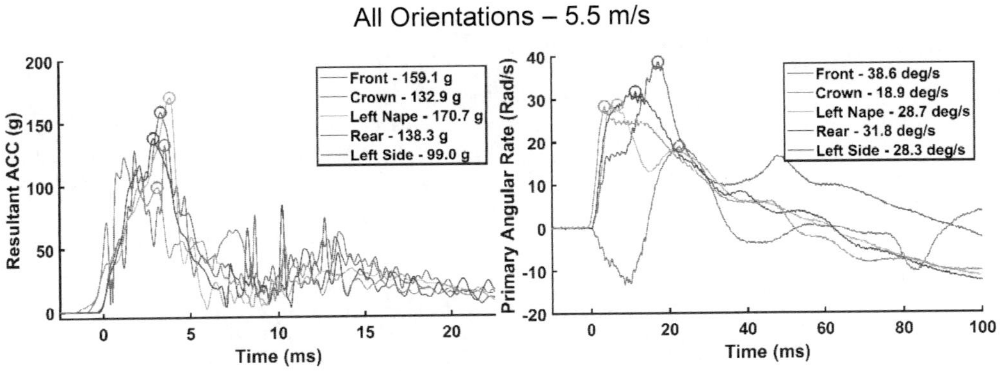

Figure 9. Examples of resultant acceleration and primary axis rotational velocity data collected from 5.5 m/s horizontal impact testing

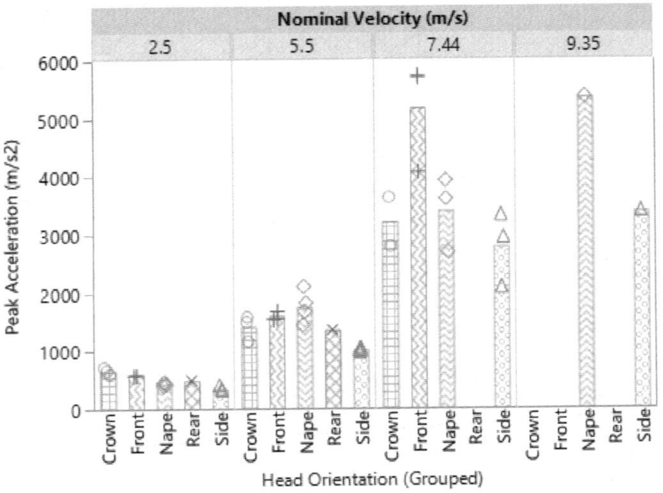

Figure 10. Peak Acceleration measured in horizontal impact testing. Bars represent average peak accelerations for each test condition, and markers represent individual test results.

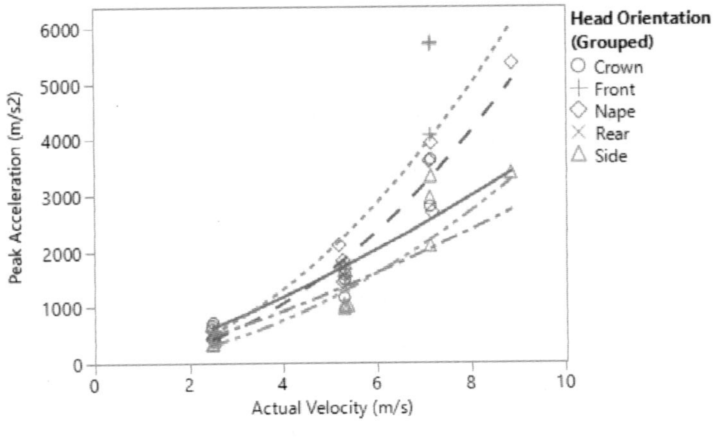

Figure 11. Acceleration data collected from the blunt impact headform at different impact velocities.

https://doi.org/10.52202/078352-0009

3.2 Ballistic Crown Testing using Air Cannon

Simulated ballistic testing was conducted to demonstrate the performance of the ballistic crown. The impact testing was conducted using a pneumatic cannon to shoot a 62 mm diameter, 67 mm long aluminium projectile, weighing 197g with a 75 mm radius of curvature tip. The projectile impacted the front ballistic crown at velocities ranging from 31 to 60 m/s. Data from 21 tests were collected and analysed to measure how the peak force changed with impact velocity.

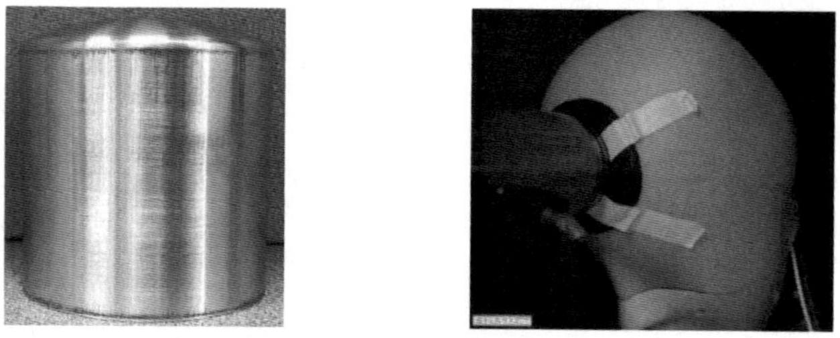

(a) (b)

Figure 12. Aluminium projectile used in air cannon testing on front ballistic crown (a). Projectile impact on front ballistic crown (b).

Results showed that the peak forces ranged from 16 to 163kN over the velocity range tested (**Figure 13**). The forces measured on the centre cap and outer ring were similar at each velocity. The coefficient of variation of the total forces ranged from 9-22% for the three nominal velocities. The test results demonstrate the ability of the CHS ballistic crown to provide both spatial and temporal measurement of the impact forces.

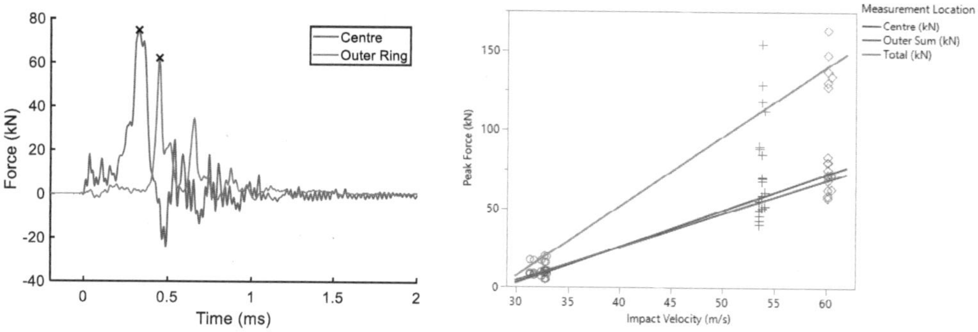

(a) (b)

Figure 13. Time-force measurement of projectile impact at 60 m/s on the ballistic crown (a). Peak forces measured at different velocities by the ballistic crown (b).

3.3 Blast Crown with External Pressure Sensor Testing

The blast crown with external pressure sensors was tested in a 91 x 91 cm advanced blast simulator [8] to evaluate the durability and response of the prototype headform, neck, and data acquisition system in a simulated blast environment. In all, 29 shock tube tests were conducted with 3 orientations of the headform (front, side, and rear facing), 2 different helmet conditions (Advanced Combat Helmet and bare), with both the new APL neck and the HIII neck. All tests were run at a nominal 1400 kPa burst pressure diaphragm, resulting in a nominal 115 kPa static peak, 7 ms positive phase duration shockwave.

External to the headform, there was a Pitot pressure sensor (PCB 113A36) to measure the incident pressure wave which was measured at 1 M sample/s. This allowed these data to be compared with data collected by the blast headform.

The data acquisition system worked well with the 9 pressure sensors monitored. The neck, CHS base and blast crown were undamaged. Data was collected at 1 M samples/s by off-board sensors and data acquisition system and compared to the data collected at 200,000 samples/s on the internal data acquisition system (both filtered with a 20 kHz 10-pole Butterworth low pass filter). There were little differences seen due to the reduced sampling rate, however, the pressure sensors in the headform registered higher peak pressures than the external pressure sensors, results that are being investigated further.

Head kinematic measurements from tests using the new, more flexible APL neck were compared to those when using the HIII neck. The flexible neck tests showed about twice the angular displacement of the HIII neck tests with a peak angular displacement time of 115 ms vs 60 ms for the HIII neck. Comparing these results to Murphy et al [2], Murphy showed that a post-mortem head and neck showed a peak rotation of 3 times that of the HIII neck, with the peak angle occurring at about 140 ms after impact. The flexible neck can be retrofitted with different springs that allow for the neck pretension and hence head and neck kinematics to be tuned to a desired level.

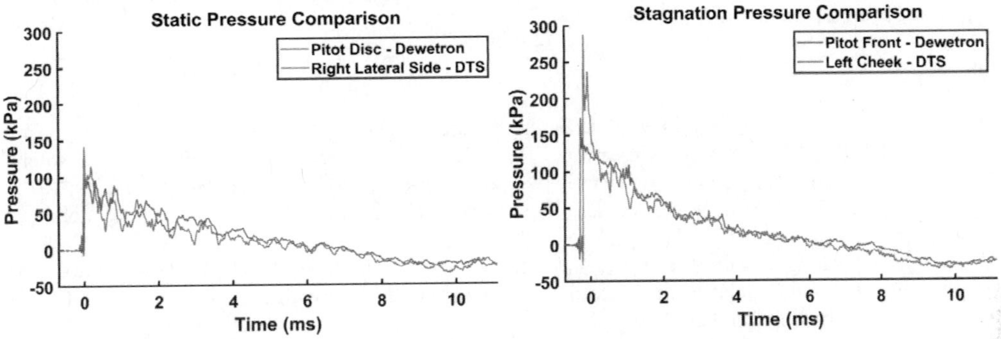

Figure 14. Static (left) and stagnation (right) pressure measured during a shock tube test with 100 kPa static overpressure wave.

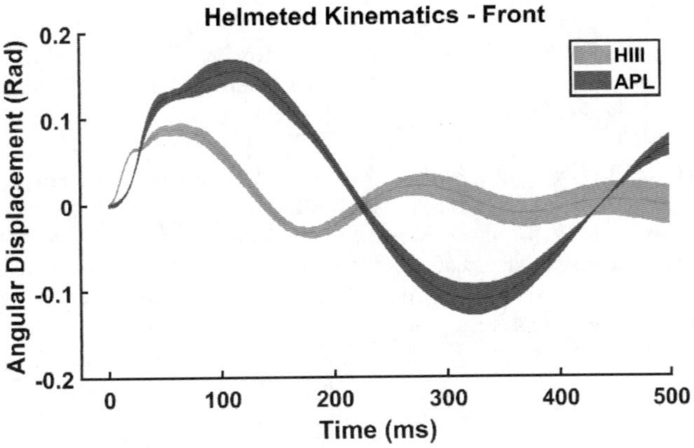

Figure 15. Angular displacement of the headform when exposed to shock tube blast overpressure wave.

4. CONCLUSIONS

A variety of headforms and test systems have been used to evaluate helmets against blunt, ballistic and blast insults. We have developed a common headform system that can be used across all three insults and provides many unique features such as:

- 3-axis accelerometers and 3-axis rotational velocity measurements for capturing 6 degrees of freedom during blunt, blast and ballistic testing
- On-board data acquisition to eliminate cabling during range testing
- A neck that is tunable and more flexible than the Hybrid III providing more human-like response to loading
- Two-zone ballistic impact force measurement to measure the peak forces found behind the helmet as well as the large, lower force surrounding area.

https://doi.org/10.52202/078352-0009

- Large number of fixed pressure sensor locations that allow for either standardized testing or focused studies in key regions.

These changes provide a wide range of new capability for helmet testing, filling measurement gaps that exist in other headform systems.

Potential future testing includes blunt impact testing at impact velocities more closely matching standardized drop tests, ballistic testing with additional threats of interest, and blast testing of helmets that were designed to protect against blast. Additionally, tuning the pretension of the neck cables would enable the surrogate neck response to more closely matches that of humans.

Acknowledgements

This research was funded by CFDRC through a US Army, CCDC Small Business Innovative Research grant. Any opinions, findings and conclusions or recommendations expressed in this material are those of the authors and do not necessarily reflect the views of the US Army.

References

[1] Title 49-Part 572, Federal Motor Vehicle Safety Standards, Section 572.33 - Neck, June 2, 2011.

[2] Murphy RJ, Wing ID, Iwaskiw AS, Armiger RS, Carneal CM, Merkle AC. "Biofidelic assessment of head-neck surrogate response to dynamic overpressure". World Congress of Biomechanics, 2014.

[3] Title 49-Part 571, Federal Motor Vehicle Safety Standards, Section 571.218 Motorcycle Helmets, October 1, 2011.

[4] National Operating Committee of Standards for Athletic Equipment, "Standard Test Method and Equipment Used in Evaluating the Performance Characteristics of Headgear/Equipment," NOCSAE Doc.001, Modified December 2019.

[5] National Operating Committee of Standards for Athletic Equipment, "Standard Pneumatic Ram Test Method and Equipment Used in Evaluating the Performance Characteristics of Protective Headgear And Faceguards," NOCSAE Doc.081, Modified September 2019.

[6] Gordon C., Blackwell C., Bradtmiller B., Parham J., Barrientos P., Paquette S., Corner B., Carson J., Venezia J., Rockwell B., Mucher M., Kristensen S., 2012 Anthropometric Survey of U.S. Army Personnel: Methods and Summary Statistics, Technical Report NATICK/TR-15/007, Dec 2014.

[7] "Review of Department of Defense Test Protocols for Combat Helmets," Committee on Review of Test Protocols Used by the DoD to Test Combat Helmets; National Research Council, National Academies Press, 2014

[8] Carneal C., Merkle A., Zinn D., Andrist J., Clark J., Alphonse V., Ritzel D., Carboni M., DeCristofano B., Cyganik J., and Maffeo M. "Development of a Laboratory Shock Tube System for Helmet Blast Overpressure Performance Assessment". Personal Armour Systems Symposium, 2016.

[9] "Blunt Impact Performance Characteristics of the Advanced Combat Helmet and the Paratrooper and Infantry Personnal Armor System for Ground Troops Helmet, USAARL Report No. 2005-12" United States Army Aeromedical Research Laboratory, August 2005.

https://doi.org/10.52202/078352-0009

Development of soil ejecta surrogate projectiles for laboratory testing of lightweight protective materials

G. Pageau[1] and **S.Ouellet[1]**

[1]*Defence R&D Canada – Valcartier, Department of National Defence, 2459, Route de la Bravoure, Québec City, Quebec G3J 1X5, Canada*
simon.ouellet@drdc-rddc.gc.ca

Abstract. This paper presents the development of Rock Simulating Projectiles (RSP) that simulate natural fragments accelerated during the explosion of a buried charge. The key features of the projectile (shape, sectional density, edges) are based on a detailed morphological study of natural fragments. Ballistic tests (V_{50} and overmatch) were conducted on a novel lightweight ballistic fabric to compare the relative penetrating capability of the proposed 2 gr RSP with 2 gr natural rocks and a 2 gr steel Right Circular Cylinder (RCC). The RSP was successfully launched with minimum yaw at low and high velocities using a 0.17 calibre gun without sabots. For the fabric tested, the lowest V_{50} value and smallest absorbed energy (from overmatch tests) were obtained with the RCC, leading to an underestimation of the actual protective capability against natural fragments. Scaling techniques, typically used for armour material with higher Areal densities (***Ad***) against FSPs and RCCs, were shown to be inappropriate for very low ***Ad*** armour materials against projectiles with a different shape. The results obtained with the RSP suggest that it is an adequate surrogate for soil ejecta, enabling a conservative assessment of the effective ballistic resistance of protective fabrics against this threat and leading to more reliable estimation of safe operating distances.

1. INTRODUCTION

In the context of asymmetric warfare, dismounted soldiers are facing a broad and complex spectrum of threats. Most current fragment protective armours were designed to defeat high-velocity fragments from air-burst warheads at long standoff distances. However, the close proximity exposure to the detonation of a buried Improvised Explosive Device (IED) leads to a very different scenario involving non-metallic natural fragments (i.e. soil ejecta), a high number of close impacts, negative impact angles and potential synergy with the blast wave [1]. While the severity of the threat may remain driven by the physical attributes of the projectiles (i.e. mass, shape, material) and their impact velocities, the specificities of the IED scenario warrants an adapted ballistic test method to evaluate potential protective strategies. Unfortunately, replicating more realistic conditions (e.g. projectile simulating soil particles, close multi-hit, etc.) has not yet been standardized in laboratory testing. In addition, the emergence of the buried IED threat has led to redefining armour coverage to include zones that were typically unprotected such as extremities, joints and urogenital areas which extends the range of materials to consider for protection. Until better-defined test methods with relevant performance metrics are implemented, the development of protective systems against the buried IED threat may not lead to the intended protection levels.

Since 2014, Defence Research & Development Canada (DRDC) has been conducting work to address this buried IED test methodology gap. Specifically, DRDC has developed complementary and rigorous test methods for full-scale explosive testing and laboratory testing of protective materials. For the full-scale tests, the method is built around test beds made of two types of narrow-graded standardized soils in order to represent two operational severity levels. The lower severity test uses a test bed made of BC 2.5-5mm crushed aggregates [2], where BC is the designation for cement concrete aggregates. This aggregate grading is meant for testing protective clothing and under garment (Tier-1 Pelvic Protection Systems (PPS)). A test bed with coarser aggregates (BC 2.5-10 mm) is used to conduct the high severity test meant for testing protective over garments (Tier-2 PPS). Well-graded soils with larger particles, such as the STANAG soil [3] specified for vehicle mine resistance testing, were found to be too severe for **PPS** testing. Because the ballistic performance of a material is sensitive to the supporting conditions (e.g. clamped or backed), a biofidelic backing replicating the "as-worn" condition and developed to test soft armour material is used to support material samples during testing [4,5]. The backing has a layered structure, ideal for recovering fragments, and comprises of an outer skin (epidermis/dermis), an under layer (hypodermis) and a soft tissue simulant. The layered backing is also used in laboratory ballistic testing of protective materials. For end-items testing (e.g. boxers, PPS), the layered backing can be easily assembled into quasi-anthropomorphic shapes. Post-test analysis of full-scale experiments is done using a CT-scanner and the *FragFinder* software [6] which enables automatic measurement of fragment properties and depth of penetration (***DoP***) in the layered backing. From these

 https://doi.org/10.52202/078352-0010

measurements and the laboratory calibration of the layered backing, a fragment's residual velocity (V_r) is estimated and used for evaluating armour system effectiveness in operational scenarios.

Using the layered backing for both laboratory and full-scale explosive test leads to coherent ballistic data sets, which can then be cross-correlated. To maintain a high level of correlation between full-scale and laboratory tests, it is necessary to use an adequate surrogate projectile and reproduce the spatial distribution of impacts on the armour. A number of laboratories [7-8] have explored using the "sand cannon" technique where multiple non-metallic projectiles are launched in a sabot to replicate the spatial and temporal distribution of impacts from soil ejecta. This method typically has a low level of reproducibility in terms of impact dispersion and velocity of the aggregate projectiles, making comparative performance assessment difficult. Multi-projectiles launchers have also been used for multi-hit testing [9-10] where the effect of near-simultaneous impacts can be evaluated. However, this method is not easy to implement due to the required specialized equipment. Using multi-hit shot-patterns is thought to be a good compromise between realism of the simulated threat, repeatability and ease of use, keeping in mind that in the context of buried IED, laboratory tests will always be limited as the combined effect of blast overpressure and ballistic impacts cannot be reproduced. The validation of the performance of protection systems against buried IEDs will always require full-scale explosive tests.

Standardized fragments simulating those from fragmenting munitions are routinely used in body armour testing [11]. The main categories are the chisel-nose fragment-simulating projectiles (FSP) and the flat-nose right circular cylinder (RCC). Both are available in a number of homologous sizes. For pre-formed fragmenting munitions, the filler fragments (RCC, spheres, cubes) are sometime directly used. The failure mechanisms of armour and the ballistic resistance are highly influenced by fragment geometry. Therefore, choosing a shape that adequately replicates the buried IED threat would appear important. Unfortunately, end items aimed at protecting against energized natural fragments still have protection levels defined using the 2 gr steel RCC or the 2.5 gr steel FSP. Soil particles are however different from the standard FSPs and RCCs. They have significantly different compositions (e.g., densities), and are often asymmetrical with numerous edges. Soil ejecta surrogates, using alternative materials, have been explored by other researchers. Glass spheres [7] have been proposed for testing lightweight fabrics, and silicon nitride balls [12] have been specified for the low-velocity testing of transparent armours. However, spheres are far from the shape of soil particles and are known to interact much differently with woven targets compared to projectiles with edges where shear cutting may dominate. Glass cannot be easily machined into complex shapes making skirted designs impossible. Cubes and parallelepipeds would potentially provide a closer match to the geometry of natural fragments, but are rarely used because their striking orientations (e.g. corners, faces and edges) and penetration capability cannot be controlled reliably. Linden [13] proposed using aluminium RCCs (3 mm and 6 mm) which led to the adoption by CEN [14] of a 4 mm diameter/length aluminium RCC (2.2 gr) for testing deminer's protection systems. An aluminium cylindrical projectile with a conical nose is also specified for testing of railway glazing against rock strike [15].

2. SURROGATE DESIGN

To support the development of a better soil ejecta surrogate projectile, a study was conducted on particle size distribution and morphology of rocks from three types of soil used in buried IED testing. The study used standard sieve analysis, 3D laser scanning and 2D image analysis. Thirty individual samples were taken from 6 different sieve opening sizes for each of the three soils (crushed granite, crushed limestone, natural granite) for 90 samples in total. The morphology parameters obtained included mass, outer surface area, minimum, maximum and average presented areas, average shape factor, and mean edge radius. The relative density with respect to water, or Specific Gravity (***SG***), was measured using a solid density tester and yielded values ranging from 2.65 to 2.72 for the three aggregate types, which is comparable to the value of 2.62 reported by Thomas [16]. The following correlations between mass (***m*** in grams), diameter (***D*** in mm), length (***L*** in mm) and elongation ratio (***L/D***) were found from the analysis:

$$\boldsymbol{m = 0.003 * D^{2.58}} \quad \textbf{(1)} \qquad\qquad \boldsymbol{\frac{L}{D} = 2.1 * D^{-0.42}} \quad \textbf{(2)}$$

Using Equation 1, the average mass for a 4.3 mm diameter (0.17 calibre) rock is 0.13 gram or 2 gr. The 4.3 mm size was chosen as it corresponds to the smallest barrel rifling available, making possible the use of skirted projectiles. It also corresponds to the upper end of the BC 2.5-5 mm aggregate grading selected for Tier 1 PPS testing making possible future performance correlations between laboratory and full-scale testing. A Rock Simulating Projectile (RSP) was designed (Figure 1) based on the standard FSP geometry and the 0.17 calibre Hornet rifling specifications. To better replicate the geometry and

https://doi.org/10.52202/078352-0010

numerous edges of natural rocks, the nose of the RSP includes four bevel planes instead of two. The planes are at an angle of 45° (version 1) instead of the standard 35° on FSPs.

The edges are also sharp with no radius, making the RSP more severe and leading to more consistent and conservative assessments of the ballistic resistance of personal armour materials. For firing at low velocity with short target distance, an un-skirted variant was also designed which can be launched using a smoothbore gas gun. For specific gravity, the closest material to natural rock is aluminium followed by soda lime glass (***SG*** of 2.5). Magnesium (***SG*** of 1.77 vs 2.7 for aluminium) was finally selected since its lower ***SG*** allows for artificially increasing the RSP length by 52% while keeping the sectional density constant (Figure 2 left). The greater elongation ratio (1.53 for magnesium vs 1.13 for aluminium) ensures better in-bore and in-flight stability (i.e. low yaw).

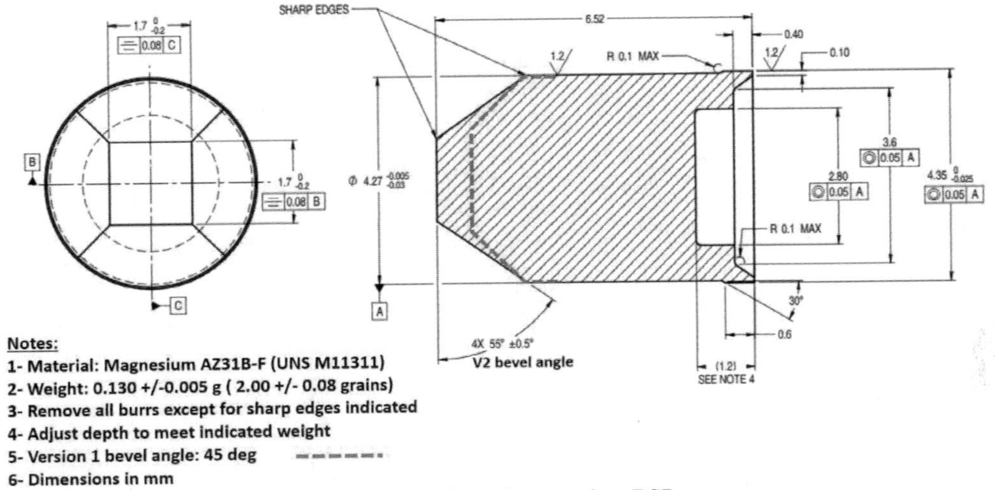

Figure 1. Drawing of magnesium RSP

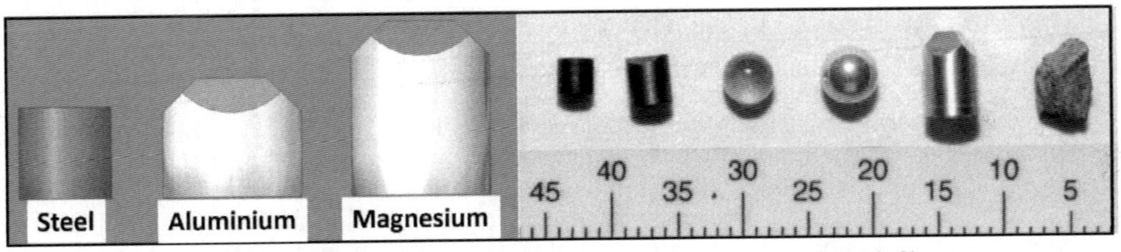

Figure 2. 2 gr steel RCC vs aluminium and magnesium RSP (left)
2 gr RCC, 3.7 gr FSP, 4 mm glass sphere, 4.5 mm aluminium sphere, 2 gr RSP, 2 gr rock (right)

A 6 gr and 16 gr variant are also proposed for representing coarser soils. The 2 gr and 16 gr masses are commonly used for testing body armour and are part of the RCC homologous family [17]. The corresponding diameters for the 6 gr and 16 gr, calculated using Equation 1, are 6.6 and 9.7 mm respectively. These values correspond closely to two standard gun calibres: the 0.270 Weatherby magnum and the 40 Smith and Wesson. The corresponding elongation ratio from Equation 2 are: 0.95 and 0.81 respectively. Although the nose geometry for the 6 gr and 16 gr variants would be the same as for the 2 gr, the 3 variants would not be fully homologous because of the different elongation ratios. This could limit scaling of ballistic data based on sectional density. To validate the integrity and flight stability of the proposed 2 gr RSP, an initial quantity of 30 units was produced. The bevel angle was subsequently changed from 45° to 55° (RSP version 2) such that the resulting edges angles are 45° as initially planned.

3. BALLISTIC TESTS: MATERIALS AND METHODS

Table 1 summarizes the four ballistics tests series conducted to validate the suitability of the proposed 2 gr RSP for replicating soil ejecta. Two types of soft armour materials were used: a 500-denier woven aramid fabric (Twaron®) typically used in fragment protective vests and a novel lightweight fabric designed for combat uniforms. Four projectiles were tested (Figure 2) namely the 4 mm glass spheres

https://doi.org/10.52202/078352-0010

(*Cospheric* precision grade soda lime), the 2 gr steel RCC, the 2 gr limestone gravel stones (rock), and the proposed 2 gr RSP machined from 6.35mm diameter magnesium alloy AZ31F rod (*Buymetal.com*). A first test series was conducted as an initial proof of concept for the RSP design (v1) where two-shots V_{50} firings could be performed against 1, 3 and 26 plies of the Twaron® fabric. The 4 mm glass spheres were only tested against the one ply Twaron® samples. Test series 2 to 4 were conducted with the 2 gr RSP (v2), 2 gr RCC and 2 gr rock against two plies of the novel fabric material selected. V_{50} tests of the novel fabric were conducted following the standard conditions (shot pattern, up-and-down method) of AEP-2920 while ballistic overmatch tests were done according to the guidelines of TOP-10-2-506 [18]. Residual velocity (***V_r***) testing was done with impact velocities (***V_i***) up to twice the V_{50}. Impact velocity was measured using a Doppler radar and two orthogonal *Photron SAZ* cameras. Rebound velocities (negative ***V_r***) were also measured for the non-penetrating impacts for better fitting the ***V_r-V_i*** data. The *ProAnalyst* image analysis software was used to compute the total projectile yaw upon target impact using the two orthogonal views. The target size for V_{50} and overmatch tests was 400 mm x 400 mm with the fabric samples stitched together at their four corners. Tests using twelve plies of the novel fabric targets (200 x 200 mm) were also conducted to measure the number of layers penetrated as a function of ***V_i*** for both the RCC and RSP. All targets were affixed with two nylon bands to the biofidelic backing.

Table 1. Ballistic tests conditions

Series	Test types	Projectiles	Target materials
1a	V_{50}	RSP 2 gr v1	Twaron® 500 den: 1, 3, 26 plies
1b	V_{50}	Sphere glass 4mm	Twaron® 500 den:1 ply
2a	V_{50} skin, ***DoP-V_i***	RSP 2 gr v2	Bare backing
2b	V_{50} &overmatch	RSP 2 gr v2	New fabric: 2 plies
2c	Perforated layers vs ***V_i***	RSP 2 gr v2	New fabric: 12 plies
3a	V_{50} skin, ***DoP-V_i***	Rock 2 gr	Bare backing
3b	V_{50} &overmatch	Rock 2 gr	New fabric: 2 plies
4a	***DoP-V_i***	RCC 2 gr (sabot)	Bare backing
4b	V_{50} & overmatch	RCC 2 gr (sabot)	New fabric: 2 plies backing & frame
4c	Perforated layers vs ***V_i***	RCC 2 gr (sabot)	New fabric: 12 plies
4d	V_{50} multi-hit	RCC 2 gr (sabot)	New fabric: 2 plies

For the V_{50} tests, the complete and partial perforation assessment was performed using three criteria: perforation of the armour sample, perforation of the backing epidermis/dermis layer and perforation of the hypodermis layer, leading to a different V_{50} value for each condition. The up-and-down procedure was driven by the perforation status of the armour to avoid removing the layered backing components after each firing. The layered configuration of the backing allowed for the precise determination of the Depth of Penetration (***DoP***) of the projectiles. This was done only after all the firings on a sample were completed. Calibration tests (***DoP*** vs ***V_i***) were conducted on the bare layered backing (series 3a and 4a) to enable the estimation of ***V_r*** from ***DoP*** data when the armour sample is perforated. The 2 gr RSP and 4 mm glass sphere were launched using a 0.17 calibre Hornet barrel chambered for receiving 0.22 calibre long rifle cartridges. Three barrels lengths were available, i.e.: 250, 400 and 660 mm for the RSP. The 250 mm barrel was used to cover the lower velocity ranges (130 to 550 m/s) while the 660 mm barrel allowed to reach up to 1450 m/s. *Hilti* 0.22 calibre single shot powder blanks (brown, green, yellow) and *Victory* crimped start blanks were used to propel the fragments. For each blank cartridge type, a velocity-distance calibration curve was generated by varying the initial position of the projectile within the barrel by steps of 12 mm up to the barrel mid-length. Projectile velocity adjustment was then done by selecting the right blank type and insertion distance.

The 2 gr rock projectiles were selected from a 15-litre container of BC 2-5mm crushed limestone aggregates. The rocks were launched using a 6.35 mm smoothbore gun tube dimensioned for firing the 16 gr steel spheres. A neoprene obturator disc was placed behind each rock to provide a gas seal and gain adequate control on impact velocity. The 2 gr RCC test series was conducted by an external laboratory using plastic sabots and short barrels which unfortunately prevented reaching the lower end of the desired velocity range. To replicate the effect of multiple close impacts (Figure 3 left), ballistic tests were also conducted using the multi-hit pattern shown in Figure 3 (test series 4d). A single-hit V_{50} value is obtained first using the AEP-2920 [11] standard shot pattern with a 64 mm spacing. A second test series is then performed where shots are placed in pairs at close distance (18 mm) from the first

https://doi.org/10.52202/078352-0010

series to generate a second V_{50} value. A third series is finally conducted using an equilateral triangle pattern for obtaining the third V_{50} value with a minimum distance between each triangle of about 50 mm.

Ballistic resistance degradation from overlapping damage may then be quantified by comparing the three V_{50} values. Although this approach only replicates the spatial and not the temporal distribution of the impact of soil ejecta, it should help identify fabric architectures and material systems that are better suited for the soil ejecta threat. From Figure 3 (centre), it can be seen that for the same projectile mass, the diameter of the RSP is 1.5 times larger than for the steel RCC and the presented area is 2.3 times larger. The larger RSP area will translate into more yarn damage. For the multi-shot pattern proposed (Figure 3 right), the RSP may lead to more overlapping damage and greater relative severity compared to the RCC. Meanwhile, the 2 gr RCC will likely be more penetrating against low thread density fabrics due to its smaller size.

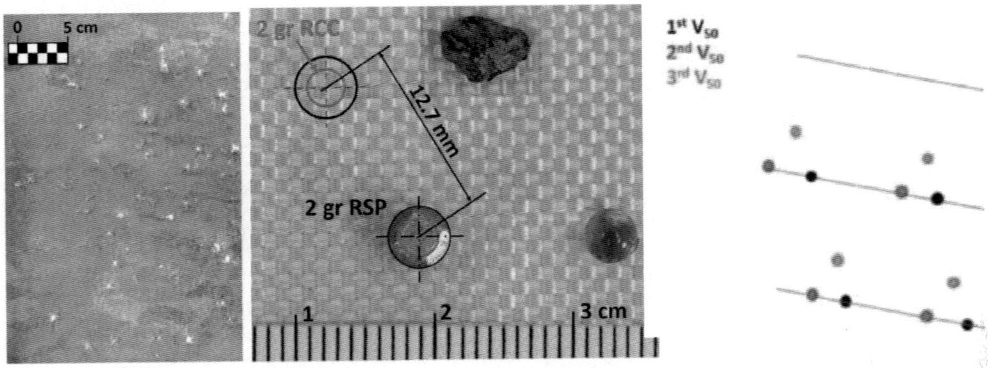

Figure 3. Typical soil ejecta spatial density (left)
RCC, RSP, Rock, and 4 mm sphere vs aramid yarn sizes (centre), multi-hit shot pattern (right)

4. BALLISTIC TESTS RESULTS

Figure 4 illustrates the level of flight stability obtained with the RSP design. The average total yaw at impact was 2.5° with 73% of the shots having yaw lower than 3° and only 5% with yaw above 5°. The yaw behaviour is quite similar to what is obtained with the 17 gr FSP, but much less than typically yaw values observed when launching the 2 gr RCC and the 2.5-gr FSP using sabots.

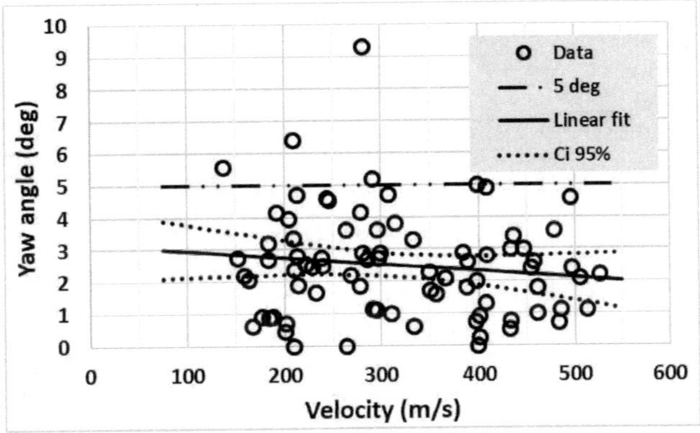

Figure 4. RSP yaw angle vs velocity

Figure 5 (left) illustrates a test where the 4 mm glass sphere penetrated the one-ply aramid fabric and remained entangled in the aramid yarn. This type of yarn pull-out did not occur with the RSP or rock projectiles as shear cutting appears to be the dominating failure mechanism. For the tests with the target made of 26 plies of Twaron®, velocities up to 1425 m/s were reached. At that point, the RSP projectiles were partly eroded with a reduction of mass of 20% and nose expansion of 5% (Figure 5 right). For all tests below 900 m/s, no erosion occurred and the projectiles remained intact. Since the 2 gr RSP is intended mainly for assessing Tier-1 PPS, erosion at very high velocities is not a concern. For the 2 gr rocks, no erosion was observed during tests at velocities around 300 m/s, but the rocks sometimes fractured into pieces upon impact. Such behaviour was also observed during full-scale buried IED trials.

https://doi.org/10.52202/078352-0010

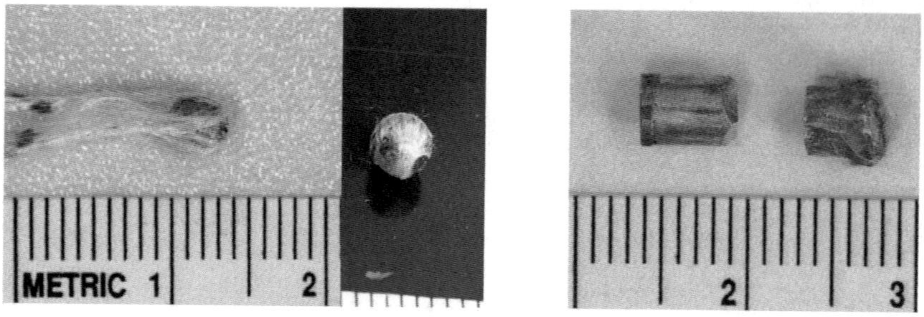

Figure 5. Twaron® yarn slippage 4mm sphere (left), recovered RSP(v1) at 500 and 1400 m/s (right)

Figure 6 shows the results obtained for test series 1 with the 4 mm glass sphere and the 2 gr RCC. Although the number of shots was limited, V_{50} values with one partial and one complete penetration within 50 m/s (2 shots V_{50}) were obtained. Results obtained previously for the same fabric against the 1 and 16 grain steel spheres and the standard FSPs and RCCs are included for comparison. To illustrate some of the scaling and testing issues involved with low areal density fabrics and non-steel fragments, the armour ***Ad*** have been normalized by the projectile sectional densities (***Sd***) as proposed by Cunniff [17]. Ballistic data for Kevlar® K706 against FSP, RCCs [19] and glass spheres [20] are also shown in Figure 5. Data from Steier [21] for Twaron® CT709 against 5.5 mm spheres of 4 densities is shown to scale well with the ***Ad/Sd*** ratio. All spheres demonstrate more severe penetrating capability than their FSP-RCC counterparts. The RSP is shown to be less penetrating than its FSP-RCC counterparts, equivalent to increasing the FSP-RCC presented areas by 39%.

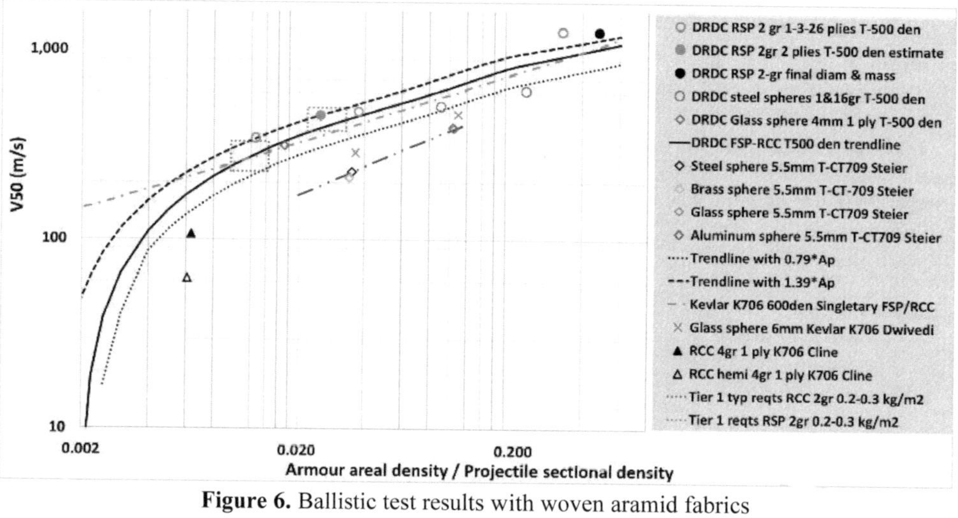

Figure 6. Ballistic test results with woven aramid fabrics

Data from Cline [22] for 1 ply Kevlar® K706 against the 4 gr RCC illustrates the effects of edge sharpness/nose radius on ballistic performance. The data does not follow the scaling trend for larger fragments and higher armour ***Ad*** for the same fabric material, suggesting that the energy absorption mechanism is changed. Figure 6 also illustrates the typical requirements (V_{50} of 230-300 m/s for 2 gr RCC) for Tier-1 PPS which are shown to increase to 375-500 m/s when scaled for the 2 gr RSP. The 2 gr RCC also appears to underestimate the protection levels of armour systems from the soil ejecta threat. The ballistic limit results obtained for the skin components of the layered backing are presented in Figure 7 (left) for the RSP and rock projectiles. The raw data was analysed using the *Stats.Blue* logistic regression calculator [23] for computing the logit penetration probability curve fits and the corresponding 95% confidence intervals. For the RCC, it was not possible to obtain a V_{50} value for the backing skin due to the lower velocity limit achievable by the firing equipment. Instead, a threshold perforation velocity $\boldsymbol{V_{th}}$ was obtained (109 m/s) by extrapolating the $\boldsymbol{V_i}$ vs ***DoP*** data (Figure 7 right). The RSP is found to be slightly more penetrating than the rock projectile with a V_{50} of 187 m/s compared to 233 m/s. For the rock projectile, a much larger standard deviation is observed which is expected since

https://doi.org/10.52202/078352-0010

each natural rock has a slightly different shape and hit the target with different orientations, making the penetration process less repeatable.

Figure 7 (right) presents the results obtained for the $\boldsymbol{V_i}$-***DoP*** calibration of the layered backing with the 2 gr RCC, RSP and rock. The three projectiles are shown to have a similar slope with the main difference being the $\boldsymbol{V_{th}}$ measured from the $\boldsymbol{V_i}$-***DoP*** fits at ***DoP* = 0**. The $\boldsymbol{V_{th}}$ values are found to be close to the V_{50} values.

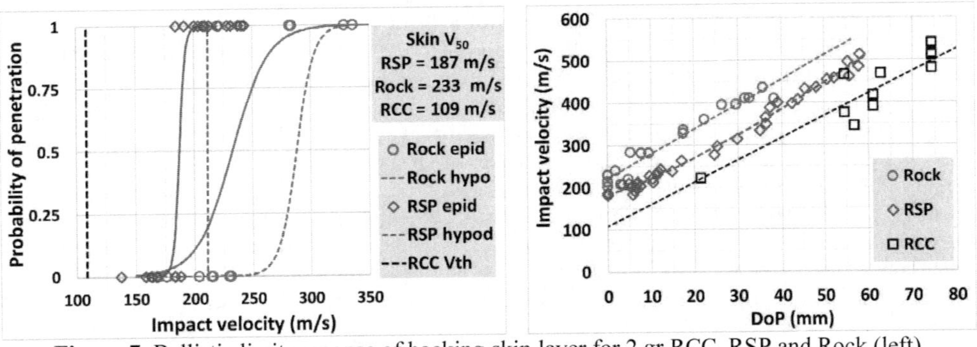

Figure 7. Ballistic limit response of backing skin layer for 2 gr RCC, RSP and Rock (left) Calibration of layered backing ($\boldsymbol{V_i}$ vs ***DoP***) for 2 gr RCC, RSP and Rock (right)

The $\boldsymbol{V_i}$-***DoP*** best fits are used in the armour overmatch test series for converting the measured ***DoP*** into $\boldsymbol{V_r}$. To further interpret the ***DoP*** results in terms of probability of incapacitation, an analysis was done using the ***DoP*** data of Breeze [24]. The data from Breeze is for ***DoP*** in goat tissues against the 0.16, 0.49 and 1.1g steel FSPs. The data was found to be best fitted by Equation 3 (R^2 of 84%) where the ***DoP*** is normalized using projectile sectional density ***Sd*** to make the correlation usable for fragments of similar shapes but made with different materials (e.g. glass, aluminium).

$$\boldsymbol{Vi = \left(\frac{DoP}{Sd}\right) * (-1176 * Sd + 108950) + 484 * Sd^{-0.4}} \qquad (3)$$

The ***DoP*** curve obtained for the RSP with the layered backing is plotted again in Figure 8, this time with ***DoPs*** in goat tissue computed from equation 3 for the 2 gr steel RCC and 2 gr RSP. The layered backing is shown to be softer than goat tissue, thus providing better resolution for determining impact velocity. From the figure, a given ***DoP*** in foam (e.g. 55 mm) can be translated into a ***DoP*** (26 mm) in goat tissue, from which the equivalent velocity of a 2 gr steel RCC having the same ***DoP*** in goat tissue can be calculated (250 m/s). This process is needed to obtain predictions for the associated probability of incapacitation ($\boldsymbol{P_i}$). $\boldsymbol{P_i}$ is calculated using the Kokinakis model [25], which was formulated for steel fragments only. The example yields a $\boldsymbol{P_i}$ of 45% (6 impacts). The black curve in Figure 8 is the result of this process for any ***DoP*** of the 2 gr RSP in the layered backing. It is also shown that no incapacitation is occurring for ***DoPs*** in the backing material less than 20 mm.

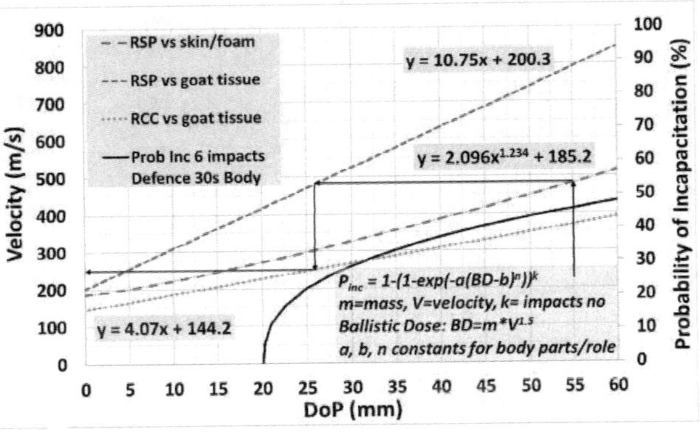

Figure 8. 2 gr RSP & RCC ***Vi*** vs ***DoP*** response for foam and goat tissue & related incapacitation

https://doi.org/10.52202/078352-0010

Figure 9 (left) presents the ballistic response curves (logit fit) of the novel fabric material against the 2 gr RCC, RSP and rock. Based on the full perforation of the skin component of the layered backing, the RCC is found to be more penetrating than the RSP.

For the RCC, the V_{50} obtained with the clamping fixture (338 m/s) is found to be much higher than that obtained with the layered backing (246 m/s). The target being allowed to deform without any restriction (air backed) and the use of the 0.5 mm aluminium witness plate which must be perforated to count as a complete penetration contribute to artificially increasing the measured ballistic resistance. For the multi-hit series with the RCC, a slight decrease of ballistic resistance was obtained for the 2nd shot V_{50} (237 m/s) with no degradation for the 3rd shot V_{50} (245 m/s) using an 18 mm triangular shot pattern. The good multi-hit performance of the novel fabric may be due to its architecture which led to very localized damage. The response of the fabric against the rock projectile demonstrates a much larger variance compared with the RSP, which was also expected given the high level of rock shape and impact angle variability. Figure 9 (right) presents the number of layers penetrated as a function of V_i for the 2 gr RCC and RSP, based on tests using samples of 12 layers of fabric. Based on a linear fit of the data, a velocity of 350 m/s would be required to penetrate two plies of fabric material, which is more than the measured V_{50} value of 291 m/s. In other words, a higher number of plies would be needed to defeat the projectile at a given impact velocity than the value estimated from semi-infinite targets. This is in agreement with the findings of Anderson [26] for metallic targets. Results for the 2 gr RCC also follow the same trend.

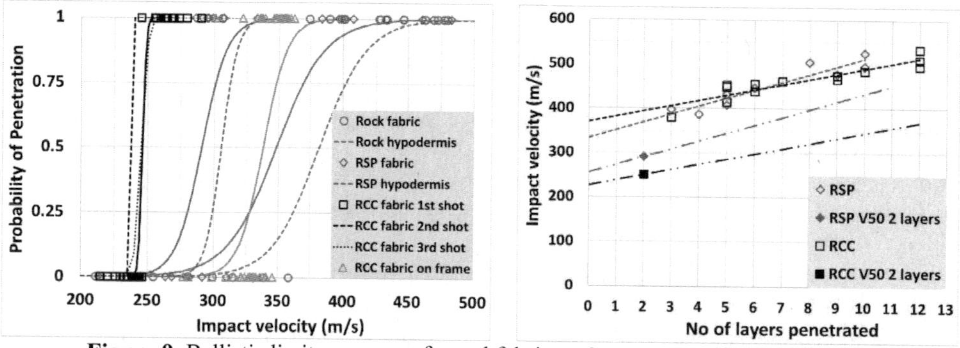

Figure 9. Ballistic limit response of novel fabric vs 2 gr RCC, RSP and Rock (left)
Number of fabric layers penetrated vs impact velocity for 2 gr RSP and RCC (right)

Figure 10 (left) presents the results obtained during the overmatch test series. The ***DoP*** data for each projectile was converted into residual velocities, from which the residual kinetic energies (***E_r***) were computed. The absorbed energies by the fabric was obtained by subtracting E_r from the initial kinetic energy of the projectile (E_i). The energy absorption ratio is given by E_a/E_i. Best fits were obtained for the energy absorption ratio data using the *Xuru* open-source software [27], which searches through more than 100 nonlinear regressions functions and ranks the resulting fits from best to worst.

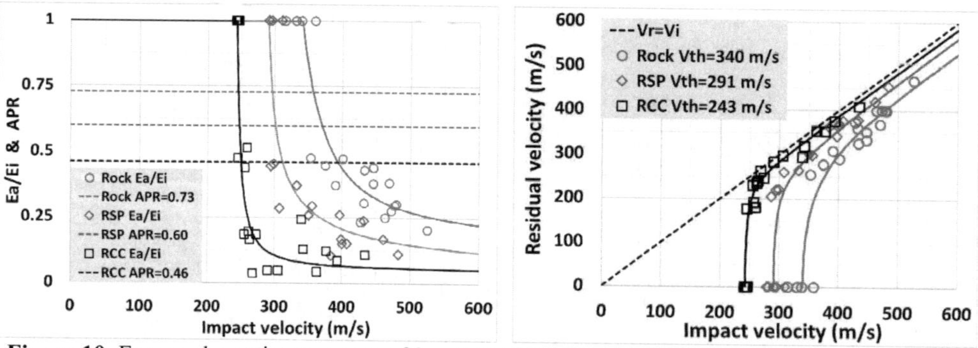

Figure 10. Energy absorption response of 2 gr RCC, RSP, and Rock (left) vs impact velocity (left)
V_r-V_i overmatch response of novel fabric for 2 gr RCC, RSP and Rock (right)

The best fit models for the energy absorption ratio were integrated over the 0-600 m/s velocity range to obtain the Armour Protection Rating (***APR***). The ***APR*** represents the average energy absorption capability of a given armour over a range of velocity. The ***APR*** captures in a single value both the

https://doi.org/10.52202/078352-0010

ballistic resistance and the overmatch response of the armour. It is similar to the *effective ballistic resistance* concept proposed by Bourget *et al.* [28]. The best fit models were converted back into V_r-V_i space (Figure 10 right), which provides key input data for vulnerability modelling tools.

Interestingly, this method of generating V_r-V_i fits was found to generate better models for the V_r-V_i data than fitting the data directly using models such as Haque and Gillepsie [29]. The obtained V_r-V_i data reproduced the typical vertical jump in V_r near V_{th} very well. Naturally, this jump is more pronounced when using the layered backing method since the projectile must penetrate both the armour and the skin component of the backing before any measurable ***DoP*** values are obtained. The ***APR*** relative ranking for the 2 gr RCC, RSP and Rock follows the one obtained from the V_{50} tests.

The operational significance of choosing the RSP instead of the RCC as a fragment surrogate for protection systems assessment can be demonstrated by analysing the fabric penetration data in the context of an explosive event producing natural fragments. First, impact velocities for a 2 gr fragment were estimated for standoff distances of 0 to 18 m from the detonation centre using the deceleration model of Thomas [15] and assuming cubic geometry and an initial velocity of 1000 m/s. The three V_r-V_i models of Figure 10 were used independently for computing corresponding V_r values. The process explained in Figure 8 was then followed to generate incapacitation predictions and the probabilities of survival (P_s =1-P_i) using the Kokinakis model [24] for the defence tactical role, 30 seconds post wounding time, with ten impacts over the entire body area. The probability of survival curves obtained are plotted relative to standoff distance for the three projectile types in Figure 11, where within a hypothetical crater radius of 1.5 m a kill probability of 100% (P_s = 0%) was assumed. For the 2 gr RCC, a safe standoff distance (P_s = 100%) of 15.9 m is obtained, which is 45% more relative to the natural rock value (11 m). For the RSP, the survival-incapacitation trendline is shown to follow much more closely to that of natural rock with a safe standoff distance of 12.6 m.

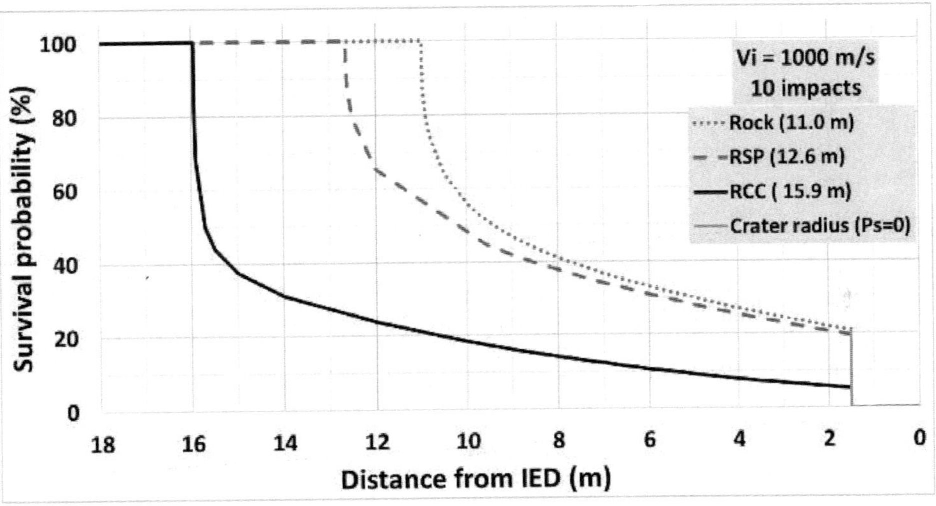

Figure 11. Estimated probability of survival vs standoff for 2 gr RCC, RSP and rock

5. CONCLUSIONS

Ballistic test results (V_{50} and overmatch) with the 2 gr Rock Simulating Projectile demonstrated that it is a suitable surrogate for soil ejecta while providing a conservative assessment of the ballistic resistance (i.e. lower V_{50}) of armour materials. The RSP can be launched in a stable manner without sabot at the low velocities required for assessing the ballistic resistance of Tier-1 personal armour systems. The results generated for the woven aramid fabric showed that the RSP V_{50} data did not follow the same scaling laws as the steel FSP and RCC data, especially for the one-ply system. Tests conducted with the 4 mm glass sphere on woven fabrics showed evidence of yarn pull-out as a dominating failure mechanism, which is different than the typical shear cutting seen with natural rocks and with the RSP. The 2 gr steel RCC was shown to be much more penetrating than the 2 gr Rock (lower V_{50} and ***APR***) leading to significant underestimation of the actual protection capability of armour systems. Additional laboratory tests (single and multi-hit) with the 2 gr RSP will be conducted to further confirm the observed trends and the scalability with the results obtained during full-scale buried IED trials. Future analyses are also planned for validating threshold and objective ballistic protection requirements (2, 6, 16 grain

https://doi.org/10.52202/078352-0010

RSPs, and impact velocities) representing the lower and upper bounds of the buried IED threat severity (levels 1 and 2).

Acknowledgments

The authors would like to acknowledge the support from the Director of Land Requirements and the Directorate of Soldier System Program Management. Special thanks to all the technical staff involved in this work, namely Mr G. Roy, M. M. Girard, and Mrs U. Gabriel from DRDC and Mr E. Fournier and Mr D. Baines from Biokinetics.

References

[1] Pageau G., Williams K., et al, Ballistic performance assessment of lightweight body armour material systems against IED threats, Proc.24th Int. Symp. on Ballistics, New Orleans, LA, USA, 2008.
[2] BNQ 2560-114, Civil Engineering Work - Aggregates, Bureau normalisation du Quebec, 2014.
[3] AEP-55 Vol 2 (Ed 2), Procedures for the evaluating the protection level of armoured vehicles - Mine threat, NATO Standardization Office, Brussels, August 2011.
[4] Ouellet S. and Pageau G, Development of a simplified torso surrogate based on selected biofidelity corridors for the assessment of the ballistic performance of soft body armor, Proc. IRCOBI conference, Athens, Greece, 2018.
[5] Ouellet S. and Pageau G, A new biofidelic backing for the evaluation of the ballistic performance of soft armour and lightweight protective fabrics, Proc. PASS2020, Copenhagen, Denmark, 2020.
[6] Gabriel, U., Pageau, G., et al, Terminal ballistics application of X-ray Computed Tomography for the analysis of fragments in collection media, Proc. Int. Symp. on Ballistics, Hyderabad, India, 2019.
[7] James G. and Hepper A., Ballistic simulation of fragmentation from buried improvised explosive devices, PASS 2014, Cambridge, United Kingdom, 8-12 September 2014.
[8] van der Jagt-Deutekom M.J., The development of a ballistic method for simulating fragments from buried explosive devices, PASS 2016, Amsterdam, The Netherlands, 19-23 September 2016.
[9] Bosik A., et al. 2002. "Initial findings on the development of test procedures for multi-hit testing of body armour", Proceedings PASS 2002, The Hague, the Netherlands.
[10] Kechagiadakis G. and Pirlot M., Development of a tool for testing PPE under near simultaneous triple impacts, Proc.11th Int. Symposium and Exhibition "Mine Action 2014", Zadar, Croatia.
[11] AEP 2920, (Ed. A, V2), Procedures for the evaluation and classification of personal armour, bullet and fragmentation threats, NATO Standardization Office, Brussels, Sept. 2016.
[12] Aldinger B.S., Evaluating the rock strike resistance of transparent armor materials, Ceramic Engineering and Science Proceedings 35(4):37-48, Dec. 2014.
[13] Liden, E., Aluminium fragment simulators for testing the effects of stone ejecta on PPE for deminers, Swedish Defence Research Agency report FOI-R-2278-SE, Feb 2007.
[14] CEN Agreement, CWA 15756, Humanitarian mine action, PPE Test and evaluation, Dec. 2007.
[15] Rail Safety and Standards Board, Standard RS 942612, Rock Strike Test
[16] Thomas J.P., Kindle C.J. et al, Modeling Soil Ejecta Threats from Buried Explosive Blasts, Proc. PASS-2016, Amsterdam, The Netherlands.
[17] Cunniff P., Variability in ballistic impact performance due to projectile physical properties and dimensions, Proc. 24th Int. Symp. on Ballistics, New Orleans, LA, Sept 2008.
[18] TOP 10-2-506, Ballistic testing of personal armour materials, US Army TECOM, 1975.
[19] Singletary J., Carbajal L., and Boogh L., Fragment simulating projectile V50 scaling rules, Proc. PASS 2008, Brussels, Belgium 2008.
[20] Dwivedi A., et al, Low velocity ballistic behavior of continuous filament knit aramid, International Journal of Impact Engineering, Vol. 96 (2016), pp. 23–34.
[21] Steier V., Carr D., et al, Effect of FSP material on the perforation of a typical body armour fabric, Proc. 28th Int. Symp. on Ballistics, Atlanta, GA, Sept 2014.
[22] Cline J., et al, The ballistic response of woven Kevlar fabric as a function of projectile sharpness, US Army Research Laboratory Report ARL-TR-8694, May 2019.
[23] Breeze J, Design validation of future ballistic neck protection through the development of novel injury models, PhD Thesis, Uni. of Birmingham 2015.
[24] http://stats.blue/Stats_Suite/logistic_regression_calculator.html
[25] Kokinakis, W. and Sperrazza J., Criteria for incapacitating soldiers with fragments and flechettes, BRL Report 1269. Ballistic Research Laboratory, USA, 1964.
[26] Anderson, C.E., Riegel, J.P., Estimate of penetration/perforation performance based on semi-infinite penetration data, Proc. 28th Int. Symp. On Ballistics, Atlanta, GA, USA, Sept. 2014.
[27] Online Nonlinear Regression Software, http://www.xuru.org/rt/NLR.asp

https://doi.org/10.52202/078352-0010

[28] Bourget D. and Pageau G., The effective ballistic resistance concept, Proc.18th Int. Symp. on Ballistics, San Antonio, Texas, USA, Nov. 1999.
[29] Haque B.Z. and Gillepsie J.W., A new penetration equation for ballistic limit analysis, J. of Thermoplastic Composite Materials, 28(7):950-972, July 2015.

 https://doi.org/10.52202/078352-0010

Ballistic Kevlar fabric with energy storage properties

Y. Chao[1], T. Bussell[2], C. Wang[1], **J. Ding[2]**,

[1] *ARC Centre of Excellence for Electromaterials Science, Intelligent Polymer Research Institute, AIIM Facility, University of Wollongong, NSW, Australia*
[2]*Land Division, Defence Science & Technology Department of Defence Melbourne, VIC, Australia (Jie.Ding@dst.defence.gov.au)*

Abstract. Shear thickening fluids (STF) are a promising component of advanced body armour materials for improved protection and flexibility. They can improve ballistic fabrics resistance to penetration by bullets, fragments or a knife without compromising weight, comfort and flexibility. Flexible body armour that allows for the wearers' maximum freedom of movement is highly desirable, especially for elbow and knee joints. Currently, batteries and power systems are required to be carried in large quantities by individual combatants due to the demands of modern warfare technology. This is a key challenge in the design of modern integrated soldier combat ensembles (SCE). It calls for the components of a SCE designed with multifunctionality. We have demonstrated success in turning STF body armour into a battery system which provides not only power but also protection against bullet impacts. Such multifunctional energy storage systems can share space and weight with existing body armour. Here, we report novel multifunctional ballistic Kevlar fabric for elbow and knee guards with energy storage properties. Lithium batteries composed of Kevlar electrodes and STF electrolyte has been developed; their electrochemical performances and protection properties against a range of kinetic impacts have been evaluated. Such batteries are capable of efficiently and safely storing power energy. It can also offer equivalent ballistic protection levels to common soft armour. The effect of shearing and bending (flexion moment and rotation) of knee and elbow joints on rheological properties of STF and electrochemical performance have been investigated.

I. INTRODUCTION

Throughout history, personal armour systems have been practical only when they have been able to provide adequate protection against prevailing threats whilst not impairing the wearer's ability to perform the tasks. Today's infantry soldier often packs more than 60 kgs and still has incomplete ballistic/blast protection. One area of increasing physical burden is in batteries to store electrical energy. Power systems carried by an individual combatant are becoming more demanding in terms of quantity and it is now a critical risk in the design of modern integrated soldier combat ensembles (SCE). It is therefore an advantage if components of a SCE are designed to perform more than one function. The development of shear thickening fluids (STF) battery offers potential to reduce the total number of systems carried by an individual combatant by combining the protective nature of body armour components and the storage of electrical energy. Extensive research is being undertaken within our group in this area [1–3]. Here we report a battery whose configuration is identical to "liquid body armour". Two basic components of "liquid body armour" were used to assemble the battery.

Adding STF into ballistic fabrics to generate "liquid body armour" has sparked significant attention, as it can improve the kinetic resistance of fabrics with more flexibility. A significant amount of research has been conducted on liquid body armour at the same areal density as conventional ballistic and stab resistant solution [4–10]. Current body armour systems are designed and implemented with a combination of different types of materials, such as ceramic, metals, polymers, ballistic fabrics and composite materials, achieving ballistic/blast protection through material layers with specific functions. The key issue is that if we provide the same level of protection to the extremities, such as arms, legs or neck as the torso, current body armour materials are far too stiff and bulky to allow wearers to move freely. Most modern body armour comprises many layers of woven Kevlar, sometimes with ceramic plates to give extra protection. For example, up to 30 or 40 layers of Kevlar are needed to offer sufficient protection; however, large number of layers of Kevlar would be too stiff and bulky for use as sleeves, trousers, and so on. Novel liquid body armour based on STFs has shown promising prospects towards improved protection and flexibility.

Shear thickening has long been a topic of interest. It is an example of a non-Newtonian fluid, often termed a dilatant fluid. At low shear rates, the fluid has low viscosity, acts as a lubricant and flows

https://doi.org/10.52202/078352-0011

easily. However, when an impact is applied (resulting in higher shear rates), the fluid adopts a solid-like state and due to a rapid increase in viscosity becomes less penetrable. More recently, the property has seen use in developing smart materials and composites [2,11–17] as their unique material properties make them ideal for many applications [2,18,19], including liquid body armour [4]. Various mechanisms have been proposed for the operation of STFs, including the formation of particle clusters by hydrodynamic lubrication forces [2,11], granular dilation [11], or impact activated solidification [13]. Simulations and experimentation have since shown that reversible shear thickening results from the formation of hydroclusters, temporary stress bearing aggregates of particles that form as a result of short range hydrodynamic lubrication forces overcoming the repulsive forces between particles during shear, as shown in Figure.1. Chains of these hydroclusters form as their numbers grow as a result of increased shear rate, resulting in larger aggregates forming which can jam flow [20]. An underlying order-disorder transition is not a requirement, but can occur in some systems prior to hydrocluster formation.

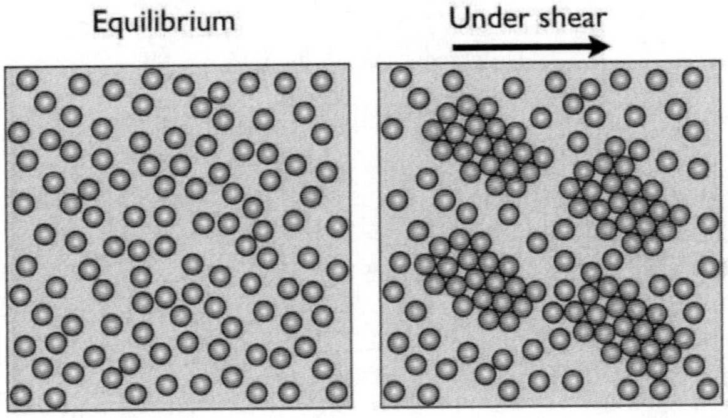

Fig.1 Schematic of particles forming hydroclusters under shear.

The ultimate goal is to create a ballistic battery that can not only share the space but also the weight of existing body armour, in order to improve the safety of batteries as well as lessen the burdens of the military personnel. This paper presents the results of work to develop materials that offer the potential of safe storage of energy, combined with an element of protection against kinetic threats. The novel lithium ion batteries were fabricated with a combination of Kevlar electrodes and STF electrolyte. To employ non-conductive Kevlar fabric as electrode substrates, silver plating process was used as a solution. After growing silver on the surface of Kevlar, high conductivity was obtained throughout the fabric. This work also investigated the effect of silver coating amount on the conductivity of Kevlar fabric, and the electrochemical performance of batteries using graphite on silver-coated Kevlar fabric (Ag-Kevlar-G). The soft-amour type of battery composed of Ag-Kevlar electrodes and STF electrolyte was also investigated.

2. EXPERIMENTAL

2.1 Preparation of silver-coated Kevlar and characterization

The Kevlar fabric was purchased from Colan Australian. Silver layer was chemically plated on the Kevlar fabric through a modified silver plating method (**Figure 1**) [20–22]. Briefly, Kevlar fabric was cut into designated sizes, e.g., small round pieces with a diameter of 15 mm using laser cutter. Acetone and Milli-Q water were successively applied to wash Kevlar pieces with the assist of ultrasonication. To ensure a good bonding with chemically grown silver layer, an air-plasma treatment was conducted to increase the surface energy of Kevlar fabric which had been dried in an oven at 70°C for 2 hour. Then pre-treated Kevlar fabrics pieces were sensitized in a solution containing 5 mg ml^{-1} $SnCl_2$ and 0.06 M HCl at room temperature for 1 h under stirring. Milli-Q water was used to thoroughly wash the sensitized fabric before it was subject to the following silver plating process.

The silver plating solution contained 12 mg ml^{-1} $AgNO_3$, 8 mg ml^{-1} NaOH and 6.25 wt% $NH_3{\cdot}H_2O$. The solution of $AgNO_3$ and NaOH produced brown precipitates immediately after the mixing, which disappeared after the addition of $NH_3{\cdot}H_2O$ solution (37%). Then the sensitized Kevlar fabrics were immersed in such sliver plating solution, followed by the quick addition of glucose solution (2.6 mg ml^{-1}) to start the dynamic chemical reaction. The reaction was kept at room temperature for 40 min

 https://doi.org/10.52202/078352-0011

under stirring. The fabric showed a colour change from yellow to metallic silver. After the final rinsing process using acetone and Milli-Q, silver-coated Kevlar fabric were obtained and denoted as Ag-Kevlar.
Field emission scanning electron microscopy (FE-SEM, JEOL JSM-7500FA) was applied to investigate the morphology of silver coating. Four-point-probe equipment was used to measure the conductivity.

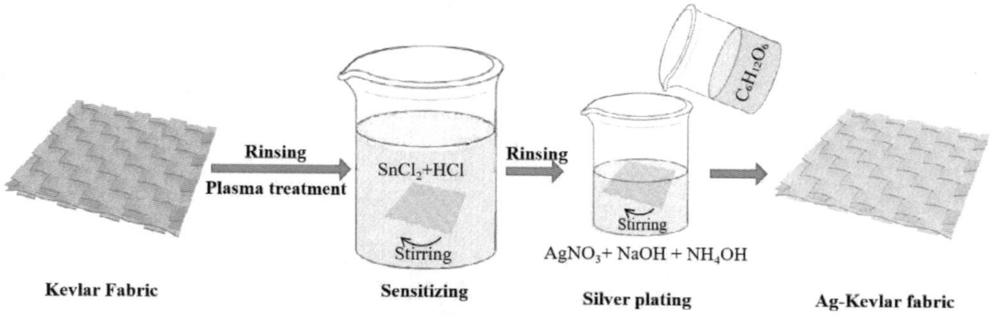

Figure 1. Schematic procedures illustrating the silver plating process on Kevlar fabric.

2.2 Preparation of Ag-Kevlar electrode

Doctor blade technique was used to coat active material slurry onto the conductive Ag-Kevlar substrate. Briefly, graphite/carbon black/PVDF (weight ratio: 8:1:1) in N-Methyl-2-pyrrolidone (NMP) was used as the slurry for anodes. Ag-Kevlar fabric was weighted prior to the coating and after the slurry-coated Ag-Kevlar was dried. The mass difference was used to calculate the weight of coated active material (graphite). The mass loading of graphite was in a range of 2-3 mg cm^{-2}. The Ag-Kevlar anodes were denoted as Ag-Kevlar-G.

2.3 Preparation of STF (shear-thickening fluids) electrolyte

The shear thickening electrolyte were prepared by mixing fume silica particles (SiO_2, S5505, Sigma-Aldrich) in a commercial lithium ion battery electrolyte (Sigma Aldrich), 1 M $LiPF_6$ in ethylene carbonate/diethyl carbonate (EC/DEC, volume ratio 1:1). SiO_2 particles were first dried in a vacuum oven at 110 °C for 24 hours and then kept in the glove box (Oxygen < 1 ppm, H_2O < 1 ppm) for 2 days before the mixing process. Based on previous results [3], 6.3 wt% SiO_2 was added into the commercial electrolyte to function as the shear-thickening electrolyte.

2.4 Electrochemical Measurement

Different combinations of electrodes and electrolyte were assembled into 2032 type coin cell batteries in the glove box (MBrau, UNIlab Plus). The Ag-Kevlar-G anode was used as working electrode with lithium foil as counter electrodes. Two type of electrolyte were used: commercial electrolyte - 1 M $LiPF_6$ in ethylene carbonate/diethyl carbonate (EC/DEC, volume ratio 1:1), and STF electrolyte. The galvanostatic charge/discharge performance of assembled batteries was investigated using a Neware testing system (Neware Electronic Co.). All the applied current density and obtained specific capacity was calculated based on the mass of active materials in Ag-Kevlar-G anodes.

2.5 Ballistic resistance testing

Ballistic resistance testing was conducted by using a gas gun (8 m). The projectile is Hornady 30 Cal.309 110 GR FML with a mass of 7.1g.

3 RESULTS AND DISCUSSIONS

3.1 Silver coated Kevlar fabric

The morphologies of blank Kevlar fabric and silver-coated Kevlar fabric (Ag-Kevlar) is displayed in Figure 2. After the sliver plating process, pristine Kevlar fabric in yellow color (Figure 2a) was

https://doi.org/10.52202/078352-0011

changed into silvery Ag-Kevlar (Figure 2b). For the sake of uniform Ag coating on Kevlar as well as the accurate mass loading on Ag-Kevlar, a laser cutter machine was used to cut the Kevlar fabric into small round pieces with a diameter of 15 mm (Figure 2c). These type of Kevlar pieces could be directly applied as electrode substrates for use in coin cells. Blank Kevlar fabric were woven with smooth fiber (Figure 2 d-e). After the silver plating process, all fibers were completely covered by silver (Figure f-g). It is also noticed that there were many small silver particle aggregates on the fiber surface, which may lead to increased surface area and has a positive effect on its performance.

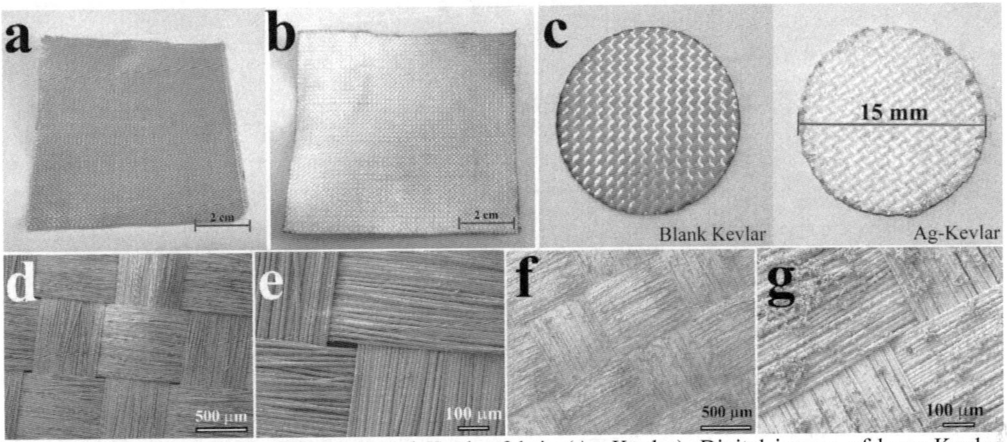

Figure 2. Demonstration of silver-coated Kevlar fabric (Ag-Kevlar). Digital images of large Kevlar fabric before (a) and after silver plating (b); (c) Digital images of the laser-dissected blank Kevlar fabric and silver coated Kevlar (Ag-Kevlar); SEM images showing the surface morphology of (d-e) blank Kevlar fabric and (f-g) Ag-Kevlar at different magnifications.

The properties of blank Kevlar and Ag-Kevlar including density, thickness, conductivity were all investigated, and shown in Table 1. Ag-Kevlar with different silver content were produced for comparison. Sample with low and high mass silver content were denoted as L-Ag-Kevlar and H-Ag-Kevlar respectively. Although Ag-Kevlar samples displayed a higher areal density than the commonly used copper foil (8.7 mg cm^{-2}) in commercial lithium ion batteries, Kevlar fabric with high strength and stress-dispersing weave pattern can well promote the ballistic resistance of batteries. The coated silver layer impacted on the conductivity of Kevlar fabric. The resistance between the double sides of Ag-Kevlar were collected for comparison and named as double-side resistance. With a silver content of 16.5%, the L-Ag-Kevlar presented a double-side resistance in the range of 0.45-0.85 ohm. While the H-Ag-Kevlar with a silver content of 34.7% displayed a high conductivity of 3.78-7.55 S m^{-1}. This good conductivity of the Ag-Kevlar samples demonstrated the potential for used as electrode substrates.

Table 1. Properties of AG-Kevlar fabric.

Sample	Blank Kevlar	L-Ag-Kevlar	H-Ag-Kevlar
Areal density (mg cm^{-2})	12.22	14.63	18.72
Sliver content	0	16.5%	34.7%
Thickness (mm)	0.165	0.182	0.2
Double-side resistance (ohm)	/	0.45-0.85	0.15-0.3
Area (cm^2)	1.766	1.766	1.766
Conductivity (S m^{-1})	/	1.21-2.29	3.78-7.55

To evaluate the performance of the above two type of Ag-Kevlar fabrics, they were all fabricated into electrodes after coating with the graphite slurry. The rate performance and the followed cycling performance with commercial liquid electrolyte are shown in Figure 3a. Clearly, H-Ag-Kevlar displayed excellent rate and cycling performance while L-Ag-Kevlar delivered slightly lower capacity during all the cycles. The discharge capacity for H-Ag-Kevlar-G and L-Ag-Kevlar-G were 547 mAh g^{-1} and 470 mAh g^{-1} at 20 mA g^{-1} (5th cycle), respectively. Even after a cycling test at 20 mA g^{-1}, H-Ag-Kevlar-G delivered a high capacity of ~559 mAh g^{-1} while it was ~500 mAh g^{-1} for L-Ag-Kevlar-G. It is interesting to notice that these two types of electrodes all delivered higher capacity at low current

https://doi.org/10.52202/078352-0011

density than the theoretical capacity of graphite. Silver can form alloys with lithium ions providing extra capacity as reported[23]. In addition, such fabric electrodes offered large surface area that promoted the diffusion and migration of electrolyte ions into active materials for realizing a high capacity. At a high current density of 100 mA g^{-1}, the discharge capacity was 352 mAh g^{-1} and 240 mAh g^{-1} for H-Ag-Kevlar-G and L-Ag-Kevlar-G electrodes, respectively. The better rate capability of H-Ag-Kevlar-G electrode can be attributed to its higher conductivity.

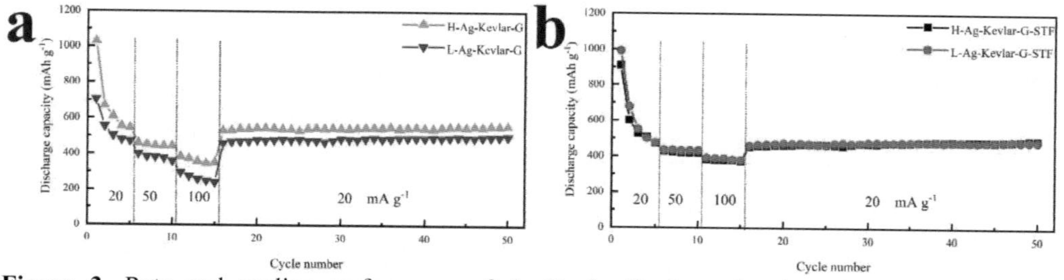

Figure 3. Rate and cycling performance of Ag-Kevlar-G electrodes in a battery system with commercial electrolyte (a) and STF electrolyte (b).

Then soft-amour type battery, Ag-Kevlar-G electrodes in combination with STF electrolyte was assembled, and the electrochemical performance is shown in Figure 3b. H-Ag-Kevlar-G and L-Ag-Kevlar-G electrodes did not show obvious difference regarding to their rate and cycling performance. The capacity for H-Ag-Kevlar-G was 472 mAh g^{-1} at 20 mA g^{-1} (5th cycle), 417 mAh g^{-1} at 50 mA g^{-1} (10th cycle) and 372 mAh g^{-1} at 100 mA g^{-1} (15th cycle); while they were 475 mAh g^{-1}, 432 mAh g^{-1} and 379 mAh g^{-1} for L-Ag-Kevlar-G electrode, respectively. After 35 cycles at 20 mA g^{-1}, both of them delivered a capacity of ~490 mAh g^{-1}, the same range as that battery system with Ag-Kevlar-G anodes and commercial liquid electrolyte. This clearly demonstrates that STF can offer similar function as commercial electrolyte for lithium ion batteries. It also proved that these Ag-Kevlar-G electrodes delivered excellent electrochemical performance in STF electrolyte. It should be pointed out that L-Ag-Kevlar-G electrodes are preferable than H-Ag-Kevlar-G to produce batteries with higher energy density due to its low weight.

3.2 Ballistic resistance testing

Ballistic resistance testing was conducted as shown in Figure 4. As Ag-Kevlar-G electrodes were developed and the electrochemical performances were tested in half cells with lithium foils as counter electrodes. It would most likely cause fire/explosion if performing the ballistic tests with the half cells due to the flammable lithium foil, which become explosive upon the contact with the moisture in air. The ballistic testing were conducted using full cells coupled with Kevlar-based cathodes and the STF electrolyte, the V50 performance of the full cell was 122 m/s. By contrast, the V50 of a full cell with commercial liquid electrolyte and electrode was 77m/s.

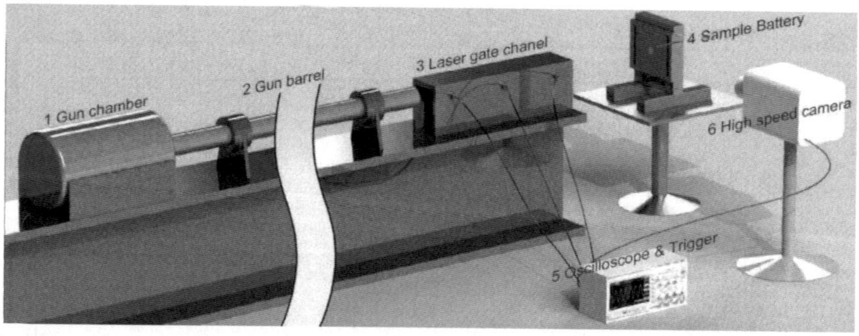

Figure 4. Schematic procedures illustrating the battery ballistic resistance test

4. CONCLUSION

This work intends to develop a ballistic-resistant soft amour battery with Kevlar electrode in combination with a STF electrolyte. With the coating of silver layer through an electroless plating

https://doi.org/10.52202/078352-0011

process, conductive Ag-Kevlar fabric was obtained and displayed low resistance, which is suitable as battery electrode substrate. The amount of sliver coating impacted on the performance of Ag-Kevlar supported graphite (Ag-Kevlar-G) anodes. The Ag-Kevlar-G with high sliver content displayed better performance than that with low sliver content (L-Ag-Kevlar-G) when using commercial liquid electrolyte. They all display stable rate capacity and cycling performance when substituted the commercial electrolyte with STF electrolyte. Take the fabrication of bullet-proof battery into consideration, thin and light L-Ag-Kevlar is preferred as electrode substrates to produce a battery with high energy density. To realize ballistic elbow and knee guards with energy storage capabilities, soft battery system composed of Ag-Kevlar electrodes and STF electrolyte need to be developed. Currently, the research work on Ag-Kevlar cathode is undergoing.

REFERENCES

[1] J. Ding, T. Tian, Q. Meng, Z. Guo, W. Li, P. Zhang, F.T. Ciacchi, J. Huang, W. Yang, Smart multifunctional fluids for lithium ion batteries: Enhanced rate performance and intrinsic mechanical protection, Sci. Rep. 3 (2013) 1–7. doi:10.1038/srep02485.

[2] J. Ding, G. Peng, K. Shu, C. Wang, T. Tian, W. Yang, Y. Zhang, G.G. Wallace, W. Li, Novel reversible and switchable electrolytes based on magneto-rheology, Sci. Rep. 5 (2015) 1–11. doi:10.1038/srep15663.

[3] J. Ding, T. Bussell, C. Wang, K. Shu, Y. Ge, BODY ARMOUR WITH POWER STORAGE CAPABILITIES, n.d.

[4] Y.S. Lee, E.D. Wetzel, N.J. Wagner, The ballistic impact characteristics of Kevlar® woven fabrics impregnated with a colloidal shear thickening fluid, J. Mater. Sci. 38 (2003) 2825–2833. doi:10.1023/A:1024424200221.

[5] E. Haro Albuja, J.A. Szpunar, A.G. Odeshi, Ballistic impact response of laminated hybrid materials made of 5086-H32 aluminum alloy, epoxy and Kevlar® fabrics impregnated with shear thickening fluid, Compos. Part A Appl. Sci. Manuf. 87 (2016) 54–65. doi:10.1016/j.compositesa.2016.04.007.

[6] Y. Park, Y.H. Kim, A.H. Baluch, C.G. Kim, Numerical simulation and empirical comparison of the high velocity impact of STF impregnated Kevlar fabric using friction effects, Compos. Struct. 125 (2015) 520–529. doi:10.1016/j.compstruct.2015.02.041.

[7] A. Haris, H.P. Lee, T.E. Tay, V.B.C. Tan, Shear thickening fluid impregnated ballistic fabric composites for shock wave mitigation, Int. J. Impact Eng. 80 (2015) 143–151. doi:10.1016/j.ijimpeng.2015.02.008.

[8] Q. He, S. Cao, Y. Wang, S. Xuan, P. Wang, X. Gong, Impact resistance of shear thickening fluid/Kevlar composite treated with shear-stiffening gel, Compos. Part A Appl. Sci. Manuf. 106 (2018) 82–90. doi:10.1016/j.compositesa.2017.12.019.

[9] X. Feng, S. Li, Y. Wang, Y. Wang, J. Liu, Effects of different silica particles on quasi-static stab resistant properties of fabrics impregnated with shear thickening fluids, Mater. Des. 64 (2014) 456–461. doi:10.1016/j.matdes.2014.06.060.

[10] S. Gürgen, M.C. Kuşhan, The stab resistance of fabrics impregnated with shear thickening fluids including various particle size of additives, Compos. Part A Appl. Sci. Manuf. 94 (2017) 50–60. doi:10.1016/j.compositesa.2016.12.019.

[11] G. Peng, Y. Ge, J. Ding, C. Wang, G.G. Wallace, W. Li, Magnetorheological technology for fabricating tunable solid electrolyte with enhanced conductivity and mechanical property, Smart Mater. Struct. 27 (2018) 035022. doi:10.1088/1361-665X/AA9F7F.

[12] G.M. Veith, B.L. Armstrong, H. Wang, S. Kalnaus, W.E. Tenhaeff, M.L. Patterson, Shear Thickening Electrolytes for High Impact Resistant Batteries, ACS Energy Lett. 2 (2017) 2084–2088. doi:10.1021/acsenergylett.7b00511.

[13] B.H. Shen, B.L. Armstrong, M. Doucet, L. Heroux, J.F. Browning, M. Agamalian, W.E. Tenhaeff, G.M. Veith, Shear Thickening Electrolyte Built from Sterically Stabilized Colloidal Particles, ACS Appl. Mater. Interfaces. 10 (2018) 9424–9434. doi:10.1021/acsami.7b19441.

[14] B.H. Shen, G.M. Veith, B.L. Armstrong, W.E. Tenhaeff, R.L. Sacci, Predictive Design of Shear-Thickening Electrolytes for Safety Considerations, J. Electrochem. Soc. 164 (2017) A2547–A2551. doi:10.1149/2.1171712jes.

[15] N.J. Wagner, J.F. Brady, Shear thickening in colloidal dispersions, Phys. Today. 62 (2009) 27–32. doi:10.1063/1.3248476.

[16] X. Cheng, J.H. McCoy, J.N. Israelachvili, I. Cohen, Imaging the microscopic structure of shear thinning and thickening colloidal suspensions, Science (80-.). 333 (2011) 1276–1279. doi:10.1126/science.1207032.

[17] S.R. Waitukaitis, H.M. Jaeger, Impact-activated solidification of dense suspensions via dynamic jamming fronts, Nature. 487 (2012) 205–209. doi:10.1038/nature11187.

[18] X. Wu, F. Zhong, Q. Yin, C. Huang, Dynamic response of shear thickening fluid under laser induced shock, Appl. Phys. Lett. 106 (2015) 071903. doi:10.1063/1.4913423.

[19] J. Yang, S. Sun, W. Li, H. Du, G. Alici, M. Nakano, Development of a linear damper working with magnetorheological shear thickening fluids, J. Intell. Mater. Syst. Struct. 26 (2015) 1811–1817. doi:10.1177/1045389X15577653.

[20] S.Q. Jiang, E. Newton, C.W.M. Yuen, C.W. Kan, Chemical silver plating and its application to textile fabric design, J. Appl. Polym. Sci. 96 (2005) 919–926. doi:10.1002/app.21541.

https://doi.org/10.52202/078352-0011

[21] Snouqiang Jiang, E. Newton, C.-W. Marcus Yuen, C.-W. Kan, Application of Chemical Silver Plating on Polyester and Cotton Blended Fabric, Text. Res. J. 77 (2007) 85–91. doi:10.1177/0040517507078739.

[22] C.W.M. Yuen, S.X. Jiang, C.W. Kan, S.K.A. Ku, P.S.R. Choi, K.P.M. Tang, S.Y. Cheng, Polyester Metallisation with Electroless Silver Plating Process, Fibers Polym. 14 (2013) 82–88. doi:10.1007/s12221-013-0082-y.

[23] M.N. Obrovac, V.L. Chevrier, Alloy Negative Electrodes for Li-Ion Batteries, Chem. Rev. 114 (2014) 11444–11502. doi:10.1021/cr500207g.

https://doi.org/10.52202/078352-0011

Numerical investigation of a shock-absorbing layer for a ballistic helmet capable of stopping rifle projectiles

A. Azevedo[1], A. Miranda-Vicario[1], F. Coghe[1] and F. Teixeira-Dias[2]

[1] *Department of Weapon Systems & Ballistics, Royal Military Academy, Brussels, Belgium, ana.ferreira@dymasec.be*
[2] *School of Engineering, University of Edinburgh, Edinburgh EH9 3JL, United Kingdom*

Abstract. Fielded ballistic helmets, both military-style and specialised police helmets, are currently unable to give appreciable protection against the AK-47 weapon system using the M43 (mild steel core) projectile. The Belgian Armed Forces (BAF) have thus a considerable interest in developing optimised systems, both for regular high-intensity conflicts, special intervention operations (e.g. Special Forces Group) and homeland operations against this wide-spread threat. Due to this capability gap, a research project was started to develop a ballistic helmet capable of stopping the Kalashnikov M43 round at muzzle velocity. The current helmet concept design consists of four layers. Whereas for the first three layers the design criteria are relatively clear and straightforward (stop the M43 Kalashnikov round with a minimum weight design) and a composite design had already been developed, designing the fourth layer poses a challenge as no internationally accepted method and threshold for behind-helmet blunt trauma (BHBT) exists. In order to reduce the risk of injury due to the local loading of the head without dramatically increasing the required standoff of the helmet, a possible approach is to maximally reduce the dynamic deflection of the helmet shell by 'pushing back' the deflection using a shock-absorbing layer. As this shock-absorbing layer is in direct contact with the head, all forces exerted on the shock-absorbing layer by the deflecting shell, will be transferred to the head. The aim of the present work was to study an innovative structure for energy absorption that minimises the likelihood of head injuries caused by a ballistic impact. The innovative helmet liner consists of an array of thermo-plastically formed cylinders manufactured by Koroyd® absorption liners. Energy is absorbed via a combination of folding and collapsing of the cylinders. The main advantage of using such liner is an optimised energy absorption for different helmet configurations and impact locations. Numerical simulations for orthogonal impacts on two different cylinder configurations of the novel liner in conjunction with the other three layers were carried out with LS-Dyna®.

1. INTRODUCTION

The most commonly found small calibre threat, both in military and in law enforcement operations, is the ubiquitous AK-47. More than 100 million AK-47s have likely been produced worldwide, and the weapon system has proliferated extensively around the world, both in the legal and in the illegal circuit. Due to this, it has often been encountered by both military and police forces [1]. Recent criminal investigations and the recent terrorist activities in Belgium and abroad have only confirmed the universal presence of the AK-47 weapon system [2-4].

A typical military-style ballistic helmet only offers protection against high-velocity fragments, which until the conflicts in Iraq and Afghanistan was also the only required level of protection as in conventional warfare, the fragmentation threat from artillery and mortar fire, is by far the most common threat. Although efforts have been made to increase the level of protection offered by current ballistic helmets, the mass constraints do not allow for any significant increase in protection by adding additional layers of protective materials [5, 6].

For continuous usage such as in a military context, the maximum allowable helmet mass is 1.8 kg, whereas for specialised helmets typical of special intervention forces, the maximum allowable mass is typically around 2.5-3.0 kg, which excludes them from being worn over extended periods. Several helmet manufacturers claim to have developed ballistic helmets offering protection against handgun and even rifle projectiles, but these helmets have never been evaluated for the risk on behind-helmet blunt trauma (BHBT) caused by the dynamic deflection of the inside of the helmet upon impact. This is due to the lack of an internationally accepted test method and associated criterion. Nevertheless, specialised helmets (generally made of titanium to reduce the dynamic deflection) offering protection against handgun threats have been developed that largely reduce the risk on BHBT, typically for use by special intervention forces.

Fielded ballistic helmets, both military-style and specialised police helmets, are currently unable to give appreciable protection against the AK-47 weapon system and the M43 projectile. The Belgian Armed Forces (BAF) have thus a considerable interest in developing optimised systems, both for regular high-intensity conflicts, special intervention operations (e.g. Special Forces Group) and homeland operations.

 https://doi.org/10.52202/078352-0012

Due to this capability gap, a research project was started within the Belgian MOD to develop a ballistic helmet capable of stopping the Kalashnikov M43 round at muzzle velocity. The final design of helmet shell would consist of four layers:

- a first ceramic layer capable of breaking and eroding the impacting projectile, especially the steel core;
- a second composite layer able to absorb the kinetic energy of the impacting projectile;
- a third stiff layer to limit the back-face deflection of the first two layers;
- and finally, a fourth layer able to absorb the shock wave of the initial impact and to provide the necessary standoff for the first three layers so that direct contact between these layers and the head is avoided.

Whereas for the first three layers the design criteria are relatively clear and straightforward (stop the M43 Kalashnikov round with a minimum weight design), designing the fourth layer poses a challenge as there is no internationally accepted method and threshold to quantify and/or evaluate behind-helmet blunt trauma (BHBT). Although several standardisation organisations have published helmet test standards with methods to assess BHBT, most of them do not specify threshold values for acceptance (or not) of a specific helmet design [7]. Those that do state a threshold value generally have no clear biomechanical basis for the specific threshold [8, 9].

A literature study was performed to identify relevant injury criteria and ways to experimentally assess the risk of injury based on these injury criteria. Looking at the different mechanisms leading to head injuries, a distinction can be made between injury mechanisms based on contact loading of the head due to the deformation and inside deflection of the helmet shell, and the associated head motion (translational, rotational, angular) [10]. Whereas the latter can be easily mitigated using the inertia of the helmet shell, the contact loading poses a significant challenge. This is mostly because light ballistic shell solutions generally have higher dynamic deflections, requiring larger helmet shell standoff in order to sufficiently decrease the risk of BHBT. A larger standoff, however, also means a larger average radius of the helmet shell, which leads to an increase in helmet weight, and a higher rotational inertia effect (which is very important from an ergonomic point of view). However, if localised loading cannot be sufficiently reduced, there is a high risk of skull fracture and injury to the brain due to dynamic wave propagation and possibly reflection.

In order to reduce the risk of injury due to local loading of the head without dramatically increasing the required standoff of the helmet, a possible approach is to maximise the reduction of the dynamic deflection of the helmet shell by 'pushing back' the helmet deflection using a shock-absorbing layer. As this shock-absorbing layer is in direct contact with the head, all forces exerted on the shock-absorbing layer by the deflecting shell will be transferred to the head. Due to natural physiological limitations, it is necessary to avoid excessive forces transferred from the shock-absorbing layer to the head. Analysing the available literature for the maximum allowable dynamic loading of the head (generally based on cadaveric experiments) a threshold seems to have been identified by several researchers for frontal head impacts of approximately 5 kN [11, 12].

In order to control the force transmitted through the shock-absorbing layer, an interesting approach is the use of crushable structures, which can be designed to crush at controllable pre-defined force levels. New shock absorbing materials based on crushable structures, such as the Koroyd material structures used for this research [13], are designed specifically for applications requiring exceptional energy absorption characteristics or excellent strength-to-density ratios.

Koroyd is a new material created by thermally welding miniature tubes together to form a whole that crushes on impact, absorbing energy in a measurable, effective way. It may look like a honeycomb, but unlike those materials it is made without glues or adhesives which can be weak or difficult to manufacture. Instead, each tube is created with a co-polymer extrusion process (see Figure 1). The inner layer is much thicker, providing the energy absorption and strength, while the outer layer is a thin membrane just a few microns thick. Its melting point is lower than that of the inner structure so, to build a sheet of the material, the tubes are stacked together and heat is applied. They then bond together across their entire length, creating a unified, consistent structure. This "sheet" of Koroyd tubes can then be bent in three dimensions, allowing it to wrap a head form with consistent thickness. Its ability to absorb varying levels of energy can be tailored by the thickness of the tube walls. Unlike EPS (expanded polystyrene) as found in many automotive helmet applications, Koroyd is also able to provide meaningful dampening for low energy impacts by elastically deforming.

The objective of this study was to perform numerical simulations for orthogonal impacts of the M43 projectile on different cylinder configurations of a novel liner concept based on the Koroyd material.

https://doi.org/10.52202/078352-0012

The different cylinder configurations were modelled in conjunction with the remaining three layers of the aforementioned advanced helmet concept. All the simulations were carried out with LS-Dyna.

The desired output of the present study was aimed at:

- evaluating the effectiveness of the two different cylinder configurations in providing force transmission at constant level between the helmet shell and the head at different impact locations.
- indicating ways in which the cylinder configurations can be optimised to reach a constant force transmission.

Figure 1. Koroyd cylinders.

2. EXAMINED CONFIGURATIONS

Contrary to other typical helmet applications (bicycle, motorcycle, skiing, etc.), the basis of the shock-absorbing layer investigated here is not a continuous layer of Koroyd, partially or completely moulded to match the inside shape of the helmet. It is instead composed of an array of short, crushable cylinders made out of Koroyd. These cylinders are attached to the inner surface of the newly developed helmet shell and support the helmet on the head. The reason for this is that ballistic impacts are more localized than typical impacts experienced during automotive or sports accidents. It can thus not be assumed that the impact load can be redistributed sufficiently fast onto the whole helmet shell (and head) to assist in absorbing and mitigating the impact energy. This also makes it a lot more difficult to 'tune' the crushing resistance of the Koroyd structure to remain under specified thresholds and keep the injury risk low.

The main aim of the research here described is to determine the optimum cylinder size and spacing. The study of the cylinder size is limited to the radius, as the height of the cylinders has already been determined [14]. Azevedo et al. [14] determined the radius and height of a stand-alone cylinder crushed by the newly developed helmet shell, based on an optimum crush strength and hence transmitted force of 5 kN. The work shown here was performed to investigate the effect of an array of multiple cylinders attached to the helmet shell.

Initially, two main configurations were considered for the array of cylinders: a regular hexagonal, close-packed spacing (configuration A), and an 'extended' hexagonal spacing (configuration B), i.e. the same hexagonal pattern as for the first configuration, with additional spacing between the cylinders.

In a hexagonal, close-packed structure, different principal directions can be identified corresponding to the most-closely packed directions (see Figure 2). These are the directions along which the units of the structure are in contact with one another in a fully dense structure. These are also the directions in which it is necessary to ensure an equal, constant inter-distance between the different units, to go from a fully dense configuration to an extended configuration.

https://doi.org/10.52202/078352-0012

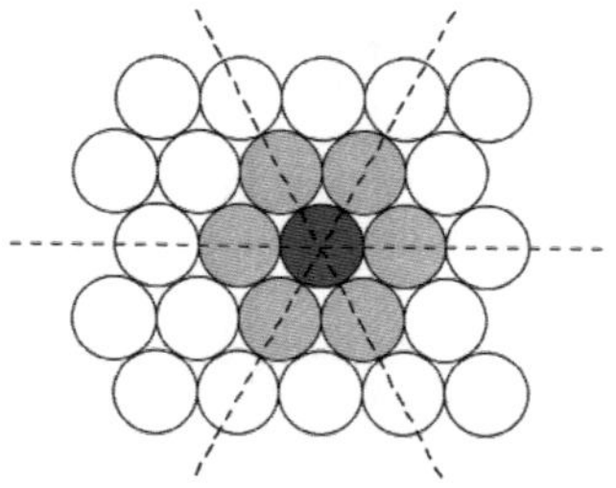

Figure 2. Schematic representation of a close packed layer of equal sized spheres/cylinders. The close packed rows (directions) are shown by the dashed lines (configuration A).

The two examined configurations are shown in Figure 3. Configuration B is basically the same as configuration A, but with a constant (non-zero) spacing along the most-closely packed directions. Figure 3 also shows the considered unit cell (dashed red line), i.e. the basic element that, if repeated, reproduces the whole structure, that is, the full array of cylinders.

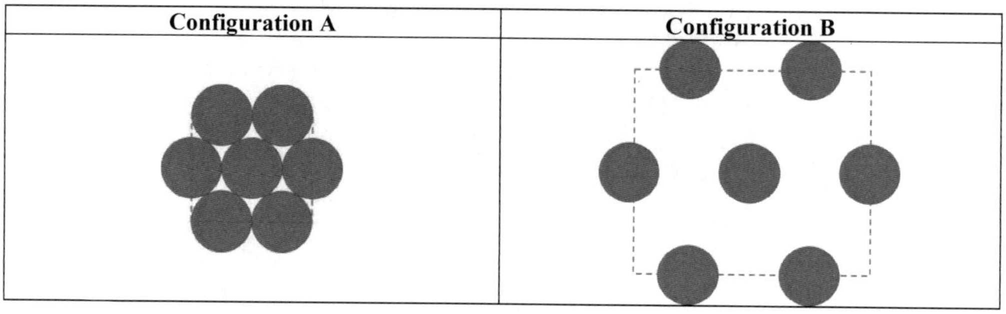

Figure 3. The two different configurations (top view): Configuration A – no space between Koroyd cylinders; Configuration B – one cylinder diameter distance (50 mm) between Koroyd cylinders.

After the initial comparison of configurations A and B, it was decided to perform further simulations using configuration B to investigate the effect of the impact location on the response of an extended structure based on configuration B. This was done as it was suspected that the transmitted force levels might be very different for an impact directly over a cylinder than for an impact in between the cylinders. The same behaviour has been observed for helmets equipped with pads [15]. Different impact locations were again selected along the most closely packed direction. Four impact locations were considered in this second phase, of which the first location corresponds to the location used for comparing configurations A and B (centre cylinder), as shown in Figure 4.

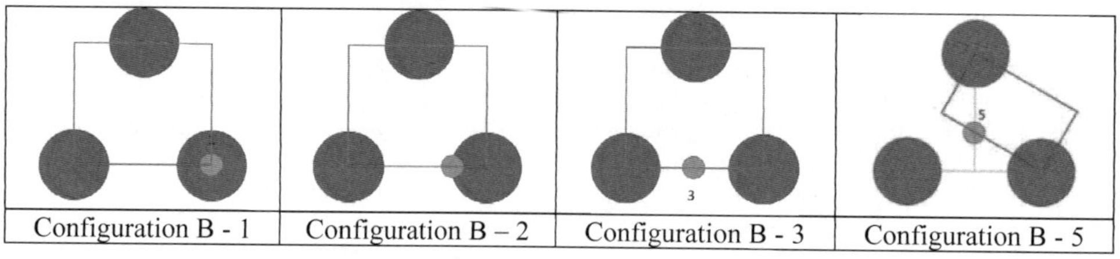

Figure 4. Different impact locations for configuration B (half model represented): (1) impact on centre; (2) on the edge, i.e. 25 mm from the centre; (3) half-way between two neighbouring cylinder; and (4) centroid of three neighbouring cylinders.

3. FINITE ELEMENT SETUP

3.1 Projectile

https://doi.org/10.52202/078352-0012

The finite element model of the projectile was previously developed and validated for different impact conditions. The model incorporates the main components of the M43 projectile: the steel jacket, the steel core, and the lead filler. An overview of the modelled components is shown in Figure 5.

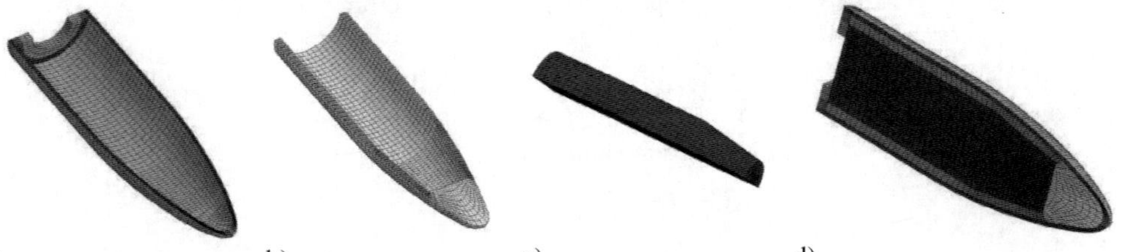

a) b) c) d)

Figure 5. Numerical model of the 7.62 x 39 mm projectile: (a) steel jacket; (b) lead filler; (c) steel core; and (d) assembled projectile.

The CAD model of the projectile was developed using the commercial package Solidworks™ and exported to the commercial HyperMesh™ software to create a hexahedral mesh. The finite element mesh was then imported to LS-Dyna™. The Johnson-Cook constitutive relation was used to simulate thermo-mechanical deformations of the steel jacket and the steel core. The lead filler of the projectile was modelled with the MAT_ISOTROPIC_ELASTIC_PLASTIC material model in LS-Dyna™ [16]. More details on the material properties and the validation of the model can be found in [17]. The impact velocity was 720 m/s for all simulations.

3.2. Helmet shell

3.2.1 Helmet shell geometry

The helmet geometry was simplified to a flat panel geometry, as this is also the way in which the experiments were performed to optimize and evaluate the ballistic shell concept [17]–[19]. Additionally, the simplification of the geometry considerably reduces the CPU time. Although both numerical and experimental results have not shown any significant difference between the local behaviour of respectively the flat panel and full helmet geometry, the global behaviour of a flat panel sample and a full helmet can be slightly different. The global behaviour is however also significant for the final load transfer between the helmet shell and the head and the movement of the head after impact on the helmet shell. The difference between the tested flat panels and the full helmet geometry can be mostly attributed to the difference in inertial effect and boundary conditions.

3.2.2 Helmet shell materials and models

The first layer of the helmet is a silicon carbide (SiC) ceramic layer, followed by a composite layer and a metallic layer, and finally the shock-absorbing layer. The material properties of the ceramic layer are those published by Miranda-Vicario et al. [17]. In order to reduce the computational time (and once this is a preliminary study) the size of the ceramic layer is smaller compare with the dimensions of the composite and the metallic layer.

The composite layer is a Dyneema™ HB80 plate (unidirectional ultra-high molecular weight polyethylene, UHMWPE), with the material properties given by Azevedo et al. [20]. The metallic plate at the back of the target is Al5754 aluminium alloy and is used to limit the back face deformation.

The shock-absorbing layer was modelled using the MAT_HONEYCOMB material model in LS-Dyna [16], which is suited for describing material with real anisotropic behaviour such as foam and honeycomb materials. All normal and shear stresses can be defined separately to describe nonlinear elastoplastic behaviour of the material. The stresses are treated as fully uncoupled. The material properties implemented in the model were based on experimental compression tests and summarised in Table 1. The stress-strain curve of the shock absorbing material is shown in Figure 7. Koroyd has an immediate loading curve meaning a large amount of energy is absorbed from the moment of

https://doi.org/10.52202/078352-0012

impact. It should be noted that the stress plateau is almost completely flat (see Figure 6). The obtained compressive force using the numerical model was also compared with experimental results for dynamic impact tests using a drop weight. The results in Figure 7 show that the model can reliably predict the actual material behaviour.

Table 1. Material properties of the shock absorbing layer [13].

Property	Value
Density [kg/m^3]	130
Young's modulus [GPa]	2.4
Poisson's ratio	0.37

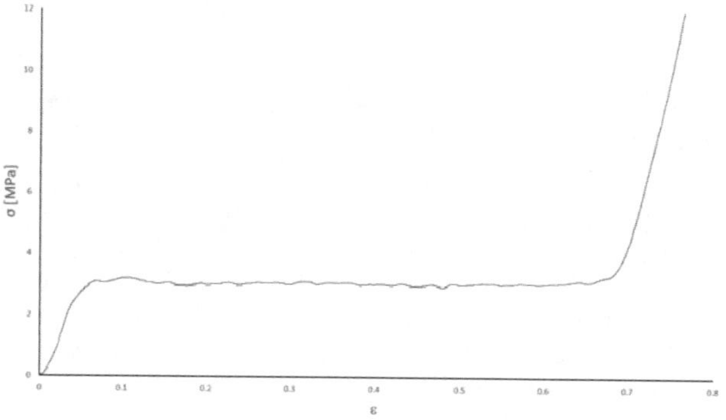

Figure 6. Stress-strain curve for the shock absorbing material (Koroyd).

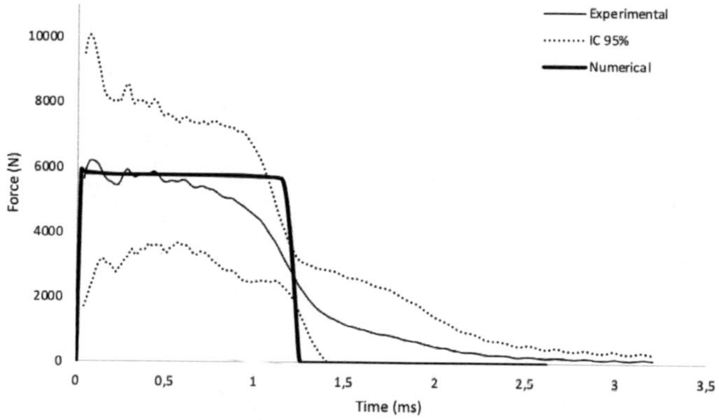

Figure 7. Comparison of the experimental (average of 3 experiments) and numerical force curves for the shock absorbing material (Koroyd) when subjected to a drop weight test.

The shock absorbing cylinder had dimensions of 30 mm length and 50 mm diameter and was modelled using eight nodded hexahedral solid elements, hence replicating exactly the experimental drop weight and impact test. A mesh convergence analysis was done on the shock absorbing material and an edge of 1 mm was chosen for the shock absorbing material.

The CONTACT_TIED_SURFACE_TO_SURFACE contact algorithm was used to simulate the interaction between the ceramic tile and the first layer of the Dyneema™ composite plate, and the interaction between the aluminium layer and the Dyneema™ plate. The element size in the impacted region was chosen to be relatively small (approximately 0.3 mm) for the composite material and for the metal layer at the back of the target. For the ceramic material, the element size was kept constant in the whole plate in order to ensure consistent crack propagation and evolution. The target was modelled with hexahedral solid elements. An example of the complete target and projectile model and discretisation is

https://doi.org/10.52202/078352-0012

shown in Figure 8. The ceramic layer was also reduced to a small plate in order to reduce considerably the computational cost.

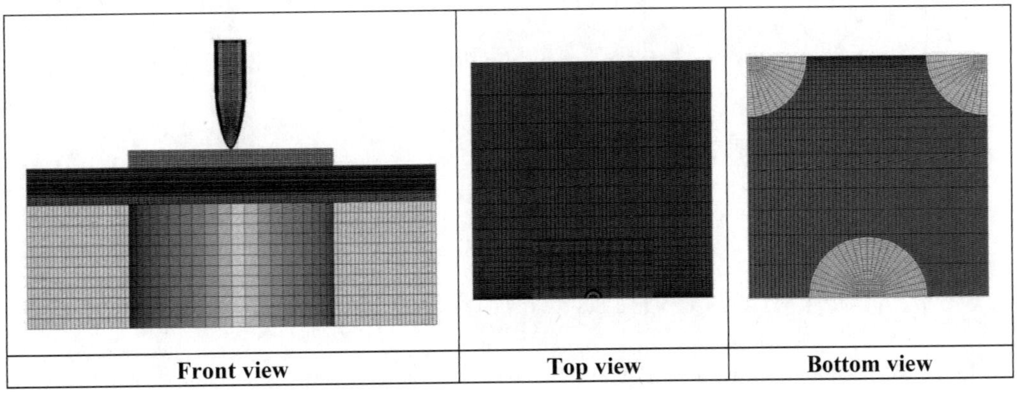

Figure 8. Different views showing the model setup and the discretisation of the target (half model) and the projectile (configuration B).

4. RESULTS AND DISCUSSION

4.1. Influence of the spacing for a centre-cylinder impact

Based on previous work by Azevedo et al. [14], 30 mm high cylinders with a 50 mm diameter were used as the basis of the evaluated configurations. The main difference between the two tested configurations A and B is the distance between the shock absorbing cylinders. In the case of configuration A, the cylinders are in contact. For configuration B, a spacing of 50 mm was selected.

The numerical setup is shown in Figure 8. Based on the unit cell shown in Figure 3, two symmetry planes were identified to minimise the required calculation time, leading to the simulation of only a quarter of the setup for an impact aimed at the centre of a cylinder (Figure 8 shows half of the model instead of one quarter for clarity). Only the nearest six neighbouring cylinders were considered to interact with the deflecting armour shell, based on the size of the actual bulges observed during experimental testing of the new helmet shell configuration. The head was considered to be a rigid wall (i.e. infinite inertia), as a worst-case scenario.

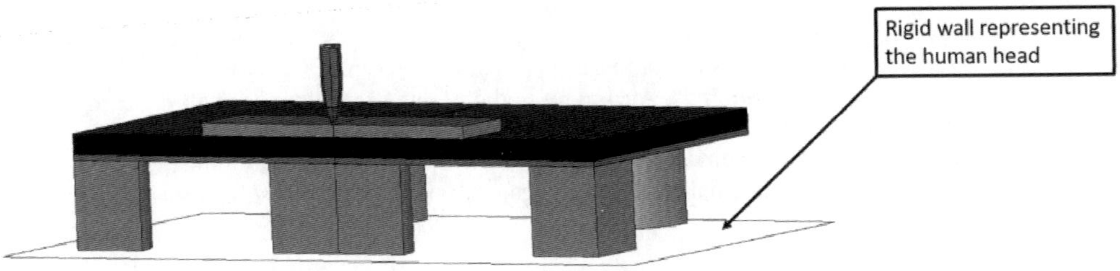

Figure 9. Illustration of the numerical model including the rigid wall representing the resistance of the human head on the inside of the helmet shell.

https://doi.org/10.52202/078352-0012

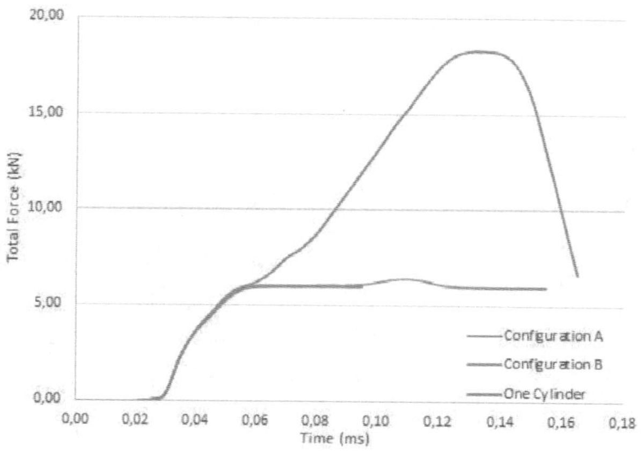

Figure 10. Total force as a function of time for configurations A and B, and a single cylinder case.

The results in Figure 10 show the total force transmitted by the Koroyd cylinders to the rigid wall for both configurations. This force is calculated using a rigid frame under the Koroyd cylinders to represent the head (see Figure 9). Configuration A shows a maximum absorbed force three times higher than configuration B. The two configurations are also compared to the force experienced by a single cylinder. As expected, configuration A shows a much higher force as the neighbouring cylinders to the central cylinder will also transmit a considerable load to the head. This is due to the fact that the bulge formed at the inside of the helmet shell is larger than the diameter of a single cylinder, and hence is in contact with several crushable cylinders. Conversely, as there is a considerable spacing between the individual cylinders in Configuration B, the load transfer in this case is limited to only the central cylinder, giving almost exactly the same force transfer as for a single cylinder.

Based on the results presented in Figure 10 and the maximum allowable dynamic loading of the head (5 kN), only configuration B seems to give reasonable results, although the transmitted force is 20% too high. This could however easily be solved by choosing a slightly smaller diameter for the Koroyd cylinders.

4.2. Influence of the impact point for an extended hexagonal configuration

Based on the results presented in Figure 10 and the maximum allowable dynamic loading of the head (5 kN), only configuration B seems to give acceptable results. Consequently, the following simulations were done with configuration B only, changing the impact location but keeping the distance between the Koroyd cylinders fixed. The four different impact locations are shown in Figure 4.

Figure 11 shows the total absorbed force for the four different impact locations as a function of time. As can be seen, configuration B-1 shows the highest total force. The remaining three configurations show a total absorbed force under 4 kN, demonstrating that the crushable cylinders are not used up to their full potential in these configurations. According to these numerical results for all different impact locations, configuration B can hence still be further optimised by changing the diameter of the Koroyd cylinder and the spacing between cylinders. In this way, the minimum standoff, and hence the mass of the helmet shell of a rifle-resistant ballistic helmet, can be obtained while still ensuring a constant maximum load transfer.

https://doi.org/10.52202/078352-0012

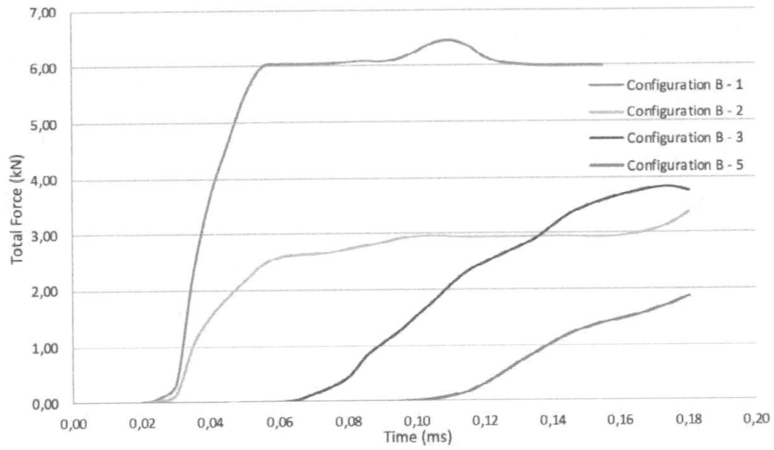

Figure 11. Total force as a function of time for the four different impact locations (configuration B).

5. CONCLUSIONS

From the simulation results, it is clear that it is possible to develop a ballistic helmet able to stop Kalashnikov ammunition keeping an acceptable BHBT.

Further research will focus on the optimisation of the shock absorbing cylinders, namely the diameter of the cylinder to have a maximum of 5 kN transmitted force, and the spacing between the cylinders in order to have a transmitted force as constant as possible, independent of the impact location. Currently only a proof of concept has been demonstrated. The next steps will include the development of a technology demonstrator and ballistic testing of the technology demonstrator under laboratory conditions. This technology demonstrator would combine the results from the different experimental and numerical approaches, in order to illustrate the available design possibilities to increase the survivability of law enforcement and military personnel during high-risk interventions (hostage rescue, high-profile arrests, forced entry, close quarters battle, etc.) This might however require the development of a new experimental setup as the testing solutions currently used in the framework of this project (US Army headform, Ballistic Load-Sensing Headform) are not suitable in their current form due to the limited size of the witness/measurement zone, compared to the expected bulge size of the helmet shell and the dimensions of the supporting array of crushable cylinder absorbing the impact (configuration B).

References

[1] B. Pauker, “Congo: On the Trail of an AK-47”, *Frontline world*, 2007.

[2] Le Parisien, “Je suis Charlie,” *Je suis Charlie*, 2015. http://atelier.leparisien.fr/sites/Je-Suis-Charlie/ (accessed Jan. 13, 2017).

[3] BBC News, “Paris attacks weapons ‘made by Zastava Arms in Serbia”, *BBC News*, Nov. 28, 2015.

[4] Associated Press in Bamako, “Two arrested in connection with Bamako hotel attack”, *The Guardian*, Nov. 27, 2015.

[5] S. G. Kulkarni, X.-L. Gao, S. E. Horner, J. Q. Zheng, and N. V. David, “Ballistic helmets - their design, materials, and performance against traumatic brain injury”, *Composite Structures*, vol. 101, pp. 313–331, Jul. 2013, doi: 10.1016/j.compstruct.2013.02.014.

[6] L. Vargas-Gonzalez, S. M. Walsh, and J. Wolbert, “Impact and ballistic response of hybridized thermoplastic laminates”, DTIC Document, 2011. Accessed: Oct. 09, 2014. [Online]. Available: http://oai.dtic.mil/oai/oai?verb=getRecord&metadataPrefix=html&identifier=ADA538498.

[7] NATO Standardization Agency, “STANAG 2920 - Ballistic test method for personal armour materials and combat clothing”, North Atlantic Treaty Organization (NATO), Jul. 2003.

[8] VPAM, “VPAM HVN 2009 Durchschusshemmender Helm mit Visier und Nackenschutz 2009,” *Vereiningung der Prufstellen fur Angriffshemmende Materialien und Konstruktionen*, 2017.

[9] M. B. Mukasey, J. L. Sedgwick, and D. W. Hagy, “Ballistic Resistance of Body Armor NIJ standard-0101.06”, *Law Enforcement Standards Laboratory of the National Bureau of Standards*, p. 89, 2008.

https://doi.org/10.52202/078352-0012

[10] T. A. Gennarelli, "Head Injuries: How to protect what", Snell HIC Conference, Milwaukee, Wisconsin, USA, May 2005.
[11] K.-U. Schmitt, P. F. Niederer, D. S. Cronin, M. H. Muser, and F. Walz, *Trauma Biomechanics: An Introduction to Injury Biomechanics*, 4th ed. Berlin Heidelberg: Springer-Verlag, 2014.
[12] D. E. Raymond, "Biomechanics of blunt ballistic temporo-pariental head impact", PhD thesis, Wayne State University, Detroit, Michigan, 2008.
[13] Koroyd, "Koroyd," 2019. https://koroyd.com (accessed Jul. 01, 2019).
[14] A. Azevedo, A. Miranda-Vicario, F. Coghe, and F. Teixeira-Dias, "Numerical evaluation of the feasibility of a novel ballistic helmet concept", In Proceedings of the 30th International Symposium on Ballistics, Long Beach, CA, Sep. 2017.
[15] C. J. Freitas, J. T. Mathis, N. Scott, R. P. Bigger, and J. MacKiewicz, "Dynamic response due to behind helmet blunt trauma measured with a human head surrogate", *Int J Med Sci*, vol. 11, no. 5, pp. 409–425, Mar. 2014, doi: 10.7150/ijms.8079.
[16] J. Hallquist, *LS-Dyna Keyword User's Manual*, vol. II. Livermore Software Technology Corporation (LSTC), 2012.
[17] A. Miranda-Vicario, A. Azevedo, F. Coghe, J. C. Matos, and M. Pirlot, "Experimental and numerical testing of different armour configurations for ballistic helmets", presented at the Personal Armour Systems Symposium (PASS), Cambridge, UK, Sep. 2014.
[18] F. Coghe, T. Vandeveld, L. Rabet, and P. Mermans, "Feasability study for an improved helmet design", in *Personal Armour Systems Symposium*, 2010, pp. 404–413.
[19] F. Coghe, T. Meunier, L. Rabet, and M. Pirlot, "Analytical and numerical modeling of an improved concept for a ballistic helmet", Freiburg – Germany, 2012.
[20] A. Azevedo, F. Teixeira-Dias, and F. Coghe, "Modeling of the ballistic behavior of Dyneema® HB26 and HB80 using LS-DYNA®", 2013.

https://doi.org/10.52202/078352-0012

The Relationship between the Shape of Backface Deformation and Behind-Armour Blunt Trauma

K. A. Rafaels[1], M. E. Lizins[2], K. L. Loftis[3], and C. A. Bir[4]

[1]*Army Futures Command, CCDC Army Research Laboratory, Aberdeen Proving Ground, MD, USA, karin.a.rafaels.civ@mail.mil*
[2]*College Qualified Leaders Program, Army Futures Command, CCDC Army Research Laboratory, Aberdeen Proving Ground, MD, USA*
[3]*Army Futures Command, CCDC, DAC, Aberdeen Proving Ground, MD, USA*
[4]*Department Chair of Biomedical Engineering, Wayne State University, Detroit, MI, USA*

Abstract. Measuring the imprint of body armour backface deformation in clay is still the most widely-used method to assess BABT. Previous efforts have demonstrated that the current metric using the maximum depth in clay may not predict injuries as well as a metric containing the velocity of the armour backface deformation; however, measuring the velocity requires upgrading facilities with expensive equipment and, even then, backface velocity is difficult to capture. This study investigates the association of thoracic injury with the depth and shape of the backface deformation in clay that is associated with the energy and energy density of the impact. This metric including depth and shape reflects the empirical evidence that deeper deformations are more likely to cause injury; but for similar depths, deformations with larger volumes are more injurious. Reconstructions of several law enforcement survivor cases of behind-armour blunt trauma to the chest were performed to obtain the backface signature in clay for these known injury events. Moulds were made of the backface signatures and then scanned to characterise the diameter, depth, volume, and surface area of the deformations. The relationship between the depth and surface area/volume ratio and the injuries from the survivor cases were determined. The depth and surface area/volume ratio showed good correlation to the injury severities reported in the survivor cases. The analysis also confirmed that even if the surface area to volume ratio between two different backface deformations were maintained, the impact that results in a deeper clay depth has a higher risk of increased injury severity. The dataset for this analysis is currently limited, but the relationship exhibits potential for an improved metric that can be easily implemented into current testing methodologies and facilities.

1. INTRODUCTION

Although many factors of the backface deformation (BFD) of body armour can contribute to injury risk, it is not feasible in the near-term to define a universal and all-encompassing injury assessment tool for behind-armour blunt trauma (BABT) since there are far too many potential locations and angles of impacts, mechanisms of thoracic injury, and various threats and protection systems that change the signature of the BFD. Therefore, simple metrics have been implemented to evaluate BABT injury risk. One of the most common metrics involves the evaluation of the BFD of the armour imprinted into a clay medium, such as in the NIJ 0101.06 standard [1]. Previous efforts have demonstrated that the current metric using the maximum depth in clay may not predict injuries as well as a metric containing the velocity of the armour backface deformation [2]. However, it is difficult to measure backface velocity. Common current techniques to measure the backface velocity of armour include: digital image correlation and tracking of unbacked armour [3-4], flash x-ray images of the armour on the backing material [5-8], high-speed video tracking and analysis of gelatin-backed armour [9-11], and inner surface measurements of thoracic rig membranes [12-14]. Each technique helps in the understanding of the backface armour response, but all have limitations. For example, unbacked armour does not demonstrate the same ballistic performance as backed armour, current flash x-ray technology has limited spatial and/or temporal resolution, high-speed video tracking is difficult to implement with hard plates that have curvature in more than one plane, and the thoracic rig membranes act as mechanical filters of the backface signature (BFS). Furthermore, these techniques to measure velocity are not commonly found in certification labs, and would add significant costs to implement. Consequently, a metric that can be determined with minimal alteration of the current clay methodology that is better correlated to injury than clay depth would be desirable to address the immediate need to evaluate newer and lighter-weight armour systems.

Previous work has investigated the relationship of the volume of the displaced clay with increased severity of injury, but it was not shown to be as well correlated as depth [2]. Even though the volume of displaced clay should be correlated to the amount of energy imparted to the body during ballistic events, the way in which that energy is deposited also affects the injury risk, e.g., the contact area and

https://doi.org/10.52202/078352-0013

sharpness of the impact [15-18]. The impression left in the clay can only provide the greatest extent of the deformation and not the evolution of the contact area over the duration of the behind-armour event. Using the area of the top of the crater (in the same plane as the original clay surface) does not adequately represent the changes in contact area against the body during the entire BABT impact event. Following the association of penetrator geometry with injury risk [17-18], then if the geometry of the backface is sharper, then the force required to penetrate soft tissue would be decreased. Ideally, all of the clay deformations would be simple, spherical geometries, so the sharpness would be calculated by obtaining the inverse of the radius of curvature at the bottom of the crater. However, it is not only the radius of curvature that affects the injury risk. For example, deformations with the same radius of curvature can have different geometries (Figure 1a) with potentially different injury risks, such as increases in volume of displaced clay. Furthermore, the geometry of the deformation can be irregular such that the radius of curvature is not the same along the bottom surface of the crater (Figure 1b). To quantify the effect of the geometry of the deformation, a surface area/volume ratio (SA/V) of the deformation was calculated since for similar depths, radii of curvature, or crater diameter, a larger surface area to volume ratio would indicate an impact with a smaller contact area.

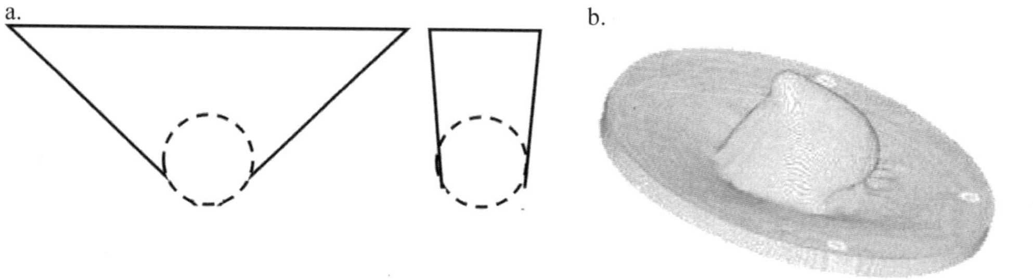

Figure 1. a. Examples of deformations with identical radii of curvature, but different crater geometries. b. An example of a deformation with an irregular geometry.

Measuring the BFS in clay is still a widely-used method to assess BABT. Despite clay depth having been shown to have only a moderate correlation to increased injury severity [2], it is still the only official metric used to assess BABT. Hence, an examination into alternative metrics that can be used with minimal modification to the clay methodology, and inclusion of parameters associated with injury risks may allow for a more immediate improvement of a test methodology to assess and/or certify the performance of body armour. Examining injuries from reconstructed cases of BABT events on clay, the relationship between depth and SA/V and the injuries from the cases will be determined.

2. METHODOLOGY

This study expanded on previous efforts to obtain and re-create cases of BABT on Roma Plastilina No. 1 clay [2, 19-20]. Similar to the study investigating the association of other clay metrics with the reconstructed cases of BABT, only handgun round impacts to the chest, defined as impact locations over the sternum or rib cage, were considered in this investigation. When available, the data previously collected in [20] was re-evaluated for this study. Additional re-creations were performed on newly-acquired cases, or for previously obtained cases if the case did not have a mould of the clay depression available, and if exemplar armour panels were available to do so. Ultimately, there were a total of 20 cases included in this analysis.

The new re-creations were performed in a similar manner to the ones previously described [20]. Prior to any data collection, approval was garnered from the Wayne State University Human Investigation Committee (HIC) and/or the University of Southern California Office for the Protection of Research Subjects (OPRS) Institutional Review Board (IRB). In summary, through surveys, interviews, medical reports, and police reports, details regarding the armour, firearm, ammunition, and range were obtained. Each incident was then re-created using the reported munition, distance, impact location, and either the same or equivalent armour system, verified by the manufacturer. Some specifics are provided in Table 1. The armour was tested on Roma Plastilina No. 1 clay conditioned in the same manner as outlined in the NIJ standard [1]. The armour was strapped to the clay with 51 mm (2-in.) wide straps, ensuring that the intended impact location was at least 76 mm (3-in.) from the border of the clay box or

https://doi.org/10.52202/078352-0013

any of the calibration indentations. Since the obliquity of the shotline was not well characterised in the collected data, all tests were performed with a 0° obliquity.

Table 1. Summary of law enforcement BABT cases

Case	Threat Caliber	Threat Type	Threat Weight (gr)	Range (m)	Armour Level (NIJ)	Armour Description
NM07^ ribs'	.45	JHP	230	1.8	IIIA	Aramid/UHMWPE
990	.40 S&W	FMJ	180	2.4	II	Aramid
3029	.357 Mag	JHP	110	4.6	II	Aramid
3138	.40 S&W	JHP	135	0.3	IIIA	Aramid
1805	.357 Mag	FMJ FN	158	0.3	II	Aramid
1753	9 mm	FMJ	124	2.4	II	Aramid
3123	.38 Sp	FMJ	158	1.5	II	Aramid/UHMWPE
2576	9 mm	JHP	115	1.8	II	Aramid/UHMWPE
2840'	9 mm	JHP	147	3.0	II	Aramid/UHMWPE w/aramid-based trauma pack
3128	9 mm	FMJ	124	0.6	II	Aramid
1716	.38 Sp	FMJ FN	158	0.9	IIIA	Aramid/UHMWPE
3166	9 mm	JHP	147	1.8	IIIA	Aramid/UHMWPE
NM07^ chest'	.45	JHP	230	1.8	IIIA	Aramid/UHMWPE w/aramid-based trauma pack
3109'	.45	FMJ	230	0.3	IIIA	Aramid
2914'	.380	FMJ	90	2.4	II	Aramid/UHMWPE
3167'	.45	FMJ	230	1.4	IIIA	Aramid
3108'	.45	FMJ	230	2.4	II	Aramid/UHMWPE w/aramid-based trauma pack
1779'	.32 Auto	FMJ	73	0.9	IIIA	Aramid
3048	.25 Auto	FMJ	50	2.4	IIIA	Aramid/UHMWPE
954	9 mm	FMJ	124	3.7	II	Aramid w/composite plate

Abbreviations: S&W – Smith and Wesson, JHP – Jacketed Hollow Point, FMJ – Full Metal Jacket, FN – Flat Nose, NIJ – National Institute of Justice, UHMWPE – Ultra-High Molecular Weight Polyethylene

'These cases were previously re-created in [20]

^NM07 was shot twice – once in the anterior chest, once more laterally on the rib cage – re-creations of both impacts were performed

A primary difference in the re-creation methodology in this study compared to the previous study was how the moulds of the BFS were made. The Polytek EasyFlo 60 (Bare Metal Foil Company; Farmington, MI) material was still used to create the mould, but the mould captured more volume than just the clay depression to include the plane of the undisturbed portion of the clay and capture the volume of any clay that was expelled from the crater, i.e., the crater lip. To identify the plane of the original clay surface, three metal washers were placed on the surface around the outside of the effective diameter of the crater. The well-mixed, two-part resin was then poured such that the height of the solution was higher than the crater lip. Figure 2 depicts the process of making the mould of the BFS, as well as an example of a casted depression.

https://doi.org/10.52202/078352-0013

a. b. c.

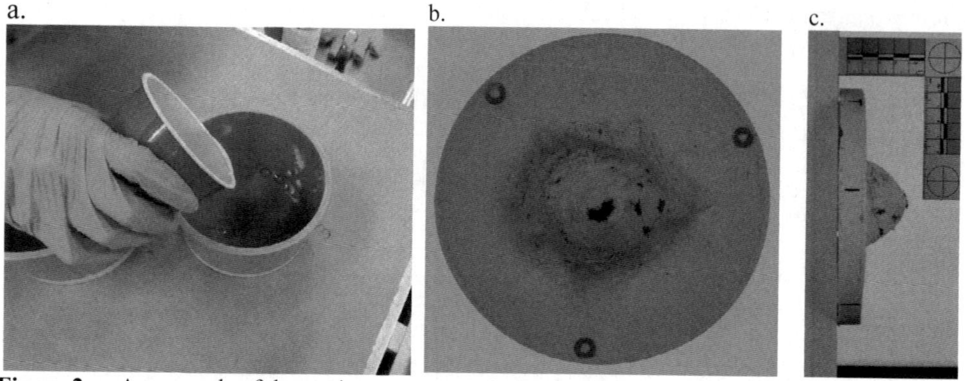

Figure 2. a. An example of the casting process to obtain the BFS of the re-created BABT cases. b. An overhead view of a mould of a BABT case created using the methodology described in this study. c. A side-view of the same mould.

All of the moulds available, both previously obtained and newly acquired, were scanned using a CT scanner (GE BrightSpeed Elite CT System, GE® Healthcare). The diameter, depth, volume, and surface area of the deformation profiles were calculated using MATLAB (MATLAB Release 2018a, The MathWorks, Inc., Natick, Massachusetts, United States). For the moulds that were previously obtained, it was assumed that the flat surface of the mould was coincident with the plane of the original clay surface. In the newly-acquired moulds, the centroids of each of the three washers were used to determine the plane of the original clay surface. The diameters, depths, volumes, and surface areas were calculated from that plane as indicated in Figure 3. If the deformation was not symmetric, the average diameter was used for further calculations.

a. b.

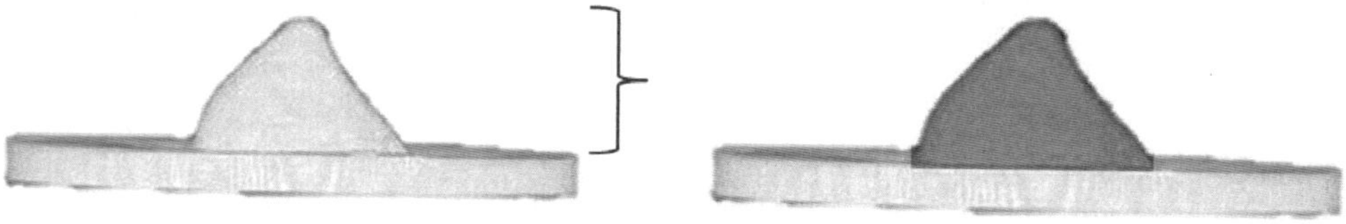

Figure 3. a. A depiction of the clay depth measurement from the new moulds obtained in this study. b. An illustration of the portion of the mould evaluated for the volume and surface area. The red-highlighted portion is the region that was analysed.

To analyse the relationship of these backface parameters to injury, the various injuries sustained by the officers had to be consolidated and described using a common scheme. The injuries were initially classified using the Abbreviated Injury Scale (AIS) [21], a commonly-used injury scoring system for traumatic injuries, but the majority of the injuries were considered AIS 1, or minor. This minor category, however, could include superficial bruises as well as fractured ribs. Consequently, similar to the previous study, an additional scoring system was applied to improve the resolution of the minor injuries [2]. Category A included minor skin injuries, while Category B included skin injuries that required wound care and/or internal injuries, e.g., rib fracture, pulmonary contusion, etc.. The categorisation of the skin injuries was confirmed by a board-certified plastic surgeon.

In a similar fashion to other injury relationships, such as the Wayne State Tolerance Curve for head injuries [22] and Bowen's curves for blast lung injuries [23], the likelihood of injury from BABT depends on multiple factors in a non-linear relationship. As in the aforementioned injury curves where the peak acceleration or pressure does not adequately describe the applied loading, the clay depth also does not accurately represent the full extent of the BFS. For this analysis, we assumed that the likelihood of increased injury severity was a function of both the depth and shape of the BFS.

Several empirical observations of how the body responds to impacts, as well as some presumed responses were considered to define the form of this relationship. Previous research has demonstrated that increases in clay depth are associated with an increased risk of injury severity [2, 20], which should be related to the delivered energy. However, other research in injury biomechanics has indicated that the shape of the delivered impact can affect levels of injury [15] and injury patterns [16]. It is expected that a BFS that has a sharper geometry would result in larger stresses and strains around the impact,

https://doi.org/10.52202/078352-0013

increasing the injury risk. Because not all of the clay signatures have simple, spherical geometries, the sharpness cannot be easily calculated, nor does it account for the rest of the surface geometry that would also affect injury risk. To quantify the effect of the shape of the deformation, SA/V of the deformation was calculated. This ratio is problematic, though, if applied across the spectrum of deformation depths, since the ratio decreases as the size increases, regardless of the shape of the deformation. Thankfully, the proportional differences between shapes remain constant for a given characteristic length, BFS depth in this study. To account for this dependence on size, the SA/V of the BFS was compared to the SA/V of a spherical cap with the same depth and diameter as the BFS. A spherical shape was selected for the comparison since that shape demonstrates the lowest SA/V. Figure 4 illustrates the parameters in the surface area and volume calculations of a spherical cap. Using half of the crater diameter for the radius of the base of the cap, *a*, and the depth for the height of the cap, *h*, the SA/V of the normalising geometry was determined (Equation 1). This form of the SA/V equation could be used since all of the BFS depths in this study were less than the diameters of the BFSs.

$$\frac{SA_{cap}=\pi(a^2+h^2)}{V_{cap}=\frac{1}{6}\pi h(3a^2+h^2)} \tag{1}$$

Equation 2 describes the comparison of the surface area to volume ratio of the BFS with the surface area to volume ratio of the spherical cap. This new ratio increases in value when the BFS becomes more spherical or less sharp of an impact, with a theoretical maximum value of 1.

$$\frac{SA}{V}_{norm} = \frac{\frac{SA_{cap}}{V_{cap}}}{\frac{SA_{BFD}}{V_{BFD}}} \tag{2}$$

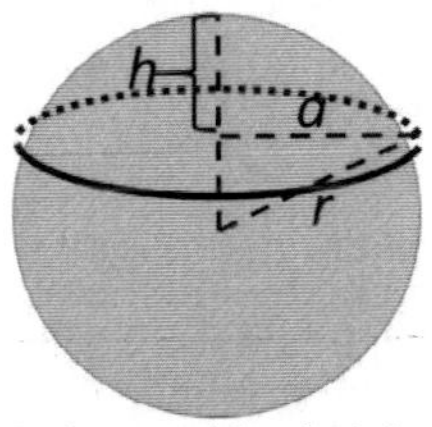

Figure 4. The surface area of a spherical cap with a height, *h*, equivalent to the clay depth, and a radius of the base of the cap, *a*, equivalent to half of the crater diameter was calculated to normalise the SA/V of the BFS.

Assuming the BFSs have primarily convex geometries in the clay, the normalised SA/V of shallower depths does not follow the same trend of describing the sharpness of the BFS for deeper depths. For example, if the depth of the BFS is small and the normalised SA/V is also small, then the surface area of the BFS must be large compared to the volume. Since the depth is small, the geometry would likely describe a relatively broad impact area. A broader impact area would spread the load across a wider area of the body which would decrease the injury risk from this type of impact relative to a more spherical impact. Likewise, when the depths increase and the normalised SA/V is small, the surface area of the BFS must again be large compared to the volume. However, in this scenario, because the depth is deep, the impact area would be narrower and more likely to be injurious. Figure 5 depicts geometries with the same SA/V, but different BFS depths to illustrate this concept.

a.

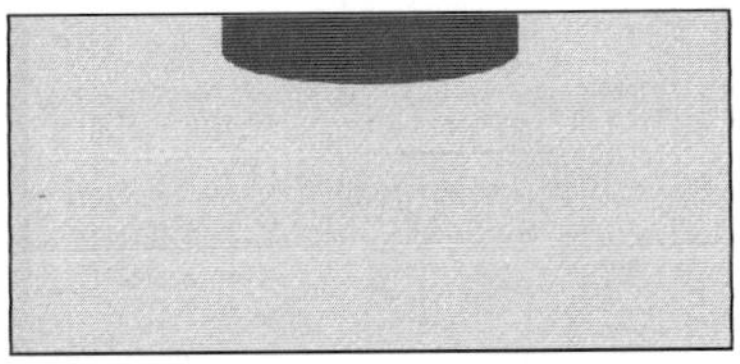

b.

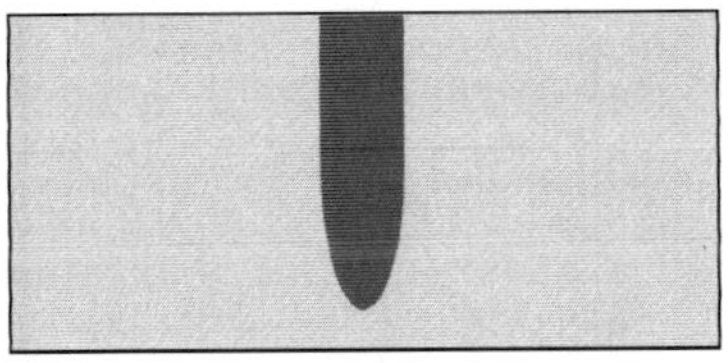

Figure 5. Comparison of similar normalised SA/V, but different BFS depths. a. An illustration of a BFS shape with a small depth and normalised SA/V. b. An illustration of a BFS shape with a large depth and small normalised SA/V.

https://doi.org/10.52202/078352-0013

The relationship between BFS depth and the normalised SA/V described in Equation 3 satisfies these changes in injury risk for the different depths and shape factor.

$$depth_{BFS} = D\left(1 + C\frac{SA}{V}_{norm}^{-b}\right) \tag{3}$$

D, *C* and *b* are constants that will be determined from the data collected from the re-created cases. As the geometry of the BFS trends towards a spherical impact, the equation will approach a constant depth, *D* + *DC*. The negative exponential constant describes the effect that in order for BFSs with small volumes, i.e., low energy, to be injurious, the BFS profile must be sharp.

A logistic regression was performed on the BABT cases re-created on clay to express the likelihood of increased injury risk in terms of the BFS depth and normalised SA/V. The logistic regression model was similar to the one described in [22], but the function used as the explanatory variable was as follows in Equation 4:

$$A = \ln\left(D\left(1 + C\frac{SA}{V}_{norm}^{-b}\right)\right) - \ln(depth_{BFS}) \tag{4}$$

Initial estimates for *D*, *C* and *b* were made based on the data distribution on the plot of BFS depth and normalised SA/V. The parameter estimates were varied until the model reached a minima with regards to the root-mean squared error and the likelihood-ratio Chi-square test statistic was significant ($p < 0.05$). Generalised R^2 was used to evaluate the goodness-of-fit. The receiver operating curve (ROC) was used to assess the predicted probabilities. A logistic regression was also performed with just the BFS depth for comparison. The logistic regression analyses were carried out using the Logistic platform of the Fit Y by X platform in JMP, version 12.0.1, statistical software (SAS Institute, Inc., Cary, NC, 2015).

3. RESULTS

A summary of the 20 cases used in this study are provided in Table 2. An injury classification of B indicates that the survivor had a skin injury that required wound care and/or had an internal injury and A indicates no medical care was required. Some of the values presented in Table 2 differ from earlier published values regarding these cases. The information presented here represents the most-up-to-date and complete data at the time of publication.

https://doi.org/10.52202/078352-0013

Table 2. Summary of the BFS in clay for the law enforcement BABT cases

Case	Injury Classification	Depth (mm)	Diameter (mm)	Volume (mL)	Surface Area (cm^2)	Normalised SA/V
NM07^ ribs'	B	46.09	87	100	62	1.08
990⁺	B	43.08	76	94	85	0.81
3029⁺	B	41.77	84	101	99	0.73
3138*	B	39.55	82	116	163	0.53
1805	B	35.46	74	93	81	0.94
1753	B	32.75	62	67	76	0.83
3123⁺	B	32.42	67	73	121	0.55
2576	B	31.42	66	52	55	0.88
2840'	B	25.26	81	44	54	0.78
3128	A	29.99	67	71	108	0.63
1716	A	27.80	76	65	148	0.41
3166	A	27.68	74	67	77	0.83
NM07^ chest'	A	27.00	93	82	151	0.49
3109'	A	23.65	98	75	230	0.31
2914'	A	23.51	67	27	28	1.06
3167'	A	21.51	75	193	312	0.69
3108'	A	21.40	98	51	111	0.48
1779'	A	21.01	66	265	409	0.76
3048	A	12.02	47	9	49	0.37
954	A	3.38	82	29	222	0.77

+These cases had an AIS2 injury associated with the BABT event, confirmed via medical records

*The survivor self-reported a pulmonary contusion (AIS2 or 3 depending on degree of lung involvement), no medical records were available to substantiate the diagnosis, the reported skin defect was severe enough to be placed in Injury Classification B

'These cases were previously re-created in [20]. The moulds from those experiments were still available so they could be scanned and analysed for this study.

^NM07 was shot twice – once in the anterior chest, once more laterally on the rib cage – re-creations of both impacts were performed

The nonlinear model fit for the 50% risk of increased severity of injury is shown in Figure 6. The injury data is plotted along with the fit to show the relationship of the Injury Classification Levels A and B to the fit. The parameters for the regression model are presented in Table 3. Table 3 also summarises the goodness-of-fit of the depth-only and nonlinear regression models. The generalised R^2 is different than the R^2 used in ordinary linear regression, so the values do not represent the proportion of variance, but it still describes a proportion in terms of the log likelihood. A higher generalised R^2 indicates a better fit. The results indicate that the goodness-of-fit and predictive capability of the model improved when using the model that included the shape parameter versus the model that only included depth.

Table 3. Model parameters and evaluation

	DF	Chi Square	p	Generalised R^2	AUC of ROC
Depth (n=20)	1	17.70	<.0001	0.64	0.96
Nonlinear Regression (n=20)	1	19.04	<.0001	0.69	0.97
Parameter Estimates					
D	10.5				
C	1.5				
-b	0.5				

https://doi.org/10.52202/078352-0013

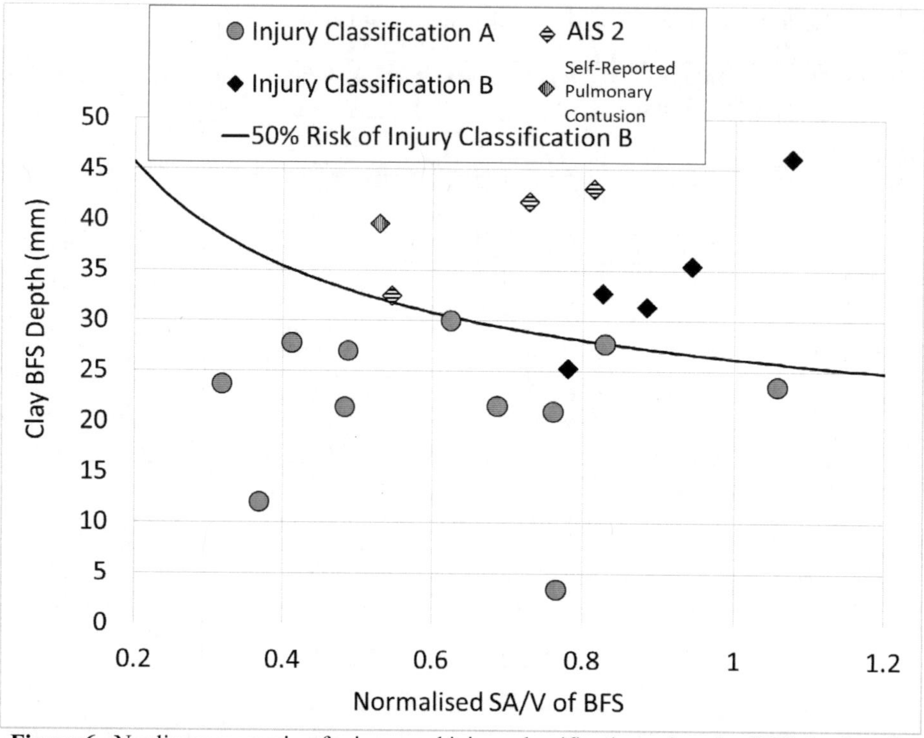

Figure 6. Nonlinear regression for increased injury classification using clay BFS depth and a normalised SA/V. The data points associated with AIS 2 injuries and the self-reported pulmonary contusion are highlighted.

4. DISCUSSION

The depth of deformation in clay has been a common method to assess BABT for decades; however, the development of the depth criterion was based on limited data using armour systems and threats that do not represent the current environment. Consequently, there is a lot of uncertainty in what the actual injury risk is for current armour systems and threats at any depth in clay. Although the depth has since been shown to have a moderate correlation to increased injury classification [2], the profile of a BFS at a certain depth can have a wide range of geometries, e.g., larger volumes, sharper shapes, etc., which can also affect the injury risk. Attempts to quantify other BFS metrics in clay have been done, but many of the studies lacked the corresponding injury data to relate the BFS parameters to injury risk [25-27]. This study is unique because it related 20 BFS profiles with associated BABT injuries in humans.

This study confirmed the notion that BFS shape along with depth affected the injury risk from BABT. Shapes that are rounded like a spherical cap have more volume of displaced clay for a particular depth than any other shape. And previous work has shown that the volume of displaced clay is related to the energy of the impact [27]. Therefore, if the volume is related to the amount of energy available to cause injury, spherical BFS shapes could deliver a threshold level of energy to the torso at smaller clay BFS depths, as indicated in Figure 6. Conversely, to reach the same energy threshold – volume of displaced clay – for sharper geometries, the clay deformation would have to be deeper. In other words, the injury relationship to geometric features of the BFS derived in this study was actually an injury assessment tool related to energy, and it has been shown that metrics associated with energy were better predictors of injury to the thorax, especially for high velocity, low mass impacts, like BABT [2, 28-29].

The fit to the nonlinear regression identified in this study describes the 50% risk of having an injury in Injury Classification B, i.e., a skin injury that required wound care and/or had an internal injury. Therefore, the injury risk described is for relatively mild injuries. However, it is expected that the general trend of the injury risk being related to the depth and shape of the BFS would be the same for serious localised injuries. The cases that had moderate injuries, AIS 2, had associated BFS depths that were generally deeper than the AIS 1 injuries. These moderate injury cases were also associated with BFSs that trended away from a spherical shape. Investigation of these profiles demonstrated irregular geometries (Figure 1b and 7), such that there were localised areas that could have relatively high impact

https://doi.org/10.52202/078352-0013

energy densities, providing more evidence that injury is associated with not just energy, but how it is delivered.

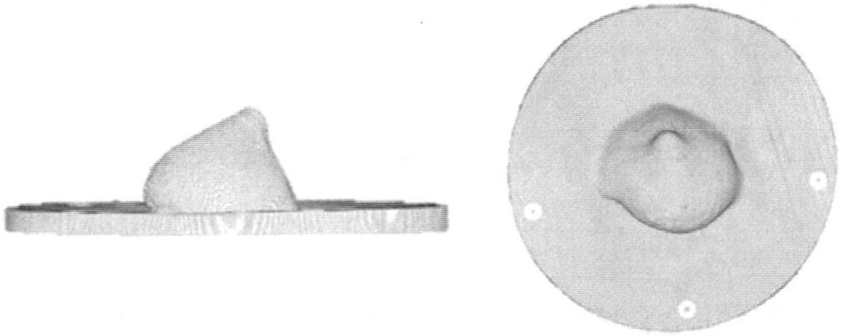

Figure 7. An example of an irregular geometry that has a region that could impact the body with a relatively high energy density.

The small number of primarily soft armour cases in this study, the unique and complex nature of each case, and the limited range of injury severities reduces the confidence in the exact probability of injury estimates of interest to armour designers and evaluators, but this analysis sets a foundation to build upon to relate the static clay BFS to the energy available to cause injury. In particular, there are uncertainties into whether this technique would apply to non-normal angles of trajectory or hard armour, as those scenarios were not directly studied, but are of interest to the community. Non-normal impacts have several aspects to consider before a robust relationship between BFS and injury can be made: how the clay deformation is affected by impact angle, and how the angle is presented to the body. For example, if the impact angle of 45° impacted the subject on the left side of the chest such that the angle was also pointing to the left, the amount of energy deposited into the body would be less than if that same angle was pointing to the right, potentially affecting injury risk. With regards to hard armor, one of the significant differences between soft and hard armour impacts is the potential for increased backface velocities for hard armour. Even though the BFS is a function of backface velocity, the BFS cannot capture, for similar deformation profiles, the rates of deformation. Previous work has indicated that the best metric for predicting increased injury was the Blunt Criterion (BC) [2], a metric that requires measuring or calculating the backface velocity of the armour, which is difficult and costly. As technology improves, it is likely it will be trivial to measure the backface velocity in the future; however, in the meantime, a metric that involves measuring the static BFS profile could be more easily implemented and may have a better relationship with injury than the current metric of BFS depth. More work needs to be done to investigate to what degree this rate parameter and non-normal impacts affect injury risk for various BABT impact conditions.

Clay is still often used to assess BABT. While alternative tools and devices have been and are being designed to improve the ability to evaluate the injury risk from BABT events, adoption of these devices has not occurred or will not likely occur quickly. Therefore, improving the metrics using the current clay methodology can more quickly address the immediate need to evaluate and optimise newer and lightweight armour systems.

5. CONCLUSION

This study has determined a relationship between the shape of the backface deformation in clay to increased injury risk from 20 re-created BABT impact events. As clay is still the most widely-used method to assess BABT, identifying a metric that can improve the ability to assess injury risk using the depression left in clay is desired. The depth and shape of the backface deformation in clay is associated with the energy and energy density of the impact, two parameters that have been shown to have a good relationship to injuries for high-rate, low-mass impacts, like BABT. The dataset for this analysis is currently limited, but the relationship exhibits potential for an improved metric that can be easily implemented into current testing methodologies and facilities.

Acknowledgments

https://doi.org/10.52202/078352-0013

This project was supported, in part, by Award No. 2011-IJ-CX-K006, awarded by the National Institute of Justice, Office of Justice Programs, U.S. Department of Justice, and Cooperative Agreement No. W911NF-17-2-0072 with the U.S. Army Research Laboratory. The opinions, findings or conclusions or recommendations expressed in this publication are those of the authors and do not necessarily reflect those of the Department of Justice or the Department of Defense. The authors would like to thank, first and foremost, all of the survivors who consented to participate in the study. We would also like to thank Erika Matheis, Rodrigo Villalta, Katherine Hewins, Nick Rowley, John Cavanaugh, and the Wayne State University Ballistics Injury Laboratory group.

References

[1] National Institute of Justice. Ballistic Resistance of Body Armor, NIJ Standard - 0101.06. (Available at: National Criminal Justice Reference System https://www.ncjrs.gov/pdffiles1/nij/223054.pdf, 2008, accessed 19 August 2011).

[2] Rafaels K.A., Loftis K.L., and Bir C.A., Can Clay Tell Us More Than Deformation?, Proceedings of the Personal Armour Systems Symposium, Washington D.C., USA, Oct. 1-5, 2018.

[3] Nader J., and Dagher H., Exp Tech, 2011; 35(2); 55-60.

[4] O'Masta MR, Compton BG, Gamble EA, Zok FW, Deshpande VS and Wadley HNG, Int. J. Impact Eng., 2015; 86, 131-144.

[5] Bass C.R., Salzar R.S., Lucas S.R., Davis M., Donnellan L., Folk B., Sanderson E., and Waclawik S., Int J Occup Saf Ergon, 2006; 12(4); 429-442.

[6] Amarilio I.B., Benes D., Asaf Z., Ya'akobovich A., Shmulevich I., Mouradjalian A., Wolf A., Grunner S., and Kluger Y., Proceedings of the Personal Armour Systems Symposium, Nuremburg, Germany, Sept. 17-21, 2012.

[7] Stuivinga M., Carton E.P, Verbeek, H.J., and van Bree J.L.M.J., Proceedings of the Personal Armour Systems Symposium, Nuremburg, Germany, Sept. 17-21, 2012.

[8] Broos J.P.F, van der Jagt-Deutekom, M., Halls V.A., and Zheng J.Q., Proceedings of the Personal Armour Systems Symposium, Nuremburg, Germany, Sept. 17-21, 2012.

[9] Metker L.W., Prather R.N., and Johnson E.M., A Method for Determining Backface Signatures of Soft Body Armors, U.S. Army Edgewood Arsenal, Technical Report, TR-75029, 1975.

[10] Mauzac O., Paquier C., Debord E., Barbillon F., Mabire P., and Jacquet J.F., Proceedings of the Personal Armour Systems Symposium, Quebec City, Canada, Sept. 14-17, 2010.

[11] Goode T., Shoemaker G., Schultz S., Peters K., and Pnakow M., Compos. Struct., 2019; 220, 687-698.

[12] Hinsley DE, Tam W, and Evison D, Behind Armour Blunt Trauma to the Thorax – Physical and Biological Models, Proceedings of the Personal Armour Systems Symposium, The Hague, Netherlands, Nov 18-22, 2002.

[13] Bourget, D, B Anctil, D Doman, and D Cronin, Development of a Surrogate Thorax for BABT Studies, Proceedings of the Personal Armour Systems Symposium, The Hague, Netherlands, Nov 18-22, 2002.

[14] Arborelius UP, Tryberg A, Gustavsson J, Malm E, Gryth D, Olsson LG, Skoglund M, Rocksen D, Proceedings of the Personal Armour Systems Symposium, Nuremberg, Germany Sept. 17-21, 2012, pp. 305-314.

[15] Cronin DS, J Biomechanics, 2007; Volume 40, Supplement 2; S48.

[16] Crandall JR, Bass CR, Duma SM, and Kuppa S. J Passeng Cars, 1998; Volume 107; 1154-1161.

[17] Shergold O and Fleck N. Proc R Soc Lond A, 2004; 460; 3037-3058.

[18] Shergold OA and Fleck NA. J Biomech Eng, 2005; 127: 838-848.

[19] Hewins K, Anctil B, Stojsih S, and Bir C, Ballistic Blunt Trauma Assessment Methodology Validation, Proceedings of the Personal Armour Systems Symposium, Nuremburg, Germany, Sept 18-21, 2012; pp. 315-323.

[20] Bir C, Lance R, Stojsih-Sherman S, and Cavanaugh J, Behind Armor Blunt Trauma: Recreation of Field Cases for the Assessment of Backface Signature Testing, Proceedings of the 30th International Symposium on Ballistics, Long Beach, CA, USA, Sept. 11-15, 2017.

[21] Association for the Advancement of Automotive Medicine. Abbreviated Injury Scale—2005 Update 2008. Chicago, IL:; 2008.

[22] Lissner HR, Lebow M, and Evans FG, Surg Gynecol Obstet, 1960; 111; 329-338.

[23] Bowen IG, Fletcher ER, and Richmond DR, Estimate of Man's Tolerance to the Direct Effects of Air Blast, Defense Atomic Support Agency, Report, DASA-2113, 1968.

[24] Bass CR, Rafaels KA, Salzar RS, J Trauma, 2008; 65; 604-615.

https://doi.org/10.52202/078352-0013

[25] Murphy MJ, Survey of the Influence of Velocity and Material Properties on the Projectile Energy/Target Hole Volume Relationship, Proceedings of the 10th International Symposium on Ballistics, San Diego, CA, USA, 1987.
[26] Broos JPF, vsn der Jagt-Deutekom M, Halls VA, and Zheng JQ, Separation Between Armour and Clay Backing during Projectile Impact, Proceedings of the Personal Armour Systems Symposium, Nuremburg, Germany, Sept 18-21, 2012.
[27] Bevan M and Luong Q, Correlation between Projectile Kinetic Energy and Displaced Clay Volume for Three Classes of Armour, Proceedings of the Personal Armour Systems Symposium, Washington, D.C., USA, Oct. 1-5, 2018.
[28] Sturdivan LM, Viano DC, and Champion HR, J Trauma, 2004; Volume 56; 651-663.
[29] Viano DC and Lau IV, J Biomechanics, 1988; Volume 21; 387-399.

https://doi.org/10.52202/078352-0013

Force Plate and Witness Material Measurement of Behind Armour Impact Forces for Different Armour Classes

M. Bevan[1], J. Clark[1], C. Peitsch[1] and Q. Luong[1]

[1]*The Johns Hopkins University Applied Physics Laboratory, Laurel MD, 20723 USA, Matthew.Bevan@jhuapl.edu*

Abstract. Initial results from experimental and analytical methods for assessing ballistic performance of three classes of armour (i.e. aramid shootpacks, ultra-high molecular weight polyethylene (UHMWPE) plates, and a multi-layered ceramic plate system) showed different forces and pressure distributions behind the armour when challenged with a matched threat. Experimental methods using a force plate and witness materials provide different perspectives of the behind armour forces than residual clay deformation alone provides. Impact plate measurements behind armour showed differences in force-time response between the different armour classes and pad thicknesses. Testing with aluminium honeycomb and tin witness materials show that residual deformations behind a shootpack were deeper and narrower than those observed behind UHMWPE and ceramic plates. These results suggest that different experimental methods and metrics used to assess armour performance may result in different criteria depending on the armour class. These results have application to developing new metrics for measuring behind armour forces that predict the potential for injury.

1. INTRODUCTION/BACKGROUND

Behind Armour Blunt Trauma (BABT) test standards have used a clay standard for decades to evaluate the potential for injury behind armour from non-penetrating projectile impacts. Initially established [1],[2],[3] for relatively low velocity (243-400 m/s), .22-.45 calibre projectiles against 7- to 12- layer woven fibre fabric armour, the clay test method has been applied to a range of projectiles ranging from 9 x 19mm NATO FMJ at 434 m/s against 22-layer Kevlar woven fibre fabric armour to NIJ Level IV projectiles at 878 m/s against a 3-layer armour system of ceramic plates, ultra-high molecular weight polyethylene (UHMWPE) fibre composite plates and Kevlar woven fabric armour.

Analysis of test results show that clay results do not necessarily parallel results gathered using other armour test methods, such as force plates or other witness materials. The development of new behind armour acceptance criteria for armour requires an understanding of how the impact force and energy are related to injury so that criteria can be developed that span numerous armour and threat combinations. Force plate and witness material responses for different classes of armour (with matching threat) are compared with previously collected clay and pressure-sensitive Fujifilm Prescale® results [4],[5]. In PASS 2014 [4], Fujifilm Prescale® was placed between armour and the clay block and the resulting pressure pattern on the Fujifilm was measured. In PASS 2018[5], the energy needed to deform clay was measured using air-cannon launched impact cones. The force plate yields information on the forces behind armour, which may be compared to the Fujifilm Prescale®, while the witness plate provides insight into the impact energy and its distribution that may be compared to clay deformation.

Different methods of behind armour impact force measurement have varying advantages and disadvantages. A force plate may have traceability of its measurement back to international standards [6], but lacks the mechanical impedance and deformation characteristics of the human body. By combining the measurement of deformation depth with the known material deformation response to impact, witness materials offer a direct measurement of the impact energy and its absorption distribution behind armour. However, while some materials have less strain hardening than others, no highly deformable, elastic/plastic materials were identified whose yielding behaviour is insensitive to strain rate. Clay has elastic/plastic/viscous behaviour and its deformation has been shown to be velocity dependent and its mechanical properties vary over time [7]. Fujifilm Prescale® is easy to use and measure, but has been shown to stretch and tear behind armour and to be strain-rate dependent [4].

2. EXPERIMENTAL METHODS

2.1 Threat-Armour Combinations Tested

Testing was conducted used three threat-armour combinations: 1) 9x19 mm FMJ NATO against an aramid Kevlar shootpack with an areal density of 5.3 kg/m^2, 2) 7.62x39 mm lead core against an

https://doi.org/10.52202/078352-0014

UHMWPE panel with an areal density of 9.8 kg/m^2, and 3) .30 calibre APM2 against a combination of Silicon Carbide (SiC) faced and UHMWPE-backed hard armour plate and aramid shootpack (SiC/shootpack) with a combined areal density of 29 kg/m^2. Each configuration was tested at two different velocity levels, the first being close to the estimated perforation velocity for that configuration, and the second being 65-80% of the V0 velocity, which corresponded to 46-65% of the perforation velocity kinetic energy.

2.2 Force Plate Design

To provide a measurement of spatial distribution of forces, the force plate design has a two-part configuration with a circular centre impact cap surrounded by a ring impact cap with an outer diameter of 152 mm. Centre impact cap diameters of 7.6 mm, 15 mm, 30 mm, and 61 mm were selected for testing. The force on the centre impact cap was measured by a single load cell while the outer ring impact cap force was measured by summing the four load cells that supported it. The five PCB 200C20 piezoelectric load cells were mounted onto a 19 mm thick stainless steel plate, and arranged as depicted in Figure 1. Data from the load cells was collected at 1 x 10^6 samples/s during impact and filtered to 50 kHz based on frequency content of signal and noise using a digital, 20-pole low pass filter. This configuration allowed measurement of the force-time distribution behind an impact as well as the force distribution.

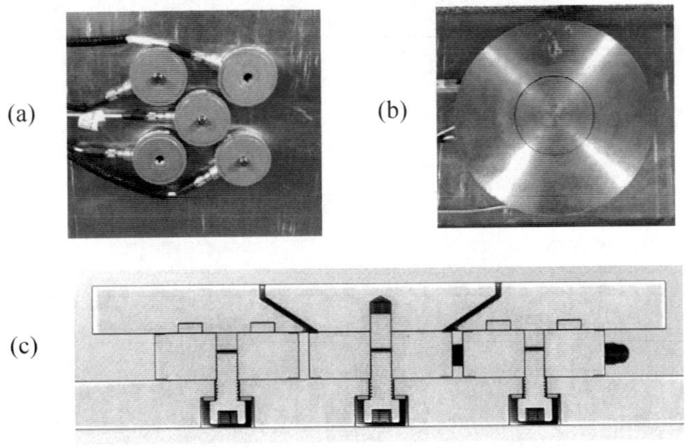

Figure 1. Force plate configuration: (a) with 5 load cells, (b) ring cap with 61 mm centre cap, (c) section view of force plate. Other impact cap diameters tested were: 7.6 mm, 15 mm, and 30 mm

Located between the back face of the armour and force plate impact caps was a sheet of neoprene (Shore-A 40 durometer), either 6 or 12.7 mm thick. The intent of adding the neoprene layer is to allow the armour to deform and to reduce the peak impact forces of the bullet.

Testing was conducted using the 3 bullet armour combinations at two impact velocities against the 4 centre caps. For the higher velocity tests, two pad thicknesses were tested.

2.3 Witness Material and Deformation Testing

An ideal witness material would have neither strain rate sensitivity nor strain hardening. Two witness materials, pure tin and an aluminium honeycomb, were selected for evaluation and testing because published data ([8] and [9]) suggested both materials exhibited elastic-plastic behaviour with little strain hardening.

2.3.1 Aluminium Honeycomb

TrussGrid (Gill Corporation) is a honeycomb-type material that was identified as a potentially suitable material for ballistic impact testing. TrussGrid is a three-dimensional aluminium honeycomb made of cross-laminated aluminium foil corrugations. Its deformation force is isotropic and uniform for 75% of the deformation range. Blocks (15x15x10 cm) with a density of 86 kg/m^3 and a crush resistance of 2.76 MPa were used for characterization and testing.

https://doi.org/10.52202/078352-0014

Three sets of tests were conducted to characterize the deformation resistance of TrussGrid. (1) compression between two flat plates on a mechanical test machine (at 2.5 mm/s) that measured deformation force and energy, (2) driving a wedge-shaped cone with either a 0.3, 0.6 or 0.9 aspect ratio into the TrussGrid at a cross-head speed of 2.5 mm/s on a mechanical test machine that measured deformation force and energy, (3) launching the same cones into the TrussGrid using an air cannon at 19-45 m/s, where a high speed camera and accelerometer measured dynamic displacement and kinematics, while later a 3 dimensional laser scanner measured residual deformation. In this fashion, the amount of force and energy needed to deform the TrussGrid could be measured across a range of velocities and indenter shapes.

The blocks used in ballistic testing were mounted on the force plate (30 mm impact cap) with double-stick tape so that the forces behind the blocks could be measured. The 3 different types of armour were mounted in front and impacted with a matching threat round at two velocities. With 3 tests at each velocity, 18 tests were conducted.

Figure 2. TrussGrid as-received (a), after deforming between two flat platens (b), and after air-cannon testing (c)

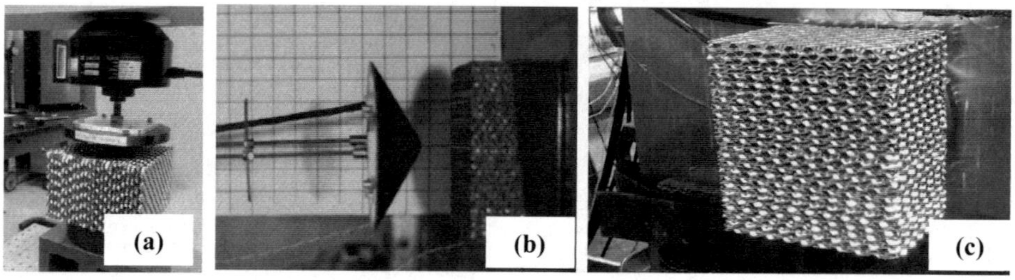

Figure 3. TrussGrid testing configurations during mechanical (a) and air cannon (b) testing and before ballistic (c) testing

2.3.2 Tin Ingot

The tin ingots used were fabricated by melting and casting pure tin pellets (99.9% pure, Rotometals inc.) into slowly cooled ingots approximately 25x76x230 mm in size. Mechancial testing was conducted using spherical indentors (32 mm and 38 mm diameter hardened steel balls) that were pressed into its surface using a mechanical test machine. The tin was loaded in a step-wise manner, measuring the increase in deformation diameter with each increase in load after the indenter was removed for each measurement.

Ballistic testing was conducted on 4 tin blocks. The tin was machined from the cast ingots into 75x75x19 mm blocks and mounted on the force plate with only the 30 mm impact cap. The three bullet-armour combinations were tested at the higher velocity and the 9 mm/shootpack combination was tested at both velocities.

2.4 Range Methods

The ballistic test velocity varied depending on the bullet and armour with the maximum velocity selected to prevent complete perforation of the armour. The bullet velocity was varied by adjusting the powder

https://doi.org/10.52202/078352-0014

loaded in the cartridge. Each shootpack received up to 9 shots while the composite panels received up to 5 shots each. Each ceramic plate was shot only once. The impact velocity was calculated based on velocity measurements made by two sets of velocity screens placed between the barrel and the armour and the distance to the target.

2.5 Post-Processing of Witness Plate Materials

The witness materials were scanned after each air cannon and ballistic test using a ROMER Absolute Laser Scanning Arm 3-D scanner. The scanner creates an array of 3-D coordinates of the surface under investigation which forms a point cloud. The point cloud was transformed into a surface using Geomagic Wrap & Geomagic Control software so that the deformation depth, diameter and displaced volume could be measured.

3.0 RESULTS

3.1 Force Plate Testing

Force-time measurements were recorded during the impact from the 5 load cells to determine the centre and outer ring forces. Example forces are shown below for the 9 mm threat impacting the shootpack at 434 m/s with 13 mm neoprene backing material (Figure 4).

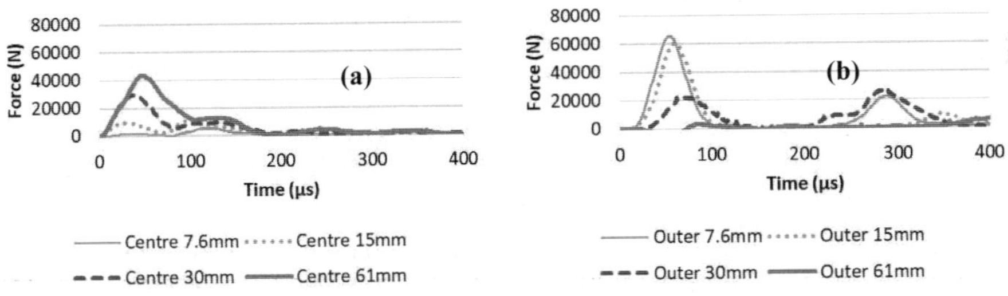

Figure 4. Centre (a) and outer (b) ring forces measured for 9 mm bullet at 434 m/s into a shootpack backed with a 13 mm neoprene pad

Impact forces were measured with time on the centre and outer rings. The outer ring force response occurs later than the centre ring response (Figure 5). Because the peak force in the centre occurred at a different time than outer peak force, the sum of peak forces (i.e. the sum of the 5 load cell measurements at the same time) is not equal to the sum of the centre and outer peak forces. Nearly all of the impact forces fall within the 61 mm cap. Unexpectedly, for many of the tests, the second peak force for the centre cap was greater than the first peak – in 8 out of 9 test conditions with the 7.6 mm cap and in 2 out of 9 conditions for the 15 mm cap. The only test condition where the first peak force was greater than the second for all tests was for the 9 mm bullet at 434 m/s (nominal) into a shootpack backed with a 6.4 mm neoprene pad.

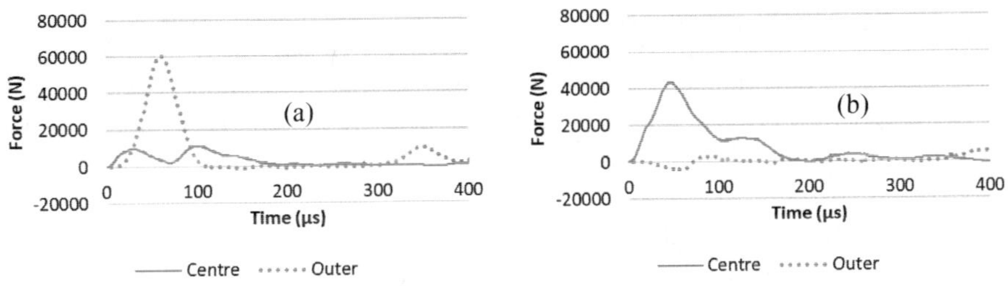

Figure 5. Centre and outer ring forces measured for 9 mm bullet at 434 m/s (nominal) into a shootpack backed with a 13 mm neoprene pad for (a) 15 mm and (b) 61 mm cap results

https://doi.org/10.52202/078352-0014

For the 9 mm threat impacting a shootpack, impact forces range from 4-13 kN for the smallest cap and 30-70 kN for the largest cap (Figure 6). The diameter where the forces are 50% of the maximum forces is estimated to be 19-23 mm. The scatter of data at the smallest cap size for the 9 mm threat may be expected since the cap is smaller than the bullet diameter and the nominal range tolerance on impact location is +/- 5 mm.

Reducing the pad thickness had a much greater effect on the impact forces than a decrease in impact kinetic energy. The double peak phenomena observed for the 7.6 and 15 mm caps nearly disappears for the thinner neoprene pad. This suggests that the thickness of the pad affects the double peak.

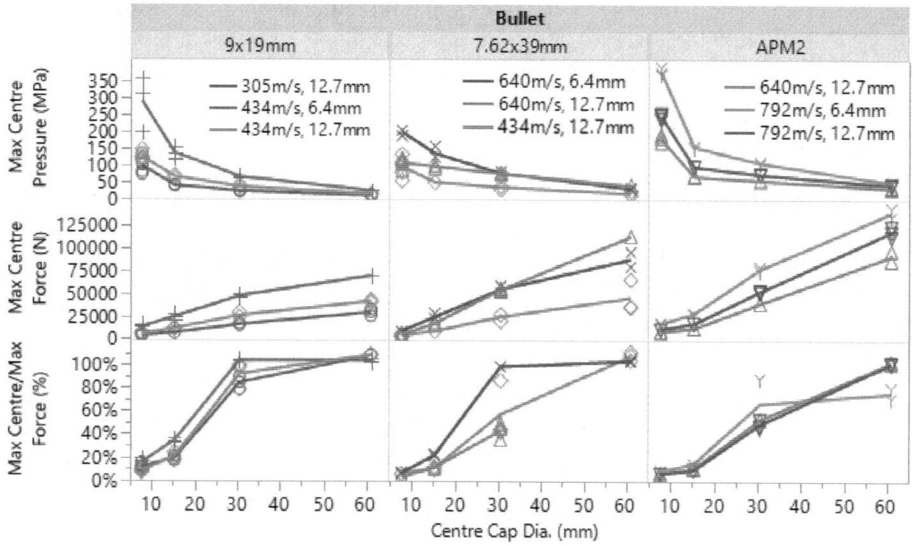

Figure 6. Force measurements for the armours and matched threats tested with various impact velocities and backing material thicknesses

The effect of the centre cap size is shown in Figure 6. While the forces on the 7.6 mm cap are relatively small, the maximum pressure (peak force divided by area of impact cap) is quite high, ranging from 106-245 MPa. Pressure drops off by about half with each doubling of diameter. The sample sizes at each test condition were limited - 1 to 3 shots.

3.2 Witness Material Testing Results

3.2.1 Aluminium Honeycomb Witness Material Results

Flat plate compression testing of three 7.5x7.5x10 cm aluminium honeycomb blocks on the square face at 2.5 mm/s showed the material, on average, deformed at 3.01 MPa, about 10% above the 2.75 MPa specified by the manufacturer. An average deforming energy of 3.01 J/cm^3 was measured in flat compression.

Mechanical testing pressing cone-shaped indenters into the square face of 15x15x10 cm aluminium honeycomb blocks at 2.5 mm/s resulted in deformation at 3.37, 3.50 and 2.90 MPa for the 0.3, 0.6, and 0.9 pitch cones respectively. Because of the elastic springback of the TrussGrid, the final pitch differed from the pitch of the cone indenter. For the 0.3 cone the final pitch was 0.29. For the 0.6 cone the final pitch was 0.55. For the 0.9 cone the final pitch was 0.79.

Overall, the deformation pressure of the cones was 3.31MPa (and a corresponding 3.31 J/cm^3 deformation energy), about 10% higher than seen in flat platen compression testing.

Air cannon testing results using cones are shown below in Table 1. The calculated deformation pressure (using the filtered peak force divided by the deformation area) was similar to that found on the mechanical test machine, with an average of 3.23 MPa. However, the energy density of the deformed area is greater than that measured in the mechanical test, potentially due to alternative damage mechanisms such as splitting of the block. Visual observation of the honeycomb blocks after testing revealed that the blocks had split along bonded layers of the honeycomb structure during impact testing.

https://doi.org/10.52202/078352-0014

This potentially created an alternative elastic deformation mechanism for storing energy in the block which could have increased the apparent energy storage density.

Ballistic testing of armour backed by the TrussGrid aluminium honeycomb was conducted, and the results are shown in the Table 2. Ballistic testing revealed similar splitting of the aluminium honeycomb observed in air-cannon testing. Both the shootpack and the UHMWPE had a conical deformation shape, while the ceramic plate/shootpack armour had a more semi-hemispherical deformation shape (Figure 7). The deformation pressure tended to increase with increasing pitch and was approximately twice the quasi-static and air cannon values. Assuming that the deformation pressure (calculated from the peak load cell force divided by deformation area) is equal to the volumetric deformation energy (in J/cm^3), it allows a lower threshold estimation of the energy to deform the aluminium honeycomb material.

Table 1. TrussGrid aluminium honeycomb data collected during air cannon testing

Test ID	Cone Pitch	Impact Velocity (m/s)	Impact Kinetic Energy (J)	Peak Force (N)	Dent Diameter (mm)	Dent Depth (mm)	Dent Pitch	Total Volume Displaced (mm^3)	Deformation Pressure (MPa)	Energy Density (J/cm^3)
AC2_053	0.6	18.6	96.5	8202.7	56.7	13.8	0.49	11057	3.25	8.73
AC2_054	0.6	27.8	216	14249	77.2	19.0	0.49	32712	3.04	6.59
AC2_055	0.6	39.2	429	22496	91.7	23.7	0.52	45295	3.41	9.47

Table 2. Data from ballistic testing against TrussGrid aluminium honeycomb

Target Material	Threat	N	Avg. Velocity (m/s)	Avg. Depth (mm)	Avg. Diameter (mm)	Avg. Pitch	Avg. Volume Displaced (mm^3)	Avg. Impact Peak Force (N)	Avg. Deformation Pressure (MPa)	Avg. Total Deformation Energy (J)
Shootpack	9mm	3	309	21.6	60.2	0.72	21549	18969	6.84	146
Shootpack	9mm	3	437	33.6	64	1.05	35327	26041	8.20	288
UHMWPE	7.62x39	3	431	11.6	94.2	0.25	29421	35907	5.15	152
UHMWPE	7.62x39	3	642	21	97.2	0.43	59402	46616	5.23	321
SiC/ Shootpack	APM2	3	642	20.3	86.1	0.47	48066	32680	5.66	271
SiC/ Shootpack	APM2	3	806	29	88.8	0.65	79019	41198	6.67	527

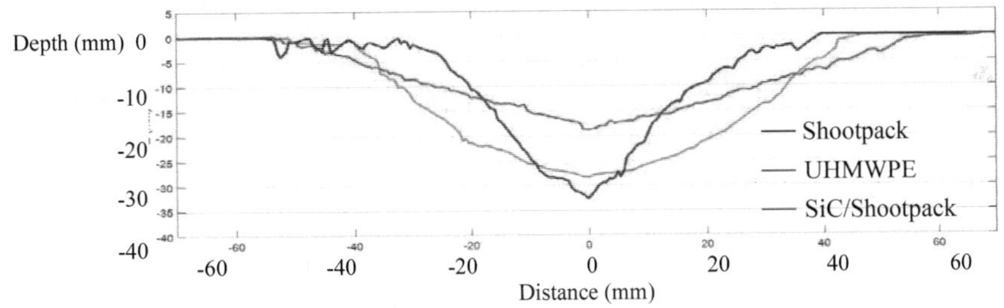

Figure 7. Comparison of TrussGrid cross-section profiles from ballistic testing at the higher velocity

3.2.2 Tin Ingot Witness Material Results

Mechanical compression testing with a 32 mm and 38 mm diameter ball was conducted on tin. The yield strength was measured to be about 42 MPa for both indentors, and the energy density of deformation was 42 J/cm^3.

The results from ballistic testing on tin are shown in Table 3. The peak force per unit area is far greater than the quasi-static results. The deformation energy is far less than measured using TrussGrid. A cross-sectional view of the tin deformation is shown in Figure 8. While the sample size was one test in each condition, the results show differences in responses.

Table 3. Ballistic test results on tin

Target Material	Threat	Velocity (m/s)	Depth (mm)	Diameter (mm)	Pitch	Volume Displaced (mm^3)	Impact Peak Force (N)	Peak Force/Area (MPa)	Total Deformation Energy (J)
Shootpack	9mm	306	2.41	19.5	0.25	364	34402	115.2	41.9

https://doi.org/10.52202/078352-0014

Shootpack	9mm	425	5.10	22.0	0.46	954	42066	110.7	105.6
UHMWPE	7.62x39	641	4.73	37.8	0.25	1739	81580	72.9	126.8
SiC/Shootpack	APM2	800	3.31	39.25	0.17	1907	107870	89.1	170.0

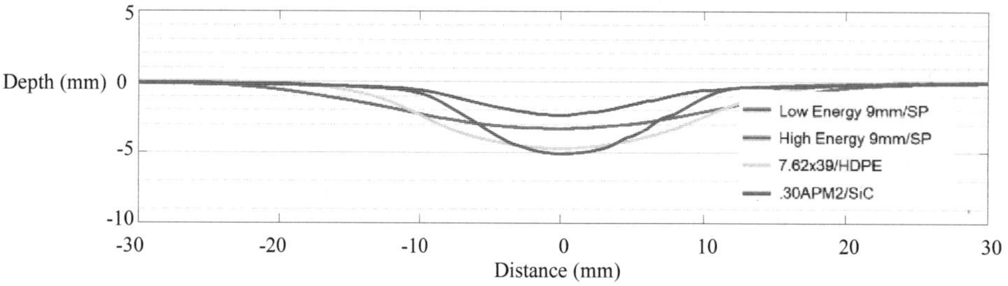

Figure 8. Cross-section of tin deformation from ballistic testing of armour

4.0 DISCUSSION

4.1 Peak Forces Behind Armour

The dynamic peak forces measured behind armour in ballistic testing using the force plate had a number of interesting trends. The centre and outer forces frequently had double and triple peaks visible in the trace. For the smallest impact caps, the second peak in time usually had a greater force than the initial peak. The outer first peak occurred noticeably later than the centre. The thicker pad produced peak forces that were less than the thinner pad. The small sample sizes show a number of trends, but limit the ability to quantify the measurement error.

For the three armour classes, at least 50% of the force falls less than 32 mm diameter centre area, with the remainder falling usually within a 61 mm diameter area. This distribution of forces may have implications in the development of force-based criteria for evaluating armour performance and predicting behind armour injury.

Three measurements, peak force, total force and deformation energy can be compared with data collected and presented at PASS 2014 [4] and PASS 2018 [5]. A model was created using the PASS 2018 data and applied to the clay deformations measured in the PASS 2014 results.

The same bullet and armour configurations for the shootpack (5.3 kg/m^2 Areal Density (AD)) and SiC/Shootpack (29 kg/m^2 AD) were used in these studies, while the UHMWPE was thicker (13.6-16.6 kg/m^2 vs. 9.8 kg/m^2 AD) in the PASS 2014 testing (Table 4). The nominal impact velocities and data fits were used to determine the Fujifilm Prescale® measurements, as well as estimates of the clay deformation. The clay energy calculations developed in PASS 2018 are compared with the TrussGrid and tin energy estimates.

Table 4. Velocities used in analysis

	7.62x39mm		9x19mm		APM2	
Measurement Method	LTM Velocity (m/s)	Muzzle Velocity (m/s)	LTM Velocity (m/s)	Muzzle Velocity (m/s)	LTM Velocity (m/s)	Muzzle Velocity (m/s)
Force Plate - 7.6 Cap, 13 mm pad	426	642	307	438	640	791
Force Plate - 7.6 Cap, 6 mm pad	NT	644	NT	439	NT	800
TrussGrid Force Plate	431	642	309	437	642	806
Tin Force Plate	NT	641	306	425	NT	800
Fujifilm Prescale & Clay	434	640	305	434	640	805

LTM – Less than Muzzle, NT – Not Tested

Peak impact pressure can be compared between the Fujifilm Prescale® and the force plate with the 7.6 mm cap and the 6 and 13 mm pads (Figure 9). In comparing them, the peak pressures follow a consistent pattern with the thickest pad (13 mm) producing the lowest peak forces and pressures while the Fujifilm Prescale®, placed directly behind the armour, produced the highest peak pressures ranging from 645-592 MPa. It is important to note that the force plate pressure was based on the average pressure over the 7.6 mm cap, whereas the Fujifilm pressure was measured at a higher spatial resolution.

https://doi.org/10.52202/078352-0014

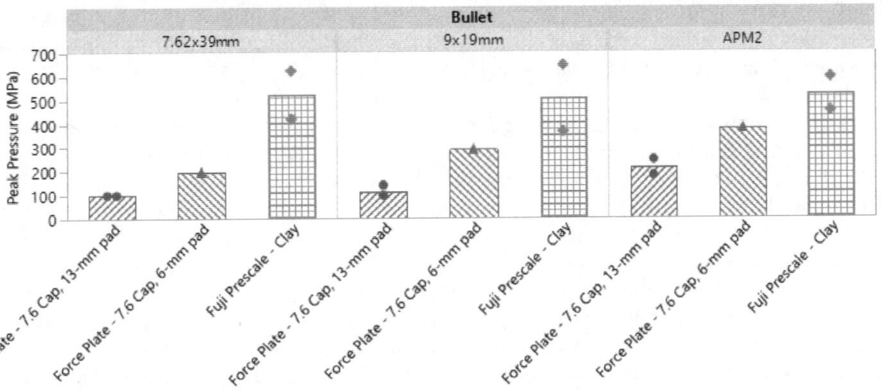

Figure 9. Peak pressure measurements from the force plate and Fujifilm Prescale® for muzzle velocity tests

The peak total force between the Fujifilm Prescale®, and force plate behind the pads and witness materials can also be compared (Figure 10). The Fujifilm Prescale®, tin and the TrussGrid underreports total force because they have a lower limit on the forces they measure. The lower limit on Fujifilm Prescale®, TrussGrid, and tin are 50 MPa [10], ~ 3 MPa and 42 MPa, respectively. The pads and witness materials are mounted on a rigid, heavy force plate that may affect the measurements, while the Fujifilm Prescale® measurements were made over clay that would resist the impacting force less. The force plate with the 6 and 13 mm pads measured the highest forces, much higher than the forces measured behind the witness plate materials and by the Fujifilm Prescale®.

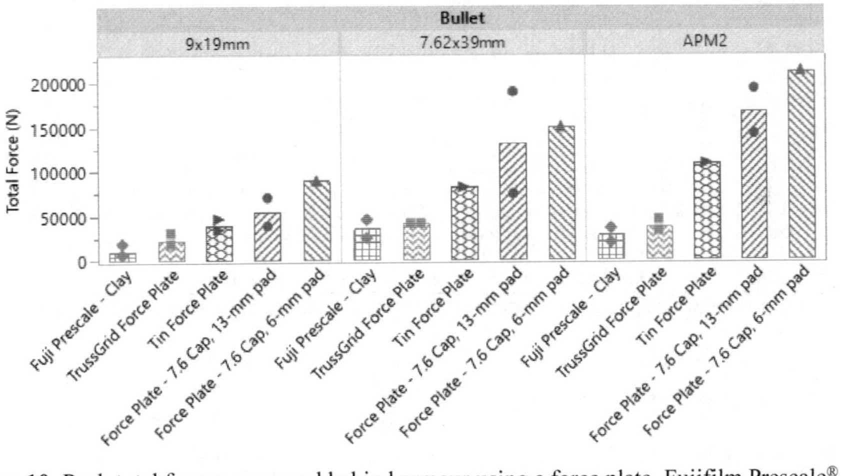

Figure 10. Peak total forces measured behind armour using a force plate, Fujifilm Prescale® and witness materials at muzzle velocity

4.2 Impact Energy

One advantage of witness materials are their ability to record the peak and distribution of energy behind the armour from the impact of the bullet. If the force displacement relationship is known, then the distribution of deformation is a direct mapping of the deformation energy distribution. However in testing armour, this deformation allows the armour to absorb more of the impact, changing the spatial-depth response of the witness plate material. The most realistic witness material response should be one that duplicates the response of human tissue at the location of impact. A stiffer material will support the armour more, result in less amour deformation and have higher forces at the interface.

The witness plate and clay measurement data provides insight into the amount and distribution of the impact energy. The maximum deformation reflects the region with the greatest energy dissipated, found in the centre of the impact. The total volume reflects the total energy while the volume/cross-

https://doi.org/10.52202/078352-0014

sectional area ratio reflects the areal energy density or kinetic energy density. If the force-deformation response did not vary with depth and velocity, the correspondence of depth with impact energy would be a simple linear relationship, but any strain-rate, strain-hardening or flow effects affects this relationship.

In this preliminary study, the sample sizes were small, making measurement of the experimental errors difficult. Quantification of the witness plate deformation energy at relevant strain rates and shapes is difficult and introduces another source of error. The measurements presented show interesting trends and highlight the challenges of using witness materials in testing and quantification of the deformation energy.

The clay deformation data collected behind the Fujifilm Prescale [4] was processed using the energy estimation methods described in [5] to estimate the energy deposited into the clay. In Figure 11, the clay energy is compared with the energy estimates for tin and TrussGrid. The energy calculated from the clay deformation is much greater than estimated from the tin and TrussGrid. There are several potential causes of this difference including the difference between deforming behaviour of the clay, tin and TrussGrid. By dividing the deformation energy by the area of deforming contact, the kinetic energy density of the impact can be compared. This comparison shows that the kinetic energy density measured by the TrussGrid and the clay behind the Fujifilm were more similar than the kinetic energy density measured behind tin. Potential reasons for this difference include the differences in yielding and backing material (clay, tin and TrussGrid) between the tests.

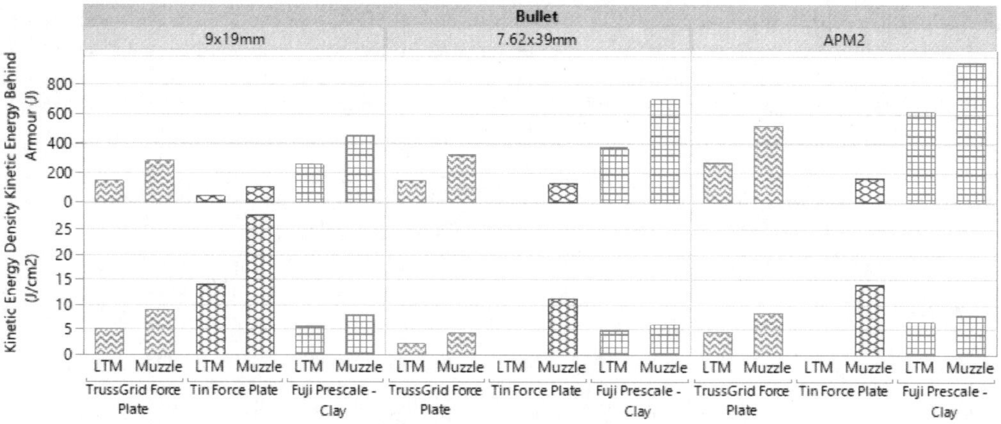

Figure 11. Behind armour total energy and kinetic energy density calculated from the clay deformation behind the Fujifilm and witness material deformation for muzzle and less-than-muzzle (LTM) impacts

5.0 SUMMARY AND FUTURE WORK

The residual impact forces, energies and their distribution behind armour were measured using force plates, Fujifilm Prescale®, and witness materials (clay, TrussGrid and tin). These measurements provide insight into the impact forces, force distribution and energy magnitudes, as well as the difficulty in quantifying these values. Because of the large deformation and complex interaction between the bullet, armour and backing material, these measurements reflect the experimental conditions tested and the limited sample sizes for each test condition. They do not provide a universal measurement of the event.

The peak pressure, total force, and total deformation energy results showed a wide range of measurements depending on the test method and bullet/armour/velocity combination. The peak pressures range from 100 MPa to 645 MPa which were measured on the shootpack while the composite and SiC/Shootpack armour combinations showed a smaller range of peak pressures. The total force measured ranged from 14.4 kN to 221 kN depending on the measurement method and bullet/armour tested. The total deformation energy ranged from 42 J to 952 J depending on the test conditions.

Future work needs to be conducted with post-mortem human tissue and matching force plate and Fujifilm testing so that these results can be grounded by understanding the potential for injury.

https://doi.org/10.52202/078352-0014

Acknowledgments

This research was funded by the US Army, PEO Soldier, under Naval Sea Systems Command (NAVSEA) Contract N00024-13-D-6400, under several Task Orders. Any opinions, findings and conclusions or recommendations expressed in this material are those of the author(s) and do not necessarily reflect the views of NAVSEA.

References

[1] Hanlon E and Gillich M., "Origin of the 44-mm Behind-Armor Blunt Trauma Standard," Military Medicine, 177, 3:333, 2012, pp 333-339.

[2] Rice K, Riley M., and Forster A. "Ballistic Resistance of Body Armor, NIJ-Standard-0101.06," National Institutes of Justice, July 2008.

[3] Prather, R, Swann, C. and Hawkins, C. "Backface Signatures of Soft Body Armors and the Associated Trauma Effects", Technical Report No. ARCSL-TR-77–55, U.S. Army Armament Research and Development Command, Aberdeen Proving Ground, Maryland, 1977

[4] Bevan M., Luong, Q., Halls V., Zheng J., "Comparison of Clay Depth Measurements to Pressure-Sensitive Film and Laser Scanning Results," 12th Personal Armour Systems Symposium (PASS), Cambridge, UK, 8-12 September 2014

[5] Bevan M., Luong, Q., "Correlation between Projectile Kinetic Energy and Displaced Clay Volume for Three Classes of Armour," 14th Personal Armour Systems Symposium (PASS), Washington DC, 1-5 October 2018

[6] "The International System of Units (SI)," NIST Special Publication 330, 2008 Edition, B. N. Taylor and A Thompson, Eds. March 2008.

[7] National Research Council. "Testing of Body Armor Materials for Use by the U.S. Army—Phase II: Letter Report." Washington, DC: The National Academies Press. https://doi.org/10.17226/12885 2010.

[8] S. Hotta, K. Matsumoto, T. Murakami, T. Narushima and C. Ouchi, "Dynamic and Static Restoration Behaviors of Pure Lead and Tin in the Ambient Temperature Range." Materials Transactions, Vol. 48, No. 10 (2007) pp. 2665 to 2673.

[9] HexWeb® Honeycomb Energy Absorption Systems Design Data HEXCEL Corporation page 5, Salt Lake City UT March 2005.

[10] Sensor Products Data Sheet "Fujifilm Prescale® Tactile Pressure Indicating Sensor Film," 24 Oct 2017

https://doi.org/10.52202/078352-0014

Numerical recreation of Police Field Cases on a human body FE model: first insights into BABT

A. Bracq[1], B. Bourel[1], R. Delille[1], C. Maréchal[1], G. Haugou[1], F. Lauro[1], S. Roth[2], O. Mauzac[3], C. Bir[4]

[1]*Laboratory LAMIH UMR CNRS 8201, University Polytechnique Hauts-de-France, 59313 Valenciennes, France*
anthony.bracq@isl.eu

[2]*University of Bourgogne Franche-Comté, UTBM, Interdisciplinary Carnot laboratory of Bourgogne UMR CNRS 6303, 90010 Belfort, France*

[3]*French Ministry of the Interior, CREL, Place Beauveau, Paris, France*

[4]*Wayne State University, 818 W. Hancock, 48201, Detroit, MI, USA*

Abstract Behind Armor Blunt Trauma (BABT) has become a topic of main importance for the law enforcement officers, soldiers and armour manufacturers. Indeed, the need for body armour weight reduction and the enhancement of projectile efficiency may result in a higher body armour deformation and therefore, an increasing risk of blunt trauma. This study focuses on the soft body armour deformation where trauma is mainly induced by the dynamic deformation of the protective system. Indeed, for the velocity range considered, it is assumed that trauma linked to shock waves may be neglected.

Three US Police field cases intend to be numerically investigated through impact simulations on a biofidelic human torso Finite Element (FE) model. These cases involve a wide range of BABT, from bruising to rib fractures and lung contusion. In order to faithfully replicate impact conditions on a human body FE model, a previously published method is used to propose a FE modelling of projectiles and body armours. It basically relies on impact events recreated on a transparent synthetic SEBS gel and dynamic gel wall displacement profile measurements.

Firstly, projectiles and protective systems are modelled through an inverse iterative approach using both experimental and numerical model of the gel block. The maximum backface deflection is used as objectives to reach in the identification procedure, along with the shape of the deformation. Secondly, impact conditions related to each field case are replicated on the torso FE model. Following observed trauma, strain, pressure fields and derived metrics are computed from the human body model. Then, comparisons of trauma and numerical metrics values are made and first conclusions are drawn. This study represents an important step along the way to a better understanding of BABT and human body-protective systems' interactions.

1. INTRODUCTION

Ballistic protective systems are in constant improvement to absorb the kinetic energy of projectiles and prevent penetration. The requirement of body armour weight reduction may cause a higher deflection of armours covering the body and consequently an increasing risk of Behind Armour Blunt Trauma (BABT) [1, 2]. Over the last decades, numerous experimental methods have been established to assess body armours. For example, it led to the well-known National Institute of Justice (NIJ) Standard-0101.06, where Roma Plastilina No 1 clay is used as target material placed behind the studied body armour [3]. Among others ballistic testing media present in the literature, the polymer SEBS (Styrene-Ethylene-Butylene-Styrene) gel exhibits advantageous properties: transparency, mechanical consistency and environmental stability [4-7]. This material is adopted in the present study to analyse non-penetrating ballistic impacts.

The use of Finite Element (FE) modelling is also considered in order to replicate impact conditions on a biofidelic human torso model. The Hermaphrodite Universal Biomechanical YX model (HUByx) is a commercially available human body model in Altair HyperWorks software packages developed by CEDREM. This model has been validated against ballistic impact replications and respective biomechanical corridors [8, 9]. Nonetheless, several papers point out the complexity of numerical modelling of ballistic impacts involving body armours [10-13]. A solution is proposed by

https://doi.org/10.52202/078352-0015

Bracq et al. [14]. It is based on an equivalent modelling of projectiles and body armours. The approach mainly relies on experimental and numerical modelling of ballistic impacts on a SEBS gel block [14].

However, results of ballistic impact simulation are meaningless without comparisons with actual field cases. A recent research work of Bir et al. reports 47 police field cases while providing useful information as body armour properties, fired ammunition and sustained injury [15].

The aim of the present paper is thus to recreate numerically some of the most relevant field cases mentioned in the study of Bir et al. [15] and correlate numerical metrics with observed injuries. For this purpose, the authors depict chosen police field cases and present the coupled experimental-numerical method to complete a FE model of the incident. Then, experimental and numerical results are provided and discussed. Finally, conclusions are drawn about the potential of FE modelling but also the requirement of statistical data. This may bring other perspectives to current standards for body armour assessment.

2. MATERIAL AND METHODS

2.1. Description of police field cases

Three police field cases are available in the present study and their data originate from the study of Bir et al. [15]. They are the main results of continuous efforts made by the IACP/DUPONT® and Safariland members to assess protective systems and potential BABT. Police officers' testimony, medical and police records as well as information regarding the manufacturer, model and threat level of the ballistic pack worn during the incident were collected. The injuries observed were classified following two injury scales. The first one noted IR and introduced by Bir et al. [15] ranks injuries from 1 to 3 according to clinical significance:
1 = minor - bruise, red mark, minor wound care for abrasions
2 = moderate - bruising with penetration (BFS), lung contusion, open wound care
3 = severe - internal injuries requiring medical intervention, advanced wound care
The Abbreviated Injury Scale (AIS) is also used to rank injuries while taking into account trauma location on the body. Table 1 outlines the collected data for every field case.

Table 1. Summary of field cases data used in this study [1].

Case	Armour	Projectile	Range [m]	Impact location	IR	AIS	Injury details
USC-990	Second Chance Ultima SMU II, Level II	Remington 40 Cal S &W 180 gr	2.4	Left flank at level of 8th rib	3	2	Broken 8th rib, contusion/ laceration of spleen, hemo peritoneum
USC-1716	PACA KSG, Level IIIA	RWS 38 special 158 gr FMJ	0.9	Upper right corner, just above trauma pack	1	1	Skin contusion
USC-3138	Point blank, CIIA-1 Level IIA	Federal Premium 40 cal, S&W, 180 gr HP	<0.3	Front upper torso, right of center	2	3	Pulmonary contusion

 https://doi.org/10.52202/078352-0015

2.2. Experimental-numerical method for impact FE modelling

In order to carry out impact modelling on HUByx and evaluate BABT, projectiles and body armours need to be modelled. The complexity in such modelling forced the authors to propose a methodology based on experimental and numerical approaches. Firstly, it relies on experimental tests on a gel block. This procedure has already been fully depicted in a previous paper [14]. Thus, only the most relevant part will be presented here.

The ballistic testing media is the polymer gel SEBS. A gel sample is created by mixing SEBS powder and mineral oil with a SEBS/oil ratio about 30/70%. Firearm projectiles are impacted on the centre of body armours using a 25 cm gel block cube as backing material. A barrel is employed to fire any projectiles. Hook and loop straps are employed to hold still body armours against the gel block surface as shown in Figure 1 (right). Moreover, the gel transparency leads to the measurement of the dynamic gel wall displacement due to armour deformation through the use of a lighting system and a high-speed camera. Images are processed to capture the gel wall displacement history using gray level thresholding. The experimental set-up dedicated to non-penetrating impacts is illustrated by Figure 1 (left). Other metrics can be deduced from gel wall displacement curves. A pseudo viscous criterion defined in the literature may be calculated from the gel wall displacement (Xmax) [16]. It is called Energy Transfer Parameter (ETP), expressed in m/s. The displaced volume (VOL) may also be computed at each time and its derivative, the volume growth rate (VGR). The maximum value of each parameter can be employed to analyse experiments.

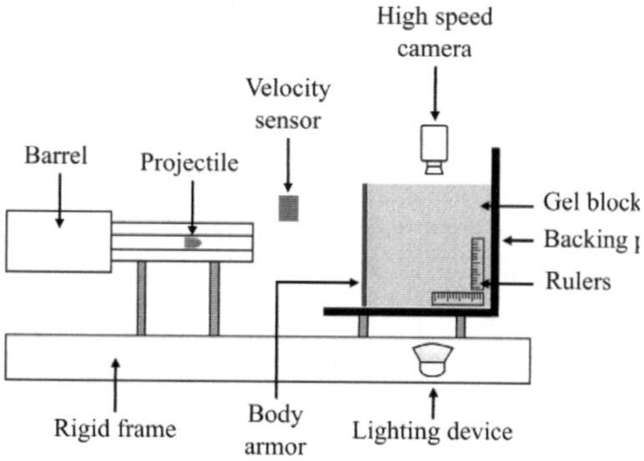

Figure 1. Experimental set-up for ballistic impact studues (left) and body armour positioning before ballistic experiments (right).

The gel wall displacement profile captures the projectile kinetic energy dissipated by both the gel block and the body armour. Therefore, the mechanical properties of the body armour can be determined if the SEBS gel material behaviour is known. Actually, a visco-hyperelastic model has been implemented for the constitutive modelling of the SEBS gel [4, 17]. An inverse methodology can be applied by modelling impact experiments on the gel block and optimizing model parameters to fit with experimental data. However, the impact modelling of a projectile on a body armour still results in complex phenomena and high computing costs.

Hence, the authors developed an equivalent FE modelling of the impact based on several assumptions. These assumptions rely upon the physical phenomena occurring before the gel wall displacement [14]. Thus, the projectile is considered as rigid in the equivalent FE model and its novel geometry is identified through the analysis of the gel wall displacement profile during an impact as well as the measurements of the actual projectile geometry after impact. By means of the length of the real deformed projectile, the equivalent rigid projectile can be meshed using the commercial software HyperMesh (Altair HyperWorks ©). The material density indicated for simulation is adjusted to obtain an equivalent mass with the actual projectile. Then the body armour is modelled at the macroscopic scale with one layer of 2D shell elements, as illustrated in Figure 2 (right). It simplifies the modelling and reduces computing costs. A simple anisotropic hyperelastic law for fabric material proposed by the

https://doi.org/10.52202/078352-0015

explicit code Radioss (LAW58) is used for the constitutive modelling of body armours. When the impact is located far from the edges of the ballistic pack, planes of symmetry are employed for impact modelling reducing the FE model to a quarter (Figure 2 (right)). Such boundary conditions can be imposed because gel block dimensions and impact location make it possible to neglect edge effects.

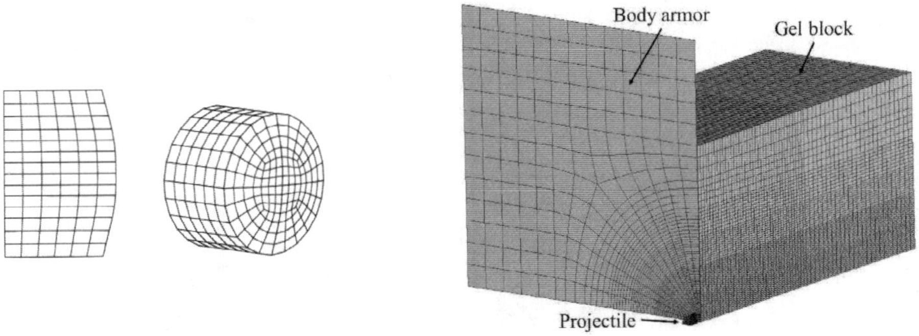

Figure 2. FE modelling of an equivalent projectile (left) and representation of the FE model developed to simulate ballistic impacts on a gel block.

An optimization procedure is carried out to identify model parameters to fit with experimental gel wall displacements and their 2D profiles. Relevant model parameters include the projectile initial velocity, the density of the body armour, its thickness and its shear modulus. Finally, a Response Surface Method proposed by the software HyperStudy is chosen to optimized parameters. Once the model parameters adjusted for each impact condition, experimental and numerical results can be compared. For instance, this procedure is applied to an impact of a 9 mm bullet on a soft body armour made of 40 layers of para-aramid Kevlar® fibres protecting a gel block. Figure 3 (left) presents the experimental and numerical profile at maximum gel wall displacement. Figure 3 (right) illustrates experimental and numerical gel wall displacements over time.

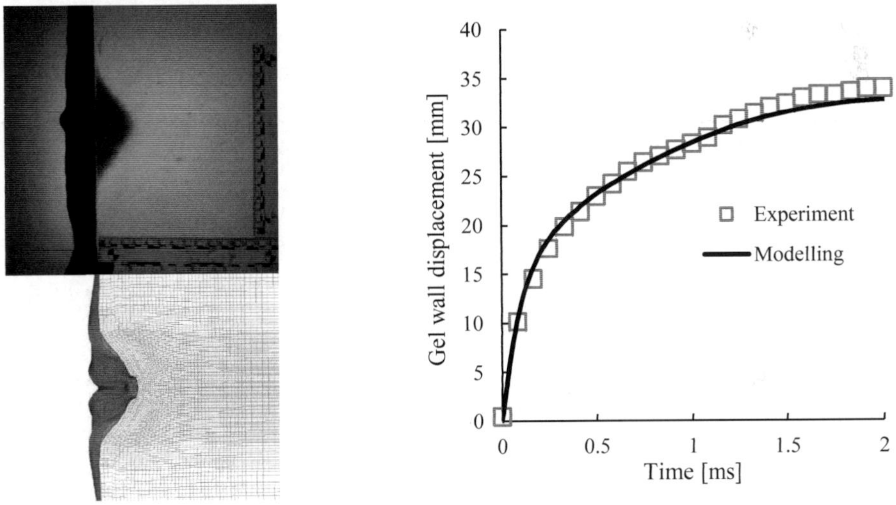

Figure 3. Experimental and numerical profile at maximum gel wall displacement (left) and experimental and numerical gel wall displacements versus time (right).

2.3. Human torso Finite Element model HUByx

The FE model HUByx is developed by means of a 3D reconstruction of the human torso geometry and corresponding CT scans image processing. A 50th percentile male subject has been created. Through the

https://doi.org/10.52202/078352-0015

HyperMesh software (Altair HyperWorks ©), a finite element model is constructed with skin and muscle, skeleton and internal organs [8, 9]. A detailed description of this FE model in terms of mesh discretization, contact interfaces and material models depicting the human body response under complex and dynamic loadings can be found through the article of Roth et al. [9]. However, this study focuses on behind armour blunt trauma such as rib fractures and lung contusion.

Once the hybrid method presented in the latter part is applied to the studied case, impact conditions can be simulated on HUByx at a considered location. Figure 4 intends to illustrate the FE model of the projectile, the body armour and HUByx. Mesh discretization procedure to model the projectile and the body armour is preserved from the identification process.

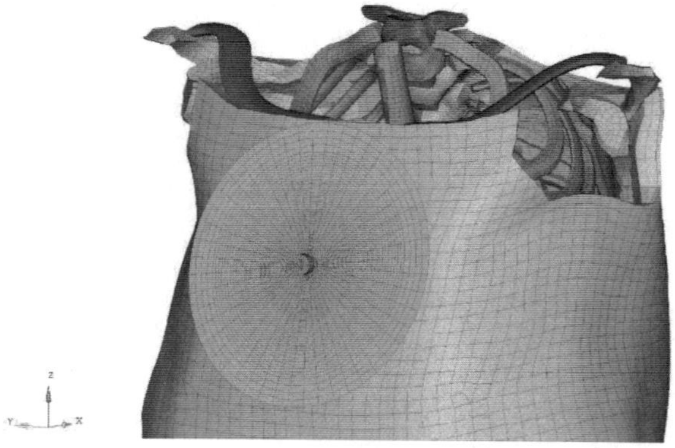

Figure 4. Representation of a FE impact modelling on HUByx [14].

To relate numerical simulations with the appearance of blunt trauma requires first of all a procedure to compute relevant numerical metrics. As suggested by the authors in a previous study [18], numerical pressure fields of the 3D elements depicting soft tissues are processed. The following softwares HyperView, HyperGraph and MatLab are combined to identify the maximum pressure over time for the body part of interest. Additional numerical metrics can be deduced from pressure time history as the pressure impulse, the peak of pressure, the duration of the pressure wave above 75% of the pressure maximum value as well as the wave duration. The risk of rib fractures is numerically accounted for by computing the maximum value of the specific energy field of the rib cortical bone. A simple average filtering method is proposed in HyperView and used to record numerical values. The following part of this paper will focus on the recreation of some Police field cases and first outcomes regarding BABT and their mitigation.

3. RESULTS AND DISCUSSION

3.1. Ballistic experiments on synthetic gel

Based on the case reports, impact conditions were faithfully reproduced on exemplar body armours. Stand-off distance, ammunition and impact location were carefully chosen. One ballistic experiment is carried out for each case. Images from the high-speed camera Phantom V1212 are saved at 20,000 frames per second with a resolution of 640x480 pixels. The image and data processing routine is used to determine the 2D gel wall displacement profile at each time step. Figure 5 to Figure 7 present a high-speed image at maximum gel wall displacement and the dynamic contour profile for each case.

Post impact projectiles' dimensions are measured such as the maximum diameter D and the thickness or width e. Table 2 summarises the impact conditions, "post-mortem" projectile dimensions and geometry for various case studies. Experimental metrics are also derived according to the gel wall deformation, as for instance the maximum gel wall displacement Xmax or the maximum displaced volume VOL (Table 3). These data added to 2D dynamic contour profiles are mandatory to apply the modelling procedure and to propose suitable ballistic simulations.

https://doi.org/10.52202/078352-0015

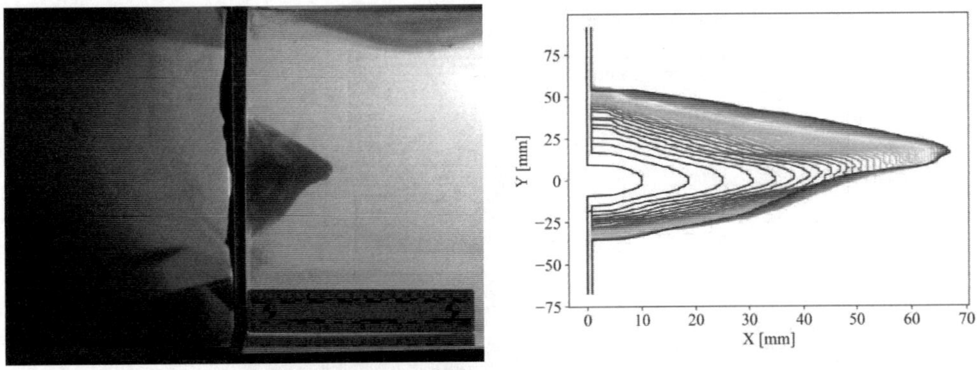

Figure 5. High-speed image at maximum gel wall displacement (left) and 2D dynamic gel wall displacement profile (right) related to the field case USC-990.

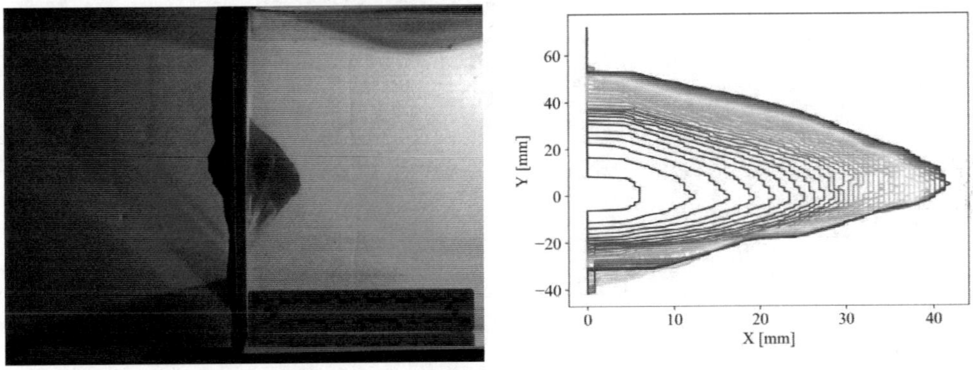

Figure 6. High-speed image at maximum gel wall displacement (left) and 2D dynamic gel wall displacement profile (right) related to the field case USC-1716.

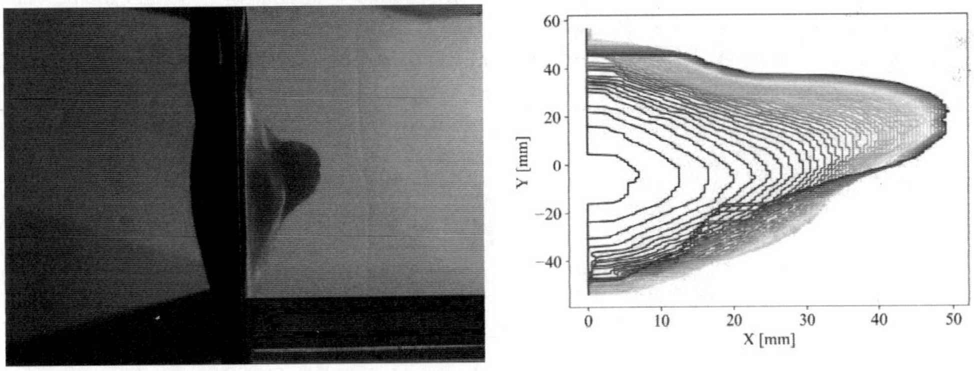

Figure 7. High-speed image at maximum gel wall displacement (left) and 2D dynamic gel wall displacement profile (right) related to the field case USC-3138.

https://doi.org/10.52202/078352-0015

Table 2. Projectile characteristics, from impact condition to post impact figures, for three field cases.

Case field	Projectile	Mass [g]	Velocity [m/s]	Post impact photograph	D [mm]	e [mm]
USC-990	.40 S&W 180 gr FAE FMJ	11.67	295.4	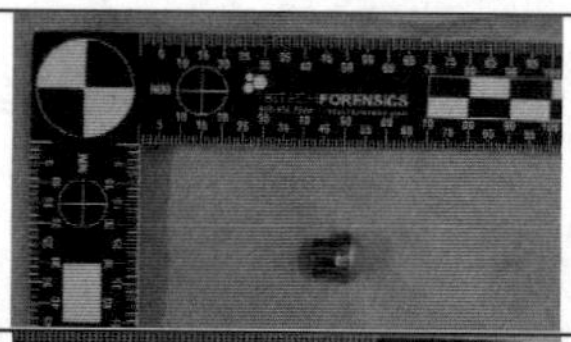	13.3	12.9
USC-1716	.38 158 gr FMJ FN GECO	10.20	262.2	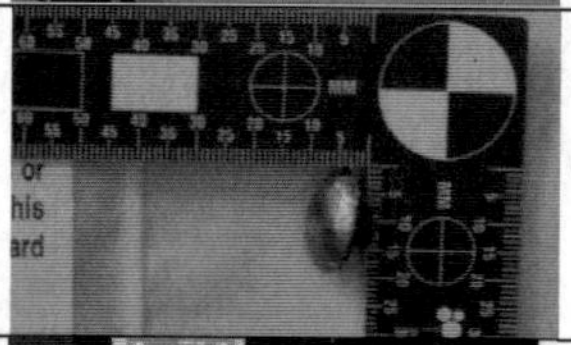	17.6	10.0
USC-3138	.40 S&W 180 gr FP JHP	11.71	297.3	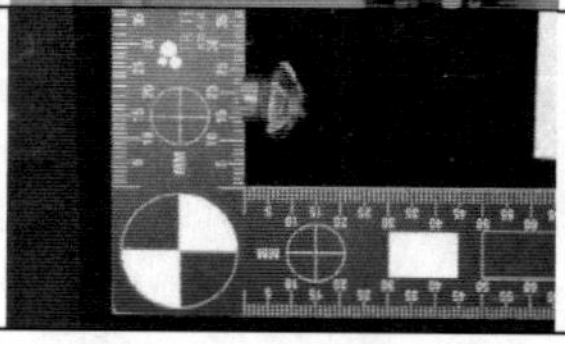	11.0	11.6

Table 3. Experimental metrics related to ballistic tests on SEBS gel block for three Police field cases.

Case field	Xmax [mm]	t_{Xmax} [ms]	ETP [m/s]	VOL [cm^3]	VGR [dm^3/s]
USC-990	67.1	2.2	15.4	185.5	168.5
USC-1716	41.9	2.1	6.3	112	102.2
USC-3138	49.6	2.9	6.5	199.5	179.1

3.2. FE modelling of impacts on synthetic gel

Numerical simulations of impacts on the gel block are firstly performed. The objective is the identification of the parameters, by an optimisation procedure, of an equivalent model of the body armour as well as the initial velocity. The boundary conditions were faithfully reproduced as experiments like presented in the previous section (2.2). Impact locations close to the body armour's edges prevent from using planes of symmetry. The equivalent projectiles forms identified from the analysis of the gel wall displacement are modelled as presented in Figure 8.

Figure 8. Equivalent projectile geometry for three field cases.

Results obtained for the three cases are presented in terms of gel wall displacement over time (Figure 9) and all the parameters are given in Table 6. A good correlation is observed for all cases. Nevertheless, some discrepancies are present for the USC-990 case due to a large wall displacement, which leads to some mesh distortion at the end of the numerical calculation.

https://doi.org/10.52202/078352-0015

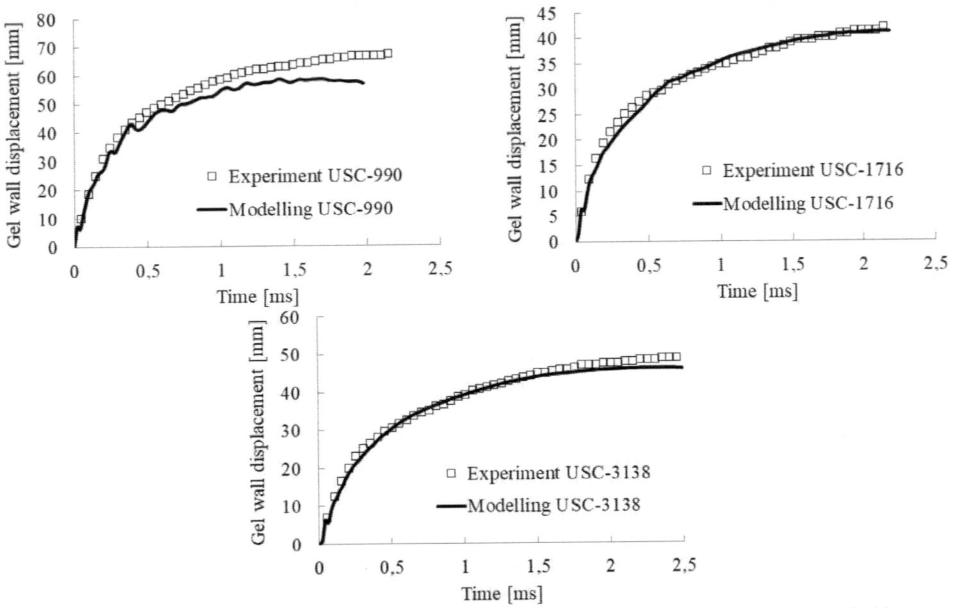

Figure 9. Numerical and experimental comparison of the gel wall displacement for three field cases.

Table 4. Optimized parameters of the projectiles / body armours equivalents models for studied cases.

	Body armour			**Projectile**
Case field	**Density [kg/m^3]**	**Shear modulus [MPa]**	**Thickness [mm]**	**Velocity [m/s]**
USC-990	1.46×10^3	0.101	2.173	241.5
USC-1716	1.16×10^3	0.101	2.710	175.7
USC-3138	2.34×10^3	0.098	1.645	154.4

Field cases are then numerically reproduced with the HUByx dummy and the equivalent body armour. Locations of bullet impacts are proposed according the documents produced from the study of Bir et al. [15] and highlighted by a red circle on Figure 10. Numerical data are collected, especially the specific energy on the ribs and the lungs peak pressure.

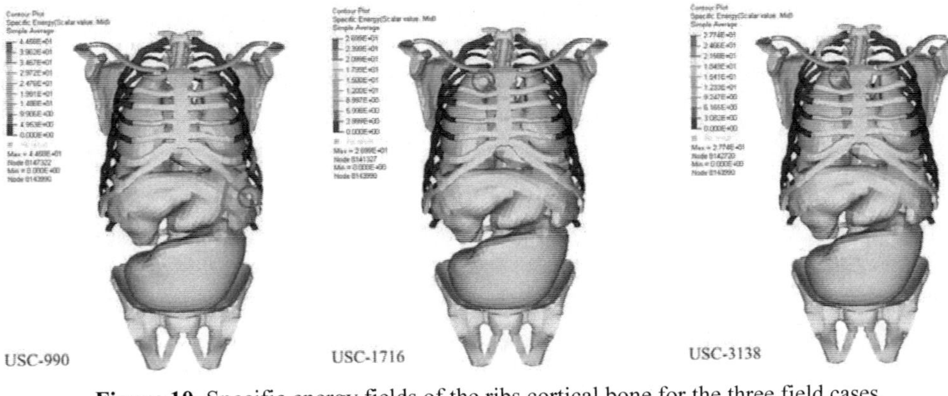

Figure 10. Specific energy fields of the ribs cortical bone for the three field cases.

A high specific energy is obtained on 8th rib for the USC-990 case and lower but similar ones for the other two cases. Concerning the pressure, the lungs are not loaded during the impact for the USC-990 case which is indicated by N.A. in the Table 5. For the USC-3138 case the maximum pressure value is higher with a shorter duration compare to the USC-1716 case. Pressure values are numerical data and probably not representative of the real ones in the human body.

https://doi.org/10.52202/078352-0015

Table 5. Results of impacts on human body FE model for three field cases.

	Ribs Specific Energy	Lungs peak pressure		
Case field	Max value [J/Kg]	Maximun [MPa]	Duration [ms]	Integral [MPa.ms]
USC-990	44.58 (8th Rib)	N.A.	N.A.	N.A.
USC-1716	26.99 (2nd Rib)	0.8897	0.8897	0.3856
USC-3138	27.74 (1st Rib)	0.9806	0.5660	0.4644

Such data lead the authors to investigate correlations with BABT. Data obtained from the HUByx simulations, in terms of ribs specific energy, could be added to the injury risk curve already created with data from 22 cases based on various projectiles, armours and impact speeds [19]. This curve gives the probability to have a rib fracture with a corresponding AIS score between 2 and 3.

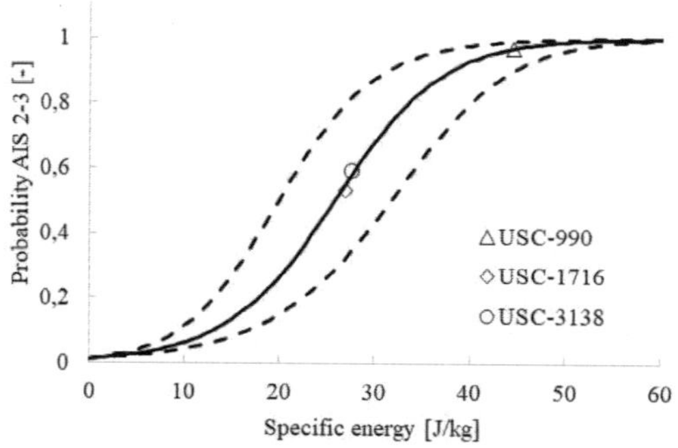

Figure 11. Injury risk curve with 95% confidence interval and injury prediction for three field cases.

The risk of rib fractures is about 97% for the USC-990, in agreement with the field case. Moreover, the same rib is broken in the simulation and in the field case. For the two other cases, the probability of rib fractures is close and about 53% and 58% for USC-1716 and USC-3138 respectively. No rib fracture is observed for these latter. For the USC-1716, the anthropometric values (1.80 m height and 88 kg weight) are higher than the 50th percentile representation of the virtual HUByx dummy. Therefore, the rib fractures probability obtained by this simulation is certainly overestimated. Unfortunately, for the USC-3138 case the anthropometric values are not given in the field case report. About the numerical pressure analysis, no data is available in the literature and thus, the probability to sustain a lung injury can't be determined. However, with these field cases, a lung injury is observed for the USC-3138 and none for the USC-1716. Data available in Table 5 don't yet enable the authors to conclude over the lung risk injury best indicator. More field cases are necessary to proceed further with this investigation.

4. CONCLUSION

The present paper intends to draw a combined experimental-numerical approach to assess the risk of behind armour blunt trauma. To obtain a representative study the authors have been encouraged to rely on police field cases available in the literature. Three relevant cases have first been recreated on a SEBS transparent gel block placed behind exemplar body armours and targeted by appropriate ammunitions. Image processing routine has provided precious measurements of the dynamic backface deformation of the ballistic pack. The experimental metrics are employed to propose a numerical modelling of the impact. Basically, an equivalent projectile/body armour FE model is identified through direct and indirect identification techniques. The objection function to minimize in the inverse iterative approach is a function of data from ballistic tests on a gel block and their numerical simulations. Comparisons of experimental and numerical gel wall displacements for recreated police field case validate the methodology. Then, impact conditions are modelled on the human FE model HUByx as reported in field cases. The simulation results as specific energy values correlate with a risk curve of rib fractures

https://doi.org/10.52202/078352-0015

according to anthropometric data and impact locations. Numerical pressure values of the impacted lung are computed and can be related to lung contusions. However, the lack of datasets prevents the authors from establishing a risk of injury curve. Law enforcement agencies and industries have to make an effort to share statistical data for further investigations in BABT.

ACKNOWLEDGEMENTS

This research is funded by the French Ministry of the Interior and is carried out within the framework of the CNRS Research Federation on Ground Transports and Mobility. in articulation with the ELSAT2020 project supported by the European Community, the French Ministry of Higher Education and Research, the Hauts de France Regional Council. The authors gratefully acknowledge the support of these institutions. Moreover, this work would not have been possible without the valuable effort of Mr. Rodrigo Villalta and Mr. Gabriel Glasser.

REFERENCES

[1] L. Cannon, 'Behind armour blunt trauma-an emerging problem', *Journal of the Royal Army Medical Corps*, vol. 147, no. 1, pp. 87–96, 2001.

[2] A. W. Carroll and C. A. Soderstrom, 'A new nonpenetrating ballistic injury.', *Annals of surgery*, vol. 188, no. 6, p. 753, 1978.

[3] J. C. Roberts, E. E. Ward, A. C. Merkle, and J. V. O'Connor, 'Assessing Behind Armor Blunt Trauma in Accordance With the National Institute of Justice Standard for Personal Body Armor Protection Using Finite Element Modeling':, *The Journal of Trauma: Injury, Infection, and Critical Care*, vol. 62, no. 5, pp. 1127–1133, May 2007, doi: 10.1097/01.ta.0000231779.99416.ee.

[4] A. Bracq *et al.*, 'Characterization of a Visco-Hyperelastic Synthetic Gel for Ballistic Impacts Assessment', in *Dynamic Behavior of Materials, Volume 1: Proceedings of the 2017 Annual Conference on Experimental and Applied Mechanics*, J. Kimberley, L. Lamberson, and S. Mates, Eds. Cham: Springer International Publishing, 2018, pp. 109–113.

[5] A. Bracq, G. Haugou, R. Delille, F. Lauro, S. Roth, and O. Mauzac, 'Experimental study of the strain rate dependence of a synthetic gel for ballistic blunt trauma assessment', *Journal of the Mechanical Behavior of Biomedical Materials*, vol. 72, pp. 138–147, Aug. 2017, doi: 10.1016/j.jmbbm.2017.04.027.

[6] O. Mauzac, C. Paquier, E. Debord, F. Barbillon, P. Mabire, and J. Jacket, 'A substitute of gelatin for the measurement of dynamic back face deformation', in *Personal Armour Systems Symposium*, Québec, Canada, 2010.

[7] R. A. Mrozek *et al.*, 'The relationship between mechanical properties and ballistic penetration depth in a viscoelastic gel', *Journal of the Mechanical Behavior of Biomedical Materials*, vol. 44, pp. 109–120, Apr. 2015, doi: 10.1016/j.jmbbm.2015.01.001.

[8] M. Bodo, A. Bracq, R. Delille, C. Marechal, and S. Roth, 'Thorax injury criteria assessment through non-lethal impact using an enhanced biomechanical model', *Journal of Mechanics in Medicine and Biology*, pp. 1–17, Oct. 2017, doi: 10.1142/S0219519417400279.

[9] S. Roth, F. Torres, P. Feuerstein, and K. Thoral-Pierre, 'Anthropometric dependence of the response of a Thorax FE model under high speed loading: Validation and real world accident replication', *Computer Methods and Programs in Biomedicine*, vol. 110, no. 2, pp. 160–170, May 2013, doi: 10.1016/j.cmpb.2012.11.004.

[10] Y. Chu, S. Min, and X. Chen, 'Numerical study of inter-yarn friction on the failure of fabrics upon ballistic impacts', *Materials & Design*, vol. 115, pp. 299–316, Feb. 2017, doi: 10.1016/j.matdes.2016.11.013.

[11] L. Gilson, L. Rabet, A. Imad, and F. Coghe, 'Experimental and numerical assessment of non-penetrating impacts on a composite protection and ballistic gelatine', *International Journal of Impact Engineering*, vol. 136, p. 103417, Feb. 2020, doi: 10.1016/j.ijimpeng.2019.103417.

[12] C. Ha-Minh, T. Kanit, F. Boussu, and A. Imad, 'Numerical multi-scale modeling for textile woven fabric against ballistic impact', *Computational Materials Science*, vol. 50, no. 7, pp. 2172–2184, May 2011, doi: 10.1016/j.commatsci.2011.02.029.

[13] A. Tabiei and G. Nilakantan, 'Ballistic Impact of Dry Woven Fabric Composites: A Review', *Applied Mechanics Reviews*, vol. 61, no. 1, pp. 1–13, 2008, doi: 10.1115/1.2821711.

[14] A. Bracq *et al.*, 'Behind armour blunt trauma assessment by means of experimental and numerical approaches', in *Personal Armour Systems Symposium*, Washington D.C, USA, 2018.

[15] C. Bir, R. Lance, S. Stojsih-Sherman, and J. Cavanaugh, 'Behind Armor Blunt Trauma: Recreation of Field Cases for the Assessment of Backface Signature Testing', in *30th International Symposium on Ballistics*, 2017, doi: 10.12783/ballistics2017/16912.

[16] O. Mauzac *et al.*, 'Comparative assessment of Behind Armour Blunt Trauma (BABT) by means of a novel transparent synthetic gel', in *Personal Armour Systems Symposium*, Germany, 2012.

[17] A. Bracq *et al.*, 'On the modeling of a visco-hyperelastic polymer gel under blunt ballistic impacts', *International Journal of Impact Engineering*, vol. 118, pp. 78–90, Aug. 2018, doi: 10.1016/j.ijimpeng.2018.04.001.

https://doi.org/10.52202/078352-0015

[18] A. Bracq *et al.*, 'Numerical Recreation of Field Cases on a Biofidelic Human FE Model Involving Deformable Less-Lethal Projectiles', *Human Factors and Mechanical Engineering for Defense and Safety*, vol. 3, no. 1, Dec. 2019, doi: 10.1007/s41314-019-0022-8.
[19] A. Bracq, 'Contribution à la prédiction du risque lésionnel thoracique lors de chocs localisés à travers la caractérisation et la modélisation d'impacts balistiques non pénétrants', PhD Thesis, University of Valenciennes and Hainaut-Cambresis, 2018.

https://doi.org/10.52202/078352-0015

Behind armour blunt trauma (BABT) indenter simulating high-velocity impacts from rifle rounds on hard body armour

J. Op 't Eynde[1,2], C. P. Eckersley[2], R. S. Salzar[3], B. D. Stemper[4], B. S. Shender[5], T. B. Bentley[6] and C. R. Bass[2]

[1]*joost.opteynde@duke.edu*
[2]*Injury Biomechanics Laboratory, Duke University,*
101 Science Drive 1427 CIEMAS BME Box 90281, Durham, NC 27708, USA
[3]*Center for Applied Biomechanics, University of Virginia,*
4040 Lewis and Clark Drive, Charlottesville, VA 22911, USA
[4]*Joint Department of Biomedical Engineering, Marquette University & Medical College of Wisconsin, 5000 W. National Ave. Research 151, Milwaukee, WI 53295, USA*
[5]*Naval Air Warfare Center Aircraft Division,*
48110 Shaw Road, Building 2187 Suite 2203, Patuxent River, MD 20670, USA
[6]*Office of Naval Research, 875 N Randolph St, Arlington, VA 22217, USA*

Abstract. Body armour provides protection against gunshot wounds for both civilian and military personnel. When body armour deforms to defeat an incoming round, the backface deformation of the armour can produce high rate loading of the thorax, injuring the ribcage and internal organs. To improve body armour design without diminishing vital protection or increasing user burden, more accurate thorax impact behaviour and injury criteria are needed, which when integrated into finite element models improve their biofidelity. In this study, a repeatable, non-destructive test setup was developed to quantify the effects of behind armour blunt trauma (BABT) in vivo using pigs or other human surrogates. Flash x-ray images of backface deformation in hard body armour provided depth and diameter measurements, which were used to create an indenter machined out of polycarbonate (mass: 0.214 kg, diameter: 100 mm). The indenter with onboard accelerometer was propelled using high pressure helium gas to deliver impacts simulating BABT. Four tests of increasing velocity (22-54 m/s) were performed on two live anesthetized pigs. Thorax impact energy varied from 52 J to 305 J, while indenter acceleration upon impact varied from 3,593 g to 26,656 g. Force was obtained by multiplying indenter mass by acceleration. Displacement during impact was obtained from double integration of the measured acceleration and verified using high speed video images. Rib fractures, liver and lung contusions occurred for all impacts above 126 J, and a 52 J impact caused broken ribs in one pig but not in the other. All rib fractures were non-displaced and did not penetrate the pleura. Observed injuries are similar to reported in-field conditions following BABT. This test methodology provides a repeatable and robust instrumented impact scenario representative of BABT.

1. INTRODUCTION

The benefits of ballistic protective body armour systems have been well characterized for both law enforcement officers and military personnel [1]. By reducing the risk of penetrating injury and lowering the energy transferred to the body, it has been estimated that wearing body armour increases the likelihood of survival for law enforcement officers shot in the torso from 32% to 80% [2]. Modern body armour can defeat incoming pistol and rifle rounds, trading energy and momentum deposition into the armour for deformation of the armour. This deformation includes direct deformation of the body armour in soft body armours and deformation with fracture in hard body armours. While protecting the user from penetrating trauma, deformation of the armour backface can cause local, high-rate loading of the underlying tissues resulting in trauma to the thoracic cage and internal organs, leading to serious injuries or even death [3-6]. These injuries are often called Behind Armor Blunt Trauma (BABT).

Due to the effectiveness of body armour in protecting the wearer from penetrating gunshot wounds, it is often worn for prolonged periods of time by law enforcement and military personnel. Because of the weight, bulk, and thermal load, it has a negative influence on the physical and psychological performance of the wearer [7-9]. Therefore, body armour should be designed to be as light as possible, while still protecting against critical threats. To aid in this design process, balancing weight

 https://doi.org/10.52202/078352-0016

and protection, computational models with finite element analysis are often used to characterize the injuries caused by BABT. These models are an increasingly important research tool to simulate scenarios that are difficult and costly to test experimentally, and to understand local tissue behaviour. For models to provide valid predictions for injuries and material behaviour, they rely on accurate mechanical properties and dynamics of the biological tissues they represent. A 2012 report by the National Research Council stated that "The fidelity of anatomical, physical, and mathematical finite-element models simulating the human thorax, heart, lungs, liver, and kidneys, is limited" and the presence of "the need for tests using human cadavers and large animal cadavers." [10]

To experimentally characterize BABT, several studies on human cadavers and animal models have been performed. Early studies on live goats with soft body armour established an injury threshold based on projectile velocity, with the notable injuries being in the lungs and ventricles when the impact was near the heart [5, 11, 12]. Later studies on pigs also indicated severe pulmonary injuries for high-energy impacts with soft body armour [6]. More recent studies evaluating hard body armour in pigs [13-15] and human cadavers [3] have found similar injury patterns. While these studies provide valuable descriptions of the physiological effect of BABT and an injury estimate in one specific condition, injury criteria for BABT are limited to the costal and sternal fracture criteria presented in Bass et al., 2006 [3]. High morbidity in BABT results from damage to the lung, heart, and liver [4, 13, 16, 17], but injury criteria for these soft tissues have not been developed. Current criteria for assessing BABT risk rely on studies determining backface deformation in clay or ballistic gelatin; neither have direct correlation to human or animal models and are insufficient for developing accurate thoracic BABT injury criteria [13, 18]. Development of injury criteria for soft tissues will aid in future body armour assessment and design.

In this study, a repeatable, non-destructive test setup was developed to quantify the effects of behind armour blunt trauma (BABT) in vivo using pigs or other human surrogates. The setup using an indenter will allow us to obtain thorax dynamics, thorax material properties, and determine injury risk curves for rib fractures, bruising, and soft tissue injuries in future studies.

2. METHODS

2.1 Indenter Development

From previous studies (Bass et al., 2006 and Sarron et al., 2000) [3, 15], flash x-ray images provided depth and diameter measurements of maximum backface deformation in hard body armour during rifle round impact with areal densities appropriate for hard body armour. These measurements were used as the model to create an indenter machined out of a polycarbonate material with a mass of 0.214 kg, a diameter of 100 mm, and a dome height of 25 mm. An accelerometer (Endevco 7270) and battery powered data acquisition system (Slice Nano, DTS) were secured inside the indenter, in order to record the acceleration without having any external attachments. A back panel sealed off the indenter and provided flight stability with carbon fibre fins.

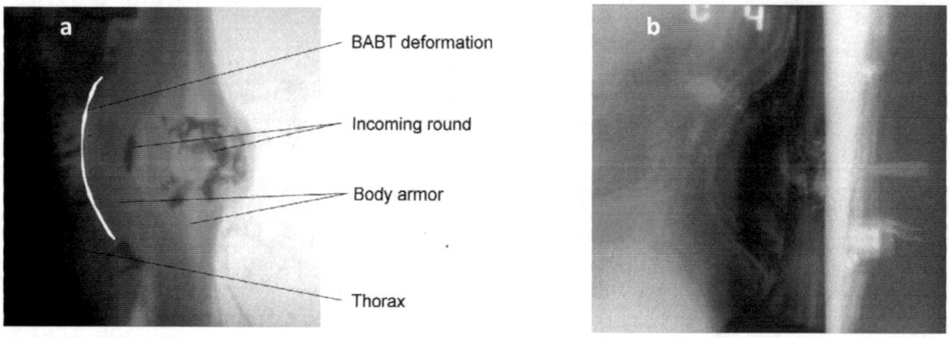

Figure 1. Maximum backface deformation profile in hard body armour from (a) Bass et al., 2006 [3] and (b) Sarron et al., 2000 [15].

https://doi.org/10.52202/078352-0016

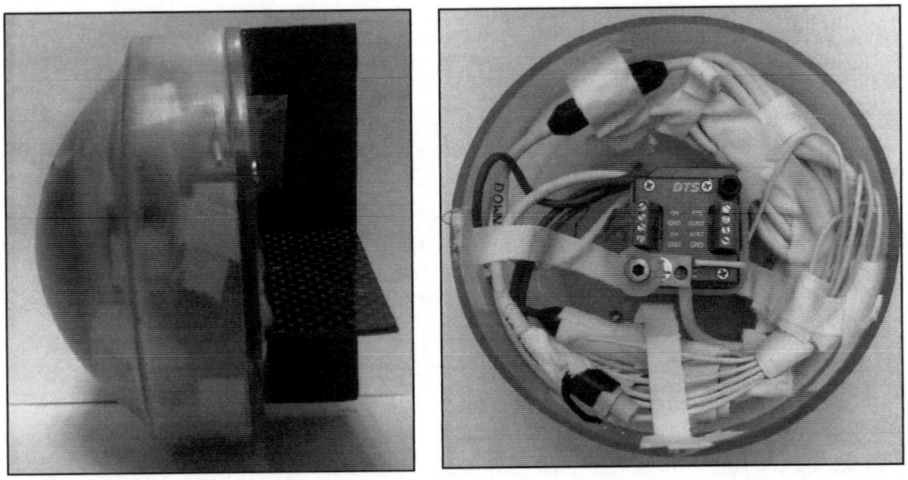

Figure 2. Polycarbonate indenter design with onboard accelerometer and data acquisition.

2.2 Impact tests

To achieve a high-speed simulated BABT, the indenter was loaded into a tightly fitted tube. The release of pressurized helium gas propelled the indenter to the target, placed at the end of the tube. Four tests of increasing velocity (22-54 m/s) were performed on each of two live anesthetized pigs, impacting the upper thorax and the lower thorax bilaterally. The two Yorkshire pigs were approximately six months old pigs with a mass of 38.9 kg and 38.3 kg. Velocities were chosen such that the kinetic energy of the indenter corresponded to realistic impact energy from a rifle round. Arborelius et al. (2012) [19] found that kinetic energy of the indenter in BABT simulation experiments most closely correlated with injury severity. The impact sites were situated above approximately the 5th and 9th rib of the animal, to achieve unobstructed impacts superficial to the lungs and liver. The animal was positioned prone with front legs forward on a lift table which was repositioned to place the desired impact site in line with the indenter (Figure 3). All procedures on these animals were approved by the Duke University Institutional Animal Care and Use Committee (IACUC). The animals were ventilated and vital signs were monitored during testing. After each impact, a 30-minute waiting period was allowed to ensure vital signs returned to a stable condition. Impacts were recorded with a high-speed video camera (Phantom V711, Vision Research) at 7500 frames per second.

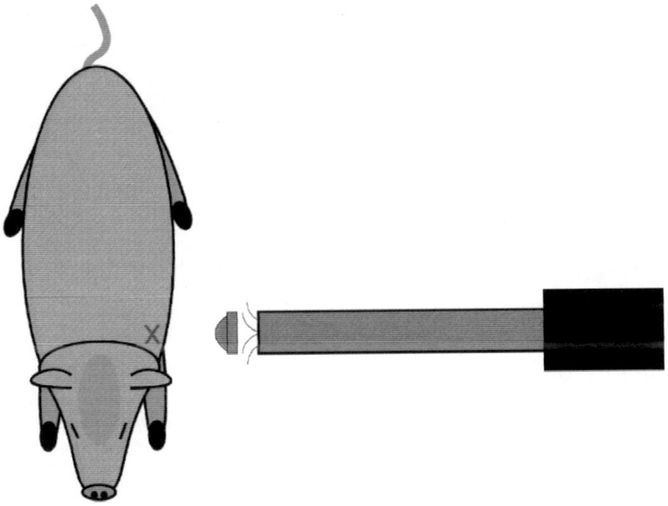

Figure 3. Test setup with pig, indenter, and launch tube.

https://doi.org/10.52202/078352-0016

2.3 Data Analysis

Indenter acceleration was recorded at 100 kHz sample frequency. Force was obtained by multiplying indenter mass by acceleration. Velocity was obtained by integrating the measured acceleration over time, and displacement during impact was obtained by integrating that velocity signal. Flight velocity was verified by high speed video analysis and matched closely the velocity obtained by integration (<1% error). Force-displacement behaviour was visualized for each impact. The impact stiffness was calculated as the steepest slope of the force displacement curve during the loading phase of the impact.

2.4 Injury assessments

After each test, the impact location was palpitated to identify any displaced rib fractures. Significant bruising at the impact location was observed and photographed. Thirty minutes after the last impact, the animal was sacrificed and a necropsy was performed. Injuries to ribcage and thorax organs were assessed and photographed during necropsy and micro CT scans were taken of the lungs, liver, and ribcage.

3. RESULTS

All four impacts were performed on each animal with a 30-minute waiting time in between. Vitals returned to a stable condition after each impact, however, there were short periods of cardiovascular instability following the two highest intensity impacts on animal #1. A time history for the heart rate and blood oxygen saturation for animal #2 is shown in Figure 6.

The results from the accelerometer data are shown in Table 1. Impact energy varied from 52 J to 306 J, by changing the incoming indenter speed from 22 m/s to 54 m/s. The peak acceleration of the indenter on impact ranged from 4,856 g to 26,656 g, resulting in a peak impact force from 7.5 kN to 55.7 kN. An example of the force-displacement curve and acceleration time history is shown in figure 4. The impact stiffness (slope of the curve) for this example is 24.6 kN/mm.

Table 1. Impact tests metrics. Each animal received an impact at four different locations on the thorax. All impacts were performed in order from low velocity to high velocity.

Animal Number	Velocity [m/s]	Energy [J]	Peak Impact Acceleration [g]	Peak Impact Force [kN]	Stiffness [kN/mm]
1	22.2	52.5	4,856	10.1	16.4
	36.9	145.1	11,283	23.6	24.6
	47.0	235.2	18,596	38.9	31.4
	53.6	305.5	26,656	55.7	36.6
2	23.1	56.6	3,593	7.5	11.0
	34.5	126.8	7,030	14.7	14.1
	37.2	147.4	7,019	14.7	15.1
	39.5	166.3	12,134	25.4	24.2

https://doi.org/10.52202/078352-0016

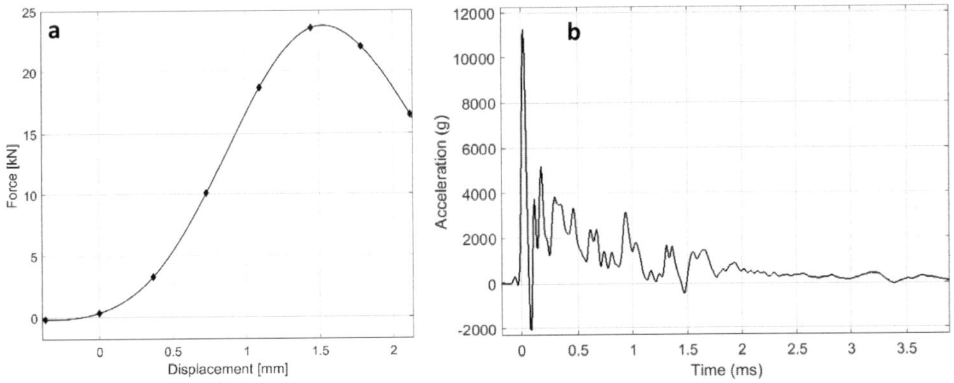

Figure 4. (a) Force-displacement graph of the initial interaction of the indenter with the thorax, smoothened using cubic spline interpolation. Measured datapoints are indicated by the markers. (b) Acceleration-time history of the same 145 J impact, unprocessed. Time = 0 was aligned with a sharp increase in acceleration experienced upon first interaction of the nose of the indenter with the thorax during impact.

Extensive bruising of the impact location occurred for each impact, similar to bruising seen in in-field BABT, but the skin remained intact. An example of the bruising can be seen in Figure 5. No rib fractures were seen for the lowest intensity test for animal #1, but a single broken rib was found under the impact site of the low intensity impact for animal #2, which was not detected before necropsy. At all other impact locations, multiple, non-displaced rib fractures were found. The impacts caused damage to the underlying internal organs. Serious lung contusions were found underneath the impact site of all but the lowest intensity impacts for both animals, and the lung exposed to 235 J and 305 J impacts was nearly entirely filled with blood. Micro CT sections of the lungs from animal #2 are shown in Figure 7. The liver of both animals, which was situated directly under the respective 145 J and 147 J impact sites, suffered contusions and lacerations, as seen in Figure 8. Following the two high intensity impacts for animal #1, there was a short period of rapid forceful muscle twitches in the impact area.

Figure 5. Bruising following 235 J (left) and 305 J (right) impacts on animal #1.

https://doi.org/10.52202/078352-0016

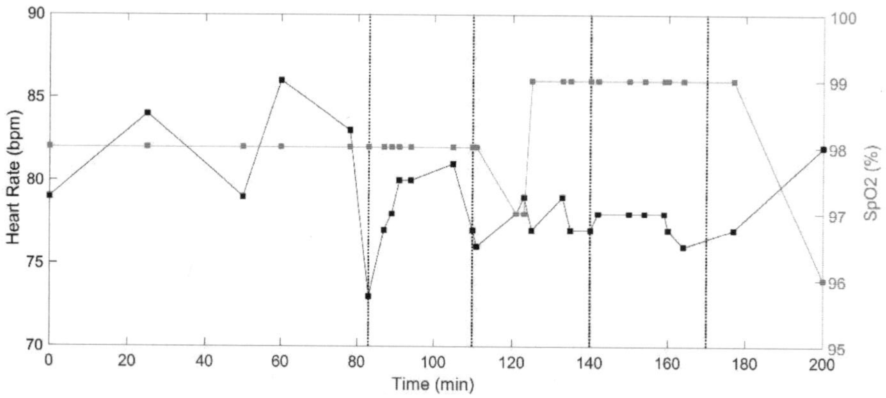

Figure 6. Heart rate and blood oxygen saturation (SpO2) for animal #2. The dotted vertical lines indicate the four impacts.

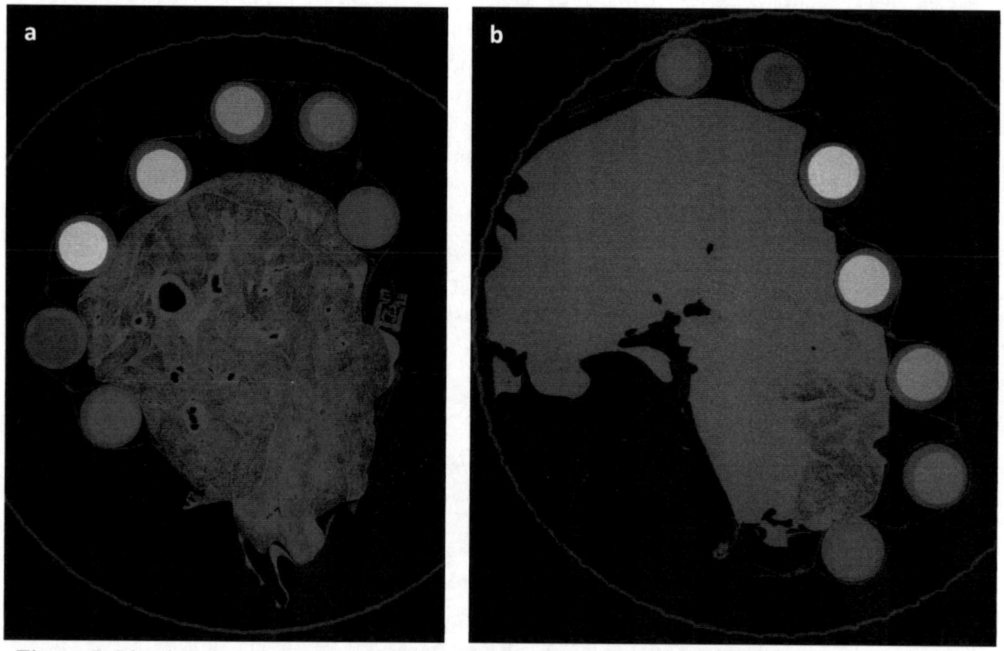

Figure 7. Blood (light grey) in between alveoli (fine, dark grey structures) of two lungs from animal #2 as seen on Micro CT scans. (a) is a slice of the right lung under the low intensity impact site and shows small areas of bleeding in an otherwise functioning lung. In (b), a section of the left lung is shown, which was exposed to two impacts at higher intensity. The majority of the alveoli in this section are filled with blood, indicating serious injury. The circles present in the scan are part of a CT phantom, included as calibration for the density of scanned materials.

https://doi.org/10.52202/078352-0016

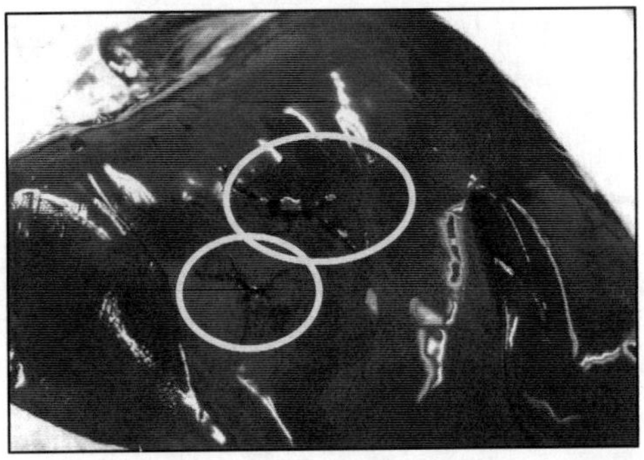

Figure 8. Liver of animal #2 after necropsy. Two lacerations on the parenchyma are examples of BABT.

4. DISCUSSION

The impact metrics measured during these tests are representative of a realistic BABT injury scenario. Bass et al. (2006) [3] measured peak impact forces in the 15k-30kN range on a cadaver sternum behind a hard body armour for a 7.62 mm round. They calculated a 50 % risk of sternum fracture at 24.9 kN. Injury severity in the current study matched expectations based on a previous indenter study by Arborelius et al. (2012) [19], as they reported an approximate impact energy threshold of 210 J for the transition from moderate to severe injuries. Park et al. (2012) [20] concluded that the transferred kinetic energy to clay from a 9 mm round impact on soft body amor was 48-100 J, and for a .44 Magnum it was 177 J. The amount of transferred energy was estimated to be 6-13% the energy of the incoming round. While no such estimates have been made for hard body armour, the results presented in this paper encompass these energy levels relevant for BABT injuries. In addition to that, the injuries observed in the animals also correspond to realistic injuries seen in modern military combat. The injuries sustained during BABT, as seen in this study, while not fatal, can be life threatening if medical treatment is not immediately available. When they are treated, they can still cause prolonged morbidity for the soldier.

In this study, only two animals were used, and four impacts were performed on each animal. While the structure of the ribcage stayed intact, fractured ribs under one impact site might have affected the integrity of the ribcage for an adjacent impact site. This is one of the limitations of the study.

The muscle twitch in the chest wall of the pig following the high-velocity impacts could be due to a reflex or local nerve damage. This behaviour is not well documented, and the response needs to be further investigated. If the behaviour is consistent, surface electromyogram electrodes could have potential as a mechanism to alert for possible BABT in military combat scenarios, where access to medical scanning equipment for evaluating these injuries might be challenging.

Future work could include expanded test conditions to allow us to find injury thresholds for the thoracic organs and develop an injury criterion. Injury criteria for these soft tissues will be instrumental in developing meaningful standards for body armour performance. In addition to BABT simulations impacting the ribcage for a whole thorax response, the impacts can be delivered directly to the organs to evaluate organ level injury criteria and mechanical behaviour. This data would be invaluable for the development of finite element models of BABT and aid in the design of new body armour.

5. CONCLUSION

A novel experimental method was developed to test the effects of behind armour blunt trauma (BABT) non-destructively. The use of an indenter created based on armour backface deformation allows for a controlled test design. The pig animal model suffered serious injuries similar to in-field reported conditions and previous experimental and theoretical studies. Rib fractures, contusions and bleeding in the lung, and liver contusions and lacerations were observed. Thorax and indenter mechanics upon

https://doi.org/10.52202/078352-0016

impact were recorded in great detail. This test methodology provides a robust instrumented impact scenario representative of BABT.

Acknowledgements

The authors gratefully acknowledge the funding and collaboration from MTEC-18-04-I-PREDICT-07, Incapacitation Prediction for Readiness in Expeditionary Domains, an Integrated Computational Tool (I-PREDICT) Thorax Model Prototype.

References

[1] Peleg, K., et al., J Am Coll Surgeons, 2006.; 202(4); pp. 643–648.
[2] LaTourrette, T., J Occup Environ Hyg, 2010; 7(10); pp. 557–562.
[3] Bass, C.R., et al., Int J Occup Saf Ergo, 2006;12(4); pp. 429–442.
[4] Cannon, L., J Roy Army Med Corps, 2001; 147(1); pp. 87-96.
[5] Carroll, A.W. and Soderstrom, C.A., Ann Surg, 1978; 188(6); pp. 753.
[6] Liden, E., et al., J Trauma, 1988; 28(1 Suppl); pp. S145-8.
[7] Larsen, B., et al., Mil Med, 2012; 177(11); pp. 1308-1315.
[8] Shanley, L.A., Slaten, B.L., and Shanley, P.S., Cloth Text Res J, 1993; 11(3); pp. 55-59.
[9] Ricciardi, R., Deuster, P.A., and Talbot, L.A., Mil Med, 2008; 173(9); pp. 817-824.
[10] National Research Council, Testing of body armor materials: Phase III. (National Academies Press, Washinton, DC, USA, 2012).
[11] Montanarelli, N., et al., Protective garments for public officials. Army Land Warfare Lab Aberdeen Proving Ground MD, 1973.
[12] Goldfarb, M.A., et al., A method for soft body armor evaluation: medical assessment. National Technical Information Service, Springfield, VA, USA, 1975.
[13] Gryth, D., et al., Mil Med, 2007; 172(10); pp. 1110-1116.
[14] Sondén, A., et al., Journal of Trauma and Acute Care Surgery, 2009; 67(6); pp. 1191-1199.
[15] Sarron, J., et al. Physiological results of NATO BABT experiments., Personal Armour Systems Symposium., Colchester, UK, 2000.
[16] Mirzeabasov, T., et al. Further investigation of modelling system for bullet-proof vests. in Personal armour systems symposium, Colchester, UK, 2000.
[17] Drobin, D., et al., J Trauma Acute Care, 2007; 63(2); pp. 405-413.
[18] Hanlon, E. and Gillich, P., Mil Med, 2012; 177(3); pp. 333–339.
[19] Arborelius, U.P., et al., Physiological Effects in Pig, and Mechanical Response in an Armour Test Rig, to Graded Chest Impacts using Behind Armour Blunt Trauma (BABT) Simulator, Personal Armour Systems Symposium, Nuremberg, Germany, 2012.
[20] Park, J., et al., Exp Mech, 2012; 52(8); pp. 1239-1250.

https://doi.org/10.52202/078352-0016

Initial injury observations in postmortem human subjects: Non-penetrating ballistic impacts over the sternum with light-weight hard armour plates

A. Iwaskiw[1], A. Wickwire[1], M. Vignos[1], C. Howes[1], N. Hahne[1], A. Injeian[1] N. Steiner[1], M. Bevan[1], E. Bar-Kochba[1], E. Mazuchowski[2], M. Clark[3], C. Carneal[1], D. Drewry[1]

[1]*The Johns Hopkins University Applied Physics Laboratory, 11100 Johns Hopkins Road, Laurel, MD 20723, Alexander.iwaskiw@jhuapl.edu*
[2]*Joint Trauma System, 3698 Chambers Pass, Joint Base San Antonio, Fort Sam Houston, TX 78234*
[3]*U.S. Army Combat Capabilities Development Command - Soldier Center / U.S. Special Operations Command, 10 General Greene Avenue, Natick, MA 01760*

Abstract Non-penetrating ballistic events, observed in personal ballistic armour use, are hypothesized to cause blunt injury. These events, termed behind armour blunt trauma (BABT), are not well understood. Early BABT studies involved soft torso armour testing on living goats, which formed the basis of armour performance requirements using a clay backface deformation metric. The lack of understanding of human injury as it is related to non-penetrating ballistic armour performance has contributed to further use of these clay standards for armour design. A more biofidelic human injury model is critical for understanding human injury risk in BABT for evaluation of modern armour systems and optimizing future armour systems.
This study measured BABT, using live ammunition and hard armour, on postmortem human subjects (PMHS) and presents initial observations for a sternum-aligned impact across relevant matched-pair clay deformations. Twelve instrumented PMHS were tested with light-weight hard armour plates along with separate matched-pair clay testing to correlate clay backface deformation to human injury. Combined injuries, coded using the abbreviated injury scale (AIS), ranged from skin damage (AIS $\leq$ 1) to multiple skeletal fractures along with lung lacerations (AIS $\leq$ 5). Further testing is ongoing to produce injury risk prediction models for the injuries seen in this study, and to develop a greater understanding of the influences of biological and experimental variation on injury risk.

1. INTRODUCTION

Modern design of personal armour using advanced materials has led to great improvements in preventing penetrating ballistic injuries [1]. With these personal armour advancements towards reducing armour weight and stopping more powerful rounds, a secondary injury modality, blunt trauma caused by armour deformation, is of increasing concern. These non-penetrating ballistic events, termed behind armour blunt trauma (BABT), while not regularly seen in combat [2] are theorized to be an emerging threat to the warfighter [3]. These BABT injuries are more common in law enforcement environments, due to the prevalence of lighter weight armour systems [4]. There is an ongoing desire in the warfighter community to reduce weight of armour systems while increasing ballistic performance [5]. However, there are concerns with reducing armour weight if it increases the prevalence and severity of BABT injuries to the warfighter community.

Early BABT research focused on soft armour systems. Notably, a series of studies in the 1970's involving soft armour on living goats, formed the basis of the original 44mm clay backface deformation (BFD) standard [6]. This standard's fidelity in assessing performance in newer armour is disputed [7] as it is not based on relevant armour systems or human injury. Studies have stated that there is a lack of human injury data in BABT events with relevant conditions [7, 8]. Continuing to define the relationship of BABT to human injury is critical to allow for continued development of new armour systems.

Classically, postmortem human surrogates (PMHS) have been used for initial evaluations of musculoskeletal injuries. Blunt impact injuries have been studied in automotive safety environments and lower speed blunt ballistic impacts [9-11] and provide a framework for assessing the injury risk. However, due to the uniquely high energy seen in terminal ballistic testing and the viscoelastic nature of the body, prior blunt injury research may fail to appropriately model the injuries hypothesized in BABT. It is also important to note that the relationship between ballistic energy and injury is highly dependent

 https://doi.org/10.52202/078352-0017

on the threat and armour pairing. While studies have been conducted evaluating BABT injury [12-15], there exists a need for more injury data in different locations and with different armour threat pairings, as evidenced by current research focus areas [8]. Additionally, these studies have not correlated injuries to the current clay BFD acceptance standard, limiting utility in validating the acceptance standard. It is critically important to identify the scope of potential BABT injuries to inform future armour acceptance criteria and design.

This study introduces an experimental protocol using instrumented PMHS, live ammunition, and light weight hard armour plates to generate BABT impacts to the sternum. Additionally, the injuries observed in this study allow for a better understanding of the potential implications of reducing the non-penetrating armour performance standards for current and future armour systems.

2. METHODS

2.1 Projectile and Armour Combination

The same projectile and armour combination was used for all tests in this study. The projectile chosen is relevant to current operational threats, while velocities were intentionally varied to achieve research goals. The armour was an ultra-high molecular weight polyethylene (UHMWPE) composite multi-curved plate with a Small Arms Protective Insert (SAPI) cut geometry, and it was purposely developed to allow for large clay BFS under specific ballistic test conditions. Each plate was shot once, along the mid-sagittal plane 100mm below the top edge. No backing material or soft armour was used between the hard armour and the impacted subjects in this study as the armour system was designed for the hard armour to be stand-alone.

2.2 Matched-Pair Clay Testing

Knowledge of associated BFS, through matched-pair testing, allows the PMHS aspect of the study to focus on relevant BFS values of interest from armour acceptance standards [6]. Prior to commencing PMHS testing in this study, 60 armour plates were tested to measure clay BFS as a function of round velocity. Testing generally followed the methods described in the National Institute of Justice (NIJ) standard 0101.03 [17]. The clay was scanned before and after each test using a coordinate measurement machine (CMM) FARO (FARO Technologies Lake Mary, Florida) laser scanner to create a point cloud representation of the clay surface. Geomagic Studio software (3D Systems, Rock Hill, South Carolina) was used to transform the point cloud into a surface model and calculate the clay deformation geometry (i.e., BFS).

Projectile velocity measurements were obtained using two sets of Oehler Research model No. 57 infrared screens with Hewlett-Packard (HP) counter chronographs (universal counters, HP model No. 53131A). Velocity was measured 4.6 m from the armour plate and then velocity loss equations were applied specific to the projectile to calculate striking velocity. Projectile velocities were tested across a range of relevant clay BFS, with a typical sample size of 3 for each velocity. The armour was placed on a clay block, with a hand rolled template used to ensure no air gaps behind the curved armour plate. The plates were oriented so that there was approximately 5° obliquity (above horizontal) to the incoming round at the specific shot location, which is consistent with current armour qualification test procedures. All firings were conducted at 7.6 m from the armour sample.

A linear fit of the clay BFS data versus the striking velocity was created. The repeatability was assessed with an r2 value equalling 0.92. The values measured in the clay BFS can be used to help describe the "dose" delivered to the PMHS for use in assessing the relationship between dose and injury for BABT.

2.3 PMHS Demographics

This study investigates the biomechanical response of instrumented PMHS and the associated characteristic injuries incurred by BABT centred over the sternum. The targeted impact location was between the level of the 3rd and 4th ribs, in the midsagittal plane. This study used twelve male, fresh-frozen, non-osteoporotic, full-body PMHS specimens. PMHS were acquired with bilateral

https://doi.org/10.52202/078352-0017

disarticulations at the humerus and femur. All PMHS were acquired under institutional review board exemption approval through Johns Hopkins Medical Institutions. Specimens were selected and acquired based on a controlled range of metrics including age, stature, weight, body mass index (BMI) and lumbar bone mineral density (BMD) based on Dual Energy X-ray Absorptiometry (DEXA), which are shown later with injury results in Table 2. Additionally, specimens with prior trauma or severe thoracic diseases requiring invasive surgical interventions were restricted from this study.

Once received, high-resolution computed tomography (CT) scans (slice thickness of 0.625 mm) were acquired for specimens at the Johns Hopkins University Applied Physics Laboratory (JHUAPL) to ensure no abnormalities or signs of trauma were apparent that could influence interpretation of injury. Specimens were stored at −20 ◦C until being thawed at 1.6 ◦C 6 days prior to testing for instrumentation and test preparation.

2.4 PMHS Instrumentation

Specimens were instrumented with high-rate biomechanical sensors in order to gain an understanding of the mechanical event and to measure potential mechanical correlates to injury. It is important to note that the instrumentation methodology was determined based on a desire to not affect injury outcomes, so key structures were left intact and minimally invasive techniques were used. While not a focus of this paper, it is important to introduce the sensors to help understand future analyses.

Instrumentation was placed on the surrounding torso skeletal system, and in key organ systems. Specific information, including location, type, and quantity of sensors are shown in Table 1. Generally, the instrumentation was grouped by skeletal motion, skeletal strain, skeletal acoustic emissions, critical organ pressure, and surface contact.

Table 1. Specimen instrumentation

Sensor Description	Location	Type	#
Rib strain	Anterior ribs 2-7, Posterior ribs 4-9	Single axis strain gauge (Kyowa KFWB-2-350)	24
Rib/Sternum Acoustic Emission	Lateral ribs 2-7, Superior manubrium	Acoustic emission sensor (Mistras Nano30)	14
Lung pressure	Left and right bronchi	Pressure transducer (Kulite XCL-100- 100A)	4
Liver pressure	Medial and lateral lobe; parenchyma	Pressure transducer (Kulite XCL-100- 100A)	4
Spleen pressure	Mid-parenchyma	Pressure transducer (Kulite XCL-100- 100A)	2
Aortic arch pressure	Aortic arch	Pressure transducer (Kulite XCL-100- 100A)	2
Left ventricle pressure	Lent ventricle	Pressure transducer (Kulite XCL-100- 100A)	2
Sternum acceleration	Superior manubrium and xiphoid (internal face)	Single-axis accelerometer (Endevco 727)	2
Spine acceleration	Co-aligned spinal level to impact location (varies)	Single-axis accelerometer (Endevco 727)	1
Surface force	On surface directly under impact, 38mm superior, left, inferior, right of impact location	Contact sensor (TekScan flexiforce 301)	5

Along with instrumentation, the specimens were outfitted with catheters and intubation equipment to perfuse the cardiovascular system and insufflate the lungs, respectively. Foley catheters were placed through the carotid arteries to the aortic arch and left ventricle of the heart, with the purpose of blocking flow out of this region of the cardiovascular system. Additionally, a Foley catheter was placed in the femoral artery. This catheter extended superior to the renal branching which allowed for isolation of the descending aorta, the aortic arch and the left ventricle. Contrast fluid was introduced to the isolated region to ensure proper installation, verified by CT. Active static pressure of 80 mmHg was introduced to the isolated heart region during the test to simulate normal cardiovascular pressure.

An intubation tube was placed through a tracheotomy used for sensor install, which was fitted to a down-regulated air compressor. Approximately 7 kPa of positive pressure was used to insufflate the lungs during the test. CT was used to ensure the lungs were fully insufflated and contacting the interior

https://doi.org/10.52202/078352-0017

surface of the anterior plural wall. A post-instrumentation, insufflated CT was taken to capture final sensor location, perfusion and insufflation.

https://doi.org/10.52202/078352-0017

2.5 PMHS BABT Experimental Setup

After instrumentation, the PMHS were transported to a NIJ certified commercial ballistic test lab. The specimens were placed on a custom mounting rig, resembling a table, and rotated upright. The specimens were strapped to the mounting rig using nylon straps around the forehead, shoulders, waist, and legs. The PMHS were supported by a seat, and the rig provided a rigid boundary condition at the back. The impact location of the specimen was in the middle of the sternum in the mid sagittal plane, specifically over the sternum between the levels of the 3rd and 4th ribs. The rig was adjusted, after insufflating the lungs, so that the impact location on the specimen was aligned with the ballistic target, and small adjustments were made to ensure the specimen was upright and symmetrical, simulating a front-on shot to the mid sternum (Figure 1).

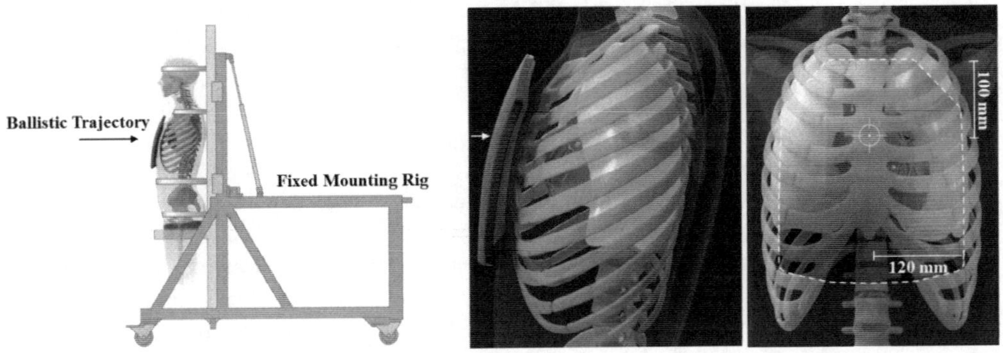

Figure 1. (Left) PMHS experimental setup. (Right) Lateral and frontal specimen and armour impact location

The armour used in the study was aligned with the shot location on the specimen using a custom carrier with holes centred on the impact location cut out to ensure no influence of the carrier to the BABT loading. The standoff of the plate, from the rear surface of the plate to the external surface of the skin, was measured using a FARO CMM. This was achieved by comparing the impact location on the PMHS before donning the armour to the impact location on the impact face of the armour after donning and subtracting out the armour thickness. Armour was donned to provide a "normal fit", which compressed parts of the PMHS but resulted in a sizable air gap between the armour and specimen at the impact location due to the curvature of the armour and curvature of the chest between the pectoral muscles. The armour standoff and obliquity was not controlled, but it was measured, spanning 3-25 mm and 5-13° above horizontal, respectively. Pretest x-rays and photos were taken of the perfused and insufflated PMHS directly before the ballistic impact to ensure proper anatomical targeting and measure upright/in situ sensor locations, shown in Figure 2.

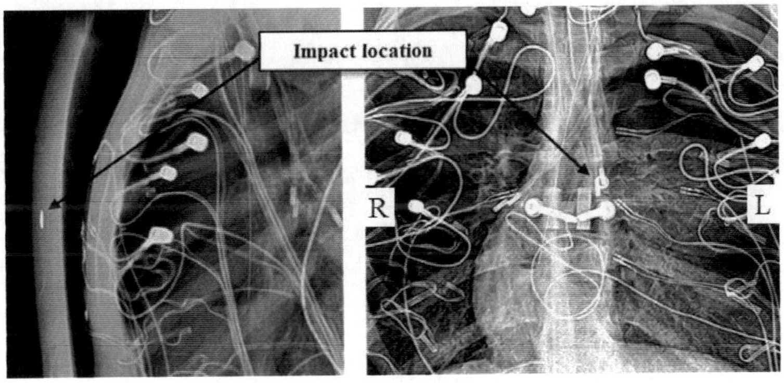

https://doi.org/10.52202/078352-0017

Figure 2. Characteristic (APL 11) specimen planar x-ray views of the sternum laterally (left) and anteroposterior of the sternum and spine (right) prior to impact.

The PMHS were then subjected to BABT loading by a single live-fire ballistic projectile across a range of relevant velocities and associated backface deformations. Projectile velocity measurements were obtained similarly to the clay matched-pair testing. A high-rate data acquisition system (Dewetron Inc., East Greenwich, RI, model DEWE-801) was used to record PMHS sensor data at 1 MHz sampling rate. Acoustic emissions sensor data was collected using a PicoScope 5000 high-rate data acquisition system (Pico Technology, Cambridgeshire, UK) at a 4.8 MHz sampling rate to accommodate higher frequencies of interest. Additionally, two high-speed camera systems (Vision Research Inc., Wayne, NJ, model Phantom v711) recorded the ballistic event at a rate of over 22,000 frames per second from both a lateral and oblique view of the specimen. All electronics systems were simultaneously triggered with the signal disruption of a frangible breakscreen by the incoming round. All instrumentation signals, accept for the acoustic emission data, were subject to a 300-kHz anti-alias filter.

2.6 Post-Test Evaluation and Forensic Analysis

Directly after the BABT impact, post-test x-rays and photos were taken, making sure not to disrupt any armour placement or insufflation. The specimens were carefully returned to a supine position and then transported back to JHUAPL for post-test CTs. Post-test CT scans were taken with insufflated lungs with armour on, armour off, and skeletal instrumentation removed. The specimen CTs were reviewed by a board-certified forensic pathologist who subsequently led an anatomical dissection of the PMHS. A detailed report, with associated pictures, for each PMHS was created to characterize and preliminarily assess severity of any injuries sustained as a result of the BABT.

3. RESULTS

3.1 Injuries

Injuries were observed, depending on their severity, in post-test x-rays, CT, and anatomical dissection, highlighted with representative images from a severe case in Figure 3. Typical post-test x-rays showed the residual armour deformation and underlying anatomical defects. CT scans were able to further identify the extent of the injury, where single slice views and maximum intensity projection images showed particularly compelling evidence of injury. Finally, an anatomical dissection was conducted to concretely evaluate the characteristics and severity of the injuries identified in the medical imaging. During this step, skeletal components were physically manipulated to assess fracture, the lungs were inflated after opening the pleura to identify lacerations, and organs were removed sequentially for further analysis. Autopsy notes were taken and integrated into injury scoring analysis using the abbreviated injury scale (AIS) [18]. AIS provides a relationship between a particular injury and probability of lethality. Associated injuries ranged from skin damage (AIS $\leq$ 1) to multiple skeletal fractures along with lung lacerations (AIS $\leq$ 5).

https://doi.org/10.52202/078352-0017

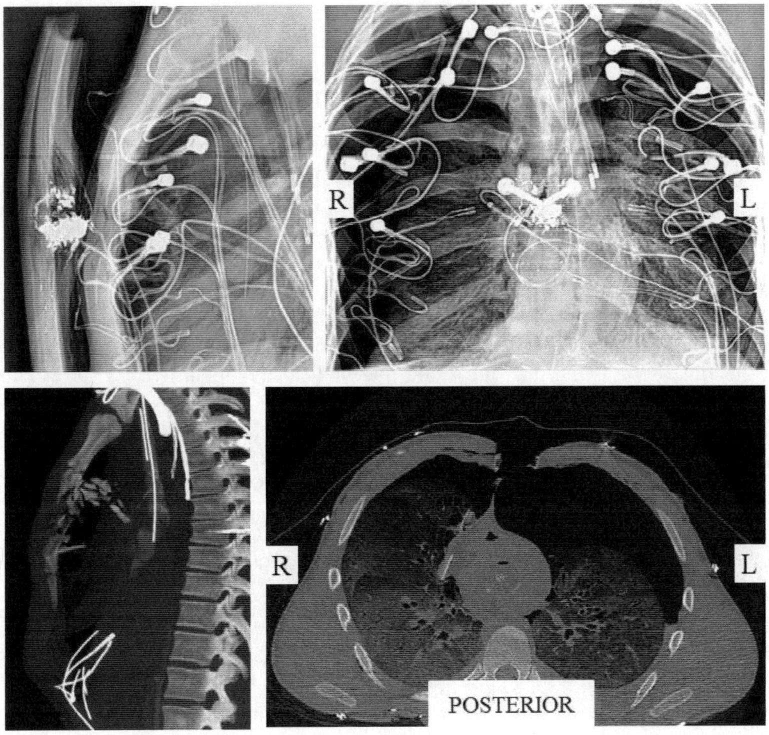

Figure 3. Characteristic (APL 11) specimen medical images after impact; (Top Left) lateral planar x-ray view of the sternum with armour on, (Top Right) anteroposterior planar x-ray of the sternum and spine with armour on, (Bottom Left) maximum intensity projection image of sternum damage in the sagittal plane with armour removed, (Bottom Right) axial CT slice at impact level with armour removed.

In order to quantify the most consequential injury, maximum abbreviated injury scale (MAIS) can be used to compare cases. An alternative way to assess the severity of a complex injury is to use the New Injury Severity Score (NISS) injury model [19]. The NISS is calculated from AIS scores and is the sum of the squares of most severe AIS coded injuries, regardless of body region. These injury outcomes can be found in Table 2. Individual injury codes can be found in Appendix 1.

Table 2. Specimen summary and injury results

APL Identifier	Age	Cause of Death	Stature (cm)	Full Body Weight (kg)	BMI (kg/m^3)	DEXA (T-Score)	MAIS	NISS
APL 01	60	Acute respiratory failure	175	71	23.03	-0.3	3	14
APL 02	64	Lung cancer	180	79	24.27	-0.4	3	17
APL 03	65	Glioblastoma multiforme	173	93	31.17	-0.8	3	14
APL 04	44	Glioblastoma multiforme	175	80	26.14	-0.5	1	2
APL 05	52	Cardiac arrest	170	67	23.02	0.5	3	17
APL 06	63	Metastatic brain cancer	178	89	28.26	0.7	3	14
APL 07	55	Sepsis	183	73	24.63	2.2	3	14
APL 08	57	Deep Vein Thrombosis	183	99	29.6	-0.5	2	9
APL 09	25	Complications: anoxic-ischemic encephalopathy	178	64	20.37	0.4	5	43
APL 10	20	Asphyxiation	178	69	21.8	-0.5	5	50
APL 11	63	Lewy Body Disease	183	100	29.8	-0.5	5	50
APL 12	62	Neoplasm of the Pancreas	178	72	22.7	0.7	5	66

https://doi.org/10.52202/078352-0017

4. DISCUSSION

4.1 PMHS Injury Outcome

Although the study is ongoing, an early identification of characteristic injuries can aid future armour design and policy. It is important to note that a majority of the clay BFS dose values used in this study are at or above the most widely adopted 44mm standard.

Thus far, four main injury presentations have emerged. The first, and least severe injury type, is a skin and soft tissue only injury. APL04 is the only result that falls into this injury outcome grouping. The AIS and NISS scores for this grouping are 1 and 2 respectively, and this outcome represents a minor injury. Future assessments planned for this study, leveraging medical experts, can further help characterize the likely clinical outcome for a living individual and the medical treatment needed for effective combat casualty care response.

The second grouping of injuries can be broadly characterized as skin and soft tissue and skeletal injuries. Five cases (APL 01, APL 03, APL 06, APL 07, APL 08) are categorized to this grouping. These cases present as skin and soft tissue injuries, crushing local fractures of the sternum (behind the impact location), and anterior fractures/separations of the ribs approximately 2-8 cm from the midline. MAIS spanning 2 to 3, and NISS spanning 9 to 14 are seen in this group. The singular MAIS 2 case injury outcome in this group, APL 08, is nearly indistinguishable from MAIS 3 of this group. The scoring distinction is that APL 08 resulted in less than 3 rib fractures, reducing its rib fracture AIS score from 3 to 2. Although in the absence of functional outcomes, the observed outcomes are assigned a serious outcome level.

The third grouping includes the same skin and soft tissue and skeletal injuries seen in the serious cases with the addition of communication with the anterior mediastinum, which could result in hemo- and pneumo- mediastinum (APL 02, APL 05). While this group does not have a higher MAIS than the serious cases, the additional clinical complication of another severe thoracic injury results in a less promising clinical outcome due to an increased risk of infection. The MAIS and NISS in this grouping are 3 and 17, respectively, with an assigned outcome level of severe.

grouping, these cases (APL 09, APL 10, APL 11, APL 12) share injuries such as skin damage, skeletal fracture, and communication with the anterior mediastinum. What differentiates these cases is the presence of damage of the underlying thoracic tissue, some suggesting pneumothorax. Accordingly, these four cases are assigned an outcome level of critical. Pneumothorax would require rapid medical intervention and is life threatening. The presence of these lung lacerations were identified through post-test x-ray and CT, where previously insufflated lungs were collapsed, and then confirmed in the anatomical dissection. Other injuries seen were flail chest rib fractures (3 adjacent ribs with 2 or more fractures), communication with the pleural cavity, laceration of the pericardial sac, and discoloration on the heart suggesting the likelihood of myocardial contusion in a living individual. These critical injuries suggest varied injuries of the thoracic organs can occur due to BFD. It was further observed that the natural positioning of key organs (e.g. anterior aspect of lung lobes positioned across mid sagittal plane directly under the impact location, mediastinum and heart directly under impact location, etc.) can influence which organs can be injured and subsequent severity. Not only does this sensitivity have implications for developing a validated human-based injury prediction model, but also for the operational medical community, as these injuries represent new challenges to combat casualty care.

The grouping of these injuries represents a pathway of likely injury outcomes. The least severe injury grouping consists of superficial injuries, and with incrementing case severity, more damage is seen deeper towards the centre of the body. However, in no way do the authors believe these categorical injury outcomes are linearly related to the dose, thusly requiring more data in each injury grouping to identify the complete injury risk spectrum [19]. The injuries seen in this study are very local and are similar to injuries seen in other similar mediastinal blunt trauma studies [11, 13], although some of the critical injuries seen in this study have not been observed before.

https://doi.org/10.52202/078352-0017

4.2 Study Limitations

This study has highlighted that, for the tested conditions, none are absent of injury, and injury severities vary greatly. It is important, however, to appropriately caveat the data as preliminary with ongoing efforts to further understand the relationship between dose and injury outcome, when dose is defined by the current clay test methodology. Additionally, due to the limitations in this model, further work must be done to understand the usability of this data in future injury risk prediction.

First, it is well known that mechanical properties of previously frozen PMHS tissue (especially soft and connective tissue) can vary from living tissue [20] and are absent of critical physiological responses. As a result, there is limited understanding of how valid soft tissue injuries are in PMHS [21]. It is important to understand the changes in tissue material properties and quality when interpreting organ injury. Non-living organs cannot accurately model lower severity organ injuries such as contusions, so living animal models may be a better model for these minor injuries. The authors believe, however, that the types of organ injuries that were observed (lacerations) and the fact that the organs of interest were properly inflated/perfused results in appropriate suggestions of potential injuries in a living human. Additionally, PMHS demographics, due to factors such as age and pathology, have been observed to contribute to a greater injury risk [22]. Ultimately, the fidelity provided by the near perfect geometrical model of a PMHS is critical in early understanding of injuries. The authors did adhere to a strict acquisition protocol, restricting age and limiting specimens with any signs of bone degeneration or disease.

Another limitation to this study is the use of a single threat/armour pairing in a limited range of dose conditions. It is very important to understand that extrapolation of this data to higher or lower doses or to other classes of armour and threats may not apply without verifying with further testing. For example, with the absence of a clothing or soft armour, the results in this study do not explain the role of soft armour in preventing BABT injuries, so further exploration is required for armour systems that include soft armour. Additionally, more information characterizing the specific contribution of the armour/threat pairing to BABT loading to the body should be generated before using this injury risk information on other cases. Other studies using mechanical surrogates can contribute greatly to the understanding of the conditions that lead to BABT injury [23, 24].

Another limitation of this study is the evaluation of a single impact location. The severe injuries seen in this study are dependent on the placement of organs with respect to the impact location. Further studies should explore key organ location and the distance/depth from potential BABT loading as well as the structural composition of the skeleton at that area [25].

Armour donning is thought to play a large role in BABT injury as "normal fit" resulted in anywhere from 3-25 mm of standoff in this study. Additionally, tissue composition and chest shape/anatomy varies within the population and could greatly affect the engagement of the armour BFD and the underlying anatomy. The authors are investigating the effect of standoff and flesh thickness, and believe that further studies should characterize the "normal fit" in military populations so that the injury risk developed in this study can be appropriately scaled to military population fit and local anatomical conditions.

5. CONCLUSION

This paper introduces an experimental methodology and initial injury results for live-fire non-penetrating ballistic impacts on instrumented PMHS. Early observations in tests show injuries ranging from minor (skin and soft tissue laceration) to critical (skin and soft tissue laceration, multiple complex skeletal fractures, lacerated lungs) that can be categorized into four distinct notional injury groupings: minor, serious, severe, and critical. The resulting injuries were generally localized to the region of armour deformation. While injury results have much value, it is important to understand that they are closely related with the conditions of this test series, and should not be extrapolated to other impact locations or armour/threat pairings without further testing. It is critical to the future use of this data for additional work to be done to understand emerging hypothesized contributors to injury risk, such as armour fit, anatomical/geometric variation, and armour type. However, the authors do believe that severe injury outcomes, as a result of BABT, are possible and that steps should be taken to ensure that future armour systems appropriately optimize the trade-offs between BABT protection and other design factors.

https://doi.org/10.52202/078352-0017

Acknowledgments

The authors would like to express their appreciation to the U.S. Special Operations Command and the US Army Combat Capabilities Development Command Soldier Center (CCDC Soldier Center) for funding this work under Contract No. is H92222-15-D-0004.

References

[1] Prat, et al., Contemporary body armor: technical data, injuries, and limits, European journal of trauma and emergency surgery 38.2 (2012); pp. 95-105.

[2] Carr, et al., Is behind armour blunt trauma a real threat to users of body armour? A systematic review, Journal of the Royal Army Medical Corps 162.1 (2016); pp. 8-11.

[3] Carroll, et al., A new nonpenetrating ballistic injury, Annals of surgery 188.6 (1978); pp. 753.

[4] Cannon L, Behind armor blunt trauma - an emerging problem, J R Army Med Corps. 2001; pp. 147:87–96.

[5] Bhatnagar, Lightweight ballistic composites: military and law-enforcement applications, Woodhead Publishing (2016).

[6] Montanarelli, et al., Protective garments for public officials. No. LWL-CR-30B73. Army Land Warfare Lab Aberdeen Proving Ground MD (1973).

[7] Hanlon, et al., Origin of the 44-mm behind-armor blunt trauma standard, Military Medicine 177.3 (2012); pp. 333-339.

[8] National Research Council, Testing of body armor materials: Phase III, National Academies Press, (2012).

[9] Kroell, et al., Impact tolerance and response of the human thorax II, SAE Transactions (1974); pp. 3724-3762.

[10] Viano,et al., A viscous tolerance criterion for soft tissue injury assessment, Journal of Biomechanics 21.5 (1988); pp. 387-399

[11] Bir, et al., Design and injury assessment criteria for blunt ballistic impacts, *Journal of Trauma and Acute Care Surgery* 57.6 (2004); pp. 1218-1224.

[12] Raymond, et al., Tolerance of the skull to blunt ballistic temporo-parietal impact, Journal of biomechanics 42.15 (2009); pp. 2479-2485.

[13] Bass, et al., Injury risk in behind armor blunt thoracic trauma, International journal of occupational safety and ergonomics 12.4 (2006); pp. 429-442.

[14] Rafaels, et al, Injuries of the head from backface deformation of ballistic protective helmets under ballistic impact, J. Forensic Sci., 60 (2015); pp. 219-225

[15] Mirzeabassov, et al., Further investigation of modelling for bullet-proof vests. In: Proceedings from the 5th Personal Armor Safety Symposium. Colchester, UK: International Personal Armour Committee (2000); pp. 211–34.

[17] Standard, N. I. J., 0101.06, Ballistic Resistance of Body Armor (2008); pp. 1-89.

[18] Gennarelli.,et al., AIS 2005, Association for the Advancement of Automotive Medicine, Barrington, IL (2005)

[19] Stevenson, et al., An overview of the injury severity score and the new injury severity score, Injury Prevention 7.1 (2001); pp. 10-13

[20] Hohmann, et al., The mechanical properties of fresh versus fresh/frozen and preserved (Thiel and Formalin) long head of biceps tendons: A cadaveric investigation, Annals of Anatomy-Anatomischer Anzeiger 221 (2019); pp. 186-191

[21] Nahum, et al., eds., Accidental injury: biomechanics and prevention, Springer Science & Business Media (2012).

[22] Laituri, et al. Derivation and evaluation of a provisional, age-dependent, AIS3+ thoracic risk curve for belted adults in frontal impacts. No. 2005-01-0297. SAE Technical Paper, 2005.

[23] Bevan. et al., Investigation of Armor and Round Effects on Water-Based Armor Assessment System Measurements, Personal Armour Systems Symposium 2016, Amsterdam, NL, (2016).

https://doi.org/10.52202/078352-0017

[24] Merkle A., et al., Assessing behind armor blunt trauma (BABT) under NIJ standard-0101.04 conditions using human torso models, Journal of Trauma and Acute Care Surgery 64.6 (2008); pp. 1555-1561

[25] Carneal C., et al., A Computational Pipeline Enabling the Generation of Multi-Organ Statistical Atlases for Improved Human Model Development, Military Health System Research Symp Kissimmee, FL (2016).

Appendix I

Table 1. Complete injury coding data

APL Identifier	AIS Coding of injury observed	
APL01	3 [450203.3]	Rib Cage; fracture(s) w/o flail; any location unilateral or bilateral ≥3
	2 [450804.2]	Sternum; fracture
	1 [410602.1], 1 [410202.1]	Skin; laceration, Skin; abrasion
APL02	3 [450203.3]	Rib Cage; fracture(s) w/o flail; any location unilateral or bilateral ≥3
	2 [442208.2]	Thoracic injury; Hemomediastinum
	2 [442209.2]	Thoracic injury; Pneumomediastinum
	2 [450804.2]	Sternum; fracture
	1 [410602.1], 1 [410202.1]	Skin; laceration, Skin; abrasion
APL03	3 [450203.3]	Rib Cage; fracture(s) w/o flail; any location unilateral or bilateral ≥3
	2 [450804.2]	Sternum; fracture
	1 [410602.1], 1 [410202.1]	Skin; laceration, Skin; abrasion
APL04	1 [410602.1], 1 [410202.1]	Skin; laceration, Skin; abrasion
APL05	3 [450203.3]	Rib Cage; fracture(s) w/o flail; any location unilateral or bilateral ≥3
	2 [442208.2]	Thoracic injury; Hemomediastinum
	2 [442209.2]	Thoracic injury; Pneumomediastinum
	2 [450804.2]	Sternum; fracture
	1 [410602.1], 1 [410202.1]	Skin; laceration, Skin; abrasion
APL 06	3 [450203.3]	Rib Cage; fracture(s) w/o flail; any location unilateral or bilateral ≥3
	2 [450804.2]	Sternum; fracture
	1 [410602.1], 1 [410202.1]	Skin; laceration, Skin; abrasion
APL 07	3 [450203.3]	Rib Cage; fracture(s) w/o flail; any location unilateral or bilateral ≥3
	2 [450804.2]	Sternum; fracture
	1 [410602.1], 1 [410202.1]	Skin; laceration, Skin; abrasion
APL 08	2 [450202.2]	Rib Cage; fracture(s) w/o flail; any location unilateral or bilateral <3
	2 [450804.2]	Sternum; fracture
	1 [410602.1], 1 [410202.1]	Skin; laceration, Skin; abrasion
APL 09	5 [442204.5]	Left chest, pneumothorax
	3 [450203.3]	Rib Cage; fractures w/o flail ≥3
	3 [441431.3]	Left lung, laceration
	2 [450804.2]	Sternum; fracture
	1 [410602.1], 1 [410202.1]	Skin; laceration, Skin; abrasion
APL 10	5 [442204.5]	Right chest, pneumothorax
	4 [415000.4]	Open ('sucking') chest wound
	3 [450203.3]	Rib Cage; fractures w/o flail ≥3
	3 [441431.3]	Right lung, laceration
	2 [450804.2]	Sternum; fracture
	1 [410602.1], 1 [410202.1]	Skin; laceration, Skin; abrasion
APL 11	5 [442204.5]	Left chest, pneumothorax
	4 [415000.4]	Open (sucking) chest wound
	3 [450212.3]	Rib Cage; fractures with flail, 3-5 ribs
	2 [450202.2]	Rib Cage; fractures w/o flail; any location unilat or bilat, 2 ribs
	3 [441431.3]	Left lung, laceration
	2 [450804.2]	Sternum; fracture
	1 [441004.1]	Heart; contusion
	1 [410602.1], 1 [410202.1]	Skin; laceration, Skin; abrasion
APL 12	5 [442204.5]	Left chest, pneumothorax
	5 [450214.5]	Bilateral flail chest
	4 [415000.4]	Open ('sucking') chest wound
	3 [441010.3]	Heart; laceration
	2 [441602.2]	Pericardium, laceration
	2 [450804.2]	Sternum; fracture
	1 [410602.1], 1 [410202.1]	Skin; laceration, Skin; abrasion

https://doi.org/10.52202/078352-0017

A computational approach to cumulative blast exposure in the brain for sequences of blast overpressure-orientation combinations

P. Matic[1] and X. Gary Tan[2]

[1]*Materials Science and Component Technology Directorate - Code 6000, Naval Research Laboratory, Washington DC 20375, USA*
peter.matic@nrl.navy.mil
[2]*Materials Science and Technology Division - Code 6300, Naval Research Laboratory, Washington DC 20375, USA*

Abstract. Blast pressure transmission into the brain, associated with traumatic brain injury (TBI), is a complex sequence of events. In this presentation (i) three-dimensional (3-D) nonlinear finite element computational simulations of the head to assess the sensitivity of internal brain pressures to individual blasts from different orientations and (ii) Monte Carlo simulations of random multiple blast sequences to assess cumulative exposure trends are discussed. The NRL high-fidelity 3-D human head finite element model is employed. This validated model, based on high resolution image data, accurately reproduces complex structures of the head and is implemented in the DoD Open Source multi-physics code CoBi. A hyper-viscoelastic model captures brain tissue behaviors with material parameters calibrated on dynamic loading data. The semi-empirical ConWep model generated transient blast from a free-field explosion in air that is nearly planar upon contact with the head model. The individual 3-D simulation results reveal that pressure patterns in the brain vary strongly with blast orientation and appear to highlight important influences of local skull curvature on blast pressure transmission. Regions of higher and lower skull curvature are associated with lesser and greater transmission of the incident blast wave. This appears to be associated with the relative phasing of the blast wave on contact with the head. Brain biomechanical injury is calculated using a spectrum of TBI thresholds associated with single and repetitive single injury. The Monte Carlo simulations, using a database constructed from the individual simulation results to analyse random sequences of blast orientations, were employed to tabulate a Cumulative Pressure Exposure Fraction (CPEF) metric versus number of blast exposures. Plots of 20 random orientation sequences of 200 blasts show the sequence dependent evolutions of the CPEF value for the brain, mean value and standard deviation CPEF convergence used to characterize early and late stage cumulative exposure.

1. INTRODUCTION

Blast pressure transmission into the brain is a complex sequence of events associated with traumatic brain injuries (TBI). In recent military operations, large numbers of service members have experienced repetitive concussion or sub-concussion exposures to explosive blast from improvised explosive devices (IEDs) [1]. Repetitive exposure to blast waves, in operational or training environments, can subject the warfighter to different ranges of pressures and incident directions of these waves. Military personnel having repeated exposures may experience neurological disorders, including concentration problems, blurred vision, irritability, headaches, sleep disorders and depression.

Current understanding of blast induced brain injury mechanisms is limited. In particular, conditions associated with cumulative head injury effects from repetitive blasts over a prolonged period of exposure are not well understood or quantified. Most effects of blast neurotrauma, resulting in structural, biochemical or electrical abnormalities in the brain, spinal cord or other nerves, are mild and difficult to detect with conventional neuroimaging. Research using controlled studies of induced mild traumatic brain injury (mTBI) is not possible with human subjects while the existing animal injury models cannot directly correlate with human injury. Three dimensional (3-D) computational analyses of blast wave biomechanics have the ability to establish relationships between the blast pressure on the head, internal biomechanical response and locations of induced brain injury. This insight can be useful in understanding injury mechanisms and designing improvement of protection systems.

The approach taken here is to quantify the biomechanical effects of individual blast events from different orientations at a fixed incident pressure using high fidelity 3-D computational simulations. The 3-D simulations track the internal response in the brain, such as stress, strain and strain rate, to each blast event and identify how the blast orientation affects the brain response. The maximum local pressures in the brain are used to track the global fraction of the brain exceeding the TBI thresholds. These detailed computational simulation results are compiled in a database covering the range of incident directions.

https://doi.org/10.52202/078352-0018

The 3-D simulation database was used to demonstrate a probabilistic Monte Carlo methodology for long random blast orientation sequences. The database is queried to calculate the cumulative exposures in the brain from the random sequences. The statistics of the internal brain pressure responses are analyzed to identify and quantify the initial variations of the mean values and standard deviation, early in the blast sequences, and the subsequent trend to convergence later in the blast sequences.

2. METHODS

2.1 Computational finite element model of human head

The 3-D human head finite element (FE) model was generated from in-vivo magnetic resonance imaging (MRI) scans with 1 mm isotropic resolution of a young adult. Figure 1 shows the human head finite element model subjected to the blast loading. The model consists of 29 material components including gray matter, white matter, ventricles, cerebrospinal fluid (CSF), skull, etc. Because of the complex geometry, a tetrahedral mesh was used for the discretization.

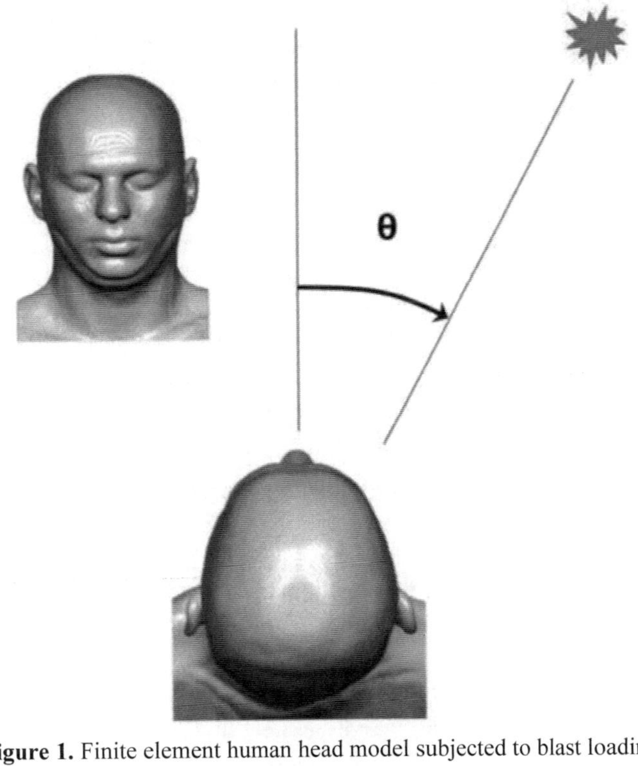

Figure 1. Finite element human head model subjected to blast loading

The multi-physics solver CoBi [2] is used for the simulation of blast-induced head biomechanics. The gray matter and white matter were modeled as hyper-viscoelastic materials. The CSF layer between the skull and brain and the ventricles inside the brain were modeled as the hyper-elastic solids with a very low shear modulus. The skull was assumed linear elastic. The full set of material models and corresponding parameters can be found in [3].

The ConWep model, incorporated in CoBi finite element solver, is used to apply the transient blast loading on the head resulting from a spherical free-field explosion in air (without the ground effect). A bare high explosive TNT charge of 1.03 kg is detonated at a distance of 2.7 m from the head surface at an angle θ relative to the middle sagittal plane of the head. The peak incident overpressure near the head is 15.0 psi (103.4 kPa). This value of overpressure is generally considered survivable in the operational environment. The angle θ ranges from 0 to 360 degrees with increments of 15 degrees, i.e. θ=0, 15, 30,…, 330, 345, 360 degrees, for a total of 24 angles.

The result of each finite element simulation with time duration of 2.5 ms produced spatially- and temporally-resolved pressure field data at each point in the 3-D brain model. These values can be compared to the minimum pressure thresholds for a critical pressure injury criterion to determine if TBI

https://doi.org/10.52202/078352-0018

has occurred. As shown in Table 1, pressure-based TBI thresholds suggested in the literature [4-6] were used to assess different levels of blast TBI. (While the pressure threshold criterion is used here, other criteria can be appled as appropriate.) The criteria are simultaneously applied on an element-by-element basis to the brain at every time step in each analysis. If an element maximum pressure exceeds a given pressure threshold at any time during the analysis, the element is considered to have been "injured". The full set of 24 blast orientation simulations generated a database containing the local pressures in each element and whether those pressures exceeded any of the four TBI thresholds.

Table 1. Pressure based thresholds of traumatic brain injury [4-6]

Blast induced TBI Type	Repetitive mTBI	Single onset mTBI	Single moderate TBI	Single intermediate TBI
Threshold (kPa)	142	173	204	235

2.2 Monte Carlo simulation of repetitive blast sequences

To conduct the Monte Carlo simulation of repetitive blast sequences, the precomputed finite element database of pressure responses in the brain containing the 24 incident blast orientations is used. Random sequences S=20, each sequence N_{tot}=200 in length, are drawn from the set of 24 blast orientation simulations. Each blast n=1, 2, 3,…, N_{tot} in a random sequence refers back to the precomputed database for dynamic pressure responses in the brain.

The pressure exposure (PE) of an element e in the finite element model is denoted as PE(e, n) and initialized to be zero, i.e. PE = 0. If the pressure in the element e equals or exceeds the specified injury threshold pressure at any time during a blast simulation, PE = 1. V(e) is the associated element volume. The cumulative pressure exposure fraction CPEF(e, N) for each element e in the brain model after N blast events ($1 \leq N \leq N_{tot}$) is defined as

$$CPEF(e, N) = \frac{\sum_{n=1}^{N} PE(e,n)}{N} \quad (1)$$

and CPEF(E_{tot}, N) for the entire brain with E_{tot} elements is defined as

$$CPEF(E_{tot}, N) = \frac{\sum_{n=1}^{N} \sum_{e=1}^{E_{tot}} PE(e,n) \times V(e)}{N \times \sum_{e=1}^{E_{tot}} V(e)} \quad (2)$$

The pressure exposure fraction (PEF) in the brain, defined for a single blast orientation, is equivalent to equation (2) with N=1 at an angle θ. Refer to [7] for further details.

3. RESULTS

The results of the individual and repetitive blast exposure analyses will be presented in terms of both local and global measures of brain dynamic response. Local measures associated with individual blasts include representative local pressure-time histories in the brain and spatial distributions of maximum pressures generated in the brain over the course of a blast analysis. Global measures associated with individual blasts include the pressure exposure fractions (PEF) versus orientation angle for a pressure injury threshold. Global measures for repetitive blast sequences are expressed in terms of the cumulative pressure exposure fraction (CPEF).

Figure 2 presents an example of data for the time history of pressure at three locations in the brain for θ=0 degrees, i.e. for a frontal blast wave, generated from an incident overpressure of 15 psi (103.42 kPa). The pressure waveform in the brain is significantly different from the incident blast wave because of wave convergence and divergence. The peak pressure at the frontal brain lobe is higher than the ConWep incident pressure and more comparable to the normally reflected pressure. The peak pressure is lower for interior brain points compared to the locations near the brain-CSF interface. Note that the peak pressure at the rear brain experiences a significant negative pressure at the early time when other parts of brain are in the positive pressures.

https://doi.org/10.52202/078352-0018

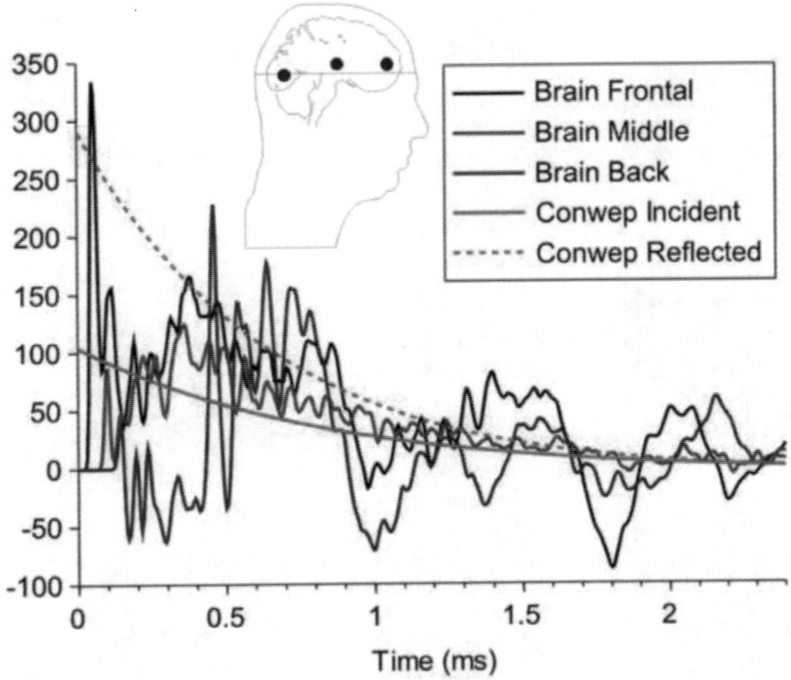

Figure 2. Time history of pressure in the brain for the frontal blast (θ=0 degrees)

The effects of different blast orientations (0, 75, 150, 180, -150 and -75 degrees) on maximum pressures experienced in the brain for a single blast are shown in Figure 3. The pressure distribution is generally symmetric for the blast orientation of 0 and 180 degrees. For 0, +/-150, and 180◦ blast orientations, the highest pressures reside in the region close to the coup and contrecoup. For +/- 75◦ blast orientation, the corresponding side of the brain incident to the blast experiences higher maximum pressures over a larger area and extending into deeper locations of the brain.

In Figure 4, the PEF parameter is plotted parametrically versus all 24 orientations and the four injury pressure thresholds for a single 15 psi (103.42 kPa) overpressure,. The plot quantifies how the PEF increases as the injury threshold decreases. It is broadly symmetric and features significant regions of both higher and lower PEF values. The slight asymmetry in the plot is caused by the natural asymmetry of the brain anatomy used to generate the model. Orientation intervals between (−120, −75) and (+75, +150) degrees have the highest CPEF values, while (−45,+45) degrees have the lowest values. The intervals between (−185, −120), (−75, −45), (+45, +75), and (+150, +180) degrees are the strong transition regions between the lowest and highest PEF values. Injury fractions of only 0.20 occur at +/−75 degrees for the higher 235 kPa injury threshold compared to 0.75 at +135 and -75 degrees for the lower 142 kPa threshold. It is clear that the side blast orientations produce much higher injury fractions in the brain compared to frontal or rear blasts.

Figure 5 shows an example of the change in PEF value relative to incident angle for the pressure injury threshold of 173 kPa in the cerebrum and cerebellum of brain, with further differentiation between grey matter (GM) and white matter (WM). The broad symmetry trend can again be observed. The PEF in the cerebrum is higher than that in cerebellum for the frontal blasts and lower for the rear blasts. Because the cerebellum is at the back of head, the PEF has the highest value in the cerebellum grey matter for the θ=180 degree blast orientations. The white matter of both cerebrum and cerebellum have a lower PEF because of the deeper interior locations compared to their respective grey matter.

It appears that the external curvature of the head, defined by the skull, is a significant factor affecting the levels of pressure infiltration into the brain. Figure 6 shows the surface curvatures of 3-D head finite element model. Both the Gaussian curvature and the mean curvature were calculated based on the triangular mesh defined by the head surface. An area on the side of the head exhibits lower curvatures compared to the forehead, back and top of the head. These low curvature areas on the right and left side of head, when aligned with the blast orientation, allow the nearly planar blast pressure wave to propagate more uniformly into the brain volume. As a result, the blasts at orientations of approximately +/-75 degrees produce the higher maximum pressures over a larger area of the brain as seen in Figure 3.

https://doi.org/10.52202/078352-0018

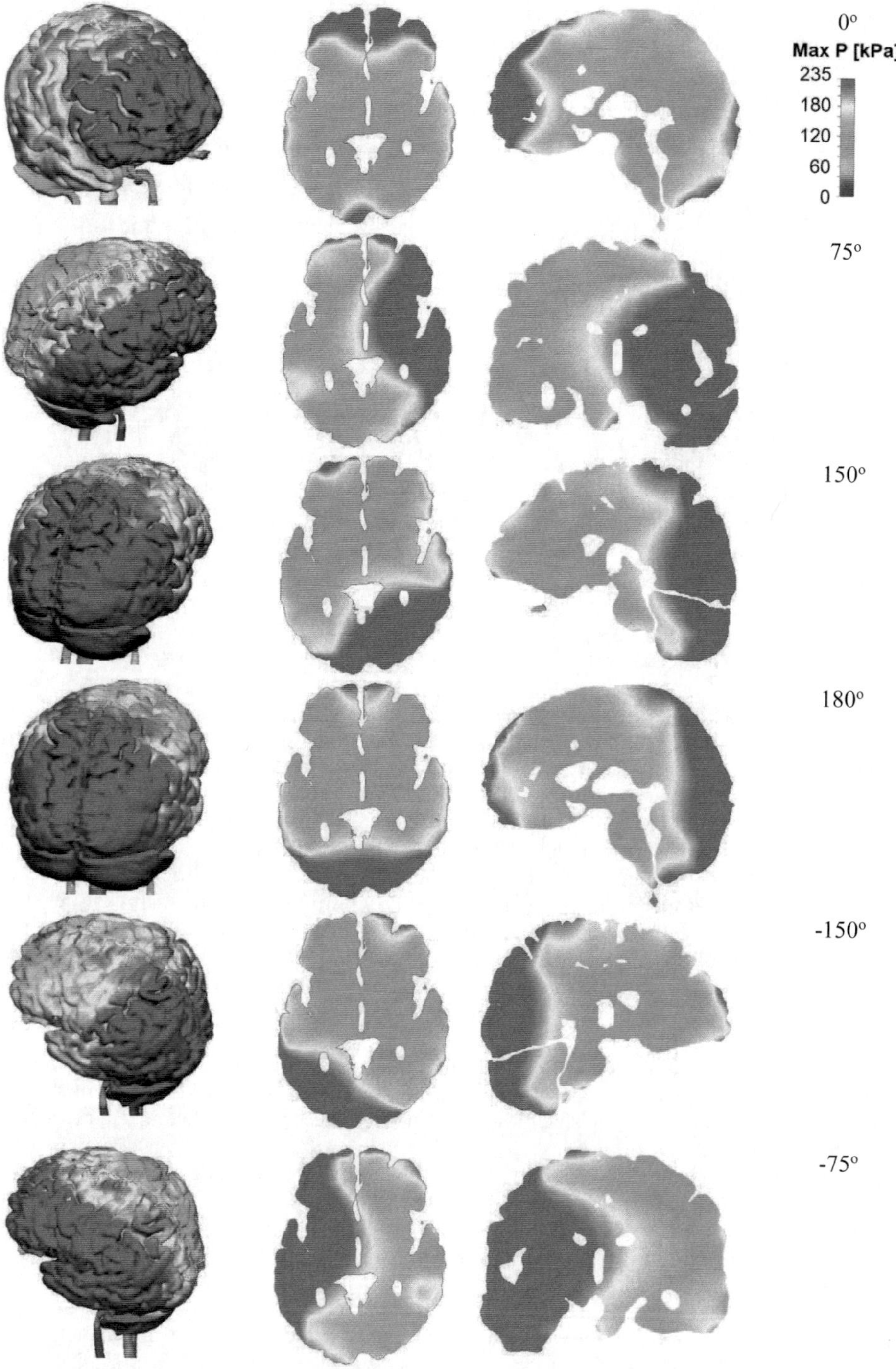

Figure 3. Maximum pressure in the brain during the blast for incident angles of 0, 75, 150, 180, -150 and -75 degrees at 15 psi (103.42 kPa) incident overpressure. Left graphic column shows contours on brain surface, middle graphic column shows contours on the transverse plane, right graphic column shows contours on the plane orthogonal to transverse plane along incident angle

https://doi.org/10.52202/078352-0018

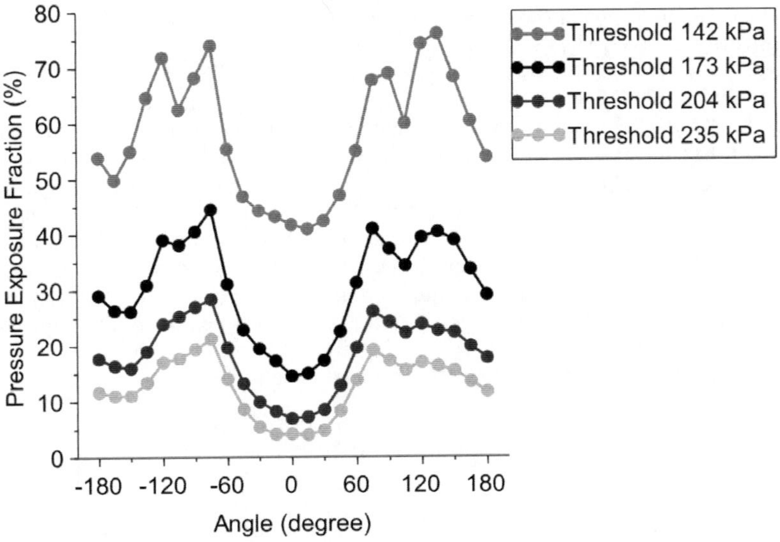

Figure 4. PEF vs. incident angle for four different pressure thresholds

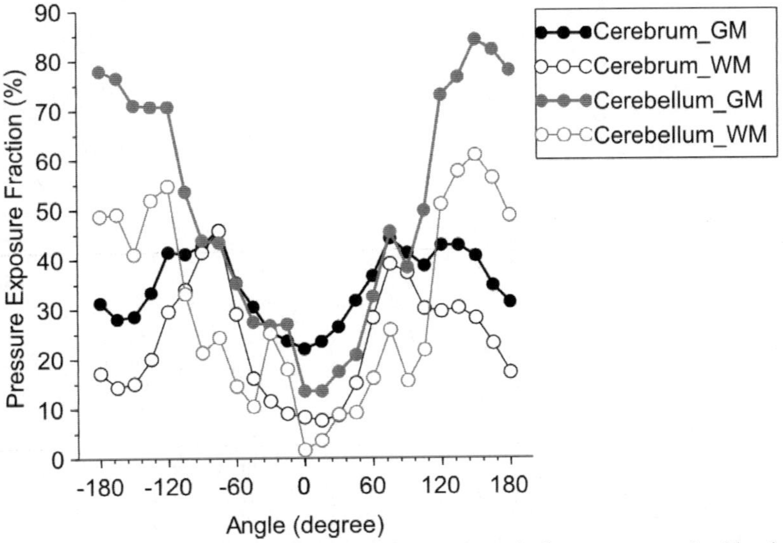

Figure 5. PEF vs. incident angle in cerebrum and cerebellum, parameterized by the grey matter (GM) and white matter (WM), for a pressure injury threshold of 173 kPa

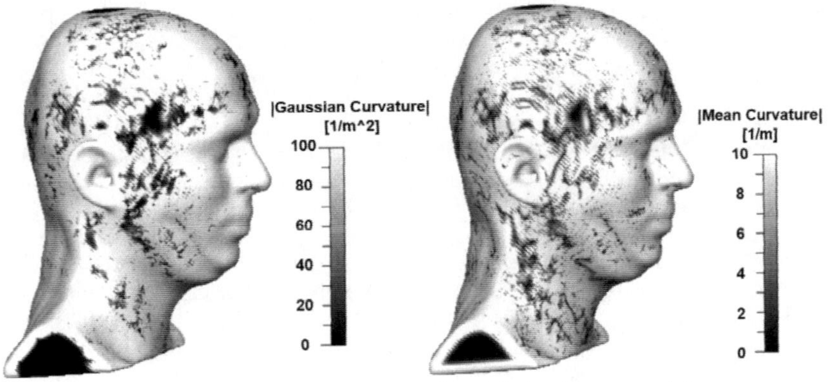

Figure 6. Surface curvatures of head finite element model

https://doi.org/10.52202/078352-0018

Using the database compiled from the individual finite element biomechanical analyses covering the 24 blast orientation angles, Monte Carlo simulations were conducted for 20 random sequences of $N_{tot} = 200$ blasts to generate cumulative response exposures statistics. Figure 7a through 7c shows the envelope curve for the 20 CPEF(E_{tot}, N) sequences exceeding the TBI thresholds of 142, 173, and 204 kPa in the brain. Also plotted are curves of the cumulative mean and cumulative standard deviation (SD), shown as red and black lines respectively. Figure 7d shows the semi-log plot of mean and SD with respect to number of blasts N. The CPEF parameter is sensitive to the direction of the blasts early in the sequence of 200 exposures, converging at approximately 40 blasts. The mean value of 0.59 is the largest, associated with the smallest TBI threshold of 142 kPa. The mean becomes much smaller, approximately 0.31 and 0.159 with the larger TBI thresholds of 173 and 204 kPa. The SD value converges at approximately 80, 70 and 40 blasts for the 142, 173 and 204 kPa injury threshold values, respectively. Analysis of the data suggests that while the mean value is associated with the scalar pressure dosing across the brain, the SD is associated with more complex geometric evolution of the CPEF patterns. When the SD converges, the patterns stabilize and do not change significantly from additional blasts.

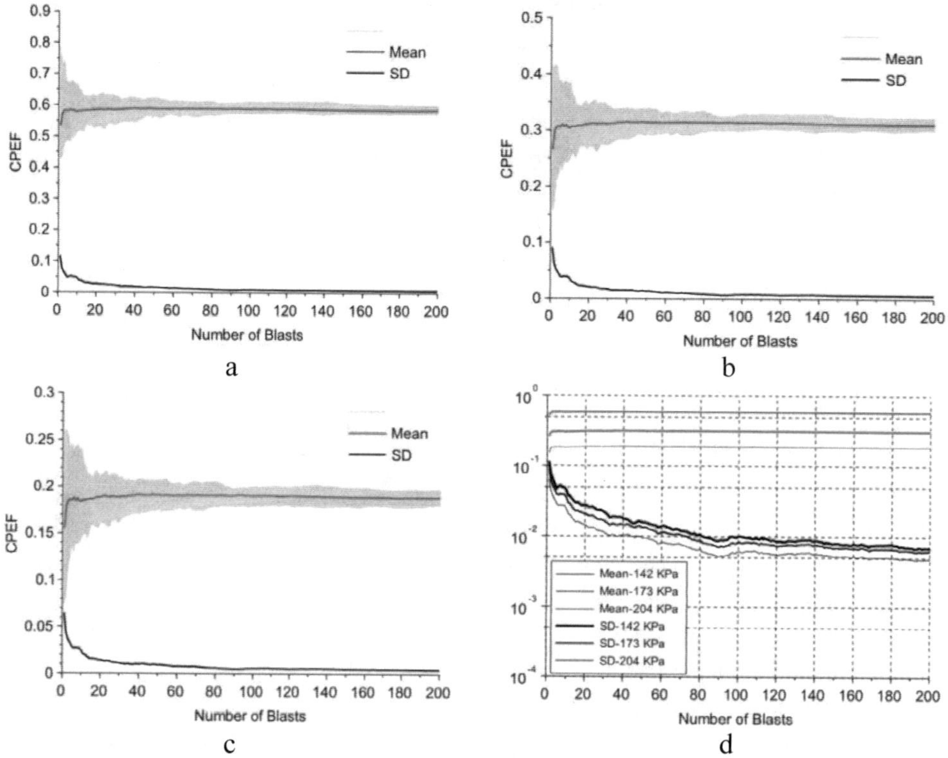

Figure 7. Three dimensional simulation Monte Carlo results for random blast sequences showing CPEF (E_{tot}, N) data for 20 blast sequences using TBI thresholds of (a) 142, (b) 173, and (c) 204 kPa. Each plot shows envelope curves (pink), cumulative mean value (red) and cumulative SD (black) for 20 blast sequences. Plot (d) is a semi-log plot of the cumulative mean and cumulative SD versus number of blasts N.

Comparison of these 3-D results with those of a prior study [7] that used a simpler transverse 2-D section along an axial plane of the 3-D human head model are useful. The 2-D study considered random blast sequences with the same 24 blast orientations while also considering four multiple blast overpressures of 7.5, 10.0, 12.5 and 15.0 psi (51.7, 68.9, 86.2 and 103.4 kPa). The effect of these overpressure-orientation combinations are shown in Figure 8 by CPEF plots for TBI thresholds of 142, 173, and 204 kPa generated from sequences of 200 random blasts. The CPEF mean values are approximately 0.17, 0.80, and 0.03 for the three injury thresholds, lower than the 3-D results due to the additional lower blast pressures. Figure 9 shows the CPEF versus number of blasts for TBI thresholds of 142, 173, and 204 kPa and a semi-log plot of mean and SD with respect to the number of blasts N. Qualitatively similar to the 3-D counterpart in Figure 7, the CPEF parameter shows the sensitivity for smaller numbers of blasts and convergence for larger numbers of blasts.

https://doi.org/10.52202/078352-0018

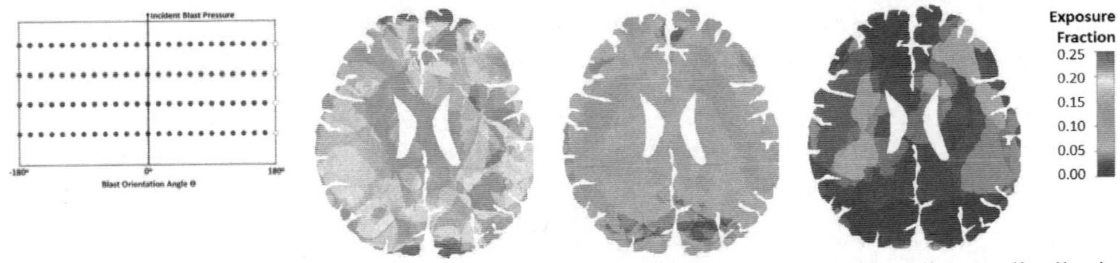

Figure 8. Monte Carlo simulations results from a two dimensions model. Uniform discrete distribution of the four blast pressures and 24 blast orientations used to construct 20 random blast sequences for N_{tot}=200, and three CPEF plots exceeding TBI thresholds of 142, 173, 204 kPa

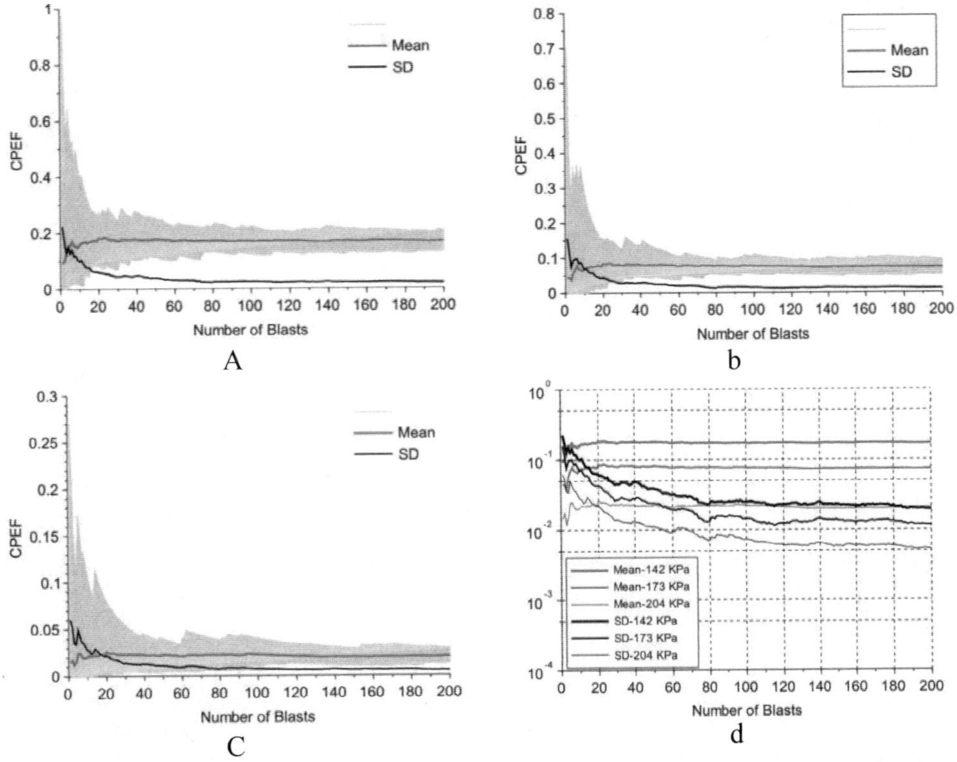

Figure 9. Two-dimensional simulation Monte Carlo results for random blast sequences showing CPEF (E_{tot}, N) data for 20 blast sequences using TBI thresholds of (a) 142, (b) 173, and (c) 204 kPa. Each plot shows envelope curves (pink), cumulative mean value (red) and cumulative SD (black) for 20 blast sequences. Plot (d) is a semi-log plot of the cumulative mean and cumulative SD versus number of blasts N.

4. DISCUSSIONS

The results presented above provide insight into the way individual blast events expose the brain to higher pressures exceeding TBI thresholds. This information is used to describe how multiple blast event exposures are accumulated by the brain. Different aspects of this and prior studies reinforce each other and quantify these trends.

From the individual blast simulations, the pressure versus time response across the brain is very heterogeneous. Pressure histories and pressure patterns vary strongly with blast orientation. Blast from the side, around +/-75 degrees, results in larger brain model injury areas than blast from other orientations. The maximum pressure encountered at each location in the brain from an individual blast creates complex patterns across the brain. These patterns include features beyond those at the typical coup and contrecoup locations associated with blast incident to the front or back of the head. For the side blast orientation the affected areas are much larger than these coup and contrecoup regions.

 https://doi.org/10.52202/078352-0018

The side blasts causing higher pressures in the brain can be related to the skull curvature. High skull curvature phases the contact of a planar or nearly planar blast wave normal to the head. The propagation of pressure into the head proceeds non-uniformly with a curved wave front, allowing a more phased dynamic response of the brain. Low skull curvature allows a normal blast wave to contact the head over a larger area nearly instantaneously, with propagation into the brain as a more linear wave front.

From the Monte Carlo studies of random blast sequences using the 3-D simulation data, the CPEF parameter tracks those portions of the brain exceeding a TBI injury threshold. CPEF tracks multiple blast exposures at each location in the brain and across the brain. It also captures the complex geometric overlay patterns that develop from different blast orientations (and overpressures), highlighting local regions of high and low cumulative exposure. The results confirm prior observations that the local low skull curvatures on the left and right sides make the brain more susceptible to blast injury. The CPEF itself is most sensitive, as shown by its fluctuations and range of responses, at lower values of N. The cumulative mean value is a more global measure of "dosing". Its convergence for the 3-D analyses at approximately 40 blasts means that in the range of 0 to 40 blasts each CPEF pattern in the brain model is likely to be significantly different and changing in geometry and in magnitude. The cumulative SDs converge between the first 40 and 80 blasts, depending on the TBI threshold, meaning that beyond this range the geometric patterns of exposure in the brain will stabilize and persist, total exposure will increase and average total exposure will remain the same.

5. CONCLUSIONS

A systematic, quantitative, and practical characterization of cumulative exposure of the 3-D brain model to repetitive blast events was developed using computational simulation of individual blast events and a Monte Carlo approach to random blast sequences. Results are generated in a computationally efficient manner for random blast sequences by constructing a detailed database, generated by a relatively small set of detailed finite element biomechanics simulations. This is followed by the Monte Carlo analysis that tracks cumulative exposure from multiple blast events utilizing the database.

The results highlight and quantify the likely role of blast orientations in random blast sequences on maximum pressure local values and spatial patterns across the brain. The associated mean values and SDs of the CPEF parameter, introduced to characterize the extent and repetition of high pressures exceeding different TBI thresholds, are tracked. Statistical convergence of the internal brain response metrics versus number of exposures quantitatively characterizes and highlights the cumulative effect in the repetitive blast environment.

These 3-D brain model results confirm the qualitative trends identified from prior analyses of a 2-D brain cross section model. 3-D simulations accounting for multiple blast pressures are planned. The significant variability in cumulative brain exposures in the early and intermediate portions of the repetitive blast sequences should be further analyzed and compared to the clinically observed range and severity of mTBI symptoms. The insights gained by this effort also complement and can maximize the utility of orientation data from blast sensors worn by the warfighter.

Acknowledgments

This study was supported by the Office of Naval Research (ONR) through the Naval Research Laboratory's Basic Research Program and the Department of Defense (DOD) High Performance Computing Modernization Program at the Army Research Laboratory, Air Force Research Laboratory and Army Engineer Research Laboratory DOD Supercomputing Resource Centers.

References

[1] Heltemes K.J., Holbrook T.L., Macgregor A.J. et al., Injury 2012; 43; 1990–1995.
[2] Tan X.G., Przekwas A.J. and Gupta R.K., Shock Waves, 2017; 27; 889–904.
[3] Tan X.G., D'Souza M.M., Khushu S., Gupta R.K., DeGiorgi V.G., et al., J Eng Sci in Medical Diagnostics and Therapy 2020; 3: 011007.
[4] Rafaels K.A., Bass C.R., Panzer M.B., et al., J Trauma Acute Care Surgery 2012; 73(4): 895–901.
[5] Salzar R.S., Treichler D., Wardlaw A. et al., J Neurotrauma 2017; 34:1589-1602.
[6] Ling G., Bandak F., Armonda R., et al., J Neurotrauma 2009; 27:815-825.
[7] Tan X.G., Matic P., Military Medicine 2020; 185:S1:214.

https://doi.org/10.52202/078352-0018

Coupling high strain rate experiments with numerical simulation for material model calibration: application to lightweight armor

H. Abdulhamid[1], J. Mespoulet[1], P. Deconinck[1] and P. Héreil[1]
[1]*Thiot Ingénierie, 830 route Nationale, 46130 Puybrun, France*
abdulhamid@thiot-ingenierie.com

Abstract. This paper presents a comprehensive mechanical study of a lightweight armor structure composed of two different kind of materials: a ceramic tile and a UHMWPE (Ultra High Molecular Weight Polyethylene) layer. The aim of this study is to provide reliable experimental data for building and validation of the whole structure under impact loading. For each material, a building block approach was used to characterize from the specimen to the structure; and a material model is fed at each step. Concerning the ceramic material, Hugoniot and spall parameters were at first identified using plate impact tests. Dynamic compression tests on Split-Hopkinson Pressure Bars (SHPB) were performed on confined specimens to analyze the effects of pressure on the damaged strength. Ballistic impacts on metal-backed ceramic tiles were performed to measure its strength in a typical shielding application. Experimental data was used to calibrate a Johnson-Holmquist (JH-2) model. For the UHMWPE, in-plane tension, and out-of-plane compression and shear were performed. This material exhibited very low shear modulus and peeling strength. Ballistic impacts have been performed on larger specimens. Based on these results, an elasto-orthotropic law with damage model was calibrated. At the end, a final validation test was performed on the armor made with the two characterized materials.

1 INTRODUCTION

Standard soldier lightweight armor (
Figure **1**) consists of a ceramic tile (2), a UHMWPE plate (3) and a polymeric foam (4). The impact fragments both the projectile (1) and the ceramic tile. The role of the UHMWPE is to decelerate and catch most of the generated fragments. The polymeric foam helps not only catching the last fragments but also reducing the impulsion transmitted to the soldier's chest.

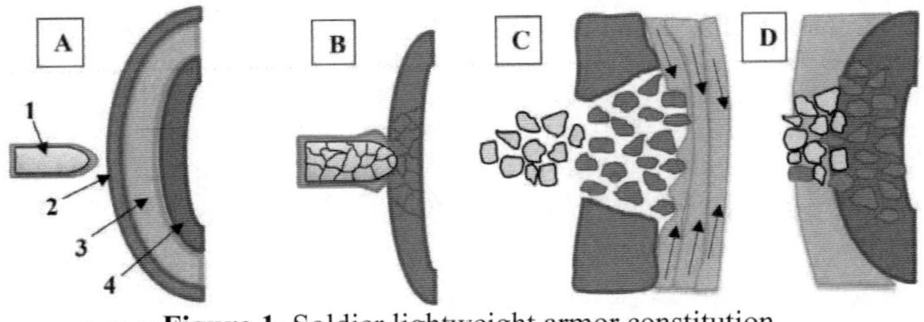

Figure 1. Soldier lightweight armor constitution

The methodology presented in this paper combines multiple experimental tests and finite element simulations in order to study the dynamic behavior of both a ceramic and UHMWPE materials. This combination helps reducing research time and cost. On one hand, experimental tests help identifying the most prominent phenomena to consider and provide data for model parameters. On the other hand, numerical simulations improve the comprehension of phenomena mechanisms especially for dynamic events that are too brief and complicated to be observed experimentally. Therefore, an iterative approach between tests and numerical simulations with a Building Block Approach (BBA - Figure 2) was conducted: studying material behavior from simple coupons to configurations with complex assemblies.

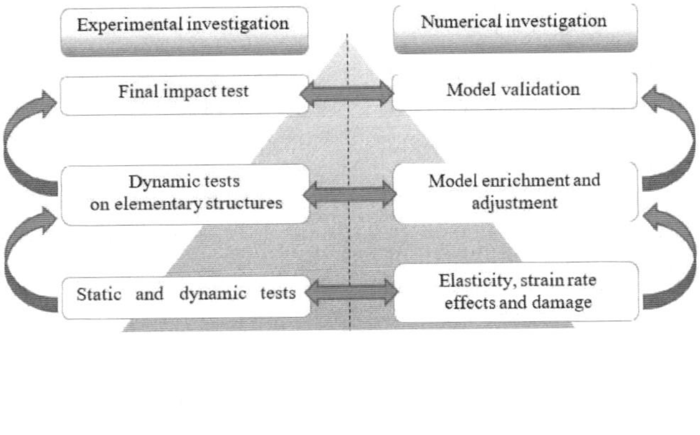

https://doi.org/10.52202/078352-0019

Figure 2. The Building Block Approach

The BBA consists in analyzing the material response from mesoscopic to structure scales. The material models in LS-DYNA were enriched throughout the tests with parameters characterizing the linear elastic response, strain rate effects and damage evolution.

For ballistic armors, and due to their weakness in tension, ceramic tiles are backed by a ductile plate to provide tensile strength to the armor configuration. In such configuration, the ceramic tiles are mainly loaded in compression and fragment the projectile while the backing prevents any penetration of the debris [1]. The design through numerical simulation of a solution to withstand a given threat remains challenging. One of the reasons is related to the difficulty to fully characterize the behavior of ceramics [2]. They show brittle failure but they can withstand important compressive load and undergo large deformation once damaged. This compressive strength depends on the confinement and is particularly important to the resistance to impact of the shield [3]. Therefore, it is important not only to measure the intact response of the ceramic but also to characterize the material in its damaged state under different confinement levels. Direct measurement of the damaged strength of brittle material is difficult to perform and indirect measurement are usually realized instead [4] [5]. A Johnson-Holmquist damage law (JH-2) [6] is calibrated. The experimental study consists in plate impact, dynamic compression of confined specimen on SHPB and impact tests. Results of numerical simulations from LS-DYNA are compared with experimental data. The purposes of the approach are both to evaluate the ability of the JH-2 to reproduce the tests and to provide additional information for the analysis of experimental data.

The use of polymer-based composite like ultra-high molecular weight polyethylene (UHMWPE) has become popular in lightweight armors. Such materials can exhibit diverse failure modes depending on the characteristics of the threats. Therefore, it is convenient to use numerical tool for the analysis and optimization of such type of armor. Many studies have been conducted on UHMWPE composite material dynamic response. For example, [7] investigated in-plane deformation behavior and [8] shock propagation in-fiber direction. Out-of-plane compressive response has been studied by [9] in static and [10] under shock loading. Furthermore, a comprehensive literature is available regarding, the ballistic response of this material [11] [12]. [12] used a numerical model combining orthotropic behavior and EOS for the simulation of ballistic response on UHMWPE plate. In this work, the mechanical response and failure mode are investigated under three types of dynamic solicitation: in-plane tension, out-of-plane compression, and out-of-plane shear. Then, impact tests are performed with spherical projectile. Experimental data is then used as input of an orthotropic material law coupled with fiber damage.

The paper is divided into three parts, the first details the characterization tests with corresponding model and the second deals with the ballistic tests and simulation. The last part is dedicated to the final validation test. Some results had to be normalized to comply with confidentiality obligations of the study.

2 MATERIAL CHARACTERIZATION

2.1 Ceramic material

The aim of this part is to calibrate a Johnson-Holmquist damage model (JH-2) that is commonly available in commercial finite element softwares. Inside LS-DYNA, the JH-2 is implemented as *MAT_110. It has been developed to model the failure of brittle materials like glass or ceramics under high strain rate. The JH-2 model consists of two main blocks. First, a linear elastic law defined by a material Equation Of State (EOS) and a shear modulus. Second, a damage evolution law which describes the failure strength of the material. The EOS is given by Equation (1).

$$P = K_1\mu + K_2\mu^2 + K_3\mu^3 + \Delta P_{n-1} \tag{1}$$

where μ is the material volumetric strain; P is the mean stress assimilated to the material pressure; K_1, K_2 and K_3 are the Hugoniot parameters which are determined from plate impact tests and ΔP_{n-1} is computed by Ls-Dyna from the amount of accumulated damage. One of the main characteristics of the JH-2 model is the pressure dependence of the material strength (

Figure 3). The material strength during damage evolution under compressive loading is defined by two curves $\sigma_i(P,\dot{\epsilon})$ describing the intact strength and $\sigma_f(P,\dot{\epsilon})$ the fractured strength and a damage variable D. The material strength is between those two curves during damage evolution and then follows σ_f once it is fully damaged.

$$\sigma = \sigma_i - D\,(\sigma_i - \sigma_f) \tag{2}$$

Where :

$$\sigma_i = \sigma_{HEL}\, A \left(\frac{P+T}{P_{HEL}}\right)^N (1 + \mathrm{C} \ln \dot{\epsilon}\,) \tag{3}$$

$$\sigma_f = \sigma_{HEL}\, B \left(\frac{P}{P_{HEL}}\right)^M (1 + \mathrm{C} \ln \dot{\epsilon}\,) \tag{4}$$

$$\Delta D = \frac{\Delta \epsilon_P}{D_1 \left(\frac{P+T}{P_{HEL}}\right)^{D_2}} \tag{5}$$

A, B, M, N, T, D1 and D2 are the material parameters. σ_{HEL} and P_{HEL} are calculated from an impact plate test data in which the HEL is reached (TI1131). Then, the value of A and N are computed so that σ_i is equal to σ_{HEL} at $P = P_{HEL}$. The damage parameters B, M, D_1 and D_2 cannot be directly measured from experiments but need to be calibrated to reproduce numerically a set of failure tests.

The ceramic material is available in blocs of 100 x 100 mm² square area with 5 or 20 mm thickness. It has then been machined to manufacture the specimens used in this study.

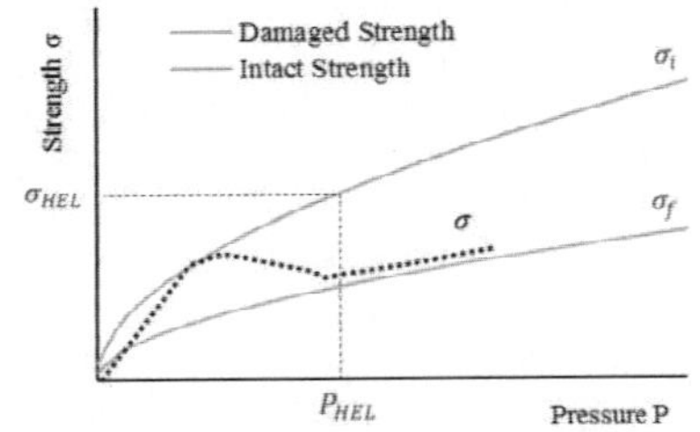

Figure 3. Description of Johnson-Holmquist damage model

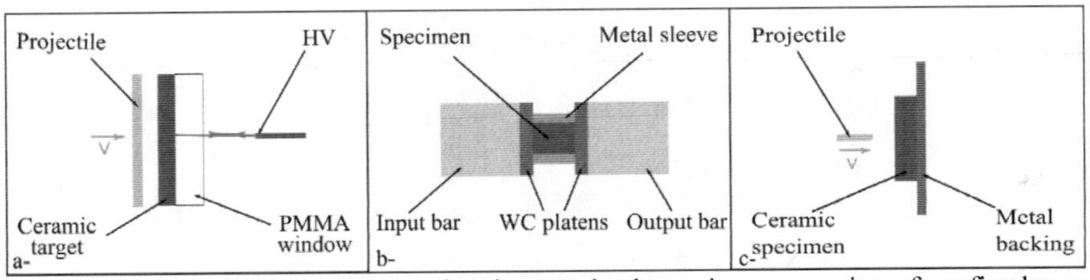

Figure 4. Schematics of experiments, a- plate impact, b- dynamic compression of confined specimens on SHPB, c-ballistic

2.1.1 *Plate impact tests*

Figure 4a shows the configuration of the plate impact tests. Copper disk projectiles are used for the experiment and PMMA windows are glued at the back of ceramic targets. All dimensions are defined so that the release stress waves do not interfere with the longitudinal stress during the period of observation. A Photonic Doppler Velocimeter (PDV) is used to monitor the normal velocity at the interface between target the PMMA. Figure 5 compares the velocity profiles recorded for three tests corresponding to different impact velocities and/or projectile thicknesses. HEL (Hugoniot Elastic Limit) state is reached only for test T1131 and the inelastic response is characterized by the ramping behavior the velocity profile. Taking into account the PMMA impedance, the HEL parameter is computed from experiment T1131 and the evolution of spall strength with impact velocity is obtained from all three curves.

Plate impact tests are simulated using a 2D-axisymmetric model. Element size of 0.05 mm are used to reproduce the shock shape of the velocity signal shown in Figure 5. Figure 6 shows the propagation of the shock wave, along y direction, inside the specimen for test T1131. The comparison of velocity profile shows good agreement with the tests. The model is able to reproduce the maximum velocity at the interface. This result shows that the elastic properties and the EOS have been well identified. Moreover, the HEL velocity for test T1131 is well predicted even though the curvature is not similar.

https://doi.org/10.52202/078352-0019

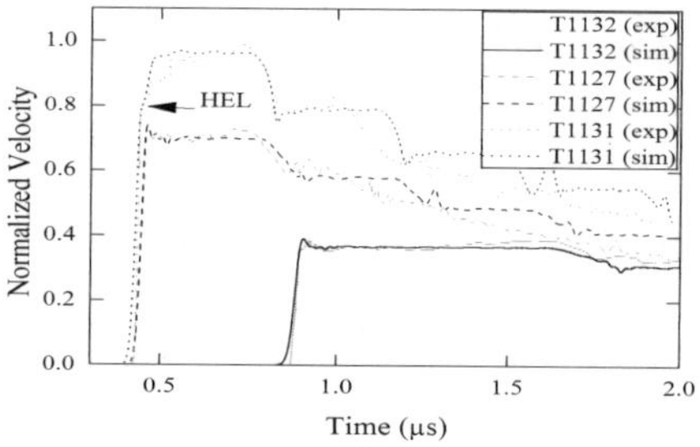

Figure 5. Velocity profiles

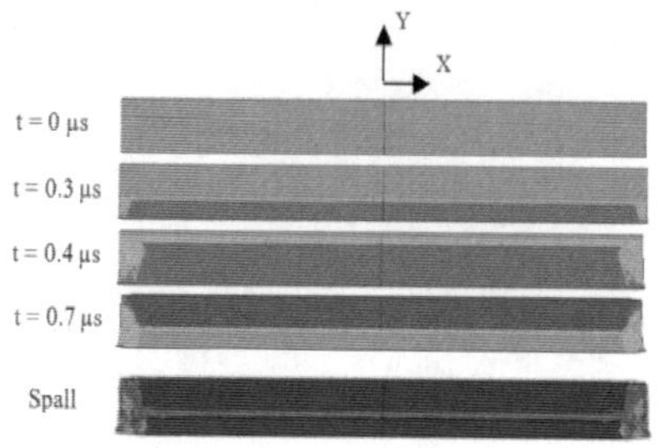

Figure 6. Simulation of plate impact tests

2.1.2 Dynamic compression of confined specimens

The configuration for SHPB tests is presented in
Figure **4**b. Cylindrical ceramic specimens are shrink-fitted inside metal sleeves of 1.5 mm thickness. Three different sleeve materials (copper, aluminum and steel) have been selected to adjust the level of confinement. Also, a copper pulse-shaper is placed between the striker and the input bars to generate an incident pulse with a ramp history.

Figure 7a and b show typical SHPB pulse signals for two tests with copper sleeve. All parameters of the tests are identical apart from the initial velocity. For comp-A test, transmitted pulse has a similar shape to incident pulse. The specimen has not been significantly damaged and only some small cracks are observed. However, for specimen comp-B, there is an abrupt discontinuity in the transmitted pulse. The ceramic has been recovered in a state of powder and the sleeve has collapsed due to buckling. Since the transmitted pulse is proportional to the load passing through the specimen, it can be noted that failure of comp-B occurs at around 60% of the maximum load passing through specimen comp-A. This result is counter-intuitive but quite repetitive and has been observed for all three types of sleeve. Furthermore, a gain in the failure load is observed as the confinement pressure increases (Figure 7c).

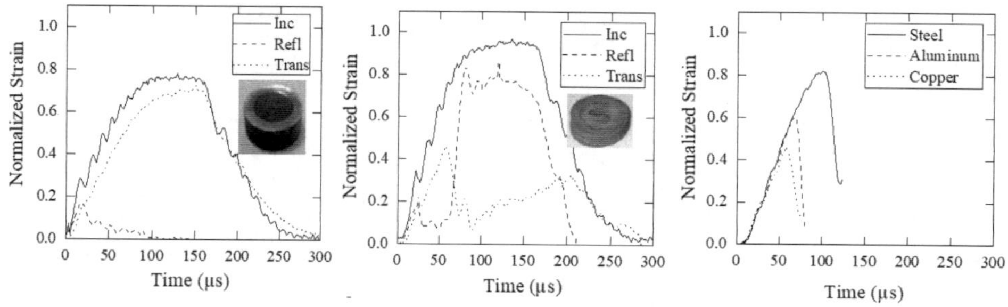

Figure 7. Strain pulses obtained from SHPB experiments, (a-) specimen comp-A, (b-) specimen comp-B, (c-) effects of confinement

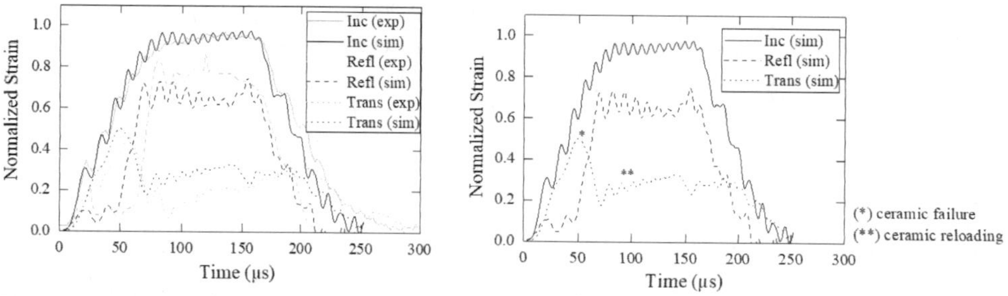

Figure 8. Comparison of SHPB pulses signal shape of specimen comp-B from numerical simulation with experimental data

The simulation of SHPB tests aims to provide additional information for the analysis of the tests results. They are simulated with 2D axisymmetric model. Apart from the ceramic and the sleeve, all the parts of the model are defined with their elastic material property. With the JH-2 parameter set initially used for the plate impact test, the model is not able to reproduce the results of comp-B test. The specimen is not damaged and the pulse signal shapes are similar to that of comp-A specimen. In order to reproduce

https://doi.org/10.52202/078352-0019

the pulse signals of comp-B specimen, the values of A and N needs to be calibrated reducing the intact strength of the ceramic. Figure 8 shows the pulse signals from the simulation. The incident pulse is well reproduced. However, reflected and transmitted signals do not match perfectly the experiment even though similar behaviors are observed. First, a linear response of the ceramic is observed until failure which is characterized by the abrupt drop in the transmitted signal. Then, the damaged ceramic is reloaded and important deformation is observed due to the buckling of the sleeve. In this test, the ceramic failed at a stress far below its HEL, therefore this experiment is complementary to the plate impact test.

2.2 UHMWPE material

The material is available as a laminate panel of UHMWPE unidirectional fibers. It is the strongest fiber available in terms of strength to weight ratio [13] and is obtained from multiple stack 0°/90° plies impregnated with thermoplastic matrix. All the tests conducted have been realized from the same batch of material.

2.2.1 Dynamic tension

Tensile tests are conducted on a 200 tons dynamic press. Tensile load is measured with a piezoelectric sensor. Displacement is measured with a laser sensor. Specimen shape is similar to conventional tensile tests with a rectangle working area of 10 x 6 mm and is loaded at 40 s^{-1} along the fiber direction (0° and 90°). Any attempts to load the 45° direction fails due to a very low in-plane shearing strength. An average 10 % increase of the strength is observed during dynamic tests compared to quasi-static data.

2.2.2 SHPB compression

Compression tests are conducted on Hopkinson bars. Compressive loading is applied in the out-of-plane (o-o-p) direction of the specimen. Eight tests are realized with three different mean strain rates: 3000, 4500 and 5600 s^{-1}. The compressive stress versus strain of the material exhibits a linear response until failure. A 20% increase of the failure strength is observed at high strain rate (Figure 9). The compression failure mode of this material is not conventional as failure is triggered by ply slipping near the edges as shown in Figure 10. Due to very low in-plane shear strength, ply cannot sustain the shear stress induced by the compressive load. Slipping is initiated near the edges since the material is less confined in this area. Similar behavior has been reported under quasi-static loading [9].

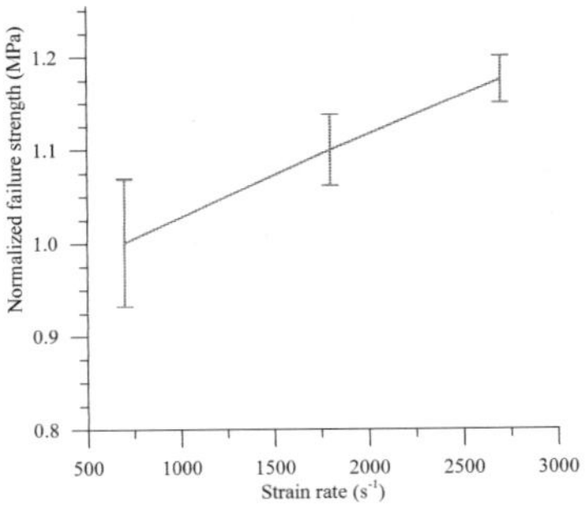

Figure 9. Strain rate effect on failure strength

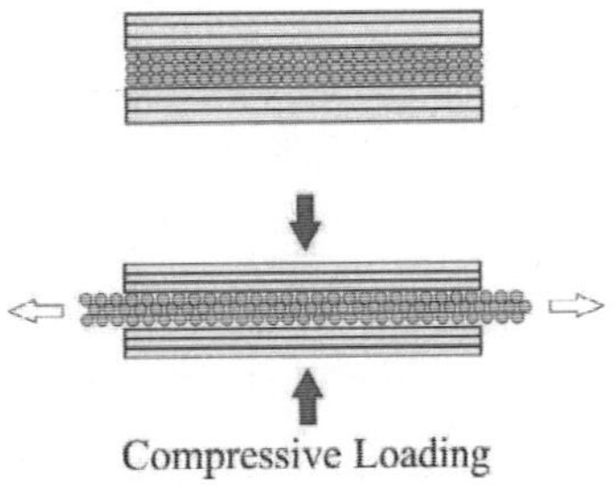

Figure 10. Failure mode under compression

2.2.3 Dynamic shear

A mean o-o-p shear response is measured with the configuration presented in Figure 11a. A cylindrical specimen is clamped on its edge and impacted by a cylindrical projectile. The diameter of the impactor is chosen to be slightly smaller than the inner diameter of the specimen clamp to generate shear stress in the area in between (Figure 11b). A load cell records the loading transmitted through the shearing of the specimen. The specimen back face velocity is measured with a PDV head and is integrated to compute the displacement δ of the specimen. The shearing angle and the shear stress are computed as:

$$\alpha = Arctan\left(\frac{\delta}{l}\right) ; \tau_{mean} = \frac{F}{\pi D e}$$

https://doi.org/10.52202/078352-0019

where D and e are the inner diameter of the clamping tool and the thickness and F the force.

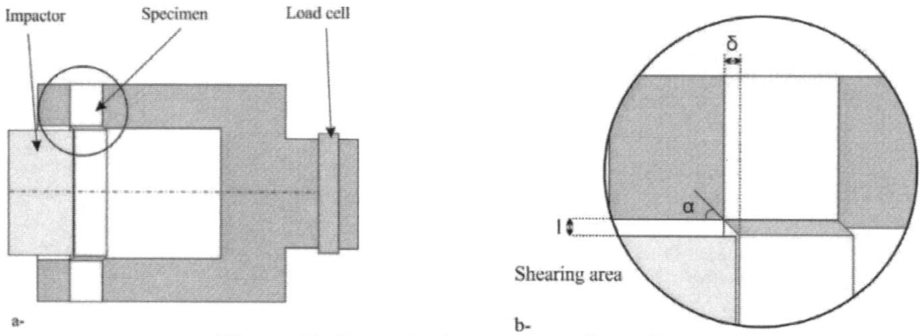

Figure 11. Dynamic shear tests configuration

Four tests are realized with different specimen thicknesses. Figure 12 shows the normalized shear stress versus shear angle curves. The strain rate ranges between 4000 and 5500 s^{-1}. The material response is quite similar for all specimen thicknesses: it confirms that the flexure contribution of the loading is negligible. The response is bilinear: a moderate slope up to 0.4 shear angle then followed by a steeper slope. The first loading phase corresponds to a pure shearing. As the displacement increases, some load is transferred as tension into the fibers which explains the increase in the slope rate.

The identification of the shear modulus G13 and G23 is realized through numerical simulations. The FE model is designed to replicate all the specificity of the testing condition (impactor, boundary condition, load cell …). A linear orthotropic model is used. The value of G13 and G23 is calibrated until the curves fit the experimental data (Figure 12). Figure 13 shows the localization of the shearing in the specimen. Identifying the shear modulus G13 and G23 can only be conducted through this method since experimental force represents a homogenized shear response along the circumference of the specimen.

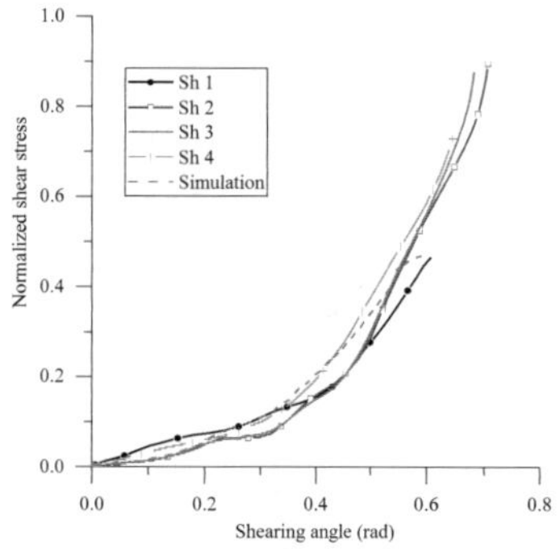

Figure 12. Normalized mean shear stress vs shearing angle

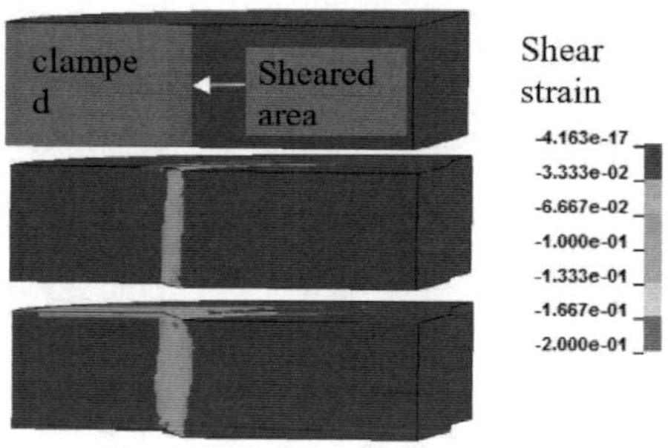

Figure 13. Simulation of the dynamic shear test

3 INTERMEDIATE IMPACT TESTS

3.1 Impact response of ceramic material

Two impact tests (bal-A and bal-B) have been performed with the configuration illustrated in Figure 4c. In both cases, a steel rod projectile of 35 mm long and 10 mm diameter impacts the target. It consists of a ceramic tile with a surface area of 100 x 100 mm^2 which is tied to an aluminum plate of 150 x 150 x 6 mm^3 dimensions at its back. Two different tile thicknesses have been tested in order to study the perforation resistance of the material: 5 mm for test bal-A and 20 mm for test bal-B. The projectile is launched with a light gas gun and the impact velocity is measured with an optical barriers system. For both tests, the impact velocity is 945 ± 10 m/s and the obliquity of the projectile, measured with an X-ray flash photography is below 2°. For specimen bal-A, a high speed camera is used to record the back side of the target, two images recorded respectively at 10 µs and 280 µs after impact are shown in Figure 14a and b. First, there is a small indentation behind the aluminum backing and later ejected

https://doi.org/10.52202/078352-0019

debris form a typical bi-conical shape. The ejection speed estimated from the camera is 750 ±10 m/s. After impact, the entire ceramic tile has been disintegrated and a hole of 18 mm diameter is present in the backing (Figure 14c). Specimen bal-B is not perforated. Increasing the ceramic thickness has stopped the projectile. The aluminum backing is severely indented and the ceramic tile is fragmented.

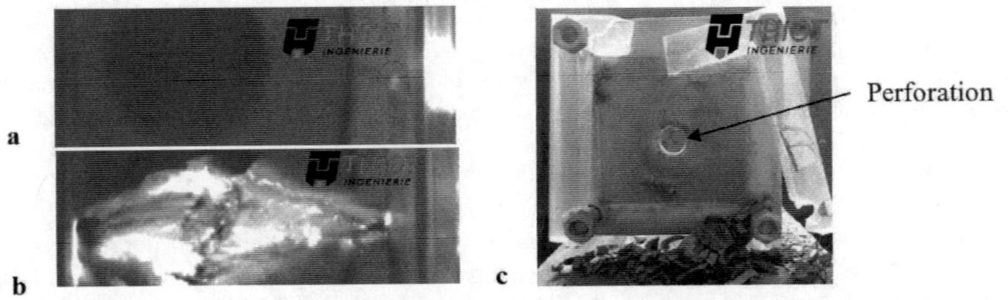

Figure 14. (a-) and (b-) Images of the side of specimen bal-A taken with high speed camera respectively at 10 μs and 280 μs after impact and (c-) perforation of the aluminum backing after the test

One of the goal of this simulation is to reproduce the velocity profile recorded with the PDV head at the backside of the backing of specimen bal-B in order to be able to analyze it. This signal is a result of both the wave propagation inside the target and its structural response. Different modelling approaches have been tested by using solid elements, SPH and hybrid Lagrange/SPH elements. Pure solid elements model gives quite good results as long as the material does not undergo large deformation (until 5-8 μs after impact). The use of SPH elements enables to go further in the calculation but the interaction of SPH elements with different material properties is quite complex. Typically, using the SPH elements interaction to represent the projectile/ceramic and ceramic/backing contacts results in inappropriate reproduction of the wave propagation inside each material and an improper contact separation when required. Also, hybrid elements generate some undesired waves when the solid elements are converted to SPH elements. A compromise is found by modelling the projectile with SPH elements and the remaining parts with solid elements. Benefitting from the symmetry of the problem, only one half of it is modelled as shown in Figure 15.

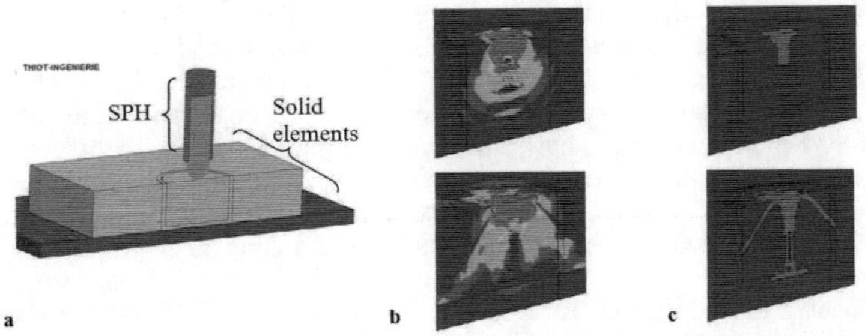

Figure 15. (a) Ballistic impact model, (b) pressure field, (c) damage behind the shock wave

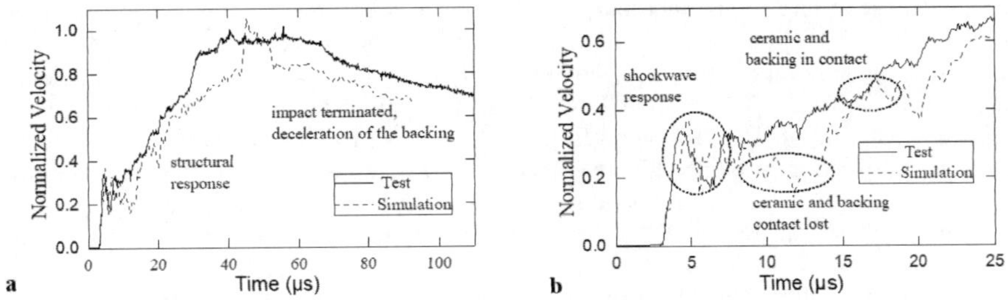

Figure 16. Comparison of the velocity profile at the backside of the backing for bal-B specimen

The projectile is modelled as elastoplastic with a failure strain of 10%. An element erosion criterion at 100% of deformation is added to the ceramic. When the projectile hits the target, two spherical shock waves are generated; one in the target and another inside the ceramic tile. As the ceramic wave propagates, it starts to partially damage the area just below the impact point. Then, the projectile

 https://doi.org/10.52202/078352-0019

disintegrates and simultaneously a network of cracks propagates inside the ceramic. Figure 16a and b compare the velocity profile obtained from the simulation with the experimental data. Figure 16a shows that the global response of the armor system is well reproduced. After the initial oscillations, a steady increase of the velocity is observed. This later characterizes the structural response of the target. As the projectile is disintegrating, the force applied to the ceramic decreases resulting in the plateau velocity observed from 35 to 70 μs. Finally, the decrease of the velocity is mainly governed by the aluminum backing. Zooming at the beginning of the curve, (Figure 16b), the initial oscillation is a result of the shock wave interaction with the backing. The gap from 9 to15 μs is due to a loss of contact at the ceramic/backing interface which may not happen in the experiment since the glue is not modelled.

3.2 Impact response of UHMWPE material

The study of the UHMWPE material is completed with impact tests. The test configuration is presented in Figure 17. A 25 mm steel sphere is launched with a gas gun. The target, a disk of 150 mm of diameter is clamped around its edge. The backface velocity is monitored with a PDV system and is observed with a high-speed camera through a mirror. Two tests are performed with different target thickness to obtain non-perforated TI1406 (Figure 19) and perforated TI1408 cases (Figure 20).

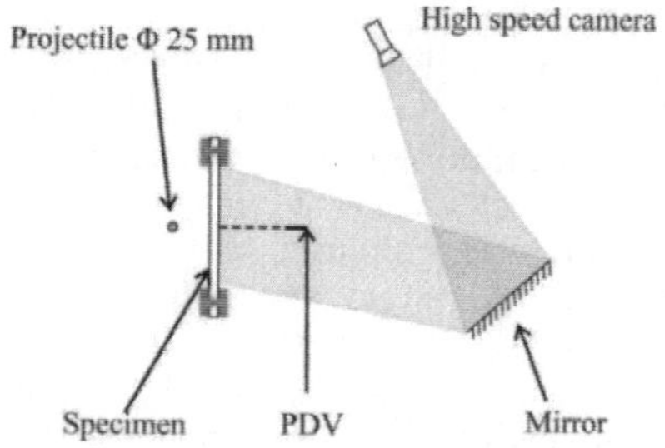

Figure 17. Ballistic tests configuration

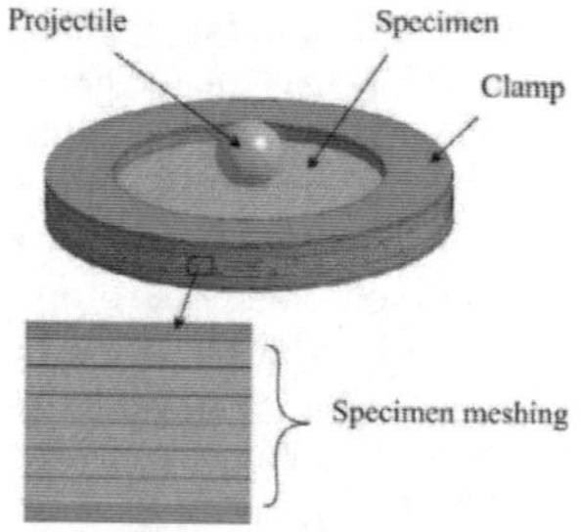

Figure 18. FE model for ballistic impact

For TI1406 specimen, one third of the plies from the front side are perforated. The projectile is recovered in between plies after impact. Important ply interface delamination is observed on the perforated plies. However, the non- perforated group of plies has been compressed and densified. Around the impact point, the nature of the polymer has also changed as if it has been heated close to its melting point. The camera images show that the target undergoes an important back face deformation. Post-mortem analysis of the recovered target shows that non-failed plies undergo large slipping which explains the non-round shape of the specimen edge after the test. For TI1408 test, a thinner target is impacted at a higher projectile velocity. The target is completely perforated. Its deformation is not as important as TI1406 however similar signs of temperature increase is found around failed plies.

Considering all the information gathered throughout the characterization tests and the ballistic impact, a FE model of the ballistic test is realized in LS-DYNA (Figure 18). The target is modelled in 3D at a pseudo-meso scale. Since representing each couple of 0°/90° plies by one element through the thickness would result in very large model, it has been decided to represent a set of 0°/90° by one element. Regarding the material law, initially, the orthotropic model developed for dynamic shear calibration is used with the tensile failure strength measured during material characterization. This version results in the perforation of both ballistic cases with a little decrease in the projectile speed. The target does not have time to deform before it is perforated. To improve the simulation results, a damage model is added to the material model using MAT221 in LS-DYNA. The parameter for damage evolution is set to reproduce the results of TI1408. This modification is motivated to account for the softening due to temperature increase before ply failure. The comparison of obtained back face velocity and projectile residual velocity with the experimental data is shown in Figure 20. The cinematics of the deformation and perforation are also coherent with the test. The target is perforated before important ply slipping. This model is then used to simulate TI1406 test and results in a partially perforated target like in the test. Furthermore, important plies slipping has occurred along the 0° and 90° as in the experiment.

4 FINAL VALIDATION

A final validation has been achieved on HERMES two-stage light-gas gun. The main goal of this test was to evaluate the prediction capabilities of the numerical developed throughout characterization and intermediate impact tests for both materials. This section only presents the impact test results as the simulations are still under progress. The target is composed of a 5mm thick ceramic tile in front of a 11mm thick UHMWPE panel. The boundary conditions are free for both the ceramic and the composite material, for representativeness of the test compare to a bullet-proof vest. The projectile is a 9.75g 7.62 lead core with copper sleeve bullet launched at 795m.s^{-1}. Two high-speed phantom cameras are used to monitor the impact: a V1840 at three quarter front view and a V2012 on one side. The images are presented in Figure 21. The post-mortem recovery of the composite panel shows that it helped catching the bullet and ceramic fragments. The ceramic tile and the UHMWPE panel thicknesses are sufficient to catch the bullet without any perforation of the composite material. These images will help evaluating the prediction capabilities of the developed numerical models.

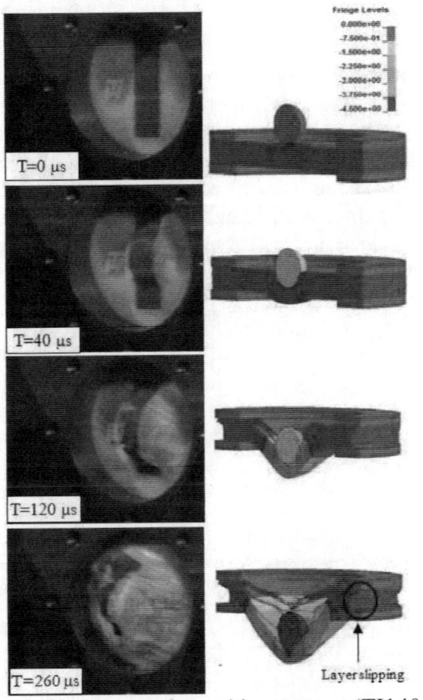

Figure 19. Non-perforated impact test (TI1406), simulation shows the z-displacement

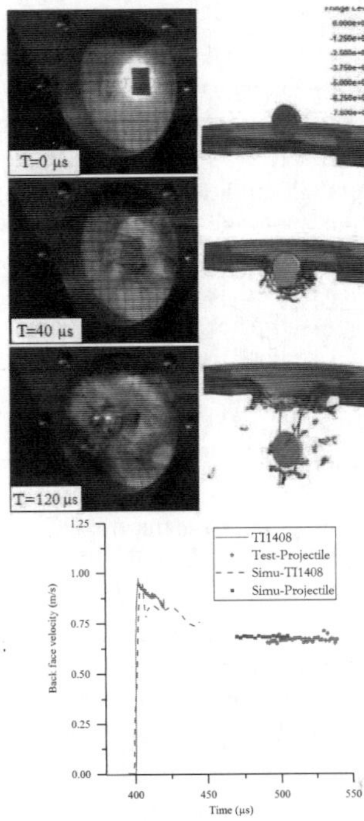

Figure 20. Perforated impact test (TI1408), simulation shows the z-displacement

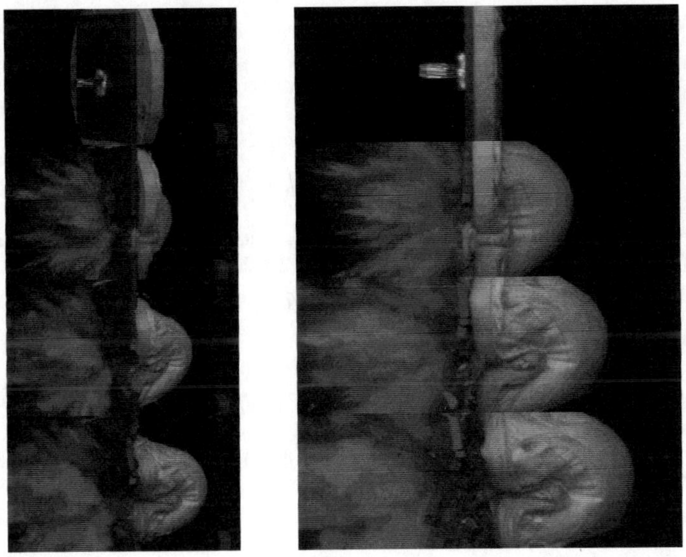

https://doi.org/10.52202/078352-0019

Figure 21. High-speed images of the final validation test. 232 μs interframe.

5 CONCLUSIONS

In conclusion, this paper has presented a characterization study for a ceramic and a composite material. The BBA approach has been established in order to use both numerical and experimental investigation.

Concerning the ceramic material, tests have been performed to identify model parameters. Numerical simulation was then used to provide complementary information and analyze the tests. EOS and HEL data of the material have been characterized and a good agreement with experiments is obtained from the simulation. However, the material damage process is complex, the study has enabled to clarify some part but there are still some questions remaining opened. Additional investigations are required towards the development of a ceramic ballistic predictive models. Overall, this project has been very enriching and will improve the way of studying the behavior of brittle materials under dynamic loading.

Concerning the UHMWPE composite, characterization tests have been conducted: in-plane tension, out-of-plane shear and compressive responses. UD fiber provides good tensile strength but the matrix low shear strength makes it prone to delamination, peeling and fiber slipping. The design of the tests has been driven by such specificity to acquire enough data to build the material model. The impact tests have highlighted the importance of ply slipping and temperature increase in the perforation mode. The simulation approach shows interesting results and the modelling scale choice represents a good compromise between calculation time and reproduction of main physical phenomena. This study should be pursued on a more detailed investigation of the failure mode of material at high temperature to confirm the modelling assumptions regarding fiber damage evolution.

At the end a final impact validation test on target made with a ceramic tile and a UHMWPE panel has been performed. This test still need to be compared with simulation results that are in progress.

Acknowledgments

The authors would like to thank the DGA Land Systems (French MoD) for financial support and all the team of Thiot Ingénierie laboratory for performing all the tests

References

[1] W. A. Gooch, "An overview of Ceramic Armor Applications," *Ceramic Transactions: Ceramic Armor Materials by Design,* vol. 134, pp. 3-22, 2001.

[2] C. E. Anderson, "A Review of Computational Ceramic Armor Modeling," *Advances in Ceramic Armor II: Ceramic Engineering and Science Proceedings,* vol. 27, pp. 1-18, 2006.

[3] H. Luo, W. Chen and A. M. Rajendran, "Dynamic Compressive Response of Damaged and Interlocked SiC-N Ceramics," *J. Am. Ceram. Soc,* vol. 89, pp. 266-273, 2005.

[4] R. Feng, G. F. Raiser and Y. M. Gupta, "Material strength and Inelastic Deformation of Silicon Carbide Under Shock Wave Compression," *J. Appl. Phys.,* vol. 83, pp. 79-86, 1997.

[5] G. R. Johnson and T. J. Holmquist, "Advances in Ceramic Armor: Ceramic Engineering and Science Proceedings," vol. 26, pp. 1-18, 2005.

[6] G. R. Johnson and T. J. Holmquist, "An improved computational constitutive model for brittle materials," *AIP Conf. Proc.,* vol. 309, p. 981, 1994.

[7] L. Govaert and P. J. Lenstra, *Cooloid Polym Sci.* , vol. 270, 1992.

[8] P. J. Hazell, G. J. Appleby-Thomas, X. Trinquant and D. J. Chapman, *J. Applied Phys.,* vol. 110, 2011.

[9] J. P. Attwood, S. N. Khaderi, K. Karthikeyan, N. A. Fleck, M. R. O'Masta, H. N. G. Wadley and V. S. Deshpande, *J. Mech. Phy. Solids,* vol. 70, no. 200, 2014.

[10] T. Lassïg, F. Bagusat, M. May and S. Hiermaier, *Int. J. Impact Eng.,* vol. 86, no. 240, 2015.

[11] M. R. O'Masta, B. G. Compton, E. A. Gamble, F. W. Zok, V. S. Deshpande and H. N. G. Wadley, *Int. J. Impact Eng.,* vol. 86, no. 131, 2015.

[12] T. Lassïg, L. Nguyen, M. May, W. Riedel, U. Heisserer, H. V. D. Werf and S. Hiermaier, *Int. J. Impact Eng.,* vol. 75, no. 110, 2015.

[13] T. Dayyoub, A. V. Maksimkin, S. Kaloshkin, E. Kolesnikov, D. Chukov, T. P. Dyachkova and I. Gutnik, "The Structure and Mechanical Properties of the UHMWPE Films Modified by the Mixture of Graphene Nanoplates with Polyaniline," *Polymers,* vol. 11, no. 23, 2019.

Development of a Physical and Mathematical Ballistic Skin Simulant

M. DeWitt[1], M. Danilich[1], K. Kong[2], M.B. Panzer[2], C. Bir[3], and B. Gillich[4]

[1]*Luna Innovations, 706 Forest St., Charlottesville, 22901 VA., USA, dewittm@lunainc.com (presenting author email address)*
[2]*University of Virginia – Center for Applied Biomechanics (UVA-CAB), 4040 Lewis and Clark Dr. Charlottesville, 22911 VA., USA*
[3]*Wayne State University, 818 W. Hancock, Detroit, 48202 MI., USA*
[4]*U.S. Army Aberdeen Test Center, 6943 Colleran Rd B400, APG, 21005-5059 MD., USA*

Abstract. Ideally, non-lethal weapons (NLWs) incapacitate or repel with a low probability of fatality or permanent injury. Rapid NLW market growth and a growing number of new kinetic weapons necessitate further relevant skin injury biomechanics research. Unfortunately, NLW development and evaluation is limited by a lack of universally accepted standard test protocols or materials. To fill this need, Luna Innovations developed TrueSkin™, a physical simulant of human skin, with a coupled Finite Element Mathematical (FEM) model of the material. (U.S. Army SBIR Phase II contract W91CRB-17-C-0032). TrueSkin comprises a proprietary nanofiber-reinforced hydrogel that mimics the architecture of human skin extracellular matrix and ultimately the skin biomechanics to failure and response to non-lethal munition impact. The physical skin simulant is designed to improve the reliability of injury risk evaluation for less-than-lethal ballistic projectiles and was iteratively designed and tested to match the material properties (ultimate tensile strength, stretch at failure) of postmortem human skin tissue. To accomplish this, the TrueSkin physical simulant was evaluated at relevant dynamic strain rates using custom equipment and protocols developed at the University of Virginia Center for Applied Biomechanics (UVA-CAB) for testing human skin tissue. Independent evaluation at HP White Laboratories validated penetration resistance of TrueSkin against established postmortem human subject (PMHS) penetration data using similar 12 gauge fin-stabilized rubber projectiles. The promising results from mechanical characterization of laboratory specimens demonstrate material property simulation capability, and comparison of penetration resistance with existing PMHS data provides initial validation of ballistic response.

1. INTRODUCTION

Non-lethal weapons (NLWs), including less-lethal kinetic energy (KE) impact munitions, are designed to incapacitate or repel targets with a low probability of fatality or permanent injury. These specialty impact munitions are typically deployed during encounters with aggressive subjects where a less-than-lethal response is needed to minimize the potential for significant injury to the target or others nearby. Single projectile KE impact munitions are typically utilized as their accuracy enables direct fire towards a specific area of the body to inflict blunt trauma with minimal risk of serious injury. Rapid market growth has created an intensely competitive effort to develop and commercialize more effective and safer NLWs. Impact munitions, such as rubber projectiles, can serve as valuable tools to military and law enforcement agencies worldwide[1]. However, concern remains regarding adverse effects that the munitions can have on potential targets when penetration through the skin occurs[2].

According to a report by Haar et al, over 2000 penetrating injuries, ranging from minor to significant, and over 50 deaths from impact munitions have occurred between the years of 1990 and 2007[3], which could increase as new projectiles are rapidly developed. Recent studies evaluated almost 1000 deployments of less-lethal kinetic energy rounds, primarily focused on 12-guage bean-bag and plastic baton rounds and found that over 80% of these deployments result in injury [4]. While most injury were contusions (51%) and abrasions (31%), fractures and other penetrating injuries were reported in these smaller caliber KE impact munitions.

As a result of these case reports potential to cause penetrating injury to the body, thorough human effects, or safety assessments, must be accomplished to define minimum engagement distances, maximum velocities, and to assess the risk of significant blunt trauma[5], [6] and projectile penetration across entire operation ranges. Numerous experimental and computational models have been developed to predict the risk of injury due to rigid projectile impacts, and many previous studies focused on safety evaluation have been used as a guide in the early stages of less-lethal projectile design[1], [7]. Unfortunately, NLW safety research and evaluation standards are limited by a lack of universally accepted test protocols or materials to determine the risk of significant injury of non-penetrating NLWs with statistical rigor to ensure safe use.

 https://doi.org/10.52202/078352-0020

The ability to test NLW safety in a controlled environment is paramount to enable a thorough biomechanical assessment of impact prior to use in the field. The risk of penetrating trauma is especially important to evaluate during safety assessments of NLWs due to the dramatic increase in injury severity that can occur when the munition penetrates into the body cavity[8]. Current methods for evaluating the injury risk, and primarily the penetration risk of NLWs, are limited. Various ballistic injury surrogates have been utilized including soap, gelatin, clay, animals, PMHS and other materials [9], [10]. Although many of these materials can be used to determine energy transfer from the munition to the tissue surrogate, they lack the visco-elastic nature of human tissue and are primarily adapted from previous methods to evaluate lethal, penetrating ballistics.

A previous study by Bir et. al, using PMHS, assessed the skin penetration of a 12 gauge, fin-stabilized, rubber projectile [2]. In this study, un-embalmed PMHS were impacted at relevant velocities by the KE impact munitions and for each impact, the energy density (Ea) was calculated and injury was determined by evaluating the underlying tissue damage. Chamois skin was also used as part of skin surrogate's development for skin penetration assessment and showed acceptable biofidelic results, however, its thickness inconsistency and reproducibility reduce its practicality [8], [11]. Efforts have been made to develop affordable synthetic skin simulant (silicone and urethane) with a stable shelf life to exhibit biomechanical failure equivalence to human skin also show little success [12].

Human dermis is a matrix of collagen (approximately 35 vol%) and elastin fibers (approximately 0.4 vol%) that are interwoven in a highly hydrated proteoglycan gel. The dermis is capable of withstanding large deformations and the strength of the skin is primarily attributed to collagen fibers, which are almost inextensible and fail at strains of 5-6% and strengths of 147-343 MPa[13]. Elastin, on the other hand, is highly deformable, and provides the dermis with its high elasticity. For accurate simulation of force loading and response to NLW impact, we pursued skin simulant methods that would recreate these strong and elastic components and simulate the tough material properties of human skin.

Using parallel biomechanical characterization of PMHS skin, the current study was conducted to develop a biomimetic human skin surrogate for use in the safety evaluation of less-lethal KE impact munitions. In order to achieve similar penetration resistance, a custom hydrated nanofiber material was devised and the tensile mechanical properties to failure were determined and compared to human skin (PMHS). Re-evaluation of a previous NLW penetration study using PMHS was accomplished to determine the penetration resistance of a 12 gauge KE impact munition over soft issue in an effort to validate the resistance of the skin surrogate developed in this study. Finally, a penetration resistance study was accomplished using the newly developed surrogate and similar munitions for direct comparison to the PMHS penetration resistance. Identification of a repeatable, cost-effective, biomimetic surrogate of human tissue to evaluate KE impact munitions will be a valuable too for munitions manufacturers during development, law enforcement agencies during munition acquisition activities, and government agencies during test standard development.

2. METHODS

2.1 PMHS and Skin Surrogate Mechanical Testing

Skin samples for mechanical characterization were excised from the back of six male PMHS at the Center for Applied Biomechanics, University of Virginia (UVA), USA. All test procedures were approved by the UVA Institutional Review Board prior to any testing and the PMHS were screened for pre-existing pathologies to avoid skin diseases that may affect skin quality. The PMHS represent an average of 57±11 years old, 178.6±3.8 cm in height and weighed 88.4±19.8 kg adult male (Table 1).

Table 1. Post-mortem human subject information (Mechanical Characterization)

Specimen ID	Age (years)	Height (cm)	Weight(kg)
795	60	175.3	83.9
757	49	185.4	122
702	42	178	86
733	74	180.3	78.9
919	59	177.8	96.6
680	58	175	63

All the PMHS were thawed for 72 hours at room temperature prior to skin excision for testing. The orientation of the skin sample was determined with reference to anatomy illustration of Langer line[14], and both parallel (0°) and perpendicular (90°) samples with respect to the Langer line were included for testing (Figure 1, right). For each PMHS, 2 different sizes of skin samples were excised on the left side of the back for uniaxial tensile to failure (static and dynamic) and stress relaxation tests. Each skin sample thickness was measured using a digital caliper prior to testing and the measured mean thickness was 3.33±0.87 mm. Further detail regarding stress relaxation testing utilized for constitutive modeling of human skin biomechanics can be found in Kong et. al [15].

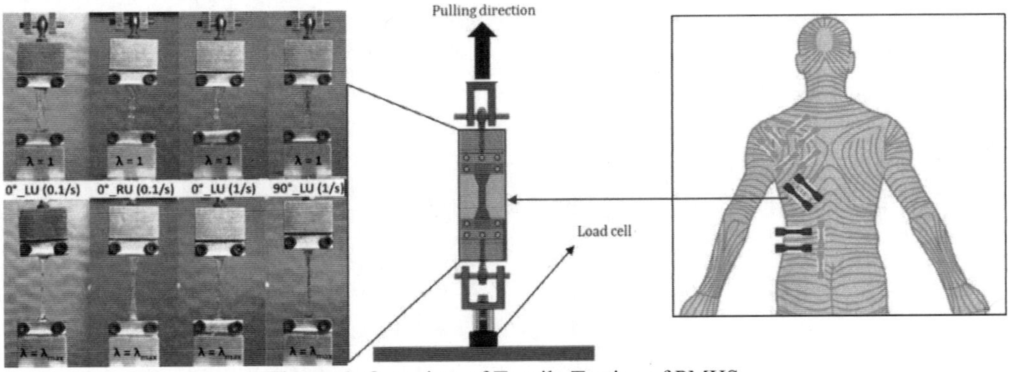

Figure 1. Overview of Tensile Testing of PMHS.

The static (1/s) tests were performed using the Instron Model 8874 servohydraulic actuated test machine (Instron, Canton, MA) and a custom-built gravity-based drop tower was used for the DT tests. Dogbone sample was clamped with 80 grit sandpapers at both ends of the custom test fixture (Figure 1, middle) to avoid slippage during testing. The test was initiated by moving the crosshead of the machine until skin failure occurred. A 1000 lbf (4.4 kN) Honeywell model 31 piezoresistive load cell (Honeywell, Charlotte, NC) was used to measure the force at 1000 and 10,000 sampling rates for ST and DT tests respectively. Similarly, test videos were recorded at 1000 fps for ST tests and 10,000 fps for DT tests using a Memrecam GX-1 high-speed camera (NAC Image Technology, Simi Valley, CA). Sharpie marks which were placed *in situ* were used for determining stretch values through video tracking software (Figure 1, left) (Tracker, ver. 4.11.0). A trigger box was utilized to activate data acquisition of the load cell and video recording simultaneously when the test was initiated. Engineering stress-stretch curve was constructed based on the measured force and the video tracked displacement. The engineering stress was calculated by dividing the measured force by the undeformed cross-section area. The stretch ratio (λ) was defined as the ratio between the current gauge length and original gauge length.

2.2 Human Skin Surrogate

Two different composite approaches were taken in an attempt to simulate human skin microstructure and response to applied loads (Figure 2). The first involved rayon microfiber-reinforcement of commercially available silicone. Early development efforts focused on recreating the fiber loading and failure mechanisms in this simulant approach (Figure 2, white box). Preliminary mechanical testing (data not shown) suggested that the microfibers were adversely impacting the mechanical properties of the samples because of nonuniform distribution and poor adhesion to silicone matrices. We transitioned surrogate development to recreation of the nanofiber structure and high hydration of human skin for optimal recreation of the viscoelasticity that dictates the mechanical properties and penetration resistance of skin. The final simulant approach therefore comprised Kevlar nanofiber (KNF) [16] reinforced poly(vinyl alcohol) hydrogel [17] (Figure 2,right). All mechanical testing (method describe above) and penetration resistance testing utilized the KNF surrogate.

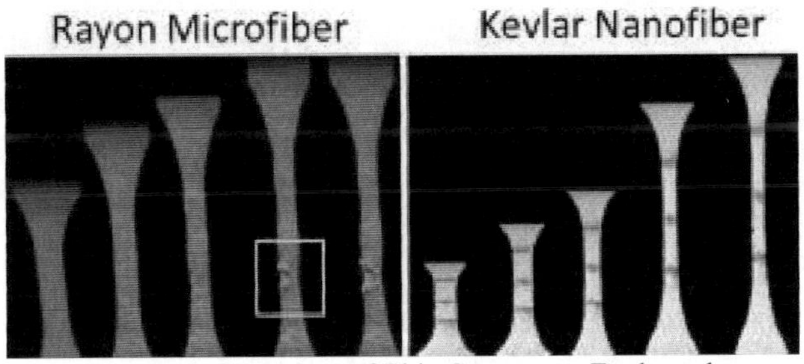

Figure 2. Overview of Skin Surrogates Evaluated.

https://doi.org/10.52202/078352-0020

2.3 Statistical Analysis of PMHS Penetration

A previous report by Bir et al [2] contained the results of ballistic testing of 8 PMHS against a 12-gauge, fin-stabilized rubber round (Figure 3). The eight cadaveric specimens, four male and four female, were procured from either the Wayne State University (WSU) Body Bequest Program or the University of Michigan (UM) Anatomical Donations Program. The cadavers were fresh, not embalmed. A summary of each cadaveric specimen is located in Table 2.

Figure 3. 12-gauge rubber round and cartridge.

Table 2. Post-mortem human subject information (Penetration Resistance)

Cadaver ID	Source	Sex	Age, yrs	Height			Weight		BMI
				cm	ft	in	kg	lbs	
31222	UM	F	58	162.5	5.331	64.0	71.21	157.0	27.0
31234	UM	M	58	175.5	5.758	69.1	57.61	127.0	18.7
562	WSU	F	72	168.0	5.512	66.1	72.57	160.0	25.7
430	WSU	M	75	174.0	5.709	68.5	84.14	185.5	27.8
31155	UM	M	76	174.0	5.709	68.5	73.03	161.0	24.1
31480	UM	M	77	172.0	5.643	67.7	78.02	172.0	26.4
545	WSU	F	78	155.0	5.085	61.0	52.16	115.0	21.7
563	WSU	F	80	164.0	5.381	64.6	68.04	150.0	25.3

Each PMHS sustained a maximum of 25 impacts consisting of shots to the anterior and posterior thorax, abdomen, and legs, for a total of 158 total impacts. The 10 specific locations of impact include 5 areas where bone lie directly under the skin (sternum, anterior on rib, anterior between ribs, posterior on ribs, scapula) and 5 fleshy areas either devoid of bone or with muscle/fatty layer before bone (liver, lateral to umbilicus (belly), posterior lower back, proximal femur, distal femur). Following each impact to a given location, a visual inspection of the injury was performed, and the wound was labeled penetrating or non-penetrating. Penetrating wounds were determined as such by evaluating whether the impactor disrupted not only the skin, but underlying tissue such as subcutaneous fat and/or muscle. Slight tearing (laceration), discoloration or marking of the skin without damage to underlying tissue was recorded and regarded as non-penetrating. Projectile velocity was measured with a single chronograph placed 22 in. from the PMHS. The mass and diameter of each round was measured and the energy density (J/cm^2) was mathematically determined.

As a validation of the skin simulant prototype, it was desired to replicate this PMHS dataset as closely as possible to see if the skin simulant would produce results similar to actual human skin. As the majority of methods for penetration currently use gelatin as backing material to simulate muscle/soft flesh [8], only the soft-backed PMHS impact locations were utilized to derive "actual human skin" expected performance, or truth set, for initial comparison and validation of the synthetic skin material. The 5 soft-tissue body locations from 8 PMHS totaled 68 shots. The data for these are plotted in Figure 4 with non-penetrations plotted as 0 and penetrations plotted as 1.

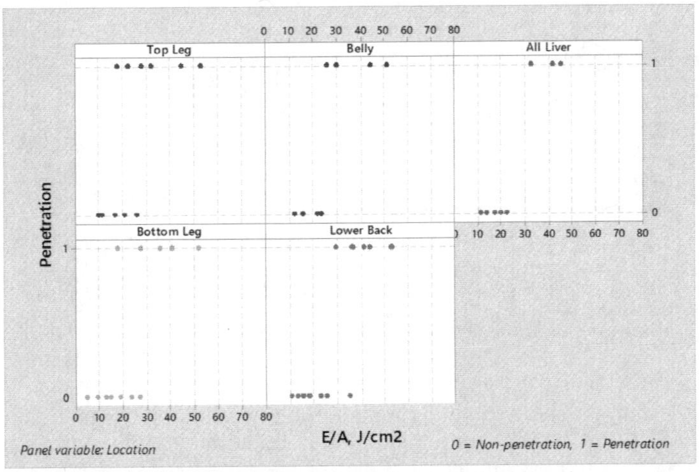

Figure 4. All data from 5 soft tissue backed locations.

To see if the five body parts produced similar results, or data from the same population, and therefore could be combined, the different populations statistical test was performed. The 0.10 alpha (or (1-alpha) = 0.90 difference confidence) was used to determine the level of significant difference. Parameter estimates using likelihood ratio analysis were calculated using three different link functions (Log Normal, Probit, Logit). A plot of the log-likelihood response curve along with lower and upper 90% confidence bounds were calculated. The log-normal link function is preferred since, unlike a probit or logit, it does not produce any negative stimulus values for the response or confidence curves.

2.4 Skin Simulant Penetration Testing

Independent penetration testing was conducted on an indoor range at HP White Laboratories (Street, Md) at ambient conditions utilizing a custom pneumatic test setup to control the KE projectile velocity. Testing was conducted using a 12 gauge less lethal rubber rocket manufactured by Defense Tech (DT3021). The test platform (Figure 5) consisted of a 20% ballistic gelatin covered by the skin surrogate, secured with a custom outer clamping ring to improve consistency of boundary conditions under applied load (impact munition force).

Figure 5. Skin Surrogate and Ballistic Gelatin utilized for penetration testing.

The test samples (Figure 5) were positioned 6.17 feet from the muzzle of the custom barrel to produce zero (0°) degree obliquity impacts (Figure 6). Each projectile was weighed prior to testing. High speed recording at 10,000 frames/sec was used to determine the striking velocity of the projectile. Final injury response criteria were selected and included: *(1)* No Injury – neither the skin simulant nor the gel block indicated irreversible damage, *(2)* Contusion – the gel block showed indication of impact but the skin simulant was not perforated, *(3)* Laceration – the skin simulant was perforated as indicated by visible light, and *(4)* Complete Penetration – the underlying gel block surface was fractured. As a validation of the skin simulant prototype, it was desired to replicate the Bir 2005 cadaveric experiment [2] as closely as possible to see if the skin simulant would produce results similar to actual human skin.

https://doi.org/10.52202/078352-0020

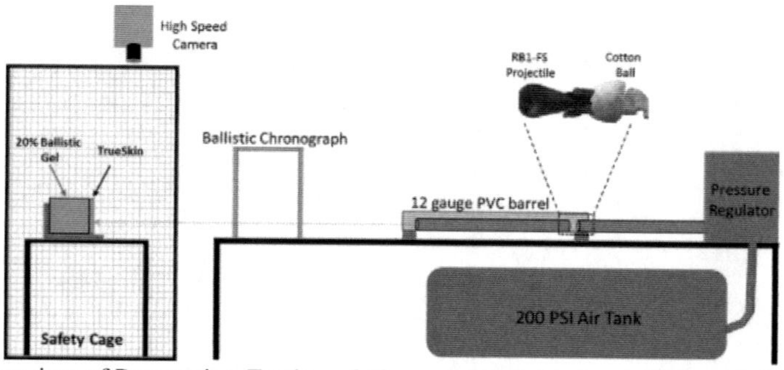

Figure 6. Overview of Penetration Testing of 12 gauge KE impact munition with Skin Surrogate.

Energy density (Ea) was calculated using projectile mass (~6g), face area(2.45 cm^2) and impact velocity. After each shot, the skin surrogate was removed from the ballistic gelatin and both the skin surrogate and the underlying gelatin were examined. The horizontal yaw was captured using a high-speed video camera mounted above the simulant. The depth of the projectile was also captured by the high-speed video camera with parallax effects accounted for in the gelatin.

3. RESULTS

3.1 Mechanical Characterization

The uniaxial tensile test matrix and average responses of the uniaxial tensile PMHS skin samples relevant to the development of a skin surrogate is summarized in Table 3. Among the 48 skin samples tested at the UVA-CAB, 5 samples slipped out of the test fixture and were permanently deformed during testing, therefore, these samples were excluded from data analysis. Additional details regarding mechanical characterization of these PMHS skin samples can be found in Kong et al [15].

Table 3. Average uniaxial tensile test results of human skin

Orientation	Strain rate (/s)	UTS (MPa)	λ_f
Parallel	1 (n=11)	**28.4**±6.3	**1.76**±0.14
	75 (n=5)	**20.6**±7.8	**1.75**±0.18
	180 (n=6)	**25.6**±5.4	**1.75**±0.07
Perpendicular	1 (n=10)	**22.6**±4.6	**1.97**±0.15
	75 (n=5)	**16.6**±5.8	**1.78**±0.12
	180 (n=6)	**20.7**±4.9	**1.81**±0.10

For the PMHS samples tested at the same strain rate, the parallel (0°-red) and perpendicular (90° - blue) stress-stretch-relationships are compared (Figure 7). The average UTS of all samples tested at all strain rates (n = 43) was 23.3 MPa and the standard deviation was 6.64 MPa. Additional efforts to perform statistical analysis between these groups was combined with the stress relaxation data (data not shown) in an effort to develop a parallel constitutive model of skin. However, the primary output of the PMHS mechanical characterization with regard to the development of an appropriate skin simulant was the UTS, stretch at failure (λ_f), and the overall stress-stretch relationship (appearance of toe region, strain-stiffening, and damage propagation) (Figure 7).

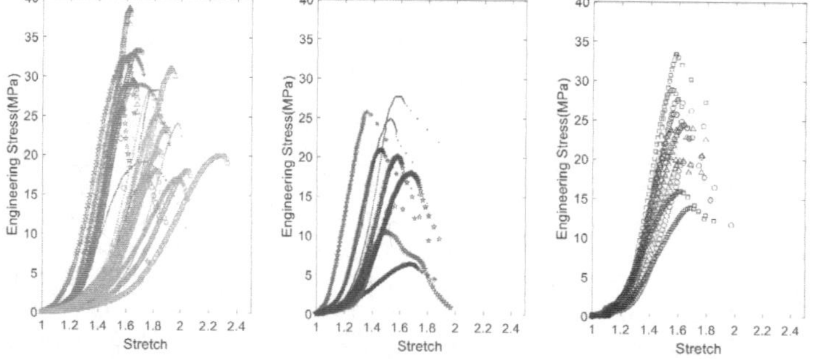

Figure 7. Stress-Stretch profiles of varied orientations (0°-red, 90°-blue) under strain rates (1,75,180/s)

As a result of iterative improvement of fabrication protocols and selection of appropriate concentrations of the PVA and KNF solutions and the ratio of the two composition, we fabricated custom skin surrogate prototypes (Figure 8 -left) that exhibited stress-stretch relationships (Figure 8-right) and mechanical properties within the range of PMHS mechanical data collected. Specifically, the skin surrogate exhibited an ultimate tensile strength of 26.97 ± 0.23 MPa and failed at a stretch ratio of 2.17 ± 0.01. An overview of the mechanical properties compared between PMHS and the skin surrogate can be seen in Table 4.

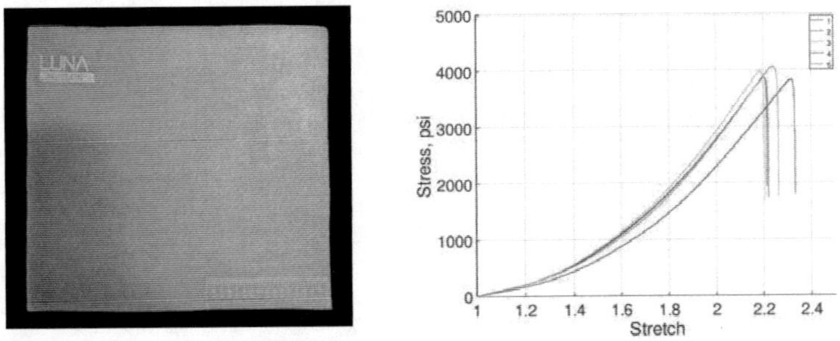

Figure 8. Final Skin Surrogate Prototype and Resulting Stress-Stretch relationship

Table 4. Average uniaxial tensile test results of human skin compared to surrogate

Material	Ultimate Tensile Strength	Stretch at Failure
PMHS Skin	23.30 ± 6.64 (MPa)	1.81 ± 0.15
Custom Nanofiber Skin Surrogate	26.97 ± 0.23 (MPa)	2.17 ± 0.01

3.2 PMHS Penetration Evaluation

Eight fresh cadaveric specimens, four male and four female, were tested for the Bir report [2]. Each PMHS sustained impacts consisting of shots to the anterior and posterior thorax, abdomen, and legs, for a total of 158 total impacts. The 10 specific locations of impact included 5 'hard-backed' areas where bone lies directly under the skin (sternum, anterior on rib, anterior between ribs, posterior on ribs, scapula) and 5 'soft tissue' areas either devoid of bone or with a muscle/fatty layer between skin and bone (liver, lateral to umbilicus (belly), posterior lower back, proximal femur, distal femur). Since the skin simulant evaluation uses gelatin as backing material, only the soft tissue PMHS locations were used.

The 5 soft tissue locations from 8 PMHS totaled 68 shots. To see if the five body parts produced similar results, or data from the same population, and therefore could be combined, the different populations statistical test was performed. The 0.10 α (or $(1-\alpha) = 0.90$ difference confidence) was used to determine the level of significant difference and identified that the lower back and belly produced different responses (Table 5 -red).

Table 5. Difference Confidence for five Soft-tissue locations

	Belly	All Liver	Bottom Leg	Top Leg
Lower Back	91.2	43.9	62.1	84.6
	Belly	0.0	85.0	86.6
		All Liver	60.2	70.5
			Bottom Leg	15.3

Deleting the Belly data resulted in 50 data points. The difference confidences are presented in Table 6, showing none exceeded the 90% difference confidence. Parameter estimates using likelihood ratio analysis were calculated using three different link functions for the combined dataset and are presented in Table 7.

Table 6. Difference Confidence for five Soft-tissue locations

	All Liver	Bottom Leg	Top Leg
Lower Back	43.9	62.1	84.6
	All Liver	60.2	70.5
		Bottom Leg	15.3

Table 7. Parameter Estimates from four soft-tissue locations combined

N	Link Function	Mu	sigma	10%	90%	L90CL Mu	U90CL Mu	L90CL sig	U90CL sig
50	Log Normal	25.43	0.2883	17.57	36.80	22.44	29.05	0.1968	0.4476
	Probit	26.57	7.13	17.43	35.70	23.53	30.23	4.84	11.11
	Logit	26.55	7.36	17.21	--	23.45	30.24	4.72	11.96

https://doi.org/10.52202/078352-0020

A plot of the log-likelihood response curve along with lower and upper 90% confidence bounds are presented in Figure 9. The log-normal link function is preferred since, unlike the other two, it does not produce any negative stimulus values for the response or confidence curves. Per Table 7 and Figure 9, in order for the skin simulant to be considered a match to the soft-tissue backed PMHS data, the surrogate should produce a mean energy density (Ea50) between 22.4 and 29.0 J/cm^2.

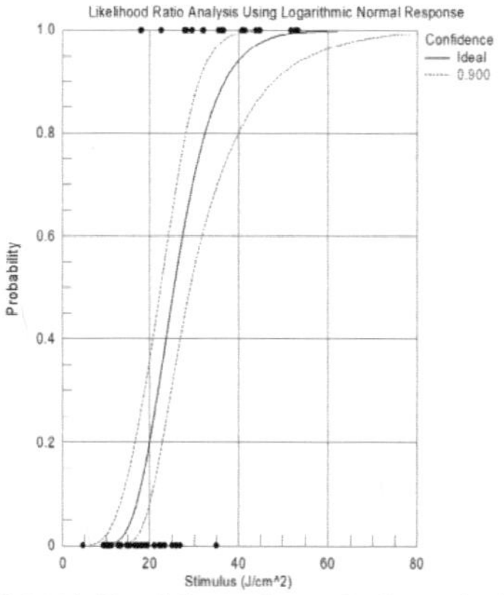

Figure 9. Response Curve and 90% Confidence bounds for soft-tissue backed PMHS locations

3.3 Surrogate Penetration Evaluation

Figure 10 displays the visual responses for skin simulant and the underlying ballistic gelatin from the first 15 shots, along with their corresponding projectile velocity and classification. 40 shots were accomplished in this study, of which 14 exhibited either high yaw (>20°) or struck the testing frame and were therefore not included into analysis. As a result, the penetration resistance of the remaining 26 shots was accomplished for comparison to the PMHS dataset (Figure 9).

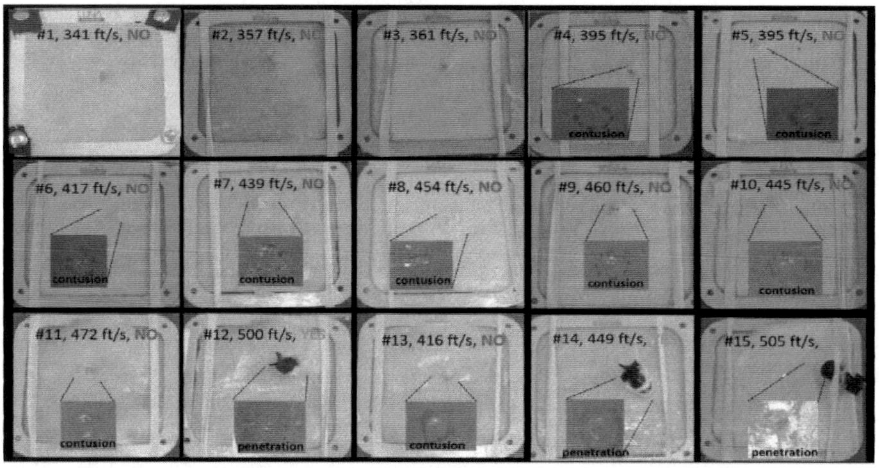

Figure 10. Visual Examination Results from Penetration Testing at HP White Laboratories

A log-normal analysis was utilized along with the likelihood ratio to determine 90% confidence intervals for the probability of penetration as it relates to the average energy per unit area, following methods utilized for PMHS testing (Figure 11). The skin simulant produced a mean energy density (Ea50) of 22.9 J/cm^2. Figure 8 presents comparison of the results from the PMHS-defined injury risk 'truth-set' and the surrogate penetration testing results accomplished at HP White Laboratories.

Table 8. Comparison of Ea50 for Soft-Tissue Backed PMHS and Skin Surrogate

Prototype	Ea50 Confidence Interval (Average) [J/cm^2]
Re-analyzed PMHS data	22.44-29.05 **(25.43)**
TrueSkin Prototype	20.71-24.75 **(22.86)**

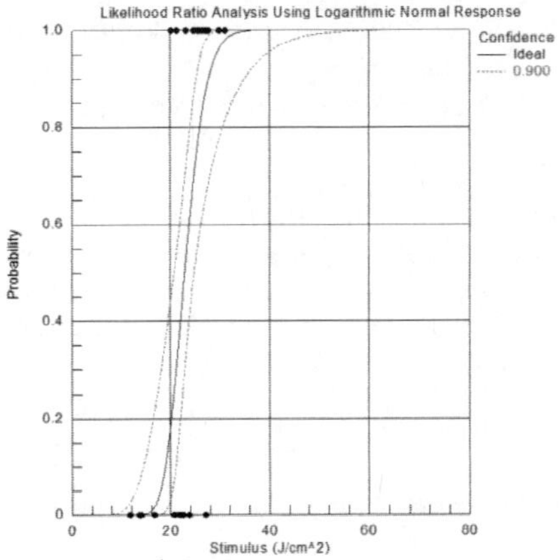

Figure 11. Response Curve and 90% Confidence bounds for Surrogate testing

4. DISCUSSION AND CONCLUSION

4.1 Discussion

A synthetic skin simulant has been developed in an effort to fulfill a need from kinetic munition developers and safety evaluators. This need simulant and set of test methods was accomplished by precisely characterizing the mechanical properties of human skin, tuning the composition of a hydrated nanofiber material, and evaluating the penetration resistance of the surrogate to 12 gauge rubber projectiles. In this study, PMHS skin samples were investigated under different strain rates and skin orientations and were utilized to generate stress-stretch profiles of human skin to failure. This data was utilized to (1) iteratively inform parallel development of a fiber-reinforced hydrated skin model by setting critical mechanical properties (UTS and λ_f) and (2) develop a constitutive model of human skin for future use in finite element modeling. While many groups have focused on characterizing the mechanical properties of human skin[18], [19], these are typically accomplished without testing to failure, a critical need to define the properties relevant to NLW testing and evaluation. Additionally, many of these studies utilize PMHS samples with excessive age[13].

While previous studies have examined the failure mechanics of animal skins, such as goat or porcine[20], [21], to failure these natural materials are known to vary from human skin biomechanics. The results of this study demonstrate how mechanical characterization of human skin (PMHS) can be accomplished to drive the design of a surrogate material that matches relevant mechanical properties, with higher reliability (lower standard deviation) and without the costs and logistical burdens associated with human subject testing.

More recent studies have attempted to standardize the evaluation of NLWs, including KE impact munitions, with the NATO Standardization Recommendation (STANREC)[22]. These recommendations include the use of test methods focused on assessing skin penetration in addition to the blunt impact effect. While numerous studies have examined the relationship between applied energy (Ea) and risk for significant injury using surrogate materials, these studies have primarily been focused on simulating and investigating viscous or blunt impact effects[3], [23]. Studies focused on penetration testing often utilize synthetic or natural materials for representing skin, such as Chamois[8]. However, these studies either utilize natural materials which can exhibit variation on par with human subjects, or utilize materials which have not been verified to match biomechanical properties of human skin.

4.2 Conclusion

The current study investigated relevant mechanical properties of human skin which was used to inform a custom hydrated-nanofiber skin simulant. The synthetic skin simulant was then utilized in a kinetic munition penetration evaluation protocol with a ballistic gelatin backing. The penetration resistance of a 12 gauge rubber projectile was determined and compared to previous PMHS penetration data [2]. The following statements highlight the takeaways from this study:

- Definition of the mechanical properties of human skin relevant to NLW penetration resistance (UTS, λ at failure) with precise characterization of PMHS evaluated at the UVA-CAB.
- Fabrication of a proprietary hydrated nanofiber skin simulant for accurate simulation of human skin mechanical properties.

https://doi.org/10.52202/078352-0020

- Evaluation of previous PMHS penetration resistance dataset and establishment of penetration likelihood over soft-tissue backed anatomies.
- Validation of physical simulant penetration performance with independent testing at HP White Laboratories.

With further refinement of test methods for penetration evaluation and validation of the simulant and methods developed in this study for additional KE impact munitions, the surrogate can provide a valuable resource for key stakeholders in the development and safety evaluation of NLWs. Data generated from this model can eventually reduce the cost and logistic burden of PMHS or animal testing and can be used to generate safety profiles of NLW or other projectiles with higher statistical confidence.

Acknowledgments

This material is based upon work supported by the U.S. Army Aberdeen Test Center under Contract No W91CRB-17-C-0032.

References

[1] J. A. Kapeles and C. A. Bir, "Human Effects Assessment of 40-mm Nonlethal Impact Munitions," Hum. Factors Mech. Eng. Def. Saf., vol. 3, no. 1, 2019.
[2] C. A. Bir, S. J. Stewart, and M. Wilhelm, "Skin Penetration Assessment of Less Lethal Kinetic Energy Munitions," J. Forensic Sci., vol. 50, no. 6, pp. 1–4, 2005.
[3] R. J. Haar, et al, "Death, injury and disability from kinetic impact projectiles in crowd-control settings: A systematic review," BMJ Open, vol. 7, no. 12. 2017.
[4] K. Hubbs and D. Klinger, "Impact munitions: Data base of use and effects," Ncj 204433, 2004.
[5] J. Pavier, et al, "On ballistic parameters of less lethal projectiles influencing the severity of thoracic blunt impacts," Comput. Methods Biomech. Biomed. Engin., vol. 18, , 2015.
[6] C. Bir and D. C. Viano, "Design and injury assessment criteria for blunt ballistic impacts," J. Trauma - Inj. Infect. Crit. Care, vol. 57, no. 6, pp. 1218–1224, 2004.
[7] C. Bir, D. Viano, and A. King, "Development of biomechanical response corridors of the thorax to blunt ballistic impacts," J. Biomech., vol. 37, no. 1, pp. 73–79, 2004.
[8] C. A. Bir, M. Resslar, and S. Stewart, "Skin penetration surrogate for the evaluation of less lethal kinetic energy munitions," Forensic Sci. Int., vol. 220, no. 1–3, pp. 126–129, 2012.
[9] A. Chanda, et al, "Experimental study on tissue phantoms to understand the effect of injury and suturing on human skin mechanical properties," Proc. Inst. Mech. Eng. 2017.
[10] M. Xiong, B. Qin, S. Wang, R. Han, and L. Zang, "Experimental impacts of less lethal rubber spheres on a skin-fat-muscle model," J. Forensic Leg. Med., vol. 67, pp. 7–14, 2019.
[11] A. Papy, et al, "Definition of a standardized skin penetration surrogate for blunt impacts," in 2012 IRCOBI Conference Proceedings - 2012, pp. 486–493.
[12] "Mechanical Characterization of Soft Tissue Simulant Materials," Adv. Exp. Mech., 2017.
[13] M. Ottenio, et al, "Strain rate and anisotropy effects on the tensile failure characteristics of human skin," J. Mech. Behav. Biomed. Mater., vol. 41, pp. 241–250, 2015.
[14] A. N. Annaidh, et al, "Mechanical properties of excised human skin," in IFMBE Proceedings, 2010, vol. 31 IFMBE, pp. 1000–1003.
[15] M. B. Panzer, K. Kong, et al, "Dynamic Mechanical Properties of Human Skin," in PASS, 2020.
[16] J. Lin, S. H. Bang, M. H. Malakooti, and H. A. Sodano, "Isolation of Aramid Nanofibers for High Strength and Toughness Polymer Nanocomposites," ACS Appl. Mater. Interfaces, 2017.
[17] S. R. Stauffer and N. A. Peppast, "Poly(vinyl alcohol) hydrogels prepared by freezing-thawing cyclic processing," Polymer (Guildf)., vol. 33, no. 18, pp. 3932–3936, 1992.
[18] A. J. Gallagher, A. Ni Anniadh, K. Kruyere, M. Ottenio, H. Xie, and M. D. Gilchrist, "Dynamic tensile properties of human skin," 2012 IRCOBI Conf., pp. 494–502, 2012.
[19] M. D. RIDGE and V. WRIGHT, "A Bio-Engineering Study of the Mechanical Properties of Human Skin in Relation to Its Structure.," Br. J. Dermatol., vol. 77, no. 12, pp. 639–649, 1965.
[20] W. L. E. Wong, et al, "Resolving the viscoelasticity and anisotropy dependence of the mechanical properties of skin from a porcine model," Biomech. Model. Mechanobiol., 2016.
[21] S. Schick, et al, "Maximum tensile stress and strain of skin of the domestic pig—differences concerning pigs from organic and non-organic farming," Int. J. Legal Med., 2019.
[22] NATO/PFP, "Risk Assessment of non-lethal Kinetic Energy projectiles," STANREC 4744
[23] A. Oukara, N. Nsiampa, C. Robbe, and A. Papy, "Injury Risk Assessment of Non-Lethal Projectile Head Impacts," Open Biomed. Eng. J., vol. 8, no. 1, pp. 75–83, 2014.

A Warrior Health Avatar for Model Based Evaluation of Personal Protective Armor against Blast and Blunt Impact Threats

RK Gupta[1], HT Garimella[2], ZJ Chen[2] and A Przekwas[2]

[1]*US DoD Blast Injury Research Program Coordinating Office, USAMRDC, 504 Scott Street, Fort Detrick, MD 21702-5000, USA,*
raj.k.gupta.civ@mail.mil
[2]*CFD Research Corporation, 701 McMillian Way NW, Huntsville, AL 35806, USA*

Abstract. In combat operations and training, military personnel may be exposed to blast waves generated by explosive devices and by heavy weapon systems such as recoilless weapons. Conventional military personal protective armor (PPA), including helmets and ceramic vest inserts, is typically designed to protect against high energy ballistic impacts. For ergonomic reasons, such armor protects against most sensitive organs (head/brain and the heart). However, blast waves and the debris load the whole body, of which only part is protected by soft armor. There is a need for user-friendly software tools for model-based design and evaluation of personalized PPA. This paper presents a novel concept of a Warrior Health Avatar (WHA) for computational model and wearable sensors-based evaluation and optimization of the PPA used in combat and military training. The WHA framework includes a fast-running blast dynamics model, subject-specific human body anatomy and PPA, biodynamics and biomechanics tools. The WHA generator, validated against the military population ANSUR II database, accounts for the anthropometric and postural variations, weapons and PPA. The fast-running tools can simulate blast and debris loads on the entire human body partially protected by the PPA as well as on blast injury vulnerable organs (head, face, ears, lung, and groin). Reduced-order models are used to simulate the dynamics of pressure load absorption and transmission through the soft and hard armor to the human body. The paper will present example simulations of IED blast loads on a human subject and several examples of blast loads on military personnel in training with heavy weapons, including recoilless shoulder-mounted rifles, mortar, and snipers. Parametric studies can be conducted to analyze and optimize military training protocols. Predicted blast loading on specific organs, brain, in particular, can be used to calculate injury criteria.

1. INTRODUCTION

Blast events accounted for nearly 70% of injuries in wounded U.S. Service members in Iraq and Afghanistan [1-3] with the blast-induced traumatic brain injury (TBI) becoming the "signature" wound inflicted by improvised explosive devices (IEDs). Although moderate and severe TBI can be easily identified and aggressively treated, concussions and mild TBIs (mTBIs) with no other visible wounds have been often overlooked [4]. While most combat-related mTBI cases are expected to recover, persistent symptoms after TBI, such as chronic dizziness, fatigue, headaches and delayed recall of memory, are common [5]. Significantly, these symptoms have been more frequently observed in Servicemembers exposed to multiple low-level blast (LLB) exposures. Moreover, it has been recently observed that military personnel involved in training with explosives such as breachers and gunners could also experience neurological conditions, mainly due to repeated exposures [7-9].

Over the last few decades, the U.S. Department of Defense (DoD) has committed substantial resources in researching ballistic and blast trauma to improve combat casualty care and personal protective equipment (PPE) [6]. The resulting improvements in the PPE and trauma care have mitigated or reduced potential blast and ballistic injury to the thorax, including lung and abdomen. However, vulnerability to face, ear, brain, groin and extremity injury still remain. Protection against blast wave TBI is particularly challenging because, despite the protective helmet, a significant part of the soldier's head and neck is still exposed to the blast. Moreover, because the blast waves traverse the entire human body, the design of an effective PPE for the warfighter should include the entire body. At the same time, any additional or reinforced PPE to be carried by a warfighter in combat conditions has to be balanced against limitations in the warfighter mobility, situational awareness, and physiological stress, including thermal loads and fatigue.

Emerging evidence suggests that Service members exposed to repetitive blast exposure from breaching and heavy weapon firing, even during training, may be susceptible to short-term neurocognitive deficits [9,10]. There is a great need for better understanding and quantification of

 https://doi.org/10.52202/078352-0021

potential injury mechanisms from repeated exposure to blast generating weapons such as artillery, mortar, recoilless rifles, sniper rifles, machine guns, heavy weapons, explosive breaching and others. In December 2017, U.S. President Donald Trump signed into law the 2018 National Defense Authorization Act, which obligated the Secretary of Defense to conduct "Longitudinal Medical Study on Blast Pressure Exposure of Members of the Armed Forces" [11]. These studies, referred to as SEC. 734, include monitoring of blast pressure exposure (Dose) and medical symptoms (Response) of Servicemembers in training. The ongoing longitudinal medical studies, focusing on the Response components, cannot provide complete answers without the precise estimation of the blast Dose. At the recent DoD scientific meeting, it was concluded that complementary experimental and computational studies should be conducted to correlate the Dose-Response effects [12]. It was also concluded that computational models of repeated low-level blast exposure are urgently needed to develop protective protocols guiding the ongoing field tests.

Mathematical models of the blast wave - human body interaction could be used for more accurate calculations and model-based interpretation of the wearable blast pressure sensor data. Both high-fidelity CFD models and calibrated fast-running models of an explosive and muzzle generated blast waves are needed to compute pressure loads on the Servicemembers exposed to the blast. The fast-running models could be used for instruction during training and can aid in rapid, real-time, calculation, collection and storage of person-specific blast loads - the Dose. There is a need for a user-friendly simulation framework for model-based development of personalized protective armor for combat and military training applications.

This paper presents a novel concept of a Warrior Health Avatar (WHA), a virtual human body for computational simulations of blast exposure events, calculation of blast loads on the entire body, evaluation of protective equipment and for development of safer military combat and training operations. The WHA framework includes a fast-running blast dynamics model, subject-specific human body anatomy, the capability to set up the human body posture, outfit the body with wearable sensors, the protective armor and weapons, immersing the body in the blast scene and graphical visualization of predicted blast loads on the body. CoBi-Blast tools are used to conduct blast loading simulations predicting space and time-resolved blast loads on the body and blast injury sensitive organs, human body biodynamics and injury biomechanics [13-17].

2. METHODS

2.1 CoBi Software Suite

CoBi is an DoD open source multiscale multiphysics C++ software tool that was developed by CFD Research. This paper presents CoBi-HBMG (***H***uman ***B***ody ***M***odel ***G***enerator) and CoBi-Blast, a blast dose estimation module of the framework. The key components of the CoBi-HBMG include: anthropometrically accurate setup of warfighter skin model, addition of clothing and protection equipment, and adjustment of posture and position. The key components of the CoBi-Blast module include: characterization of weapon specific blast signature, visual setup for the generation of a warfighter body and the blast exposure scene, and a fast running solver for calculation of blast loads on the entire human body and on injury sensitive organs (face, head, thorax, ears, eyes, neck, groin and others). CoBi-FEM, as the name indicates, is a finite element module in the CoBi framework and is used for a wide-variety of multiscale multiphysics problems.

2.2 Blast Exposure Simulation Framework

Comprehensive computational analysis of human body blast injury involves three major components: 1) generation of a virtual blast scene involving the explosive charge with one or several humans, 2) simulation of blast wave interaction with the human body, and 3) simulation of the injury biomechanics to blast sensitive organs such as the brain and lungs. Figure 1 schematically presents the overall architecture and main components of the CoBi-Blast tools used in this study.

https://doi.org/10.52202/078352-0021

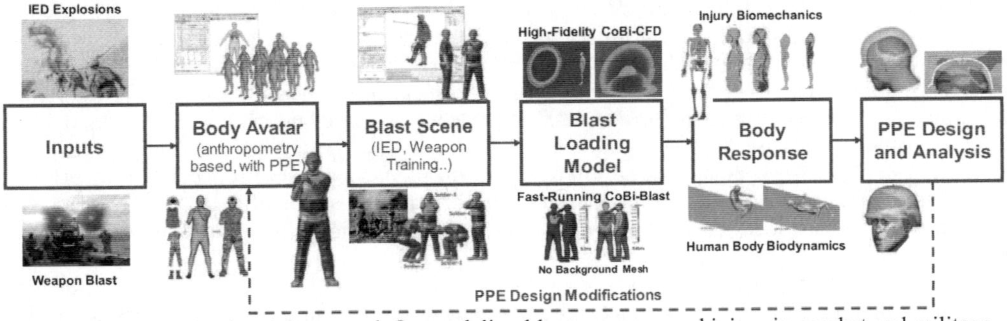

Figure 1. A computational framework for modeling blast exposure and injury in combat and military trading scenarios

The central element of the framework is the parametric anatomical model of a human body, here referred to as the Warrior Avatar, which can be personalized based on subject-specific anthropometric characteristics, outfitted with the protective equipment, articulated and placed in a virtual blast scene. The exterior surface ("skin") and the internal anatomy geometry of the avatar are used to generate corresponding computational meshes for blast loading and injury biomechanics simulations, respectively. CoBi framework enables both high-fidelity CFD and reduced-order fast running simulations of blast dynamics, blast-body interaction and calculation of time-space resolved blast loads on the entire surface of the human body and on the body armor. The blast loading results are used as input boundary conditions for the subsequent simulation of human body biodynamics and injury biomechanics. A more detailed description of the CoBi-Blast framework, validation and military application results can be found in authors' previous publications [6,13-17].

2.3 Generation of Warrior Avatars (CoBi-HBMG)

Computational modeling of a human body injury biomechanics is typically conducted using a "standard" human body anatomy obtained from databases such as Visible Human or Zygote. These databases comprise of carefully constructed mesh models representing a 50th percentile human (e.g., skin and internal anatomy models such as Total Human Model for Safety (THUMS) or Global Human Body Models Consortium (GHBMC) model [18-20]). To simulate military Servicemembers' injury biomechanics and protection, we have developed a framework for an anthropometry-based generation of human body anatomy and calibrated the model on the U.S. Army Anthropometric Survey (ANSUR II) database of body scans of male and female Servicemembers [21-23]. This human body generator can be used to construct human avatars including internal anatomy, and body exterior such as clothing and the protective armor. Figure 2 shows the subject-specific warrior avatar generation steps. The anthropometric data (weight, height, extremity lengths, chest breadth, chest depth, and others) are used as inputs to the principal component analysis (PCA) tools to generate the subject-specific or population of skin models and anatomical geometry of major internal organs (skeleton, brain, lungs, intestine, muscles adipose and others). The human body skin is then used to generate the warrior avatar by incorporating combat clothing and armor and articulating the body to the desired posture. The model setup step is the generation of the blast scene involving one or several avatars and placing the IED charge or the blast generating heavy weapon, e.g., standing gunner team with shoulder mounted recoilless rifle as shown in Figure 2.

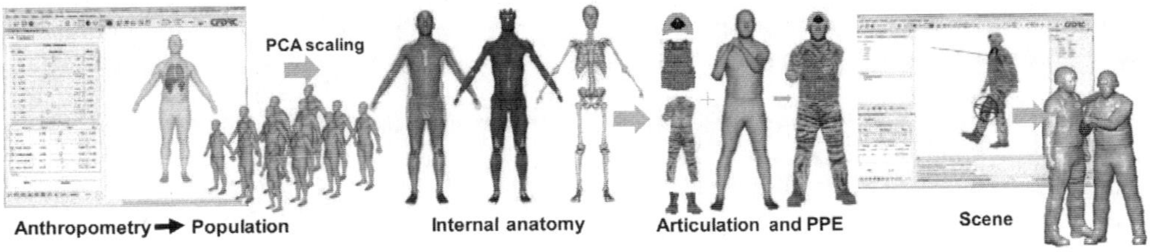

Figure 2. Generation of the warrior avatar and the blast exposure in military trading scene

2.4 Models of Blast Exposure and Injury Biomechanics

https://doi.org/10.52202/078352-0021

Computational modeling of the blast exposure event can be conducted by using either the high-fidelity CFD type software [13-15] or reduced order fast running models [16]. CFD models are more accurate but require a 3D volume computational mesh for the entire blast scene with fine mesh to accurately resolve the propagating blast wave. In our high-fidelity blast exposure studies, we used CoBi multiscale tools and the adaptive Cartesian mesh with mesh refinement around the human body. In contrast, CoBi-Blast fast-running tools use analytical models to simulate the blast wave dynamics and require only the surface mesh of human body avatars or any of the surrounding shock wave reflecting surfaces. The second approach can be used for parametric simulations of blast exposure and for an "inverse problem" reconstruction of blast loads on the entire warfighter body using data from wearable pressure sensors [16].

In the milliseconds long period of the blast wave interaction with the human body, we assume the body as the rigid object with appropriate wave reflection properties on the body's surface. The recorded pressure traces on the human body surface are then used as loading conditions for subsequent simulations (sequential approach) of injury biomechanics. The individual segmented skin regions (triangulated meshes) are used to record and store the spatiotemporally varying loading conditions associated with corresponding organ models. For the biomechanics simulations, we have used the multiphysics code CoBi-FEM (finite element module in CoBi). The anatomy geometry was used to generate hexahedral mesh using an in-house meshing tool. The different material models used to describe the different body regions can be obtained from the literature [14]. An explicit FE solver was used with a reduced integration brick element and a judicious choice of hourglass coefficients. This mesh element description reduces the computational cost, increases the stability, and minimizes the artificial stiffening.

Overall, the above approach leads to a well-defined sequential multiscale modeling framework, for analyzing the effects of PPE under LLB exposures, where multiple length (from meters for a blast scene, to cm for the brain, to µm for neurons and axons, to nm for axonal cytoskeleton and neuronal synapses) and time scales (from µsec for blast-wave transition over the head, to msec for brain biomechanical responses, to min/hr/days for secondary injury and repair cascade) can be accommodated [17].

The next section presents some of our previous results on human body IED blast exposure; and our new simulations demonstrating the effects of armor and body posture on blast exposure during military training with heavy weapons (using CoBi-Blast fast running tools).

3. RESULTS

3.1 Simulation of IED blast loads on a human body

In our blast exposure simulation protocols, we first use CoBi high-fidelity CFD simulations for detailed human body blast load analysis and assessment of the accuracy of the reduced order fast running simulations using CoBi-Blast. CoBi tools have been validated on range of experimental test cases of free field blast exposure of a human surrogate and exposure of human head physical surrogates in shock tubes [13, 14, 27-29]. Figure 3 shows simulation results for a case of a human body frontal exposure to an experimental explosion of 5lb C4 located at 233 cm away from the human body and 127 cm above the ground. For computational efficiency, a high resolution 1D spherical detonation and blast wave dynamics CFD model results, Figure 3a, were used to establish initial conditions (at t=0.45 ms after the explosion) for the 3D model of blast wave interaction with standing human body, Figure 3b-d. The recorded blast loads on the entire human body were then used as a dynamic pressure boundary conditions for modeling pressure wave propagation inside the body. Figure 3e shows a time sequence of pressure loads on the skin as the wave propagates. Note that the simulation starts at t=0 ms at the instance when the blast wave touches the thorax. The same pressure loading conditions can also be used for modeling human body biodynamics of translocation in air and impact on the ground [17, 37]. As presented below, the human body model can be enhanced by including the body armor, and by adjusting body posture and the orientation relative to the explosive location.

https://doi.org/10.52202/078352-0021

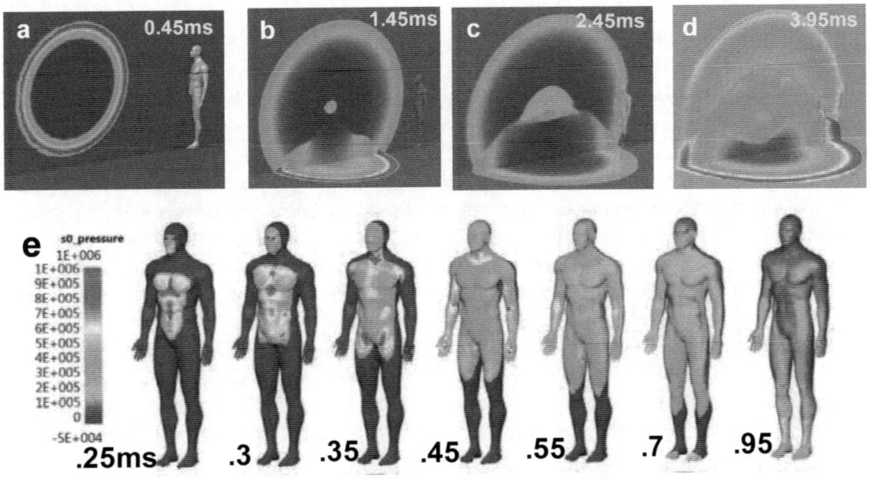

Figure 3. a-d): CFD simulations of a propagating blast wave after explosion above the ground and blast wave interaction with a human body; e): FEM simulation of pressure wave transmission through the human body.

3.2 Simulation of brain injury biomechanics

Protection against blast-induced traumatic brain injury is the most pacing challenge in the ongoing military medical resketch. Computational models of brain injury biomechanics have been used to analyze potential injury mechanisms and to analyze the role of protective head gear including the helmet, eye wear and the hearing protection devices. Figure 4 shows more detailed simulations of the human head and brain response to the blast wave exposure described above. Detailed head anatomy, including skin/scalp, cranium, cerebrospinal fluid (CSF) and brain is simulated. Figure 4a shows three-time instances of the pressure fields in the head central sagittal plane. Figure 4b shows the pressure time traces at three locations in the brain mid-plane (fm-front, mm-middle, and rm rear). Note that the first pressure pulse at the rear of the brain (contrecoup effect) is higher that the initial pressure pulse in the front (coup effect). Figure 4c shows the maximum strain and maximum stain rate in the transverse mid plane of the head. As often reported in the preclinical and human imaging data, our simulations show that most susceptible regions for the brain injury are located in the brain cortical layers, sulci and gyri and the brain-CSF interface.

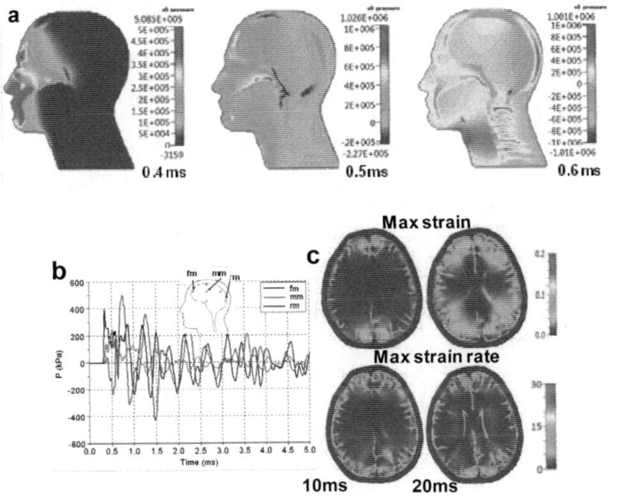

Figure 4. FEM simulations of head/brain biomechanics: CFD simulations of a propagating blast wave after explosion above the ground and blast wave interaction with a human body; e): FEM simulation of pressure wave transmission through the human body.

https://doi.org/10.52202/078352-0021

3.2 Computational Evaluation of the Head PPE in Blast Exposures

The CoBi framework has been used for evaluation of military PPE against blast injury, including: combat helmets, protective chest vests, helmet retention system, helmet suspension pads, combat boots, protective eye wear and hearing protection devices [29-32]. It is difficult to design PPE against highly variable IED blast exposures but much more feasible against blast exposures during military training. Computational analysis of the protection effectiveness of military helmets against the primary blast wave can be conducted using the sequential process presented above. The CFD blast wave loading on the entire helmeted head, Figure 5a, is used for modeling the dynamic response of the helmet shell, the suspension pads and the skull scalp. Simulations have been performed with and without the suspension foam pads. Figure 5b shows three instances of a frontal blast wave propagating over the helmet and between the helmet-head space, the effect known as the "under-wash". The under-wash wave can reflect from the posterior section of the inner helmet wall (Figure 5b, time t_3) and may impact the head as a much stronger wave than the primary incident wave. Proper design and distribution of the helmet pads can significantly attenuate the under-wash effect. The military helmet can also affect pressure loads on the unprotected parts of the face. Figure 5c presents the reflected pressure fields on the human eyes below the helmet rim, in the nose plane, and in the ear lobe space. The well-known extended helmet standoff distance around the ear lobes is partly responsible for reflecting the frontal incident blast wave from the inner helmet shell into the ear lobe and ear canal, Figure 5c.

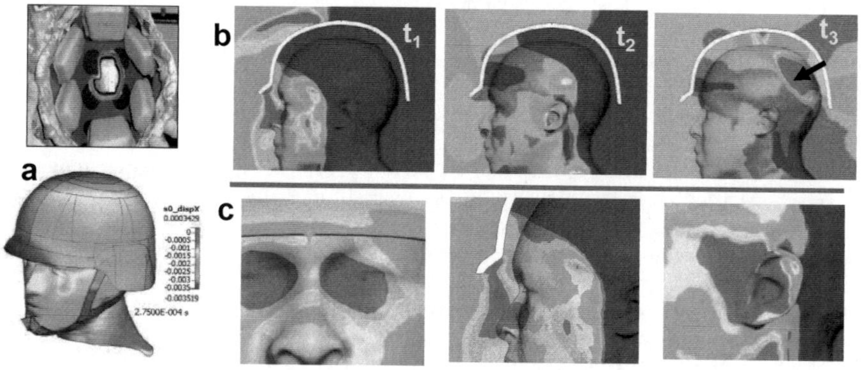

Figure 5. Simulations of a frontal blast wave interaction with a helmeted head, a) military helmet pads and a helmeted human head model; b) blast wave propagation around the helmet without pads, c) pressure loads around the eyes, nose and the ear.

3.3 Simulation of Blast Exposure in Military Training

Here, CoBi-Blast tools are used to demonstrate and validate the applicability of reduced order fast-running blast analysis tools in simulating the blast loads during IED explosions and military training of heavy weapon systems. Furthermore, CoBi-HBMG tools were leveraged for a rapid and realistic reconstruction of training scenes from field images. Figure 6a shows some of the field scenes that are reconstructed from heavy weapon training scenarios. In addition, it also shows the simulation results for back-blast exposure from Carl Gustav weapon system. Free-field sensor data was used to calibrate (i.e., estimate the equivalent blast kernel properties such charge mass and location using CoBi-Blast inverse solver tools) and validate the model. Figure 6b shows some of the initial validation of the model predictions in comparison to the field data [24]. For free-field sensors 1 and 2, the simulation predictions match well with the experimental data. For sensor 3, there are some difference in the second blast peak (due to ground reflection). This could be due to the ground reflection properties (surface type). The back-blast simulations using the CoBi-Blast tools were completed in ~10 mins. on a PC.

https://doi.org/10.52202/078352-0021

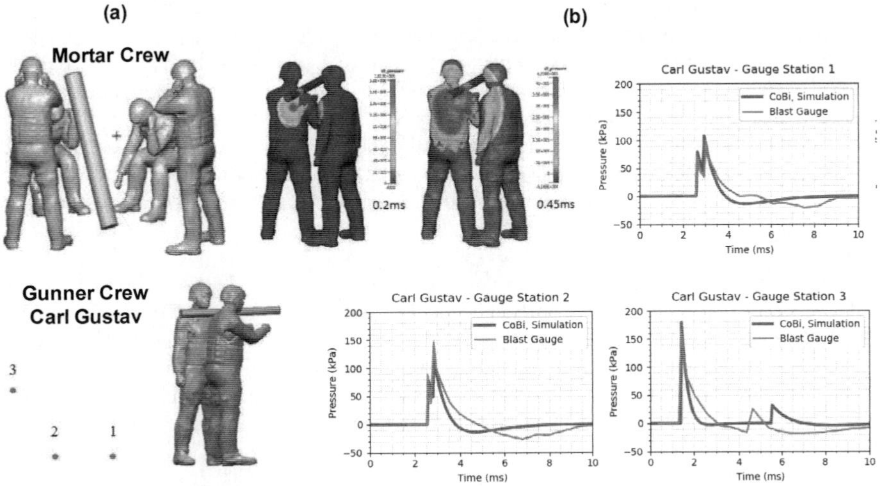

Figure 6. a) Heavy weapon training scenes reconstructed using CoBi human body model generator tools. (b) Simulation of weapon induced blast exposure using CoBi-Blast fast-running tools; and preliminary validation of the Carl Gustav back-blast training scenario.

4. DISCUSSION AND CONCLUSIONS

Recent retrospective studies of injury patterns of U.S. military personnel in combat and military heavy weapons training have considerably contributed to the awareness of the blast injury [1,2, 7-9]. Concurrently, computational studies of human body exposure to blast waves, corresponding body biodynamic and biomechanical responses helped us better understand the blast injury mechanisms of the brain, lungs, spine, and other organs [13, 14, 16,17, 25-32, 33-36]. The major limitation of these and other published studies was the use of a standard size of the human body, typically 50 percental male, represented as a "bare skin" surface geometry. In this paper, we introduce a novel concept of a Warrior Avatar, a virtual scalable human body for computational simulations of blast exposure, injury biomechanics and evaluation of protective equipment. In our CoBi Warrior Avatar framework, the male or female virtual anatomical body, with both the skin and main components of internal anatomy, is generated using subject-specific anthropometric measurements. The body skin model can be also be generated directly by fusing the subject specific skin scan and the facial de-identification step. To conduct the blast wave exposure simulations, the avatar skin model is outfitted with a military combat uniform, personal protective armor, weapons and is articulated to the desired position. We envision that ultimately every Servicemember will have his/her own Avatar, comprising not only the anatomical but also physiological and fitness models. A Servicemember equipped with mobile computing platform and set of wearable sensors could activate the data collection stream during specific events, e.g., weapon training. Such data could then be used for longitudinal monotasking of his/her blast exposures.

At present, it is not possible to quantify blast loading to the warfighter body exposed to an IED blast due to lack of precise forensic data such as body position and orientation relative to the IED and the effects for the surrounding structures. Even if the wearable pressure sensor data were available, it is difficult to reconstruct the blast loads to specific organs. As demonstrated in this paper, computational tools developed by our team can make this task easier. It is certainly possible to reconstruct blast loads on the whole body of a warfighter involved in military heavy weapons training using a photograph of the blast scene, avatars of Servicemembers involved in the blast scene, wearable pressure sensors data and, as described above, the weapon blast signature model [16]. This paper described and demonstrated two modeling approaches for the dynamic reconstruction of the blast exposure scene and calculations of blast loads on the exposed warfighters' bodies: high fidelity 3D CFD models and reduced order fast running models. Each approach has advantages and limitations related to accuracy and computational costs. The main advantages of fast running models are the ease of model setup (no 3D computational mesh needed), the use of the inverse solver problem to reconstruct the whole human body blast loads directly from wearable pressure sensors and the ability to conduct large number of parametric simulations to evaluate the effectiveness of the protective armor and safety protocols.

 https://doi.org/10.52202/078352-0021

Because of the high variability of IED weapons, detonation scenes and corresponding Servicemembers' positions, the design of the personal protective armor is very challenging. This paper has demonstrated that the conventional combat helmets do not protect the exposed part of the head and may even amplify loads to parts of the head, such as eyes and ears. Such injuries can be significantly attenuated by wearing eye ware or helmet integrated visors. Protection of Servicemembers for blast injury during heavy weapons training is much more feasible and is currently under development and evaluation in the U.S. Various protection methods are being explored including hard and soft body armor, improved head, neck and chest protection, use of lower power weapons rounds for training, optimized body postures and other safety protocols.

This study demonstrated the capability to compute the blast loads on a warfighter body from IEDs and during weapons training but also identified needs for future developments. There is a need for more precise specification of blast "dose" parameters on specific organs, in a mechanistic model for calculation of the cumulative "dose" from repeated exposures, for longitudinal monitoring of warfighter post-exposure physiological and cognitive "responses." Such a holistic approach for model-based personalized "dose-response" analysis, currently pursued in the U.S., will have immense benefits in warfighter safety and health.

Disclaimer

The opinions and assertions contained herein are the private views of the author/s and are not to be construed as official or reflecting the views of the Department of the Army or the Department of Defense

Acknowledgments

The authors would like to express their appreciation to several individuals who contributed to CoBi code and human body avatar development and for helpful discussions warfighter protection in military training including: Dr. Gary Tan, at the US NRL, Dr. Walter Carr and Dr. Gary Kamimori at WRAIR, and several scientists at CFDRC: Dr. Vincent Harrand, Dr. Alex Zhao, Mr. Timothy Zehnbauer and Mr. Keith Sedberry.

References

[1] Eskridge S.L., MacEra C.A, Galarneau M.R., et al., Injuries from combat explosions in Iraq: Injury type, location, and severity. Injury, 2012; 43(10): 1678–82.
[2] DePalma R.G, Hoffman S.W., Combat blast related traumatic brain injury (TBI): Decade of recognition; promise of progress, Behav. Brain Res. 2018, 340;102–105
[3] Magnuson J, Ling G. Explosive blast mild traumatic brain injury. In: Gorbunov N, editor. Traumatic Brain Injury-Pathobiology, Advanced Diagnostics and Acute Management. London: IntechOpen, 2018; pp. 39–48.
[4] Brix K.A., Brody D.L., Grimes J.B., Yitzhak A. et al., Summary of Working Groups and Expert Panel Findings and Recommendations. J. Neurotrauma, 2017; 34, S-18.
[5] Heltemes K.J, Holbrook T.L., MacGregor A.J., Galarneau M.R. Blast-related mild traumatic brain injury is associated with a decline in self-rated health amongst US military personnel. Injury, 2012; 43(12): 1990–5
[6] Elsayed N. and Atkins J., Explosion and Blast-Related Injuries, Elsevier Academic Press, London, June 2008.
[7] Carr W., et al., Repeated Low-Level Blast Exposure: A Descriptive Human Subjects Study, Military Medicine, 2016; 181, 5:28-39
[8] Kamimori G.H., Reilly L.A., LaValle C.R. & Silva U.B.O.D. Occupational overpressure exposure of breachers and military personnel. Shock Waves, 2017; 27, 837–847.
[9] Belding J.N., et al., Blast Exposure and Risk of Recurrent Occupational Overpressure Exposure Predict Deployment TBIs. Military Medicine, 2019; Oct 28, ahead of print.
[10] LaValle C.R., et al., Neurocognitive Performance Deficits Related to Immediate and Acute Blast Overpressure Exposure. Front. Neurol. 2019; 10:949
[11] 115th Congress (2017-2018), "Longitudinal Medical Study on Blast Pressure Exposure of Members of the Armed Forces," Sec. 734, H.R.2810, National Defense Authorization Act for FY 2018.
[12] The Neurological Effects of Repeated Exposure to Military Occupational Blast, Proc. 7th Annal Department of Defense State-of-the-Science Meeting, March 12-14, 2018, Washington DC, USA, Publ. RAND Corporation, Santa Monica, CA.

https://doi.org/10.52202/078352-0021

[13] Gupta R.K., Przekwas A. Mathematical models of blast-induced TBI: current status, challenges, and prospects. Frontiers in Neurol. 2013; 4, 59.

[14] Tan X.G., Przekwas A.J., Gupta R.K. Computational Modeling of Blast Wave Interaction with a Human Body and Assessment of Traumatic Brain Injury. Shock Waves, 2017; v27, 889–904

[15] Chen Z.J., Przekwas A.J. A coupled pressure-based computational method for incompressible and compressible flows. J Computational Physics, 2010; v229(24), 9150-65

[16] Przekwas, A. Garimella, H.T, Chen Z.J., Zehnbauer T., Gupta R.K., Carr, W.S., Kamimori, G.H. Fast Running Tools for Personalized Monitoring of Blast Exposure in Military Training and Operations, Military Medicine, Jan. 2020 submitted.

[17] Przekwas A., et al., Biomechanics of blast TBI with time-resolved consecutive primary, secondary, and tertiary loads. Military Medicine. 2019; v184, 195–205,

[18] Spitzer, V., Ackerman M., Scherzinger A. and Whitlock D. The visible human male: a technical report. J. Am. Med. Inform. Assoc. 1996; 3:118–130.

[19] Iwamoto M.Y., et al., Development of a finite element model of the Total Human Model for Safety (THUMS) and application to injury reconstruction. Proceedings of the International Research Council on the Biomechanics of Injury (IRCOBI). Sept. 18–20, 2002. Munich, Germany, 1–12

[20] Gayzik F.S., et al., Development of a Full Body CAD Dataset for Computational Modeling: A Multi-modality Approach, Annals of Biomed. Eng. 2011; v39(10), 2568–83

[21] Wilkerson P., Zhou X., Przekwas A., Buhrman J. et al., Virtual Body Generator for Anthropometry and Physiology Based Modeling, SAE Technical Paper 2009-01-2280, 2009, https://doi.org/10.4271/2009-01-2280.

[22] Zhou X., Sun K., Roos P., Li P., Corner B. Anthropometry Model Generation Based on ANSUR II Database. Int J Digit Hum. 2016; v1(4):321–43

[23] Roos P.E., Vasavada A., Zheng L., Zhou X. Neck musculoskeletal model generation through anthropometric scaling. PLoS ONE, 2020; 15(1):e0219954.

[24] Wiri S., Ritter A.C., Bailie J.M., Needham C. and Duckworth J.L. Computational modeling of blast exposure associated with recoilless weapons combat training. Shock Waves, 2017; 27(6): 849-862.

[25] Garimella, H.T., & Kraft, R.H. Modeling the mechanics of axonal fiber tracts using the embedded finite element method. Int. J. Numerical Methods in Biomedical Engineering, 2017; 33(5), e2823.

[26] Garimella, H.T., Kraft, R.H., & Przekwas, A.J. Do blast induced skull flexures result in axonal deformation? PloS one, 2018; 13(3).

[27] Tan, X.G., Kannan R., Przekwas A.J., et al., An enhanced articulated human body model under C4 blast loadings, IMECE 2012-89067, Proc. ASME IMECE, Nov 9-15, 2012, Houston, TX, USA

[28] Tan X.G., Przekwas A.J., Long J.B. Validations of Virtual Animal Model for Investigation of Shock/Blast Wave TBI, IMECE 2013-64587, Proceedings of the ASME IMECE, Nov 15-21, 2013, San Diego, CA, USA

[28] Tan X.G., et al., Computational Modeling of Blunt Impact to Head and Correlation of Biomechanical Measures with Medical Images ASME J of Medical Diagnostics. 2020; 3(1): 011007-1-13

[30] Przekwas A., et al., Integrated Experimental and Computational Framework for the Development and Validation of Blast Wave Brain Biomechanics and Helmet Protection, Proc. HFM-207 NATO Symposium on a Survey of Blast Injury Across the Full Landscape of Military Science, Halifax NS, Canada, Oct. 3-5, 2011

[31] Przekwas A, et al., Computational Modeling of Helmet Structural Dynamics During Blunt Impacts, ASME 2009 Int. Mech. Eng. Congress and Exp., Nov 13–19, 2009, Lake Buena Vista, Florida, USA

[32] Phelps S., Przekwas A., McEntire J, Chancey C. Biomechanics and Physiology Based Modeling of Military Helmet Protection against Blast Wave Traumatic Brain Injury, NATO Int. Personal Armour Committee Workshop, Paris, Ecole du Val-de-Grace, 21-22 April 2009

[33] Przekwas A. Multiscale Computational Modeling of Blast Injuries. in Explosion and Blast Injuries, Ed. by Elsayed E. and Atkins J., Elsevier Academic Press, London, June 2008.

[34] Kannan R. and Przekwas A. A computational model to detect and quantify a primary blast lung injury using near-infrared optical tomography. Int. J. Num. Meth. Biomed. Eng. 2011; v27(1) 13-28

[35] Gupta R.K., Tan X.G., Somayaji M.R. & Przekwas A.J. Multiscale Modelling of Blast-Induced TBI Mechanobiology - From Body to Neuron to Molecule. 2017; Def. Life Sci. J. 2, 3–13.

[36] Przekwas A., Somayaji M.R. & Gupta R.K. Synaptic Mechanisms of Blast-Induced Brain Injury, Front. Neurol. 2016; 7, 2.

[37] Tan X.G., and Przekwas, A. A computational model for articulated human body dynamics. Int. J. of Human Factors Model. & Simulation, 2011;v2, 1/2, 85-110.

 https://doi.org/10.52202/078352-0021

Hard Armour Trade Space Analysis

A. Moser[1] and A. Geltmacher[1]

[1]*U.S. Naval Research Laboratory, 4555 Overlook Ave., S.W., Washington, DC 20375*
Alex.moser@nrl.navy.mil

Abstract. The U.S. Marine Corps primarily fields a single type of hard armour system called the Enhanced Small-Arms Protective Insert (ESAPI). The existing ESAPI plates are heavy and contribute to increased load bearing injuries and decreased mobility and survivability. These plates are engineered to stop specific threats at the designated threats' muzzle velocity. The primary and most severe threat for which the ESAPI is rated is rarely found in operation by either allied or adversarial forces. Also, the specific threats in most combat situations are predominantly and significantly under the threat's rated muzzle velocity. Thus, in 90% of battlefield scenarios these plates are overrated for penetration performance but significantly underperform with respect to mobility and overall survivability. As part of the Marine Corps mission, an attempt is being made to maintain adequate ballistic performance while decreasing plate weight, mobility, and overall survivability. The Hard Armour Trade Space (HATS) analysis evaluated ballistic performance information from armour vendor web sites and vendor proprietary information provided to the U.S. Naval Research Laboratory. Most vendor proprietary information typically consisted of ballistic test reports from a certified commercial ballistic test laboratory. The HATS analysis system consists of a database of 40 armour manufacturers, 234 hard armour plates, 42 prevalent ballistic threats, ballistic test data from threats for which each plate was tested, and scripts to perform analysis on the commercial-off-the-shelf plates in the database. The database provides over 1000 plate/threat performance combinations. The HATS analysis resulted in contribution to the ballistic requirements for a significant new lightweight armour plate procurement of over $260M for improved overall survivability.

1. Background

Dismounted ground troops carry 41kg (90lbs.) to 64kg (140 lbs.) or more in combat [1]. Such loads cause the Warfighter to sustain injuries from dismounting vehicles, even before entering contested spaces. With fatigue, cognitive abilities decline [2], as well. The excessive weight also decreases mobility. These factors, in combination, decrease over-all survivability. Thus, to decrease Warfighter load one approach is to decrease the weight of body armour.

Progress in armour development has progressively slowed over several decades, primarily from difficulties in finding materials with enhanced ballistic performance. An overall trend described by Dr. James Zheng, formerly of US Army/PEO Soldier, indicated improvements had approached an asymptotic limit as shown in Figure 1 [3]. Because of this, armed services development centres have recently focused on finding armour solutions that fit the threat scenario within specific operational environments. This typically reduces the weight requirement of armour, since the currently fielded US military body armour is required to stop a threat rarely seen in conflicted spaces. However, the trade-off between armour weight and survivability has previously been unquantified, and there had been a reluctance to transition to lighter weight body armour with reduced ballistic specification. A ballistic test is a clear metric on armour performance and is easily quantified through values such as V_{50} result. In contrast, survivability is less quantifiable, but is thought to be a function of mobility and, ultimately, body armour weight. Recently, progress has been made to better understand the link between armour weight, mobility, and survivability [4], and the results from these simulations show a strong correlation between weight and survivability.

https://doi.org/10.52202/078352-0022

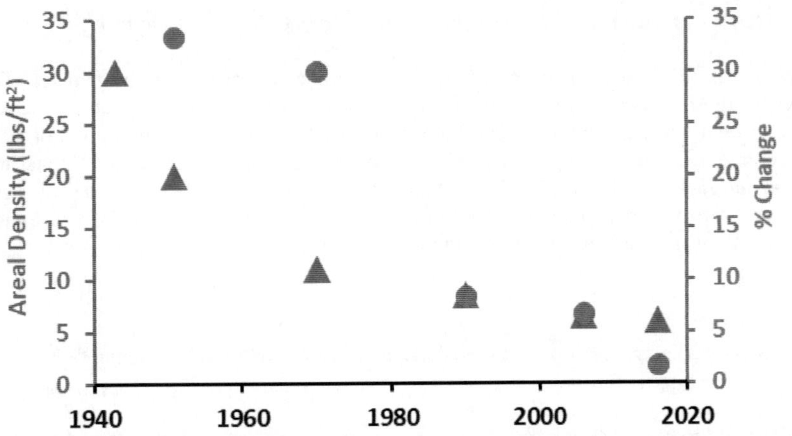

Figure 1. Plot of armour areal density as a function of development time reproduced from Fish et al [3]. The data indicates recent armour developments have been minimal.

An experimental study demonstrated a strong correlation between armour weight and mobility [5], in which a group of Marines were tasked to traverse an obstacle course carrying varying loads and in a rested and fatigued state. However, one must interpret these results carefully.

The plot in Figure 2 shows soldier's completion time through a typical military obstacle course used in training as a function of various armour configurations; the heaviest setup on the left side and the lightest on the right. One can see, for soldier's running a course in a rested state (blue) the reduction in time is less than 20% for a reduction in armour weight approaching 100%.

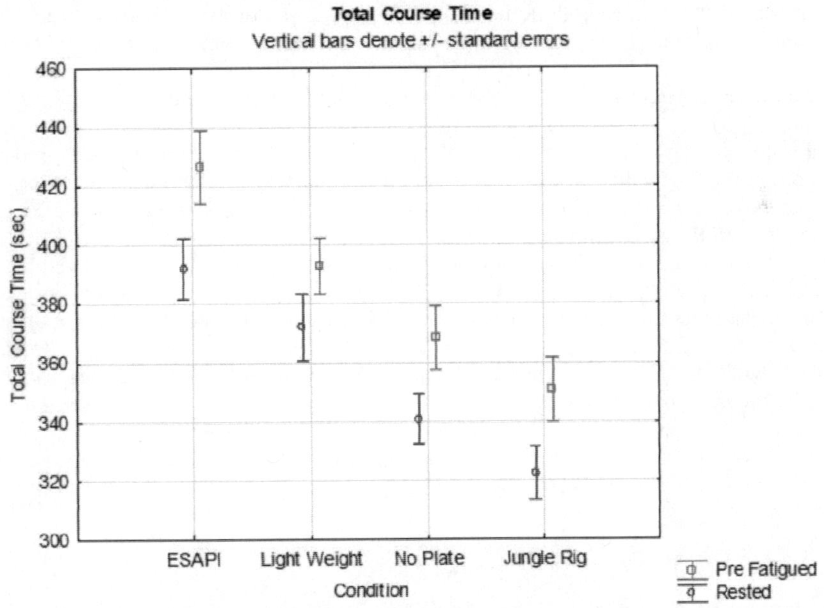

Figure 2. Results from a mobility test showing the importance a few shed kg has on mobility, and ultimately, survivability.

The trade-off between ballistic protection and mobility might lead one to believe reduction of armour weight is not worth the increased mobility. However, the most accurate metric is the total weight the soldier must carry, including the soldier's own body weight, compared to the mobility represented as time to course completion. Thus, a soldier weighing approximately 86kg (190 lbs.), carrying a minimal 90 lb load, including the full set of body armour results in a total weight of 127kg (280lbs.). Without the body armour or body armour vest the total weight is approximately 118kg (260 lbs.). This represents a weight decrease of only 7% yet the reduction in obstacle course time is approximately 17%.

https://doi.org/10.52202/078352-0022

Thus, mobility is greatly enhanced for every ounce of mass shed and is likely to improve survivability significantly.

An endeavour was initiated to develop a hard armour trade scape analysis system (HATS) to determine the weight requirement for hard armour plates based on a set of threat, each at a specified velocity or distance down range from the muzzle. Other analytical models exist, but are not based on a large pool of experimental data [6,7,8]. The process used and subsequently described consisted of amassing known manufacturer hard armour plate data into a database and applying an analysis method which could capture intrinsic characteristics common among the known plates to determine the minimum required armour weight based on a user input threat set.

2. Database

A database was constructed using a well-established database development environment [9]. The criteria for entry into the database was as follows:

1) The company providing the armour was a U.S. company or resided in a country considered a close ally of the U.S.
2) The company of criteria (1) had to be a manufacturer of the plates and not merely a vendor/reseller of the plates
3) An exception to criteria (2) was if the vendor/reseller had exclusive rights to sell the plate from the manufacturer. In essence, the plate was specified/designed by the vendor and made by someone else.
4) An exception to criteria (3) was if the manufacturer was not a U.S. company or resided in a country considered an ally of the U.S.
5) The plates were required to be of nominal areal dimensions of a standard medium sized ESAPI plate.

Armour input into the database typically consisted of an array of plates comprised of various designs and constructions, but the most prevalent consisted of either all steel plates, all polymer (Dyneema or Spectra or similar variant), all ceramic plates, or a combination of ceramic and polymer. Differing covered and foams were incorporated into plates, but these variants contributed little to ballistic performance.

Most armour plates had associated detailed ballistic test reports either obtained directly on-line or through proprietary agreement with the company providing the information. Excluding some minor variance in the data, most were obtained with the use of a soft armour surrogate as a backing material in front of a standard Roma Plastilina.

The database structure is shown in Figure 3 and consists of 15 relational database tables containing a total of 110 fields and thousands of records (2033 records). It contains personal armour plates from 38 U.S. and allied armour manufacturers down selected from an initial field of 108 vendors/manufacturers. It contains a total of 229 unique armour systems ranging from NIJ level III through IV+, and included some plates with capability to stop special threats. The tables were linked together using a variety of rules and/or programmatic processes.

https://doi.org/10.52202/078352-0022

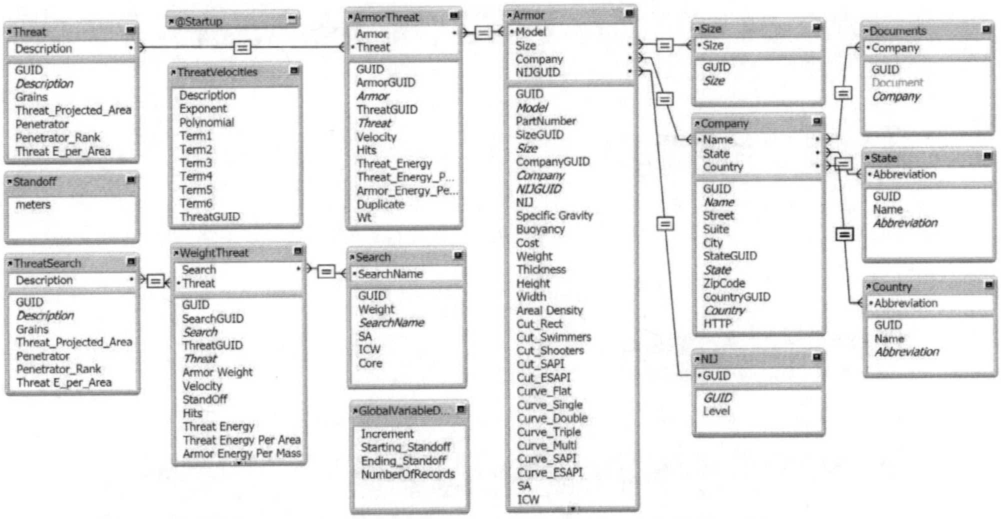

Figure 3. Trade space analysis database tables, fields, and relationships.

Armour plate weights ranged from 0.744kg (1.64 lbs.) to 4.45kg (9.8 lbs.) for medium sized plates, thicknesses from 4.8mm (0.19in.) to 33mm (1.3in). It consisted of 38 different threat types, from shotgun slugs, through NIJ level IV+ threats[1]. Since each plate was typically tested for multiple threats, the database contained ballistic test data from over 1086 armour/threat combinations.

The front-end user input section of the database is comprised of a search/analysis capability, is described in detail later, and is shown schematically in Figure 4. The user can create and save a set of search criteria based on a set of threats and associated velocity/stand-offs from muzzle. The search can be for plates known to defeat all threats in the threat set at or about the designated velocity/stand-off or those inferred to defeat the threats based on analysis of each plate's ballistic data and other parameters.

[1] NIJ IV+ is not a standard NIJ level, but presented by several vendors to represent protection against not only M2 AP threat, but other threats such as the 7.62x39 BZ API and 7.62x54R B32 API.

https://doi.org/10.52202/078352-0022

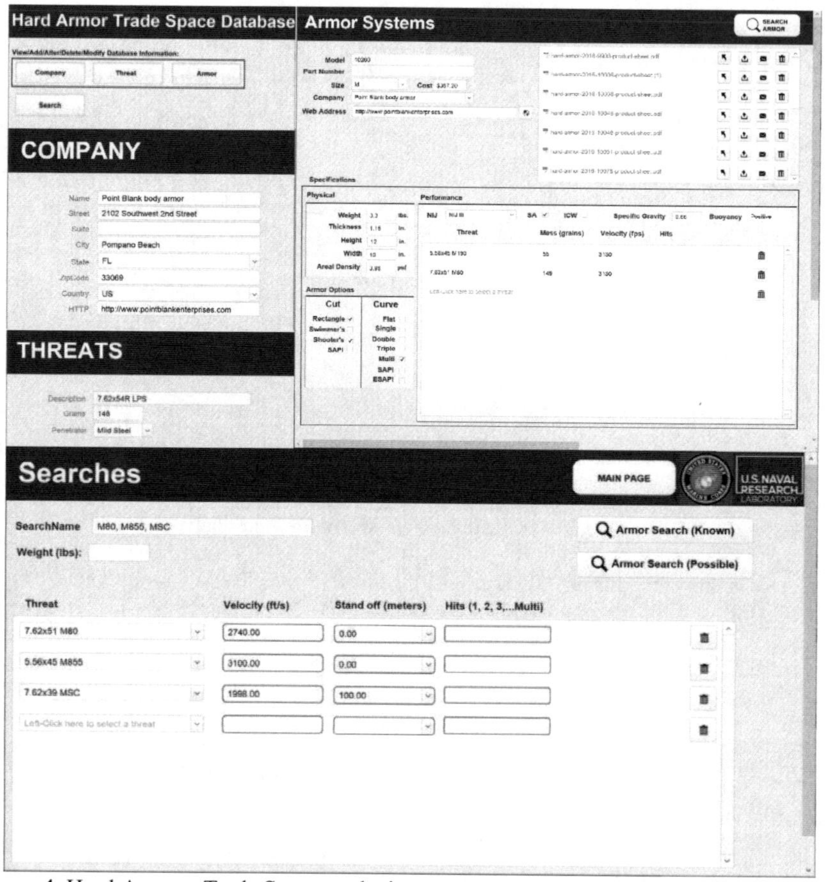

Figure 4. Hard Armour Trade Space analysis system comprised of a database built from a common database development system. Shown (upper left) the main page, the page for Company input, and Threat input, (upper right) Armour system input, and (lower) the search page for finding the required armour.

The result of the search is either a table of plates known to defeat the most aggressive threat in the threat set or a table of plates likely to defeat the most aggressive threats in the threat set and also a minimum medium ESAPI format plate weight required to defeat the most aggressive threat in the threat set. Further filtering and sorting can be applied to the tables upon user input to target specific desired attributes such as plate total weight range, buoyancy characteristic, and specific armour plate cut.

The results obtain for plate systems known to defeat specific threats is straightforward and requires a typical database record lookup. However, determination of the armour systems inferred to defeat the user input threat set and the minimum armour plate weight required is more involved.

As seen in Figure 5 for a single threat at muzzle velocity, without adequate transformation of the data, no clear determination of armour performance is possible. The upper left plot in Figure 5 shows armour weight as a function of threat velocity producing a scatter plot without any significant trend. Similarly, the remaining representations of the data within the database, armour areal density as a function of threat velocity, energy, or projected energy (as shown in Figure 6) produces similar ambiguous results.

The plots in Figure 5 all use extrinsic parameters for the data representation of the armour systems. Ultimately, transformation of the data to an intrinsic parameter similar to toughness would enable comparison of each armour system relative to another.

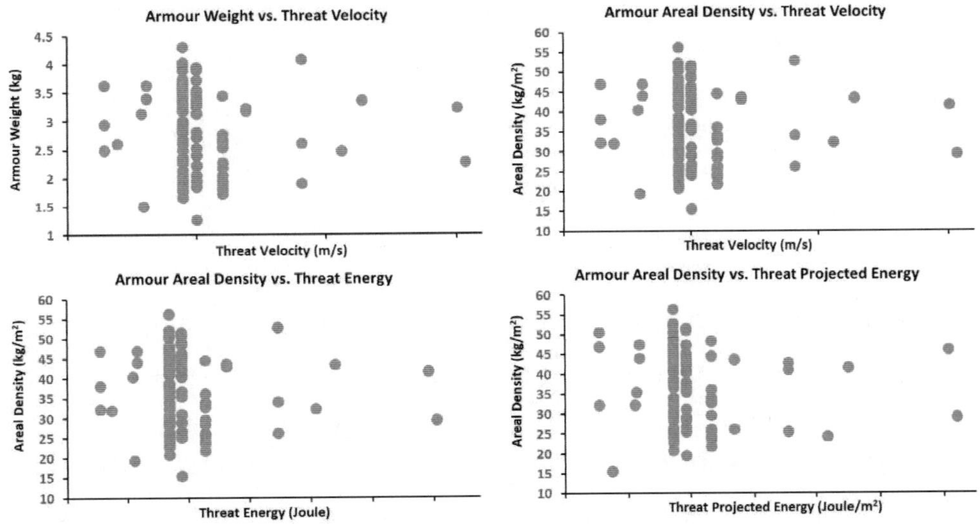

Figure 5. Representations of armour performance against a specific rifle threat. Each horizontal axis represents similar range but with differing units. These data representations make it difficult to infer any useful information.

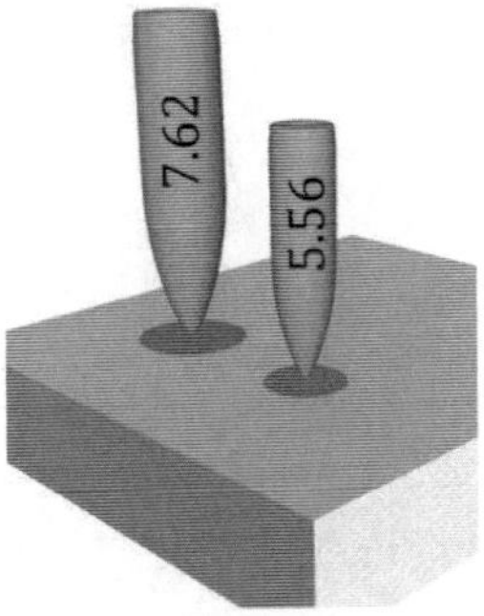

Figure 6. Schematic showing projected cross-sectional area of two different threats onto armour front surface.

The area projected by the threat onto the armour defines the armour volume in which the threat energy is absorbed. This assumption is justified because the greatest strains to the armour, greatest surface area transformation of the ceramic strike face and, ultimately, the greatest percentage of the threats kinetic energy is absorbed in this volume. If one represents the data as a plate energy absorbed per unit volume, this enables comparison across plate systems having undergone ballistic tests with differing threats. This metric is, effectively, a per-mass representation of toughness. To create this representation of the data, the threat projected energy (Joule/cm^2) is divided by the plate areal density (gm/cm^2) to obtain the effective per- mass 'toughness' of the armour system (J/gm). This representation of the armour systems against a specific threat is shown in the left-hand plot of Figure 7, and follows a $y = x^{-n}$ relationship. The inverse relationship is shown on the right-hand side of Figure 7 and follows a $y = m\ x$ relationship. This representation allows us to find the armour with the lowest areal density; lowest values along the y-axis and the armour with the greatest mass- toughness (lowest values along the x-axis).

https://doi.org/10.52202/078352-0022

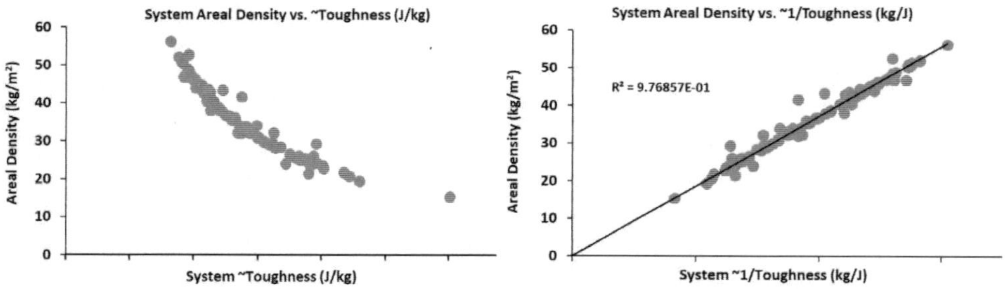

Figure 7. Data from Figure 6 transformed to show armour performance trend. (left) Data from the lower right plot of Figure 6 normalized by dividing the threat projected energy by the area density to produce an armour 'toughness' metric. (right) the same data from the plot on the left but representing the inverse 'toughness'.

Interestingly, one observes a plate that seems to significantly outperform others. This armour system with the lowest areal density and greatest toughness is separated from the other performers. Further investigation revealed the plate's ceramic component was significantly smaller than that of a standard medium ESAPI plate and was backed by a standard sized polymer backing. Thus, one must still carefully screen outliers because they can distort the desired result. In this case, the manufacturer represented the armour as a standard size plate, when, in fact, functionally this was not the case.

Figure 8 demonstrates armour systems perform differently against different threats. Each threat's performance is plotted against the plates with which it was tested. The most severe threats are those with the largest slope.

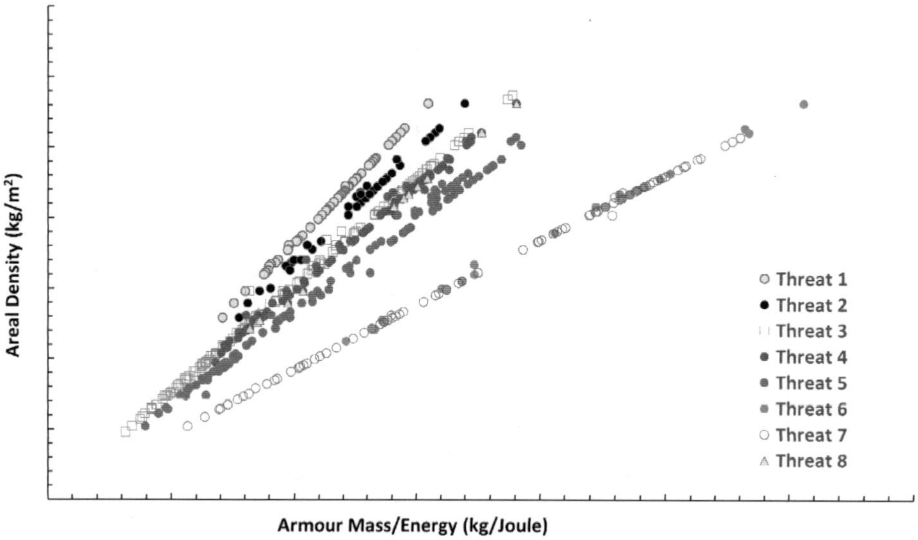

Figure 8. Plot of armour areal density vs the armour mass per unit energy which the armour system is known to absorb based on ballistic test reports. Each point represents a specific armour system/threat pair for which the armour was tested.

3. Results

One can use the database/analysis system to determine the weight requirement for a plate with specific ballistic performance characteristics. Recently, the United States Marine Corps had a requirement for a light weight plate able to stop four different threats, each at a different velocity. The number of ballistic tests to determine the required weight would be time consuming and costly, and since the tests would likely be from a limited lot of plates, the results would be less certain than those from a source utilizing a large dataset of armour plates.

Analysis to determine a feasible plate weight based on USMC threat requirements was quickly performed using the HATS tool. The results of the analysis informed the threshold and objective areal density requirements of the Lightweight Plate Performance Specification. A solicitation was published to which several vendors submitted proposal packages. Each submission was evaluated individually and independently for ballistic and non-ballistic conformance to the requirements by a Source Selection Evaluation Board. The Protective Group, a Point Blank Company, was awarded the contract to produce the USMC Lightweight Plate [10].

4. Conclusion

ESAPI plate system is capable of stopping severe threats typically not found on the battlefield and, as a result, is much heavier than needed for survivability optimization. A lighter armour system, capable of stopping more common and less severe battlefield threats, significantly improves Warfighter mobility and survivability. Without extensive knowledge of the existing armour plate data from a large vendor dataset, determination of the optimal armour weight and ballistic performance is time consuming and expensive.

Using the Hard Armour Trade Space analysis system, armour system performance and weight were tailored for a balance between ballistic performance and mobility to optimize overall survivability for a specific mission and operation theatre. This system has taken armour system physical parameter information and ballistic performance data and transformed it into a more intrinsic form of energy-per-unit-mass or the inverse. This allows for evaluation of plate performance against threats for which it was not ballistically tested and enables determination of a potential best performing plate.

The resulting lightweight plate solicitation by the USMC is one example of armour optimization, but each specific mission could have different requirements and armour optimal weight. The HATS system could find use by other organizations for each of their mission requirements in the future.

Acknowledgments

Funding for this work was provided by the United States Marine Corps System Command PM ICE.

References

[1] Scharre P., Fish L., A Strategy for Enhancing Warfighter Survivability, Center for a New American Security, Washington, DC, 2018, pp. 2.
[2] Orr R., Soldier Load Carriage: A Risk Management Approach, The University of Queensland, St. Lucia, Australia, 2012, pp. 77.
[3] Fish L., Scharre P., The Soldier's Heavy Load, Center for a New American Security, Washington, DC, 2018, pp.15.
[4] Thompson C., Paying for weight in blood: An analysis of weight and protection level of a combat load during tactical operations, Naval Postgraduate School, Monterey, CA, 2019, pp. 1-81.
[5] Pierce N, Marine Corps System Command public release document, Mobility trial comparison of ESAPI to Light-Weight Plates, US Marine Corps Systems Command, Quantico, VA, 2018
[6] Wang B., Lu G., On the optimisation of two-component plates against ballistic impact, J Mats Process. Tech., 1996; 57; 141-145.
[7] Florence A., Aherns T., Interaction of Projectiles and Composite Armor, Stanford Research Institute, Menlo Park, CA, 1967, pp. 1-125.
[8] Florence A., Interaction of Projectiles and Composite Armor: Part II, Stanford Research Institute, Menlo Park, CA,1969, pp. 1-64.
[9] FileMaker Pro, Claris International, an Apple Subsidiary
[10]https://www.pointblankenterprises.com/news/Awarded-Lightweight-Body-Armor-Insert-MARCORSYSCOM-Contract.html, Point Blank Enterprises Awarded Lightweight Body Armor Insert Contract by the United States Marine Corps Systems Command (MARCORSYSCOM), 2019

https://doi.org/10.52202/078352-0022

Helmet Blast Attenuation Performance

V. Alphonse[1], Q. Luong[1], M. Tumperi[1], C. Schuman[1], S. Herman[1], J. Clark[1], and M. Maffeo[2]

[1]The Johns Hopkins University Applied Physics Laboratory, 11100 Johns Hopkins Road, Laurel, MD, 20723, USA
vanessa.alphonse@jhuapl.edu
[2]The U.S. Army Combat Capabilities Development Command Soldier Center, General Green Avenue, Natick, MA, 01760, USA

Abstract. The Blast Overpressure Simulation System (BOSS), an advanced blast simulator, and the Human Surrogate Headform Model (HSHM), an instrumented test device, developed at the Johns Hopkins University Applied Physics Laboratory were previously used to establish requirements for a standardized blast test methodology for headborne protective equipment. This methodology was used herein to test and evaluate the blast attenuating performance of eight currently fielded and historic helmet systems. Specifically, tests were designed to examine the effects of helmet geometry, material, and suspension system (e.g. pad vs. sling). The HSHM was mounted on a Hybrid III neck, outfitted with each helmet system, and exposed to a blast overpressure pulse inside the BOSS using a face-on configuration. Overpressure was measured along the walls of the BOSS as well as on the surface of the headform at 18 locations. Data from an accelerometer and angular rate sensor inside the headform, combined with high speed video of each event, was recorded to quantify the kinematic response of the head/neck. Characteristic shock wave metrics including peak overpressure magnitude, positive phase duration, and positive phase impulse were calculated for each sensor location for every test, and used for a comparative analysis across helmet systems. Of the helmet variables examined, changing the suspension system resulted in the greatest differences in response. Helmet geometry was also a large driver in recorded differences, while helmet shell material showed the weakest contribution to differentiating between helmet systems. This test series demonstrated the ability of the previously-developed test methodology to characterize and distinguish between helmets when considering differences in geometry, material, and suspension system. Recommended future work to examine alternate orientations will elucidate a greater understanding and sensitivity of these effects on overall helmet blast attenuating performance.

1. INTRODUCTION

Currently, there is no standard test method or criteria for evaluating the blast attenuating performance of headborne personal protective equipment (PPE). Prior work conducted by the Johns Hopkins University Applied Physics Laboratory (JHU/APL) and the U.S. Army Combat Capabilities Development Command (CCDC) Soldier Center resulted in the development of a robust laboratory test methodology for evaluating headborne PPE in blast overpressure conditions[1]. The primary focus of the prior work was developing and validating highly repeatable laboratory equipment (e.g. shock tube and instrumented headform) and analytical methods to induce a planar shock wave and assess the kinetic and kinematic response of the head. The purpose of the current study is to use this methodology to assess the blast attenuating performance and examine the effects of various design elements of 8 combat helmet systems that had a variety of shell geometries, shell materials, and suspension system types.

2. METHODS

All blast tests were conducted using the Blast Overpressure Simulation System (BOSS) (Figure 1) located at JHU/APL and the Basic Human Surrogate Headform (HSHM/B) (Figure 2), as described in [1]. Briefly, the headform was mounted on a Hybrid III neck, placed within the test region of the 91 cm x 91 cm BOSS, and exposed to a shock wave created using an acetate plastic membrane and compressed air gas. The HSHM/B has 18 surface pressure measurement locations as well as an internal triaxial accelerometer and angular rate sensor.

https://doi.org/10.52202/078352-0023

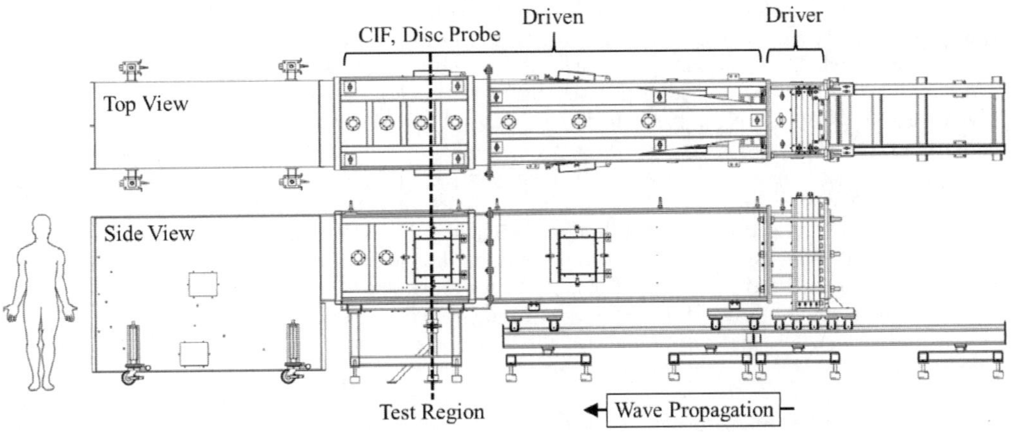

Figure 1. Blast Overpressure Simulation System (BOSS). Note that the CIF and Disc Probe sensors align with the dotted line and are located inside the BOSS; on the wall and in the flow, respectively.

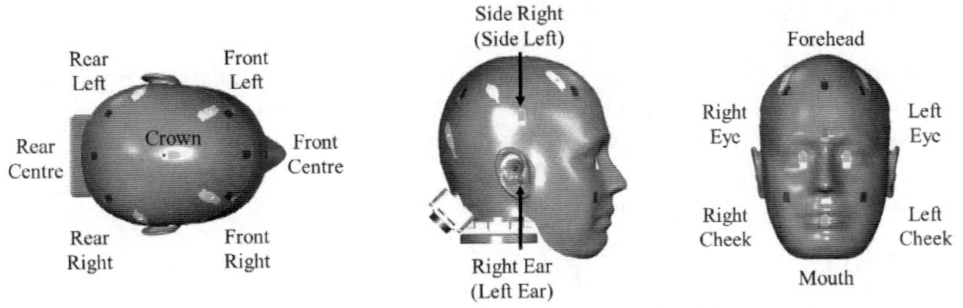

Figure 2. Basic Human Surrogate Headform Model (HSHM/B) with each sensor location labelled.

All tests in the current study were conducted with a forward-facing orientation to simulate an eye-level blast; however a subset of the helmets tested had previously been evaluated in 3 other orientations (side, rear, and inclined). Table 1 shows the test matrix. Each helmet system was tested at least 10 times.

Table 1. Test matrix of eight helmet systems.

Identifier	Shell Geometry	Shell Material	Suspension System	Fielded Status	Sample Size
No Helmet	-	-	-	-	24
A1P	A	1	Pad	Fielded	11
A1S	A	1	Sling	No*	10
B1P	B	1	Pad	No*	10
B1S	B	1	Sling	Historic	10, 3**
B2S	B	2	Sling	No	10, 3**
C3S	C	3	Sling	Historic	10
D1H	D	1	Honeycomb	Historic	10
E4P	E	4	Pad	Fielded	14

*These two configurations are not outfitted as fielded; rather, the fielded suspension systems were swapped between the shell geometries specifically to examine the effect of suspension system in the alternate shell geometry.
**The additional three tests were conducted with an updated fitting protocol for the helmet on the headform.

This study assessed the performance of eight helmet configurations consisting of combat helmets from multiple countries and from multiple generations. There were 5 unique helmet shell designs; geometry A corresponded to a tactical cut, B corresponded to a full cut, C corresponded to a historic cut,

https://doi.org/10.52202/078352-0023

D corresponded to a single-size full cut, and E corresponded to a high cut helmet. Geometries B, C, and D have a brim on the front of the helmet. Geometry E has a high cut around the ear; all other geometries cover the ear. Helmet shells were constructed of 1) aramid, 2) sheet moulded compound (SMC) consisting of fiberglass-reinforced thermoset plastic, 3) steel, and 4) ultra-high molecular weight polyethylene (UHMWPE). Note, the SMC helmet was not a fielded helmet design, but included in this study to assess the effects of material for a given shell shape and suspension system combinations. Suspension systems ranged from polyurethane foam pads (P) to fabric slings (S) and plastic honeycomb (H) designs. Non-standard suspension system configurations were tested to assess the effect of suspension system for a given shell geometry and material. A1S represents a tactical cut helmet retrofitted with a sling suspension system from a full cut helmet, and B1P represents a full cut helmet retrofitted with a pad suspension system from a tactical cut helmet. Table 2 shows midsagittal view (via computed tomography) of each of the helmets outfitting on the headform.

3. RESULTS

The metric of peak overpressure showed the greatest sensitivity to changes in helmet systems. There was less variation in positive phase duration and positive phase impulse compared to peak overpressure measurements. While positive phase duration and positive phase impulse were analysed, the following presentation of results focused on pressure versus time data traces and peak overpressure measurements.

3.1 Evaluation of Input Conditions

It is of utmost importance that any test methodology intended to evaluate performance is repeatable. Examining the sensor responses of the BOSS sensors (downstream and around the test object) is one way to examine the repeatability and reproducibility of the input conditions over a test series.

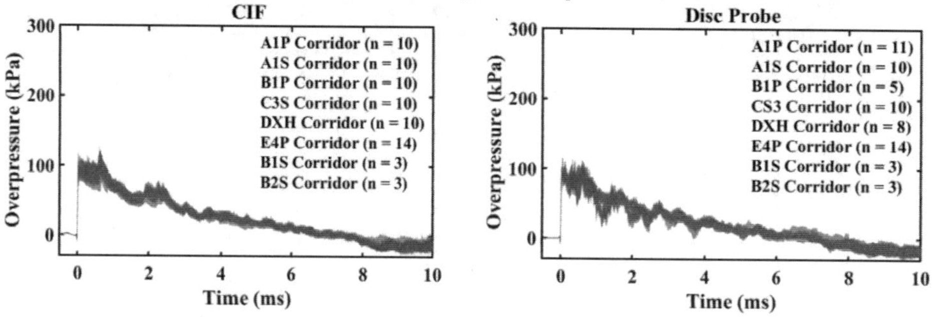

Figure 3 shows the response corridors for the Disc Probe and the CIF pressure sensor locations within the BOSS (Figure 1). The Disc Probe measures static overpressure and is located alongside the headform in the fluid flow. The CIF pressure sensor also measures static overpressure, but is located flush with the inner top surface of the BOSS. These two sensors have been selected to assess the repeatability of the system. Other sensors downstream of the headform can also be used, and show a similar result as the Disc Probe and CIF sensors. Note that in these images, all of the corridors, which represent average ± 1 standard deviation of each test configuration, are plotted simultaneously, and that a dark colour indicates an overlapping region. The lack of drastic differences in these corridors indicates repeatability. The tests shown in these and the following figures were randomized and conducted over the course of three months; the tight corridors therefore also indicate reproducibility over time. The figure legends are included to show the number of tests included within each corridor.

https://doi.org/10.52202/078352-0023

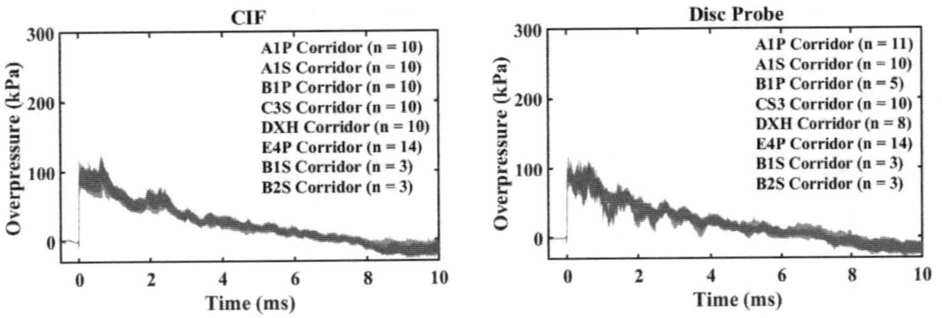

Figure 3. Response corridors at two locations in the BOSS for all test conditions, overlaid.

3.2 Effect of Geometry

In order to isolate and assess the effect of helmet shell geometry on blast attenuating performance, an identical 7-pad suspension system and identical fabric sling suspension system were installed and tested in two different aramid helmet shells, one with a tactical cut and one with a full cut. The effect of geometry was assessed through two comparisons: A1P v. B1P, and A1S v. B1S (A1P and B1S represent fielded configurations). The results can be more broadly interpreted and applied in common helmet designs by evaluating the effect of helmet shell geometry with both pad and sling suspension systems, rather than one or the other. Peak overpressures for the headform sensors are shown in

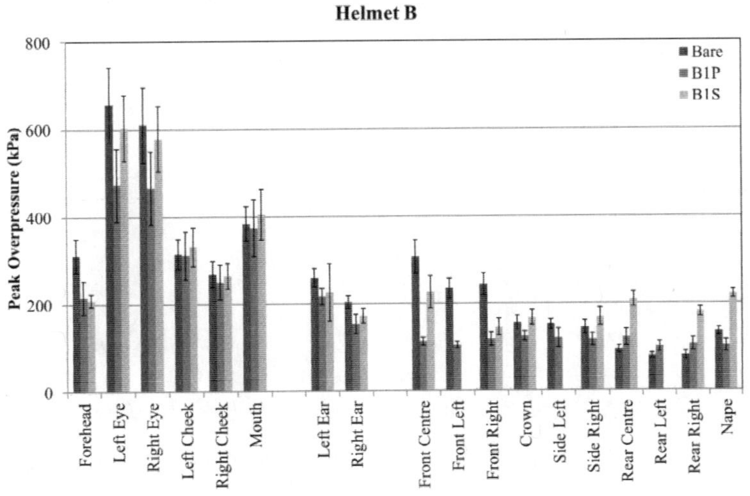

Figure 4, and corridor (average ± 1 standard deviation) responses for select headform sensors are shown in Figure 5. Peak pressures were generally similar for the two helmet geometries, except for at the forehead and eye pressure sensors, where geometry B (full cut, with brim) generally had lower peak pressures that geometry A (tactical cut, no brim). Corridor responses showed some differences in the shape of the pressure versus time response between the two helmet shapes.

3.3 Effect of Suspension System

https://doi.org/10.52202/078352-0023

The effect of suspension system (pad versus sling) was assessed through two comparisons: A1P v. A1S, and B1P v. B1S. Peak overpressures for the headform sensors are shown in

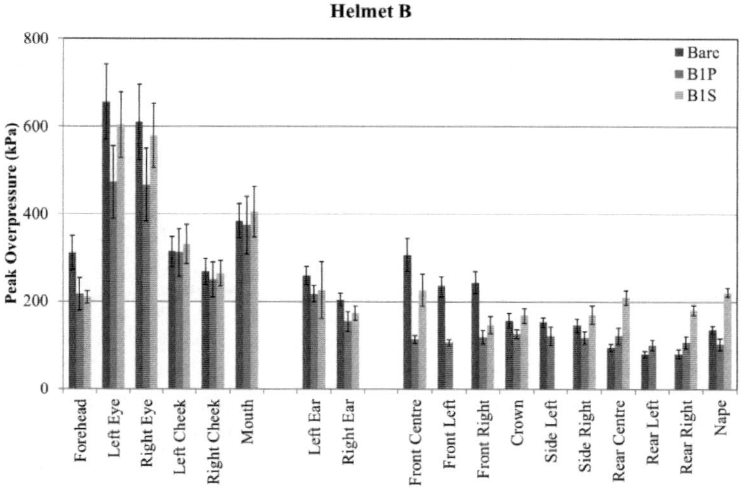

Figure 4.

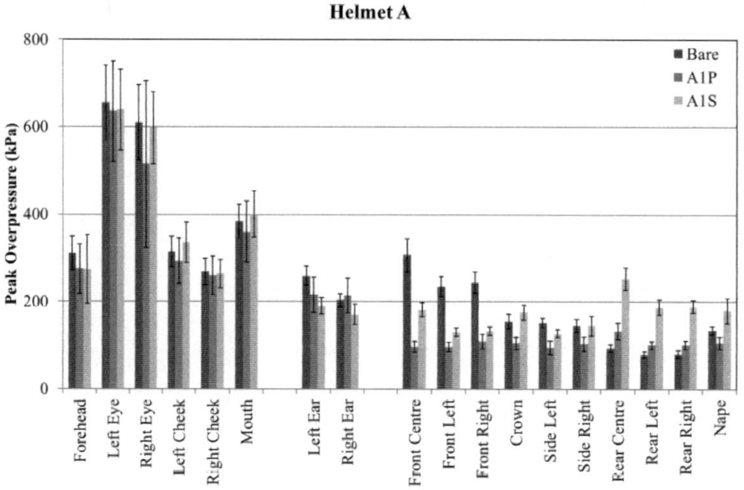

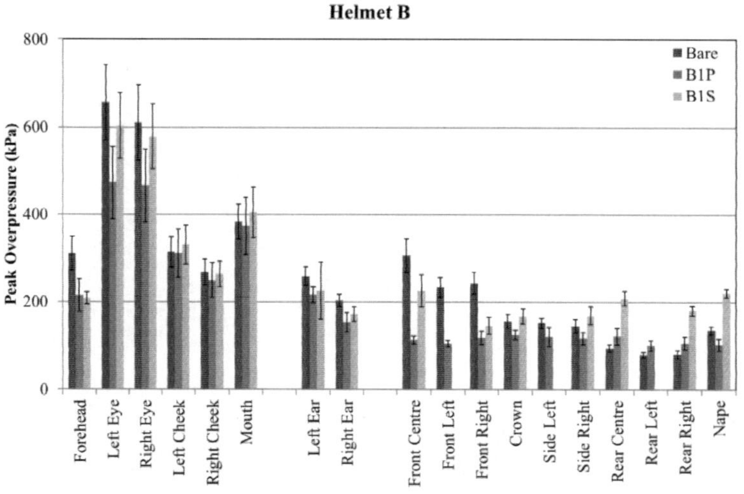

https://doi.org/10.52202/078352-0023

Figure 4. Peak overpressure measured on the headform for two helmet systems outfitted with either a pad (P) or sling (S) suspension system. Sensor locations are grouped by face (left), ear (centre), and underneath helmet (right) in the bar charts.

Corridor (average ± 1 standard deviation) responses for select headform sensors are shown in Figure 5. Figure 5 also shows the effect of helmet fit for the B1S helmet system. When the helmet systems were initially CT-scanned, the B1S and B2S helmets were fitting closer to the head than the other helmets; the suspension systems were adjusted to provide a more consistent fit, and re-tested as such. While little or no difference was observed in the facial and ear pressure sensors between pad and sling tests, pressure sensors underneath the helmet generally showed lower peak pressures for the pad suspension system and corridor responses were characteristically different between pad and sling.

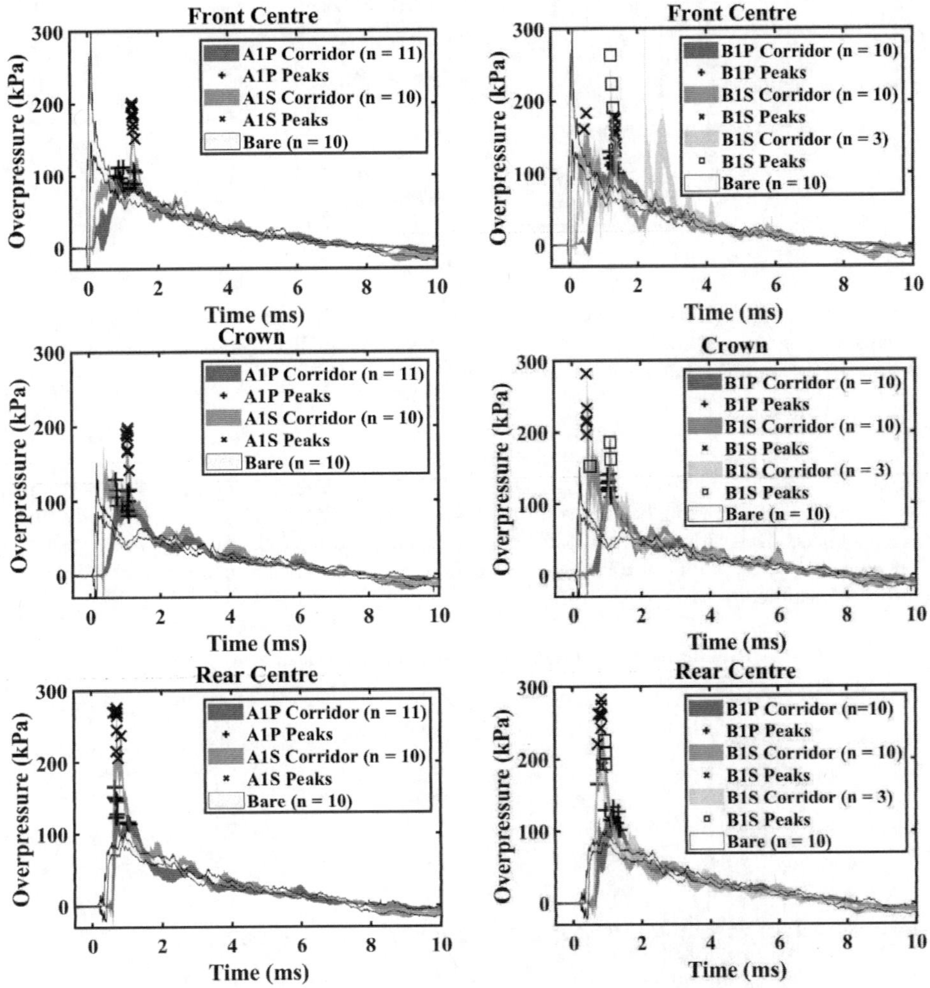

Figure 5. Overpressure corridors for select sensor locations for two helmet systems outfitted with either a pad (P) or sling (S) suspension system. Peak overpressure values for individual tests are marked with a '+', 'x', or '□'. Bare headform corridors included for comparison.

3.4 Effect of Material

The effect of material was assessed through one comparison: B1S v. B2S. Figure 6 illustrates the effect of material for a select number of sensors on the headform. Some slight differences can be observed between the B1S (Aramid) and B2S (plastic) helmets.

https://doi.org/10.52202/078352-0023

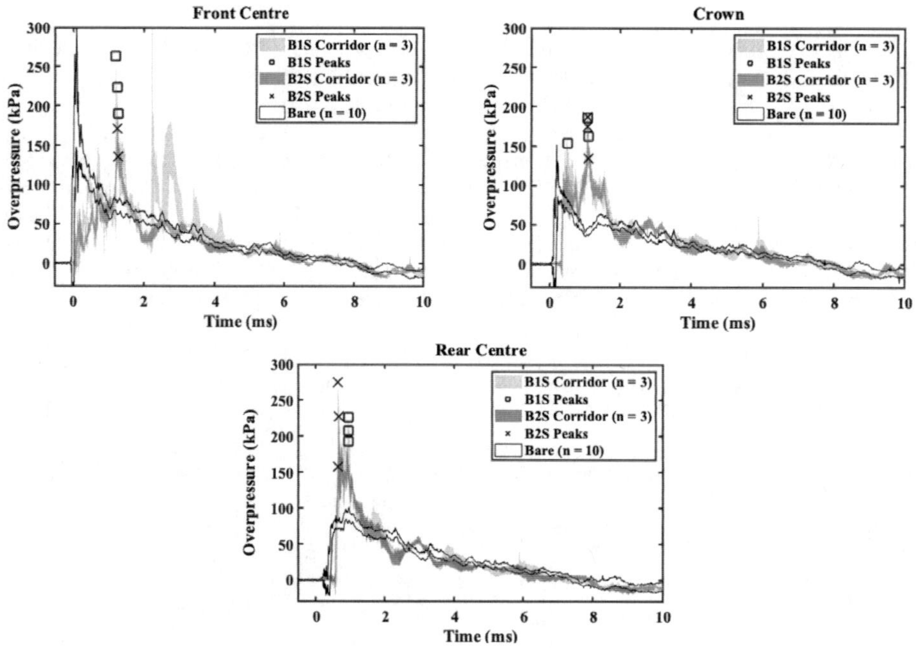

Figure 6. Overpressure corridors for select sensor locations for two helmet systems that have the same geometry and suspension system. Peak overpressure values for individual tests are marked with an 'x' or '□'. (Left) updated helmet fit; (right) original helmet fit.

3.5 Fielded v. Historic

The currently fielded helmets were compared to the historic helmets. Figure 7 shows the peak overpressures measured across the headform for the historic (B1S, C3S, DXH) versus currently fielded (A1P, E4P) helmets. Data from a bare headform are also included for reference. Figure 8 shows the corridor responses of the Front Centre Pad sensor location. Some differences can be observed between the two groups of helmets.

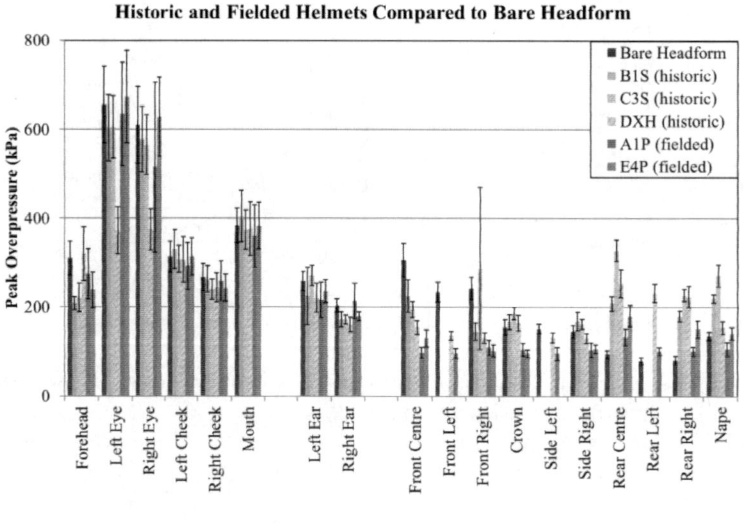

Figure 7. Peak overpressure measured on the headform for two historic helmet systems (B1S, C3S, DXH) and three fielded helmet systems (A1P, E4P), compared to a bare headform.

https://doi.org/10.52202/078352-0023

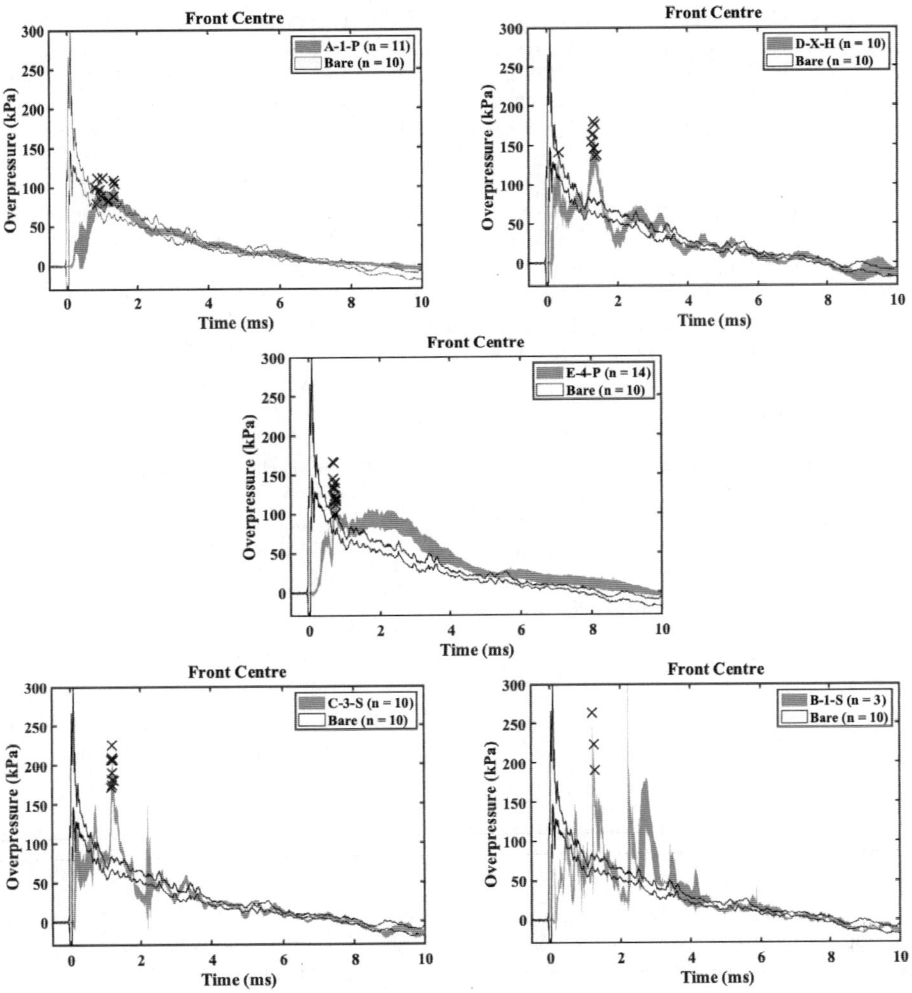

Figure 8. Overpressure corridors for a single sensor location for five historic or fielded helmet systems. Peak overpressure values for individual tests are marked with an 'x'.

4. DISCUSSION AND CONCLUSIONS

4.1 Effects of Geometry, Material, and Suspension System

The suspension system used showed the greatest difference across the helmets tested. A pad suspension system utilizes a series of foam pads that can be attached to the inside the helmet shell (often with Velcro); a sling suspension system utilizes a separate inner lining that suspends the helmet shell above the wearer's head. Either suspension system allows for standoff between the wearer's head and the helmet shell. Where the pad system directly contacts the wearer's head, a sling suspension allows air to pass through the space between the head and the helmet shell. This air gap in the standoff space presents an opening for the shock wave to propagate, and could be a factor in generally greater peak pressure readings observed for the sling versus pad suspension system.

Although there are manufacturer suggestions for the location of the pads within the helmet, an individual may choose to place the pads in slightly different locations. Similarly, a wearer can adjust the straps of the sling suspension system to provide an individualized fit beyond the manufacturer's suggestions. All of the helmets tested in this series followed manufacturer suggestions for the pad or sling suspension systems based on the size of the HSHM/B headform.

https://doi.org/10.52202/078352-0023

Generally similar trends exist across Helmet A (tactical cut) and Helmet B (full cut) compared to the bare headform; however, Helmet B shows a marked difference at the Forehead and Eye sensors compared to Helmet A, likely due to the addition of the brim on the front of Helmet B. Local geometry of the helmet and its proximity to facial structures have shown a variety of results. For example, the addition of a brim on the front of the helmet, compared to a brimless helmet, showed a marked difference (decrease) in the pressure measured at the eye.

While helmet shell materials can affect pressure propagation, the differences observed in this study between Aramid and SMC were small compared to the effects of suspension system and geometry. It is possible other helmet shell materials, including layering of various materials with significant impedance mismatch, could potentially have a greater effect on reducing pressure propagation than measured herein. Future work to specifically examine this could prove or disprove this, as the results of the current study show minimal differences due to helmet shell material.

4.2 Fielded v. Historic

Historic and currently fielded helmets showed different responses; however the differences vary by sensor location and are not consistent for all currently fielded versus all historic helmets. While assessing the evolution of combat helmets and their blast attenuating performance may be interesting, evaluating individual helmet designs on their merits may be more beneficial to informing helmet improvements.

4.3 Evaluation of the Test Methodology

The fact that various helmets illustrated markedly different responses, even in a single orientation, indicates that the test methodology used in this study was able to distinguish between different helmets, especially with respect to peak overpressure measurement. Additionally, the test methodology has proven to be robust, repeatable, and reproducible over several years of operation and over a thousand tests. The shock wave profile generated in the BOSS has been shown to be operationally relevant [2]. While this study was limited to testing in a single orientation, forward facing the blast, the system is capable of and has been used extensively to test other operationally relevant orientations such as rearward, side, and inclined [1].

Prior work developed a formal methodology detailing the required test equipment as well as specific step-by-step instructions to conduct and analyse tests. This methodology was used for the tests described herein and could serve as the basis for a future standardized helmet blast test protocol. In order to implement this methodology more globally, slight modifications or adaptations of the methodology would be needed to address differences between individual shock tubes and Advanced Blast Simulators (like the BOSS); however, the specific test equipment used to conduct the tests herein (e.g., the HSHM/B) could be used for a direct comparison of performance across specific test equipment to ensure matched results. Once matched results are confirmed, equipment-specific modifications or adaptations can be included in the standardized methodology to account for testing at different facilities.

4.3.1 Standardising Helmet Fit in Methodology

While testing helmets with different shapes and suspension systems, the need for standardising the specific placement of the helmet on the headform became apparent. Initial tests conducted to develop the methodology used helmet systems that were relatively similar in shape [1]. However; evolving helmet designs may lead to drastically different shell and suspension system designs, complicating the direct comparison across different helmets.

While external measurements of helmet symmetry and brim height with respect to landmarks on the headform are typically used during helmet testing, this study utilized an additional tool to assess and help properly adjust helmet fit. Computed tomography (CT) scans were taken of the headform outfitted with each helmet as it would be placed during testing (Table 2). While the standoff distance between the headform and the shell of the helmet is fairly fixed for helmets with pad suspension systems, for the helmets with sling suspension systems, the standoff could easily be adjusted and significantly altered by tightening or loosening the straps and/or bands. In addition to informing helmet fitting procedures, the CT scan data provided insight into factors, such as standoff and locations of pads, straps, and other materials near each headform pressure sensor, that may have affected the test results.

https://doi.org/10.52202/078352-0023

Table 2. Midsagittal slice from CT scans of the helmets outfitted on the HSHM/B. All tests were conducted using a Hybrid III neck; CT scans utilized a shorter neck that fit within the bore of the CT.

A1P	A1S
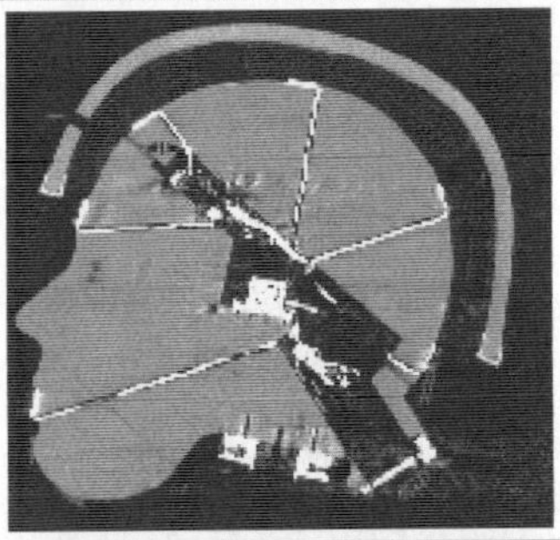	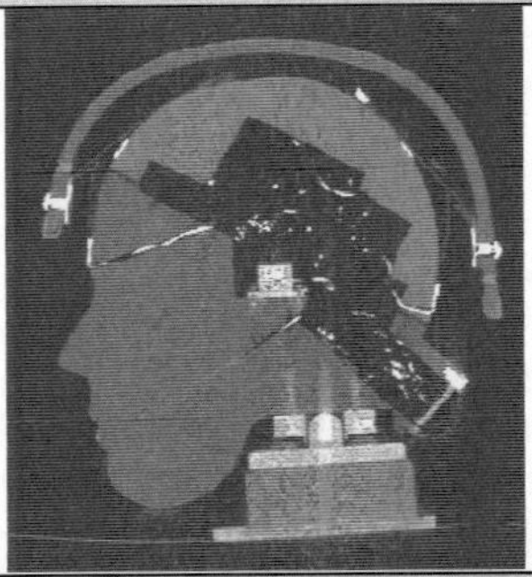
B1P	**B1S**
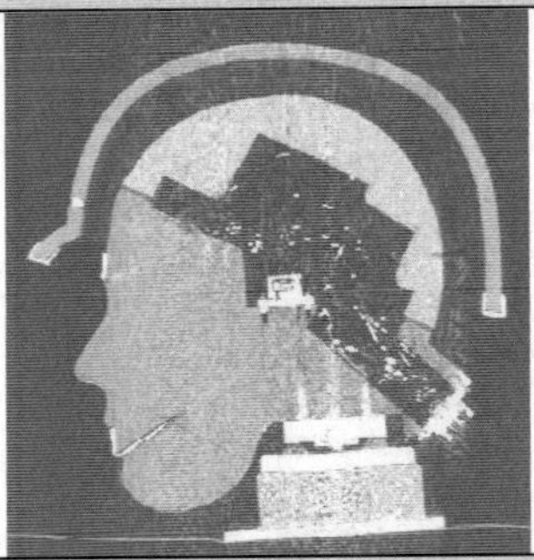	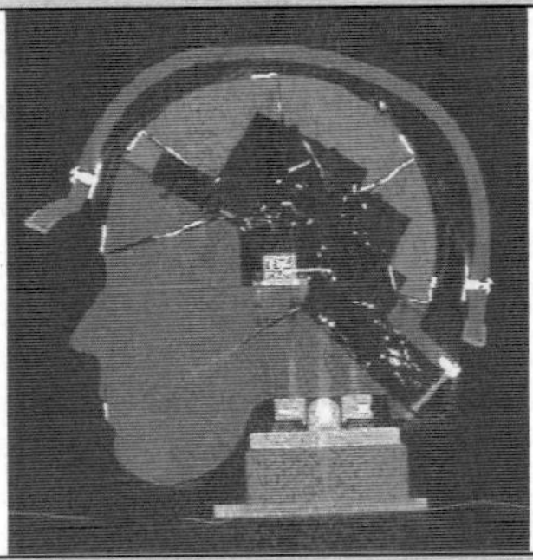
D1H	**E4P**
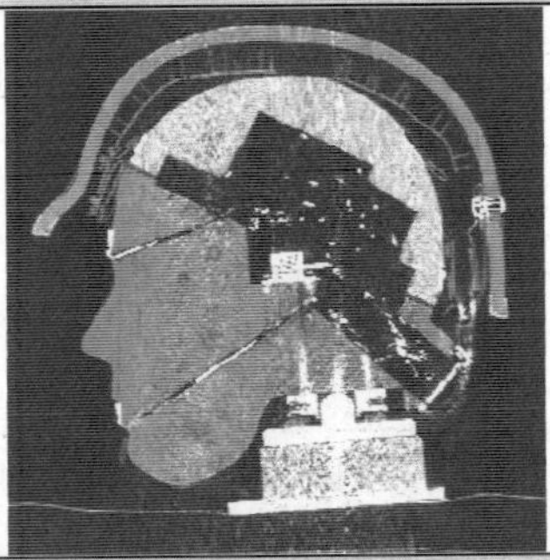	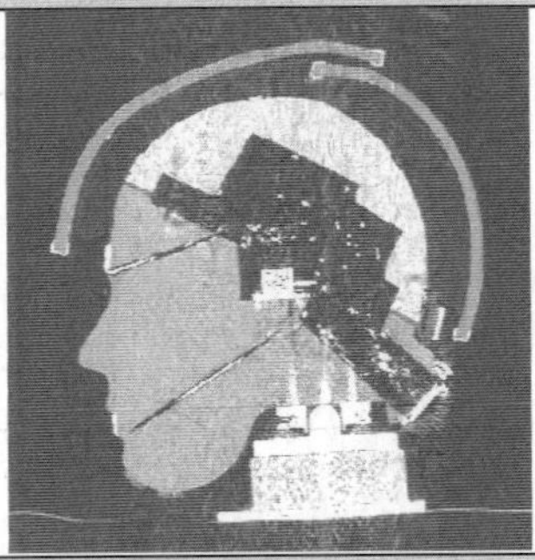
B2S	**C3S**
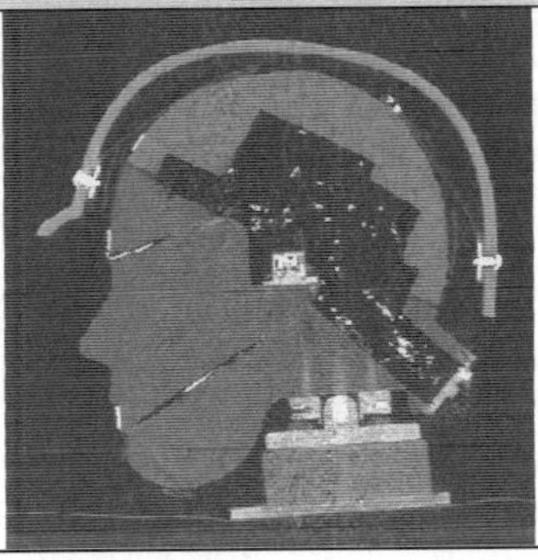	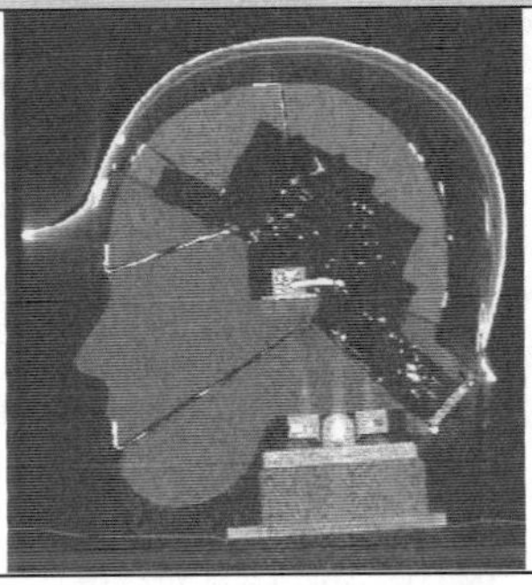

4.4 Limitations and Future Work

The current study demonstrates a test methodology that is able to discriminate differences in helmet design (geometry, material, suspension system), especially in the magnitude and time of peak overpressure measured at various locations around the headform. Depending on sensor location, the

https://doi.org/10.52202/078352-0023

addition of the helmet could increase or decrease the peak overpressure magnitude. Due to the current limitations of injury research, the utility of the current work is limited to comparative analysis. Future advancements to establish injury correlations would enable the current methodology to be used for an injury-based blast helmet requirement. Although lower pressure may be 'better' than higher pressure, there are no quantitative metrics at this time that can be used to identify injury risk. Similarly, the magnitude of the increase or decrease in pressure observed in the sensors in the current study may not be great enough to distinguish between the sensitivity of a protective system to actually 'protect'. Further work to connect field, laboratory, and clinical outcomes is needed to make this further assessment.

The HSHM/B used for this methodology included 18 surface pressure measurement locations. As the design of helmet systems evolves over time, the equipment used to test, evaluate, and compare headborne protective equipment should enable a fair assessment for designs that are drastically different. This study illustrated a marked difference in the response between a pad and a sling suspension system at the sensor locations examined. All of the HSHM/B pressure sensors covered by the helmet were located underneath a helmet pad for the A1P and B1P test configuration, except for the Side Left Pad and the Side Right Pad sensors. The location of a pad near or directly over a pressure sensor will affect the response; ensuring modularity of the test equipment used will be critical to providing an un-biased comparison that is less prone to manipulation. Increasing the number, location (e.g., especially between pads or other suspension design components), and modularity of the sensors on a piece of test equipment can be a useful first step to this end [3]. Furthermore, sensors placed inside the headform (i.e., brain) could inform how the shock wave transmits through the skull. JHU/APL developed a headform that includes four intracranial pressure sensors for this purpose. This headform is referred to as the HSHM/E (enhanced), and has a reduced number of sensors on the surface of the headform to account for the additional sensors in the brain. The HSHM/B was used in the current study in order to capture more measurements around the surface of the headform for the variety of helmets tested. Future tests repeated with the HSHM/E will yield additional knowledge of the shock wave transmission.

It should be noted that helmet performance can be drastically affected by the fit or placement of the helmet, as illustrated with the B1S and B2S tests. Although the tests in this series standardized the fit of the helmet on the headform, it is possible that an individual user may make minor modifications beyond the manufacturer's recommendations in order to obtain a more comfortable fit. It is possible these slight changes might affect the performance of the helmet. It is critical, therefore, to examine the effects of minor differences in fit; a computational model may provide valuable insight to minor variations in helmet fit more quickly than experimental testing. Results from a computational sensitivity study could inform a down-selected set of subsequent experimental tests for validation.

Lastly, all of the tests reported in this paper were conducted with a single orientation, facing the blast. Previous work showed notable differences in responses at various orientations, especially for side-on exposures to a blast [1]. A comprehensive comparative evaluation of helmet performance should therefore include tests with multiple orientations in order to elucidate orientation-specific results.

Acknowledgments

The authors would like to thank the United States Army Combat Capabilities Development Command Soldier Center for sponsoring this effort. This material is based upon work supported by the United States Army Combat Capabilities Development Command Soldier Center under NAVAL SEA SYSTEMS COMMAND Contract No. N00024-13-D-6400, Task Order #VKW03. Any opinions, findings and conclusions or recommendations expressed in this material are those of the author(s) and do not necessarily reflect the views of the NAVAL SEA SYSTEMS COMMAND (NAVSEA).

References

[1] Alphonse V, Carneal C, Luong Q, Clark J, Andrist J, Townsend K, Maffeo M, Carboni M, and Cyganik J. Effect of Helmet and Eyewear on Headform Kinematic Response to Primary Blast Overpressure Exposure. *Personal Armour Systems Symposium, 2018.*

[2] Carneal C, Merkle A, Zinn D, Andrist J, Clark J, Alphonse V, Ritzel D, Carboni M, DeCristofano B, Cyganik J, and Maffeo M. Development of a laboratory shock tube system for helmet blast overpressure performance assessment. *Personal Armour Systems Symposium, 2016.*

[3] Bevan M, Clark J, Hrivnak J, Herchek J, Alphonse V, Luong Q, Sedberry K. Common helmet test system for blast, blunt, and ballistic testing. *Personal Armour Systems Symposium, 2020.*

https://doi.org/10.52202/078352-0023

Comparison of Pressure Attenuation Performance of Bomb Suit Designs during Free-Field Blasts using an Advanced Human Surrogate

M. Vignos[1], Q. Luong[1], J. Clark[1], C. Schuman[1], V. Alphonse[1], J. Gipple[1], C. Carneal[1], R. Schott[2], J. Gardner[2], E. Wilson[2], M. Maffeo[3], and M. Zielinski[3]

[1]The Johns Hopkins University Applied Physics Laboratory, 11100 Johns Hopkins Road, Laurel, MD, 20723, USA
mike.vignos@jhuapl.edu
[2]U.S. Army Aberdeen Test Center, 6943 Colleran Rd, Aberdeen Proving Ground, MD, 21005, USA
[3]U.S. Army Combat Capabilities Development Command-Soldier Center, 10 General Greene Avenue, Natick, MA, 01760, USA

Abstract. Assessing blast overpressure attenuation and mechanical response of the body are important for evaluating protection of bomb suits. However, there is limited information and no standard test methodology for overpressure attenuation of existing suits. Additionally, existing literature has primarily used an automotive test device (50th percentile male Hybrid III, or HIII) with added surface pressure sensors, which limits our understanding of pressure experienced by internal organs of interest. Thus, the objective of this study was to investigate overpressure attenuation and the mechanical response of the body for current bomb suits using an advanced human surrogate.

Nine free-field blasts were performed using a 4.5kg spherical C4 charge positioned at mid-sternum height (1.37m) with a 1.83m standoff. The surrogate tested was a combination of a HIII head, a surrogate neck, an advanced human surrogate torso, and HIII legs. The torso was constructed from biosimulant materials representing a skeletal system, organs, flesh, and skin. Pressure sensors were embedded in major organs and on the skin surface. An accelerometer and a custom displacement sensor were mounted to the sternum. For each test, the surrogate was dressed with one of four bomb suit designs and rigidly mounted to a steel fixture at the waist to maintain a front-facing, standing position.

Reference pressures were similar across tests. Sternum acceleration and velocity differed between suits, but sternum compression remained relatively similar. Peak pressures varied across suits, with a significantly higher lung pressure with suit B than suit A. Peak surface pressures were significantly different from internal pressures. These findings suggest the suits tested provide varying levels of protection. However, further work is needed to relate these biomechanical metrics to risk of injury. This study supports previous results showing that internal pressure differs from surface pressure, indicating the benefits of using advanced surrogates in assessing bomb suit performance.

1. INTRODUCTION

Design of bomb suits to support an explosive ordnance disposal (EOD) technician during improvised explosive device (IED) defeat missions requires a complex balance of protection and performance. The bomb suit must protect against the imminent threats of blast overpressure, fragmentation, and ballistic impacts, while also allowing the EOD technician the range of motion, visibility, and dexterity needed to complete the mission. This complex design space has led to multiple test methodologies [1-3] and associated standards to verify that bomb suits meet the needs of EOD technicians. However, an official test methodology for blast overpressure attenuation performance of bomb suits does not exist.

One challenge in developing standards for assessing bomb suit blast attenuation performance is linking test methodologies to risk of primary blast-induced injuries. Current test methodologies commonly use an automotive anthropometric test device (i.e. a 50th percentile male Hybrid III) that has pressure sensors added to its outer surface to assess blast attenuation performance of a bomb suit [1-3]. While this provides a reasonable assessment for a relative comparison across bomb suit designs, there is limited data correlating this test methodology to risk of primary blast injuries. Additionally, this modified Hybrid III only provides an assessment of pressure experienced at the surface of the test device. This limits our understanding of the pressure experienced by internal organs of interest, such as the lungs, which has previously been associated with a degree of pulmonary contusion in an ovine model [4].

In previous work, an advanced surrogate system of the human torso, referred to as the Human Surrogate Torso Model (HSTM), was developed to assess blast overpressure attenuation performance of personal protective equipment [5,6]. This surrogate system was designed to measure a range of biomechanical metrics that are potentially linked to risk of blast-induced injury, including surface torso pressure, internal organ pressure, and skeletal kinematics. In working towards a robust standard test

https://doi.org/10.52202/078352-0024

methodology, an advanced surrogate system that incorporates a wider array of sensing modalities may provide a more complete understanding of blast attenuation performance of bomb suits. Thus, the objectives of this study were (1) to evaluate the feasibility of using an advanced surrogate system to assess blast overpressure attenuation of bomb suits and (2) to perform a baseline assessment of the blast attenuation performance of current bomb suit designs with this advanced surrogate system.

2. METHODS

2.1 Advanced Blast Surrogate System

The Human Surrogate Torso Model (HSTM) was developed in prior efforts as a physical test device to assess the response of the human torso to high-rate (i.e. blast and ballistic) loading with varying personal protective equipment configurations [5,6]. The HSTM is representative of the human torso's form factor, structure, and material response, and enables repeatable dynamic measurements both on the surface and inside the torso (Figure 1). The HSTM is constructed with biosimulant materials representing a skeletal structure, major organs, mediastinum, flesh, and skin. Previous versions of this surrogate were application specific and, thus, the organs represented and the types of sensors used within the HSTM differed depending on the injuries of greatest concern. The organs represented in this version of the HSTM consisted of the left and right lungs, heart, liver, stomach, and intestinal mass. The heart, liver, stomach, and intestines are fabricated from the same silicone-based material, which was previously selected based on matched-pair testing to human tissues [7-10]. For the lungs, glass microspheres were spun cast into the silicone-based material to reduce its density and bulk modulus to better match an air-filled human lung than the standard silicone material [7,9]. Pressure sensors were embedded in the left lung and heart (EPIH, TE Connectivity, Schaffhausen, Switzerland), as well as on the skin surface (Model F, Honeywell, Charlotte, North Carolina). The placement of the four skin surface sensors was selected to be comparable to the sensor placement of a Blast Test Device (i.e. front, left, right, and back), a system that is commonly used to assess reference pressure during live-fire blast events. An accelerometer (7270A, Endevco, Sunnyvale, California) and a custom displacement sensor were mounted to the sternum.

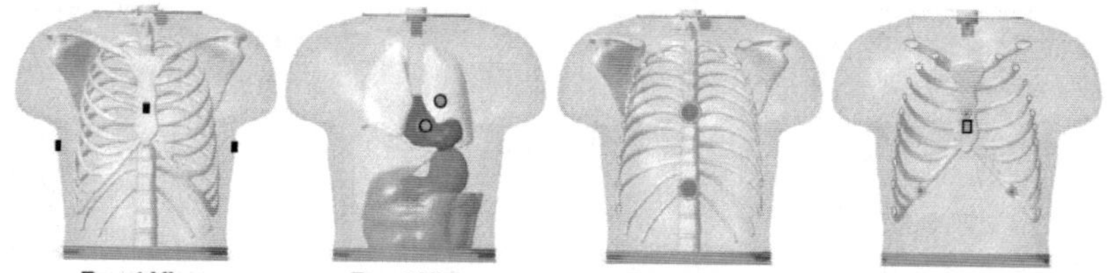

Figure 1. Sensing modalities within the HSTM consist of: (1) four external surface pressure sensors, (2) two internal pressures sensors embedded in the left lung and heart, (3) two EMF emitter coils mounted on the spine, (4) one accelerometer on the sternum and three EMF receiver coils with one mounted on the sternum and two mounted on the junction of left and right ribs 7 and 8. Together, the EMF emitter and receiver coils allow for measurement of the displacement of the skeletal system.

Previous tests utilised the HSTM independent of other components of the body to study the isolated response of the torso to blast overpressure. However, to allow for proper fit of the bomb suits tested in this study, a custom neck was designed and built to attach a Hybrid III head to the HSTM. The design of the neck was constrained by the existing HSTM connector, which is located on the top plate of the HSTM. This required the neck to be offset from the top of the HSTM to allow room for the connector. The neck component consisted of three layers, including an aluminium baseplate to attach to the top of the HSTM, a layer of silicone, and an aluminium top plate to attach to the Hybrid III head (Figure 2). This neck was then mounted to the top of the HSTM using aluminium standoffs. This allowed for physiological positioning of the Hybrid III head relative to the HSTM and proper fit of the tested bomb suit helmets.

In order to allow the bomb suits to fit on the surrogate system as specified by the manufacturer, the baseplate of the HSTM was also modified to allow for attachment to a standing Hybrid III pelvis and legs. Custom components were designed and fabricated to attach to currently existing threaded mounts

https://doi.org/10.52202/078352-0024

within the Hybrid III pelvis. This combined surrogate system was then rigidly mounted at the waist to a steel blast test stand with the HSTM mounted to the top of the stand and the Hybrid III pelvis mounted to the bottom (Figure 3). Use of the blast test stand with the HSTM and Hybrid III pelvis allowed for the bomb suit trousers to be worn as specified by the manufacturer, with the straps of the trousers tightly fitting over the shoulders of the HSTM. Additionally, this test stand allowed the surrogate system to maintain a front-facing, standing posture during tests.

Figure 2. Custom neck used to connect the Hybrid III head to the HSTM. This silicone neck was mounted to the top of the HSTM using standoff mounts to allow for room for the sensor connector.

Figure 3. Surrogate system mounted to the steel blast test stand. The stand included a plate to mount the HSTM and the Hybrid III pelvis and a head support to prevent excessive torso hyperextension.

2.2 Bomb Suit Tests

The bomb suit test series consisted of 2 calibration tests followed by 9 assessment tests. The calibration tests were used to determine the correct standoff for the assessment tests. Prior to the calibration tests, it was determined that a 4.54 kg spherical charge of C4 suspended from 1.37 m would be used for each blast test. The charge height was selected to prevent the Mach stem from passing through the HSTM. The charge weight was optimised based on the predicted blast overpressure and estimated size of the fireball. The Conventional Weapons Effects (ConWep) software [11] was used to determine that this charge weight would produce a blast overpressure at a 1.52 m standoff that is comparable to the overpressure achieved with the current National Institute of Justice (NIJ) standard for bomb suit blast integrity testing (0.567 kg C4 at 0.6 m standoff) [12]. Thus, the target standoff was set at 1.52 m. However, during the calibration tests it was discovered that the fireball was larger than anticipated. This led the final standoff to be set at 1.83 m, given that this distance resulted in an acceptable blast overpressure and minimised the risk of damage to the surrogate system when protected by the EOD suits.

Four bomb suit designs were used in this test series (referred to as suits A, B, C, and D) with each suit design tested a different number of times (n = 4, 3, 1, and 1, respectively). The number of tests for each design was set based on availability of suits for testing. The primary functional differences across these bomb suit designs were variations in the frontal thoracic layup (Figure 4). Suit D had a frontal thoracic layup that consisted of a hard armour plate over a soft armour package with a thin layer of foam present between the soft armour and the torso. The other three suits (A, B, and C) had a frontal thoracic layup that consisted of at least two rigid ballistic plates separated by thin foam, a soft armour package, and a layer of thick foam between the soft armour and the torso. This layup provided a substantial additional layer of foam between the ballistic protection and the torso, in comparison to Suit D.

 https://doi.org/10.52202/078352-0024

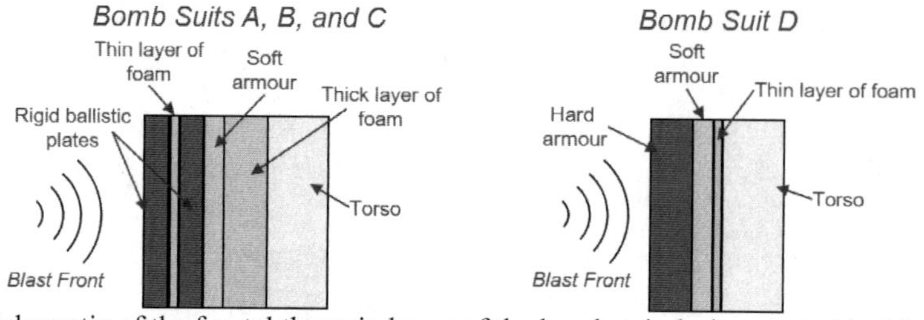

Figure 4. Schematic of the frontal thoracic layup of the bomb suit designs tested in this study. This drawing is not to scale and, thus, does not reflect exact relative thicknesses.

Each blast test included the surrogate system and two reference pressure systems. Reference pressure systems consisted of a pitot-static probe (Model 113A36/061A50, PCB Piezotronics. Depew, NY) and a blast test device (BTD) [13]. The pitot-static probe allowed for measurement of both the stagnation pressure and static overpressure. The BTD included four pressure sensors that were mounted flush to the front, left, right, and rear surfaces and allowed for an assessment of the stagnation and static overpressure that an unprotected individual would experience.

The test devices were positioned such that a reference point on each device was at the charge height of 1.37 m and at a 1.83 m standoff relative to the centre of the spherical charge. Reference points consisted of the front surface pressure sensor on the HSTM and the BTD and the side pressure sensor on the pitot-static probes. To ensure consistent standoff, the distance from the charge suspension line to the reference point on each test device was measured and adjusted as needed prior to each test. Once standoff was achieved, the HSTM was dressed with the bomb suit.

Cables for the test devices was routed from the blast pad into a steel blast shelter, which housed all data acquisition systems. For the HSTM, cables were first run from the surrogate into a set of buffer boards. These buffer boards provided signal conditioning and amplification needed for the sensors within the HSTM. The buffer boards were housed in aluminium boxes that provided blast protection and isolation from vibrations. Cables were then run from the buffer boards into the blast shelter, where it interfaced with the data acquisition systems. Where reasonable, cables were wrapped with Kevlar fabric to provide further protection from fragmentation during tests.

2.3 Data Processing

All data were recorded in the J211 coordinate system [14] and filtered using previously established protocols for free-field blast testing (**Error! Reference source not found.**). HSTM sternum velocity was then computed using numerical integration of the filtered sternum acceleration. Relevant metrics of interest were then computed for the HSTM and the reference pressure systems (**Error! Reference source not found.**). Sternum viscous criterion was computed using sternum displacement and accelerometer-derived velocity (Equation 1, [15]).

$$VC = \frac{y(t)\dot{y}(t)}{D}, \tag{1}$$

where $y(t)$ and $\dot{y}(t)$ are the sternum displacement and velocity, respectively, and D is the chest depth. The peak viscous criterion was then computed for each test. This metric has been shown to relate to the risk of severe chest injury (i.e. Abbreviated Injury Score ≥ 4) during moderate velocity, long duration blunt impacts (e.g. pendulum impacts), with the injury threshold typically set at a viscous criterion of 1.0 m/s [15]. However, this threshold has not been validated for blast induced chest injuries.

Table 1. Sampling frequency and filtering scheme used for each signal collected in this study.

Signal	Sampling Frequency (kHz)	Filtering Scheme
Pitot-static probe Pressure	400	40 kHz low-pass hardware filter
BTD Pressure	400	40 kHz low-pass hardware filter
HSTM Pressure	1000	60 Hz, 20-pole band-stop filter and 10 kHz, 20-pole low-pass filter
HSTM Sternum Displacement	1000	CFC60 filter [14]
HSTM Sternum Acceleration	1000	60 Hz, 20-pole band-stop filter and a 30 Hz high-pass, 10 kHz low-pass, 20-pole band-pass filter

https://doi.org/10.52202/078352-0024

Table 2. Metrics computed from the HSTM response and the reference pressure systems.

Metric	Definition
Peak Pressure	Maximum positive pressure
Start of Pressure Pulse	Time at which the pressure reached 10% of the peak value prior to the peak
End of Pressure Pulse	Time at which the pressure first dropped to 10% of the peak value following the peak
Pulse Duration	Length of time between the peak start and end
Pressure Positive Phase Impulse	Integral of pressure with respect to time from the pulse start to end
Peak Sternum Kinematics (Acceleration, Velocity, and Displacement)	Minimum value of signal, which corresponded with the maximum compressive sternum kinematics

2.4 Statistical Analysis

Peak stagnation pressure, static overpressure, and rear pressure were sorted based on suit design worn by the HSTM for each test. Unpaired t-tests with a Bonferroni-Holm correction for multiple comparisons were then used to assess differences in peak reference pressure experienced by each suit design. For suit designs tested multiple times, an unpaired t-test with a Bonferroni-Holm correction for multiple comparisons was performed for each of the computed metrics (i.e. peaks, pulse durations, pressure positive phase impulses, and viscous criterion) to determine statistically significant differences in blast attenuation performance across bomb suit designs. Significance was set at $p < 0.05$ for all comparisons.

3. RESULTS

3.1 Reference Pressures

Peak stagnation and static overpressure were not statistically different across all tests (Table 3). Additionally, peak reference pressures when separated by suit design worn by HSTM were also comparable (Figure 5). This indicates that the blast pressure dose experienced by each suit was similar and, thus, allowed for a valid comparison across suit designs.

The BTD peak stagnation pressure measured with the front pressure sensor was substantially higher than the peak stagnation pressure measured with the pitot-static probe (Table 4). However, the BTD left and right peak pressures were comparable to the static overpressure measured with the pitot-static probes.

Table 3. Peak reference pressures (mean ± standard deviation) across all nine tests. Table includes peak unfiltered and filtered pressures measured with the pitot-static probe.

	Stagnation Pressure (MPa)	Static Overpressure (MPa)
Unfiltered Pitot-Static Probe	5.66 ± 0.61	1.03 ± 0.090
Filtered Pitot-Static Probe	4.05 ± 0.26	0.862 ± 0.056

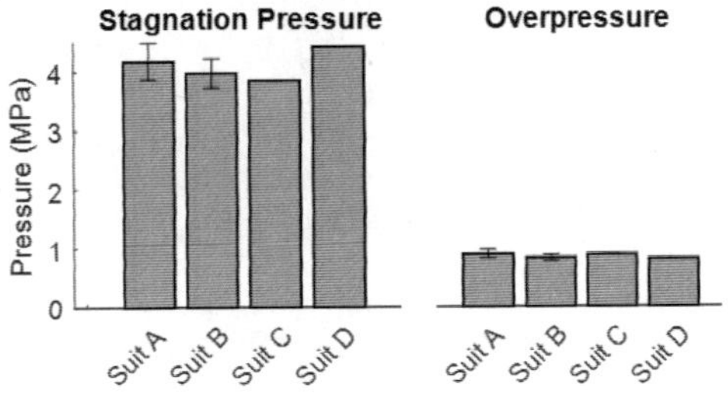

Figure 5. Bar plots show peak stagnation pressure and static overpressure (mean ± standard deviation) measured with the pitot-static probe for each suit. No significant differences were observed across suits, indicating that each suit experienced comparable blast pressure doses across each test.

Table 4. Peak pressure (mean ± standard deviation) measured with the front, left, and right pressure sensors of the BTD across all tests.

	Front (MPa)	Left (MPa)	Right (MPa)
BTD Peak Pressure	5.01 ± 0.60	0.903 ± 0.21	0.814 ± 0.25

https://doi.org/10.52202/078352-0024

3.2. Bomb Suit Performance

The HSTM sternum kinematics differed significantly across bomb suit designs. Sternum accelerations that occurred with suit D were substantially different from the accelerations with suits A, B, and C (Figure 6). Peak sternum accelerations were greater for suit D than suits A, B, and C and significantly greater for suit B than suit A (Figure 7, $p = 0.049$). This relationship remained consistent for sternum velocity with a greater peak velocity observed for suit B than suit A ($p = 0.006$). However, there were no significant differences observed in sternum displacement across bomb suit designs (Figure 6 and Figure 7).

Previously, a sternum viscous criterion threshold of 1.0 m/s has been defined as the cut-off for the risk of severe chest injury. In this study, the peak viscous criterion were well beneath this previously defined threshold for all suit designs (**Error! Reference source not found.**Figure 8). However, there was a significant difference in peak viscous criterion between suit A and B ($p = 0.03$).

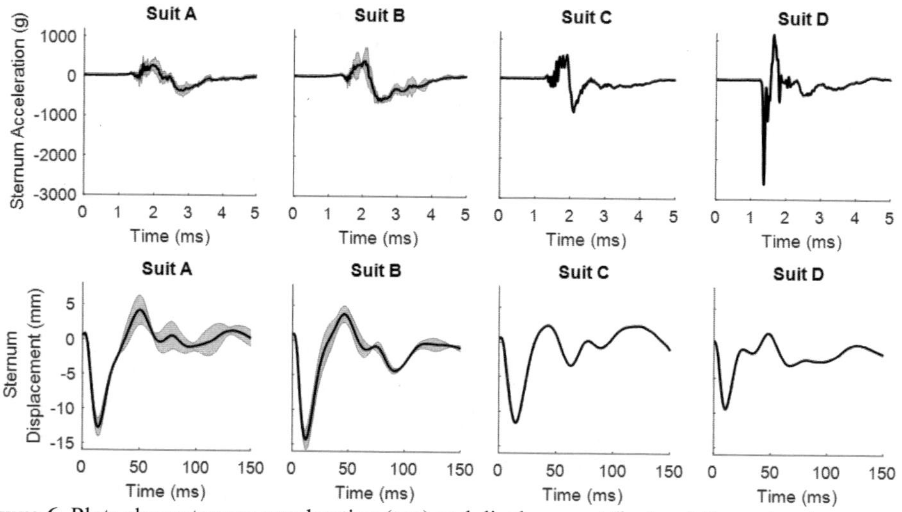

Figure 6. Plots show sternum acceleration (top) and displacement (bottom) for each suit measured across all tests (mean ± standard deviation). For suit designs tested once, the time history shown is for the single test. Negative values indicate the sternum moving towards the spine.

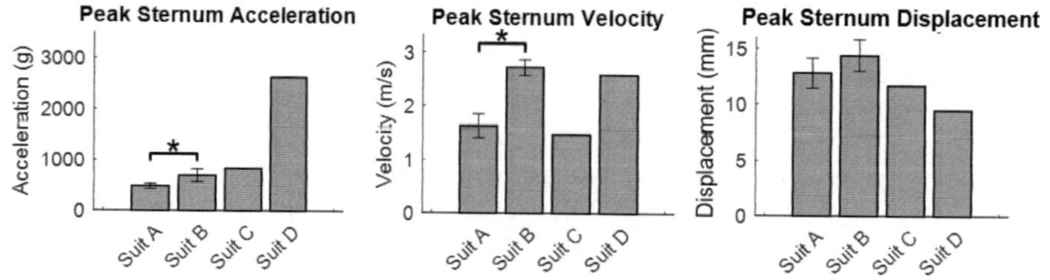

Figure 7. Bar plots show the peak sternum acceleration (left), velocity (middle), and displacement (right) across all tests for each suit design. * indicates a significant difference across suit designs peak.

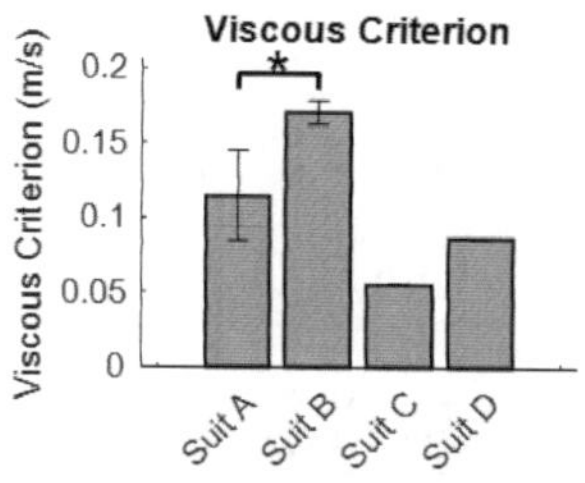

Figure 8. Peak torso viscous criterion (mean ± standard deviation) for each suit design. * indicates a significantly different peak viscous criterion between suit design.

https://doi.org/10.52202/078352-0024

Surface pressures measured with the HSTM also differed substantially across bomb suit designs (**Error! Reference source not found.**). Each of the bomb suit designs resulted in a significant pressure attenuation relative to the unprotected configuration, based on a comparison of the BTD front pressure to the HSTM front surface pressure across each test (**Error! Reference source not found.**). However, this pressure attenuation performance varied across suit designs with a greater peak front pressure for suit D than suits A, B, and C. Additionally, suits C and D exhibited greater peak left surface pressures than suits A and B and suit D exhibited a greater peak right surface pressure than the other suits (**Error! Reference source not found.**).

Lung and heart pressures measured with the HSTM followed a similar trend to the front surface pressure. Greater peak lung pressures occurred with suit D than suits A, B, and C and significantly greater peak lung pressures occurred with suit B than suit A ($p = 0.03$). No significant differences were observed in peak heart pressure. Significant differences were also observed between the surface and internal pressures measured with the HSTM (**Error! Reference source not found.**). Across all tests, the peak lung pressure was significantly greater than the front surface pressure, while the peak heart pressure was significantly lower than the front surface pressure ($p < 0.001$ and $p = 0.007$, respectively).

No significant differences were observed in the pressure positive phase impulse for the front surface, lung, and heart across the bomb suit designs (Figure 12). It is important to note that the threshold for injury based on these metrics has not been defined for an armoured configuration. Therefore, it is unclear how the magnitude of the pressure impulse relates to risk of injury across suit designs.

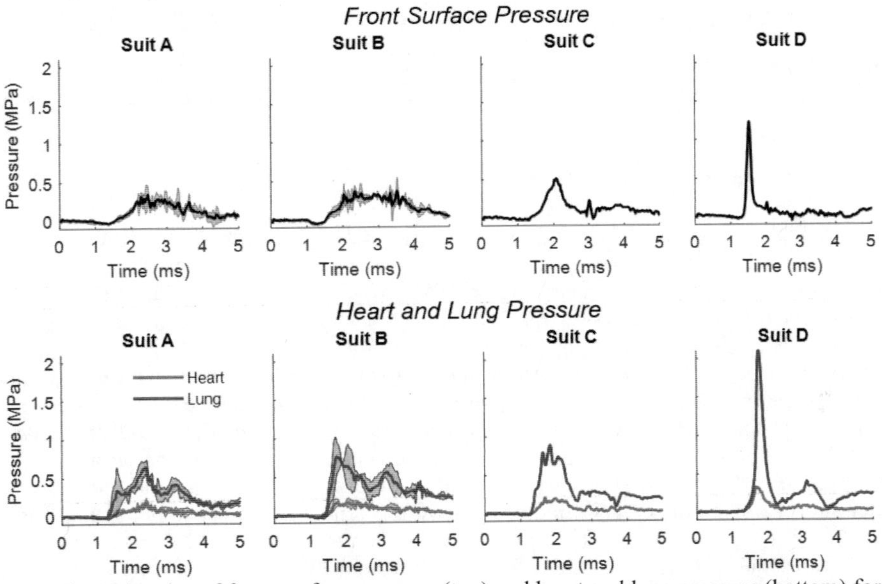

Figure 9. Time histories of front surface pressure (top) and heart and lung pressure (bottom) for each suit measured across all tests (mean ± standard deviation). For suit designs tested once, the time history is shown for the single test.

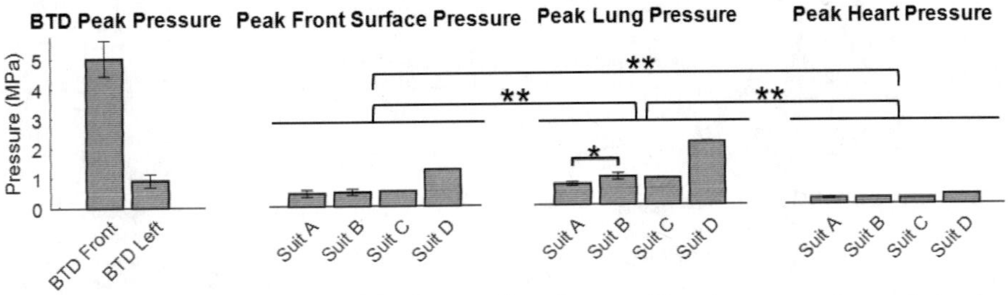

Figure 10. Bar plots show peak BTD, front surface, lung, and heart pressures (mean ± standard deviation) measured across all tests for each suit design. * indicates a significantly different peak metric across suit design. ** indicates a significantly different pressure across locations.

https://doi.org/10.52202/078352-0024

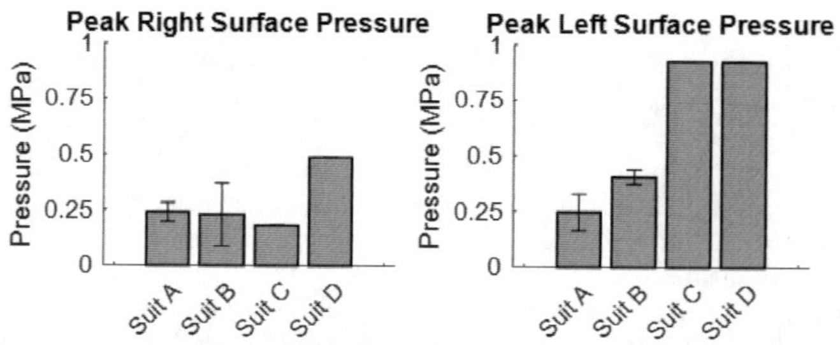

Figure 11. Bar plots show peak right and left surface pressures (mean ± standard deviation) measured across all tests for each suit design. No significant differences were observed for these pressure metrics.

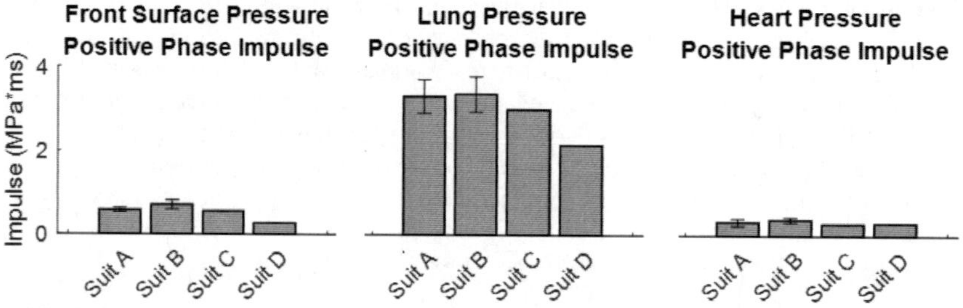

Figure 12. Bar plots show front surface (left), lung (middle), and heart (right) primary pressure impulse (mean ± standard deviation) for each suit design. No significant differences were observed.

4. DISCUSSIONS AND CONCLUSIONS

This study sought to evaluate the feasibility of using an advanced surrogate system to assess blast overpressure attenuation of bomb suits and to perform a baseline assessment of current bomb suit designs. The findings of this study indicate the HSTM provides a repeatable assessment of bomb suit performance, given the relatively low variability measured within a given bomb suit design. This is similar to previous assessments of the repeatability of the HSTM during ballistic impacts [16]. Additionally, this study suggests that the test methodology used can detect relative differences in bomb suit blast attenuation performance. It was observed that suit D exhibited the greatest sternum accelerations, but similar sternum velocities, and displacements when comparing to the other suit designs (**Error! Reference source not found.**). Additionally, suit D resulted in the greatest surface and internal pressures experienced by the HSTM, but also exhibited the shortest peak pressure duration (**Error! Reference source not found.**). These findings suggest a potential difference in protection from blast-induced injuries across the bomb suit designs. However, future work is needed to directly link these metrics to risk of injury.

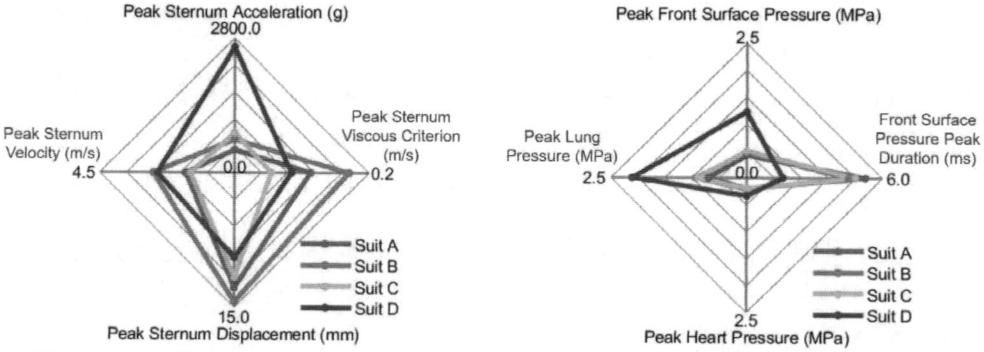

Figure 13. (Left) Plot shows the average peak sternum acceleration, velocity, displacement, and viscous criterion for each suit design. (Right) Plot shows the average peak front surface, lung, and heart pressure and the average front surface pressure duration for each suit design.

The second aim of this study was to assess the use of an advanced blast test surrogate for comparing bomb suit performance. In this test series, the HSTM exhibited excellent repeatability across the same

https://doi.org/10.52202/078352-0024

suit design and good robustness with no sensors damaged. The HSTM also measured a significant difference between surface and internal pressures. This suggests that it may be important to use an advanced surrogate system, with the ability to measure pressure experienced by organs of interest, when assessing blast attenuation performance of bomb suits. The benefit of this blast surrogate was further supported by the differences observed in the measured lung and heart pressures. This finding indicates that the HSTM may be useful in detecting relative differences in injury to the internal organs.

Differences in suit performance were potentially influenced by the frontal thoracic layup of each suit design. As previously discussed, suit D was designed with only a thin layer of foam present between the hard armour plate and the surface of the HSTM. Suits A, B, and C were designed with a much thicker layer of foam between the ballistic protection and the HSTM surface. The results of this study suggest that this additional layer of foam is critical for improving the blast attenuation performance of a given suit design and mitigating the pressure applied to the surface of the body and to internal organs. Previous studies have shown that addition of a foam layer between a hard armour plate and the torso, resulting in an impedance to transmitted blast stress waves, can significantly reduce the risk of lung injury in an animal model [17-20]. The results of this study support these previous findings and provide a first step in identifying metrics that may be useful in detecting lung injury.

As previously mentioned, measurements made with the HSTM in an armoured configuration have not been directly linked to risk of blast-induced injury. However, previous work has compared the peak lung pressures measured with the HSTM when wearing hard armour plates to lung injury scores assessed in an ovine model [4]. In this previous study, blast testing was performed in a relatively small room with rigid walls using both an armoured HSTM and an armoured ovine model. The HSTM and ovine testing did not occur at the same time, but the testing configurations and charge weights were matched between the studies. In this test series, the ovine lung injury scored at an average of 31.9 based on a modified version of the Blast Injury Scoring System [21,22] (i.e. moderate lung injury) and the HSTM measured an average peak lung pressure of 0.625 MPa with a charge weight of 1.25 kg. In the current study, the peak lung pressures measured with the HSTM were greater than 0.625 MPa for each bomb suit design. While this may indicate a risk of lung injury for the bomb suits tested, it is important to note that attempts in this previous study to use the HSTM lung pressure to directly predict lung injury score were inconclusive. This may indicate that peak pressure alone is unable to predict lung injury and that further investigation is needed to determine a better injury prediction metric. Additionally, it is unknown how ovine lung injury scores relate to lung injury risk in humans, which limits the translation of these findings to the EOD technician.

While use of the HSTM and the associated test methodology allowed for a comparison across bomb suit designs, some limitations were noted in this study. The primary limitation of the test methodology is the mounting fixture used for the surrogate system. In the current design, the bottom of the HSTM and the top of the Hybrid III pelvis were rigidly mounted to a steel blast support system. While this mounting technique provided a stable support and helped limit variability across tests, it also resulted in an approximation of the worst-case scenario in regards to mechanical loading experienced by the skeletal system and internal organs of the surrogate. This approach was necessary due to the cables required for the HSTM. However, recent updates in miniaturised data acquisition systems have allowed for the potential to develop a wireless HSTM. Moving to a wireless surrogate will allow for modifications to this mounting design that result in a more realistic motion of the surrogate and a more accurate assessment of the blast attenuation performance across bomb suits. However, any updates to the test methods will need to maintain a high level of repeatability across tests to allow for an accurate comparison of suit designs. The primary limitation with the HSTM is the current inability to directly link the response of this surrogate system to human injury. Currently, the HSTM is best utilised in comparative assessments of different loading conditions or different armour systems, as was performed in this study. While there is some data linking the HSTM pressure response to risk of lung injury in an ovine model, future experiments are needed to develop a method to predict risk of blast-induced injury for humans with this advanced surrogate. The primary limitation in comparing performance across bomb suit designs was the limited sample size, which prevented a robust statistical comparison of some of the results. This was primarily due to difficulty in obtaining multiple bomb suits for each of the designs. However, the limited number of bomb suits did not detract from demonstrating the feasibility of using the HSTM with this new test methodology. In future studies that are primarily focused on comparing bomb suit designs, performing at least three tests with a given suit design is recommended.

In conclusion, this study achieved the proposed objectives of developing a bomb suit test methodology using an advanced blast surrogate and demonstrating its use for comparing across suit designs. The first key finding of this study was that the advanced blast surrogate allowed for a repeatable and reliable framework for bomb suit assessment and provided more information related to the human body response than previous test methodologies. The second key finding was that suits with thicker layers

https://doi.org/10.52202/078352-0024

of foam between the hard armour and the torso exhibited lower peak sternum accelerations and lower peak surface and internal pressures, as compared to the suit with a thin layer of foam. These results suggest potential differences in the injury mitigation performance of the tested bomb suits, but future work is needed to understand the types of injuries that may have occurred in these free-field blast tests.

Acknowledgements

The authors would like to thank the United States Army Combat Capabilities Development Command-Soldier Center for sponsoring this effort. This material is based on work supported by the United States Army Combat Capabilities Development Command Soldier Center under NAVAL SEA SYSTEMS COMMAND (NAVSEA) Contract No. N00024-13-D-6400, Task Order #VKW03. Any opinions, findings and conclusions or recommendations expressed in this material are those of the author(s) and do not necessarily reflect the views of NAVSEA.

References

[1] Bass C., Davis M., Rafaels K., et al., A Methodology for Assessing Blast Protection in Explosive Ordnance Disposal Bomb Suits, *Int. J. Occup. Saf. Ergon.*, 2005; 11(4); 347-361.
[2] Dionne J.P., Levine J., and Makris A., Acceleration-Based Methodology to Assess the Blast Mitigation Performance of Explosive Ordnance Disposal Helmets, *Shock Waves*, 2018; 28(1); 5-18.
[3] Dionne J.P., Levine J., and Makris A., Investigating Bomb Suit Blast Overpressure Test Methodologies, Homeland Security and Public Safety: Research, Applications and Standards, P. Mattson, and J. Marshall, eds. (ASTM International, West Conshohocken, PA, 2019; 216–236.
[4] Carneal C., Merkle A., et al. Influence of Armor on Thoraco-Abdominal Organ Injury Patterns During Complex Blast Loading, *Personal Armour Systems Symposium*, Cambridge, United Kingdom, 2014.
[5] Wickwire, A., Carneal, C., et al. Effect of Torso Armor on Surface and Internal Pressure Response of a Human Surrogate, *Personal Armour Systems Symposium*, Cambridge, United Kingdom, 2014.
[6] Merkle A., Roberts J., et al., Evaluation of the Human Surrogate Torso Model Response to Ideal and Complex Blast Loading Conditions, *Personal Armour Systems Symposium*. Quebec, Canada, 2010.
[7] Biermann P., Ward E., Cain R., et al., Development of a Physical Human Surrogate Torso Model for Ballistic Impact and Blast, *J. Adv. Mater.*, 2006; 38(1); 3-12.
[8] Yen, R., Fung, Y., Ho, H., and Buttermen, G., Speed of Stress Wave Propagation in Lung, *J. Appl. Physiol.*, 1986; 701–705.
[9] Saraf, H., Ramesh, K. T., Lennon, A. M., et al., Mechanical Properties of Soft Human Tissues under Dynamic Loading, *J. Biomech.*, 2007; 40(9); 1960-1967.
[10] Merkle, A., Roberts, J., et al., Evaluation of an Instrumented Human Surrogate Torso Model in Open Field Blast Loading, *ASME IMECE Proceedings. 2009*.
[11] Hyde, D., ConWep–Conventional Weapons Effects, *US Army Eng. Waterw. Exp. Stn.* 1992.
[12] National Institute of Justice, Public Safety Bomb Suit Standard NIJ Standard-0117.01, April 2016.
[13] MacFadden L.N., Chan P.C., Ho K.H.H., and Stuhmiller, J.H., A Model for Predicting Primary Blast Lung Injury, *J. Trauma Acute Care Surg.*, 2012; 73(5); 1121-1129.
[14] SAE International (SAE), SAE J211: Instrumentation for Impact Test, 2014.
[15] Viano D.C., and Lau I.V., A Viscous Tolerance Criterion for Soft Tissue Injury Assessment, *J. Biomech.*; 1988; 21(5); 387-399.
[16] Roberts J.C., Merkle A.C., Biermann P.J., et al., Computational and Experimental Models of the Human Torso for Non-Penetrating Ballistic Impact, *J. Biomech.*, 2007; 40(1); 125-136.
[17] Cooper G.J., Protection of the Lung from Blast Overpressure by Thoracic Stress Wave Decouplers. *J Trauma*, 1996; 40(3S); 105-110.
[18] Cooper G.J., Townend D.J., Cater S.R., et al., The role of stress waves in thoracic visceral injury from blast waves. *J Biomech.,* 1991;24(5):273-285.
[19] Jonsson A., Experimental investigations on the mechanism of lung injury in blast and impact exposure (Dissertation). Linkoping University, Sweden, 1979.
[20] Sedman A. and Hepper A., Protection of the lung from blast overpressure by stress wave decouplers, buffer plates or sandwich panels. *J R Army Med Corps*, 2019;16(5);22-26.
[21] Carneal C., Merkle A., et al. Thoraco-Abdominal Organ Injury Response Trends due to Complex Blast Loading, *Personal Armour Systems Symposium*, Nuremberg, Germany, 2012.
[22] Yelverton, J. T., Pathology scoring system for blast injuries. *J. Trauma,* 1996; 40(3 Suppl); S111-115.

https://doi.org/10.52202/078352-0024

Human factors and traumatic injury considerations associated with small changes in combat helmet mass

S. Laing[1], J. Dutschke[2], S. Doecke[2] and S. Davis[1]

[1]*Land Division, Defence Science and Technology Group, 506 Lorimer St, Fishermans Bend, Melbourne, VIC 3207, Australia, sheridan.laing@dst.defence.gov.au*
[2]*Centre for Automotive Safety Research, The University of Adelaide, South Australia 5005, Australia*

Abstract. Reducing the mass of combat helmets will benefit the user's comfort and mobility and decrease the risk of musculoskeletal injury. Technological advancements have resulted in small incremental reductions in helmet mass. However, there is limited data quantifying the potential user benefits associated with these small mass reductions. This research considered both human factors and traumatic injury potential associated with 200 g changes in helmet mass. The human factors assessment comprised three helmets of different masses (1.0, 1.2 and 1.4 kg) worn by 20 Army qualified riflemen and applied a within-subject design. A carbon 'bump' helmet was modified to allow mass to be discretely added to the crown of the helmet while participants remained blinded to the mass condition. After completing two hours of representative military tasks, a questionnaire concerning the human factors effects of helmet mass was completed. The risk of neck injury during frontal vehicle impacts was determined via validated male and female MADYMO models for five different helmet masses ranging from 0.6 to 1.4 kg in 200 g intervals and for a no helmet condition. The helmet moment of inertia was scaled appropriately with the mass and the crash pulse consistent across all simulations. The feedback from participants suggested a 200 g helmet mass reduction is likely to a have a small positive impact on human factors. Similarly, the neck injury risk during frontal impacts is only slightly reduced with a 200 g change in mass. This reduction in injury risk was equivalent to a reduction in impact speed of between 1.3 and 5.2 km/h for males and 2.7 and 5.3 km/h for females for driving speeds 20 to 80 km/h. Although a 200 g reduction in helmet mass was found to induce only small effects, any reduction is head supported mass is seen as beneficial.

1. INTRODUCTION

1.1 Human factors considerations

The addition of head supported mass (HSM) causes the centre of mass (COM) of the head/HSM system to move away from the head's natural COM as well as increasing the head/HSM system mass moment of inertia (MMOI) and the resultant torque on the spine when moving the head. Ivancevic and Beagley provided three basic ergonomic recommendations for choice of HSM [1]. These were to choose an HSM with the smallest mass, with the mass most aligned to the head's natural COM, and with the mass closest to the head in terms of smallest diameter to limit MMOI [1].

The increased muscle activity required to support HSM as well as to maintain the head in an upright position to counter the effects of head-helmet system COM changes induced by the HSM can cause muscle fatigue [2-4]. Such muscle fatigue can cause pain and soreness which can cause distraction and impact operational performance [5]. The strain and increase in fatigue associated with increased HSM has been found to cause reduced efficiency, lower mental concentration and to increase the potential to make mistakes [6], as well as a decrease in the speed and number of head scanning movements [7]. Increased MMOI causes similar issues relating to neck muscle fatigue and the indirect effects these can have on operational performance. Physical discomfort has been found to be affected by HSM with a linear relationship between reports of discomfort and an increase in HSM [7, 8]. Discomfort can cause headaches and distraction which can subsequently affect operational performance [6, 9, 10]. MMOI arguably has the most direct effect on operational performance. Increased MMOI decreases head angular accelerations (for starting and stopping motion) and can therefore cause delayed head movements which can result in deficiencies in performance of tasks including those requiring efficient tracking and sighting of a moving target [11].

The mass of a helmet system has been proposed as a factor that contributes to overall user acceptance [12]. User acceptance is a multifaceted attribute and failure to achieve user acceptance can result in misuse and disuse of an item of equipment [6]. One of the key issues is that there is very limited data on limits of HSM and so it is not known at what mass the effects begin or at which they are exacerbated. There is also limited data on the difference that small changes in mass can make.

 https://doi.org/10.52202/078352-0025

1.2 Injury considerations

The optimum position for any HSM is on the vertical (superior-inferior) axis connecting the COM of the head and the COM of the body. Any shift away from this axis increases the strain and fatigue on the neck muscles as they work to keep the head in an upright position [4]. The detriment of increased HSM on muscle fatigue and injury is greater when the COM of the HSM is further away from this vertical axis [13, 14]. Additional mass can lead to neck muscle fatigue which can reduce the protective efficiency of the musculature. It is also understood that repetitive loading can lead to intervertebral disc degeneration and that increased loads exacerbate this degeneration. However, these effects have not been quantified for small changes in HSM. The chronic neck injury risk associated with 3 kg and 5 kg HSM has been quantified using musculoskeletal modelling [13, 14]. Small reductions in the intervertebral stresses and repetitive injury risk were found for the 2 kg mass reduction when HSM has no anterior COM component. Hence, it is likely that these differences would be negligible when considering a small (i.e. 200 g) mass reduction. A large anterior COM component is expected when anteriorly-positioned helmet ancillaries such as night vision goggles (NVGs) are used. This study considers helmets alone, thus a small, or no, anterior COM component is of interest.

During traumatic injury events, such as falls or vehicle accidents, the addition of HSM can exacerbate neck injuries due to the increased MMOI. A limited number of studies have suggested that HSM may contribute to increased risk of neck injury in the case of an automotive frontal crash [15-17]. Consequently, minimising the total head supported mass may be an important strategy for minimising neck injury risk during a crash. However, although Merkle et al. showed that helmet mass was positively correlated with upper neck shear and tensile forces, they also showed that upper neck flexion, lower neck tension and lower neck shear were not significantly correlated to mass [17]. Doczy et al. showed that the effect of helmet mass was less important than impact deceleration [16] and Manoogian et al. suggested that an increase in mass, provided it was placed in the correct location, could actually reduce the neck injury risk [15]. However, if it was placed in the wrong location, it could increase the neck injury risk [15]. Merkle et al. utilised mass deltas of 900 g [17] and the mass deltas adopted by Manoogian et al. were not reported [15]. Although Doczy et al. adopted smaller mass deltas (227 to 460 g) the study was limited to sub-injurious loading due to the use of volunteer participants [16]. Thus, the influence of small HSM deltas on injury potential during frontal impacts remains unknown.

1.3 Aims

Technological advancements have resulted in small incremental reductions in helmet mass. Although reducing the mass of combat helmets has recognised benefits with regards to user performance and decreased risk of chronic and traumatic injuries, there is limited data quantifying the potential user benefits associated with small mass reductions. The aim of this research was to quantify changes in both user acceptance and traumatic injury risks associated with small (i.e. 200 g) changes in helmet mass.

2. METHODS

2.1 Human factors

Twenty Australian Army riflemen participated in the study. Three different helmet masses were employed. Each participant conducted one helmet condition on each trial day with the order of conditions balanced across the trial days. Each trial day comprised of the following activities:

- Two hours of representative military tasks:
    - 5 km pack march, into contact, drop pack
    - Section attack with blanks
    - Urban Assault with blanks
    - Obstacle course (comprising overs and unders, traverse ropes, monkey bars, balance walk, three tier tower, rope swing over, horizontal log walk, tunnel crawl, ten foot wall, leopard crawl, 'A' frame, burma bridge, cargo net crawl, balance cargo/net walk, and swinging bridge)
- Administration of questionnaire.

https://doi.org/10.52202/078352-0025

A focus group was conducted at the end of the final trial day. During the trial days, participants wore combat uniform, boots, Australian body armour carrier with training soft armour and plates (plates were removed for the obstacle course as per standard operating procedures). Two gunners wore belt rigs in addition to the body armour system vest. Large field packs were worn during the pack march and cached prior to the other activities. Hearing protection was worn during the section attack and urban assault.

A carbon "bump" helmet had been modified to allow the addition of a mass adjustment rig (Figure 1). This rig, and specially designed helmet cover, allowed the mass of the helmet to be altered without participants being aware of the mass and with all other features of the helmet remaining the same. The mass was added centrally and vertically over the mid-point of the helmet, this would have the effect of slightly raising the COM of the system vertically but not affect the anterior-posterior or left-right positioning of the COM. Thus, the helmet mass was the main inertial variable to change across the conditions.

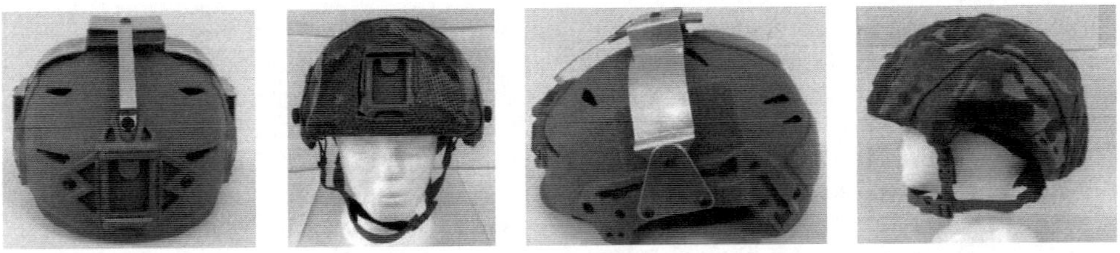

Figure 1. Trial helmets – mass adjustment rig and helmet cover

The base system was 1.0 kg, comprising the helmet, mass adjustment rig and helmet cover. Masses were added to produce the 1.2 kg and 1.4 kg conditions. Thus, the three helmet mass conditions were: 1.0 kg, 1.2 kg and 1.4 kg. Small system mass variations were expected due to variability in manufacturing of the helmet shell, mass adjustment rig and helmet covers. All test systems were weighed and found to be within 20 g of the target mass condition. The total HSM generally worn by the participants depended on the helmet size and which ancillaries were used. Based on these factors, the total HSM that the participants were accustomed to was likely between 1.2 kg and 2.2 kg, and thus encompassed the two heaviest test systems in this study.

A questionnaire was administered at the end of the two-hour trial. Six questions on discomfort, mass, mass distribution, difficulty in keeping head/chin up and overall user acceptance were addressed:

- The weight of the helmet was… (5-pt acceptability scale)
- The weight distribution of the helmet was… (5-pt acceptability scale)
- Did you experience discomfort to your head/neck whilst wearing the helmet? (Binary 'yes/no' response)
- The weight of the helmet made it difficult to keep my chin/head up when in prone… (5-pt agreement scale)
- Do you feel that the helmet slowed your head movement? (Binary 'yes/no' response)
- Overall, how acceptable was the helmet (as it is) to you as a user? (5-pt acceptability scale).

Given the low participant numbers, the alpha level was set at $\alpha = 0.10$ for all tests, where $p < 0.10$ was suggestive of a significant trend. Where questions used a 5-point acceptability scale or a 5-point agreement scale, non-parametric Friedman tests were conducted. Post-hoc analyses were conducted using Wilcoxon signed ranks when the Friedman tests indicated significance. The effect size was calculated using the Probability of Superiority, A. The A value is expressed as a percentage and indicates the probability that a person randomly chosen from Group 2 will have a higher score than a randomly chosen person from Group 1. Values of A equal to or greater than 56%, 64% and 71% were interpreted as small, medium and large effect sizes respectively [18].

Where questions used a binary 'yes/no' response, a Chi square test was conducted. The effect size for the binary response questions was calculated using the Odds Ratio (OR). The OR indicates the odds that someone with a given response came from a particular treatment group. OR values equal to or greater than 1.68, 3.47 and 6.71 were interpreted as small, medium and large effect sizes respectively [19].

https://doi.org/10.52202/078352-0025

2.2 Neck injury risks while driving

2.2.1 Previous sled tests and models

The influence of additional helmet mass while driving was assessed using previously developed and validated MADYMO models of 50th percentile male and 5th percentile female Hybrid III anthropomorphic test devices (ATDs). The models were validated from a previous study comprising 13 sled tests with three Hybrid IIIs per test (two 50th males and one 5th female). The tests were conducted at four different impulses, varied HSM conditions (i.e. no helmet, two varieties of helmet and helmets with NVGs in different configurations) and with two different seats from Australian military vehicles. Initial testing showed significant rotation of the helmet on the head of the ATD during the impact, especially for the female ATD. To eliminate this motion and ensure a 'worst-case' rigidly coupled helmet mass, the helmets were subsequently bolted to the existing threads of the ATD heads.

A model of each of the seats was constructed in MADYMO. The surfaces of the models were captured with a coordinate measuring machine and were based on the geometry of the seats after they were installed on the sled but before they were used in any tests. To reduce the geometric complexity, only the centreline of the seats was modelled. The seats were then modelled with flat horizontal planes. The positions of the elements of the restraint systems were positioned in MADYMO at the locations recorded with the coordinate measuring machine. A finite element belt type was used where the belt and the ATD came in to contact. Line elements were used for the other parts of the belt that did not contact the ATD. The dimensions of the belts and their positions were also based on the measurements made with the coordinate measuring machine, while taking note of the belt routing around the ATD prior to the tests. The seat belt was modelled in two parts, a sash portion and a lap portion, tied at the buckle. The ATD was positioned in each seat to reflect the median position of the ATD across all tests (Figure 2). The body armour was modelled with multiple ellipsoids to represent the contoured front and back plates, with particular attention paid the areas where armour would come into contact with the belt or seat. The front and back plates were connected with a translational joint to model the straps. A helmet was modelled as a point mass with finite moments of inertia. This point mass was positioned at the location that represented the centre of gravity of the helmet. The point mass has several advantages in that it is simple to model, and that it represents the bolted helmet that was used in the sled tests. For the purposes of visualising the helmet in a model, an ellipsoid of no mass was added over the head. The MMOI of the helmet was based on that previously measured for a helmet of similar cut and was scaled to reflect the mass of the helmet under test.

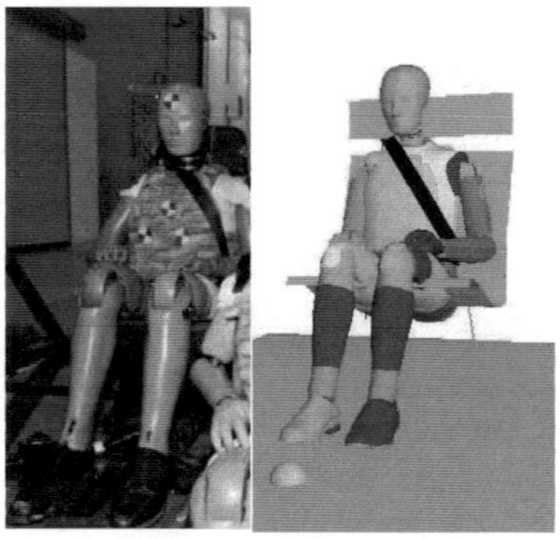

Figure 2. 50th percentile male ATD in seat with no helmet and corresponding MADYMO model

The models were adjusted to improve their concordance with the sled test result where no helmet was worn. The adjustments were limited to the gap between the seatbelt and the ATD, the length of seatbelt stored in the retractor, and the stiffness of the seatbelt. The length of seatbelt left in the retractor was adjusted until the payout of the seatbelt in the model closely resembled the payout of the seatbelt in the tests. Finally, the stiffness of the seatbelt was adjusted to provide a good fit to the apparent stiffness in the test results. It was found that this adjustment had a relatively small effect on ATD response. Three ATD kinematic variables were examined: the peak head acceleration, peak chest acceleration and peak pelvis acceleration; agreement between the sled tests and simulations in these variables indicated model validity.

https://doi.org/10.52202/078352-0025

MADYMO simulations were completed to quantify the risk of neck injury with increasing impact velocities. Simulations were performed with each the 50th percentile male and 5th percentile female ATDs with the 1.2 kg helmet and 90 ms impact duration held constant and with varying impact speeds. These speeds were 20, 40, 50, 60, and 80 km/h. The results from these simulations were used to calculate regression equations for the neck injury metrics with varying impact speed (given a 1.2 kg helmet condition and 90 ms impact duration)(Table 1). Based on results of previous testing, the three most likely injury modes are lower neck (LN) flexion, upper neck (UN) flexion and the UN combined tension-flexion (N_{TF}); thus, these metrics are the focus of the current study.

Table 1. Previously developed linear regression equations describing neck injury metrics as a function of change in speed between 20 and 80 km/h, for 50th percentile male and 5th percentile female with the 1.2 kg helmet, with 90 ms impact duration

Regression coefficients	**LN Flexion (Nm)**		**UN Flexion (Nm)**		**UN N_{TF}**	
	Male	**Female**	**Male**	**Female**	**Male**	**Female**
a	6.76	6.20	1.61	2.09	0.012	0.03
b	83.16	30.48	62.07	5.80	0.04	-0.05
R^2	0.87	0.96	0.76	0.90	0.92	0.96

2.2.2 Current study

The impulse, helmet (1.2 kg condition) and seat used in the current study were included in the previous sled tests (13 male ATD and 6 female ATD runs) and subsequent MADYMO modelling. The seats were from tactical training vehicles which featured a standard 3-point harness comprising a lap and sash belt. To investigate the influence of small changes in helmet mass, five simulations were performed for each ATD with varying helmet mass and an additional test with no helmet worn. Helmet masses varied at 200 g intervals and comprised 0.6 kg, 0.8 kg, 1.0 kg, 1.2 kg and 1.4 kg. For the helmet tests, the COM was kept consistent for each test, with the value as previously determined for the 1.2 kg helmet. The MMOI was scaled according to the helmet mass. All simulations represented a 40 km/h impact with 90 ms duration 12.4 g deceleration pulse.

The neck injury metrics calculated were the UN tension, UN flexion, UN N_{TF}, LN tension, LN flexion and LN N_{TF}. Internationally-used thresholds exist for the male and female ATD for each metric, where exceeding the threshold generally represents a low (i.e. $\leq 5\%$) risk of a serious injury. Examples of serious injuries to the cervical spine include intervertebral disc rupture, major compression fracture of the vertebral body (no cord involvement) or spinal cord contusion with transient neurological symptoms.

3. RESULTS

3.1 Human factors

The human factors study results showed the following general tendencies: discomfort increased with mass, difficulty in keeping head/chin up when in prone increased with mass, acceptability of helmet weight decreased as mass increased, acceptability of helmet weight distribution decreased as mass increased, perceived speed of head movement decreased as mass increased and overall acceptability of helmet decreased as mass increased (Figure 3, Figure 4 and Figure 5). However, statistically significant differences were only reported for difficulty in keeping head/chin up when in prone, for the perceived weight of the helmet and for the perceived weight distribution of the helmet. Furthermore, whilst significant differences were reported between the 1.0 kg and 1.4 kg condition for these three parameters, only the difficulty in keeping head/chin up when in prone was significantly different at $\alpha = 0.10$ between 1.0 kg and 1.2 kg, and 1.0 kg and 1.4 kg. This demonstrates that a 400 g HSM difference had an effect on all three factors but 200 g only had a significant effect on difficulty in keeping head/chin up.

Only the 400 g mass difference resulted in medium or large observed effects, however small effects were observed between the 200 g conditions for most questions. The greatest effect size (large effect) was seen between the 1.0 kg and 1.4 kg conditions for the feeling of slowed head movement. Small effects were also observed between the 1.0 kg and 1.2 kg and the 1.0 kg and 1.4 kg conditions for experiencing discomfort, as well as between the 1.2 kg and 1.4 kg and the 1.0 kg and 1.4 kg conditions for overall acceptability.

https://doi.org/10.52202/078352-0025

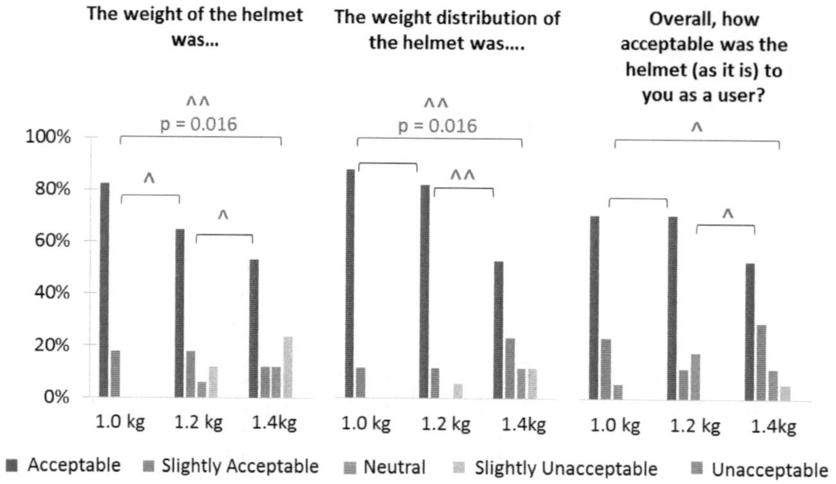

Figure 3. Results for questions answered using a 5-pt acceptability scale. Significance (p) only shown when p < 0.10, effect size, *A*, indicated if the requirements for small (^), medium (^^) or large (^^^) effects were met.

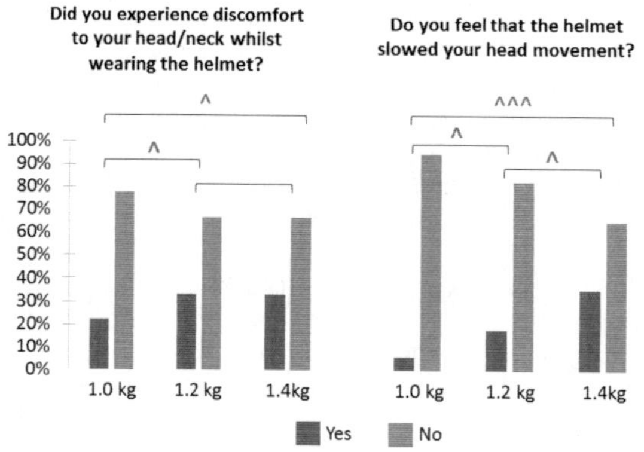

Figure 4. Results for questions answered using binary yes/no response. Significance (p) only shown when p < 0.10, effect size, OR, indicated if the requirements for small (^), medium (^^) or large (^^^) effects were met.

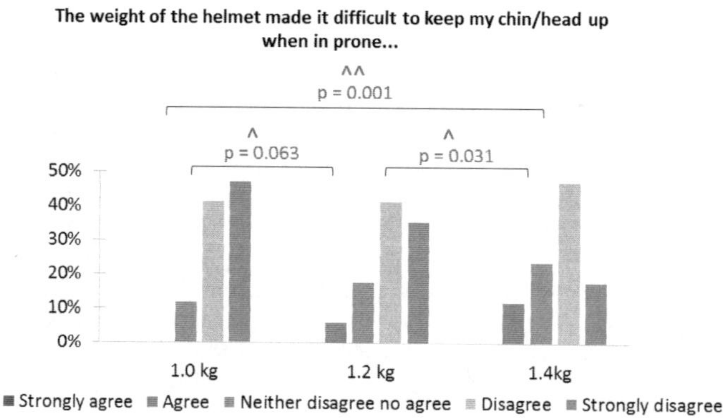

Figure 5. Results for the question answered using a 5-pt agreement scale. Significance (p) only shown when p < 0.10, effect size, *A*, indicated if the requirements for small (^), medium (^^) or large (^^^) effects were met.

https://doi.org/10.52202/078352-0025

Regarding task performance, a small number of participants indicated that the helmet interfered with their ability to perform a pack march in the 1.2 kg and 1.4 kg condition but no statistically significant differences were identified. The qualitative comments indicated that the weight of the helmet made it difficult to keep the head up and there was a need to look down to obtain relief. The results of the rating scale data indicated that the weight of the helmets used in the trial did not appear to cause issues with performing a section attack, an urban assault or shooting. However, the qualitative comments indicated that some participants felt strain on their neck, particularly during pack marching, urban assault and section attack.

During the military tasks, the participants had been blinded to the helmet mass conditions. At the outset of the focus group, which took place at the end of the last trial day, participants were told that there had been three conditions. When asked what they thought the weight of the lightest helmet was, responses ranged from 300 g to 1 kg. When asked what they thought that the weight of the heaviest helmet was, responses ranged from 1.3 kg to 2 kg. When asked what they thought the weight difference was between each condition, there was general agreement that there was between a 50 g and 100 g difference.

3.2 Mass changes as equivalent speed reductions

The most likely injury modes were previously identified as LN flexion, UN flexion and the UN N_{TF}. Indeed for all simulations performed in the current test series, the only injury thresholds to be exceeded were these three injury metrics for the 5th percentile female and the LN flexion alone for the 50th percentile male. The LN flexion threshold is exceeded at all HSM conditions for the 5th percentile female and the no helmet condition is at 96% of the threshold (Figure 6). Comparatively, only the 1.2 and 1.4 kg conditions exceeded the LN flexion threshold for the 50th percentile male. The UN flexion and N_{TF} thresholds for the 5th percentile female were exceeded for HSM conditions of 1.0 kg and greater; the UN N_{TF} was at the threshold for 0.8 kg HSM (Figure 6).

For the 50th percentile male, the LN flexion threshold was exceeded for the 1.2 kg HSM condition but not the 1.0 kg condition (Figure 6). There were no other injury thresholds exceeded at the 1.2 kg HSM mass that was not exceeded at the 1.0 kg mass. Linear regression analysis was completed to predict the injury metric as a function of helmet mass for the given impact conditions (Table 2) as per Equation 1.

$$\textit{Injury metric} = a(\textit{helmet mass in kg}) + b \tag{1}$$

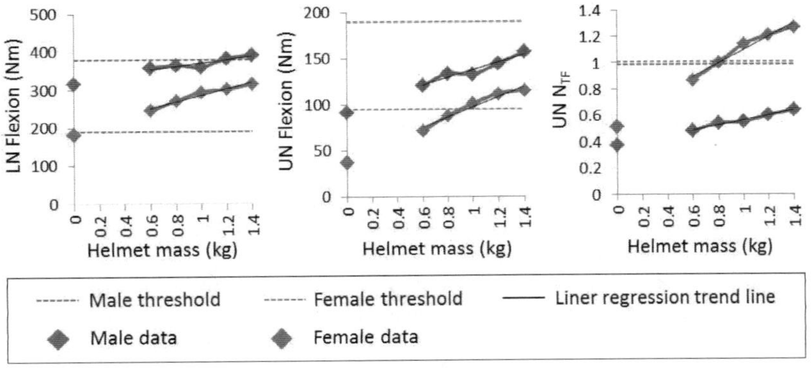

Figure 6. Influence of helmet mass on the injury metrics of LN flexion, UN flexion and UN N_{TF}. The 0 kg (no Helmet) data was not included in the linear regression due to the varied COM properties of the test compared to the helmeted tests. Note the 50th percentile male and 5th percentile female N_{TF} thresholds are both 1.0.

Table 2. Regression coefficients (a and b) and the coefficient of determination (R^2) for the three injury metrics of LN flexion, UN flexion and UN N_{TF} as a function of HSM (kg)

Regression coefficients	LN Flexion (Nm) Male	LN Flexion (Nm) Female	UN Flexion (Nm) Male	UN Flexion (Nm) Female	UN N_{TF} Male	UN N_{TF} Female
a	43.0	84.5	41.6	55.9	0.19	0.50
b	328.7	201.8	96.1	41.6	0.37	0.59
R^2	0.82	0.96	0.93	0.96	0.97	0.97

https://doi.org/10.52202/078352-0025

Regression coefficient *a* indicates the change in the injury metric for each change in helmet mass of 1 kg. The LN flexion, UN flexion and UN N_{TF} data increases linearly with increasing HSM ($R^2 \geq 0.82$) for both the 50th percentile male and 5th percentile female. The regression results were used to calculate generalisable data regarding the influence of changing helmet mass.

Simultaneous solving of the equations of Table 1 and Table 2 permits the expression of the change or delta (Δ) helmet mass as an equivalent impact speed delta (Table 3). That is, for a 200 g reduction in helmet mass, the reduced risk of LN flexion injury for a midsized male was equivalent to that of travelling 1.3 km/h slower. Equally, with a 200 g lighter helmet, a midsized male could travel 1.3 km/h faster than otherwise with an equivalent risk of LN flexion injury. The change in HSM had a greater influence on the UN flexion and N_{TF} metrics, with a 200 g mass delta equivalent to speed deltas of 5.2 and 3.2 km/h respectively. Considering the LN flexion (i.e. the most likely injury mode) even a large mass change of 1.0 kg corresponded to a relatively small equivalent change in speed (6.4 km/h) for the same injury risk for a midsized male. Changing helmet mass had a slightly greater influence on neck injury risks for 5th percentile female than 50th percentile male. A helmet mass change of 200 g was equivalent to a change in speed of 2.7 to 5.3 km/h for the three injury metrics considered. Considering LN flexion, a large mass delta of 1.0 kg corresponded to a substantial equivalent speed delta of 26.7 km/h.

Table 3. The change in the injury metrics and equivalent changes in driving speed associated with helmet mass changes of 1 kg and 200 g for midsized male and small female occupants

Helmet mass	Δ Injury metric LN flexion (Nm)	UN Flexion (Nm)	UN N_{TF}	Equivalent Δ speed (km/h) LN flexion	UN Flexion	UN N_{TF}
			Midsize male			
Δ 1kg	43.0	41.6	0.19	6.4	25.9	15.8
Δ 200g	8.6	8.3	0.04	1.3	5.2	3.2
			Small female			
Δ 1kg	84.5	55.9	0.51	13.6	26.7	16.8
Δ 200g	16.9	11.2	0.10	2.7	5.3	3.4

4. DISCUSSION

This research aimed to quantify the changes in user acceptance and traumatic injury risks associated with small (i.e. 200 g) changes in helmet mass. Although other studies have investigated human factors and injury considerations for helmet masses which differ by larger amounts, decisions regarding helmet procurement are generally concerned with small differences between tendered items. Thus, these studies provide useful evidence with which helmets of plausibly different masses can be assessed and compared.

A 200 g change in helmet mass was found to have a significant effect on the difficulty of keeping the head up when in a prone position. No other significant differences were found due to 200 g mass changes and all effect sizes were small. A larger mass delta of 400 g caused significant differences in the acceptability of the helmet weight and weight distribution and the difficulty of keeping the head up with small to large effects found across all questions.

The study was limited to a sample of 20 participants and each helmet condition was only worn for 2 hours. It is likely that the small effects of the additional mass that were identified following 2 hours of activities would become larger, the longer the heavier helmet was worn. The results are, therefore, indicative and do not provide conclusive data on the implications of a 200 g HSM delta as worn by a large population or on an ongoing basis. As per the ergonomic recommendations in Ivancevic and Beagley, an HSM with the smallest mass is preferable [1]. The human factors study addressed the basic mass of the helmet; however the helmet also functions as a platform for mounting ancillaries such as NVGs, counterweights, torches, and cameras, which would substantially increase the total HSM. At higher masses and varied head-HSM system COMs, a reduction in mass of 200 g would potentially yield greater human factors benefits.

HSM reductions of 200 g were found to reduce the risk of injury equivalent to quite small reductions in driving speed of 5.3 km/h or less for both midsized males and small females. For midsized males, the only injury threshold exceeded was LN flexion at the 1.2 and 1.4 kg mass conditions. The changing HSM had slightly greater effect on the 5th percentile female injury metrics;

https://doi.org/10.52202/078352-0025

the UN flexion and UN N_{TF} thresholds were exceeded for helmet masses 1 kg and greater and the LN flexion threshold was exceeded for all helmet conditions assessed. However, the no helmet condition represented 96% of the threshold value, indicating injury risk for small females at the tested impact condition even without HSM. The aforementioned injury predictions are limited to the impact conditions modelled in the simulations. This impulse was derived from publically available crashworthiness data of the Ford F150 truck. The speed equivalencies for changing helmet mass are limited to driving speeds of 20 to 80 km/h and impacts with a deceleration duration of 90 ms. The deceleration duration is affected by the stiffness and mass of the vehicle, whereby vehicles with greater stiffness during impacts, or lighter mass, have shorter impact durations.

Frontal vehicle impacts cause forward flexion of the neck. Therefore, a COM that is more anterior and superior can exacerbate the flexion and tension forces. However, helmets alone can result in the COM moving posterior to the natural COM of the head. Thus, equivalent rearward vehicle impacts may be the more dangerous impact scenario for this HSM condition. Further, the neck is less tolerant to extension than flexion bending moments. However, rearward vehicles impacts are generally at lower impact speeds than frontal impacts.

The injury analyses assessed the neck injury risk with helmets alone and in a specific scenario. If the helmet was used as part of a system including NVGs, it would have the effect of moving the COM anteriorly. The anterior translation of the head-helmet COM would likely exacerbate the chronic injury risks [13, 14], traumatic injury risks during frontal impacts and potentially the traumatic injury risks during events with lower peak decelerations e.g. falls.

This study was funded by the Australian Department of Defence, as part of an ongoing effort to improve the fit, form, and function of equipment to accommodate and protect Australian soldiers.

5. CONCLUSIONS

Human factors data indicates that 200 g reductions in HSM resulted in small improvements in the human factors areas considered in this report. The traumatic neck injury risk while driving was slightly reduced with 200 g HSM reductions, equivalent to driving 5 km/h slower or less. This report addressed one type of traumatic injury and participants only worn the helmets for 2 hrs. The results and conclusions of this report are, therefore, bounded by this scope. Small mass differences may have greater impacts during other traumatic or chronic injury modes and on other human factors considerations. Further, a 200 g HSM delta may have a greater effect when the effect of wearing the helmet is an anterior shift of the head-HSM COM or when the total system mass is greater. From both a human factors and injury perspective, any reduction in mass is seen as beneficial.

Acknowledgements

The authors would like to acknowledge the contributions of Dr Steve Milanese and Mr Alistair Furnell in the support to user trial planning and review of trial outcomes.

References

[1] Ivancevic V. and Beagley N., 2004, Determining the acceptable limits of head mounted loads, Land Operations Division, DSTO, Report number: DSTO-TR-1577.

[2] Petrofsky J.S. and Phillips C.A., 1982, The strength-endurance relationship in skeletal muscle: Its application to helmet design, Aviation, Space, and Environmental Medicine, vol. 53, no. 4, pp. 365-369.

[3] Milanese S. and Steele E., 2002, Investigation of issues associated with the use of head supported mass systems for the combat soldier, Centre for Allied Health Research, University of South Australia, Australia.

[4] Kulkarni S.G., Gao X.L., Horner S.E, Zheng J.Q. and David N.V., 2013, Ballistic helmets – Their design, materials, and performance against traumatic brain injury, Composite Structures, vol. 101, pp. 313-331.

[5] Harms-Ringdahl K., Linder J., Spångberg C. and Burton R.R., 1999, Biomechanical considerations in the development of cervical spine pathologies, in Burton, R.R. (ed.) Cervical spinal injury from repeated exposures to sustained acceleration, NATO Research and Technology Organization (RTO) Technical Report (RTO-TR-4) Neuilly-Sur-Sein Cedex, France.

[6] Rash C.E., McLean W.E., Mora J.C., Ledford M.H., Mozo B.T., Licina J.R. and McEntire B.J., 1998, Design issues for helmet-mounted display systems for rotary-wing aviation', U.S. Army Aeromedical Research Laboratory, Fort Rucker, Alabama. Report number: USAARL 98-32.

 https://doi.org/10.52202/078352-0025

[7] Tack D.W., Nakaza E.T., McEachern A. and Marrao C., 2006, Investigation of the preferred mass properties for infantry headwear systems, Human Systems Inc., Guelph, Ontario, Canada. Report number: DRDC Toronto CR-2005-230.
[8] Barker D.J. and Albery C., 2010, Neck fatigue and comfort effects due to the extended wear of law enforcement representative head-borne personal protective equipment, Research and Technology Directorate, Edgewood Chemical Biological Center, Aberdeen Proving Ground, Maryland, U.S. Report number: ECBC-TR-825.
[9] McKenzie D.M., 1969, A human engineering evaluation of the combat vehicle crewmans helmet T56-6, Human Engineering Laboratories, Aberdeen Research & Development Center, Aberdeen Proving Ground, Maryland. Report number: Technical Memorandum 10-69.
[10] Van den Oord M.H.A.H., Steinman Y., Sluiter J.K. and Frings-Dresen M.H.W., 2012, The effect of an optimised helmet fit on neck load and neck pain during military helicopter flights, Applied Ergonomics, vol. 43, pp. 958-964.
[11] Carey M.E., Herz M., Corner B., McEntire J., Malabarba D., Paquette S and Sampson J.B., 2000, Ballistic helmets and aspects of their design, Neurosurgery, vol. 47, no. 3, pp. 678-689.
[12] Hickling E.M., 1986, Factors affecting the acceptability of head protection at work, Journal of Occupational Accidents, vol. 8, pp. 193 – 206.
[13] Eckersley C., Cox C., Ortiz-Paparoni M., Lutz R., Sell T., Bass C., 2018, A real pain in the neck: design limits on magnitude and location of head supported mass, Presented at the Personal Armour Systems Symposium, 1st – 5th Oct, Washington DC, USA
[14] Cox C., Eckersley C., Ortiz-Paparoni M., Schmidt A., Shridharani J., Salzar R., Bass C., 2018, Men and women and helmets and necks, Presented at the Personal Armour Systems Symposium, 1st – 5th Oct, Washington DC, USA,
[15] Manoogian S.J., Kennedy E.A., Wilson K.A., Duma S.M., and Alem N.M., 2006, Predicting neck injuries due to head-supported mass. Aviat Space Environ Med, vol. 77, no. 5, 509-514.
[16] Doczy E., Mosher S., and Buhrman J., 2004, The effects of variable helmet weight and subject bracing on neck loading during frontal-Gx impact. Paper presented at the Forty-second Annual SAFE Association Symposium, 27th-29th Sept, Salt Lake City, UT, USA
[17] Merkle A.C., Kleinberger M., and Uy O.M., 2005, The effects of head-supported mass on the risk of neck injury in army personnel. Johns Hopkins APL technical digest, vol. 26, no. 1, 75-83.
[18] Vargha A., and Delaney H.D., 2000, A critique and improvement of the CL common language effect size statistics of McGraw and Wong, Journal of Educational and Behavioral Statistics, vol. 25, no. 2, pp. 101–132.
[19] Chen H., Cohen P., and Chen S., 2010, How big is a big odds ratio? Interpreting the magnitudes of odds ratios in epidemiological studies, Communications in Statistics—Simulation and Computation, vol. 39, no. 4, pp. 860-864.

https://doi.org/10.52202/078352-0025

Scaling of animal and PMHS thoracic BABT data to live human data

Bourget, D.

Defence Research and Development Canada – Valcartier Research Center, 2459 Pie-XI Blvd North
Québec (Québec) G3J 1X5, Canada, Daniel.Bourget@drdc-rddc.gc.ca

Abstract. The reduction of mass and the increase of soldier comfort and performance of armour systems are closely related and necessitates the optimization of those systems. Lighter weight materials and improved armour fabrication techniques are key to this objective. Another aspect of body armour optimization process concerns our knowledge of appropriate threshold to guide body armour design. Indeed, tests to assess the appropriate thresholds of injury are frequently done using either animals (either live or deceased) or PMHS (Post Mortem Human Subjects). We are assuming that the dynamic response and injury sensitivity of those surrogates is one to one compared to live human being. It has been shown in the past that it is not necessarily the case ([1], [2] and [3]). This paper investigates open literature data on the subject and determines the expected relationship between porcine (live or dead) and PMHS to live human. The relationship is then used to discuss and assess why, for similar impact loadings, experimental data on animals are resulting in injuries while field data are not showing the injurious response for live humans. As many available datasets concern blunt projectile impacts, the applicability of using those data for BABT studies is discussed.

1. INTRODUCTION

One aspect of body armour optimization process concerns our knowledge of appropriate threshold to guide body armour design. This knowledge can only be acquired through the use of Post-Mortem Human Subjects (PMHS) and live or post-mortem animals as a proxy to the response of live human subjects (LHS). This estimation is usually made by assuming that there is a one on one relation between PMHS and live or post-mortem animals to LHS. The present study uses available PMHS, Live Porcine Subject (LPS) and Post-Mortem Porcine Subjects (PMPS) data low mass high velocity and high mass-low velocity thoracic impacts to estimate the response of LHS. The range of impact conditions is selected to cover soft and hard thoracic armour BABT conditions. First a literature review is done to determine which data are available and how the automotive industry transfers data from animal to human subjects. Then, the available data are analysed and trends of selected dependent data for various relevant independent data are presented and discussed. Based on the relationships found, an estimate of LHS response is produced. Because the majority of porcine and PMHS thoracic response data used to assess LHS response concern blunt projectile impacts, a discussion on their validity for BABT studies is presented.

2. LITERATURE REVIEW

A literature review was made to determine the state of knowledge related to comparative impact biodynamic and associated injuries data for thoracic impacts on animals (live or post-mortem) and humans (live and post mortem). The results are presented in the following sub-sections:

2.1 Cardio-Pulmonary Resuscitation (CPR)

Although related to low rate impacts (0.25 to 2.4 m/s), data gathered using instrumented CRP defibrillators are particularly important as they are the only data set that relate live human response to PMHS and porcine response. Reference [1] to [6] have reported on measures of force-displacement-time data for a wide variety of subject mass and age. In particular they have come up with estimations of the stiffness and damping of anterior-posterior (AP) thoracic response to CRP maneuvers and scaled these data to pediatric size humans.

In [4], the authors measures CPR maneuvers response for live human (n = 91 adult, 61 males, average age = 70 years) compared to PMHS (n = 13, 6 males, average age = 71.2 years). They compared their force-displacement results to results from [5] and [6] (11 and 16 live human subjects). The area of load application was different for each dataset and no clear trend were seen related to thoracic stiffness. The authors found that the force-displacement response of the PMHS is statistically higher than for live human subjects, i.e. PMHS thoraces are stiffer than live human thoraces.

 https://doi.org/10.52202/078352-0026

On his side, Neurauter [3] measured stiffness and damping for live humans (90 patients, 18 to 92 years old) and 14 live porcine subjects (a group of 7 weighing 23-30 kg and a group of 7 weighing 34-46 kg) using an instrumented CPR defibrillator. Comparison of the human and animal data revealed similar chest stiffness and viscosity values at the beginning of the CPR episodes for 15 mm chest deformation, but at higher deformation (30 mm and 50 mm) porcine thoracic stiffness was significantly higher. Mean and median force-displacement response curve for the live pigs and live humans were compared and found to be similar.

2.2 High mass-low velocity impacts

In [7] and [8], Viano studied the response to low velocity (8.1 to 12 m/s), high mass (21 kg) thoracic impacts on live (7 sibling subjects, mass = 50.9 to 70.0 kg and 5 other subjects, mass = 49.9 to 59.0 kg) and post-mortem (4 sibling subjects, mass = 48.2 to 68.2) porcine subjects and compared these results to similar impacts on PMHS (Neathery [9]). Higher forces were measured at approximately mid-deflection for the post-mortem porcine subjects compared to the live porcine subjects but at higher and lower deflections, forces were essentially the same. Maximum compression, peak force and impact velocity were shown to be correlated with the AIS injury score and the number of rib fracture. Figure 24 of [8] shows the relation between the AIS injury level and the maximum thoracic compression for post-mortem porcine subjects, live porcine subjects and PMHS. The post-mortem porcine subjects' presents higher AIS scores compared to live porcine subjects and PMHS values are shown to be lower than those two.

In [10], Yaek did a validation study of the scaling process used in the car crash industry for pediatric aged subjects. Her approach was as follows:

a) Evaluate the geometry/material characteristics of porcine subjects and compare them to target pediatric human subjects.
b) Select representative porcine subjects representative of the 50th percentile human male subject as well as 10 years old, 6 years old and 3 years old human subjects and execute a series of pendulum tests to measure their response to lateral abdominal and lateral thoracic impacts. Impact mass and velocity were scaled to fit the scaled values used in the car crash industry
c) Define response corridors and compare them to: scaled down 50th percentile equivalent porcine subject and scaled down 50th percentile human subject

Her results indicated that the scaling process used for abdominal and thoracic side impacts is valid. She also demonstrated that the response corridor for porcine subjects and human subjects of the same total body mass are the same.

2.3 Low mass-high velocity impacts

Bir [11] executed 15 impacts on 7 PMHS thoraces. Impacts were located above the mid-sternum. The projectiles were: 140-g rigid PVC cylinders that impacted the subjects at 20 m/s (5 shots) and 40 m/s (5 shots) and 60-g rigid PVC cylinders that impacted the subjects at 60 m/s (5 shots). All projectiles were 37 mm in diameter. Their characteristics corresponded to typical KENLW (Kinetic Energy Non-Lethal Weapon) projectiles. Response corridors were determined along with the maximum thoracic compression (Cmax), the viscous criterion (VCmax), the blunt criteria (BC), AIS injury scores and number of rib fracture.

Prat and his colleagues, [12] to [15], presented results of a series of blunt impact experiment involving PMHS (n = 12, 5 males, 21 valid impacts, mass = 46 to 101 kg, average mass = 68.8 kg) and live porcine subjects (n = 19, average mass = 49 ± 1.5 kg) against two commercial KENLW projectiles weighing 31 g and 61 g (40 mm diameter). Both had a foam nose and a hard plastic pellet. Results presented included maximum thoracic deformation, Cmax, VCmax, BC and number of rib fracture. Impact location on the PMHS was the 4th left or right rib at the mid-clavicular line. On the porcine subjects, impact location was situated at the 7th right rib equidistant from the sternum and the spine. The conclusion of the principal document of this series [12] is as follows:

To conclude, both pigs and PMHS represent good surrogates for the human adult, which is the subject of interest for ballistic forensic assessment. Though PMHS provide good anatomic thoracic wall conformation, pigs allow for the study of pathophysiological responses to the impact. Because

https://doi.org/10.52202/078352-0026

only one type of impact was used in this study, we cannot build on acute correlation between pigs and PMHS for the thoracic wall response, but we can affirm that, under the same threat:

- *The motion of the pig's chest is greater than that of the PMHS*
- *Severity of the impact for a given Blunt Criterion is always higher for PMHS than pigs*
- *The bone in the porcine model is more elastic and less brittle than older PMHS bone.*

Pavier [14] did a series of test against porcine eviscerated hemi thoraces of 80 kg subjects using 2 short (42-g) and 2 long (78-g) projectiles made of two proprietary foam nose material. All projectile were 30 mm diameter. Dynamics of the thoracic deformation was measured and included the maximum thoracic deformation (Dmax). The authors demonstrate a relationship between the measured Dmax and the initial momentum of the projectile and showed similar relation using live porcine thoracic deformation from [12].

Finally, to study commotio cordis, Dau [16] did a series of experiments with an instrumented Lacrosse balls (64.7 mm diameter, 188.4 g for the 2 lowest impact speed and 214.5 g for the 2 highest impact speed) impacting at 13.4, 17.9, 22.4 and 26.8 m/s against:

- PMHS thoraces (7 subjects, 1 male, weighing 44.9 to 71.2 kg, mean 58.9 kg) on which a total of 23 impacts were done. PMHS thoraces were tested until injury occurred. As a result, between 2 and 8 impacts per PMHS were done.
- Live porcine subject thoraces (n = 17, weighing 21 to 45 kg, mean 32.9 kg) for a total of 187 impacts. Porcine thoraces were tested until normal cardiac rhythm could not be reached between shots. As a result, between 9 and 12 impacts per porcine subjects were done.

All impacts occurred over the center of the cardiac silhouette. For the porcine subjects, only the consolidated ventricular fibrillation risk (VF) function against impact characteristics and average thoracic response corridors are presented, while for the PMHS, detailed Cmax, VCmax and peak force data as well as average thoracic response corridors are presented. Comparison between live porcine and PMHS response shows that for low impact velocities (13.4 and 17.9 m/s), response are similar, while for the two highest impact velocities (22.4 and 26.8 m/s), the peak forces and peak deflections of the porcine subjects are higher than for the PMHS.

2.4 Summary of findings

This literature review indicates that a relationship between animal (live or post-mortem) and PMHS response to thoracic impact exits. The exact nature of the relationship is not clear although some trends can be underlined:

- Yaek [10], clearly demonstrate that PMHS and live porcine response are comparable for subjects with the same total body mass under the same impact conditions. She also validated the use of scaling laws to scale thoracic response for different subject size. Unfortunately, these demonstrations were done for side impacts and for high mass-low impact velocity (~10 m/s) impacts.
- For lower mass-higher impact velocity impact conditions, Prat [12], conclusion mentions that the motion of the pig's chest is greater than that of the PMHS. This seems contrary to Yaek conclusion. The difference can be explained if we consider that the range of mass of the porcine subjects (49 ± 1.5 kg) is quite different from the range of mass of the PMHS (46 to 101 kg). Scaling of thoracic response to body mass might reduce the observed difference.
- Still for low mass-high impact velocity impact conditions, Dau [16] shows that as the projectile impact velocity increases, the discrepancy between the PHMS and porcine response increases, porcine thoracic dynamic being higher. This fits well the conclusion from [12], but again, the difference between the PMHS body mass (44.9 to 71.2 kg) and the porcine subject's body mass (21 to 45 kg) might explain the observed difference.
- In [3], Neurauter compared live porcine subject to live human thoracic stiffness. Response for both species were found to be different for larger deformation (live porcine thoraces being stiffer than live human thoraces). Porcine subject's mass range (23-46 kg) was most likely lower than that of the 90 human patients (mass data not available). These data were generated for very low loading rates.
- Only CPR data from [4] compares thoracic response of live humans to PMHS. In this case, subject's mass data is not available. The authors observed that PMHS thoraces are stiffer than live human thoraces. Similar to above, these data were generated for very low loading rates.

https://doi.org/10.52202/078352-0026

If a relationship between live or post-mortem porcine thoracic response and PMHS/live human thoracic response exists, clearly the rate of application of the force and the subjects mass have to be considered.

3. OPEN LITERATURE DATA ANALYSIS

Analysis of the above mentioned open literature data is presented in the following sections. For the response related dependent variables VCmax (m/s), Cmax (mm/mm) and Dmax (mm), the independent variables studied are: Energy (E, in J), Energy per unit area for the impactor (E/A, in J/cm^2), Impulse (I, in Ns), Impulse per unit area for the impactor (I/A, in Ns/cm^2), Impact velocity (V, in m/s), Blunt Criterion (BC[1]) and mV/W/D (projectile momentum, in Ns scaled by animal mass W, in kg and projectile diameter D, in cm). For all variables, the projectile mass (m) is in kg. The independent variable mV/W/D is used here since it was used in [18] as a preliminary variable and enables scaling to the impact to the animal size and impact surface.

For the injury severity related data, dependent variables # Rib Fx (number of rib fractures) and # Rib Fx/A (# Rib Fx scaled to the projectile impact area, in cm^2) are used. The independent variables studied are, with the same units as above: E, E/A, I, I/A, V, BC, mV/W/D, VCmax and Cmax.

For all curve discussed the coefficient of determination (R^2) is calculated and presented in the figures and an ANOVA (ANalysis Of VAriance) is done on the fitted equation parameters to assess if the curves for the different subject groups (PMHS, PMPS and LPS) are statistically different. The data fitting process was done using MS Excel LINEST function from which the standard error estimate can also be found. Statistically significant difference is reached if p-value of the test is less than 0.05.

3.1 Dynamic response related dependent variables correlation

Error! Reference source not found. presents VCmax versus projectile momentum data. Data used for LPS are from [12] and data used for PMHS are from [12], [11] and [16]. Consequently, all data used are for low mass-high velocity impacts. No VCmax data are available for PMPS. It can be observed that response for LPS is higher than for PMHS, but an ANOVA of the equation parameters shows that they are not statistically significant, i.e. both curves are essentially the same (equation: $y = Ax^B$, for A, p =0.162, for B, p = 0.238). This is due to the large scatter of the PMHS data. Therefore, for the same projectile momentum, LPS present similar VCmax values compared to PMHS.

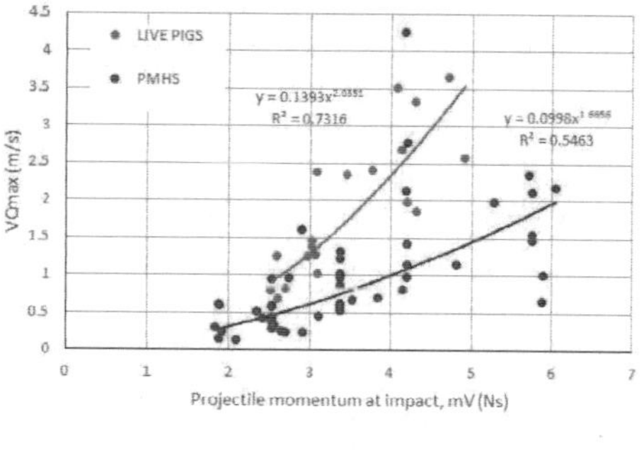

Figure 1 – VCmax (m/s) versus projectile momentum (Ns) for live porcine subjects and for PMHS

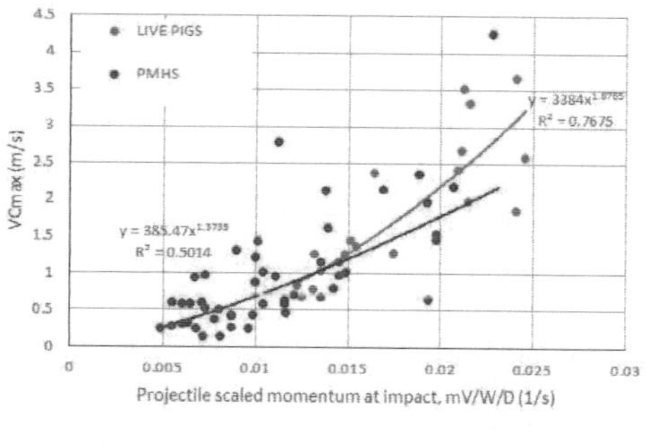

Figure 2 – VCmax versus mV/W/D (1/s) for live porcine subjects and for PMHS

Error! Reference source not found. presents VCmax versus mV/W/D. Data used for LPS are from [12], data used for PMHS are from [12], [11] and [16]. Consequently, all data used are for low mass-high velocity impacts. No VCmax data are available for PMPS. The relationship between VCmax and BC (not shown) did not produced good correlation. Animal mass and projectile diameter scaling of the projectile momentum data have not changed the outcome: both curves are essentially the same (equation: $y = Ax^B$, for A, p =0.061, for B, p = 0.058). The same conclusion can be given for VCmax versus BC data (not presented here).

---

[1] Blunt Criterion, BC is defined in [17] as: $BC = \ln(E/(W^{1/3} \times T \times D))$, where E is the projectile energy in J, W is the animal mass in kg, T is the thoracic wall thickness in cm and D is the projectile diameter in cm. The thoracic wall thickness is estimated in [17] using the following formulas: $T = k \times W^{1/3}$ where k = 0.751 for pigs, k = 0.593 for women and k = 0.711 for men.

https://doi.org/10.52202/078352-0026

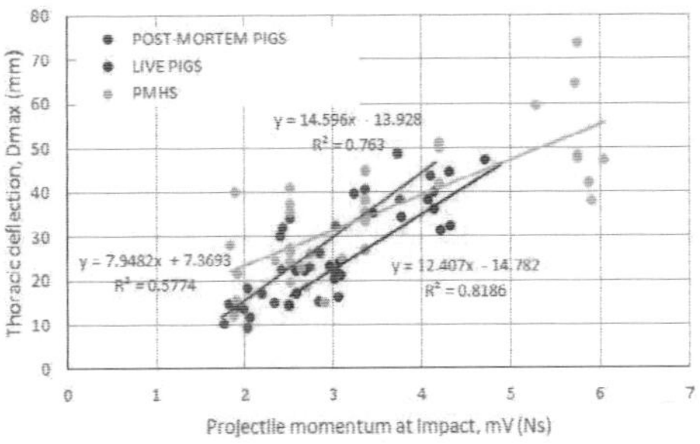

Figure 3 – Maximal thoracic deformation versus initial projectile impulse for LPS and PMPS, reproduced from [14], with PMHS ([12], [11] and [16]) data overlaid.

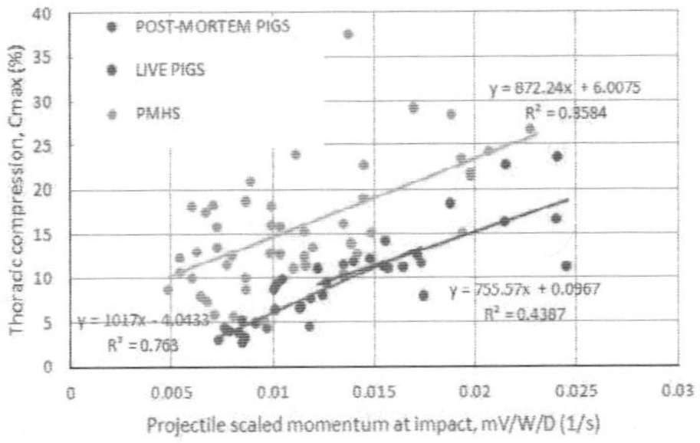

Figure 4 – Cmax versus mV/W/D for LPS, PMPS and PMHS. Data for LPS are from [12], data for PMHS are from [12] and [11] and finally, data for PMPS are from [14].

Pavier [14] shows that for PMPS and LPS, the projectile initial momentum correlates to the maximal thoracic deformation, Dmax. The figure reproduced from [14] is presented in Figure 3 with PMHS data from [12], [11] and [16] overlaid on top. The average body mass of the LPS in [12] is 49 kg and it is 80 kg for PMPS in [14] while for PMHS, it is 69.3 kg in [12], 79.3 kg in [11] and 58.8 kg for [16]. Despite the different average body mass of the datasets and the large variation of the PMHS data, analysis shows a statistically significant difference between both types of porcine subjects and PMHS (PMHS vs LPS: $y = Mx + B$, for M, $p < 0.01$, for B, $p < 0.01$), while it shows no statistically significant difference between LPS and PMPS (PMPS versus LPS: $y = Mx + B$, for M, $p = 0.170$, for B, $p = 0.452$). Therefore, under the same projectile momentum, PMHS thoraces deforms differently than porcine subjects, either live or post-mortem. Another ANOVA (correlation not shown here) shows that thoracic compression (Cmax) versus projectile momentum curves are not statistically significant between PMHS, PMPS and LPS. Scaling of the projectile momentum with the subjects' body mass and projectile diameter (mV/W/D), Figure 4 shows that maximal thoracic compression curves are not statistically significant between PMHS, PMPS and LPS (PMHS vs PMPS: $y = Mx + B$, for M, $p = 0.245$). Again, this is due to the large scatter of the PMHS data.

For VCmax and Cmax data, attempts were made to reduce variability of the PMHS data by using only one dataset at a time (either [12] or [11] or [16]) to see if significant difference between PMHS and porcine data fit could be observed. No significant differences were observed. In conclusion, comparison of thoracic response data for a variety of blunt impact conditions shows that:

- VCmax values for thoracic impacts on LPS and PMHS are similar (not statistically different) for a given scaled (mV/W/D or BC) or unscaled (mV) projectile impulse
- Cmax values for thoracic impacts on LPS, PMPS and PMHS are similar (not statistically different) for a given scaled (mV/W/D or BC) or unscaled (mV) projectile impulse.
- Post-mortem human subject Dmax is different of porcine subjects (live or post-mortem) Dmax. No difference is found between live and post-mortem porcine subject responses.

3.2 Injury severity related dependent variables correlation

Figure 5 presents number of rib fracture versus Cmax data for LPS, PMPS and PMHS. Data for LPS are from [12] and [8], data for PMHS are from [12] and [11] and finally, data for PMPS are from [14] and [8]. The data cover a wide range for impact conditions covering low mass-high velocity impacts and high mass-low velocity impacts. Direct comparison between curves shows they are all statistically different. Notice also that the R^2 for PMPS (0.97) and LPS (0.71) are high, whereas it is low (0.52) for PMHS.

https://doi.org/10.52202/078352-0026

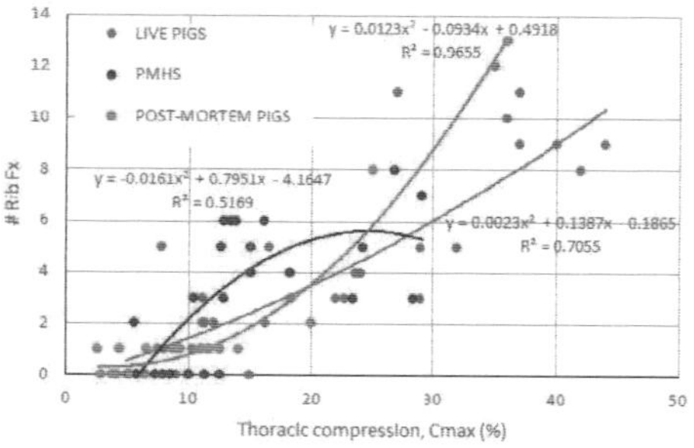

Figure 5 – Number of rib fracture versus Cmax for live and post-mortem porcine subjects and for PMHS

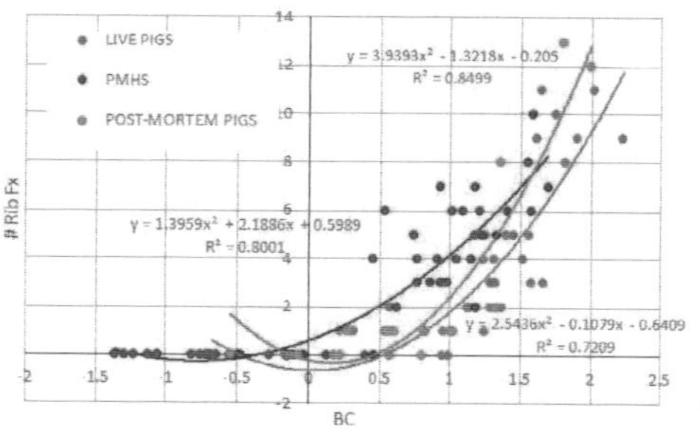

Figure 6 – Number of rib fracture versus BC for live and post-mortem porcine subjects and for PMHS

Figure 6 presents number of rib fracture versus BC for LPS, PMPS and PMHS. Data used for live pig are from [12] and [8], data used for PMHS are from [12], [11] and [16] and finally, data for post-mortem porcine subjects are from [14] and [8]. Again, impact conditions cover low mass-high velocity impacts and high mass-low velocity impacts. Direct comparison between curves shows that PMHS curve is statistically different from both, PMPS and LPS curves and the difference between PMPS and LPS curves is also statistically different. Notice also that the R^2 for each curve is high (0.72 for LPS, 0.80 for PMHS and 0.85 for PMPS).

To conclude, comparison of injury severity (number of rib fracture) for a large variety of blunt impact conditions shows that:

- The BC independent variable enables significant discrimination between PMHS, PMPS and LPS injury response. Also, for each subject types, high correlation is found with BC.
- The Cmax independent variable enables significant discrimination between PMHS, PMPS and LPS injury response. Also, for porcine subjects, high correlation is found with Cmax, but it is not the case for PMHS.
- No correlation is found between the number of rib fracture and VCmax.

4. DISCUSSION

For a given projectile momentum (scaled or unscaled to animal mass), scaled dynamic thoracic response (Cmax and VCmax) are the same between human and porcine subjects. This is due, in part to the large variability of the PMHS response as it can be seen in **Error! Reference source not found.** to Figure 4. Normalisation of the data presented above for a given body mass (50th percentile male, 77.7 kg) following the process detailed in [19], [20] and [21] results in slight decrease of the PMHS response variability. For Dmax, Cmax and VCmax, use of normalisation does not improve the correlation coefficient of the different thoracic response variables for the different subjects, nor does it enable differentiation between PMHS, PMPS and LPS responses. In Yaek [10], dynamic thoracic response of PMHS is found to be the same as the dynamic thoracic response of LPS when data are scaled to the same body mass. Similarly here, it is observed that the dynamic thoracic response of PMHS cannot be differentiated from porcine dynamic thoracic response. Part of the observed PMHS data variation is due to the different projectile compliance used in [11], [12] and [16] and also variation in impact location on the subjects' thorax, which also results in varying compliance. Nevertheless, the current analysis of the dynamic response data seems to support the observations made by Yaek [10] for high mass-low velocity impacts and expands the applicability of Yaek's conclusion to low mass-high velocity data.

Thoracic injury severity response (# of rib fracture) is found to correlate with BC. Furthermore, distinct responses can be observed for PMHS versus PMPS versus LPS. It is found that for the same impact severity, PMHSs' show higher number of rib fracture compared to porcine subjects. Also, PMPSs' show higher number of rib fracture than LPSs' for the same insult. As expected, post-mortem subjects (human or porcine) are more sensitive to impacts than live subjects. In [8], Viano showed that it is the case for high mass-low velocity impacts and in [12], Prat reached a similar conclusion for low mass-high velocity impacts. Results shown here seem to show that it is also the case for a wider range of impact conditions. An estimate of impact sensitivity for LHS is shown in Figure 7. It is calculated assuming that the difference between post-mortem and live porcine subjects is the same as the difference between PMHS

https://doi.org/10.52202/078352-0026

and LHS. An analysis of Figure 7 shows that the expected number of rib fracture (and therefore the severity of the injury) for a LHS is higher than for either a PMPS or a LPS. For BC values higher than 1.56, the expected severity of injury for LHS is lower than for PMPS and LPS. An impact with a BC value of 1.56 is quite severe as it is is higher than most BABT related impact severity (see next section).

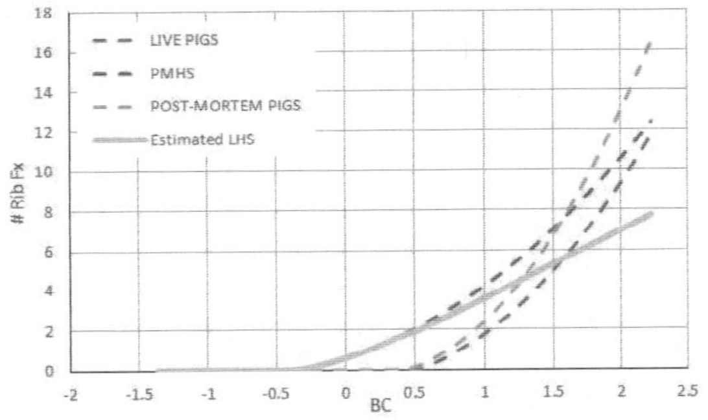

Figure 7 – Estimate of # of rib fractures versus BC for LHS (= PMHS – (PMPS-LPS)). LPS, PMHS and PMPS curves are the same as in Figure 6 with negative # of rib fracture values removed. LHS curve equation is $0.2765x^2 + 2.6194x + 0.6385$

The relationship shown above for LHS does not explain why, for similar impact loadings, experimental behind armour data on animals are resulting in injuries while operational data are not showing injurious response. The assumption made above might not be true. In order to assess the truthfulness of that assumption, thoracic blunt impact injury data from theaters of operation would be necessary.

4.1 Severity of KENLW impacts relative to BABT related impacts

The conclusions made above are mostly based on data related to KENLW projectile impacts. Although BABT and KENLW projectile impacts are both considered as blunt impacts, they are not necessarily equivalent. Literature contains a large number of thoracic BABT test results that involve the use of specific alumina and UHMWPE armours against a variety of projectiles, including 12.7 mm Ball, 7.62 mm Ball and 5.56 mm Ball projectiles, [22] to [33]. As these tests involve the use of LPS some comparison can be drawn at least on the severity of the impacts generated in both scenarios. To enable comparison, it is necessary to translate the complex projectile-armour interaction into an equivalent projectile. Assuming conservation of momentum just before and just after impact on the armour and assuming that the armour surface pushed against the thorax corresponds to the surface of the bruise left on the animal, it is possible to calculate an effective mass, an effective velocity and an effective BC value for a given BABT event. Figure 8 presents effective BC versus effective mass for BABT ([22] to [33]) and KENLW ([11], [12], [16]) events. The use of BC to assess impact severity accounts for both, the severity of the impact and the size of the animal and therefore enables a direct comparison of impact severity.

It can be seen that the severity level (BC value) for each type of event is generally equivalent and it is lower than BC = 2.0. The impact with a BC value of 2.85 was obtained for a very light animal (20 kg only). The effective projectile mass range for BABT experiments is much larger than for KENLW impacts, i.e. impacts should be done with heavier blunt projectile to cover the full range. Finally, severity of impact for typical 7.62 mm Ball round corresponds to the upper end of KENLW impact severity, i.e. impacts with typical KENLW projectiles should be done at higher velocity to reach the level required to simulate BABT impacts.

5. CONCLUSIONS

Dynamic and injurious response of PMHS and porcine thoraces to blunt impacts for a wide range of impact conditions were studied with the objective to determine the expected thoracic response of live human subjects. The study showed that dynamic thoracic response of PMHS and porcine subjects are difficult to differentiate due to the large variations in PMHS response data available. The injury severity for PMHS, post-mortem porcine subjects and live porcine subjects with respect to impact BC was shown to be statistically different. This enables the estimation of the thoracic response of live humans by assuming that the difference between post-mortem and live porcine subjects is the same as the difference

https://doi.org/10.52202/078352-0026

between PMHS and live human subjects. It is predicted that LHS would have more severe injuries than PMPS and LPS for similar impact for a wide range of impact severity. Clearly, BABT related injury data from the field must be used to verify the assumptions used to assess live human subject response. In addition, it is demonstrated that the range of blunt impact tests on animal and PMHS test should be expanded to cover the full range of BABT injury severity and that those tests should be done against animals with the same body mass as the subjects to be protected. The data presented cannot explain the observed difference in BABT injury severity between laboratory animals and humans.

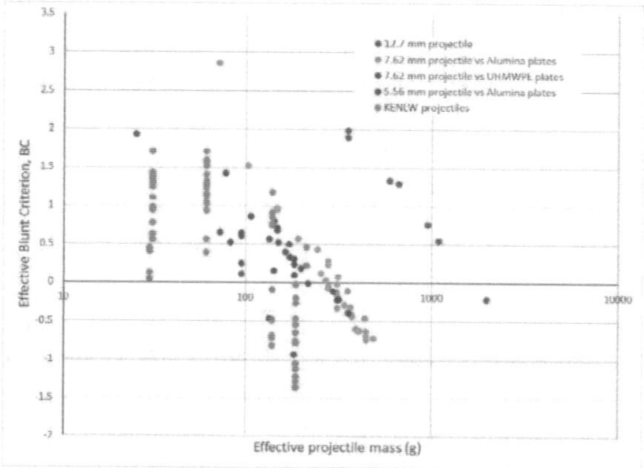

Figure 8 – Estimate of effective BC versus effective impact mass on the thorax of LPS and PMHS for KENLW projectile and BABT events

REFERENCES

[1] Maltese, M.R., Arbogast, K.B., Wang, Z. & Craig, M.T., Scaling Methods Applied to Thoracic Force Displacement Characteristics Derived from Cardiopulmonary Resuscitation, National Highway Traffic Safety Administration, 22nd International Technical Conference on the Enhanced Safety of Vehicles, 13-16 June 2011, Washington, D.C., USA.

[2] Maltese, M.R., Arbogast, K.B., Nadkarni, V., Berg, R., Balasubramanian, S., Seacrist, T., Kent, R.W., Parent, D.P., Craig, M. and Ridella, S.A., Incorporation of CPR Data into ATD Chest Impact Response Requirements, Ann Adv Automot Med. 2010 Jan; 54: 79–88.

[3] Neurauter, A., & al. Comparison of mechanical characteristics of the human and porcine chest during cardiopulmonary resuscitation, Resuscitation, Vol. 80, 2009, pp 663-469

[4] Arbogast, K.B., Maltese, M.R., Nadkarni, V.M., Steen, P.A. & Nysaether, J.B., Anterior-Posterior Thoracic Force-Deflection Characteristics Measured During Cardiopulmonary Resuscitation: Comparison to Post-Mortem Human Subject Data, Stapp Car Crash Journal, Vol. 50, November 2006, pp. 131-145

[5] Gruben, K., Guerci, A., Halperin, H., Popel, A. & Tsitlik, J., Sternal force-displacement relationship during cardiopulmonary resuscitation, Journal of Biomechanical Engineering, Vol. 115, 1993, pp. 195-201.

[6] Tsitlik, J., Weisfeldt, M., Chandra, N., Effron, M. Halperin, H. & Levin, H., Elastic Properties of the human chest during cardiopulmonary resuscitation, Crit. Car. Med., Vol. 11, No. 9, 1983, pp. 685-692

[7] Viano, D.C., Warner, C.Y., Thoracic Impact Response of Live Porcine Subjects, Paper 760823, Proceedings of the 20th Stapp Car Crash Conference, New York, Society of Automotive Engineers, Inc., 1976

[8] Viano, D.C., Kroell, C.K. and Warner, C.Y., Comparative Thoracic Impact Response of Living and Sacrificed Porcine Siblings, SAE Technical Paper 770930, 1977.

[9] Neathery, R.F., Kroell, C.K. and Mertz,, H.J., Prediction of Thoracic Injury from Dummy Response, Paper 751151, Proceedings of the 19th Stapp Car Crash Conference, New York, Society of Automotive Engineers, Inc., November 1975

https://doi.org/10.52202/078352-0026

[10] Yaek, J., Biofidelity Assessment Of 6-Year-Old Anthropometric Test Devices (ATDs) And Scaling Laws In Lateral Impact, Wayne State University Doctoral Dissertation, Paper 1902, Jan. 2017.

[11] Bir, C. A., The Evaluation of Blunt Ballistic Impacts of the Thorax, Wayne State University Doctoral Dissertation, 2000

[12] Prat, N., Rongerias, F., de Freminville, H., Magnan, P., Debord, E., Fusai, T., Destombe, C., Sarron, J.C., and Voiglio, E.J., Comparison of thoracic wall behavior in large animals and human cadavers submitted to an identical ballistic blunt thoracic trauma, Forensic Science International, 222 (2010)

[13] Prat, N., Rongerias, F., Voiglio, E., Magnan, P., Destombe, C., Debord, E., Barbillon, F., Fusai, T. & Sarron, J.C., Intrathoracic Pressure Impulse Predicts Pulmonary Contusion Volume in Ballistic Blunt Thoracic Trauma, The Journal of Trauma, Vol. 69, No. 4, October 2010.

[14] Pavier, J., Langlet, A., Eches, N., Prat, N., Bailly, P. and Jacquet, J.F., Experimental study of the coupling parameters influencing the terminal effects of thoracic blunt ballistic trauma, Forensic Science International, 252, 2015, pp. 39-51

[15] Pavier, J., Langlet, A., Eches, N., and Jacquet, J.F., On ballistic parameters of less lethal projectiles influencing the severity of thoracic blunt impacts, Computer Methods in Biomechanics and Biomedical Engineering, Vol. 18, No. 2, 2015, pp. 192-200.

[16] Dau, N., Development of a biomechanical surrogate for the evaluation of commotio cordis protection, Wayne State University Dissertations, Paper 407, 2012

[17] Sturdivan, L.M., Viano, D.C., Champion, H.R., 'Analysis of Injury to Assess Chest and Abdominal Injury Risks in Blunt and Ballistic Impacts', Journal of Trauma, Volume 56, No. 3, March 2004.

[18] Clare, V.R., Lewis, J.H., Mickiewicz, A.P. and Sturdivan, L.M., Blunt Trauma Data Correlation, Report number EB-TR-73016, Edgewood Arsenal, Maryland, USA, May 1975

[19] Mertz, H.J., Irwin, A.L., Melvin, J.W., Stalkner, R.L., Beebe, M.S., Size, Weight and Biomechanical Impact Response Requirements for Adult Size Small Female and Large Male Dummies, Society of Automotive Engineers, Paper 890756, 1989

[20] Irwin, A.L., Mertz, H.J., Elhagediah, A.M., Moss, S., Guidelines for Assessing the Biofidelity of Side Impact Dummies of Various Sizes and Ages, Stapp Car Crash Journal, Vol. 46, November 2002, pp. 297-319.

[21] ISO/TR 12350:2013, Road vehicles – Injury risk curves for the evaluation of occupant protection in side impact tests, Edition 2, October 2013.

[22] Gryth, D., Maj. 'Hemodynamic, Respiratory and Neurophysiological Reactions after High-Velocity Behind Armor Blunt Trauma', Thesis for Doctoral Degree, Karolinska Institutet, Sweden, 2007.

[23] Drobin, D., Gryth, D. Maj., Persson, J.K.E., Rocksén, D., Arborelius, U.P., Olsson, L-G., Bursell, J. and Kjellström, B.T., 'Electroencephalogram, Circulation, and Lung Function After High-Velocity Behind Armor Blunt Trauma', Journal of Trauma, Vol. 63, No. 2, pp 405-413, August 2007.

[24] Gryth, D. Maj., Rocksén, D., Persson, J.K.E., Arborelius, U.P., Drobin, D., Bursell, J. and Olsson, L-G., 'Severe Lung Contosion and Death after High-Velocity Behind-Armor Blunt Trauma: Relation to Protection Level', Military Medicine, Vol. 172, pp 1110-1116, October 2007.

[25] Gryth, D. Maj., Rocksén, D., Arborelius, U.P., Drobin, D., Persson, J.K.E., Sondén, A., Bursell, J., Olsson, L-G., Kjellström, B.T., Bilateral vagotomy inhibits apnea and attenuates other physiological responses after blunt chest trauma, J Trauma, Vol. 64, No. 6, Jun 2008, pp 1420-1426.

[26] Rocksén, D., Gryth, D., Druid, H., Gustavsson, J. and Arborelius, U.P., 'Pathophysiological effects and changes in potrassium, ionised calcium, glucose and haemoglobin early after severe blunt chest trauma', Injury, Int. J. Care Injured, 43 (2012), 632-637

[27] Gryth, D., Rocksén, D., Drobin, D., Druid, H., Weitzberg, E., Bursell, J., Olsson, L-G., Arborelius, U.P., Effects of fluid resuscitation with hypertonic saline dextrane or Rigner's acetate after nonhemorrhagic shock caused by pulmonary contusion, J Trauma, Vol. 69, No. 4, Oct. 2010, pp 741-748

[28] Sondén, A., Rocksén, D., Riddez, L., Davidson, J., Persson, J.K., Gryth, D., Bursell, J. and Arborelius, U.P., 'Trauma Attenuating Backing Improves Protection Against Behind Armor Blunt Trauma', The Journal of Trauma, Vol. 67, No. 6, December 2009

https://doi.org/10.52202/078352-0026

[29] Riddez, L., Rocksén, D., Dondén, A., Persson, J.K., Gryth, D., Bursell, J. & Arborelius, U.P., 'Increased Protection Against Behind Armour Blunt Trauma using Trauma Attenuating Backing (TAB)', Proceedings of the Personal Armour System Symposium 2006, Leeds, UK, 19-22 September 2006.

[30] Johansen - Trial Report 02/97 - Behind Armour Blunt Trauma, Danish Army Combat School, Trials and Safety Branch, 24-28 February 1997

[31] Sarron, J.C., Destombe, C., Da Cunha, J., Martinez, Vassout, P., Magnan, P., Blessures thoraciques par balle de guerre sous protection balistique individuelle – Étude comparative de trois plaques de protection – Essais conduits à Oksbøl (Danemark) par le groupe de travail TG001 du groupe OTAN HFM 024, PEA 980823, DGA/DSP/STTC/DT-SH, décembre 2000

[32] Sarron, J.C., Destombe, C., Da Cunha, J., Martinez, Vassout, P., Lésions Thoraciques fermées par balles de guerre – Étude comparative de deux plaques de protection : céramique et polyéthylène, PEA 980823, Contrats DGA 98028, 9810075 et 98040, septembre 1999

[33] Sarron, J.C., Destombe, C., Da Cunha, J., Morin, Chene, Vassout, P., Magnan P., Gravité des blessures non-penetrantes du thorax protégé par in gilet pare-balles en fonction de l'énergie d'impact d'une munition de calibre 7.62 – Étude de seuils lésionnels, PEA 980823, DGA/DSP/STTC/DT-SH, février 2003

https://doi.org/10.52202/078352-0026

Assessment of Head Injuries: Blunt versus Penetrating

K. Loftis[1], E. Matheis[2], and K. Rafaels[3]

[1]*Army Futures Command, CCDC, Data and Analysis Center, Aberdeen Proving Ground, MD, USA, kathryn.l.loftis2.civ@mail.mil*
[2]*Bennett Aerospace Inc., Contractor to Army Research Laboratory, WMRD, Aberdeen Proving Ground, MD, USA*
[3]*Army Futures Command, CCDC, Army Research Laboratory, WMRD, Aberdeen Proving Ground, MD USA*

Abstract. Current personal armour systems, such as helmets, are optimized for penetration protection to the vital areas of the body – most notably the head and torso. However, even if the helmet prevents penetrating injuries, impacts that produce behind-helmet blunt trauma (BHBT) can still occur. With current armour solutions, this presents a design trade-off – prevent penetrations which may result in high-rate blunt injuries, or limit blunt injuries while decreasing the penetration performance. As a first look at the mortality differences between blunt and penetrating head injuries, this study utilized a large civilian trauma hospital dataset (National Trauma Databank 2013-2015) to determine injury trends. While these civilian injuries are not specific to gunshot wounds or BHBT, the analyses did allow a direct comparison between head injuries categorized with either blunt or penetrating mechanism in a large dataset where mortality could be calculated. Examining patients with only head injuries (n=92,989 patients), 83,631 had blunt and 5,687 had penetrating mechanism injuries (3,671 patients had unspecified mechanisms). The mortality rate of those with blunt mechanism head injuries was 3% while those with penetrating mechanism head injuries was 54%, which was a statistically significant difference. Additionally there were differences in the types of head injuries between blunt and penetrating trauma (3% vs. 10% involved skull fracture without focal brain injury, 31% vs. 19% involved focal brain injury without skull fracture, and 16% vs. 52% had both skull fracture and brain injury, respectively). Cases of penetrating mechanism head injuries were more likely to involve co-occurring injuries to the skull and brain, i.e., increased injury severity. Thus, this analysis implies that preventing penetrating mechanism injuries with a helmet, even if blunt injuries occur, should increase survival for a person. However, with regards to BHBT, this assumes that the deformation shape and rate would only generate a blunt impact and injury to the head versus stopping the threat but still resulting in a penetrating impact and injury. Hence, BHBT metrics should consider shape and not just depth to evaluate significant injury risk. Additional research should be done investigating injury trends specific to gunshot wounds and blunt trauma to the head within a military dataset, likely through a case analysis comparison given the small number of cases available.

1. INTRODUCTION

Ballistically-rated personal protective equipment (PPE), such as a helmet, is designed to defeat fragments and bullets by absorbing and dissipating their kinetic energy without allowing the projectile to completely penetrate the armour. When a helmet is worn for protection, it is designed to prevent the penetrating threat, but in return may result in a blunt head injury following backface deformation of the helmet material. In such a case, the helmet has exchanged a penetrating head injury for the solider with a blunt head injury. There has been much debate over the injury implications for behind-armour blunt trauma, so it is relevant to look into the injury outcomes and patterns between penetrating and blunt trauma to the head. With current armour solutions and the goal to reduce the weight of body armour, this presents a design trade-off – prevent penetrations which may result in high-rate blunt injuries, or limit blunt injuries while decreasing the penetration performance [1].

There is currently a poor understanding of the implications for outcome between penetrating and blunt head injuries, which need to be understood to make future decisions about protection and allowable injury risk. As an example of the trade-off, maybe a simple linear skull fracture is an appropriate acceptable injury level if the helmet is protecting against a catastrophic penetrating skull and brain injury. Current laboratory testing can provide key outcome information for head impacts relating to helmets when investigating skull fracture. It is much more challenging to replicate brain injury within

 https://doi.org/10.52202/078352-0027

the laboratory setting, as it is typically an injury process which requires a living subject to propagate the bleeding and properly diagnose the patient based on neurological exams [2-4]. Alternatively, skull fracture can be generated in the laboratory, both with penetrating and blunt threats and specific fracture patterns are seen between the two mechanisms. The relationship between skull fracture patterns and brain injuries is currently poorly understood and documented.

Curently there is limited access to military medical records that contain enough detailed information to perform large sample size statistics on head injuries. Therefore, this study utilized the National Trauma Databank Research Dataset (NTDB RDS) to obtain injury and outcome information for civilian hospital patients with blunt or penetrating head injuries. The goal of this work was to eludicate injury and outcome pattern differences between penetrating and blunt head injuries, investigating the incidence of underlying brain injury associated with each. While this dataset contains a broader scope of head injuries than that just focused on behind-helmet blunt trauma (BHBT), it provided a large injury sample size to investigate the overall relationship between head injury mechanism (penetrating or blunt), associated injury patterns (skull fracture and brain injury), as well as outcome (mortality).

2. METHODS

Utilizing the NTDB RDS from 2013 through 2015 (American College of Surgeons), all patients with demographic information were initially selected. Data was imported into JMP®14 software (SAS, Cary, NC, USA) using a systematic approach where the International Classification of Diseases, ninth revision (ICD-9-CM) codes and other pertinent information provided by NTDB were joined in the order of 1) external cause of injury codes (RDS_ECODES), 2) diagnosis codes (RDS_DCODES), 3) demographic information (RDS_DEMO), 4) emergency department information (RDS_ED), and 5) hospital stay information (RDS_DISCHARGE) matching each specific patient when joining. A filtering process with exclusion criteria based on demographic information, external cause of injury codes, relevant military age, and Abbreviated Injury Scale (AIS) chapters matched to ICD-9-CM diagnosis was implemented (Figure 1). The total number of patients and their injury count was tabulated for each step, using the unique patient code. Initially there were 2,607,945 patients in the dataset from 2013-2015.

In this way, only patients with isolated head injuries were kept for analysis, as previous research has shown that a patient with injuries to multiple body regions has a higher likelihood of fatality [5] and the concomitant injuries may have influenced patient outcome. Once the head-injury only population was identified, the "external cause of injury codes" (RDS_ECODES) were used to distinguish between patients with penetrating and blunt mechanisms. Other injury mechanism types were excluded for this analysis. Additionally, analysis was performed to determine whether each patient had a skull fracture, brain injury, or both, to investigate the mortality rate for each group. For the purposes of this study, a head injury with a penetrating mechanism will be termed a penetrating injury, whether or not the injury involved penetration into the skull. Likewise, a head injury with a blunt mechanism will be termed a blunt injury for this paper, even if the skin and skull are broken. More details about each data filtering step are provided in the following paragraphs.

External Cause of Injury/Age Sorting

Patients with a lack of demographic information were excluded and then the remaining were filtered through the external cause of injury mechanisms. To evaluate injuries most relevant to the research question, only external cause codes associated with penetrating or blunt force trauma were included. These injury mechanisms were identified by using ICD-9-CM external cause of injury codes E2-20 and E800-999. Within that set, adverse outcome due to accidental poisoning (E850-869), surgical complications (E870-879), fire and flames (E890-E899), natural and environmental factors (E900-908, except 908.1-908.3, 908.8, 909.2, 909.3), suffocation or submersion (E910-913), other accidents (E924-928), adverse effects (E930-949), suicide by other than physical trauma (E950-952, E953-954,E958.1-959), and other inflicted injurious events (E961-962, 968, 968.3,972,977,979.3, 980-984, 988.1-988.4 ,988.7) were also excluded. Patients were then categorized by the injury type, either blunt, penetrating, or other/unspecified. Blunt events were mechanisms such as a fall, motor vehicle accident, or struck by/against an object. Penetrating mechanisms included firearms and cut or pierce events. Event mechanisms are characterized by the primary event and it is to be noted that there may be penetrating mechanisms in the blunt category and vice versa. For example, a person may be in a car

https://doi.org/10.52202/078352-0027

accident, which would be characterized as a blunt mechanism, but then possibly have cut or pierce injuries from broken glass, which would not be characterized as penetrating in this database.

After external cause and mechanism sorting of population, the dataset was then filtered by a military relevant age, including only ages 17-55 in the dataset.

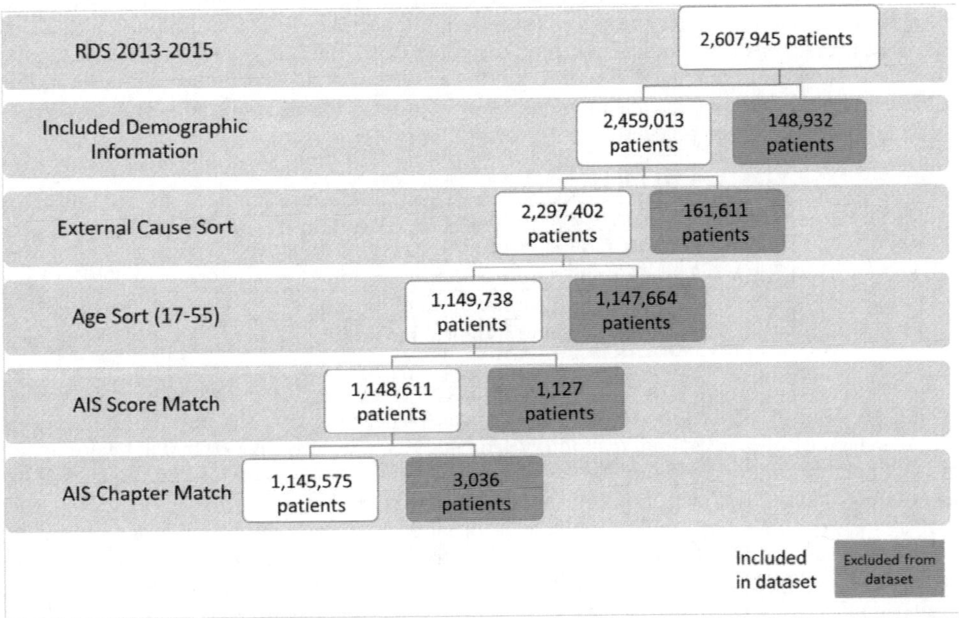

Figure 1: Filtering process from NTDB RDS 2013-2015 with patient quantities for each step in the filtering process. Totals presented in the gray boxes indicate the data that was excluded from the dataset through each step in the process, while the white boxes represent the included data.

Head Injury Population/AIS Sorting

NTDB uses ICD-9-CM as the main injury coding component within the database. ICD-9-CM groups the head, neck, and face together, making it difficult to separate out head injuries. Therefore, ICD-9-CM diagnosis codes (D codes) were mapped to AIS scores [6,7] using a protocol that was developed by certified coders of both ICD and AIS to isolate head injuries of interest. This allowed for separation of scalp, brain, and skull injuries from face and neck injuries through AIS chapter classification. Although the map is robust, there was no match in AIS for some ICD-9-CM D codes. For example, codes that included multiple body regions or multiple injuries within one code, such as fractures involving skull or face with other bones (804) had no equivalent AIS score. Therefore, incidents and injuries that did not match with any AIS codes or chapters were excluded from this analysis (Figure 1). After mapping the injuries to AIS chapters, it was possible to select only the injuries that were associated with the head (i.e., AIS Chapter 1). From this cohort a head-injury-only population was determined by totaling the injuries within each region and selecting patients with chapter 1 injuries only.

Level of Medical Care

Patients were either admitted to the emergency department, hospital, or both and both the emergency department cases and hospital admissions were included in the dataset. Patients that were either not applicable (BIU1) or not known/not recorded in the hospital disposition (HOSPDISP) were excluded from this analysis.

Specific head injury designations

The ICD-9-CM codes were used to classify the specific head injuries in the patients, in particular, skull fracture and focal brain injury. Using the diagnosis code (D-code) descriptions for presence of skull fracture, the patient was categorized as either having or not having a skull fracture. In addition to D-codes describing any type of skull fracture, skull fracture was also implied for any ICD code that had an open brain injury, even if skull fracture was not specifically described in the D-code (i.e., 851.1, 851.3, 851.5, 851.7, 851.9, 852.1, 852.3, 852.5, 853.1, and 854.1). Similarly, D-code descriptions were

https://doi.org/10.52202/078352-0027

used to categorize patients with focal brain injuries. For the purposes of this study, the definition of focal brain injury included the brain and surrounding internal soft tissue structures. D-codes such as cerebral contusion, injury to cranial nerves, cerebellum/brain stem laceration, and subdural, subarachnoid or extradural haemorrhages were categorised as a focal brain injury. Any crush, unspecific blood vessel injury, superficial wound, skull fracture only, or unspecified brain injury were not included in the focal brain injury category for this study. Concussion was also excluded because of the inconsistencies with diagnosis and notations within NTDB. Because some of the head injury patients had both skull fracture and focal brain injury, each patient was then further separated by the groups described in Table 1 to elucidate possible trends within the head injury population and for a better understanding of coexisting injuries.

Table 1: Head Injury Group Definitions

Group	**Head injury category**
None	No skull fracture and no focal brain injury
SF	Skull fracture, no focal brain injury
FB	Focal brain injury, no skull fracture
Both	Skull fracture and focal brain injury

Mortality was used as a means to investigate the differences in patient outcomes between blunt and penetrating mechanism populations using the hospital disposition code (expired). To calculate mortality, the number of patients within an injury grouping that died was divided by the total number of patients within that injury grouping. Prevalence of patients with each injury type was calculated by dividing the population by primary mechanism (blunt/penetrating). Further comparative numbers and percentages were calculated by dividing the number of interest by the total number in the group. In this way, a cross-tabulation table was created for both blunt and penetrating trauma providing the number of patients within each head injury category shown in Table 1. To determine if there was statistical significance between the mortality rates between blunt and penetrating mechanisms of injuries , a two-tailed z-score was calculated and evaluated at 0.05 for significance.

3. RESULTS

Using 2013-2015 NTDB RDS after applying all the exclusion criteria there were 83,631 patients with blunt mechanism head injuries and 5,687 patients with penetrating mechanism head injuries. Each of these groups were further broken down to investigate trends between skull fracture and focal brain injury.

Blunt mechanism head injuries

Table 2 shows the results of the blunt mechanism head injury analysis, where the majority of patients didn't have skull fracture or brain injury due to the blunt mechanism (49%, 41,264/83,631) and overall mortality was 3% (2,304/83,631). Comparing mortality rates between those with only a skull fracture (3%, 76/2,723), those with only a focal brain injury (4%, 1,064/26,313), and those with both a skull fracture and brain injury (6%, 833/13,331), there is a slight increase in mortality for the combined head injury group. While those with a blunt mechanism only sustained a skull fracture in 19% of patients, the percentage with a brain injury was closer to even at 47%. Blunt mechanism injury patients with a skull fracture had a higher incidence of associated brain injury at 83% (13,331/16,054), compared to those without skull fracture with a brain injury incidence of 39% (26,313/67,597).

https://doi.org/10.52202/078352-0027

Table 2. 2013-2015 NTDB RDS patients with isolated blunt mechanism head injuries, where a) number of patients that died in each injury category (red) and the total in that category and b) the percent of patients in that injury category that died (red) and the percent of patients from the blunt mechanism injury population that fell into that injury category.

2013-2015 NTBD RDS Blunt Mechanism Head Injuries and Deaths

	No Skull Fx	Skull Fx	Total
No Focal Brain Inj	331 **4,1264**	76 **2,723**	407 43,987
Focal Brain Inj	1,064 **26,313**	833 **13,331**	1,897 39,644
Total	1,395 67,597	909 16,054	2,304 83,631

	No Skull Fx	Skull Fx	Total
No Focal Brain Inj	1% **49%**	3% **3%**	1% 53%
Focal Brain Inj	4% **31%**	6% **16%**	5% 47%
Total	2% 81%	6% 19%	3% 100%

Penetrating mechanism head injuries

For the 5,687 patients with penetrating mechanism head injury, the overall mortality was 54% (3,066/5,687), as shown in Table 3. The majority of patients had both skull fracture and brain injury, 52% (2,979/5,687), with the smallest percentage of the group only having a skull fracture (10%, 573/5,687). Comparing mortality rates between those with only a skull fracture (57%, 325/573), those with only a focal brain injury (59%, 618/1,054), and those with both a skull fracture and brain injury (67%, 1984/2979), the combined head injury group had the highest mortality. Patients with a skull fracture had a higher mortality rate (66%, 2,339/3,552) than those without a skull fracture (35%, 757/2,135). Penetrating mechanism injury patients with a focal brain injury also had a higher mortality rate (65%, 2,602/4,033) than those without (30%, 494/1,654).

Table 3. 2013-2015 NTDB RDS patients with isolated penetrating mechanism head injuries, where a) number of patients that died in each injury category (red) and the total in that category and b) the percent of patients in that injury category that died (red) and the percent of patients from the penetrating mechanism injury population that fell into that injury category.

2013-2015 NTDB RDS Penetrating Mechanism Head Injuries and Deaths

	No Skull Fx	Skull Fx	Total
No Focal Brain Inj	139 **1,081**	325 **573**	494 1,654
Focal Brain Inj	618 **1,054**	1,984 **2,979**	2,602 4,033
Total	757 2,135	2,339 3,552	3,096 5,687

	No Skull Fx	Skull Fx	Total
No Focal Brain Inj	13% **19%**	57% **10%**	30% 29%
Focal Brain Inj	59% **19%**	67% **52%**	65% 71%
Total	35% 38%	66% 62%	54% 100%

Blunt and Penetrating Mortality Comparison

A z-score for a proportion comparison between the overall mortality rate of blunt or penetrating mechanism injuries was calculated. The value of z is -157.0404 which equates to a resultantant p value < .00001. Significance was calculated at $p<0.05$. Therefore, the mortality rate of those with penetrating head injuries was significantly different than the mortality associated with blunt head injuries.

https://doi.org/10.52202/078352-0027

4. DISCUSSION

While this research would ideally be performed using a military dataset to more closely match blunt and penetrating mechanisms that are pertinent to the US military, that was not possible given the limited data accessability and the small numbers in military injury data versus the large civilian hospital dataset that is available through NTDB. Within each injury mechanism category, regardless of whether the source of data is military or civilian based, there can be a wide range of injury mechanisms that may demonstrate their own unique injury patterns and outcomes. For example, the penetrating injury category includes mechanisms of gunshot wounds and stab wounds. Similiarly, blunt mechanisms vary between falls, motor vehicle crashes, and sports injuries. However, at this early state of understanding the pathophysiologic differences between blunt and penetrating mechanism head injuries, describing the general outcomes across the wide spectrum of insults within the blunt and penetrating injury mechanisms acts an an initial survey of the unique injury patterns and outcomes present within each mechanism to help inform future research directions for more specific mechanisms. Consequently, considering the shear number of incidents and many types of injury mechanisms included in this civilian dataset, analysis of the NTDB data was suitable for this inital inquiry, especially after applying the proper exclusions to investigate specific head injury diagnoses [8]. Future work will be completed with more specific military datasets to ensure the data trends are similar and to investigate more into specific injury patterns and cases related to gunshot wounds and behind-helmet-blunt-trauma. The results presented in this paper agree with previous smaller comparisons in mortality between penetrating and blunt mechanisms of injuries to the head, and further analysis into specifics within mechanisms is suggested [9].

Given that skeletal injury can more easily be created and measured in the laboratory versus brain injury, which requires living tissue, it was important to investigate the relationship between skull fracture and brain injury. While more work can be done to investigate specific linked injuries, this study provides the first documentation of the relationship between skull fracture and brain injury for both blunt and penetrating threats. Results showed that combined head injuries (those including both skull fracture and brain injury) were more frequent with penetrating threats, encompassing 52% of the penetrating dataset but only 16% of the blunt dataset. For the blunt mechanism, there were more patients with only skull fracture (2,723) compared to the penetrating mechanism dataset (573), but the percentage of blunt mechanism skull fracture only patients (3%) was still less than the penetrating mechanism skull fracture only patients (10%). Former research has shown that occult brain injuries can go undocumented in Emergency Department (ED) records when the patient dies before being admitted to the hospital, so additional investigation into the patients with skull fractures but no brain injuries was undertaken considering how high the mortality rate was for this group in the penetrating injury mechanism. After examining that count, 261 of the patients that died from penetrating mechanism injuries that were reported as only sustaining skull fracture (out of the 325 listed) were deaths reported in the ED instead of the hospital. This is likely due to under-reporting of the focal brain injuries for this group since the patient often expires before more definitive studies to identify brain injury can be performed [10]. Given this information, if proper injury documentation were available, one would expect the skull-fracture-only mortality rate to drop, with a corresponding increase in mortality rate and count for the skull fracture and focal brain injury group of patients.

This particular study excluded concussion-type injuries because they are not “focal” brain injuries. These injuries are typically associated with blast or blunt trauma to the head, but due to complications in diagnosing this injury, especially in patients that don’t survive, it was excluded from analysis for this first investigation into head injuries [3]. This exclusion made up the majority of patients that did not have skull fracture or focal brain injury. Future research with this data will investigate the breakdown of injury types within the focal brain injury and skull fracture categories to further elucidate injury patterns, especially linking skull fracture with the most frequently occuring focal brain injuries, as this is pertinent to injury risk predictions being developed in the laboratory.

For patients with a skull fracture, those resulting from a blunt mechanism accounted for only 19% of the dataset, while those resulting from a penetrating mechanism accounted for 62% of that dataset. As one would expect from penetrating trauma, skull fracture was a more prominent injury result. Overall between both datasets, skull fracture without brain injury was infrequent compared to the alternate injury counts. Future analysis will be undertaken to better understand if there is skull fracture pattern

https://doi.org/10.52202/078352-0027

deliniation between these groups, with the hypothesis that simple linear skull fractures are less likely to be associated with significant underlying brain injury compared to complex skull fractures.

In every injury category between blunt and penetrating mechanisms, the penetrating trauma resulted in higher mortality for the patients. This was most pronounced for patients with both a skull fracture and a brain injury, where the blunt mechanism had a mortality of 6% and the penetrating mechanism had a mortality of 67%. Additionally, the blunt mechanism cases had more patients without skull fracture (81% versus 19%) and more patients without focal brain injury (53% versus 47%), whereas the penetrating mechanism cases had much higher percentages of patients with skull fracture (62% versus 38%) and focal brain injury (71% versus 29%). This signifies that for the civilian dataset overall, blunt trauma results in fewer skull fractures and brain injuries compared to penetrating trauma and when they do occur with blunt trauma, the patients are more likely to survive. Although this dataset did not contain BHBT events, the analysis suggests that if a helmet protects the wearer from a penetrating insult, it could improve the wearers chance of survival and reduce the risk of serious injury, even if it still resulted in a blunt injury. This assumes that if the helmet stopped the threat from entering the head, the speed and shape of the back face deformation would not result in a penetrating injury. Therefore, when considering helmet backface measures and their relationship with injury, the velocity, size, and shape of the deformation should be taken into consideration, not just deformation depth or whether or not the threat perforated the helmet. Future work will investigate the mortality risk of non-penetrating BHBT impacts, specifically to investigate the comparative risk between high-rate, focal blunt injuries and penetrating trauma to verify the analysis of the civilian data for military populations.

5. CONCLUSION

Body armour research currently involves a trade-off between preventing penetrations but possibly inducing blunt injury due to back face deformations. This study investigated the differences in head injuries and mortality between blunt and penetrating mechanisms within a civilian trauma hospital dataset (NTDB RDS 2013-2015). The mortality rate of those with blunt impact head injuries was 3% while those with penetrating impact head injuries was 54%, which was a stastistically significant difference. Additionally there were differences in the types of head injuries between blunt and penetrating mechanisms (3% vs. 10% involved skull fracture without focal brain injury, 31% vs. 19% involved focal brain injury without skull fracture, and 16% vs. 52% had both skull fracture and brain injury, respectively). Cases of penetrating mechanism head injuries were more likely to involve co-occurring injuries to the skull and brain, i.e., increased injury severity. Thus, this analysis implies that helmets designed to prevent penetrating mechanisms should increase survival of the wearer, even if there is a risk of resulting blunt injury. However, with regards to BHBT, this assumes that the deformation shape and rate would only generate a blunt injury to the head versus stopping the threat but still resulting in penetrating injury. Hence, BHBT metrics should consider shape and not just depth to evaluate significant injury risk. Additional research should be done to link these civilian injury trends to more specific injury patterns specifically associated with penetrating gunshot wounds and high-rate blunt head impacts.

Acknowledgements

The research reported in this document was performed in connection with contract/instrument W911QX-16-D-0014 with the U.S. Army Research Laboratory. The views and conclusions contained in this document are those of the authors and should not be interpreted as presenting the official policies or position, either expressed or implied, of the U.S. Army Research Laboratory or the U.S. Government unless so designated by other authorized documents. Citation of manufacturer's or trade names does not constitute an official endorsement or approval of the use thereof. The U.S. Government is authorized to reproduce and distribute reprints for Government purposes notwithstanding any copyright notation hereon.

References

[1] U. S. Government Accountability Office. Personal protective equipment: Army and Marine Corps are Pursuing Efforts to Reduce the Weight of Items Worn or Carried in Combat. United States Government Accountability Office, Report to Congressional Committees, GAO-17-431, May 2017.

https://doi.org/10.52202/078352-0027

[2] Jennett, B., Epidemiology of head injury. Journal of Neurology, Neurosurgery, and Psychiatry, 1996. 60(4): p. 362-369.

[3] Langlois, J.A., W. Rutland-Brown, and M.M. Wald, The epidemiology and impact of traumatic brain injury: a brief overview. The Journal of head trauma rehabilitation, 2006. 21(5): p. 375-378.

[4] Schreiber, M.A., et al., Determinants of mortality in patients with severe blunt head injury. Archives of Surgery, 2002. 137(3): p. 285-290.

[5] Baker SP, O'Neill B, Haddon W, Long WB. The injury severity score: a method for describing patients with multiple injuries and evaluating emergency care. J Trauma; Vol 14, pp187-196, 1974.

[6] Association for the Advancement of Automotive Medicine, The Abbreviated Injury Scale 2005 Update 2008, AAAM, Des Plaines Illinois, 2008.

[7] Loftis, K.L., Price, J.P., Gillich, P.J., Cookman K.J., Brammer, T.S., Barnes, J., Graymire, V., Nayduch, D.A., Read-Allsopp, C., Baus, K., Stanley P.A., Brennan, M. Development of an expert based ICD-9-CM and ICD-10-CM map to AIS 2005 Update 2008. Traffic Injury Prevention 2016, 17:sup1, 1-5

[8] Haider, A.H., et al., Influence of the National Trauma Data Bank on the study of trauma outcomes: is it time to set research best practices to further enhance its impact? Journal of the American College of Surgeons, 2012. 214(5): p. 756-768.

[9] Demetriades, D., Kuncir, E., Murray, J., Velmahos, G., Rhee, P., Chan, L. Mortality Prediction of Head Abbreviated Injury Score and Glasgow Coma Scale: Analysis of 7,764 Head Injuries. J Am Coll Surg, 2004. 199(2): p. 216-222.

[10] Matheis, E., Loftis, K., Barrett, C., Saillant N., Moore E., Rafaels, K. Under-reporting of focal brain injuries in emergency department records versus hospital inpatient records within the National Trauma Databank, in draft.

https://doi.org/10.52202/078352-0027

Fragment Penetrating Injury and Light-Weight Protection of the Lower Leg

T-T. N. Nguyen[1], D. Carpanen[1], G. Meek[1], I. Rankin[1], A. Ramasamy[1,2,3], J. Breeze[2], W. G. Proud[4], J. Clasper[1] and S. Masouros[1]

[1]*Department of Bioengineering, Imperial College London, London, United Kingdom*
thuy-tien.nguyen08@imperial.ac.uk
[2]*Royal Centre for Defence Medicine, Queen Elizabeth Hospital Birmingham, United Kingdom*
[3]*Academic Department of Trauma and Orthopaedics, Queen Elizabeth Hospital Birmingham, United Kingdom*
[4]*Institute of Shock Physics, Imperial College London, London, United Kingdom*

Abstract. Explosive devices (EDs) are a major threat in both contemporary battlefields and terrorist attacks. EDs are designed to energise fragments aimed at causing injuries that are life-threatening and associated with poor clinical and functional outcomes. The extremities are the most affected body regions, with the tibia having the highest rate of occurrence. Current personal protective equipment offers little protection against blast fragments to the extremities, especially the lower leg. This study quantified the risk of penetrating injury to the tibia as well as tested the potential of a light-weight hybrid liner for protecting it. A 32-mm-bore gas gun was used to launch 0.78 g cylindrical carbon-steel fragment-simulating projectiles (FSPs) over a range of impact velocities up to 600 m/s. Recovered tibia samples underwent radiography and dissection to quantify the injury outcome. Fracture-risk curves were developed for different severities of fracture based on the modified Winquist-Hansen (WH) classification. The ovine tibia was used as a surrogate for the human tibia. The cortical thickness ratio was used as the scaling factor for impact velocity from the ovine results to the human. For assessment of protection, various common ballistic materials such as silk, Twaron®, Kevlar®, and Dyneema® were tested in a single-layer configuration with 20% ballistic gelatine – a soft tissue simulant – as the backing material. At least 9 shots were conducted for each material. Based on their performance, the top three materials were combined into a 20x20 cm hybrid panel; 6 samples of this panel were tested using the same protocol. The scaled impact velocity at 50% risk (± 95% confidence intervals) for EF1+, EF2+, EF3+, and EF4+ fractures to the posterior surface of human tibia – using the modified WH classification – was 260 ± 0, 364 ± 35, 400 ± 34, and 491 ± 48 m/s, respectively. The resultant light-weight hybrid panel was shown to provide meaningful protection to the lower leg, reducing the predicted fracture severity of the tibia by at least one category in the WH classification.

1. INTRODUCTION

Explosive devices are commonly used in both modern warfare and terrorist attacks. These can cause severe damage to the human body from pressure effects such as blast lung injury, blunt and penetrating trauma, and traumatic amputation of the extremities, [1]–[3]. Of these the most common wounding mechanism is penetration of the body by blast fragments [4], [5]. Blast fragments can be glass, masonry, soil ejecta to foreign bone fragments, and objects purposely included in explosive devices such as bolts and nails [6], [7]. They are projected by the explosion with an initial speed of the order of 1000 m/s which, due to their small mass and irregular shape, quickly decelerate to 600 m/s or less as observed in survival casualties reaching surgery [8]. Extremities, especially the tibia, are the most frequently impacted body regions [9], [10], whose resulted wounds often associate with high risk of amputation, infections and slow recovery [11], [12].

The current personal protective equipment (PPE) focuses on stopping bullets to the thorax and abdomen area for essential coverage of vital organs [13]; these hard body armours provide excellent protection, but can be heavy, cumbersome and poorly fit. The UK Armed Forces also employ an elastic silk soft body armour Tier 1 shorts for the pelvic region, which has been shown to reduce ingress of debris and improve the injury outcome to this area [14], [15]. These shorts, however, only cover the pelvis and upper leg.

The aim of this work was to quantify the risk of fracture to the tibia by a small metal blast fragment. A second aim was to design and test a proof of concept with a light-weight hybrid liner against the same threat and predict the resultant improvement in ballistic protection to the lower leg based on the quantified risk of fracture to the tibia.

 https://doi.org/10.52202/078352-0028

2. MATERIALS AND METHOD

Impact tests were performed using a stainless-steel 32-mm-bore gas gun system (Figure 1a) as described by Nguyen *et al.* [16]. The fragment-simulating projectile (FSP) chosen for this study was a 4.5 mm wide, 0.78 g carbon-steel cylinder designed with the same ratio as recommended by the AEP-2920 NATO Standard [17]. This choice of FSP was based on the study by Breeze *et al.* [18], which found that the most common shape of small metal fragments recovered from blast penetrating injuries was cylindrical with a mean mass of 0.78 g. The test chamber of the system is compatible for studying the fracture to tibia as well as testing the performance of ballistic materials. High-speed photography was used to record the event and to estimate the impact velocity of the FSP.

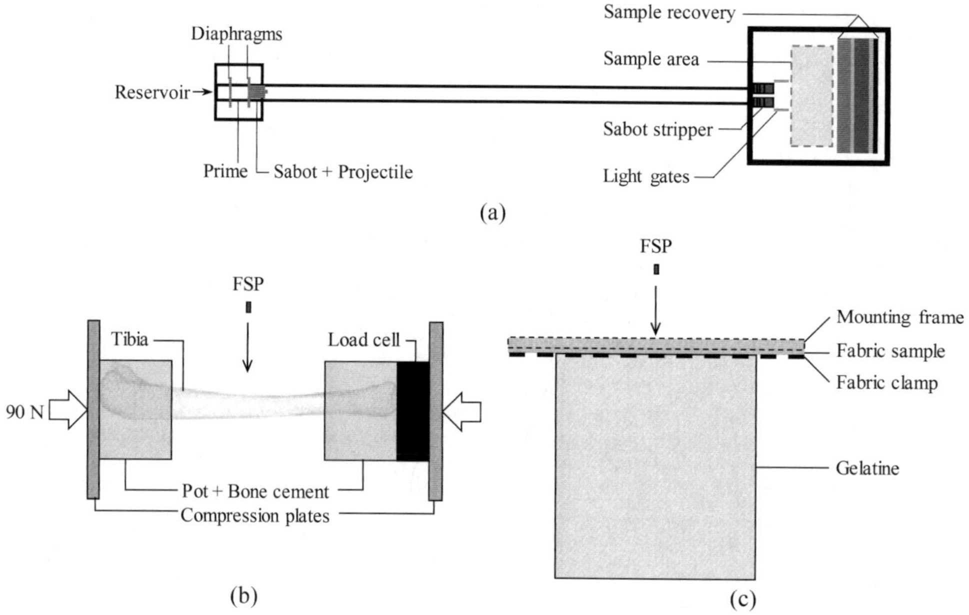

Figure 1. (a) Schematic for the set-up of the gas gun system. (b) The ovine-tibia samples were potted with bone cement and compressed by 90 N to simulate standing gait. (c) the set-up for assessment of ballistic materials; these was clamped onto a frame and put in contact with the ballistic gelatine backing.

2.1 The injury model

The animal model described by Nguyen *et al.* [19] was used to investigate the fracture pattern against impact velocity, where the tibia from skeletally mature sheep was used as a surrogate for the human tibia. The specimens were stored at -20°C for a maximum of three months after all the soft tissue had been removed leaving only the periosteum intact. Before the impact testing, the sample was thoroughly thawed and carefully potted at the two ends in acrylic resin (Figure 1b). Prior to mounting in the test chamber, the sample was compressed with a 90 N load, which is the body weight of a 5-year-old boy, to simulate the standing posture [19]. A 6-axis load cell was used to monitor the compression. The sample was kept moist with spray water during the preparation process.

The posterior aspect of the tibia was chosen as the impacted location. According to Nguyen *et al.* [16], the effect of the soft tissue surrounding the tibia, whose average thickness is 85 mm [20]–[22] for the human posterior mid lower leg, can be considered separately, hence no soft-tissue simulant was used in this model. The prepared sample was aligned in the test chamber with the aid of a laser diode so that impact was accurately aimed at its posterior surface of the mid-diaphysis. Each tibia underwent radiographic scanning using a mini C-arm before and after each test for fracture detection and classification. For optimising the number of samples required, those with no fracture after a test were tested once more at a higher impact velocity where fracture occurred. The resulted fractures were scored by three independent orthopaedic surgeons according to the modified Winquist-Hansen (mWH) classification [23], [24]. A survival analysis using the lognormal regression model was performed to obtain the fracture risk curve of the tibia at different mWH fracture severities; the data with the fracture

https://doi.org/10.52202/078352-0028

severity of interest were classified as left-censored, and those without were classified as right-censored. For samples that were re-tested, interval censoring was applied.

The fracture risk curves obtained from the ovine injury model were subsequently scaled for the human tibia, taking into account the effect of the soft tissue, using:

$$v_{impact}^{(human)} = \sqrt{\left(2.5 \times v_{impact}^{(ovine)}\right)^2 + 210^2}$$

where $v_{impact}^{(human)}$ is the scaled FSP impact velocity at the posterior surface of the lower leg for human, $v_{impact}^{(ovine)}$ is the value of the predictor variable from the survivability analysis of the ovine data, 2.5 is the human-to-ovine cortical thickness ratio and is the scaling parameter proposed and validated by Nguyen *et al.* [19], and 210 (m/s) is the FSP velocity required to penetrate the ballistic gelatine to the thickness of the soft tissue at the posterior side of the mid lower leg [25].

2.2 The hybrid panel

Seven commercially available ballistic fabrics were examined to choose the materials for the light-weight hybrid panel. These materials (Figure 2) are Twaron® plain woven aramids (synthetic aromatic polyamides) of two different areal densities, Kevlar® plain woven aramid, Kevlar® knitted aramid, Kevlar® felt aramid, Dyneema® knitted HPPE (high-performance polyethylene), and Kevlar® plain woven aramid laminated with a highly strain-rate sensitive polymer with a range of functional geometries. The two knitted materials were carefully clamped at all sides, in a neutral stretch state, to a 100 × 100 mm test area. The other materials were cut into 400 × 400 mm squares with all sides rolled up into a 100 × 100 mm taut test area.

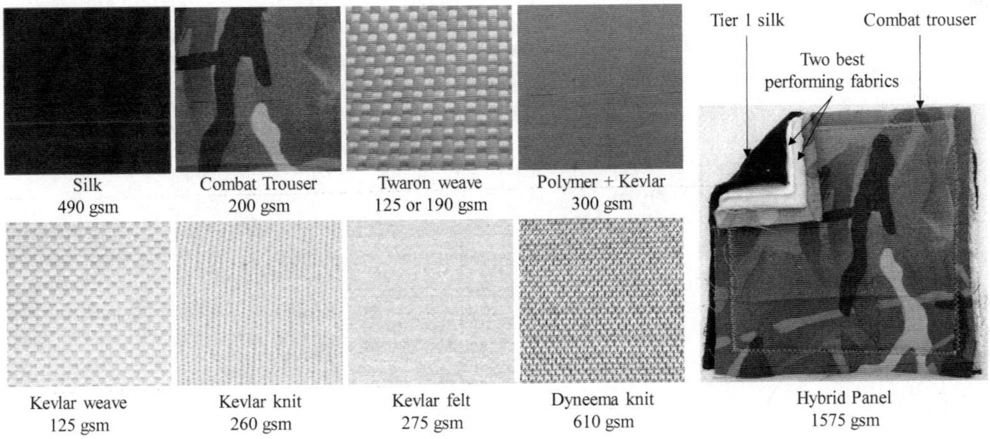

Figure 2. Ballistic materials investigated in the study

For all impact tests, a soft-tissue simulant was placed against the inner most surface of the sample to create a biofidelic boundary condition. The tissue simulant was made from type A 300-bloom ballistic gelatine of 20% by weight [16], [26]. After each impact, the fabric material was checked for deformation and the resulted depth of penetration (DoP) in the gelatine tissue simulant was measured.

A survival analysis using the Weibull regression model was performed with the FSP impact velocity as the predictor variable. Produced velocities with 50% risk (V_{50}) of (a) material perforation, (b) any penetration in the ballistic gelatine, and (c) more-than-25-mm-deep penetration in the ballistic gelatine were used to assess the ballistic performance of the seven commercial fabrics and the hybrid panel.

The light-weight hybrid panel proposed here was constructed from the two best performing materials (out of the seven tested) with the silk material from the Tier 1 shorts at the inner most surface and the fabric from standard combat trousers at the outer most surface (Figure 2). Materials from the Tier 1 shorts and combat trousers were included as the panel is expected to be used as an additional liner in the current combat clothing. The panel was tested in a 200 × 200 mm test area – a practical size for a trouser liner – where all layers were attached together along the edges. As a comparison, additional tests were performed, in the same set-up, on panels with silk and combat trousers, and on panels of combat

 https://doi.org/10.52202/078352-0028

trousers only in order to gauge the protection offered by the current PPE for the upper leg, and the lower leg respectively.

3. RESULTS & DISCUSSION

3.1 Risk of fracture to the tibial posterior mid-diaphysis

Thirty-five impact tests on the posterior cortex of the ovine tibia were performed and resulted in six cases with no bony fracture (F0 of mWH classification), nine fractures with less than 25% comminution (EF1 of mWH classification), three fractures with between 25% to 50% comminution (EF2 of mWH classification), seven fractures with between 50% to 75% comminution (EF2 of mWH classification), and ten fractures with more than 75% comminution (EF4 of mWH classification). The obtained fracture-risk curves for the ovine model (Figure 3) show that the V_{50} for EF1+ (EF1 or more severe fracture), EF2+ (EF2 or more severe fracture), EF3+ (EF3 or more severe fracture), and EF4 is respectively 86, 159, 186, and 238 m/s. After scaling and accounting for the kinetic energy absorbed by the soft tissue, the V_{50} risk of fracture to an adult lower leg for the fracture groups mentioned above is 300, 449, 509, and 632 m/s respectively.

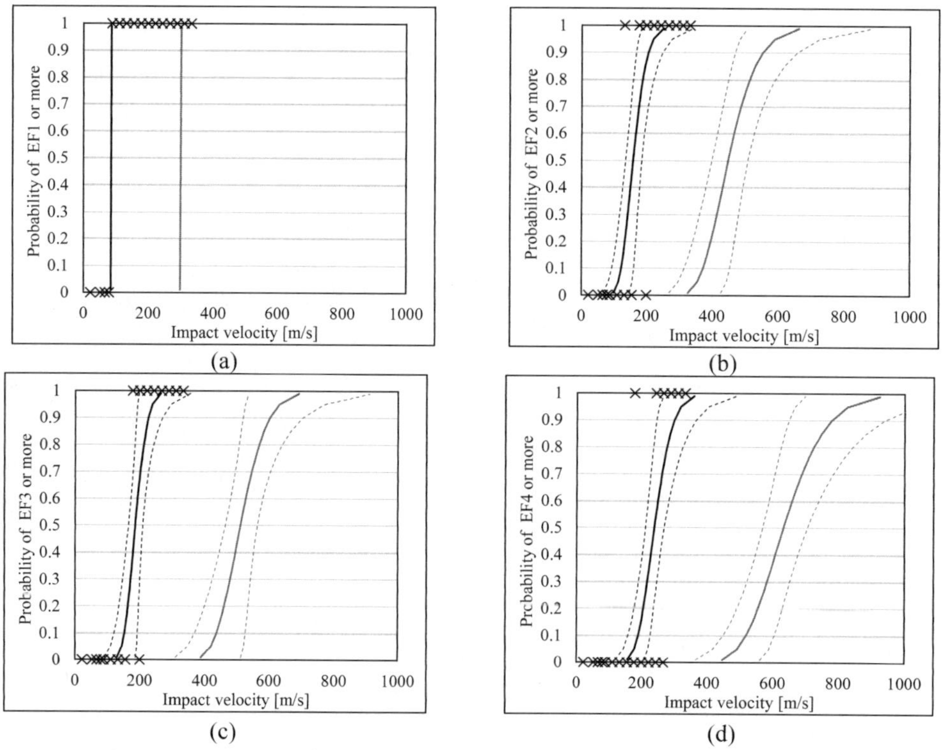

Figure 3. Fracture-risk curves for (a) EF1+, (b) EF2+, (c) EF3+, and (d) EF4 modified Winquist-Hansen fracture types of the experimental ovine model (black) and scaled human model (blue). The dashed lines indicate 95% confidence intervals.

The values of V_{50} for the posterior surface of the lower leg reported here are smaller than those for the anterior surface reported by Nguyen *et al.* [19]. This suggests that the posterior cortex of the tibia is more susceptible to injury by the FSP despite having a thicker soft-tissue layer. Since the anterior cortex is thicker than the posterior cortex, it implies that the cortical thickness is the most important factor in the risk of fracture of the tibia specifically, and by extrapolation likely of all long bones. It needs to be noted that the scaling parameter used here of 2.5 was obtained in a previous study for the anterior cortex whose cortical thickness is different. The cortical thickness ratio between human and ovine should be obtained specifically for the posterior cortex for a more accurate scaling.

https://doi.org/10.52202/078352-0028

3.2 Performance of the light-weight hybrid panel

Table 1 summarises the V_{50} values for fabric perforation, any gelatine penetration and more-than-25-mm-deep gelatine penetration of the seven commercial materials tested in single-layer construction. Across these three categories, Dyneema® gave the best performance, likely due to its high areal density (610 gsm). On the other hand, despite having the second highest areal density (300 gsm), the strain-rate sensitive polymer laminated Kevlar® plain weave showed the least protection, worse than the thinner and lighter original Kevlar® plain weave (125 gsm). This is likely because the lamination fixes the woven architecture in place and limits the movement of the primary strands, thus reducing the inter-yarn friction and the energy dissipation.

Kevlar® felt and Kevlar® knit are respectively the second and third best performing materials. Their areal densities are similar (275 and 260 gsm) and they have similar V_{50} for gelatine penetration of > 25 mm. Being more elastic, the knitted aramid has slightly higherV_{50} for material perforation (10%), thus is less susceptible to damage by an FSP than the felt aramid, but had a lower V_{50} for any penetration in gelatine (20%), thus offering less protection for soft tissue than the felt aramid. Between the two weaves of Twaron® aramid, the one with higher areal density expectedly showed better performance.

Table 1. The assessed ballistic materials

V_{50} [m/s] / Material	Fabric perforation	Any gelatine penetration	Gelatine penetration of > 25 mm
Hybrid panel, 1575 gsm	298	290	298
Dyneema® HPPE knit, 610 gsm	245	218	249
Kevlar® felt, 275 gsm	169	150	178
Kevlar® knit, 260 gsm	188	121	174
Twaron® plain weave, 190 gsm	157	112	186
Kevlar® plain weave, 125 gsm	148	146	148
Twaron® plain weave, 125 gsm	133	124	157
Strain-rate sensitive polymer + Kevlar® plain weave laminated, 300 gsm	< 90	< 90	136

As the two best performing fabrics in single layer configuration, Dyneema® knit and Kevlar® felt were chosen to construct the hybrid panel. The different layers of the panel, from outer to inner, were light combat trousers (the default outermost layer) – Kevlar® felt – Dyneema® knit – Tier 1 silk (Figure 2). The less elastic Kevlar® felt was put outside the knitted Dyneema® HPPE to eliminate the momentum of the FSP and reduce deformation to the inner more elastic layers. The V_{50} values for fabric perforation, any gelatine penetration and more-than-25-mm-deep gelatine penetration of the hybrid panel were 298 ± 15 m/s, 290 ± 12 m/s, and 298 ± 15 m/s respectively.

Figure 4 shows the trends in DoP in gelatine and the impact velocity of the FSP for the hybrid panel (all four layers), silk and trousers fabrics only (current protection for the upper leg), trousers fabric only (current protection for the lower leg), and ballistic gelatine only (no protection). The relationship between the DoP and the FSP impact velocity for ballistic gelatine only was obtained from a previous study by Nguyen *et al.* [16]. Up to 250 m/s, the panel could stop all penetration to the soft-tissue simulant and provide a DoP reduction up to 91% compared to existing PPE. Between 250 m/s and 300 m/s, the panel started getting perforated but could still reduce the DoP by 85-90%. Between 300 m/s and 500 m/s, the protective ability dropped and the DoP reduction was between 7% and 30%. Impact tests were not performed beyond 500 m/s as this was at the protection limit of the panel.

 https://doi.org/10.52202/078352-0028

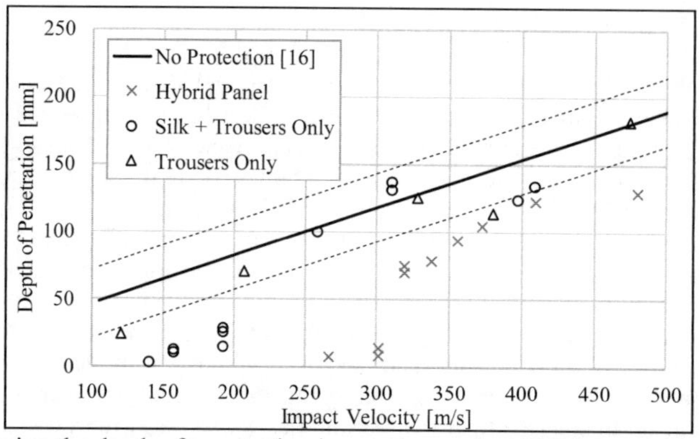

Figure 4. Comparing the depth of penetration in gelatine against FSP impact velocity between the hybrid panel (red cross), silk and trousers fabric only (black circle), trousers fabric only (black triangle), and no protection (solid line; dashed lines indicate 95% confidence intervals; obtained from Nguyen *et al.* [16]).

From Figure 4, it was estimated that below an impact velocity of 300 m/s, the hybrid panel can absorb most of the kinetic energy of the FSP as evidenced by no or almost no penetration in the gelatine; above 300 m/s and up to 480 m/s, the energy absorbed by the fabric panel has no apparent trend with impact velocity and is calculated to be 32 ± 7 J. This information was used to predict the risk of fracture to the tibia with the protection of the hybrid panel (Figure 5), assuming that the energy absorbed by the fabric panel remains constant (32 ± 7 J) above the impact velocity of 480 m/s. The equation used to predict these risk curves was:

$$v_{resultant} = \sqrt{v_{initial}^2 + 300^2}\ , \text{ for } v_{initial} \leq 300$$

$$v_{resultant} = \sqrt{v_{initial}^2 + \frac{2}{m_{FSP}} \times 32}\ , \text{ for } v_{initial} > 300.$$

The prediction shows that the light-weight hybrid panel reduces the resultant fracture by at least one injury severity classification.

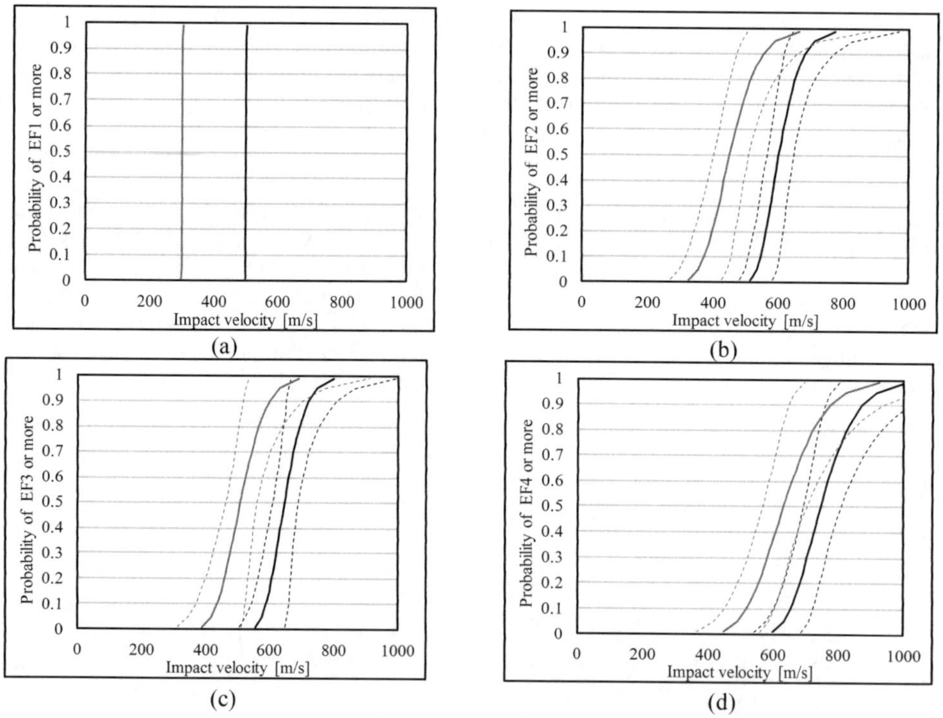

https://doi.org/10.52202/078352-0028

Figure 5. Predicted fracture risk curves for (a) EF1+, (b) EF2+, (c) EF3+, and (d) EF4 modified Winquist-Hansen fracture types of the scaled human model with the hybrid panel (black) in comparison to no protection (blue). The dashed lines indicate 95% confidence intervals.

The results reported here are specific for the FSP used; a more thorough investigation needs to be carried out by repeating the experiment with other FSPs such as the traditional 1.10-g chisel-nosed cylinder and ball bearings. Due to the availability of supply, the tested ballistic fabrics had varied areal densities. Therefore, the comparison result was solely to choose between the available options to construct the hybrid panel. This result cannot be used to conclude whether one type of fabric is better than the other; further studies with fabrics of the same areal density are needed for that purpose. In addition, choosing the two best performing fabrics to include in the hybrid panel is just one combination option. More tests need to be carried out where other combinations of fabrics are considered to achieve even better protection. Finally, the protective ability of the hybrid panel, in terms of fracture risk, reported here was estimated based on the gelatine backing. The extrapolation of this prediction to a specific body part, such as the tibia, needs to be validated by conducting experiments using specimens of the specific body part. Such validation will render the method of risk prediction presented here a useful and practical assessment for the performance of ballistic fabrics.

4. CONCLUSIONS

This study reports the use of a gas gun system to quantify the risk of fracture at the posterior cortex of the tibia. An ovine model was used and scaled to the human using the ratio of the local cortical thickness and by accounting for the thickness of the outer soft tissue. The methodology presented here may be applied to quantify the risk of injury due to an FSP on a variety of biological tissues.

This study also produced and tested a proof of concept for a light-weight hybrid liner for combat clothing using one layer of commercially available ballistic fabrics. The results show a good improvement in the depth of penetration of the soft-tissue simulant suggesting a meaningful reduction in the risk of bone fracture, which would be beneficial to dismounted soldiers. Work is being carried out to investigate the performance of the hybrid panel with the tibia itself as the backing material and at the anteromedial aspect of the lower leg.

Acknowledgments

This work was conducted in the Royal British Legion Centre for Blast Injury Studies at Imperial College London. The authors would like to acknowledge the financial support of the Royal British Legion. We wish to thank DSM, DuPont, Rheon Labs, and Tejin for the provision of samples. We also wish to thank Mr. Satpal Sangha, Dr. Nicolas Newell, and Dr. Grigorios Grigoriadis for their technical support.

References

[1] H. R. Champion, J. B. Holcomb, and L. A. Young, "Injuries from explosions: Physics, biophysics, pathology, and required research focus," *J. Trauma - Inj. Infect. Crit. Care*, vol. 66, no. 5, pp. 1468–1477, May 2009, doi: 10.1097/TA.0b013e3181a27e7f.

[2] A. Ramasamy, A. Hughes, N. Carter, and J. Kendrew, "The effects of explosion on the musculoskeletal system," *Trauma*, vol. 15, no. 2, pp. 128–139, 2013, doi: 10.1177/1460408613484683.

[3] D. S. Edwards, L. McMenemy, S. A. Stapley, H. D. L. Patel, and J. C. Clasper, "40 years of terrorist bombings-A meta-analysis of the casualty and injury profile," *Injury*, vol. 47, no. 3, pp. 646–652, 2016, doi: 10.1016/j.injury.2015.12.021.

[4] J. M. Ryan, G. J. Cooper, I. R. Haywood, and S. M. Milner, "Field surgery on a future conventional battlefield: Strategy and wound management," *Ann. R. Coll. Surg. Engl.*, vol. 73, no. 1, pp. 13–20, 1991.

[5] D. C. Covey and J. Ficke, "Blast and fragment injuries of the musculoskeletal system," *Orthop. Disasters Orthop. Inj. Nat. Disasters Mass Casualty Events*, pp. 269–280, 2016, doi: 10.1007/978-3-662-48950-5_25.

[6] E. A. Dick *et al.*, "Bomb blast imaging: bringing order to chaos," *Clin. Radiol.*, vol. 73, no. 6, pp. 509–516, 2018, doi: 10.1016/j.crad.2017.12.001.

[7] M. Kalem and N. Ercan, "Where is the fracture? Penetrating injury with a foreign bone," *Acta*

https://doi.org/10.52202/078352-0028

Orthop. Traumatol. Turc., vol. 52, no. 4, pp. 320–322, 2018, doi: 10.1016/j.aott.2018.01.004.
[8] G. W. Bowyer, "Management of small fragment wounds: experience from the Afghan border.," *J. Trauma*, vol. 40, no. 3 Suppl, pp. S170–S172, 1996.
[9] Y. A. Weil, K. Petrov, M. Liebergall, Y. Mintz, and R. Mosheiff, "Long bone fractures caused by penetrating injuries in terrorists attacks," *J. Trauma - Inj. Infect. Crit. Care*, vol. 62, no. 4, pp. 909–912, 2007, doi: 10.1097/01.ta.0000197178.90194.3e.
[10] B. D. Owens, J. F. Kragh, J. C. Wenke, J. Macaitis, C. E. Wade, and J. B. Holcomb, "Combat Wounds in Operation Iraqi Freedom and Operation Enduring Freedom," *J. Trauma Inj. Infect. Crit. Care*, vol. 64, no. 2, pp. 295–299, 2008, doi: 10.1097/TA.0b013e318163b875.
[11] M. Khatod, M. J. Botte, D. B. Hoyt, R. S. Meyer, J. M. Smith, and W. H. Akeson, "Outcomes in open tibia fractures: Relationship between delay in treatment and infection," *J. Trauma*, vol. 55, no. 5, pp. 949–954, 2003, doi: 10.1097/01.TA.0000092685.80435.63.
[12] N. Enninghorst, D. McDougall, J. J. Hunt, and Z. J. Balogh, "Open tibia fractures: Timely debridement leaves injury severity as the only determinant of poor outcome," *J. Trauma - Inj. Infect. Crit. Care*, vol. 70, no. 2, pp. 352–357, 2011, doi: 10.1097/TA.0b013e31820b4285.
[13] J. Breeze, E. A. Lewis, R. Fryer, A. E. Hepper, P. F. Mahoney, and J. C. Clasper, "Defining the essential anatomical coverage provided by military body armour against high energy projectiles," *J. R. Army Med. Corps*, vol. 162, no. 4, pp. 284–290, 2016, doi: 10.1136/jramc-2015-000431.
[14] J. Breeze, L. S. Allanson-Bailey, A. E. Hepper, and M. J. Midwinter, "Demonstrating the effectiveness of body armour: A pilot prospective computerised surface wound mapping trial performed at the role 3 hospital in Afghanistan," *J. R. Army Med. Corps*, vol. 161, no. 1, pp. 36–41, 2015, doi: 10.1136/jramc-2014-000249.
[15] E. A. Lewis, M. A. Pigott, A. Randall, and A. E. Hepper, "The development and introduction of ballistic protection of the external genitalia and perineum," *J. R. Army Med. Corps*, vol. 159, no. Supp I, pp. i15–i17, 2013, doi: 10.1136/jramc-2013-000026.
[16] T.-T. N. Nguyen *et al.*, "Fragment penetrating injury to long bones," in *Proceedings of Personal Armour Systems Symposium 2018*, 2018, vol. 1979, pp. 312–321, doi: 10.1063/1.5044868.
[17] NATO, "NATO STANDARD AEP-2920 PROCEDURES FOR THE EVALUATION AND CLASSIFICATION OF PERSONAL ARMOUR," NATO Standardization Office, 2016.
[18] J. Breeze *et al.*, "Characterisation of explosive fragments injuring the neck," *Br. J. Oral Maxillofac. Surg.*, vol. 51, no. 8, pp. e263–e266, 2013, doi: 10.1016/j.bjoms.2013.08.005.
[19] T.-T. N. Nguyen *et al.*, "The risk of fracture to the tibia from a fragment simulating projectile," *J. Mech. Behav. Biomed. Mater.*, vol. 102, Feb. 2020, doi: 10.1016/j.jmbbm.2019.103525.
[20] P. Nande, V. Mudafale, and S. Vali, "Anthropometric Profile of Female and Male Players Engaged in Different Sports Disciplines," *Internet J. Nutr. Wellness*, vol. 8, no. 1, 2008.
[21] M. Bucar *et al.*, "Diferencias Morfológicas Bilaterales de Gimnastas de Nivel Superior Morphologic Bilateral Differences of Top Level Gymnasts Diferencias Morfológicas Bilaterales de Gimnastas de Nivel Superior," *Artic. Int. J. Morphol.*, vol. 30, no. 1, pp. 110–114, 2012, doi: 10.4067/S0717-95022012000100019.
[22] A. L. Claessens *et al.*, "Anthropometric characteristics of outstanding male and female gymnasts," *J. Sports Sci.*, vol. 9, no. 1, pp. 53–74, 1991, doi: 10.1080/02640419108729855.
[23] R. A. Winquist and S. T. Hansen, "Comminuted fractures of the femoral shaft treated by intramedullary nailing.," *Orthop. Clin. North Am.*, vol. 11, no. 3, pp. 633–48, Jul. 1980.
[24] S. A. Brito, Z. Gugala, A. Tan, and R. W. Lindsey, "Statistical validity and clinical merits of a new civilian gunshot injury classification trauma," *Clin. Orthop. Relat. Res.*, vol. 471, no. 12, pp. 3981–3987, 2013, doi: 10.1007/s11999-013-2953-3.
[25] L. Cristofolini and M. Viceconti, "Mechanical validation of whole bone composite tibia models," *J. Biomech.*, vol. 33, no. 3, pp. 279–288, Mar. 2000, doi: 10.1016/S0021-9290(99)00186-4.
[26] J. Breeze, N. Hunt, I. Gibb, G. James, A. Hepper, and J. Clasper, "Experimental penetration of fragment simulating projectiles into porcine tissues compared with simulants," *J. Forensic Leg. Med.*, vol. 20, no. 4, pp. 296–299, 2013, doi: 10.1016/j.jflm.2012.12.007.

https://doi.org/10.52202/078352-0028

Chest wall velocity and intra-thoracic pressure impulse as relevant parameters in predicting thoracic injury for short-duration blast wave

J. Boutillier[1], N. Prat[2], S. De Mezzo[1], P. Magnan[1] and P. Naz[1]

[1]French-German Research Institute of Saint-Louis (ISL), 5 rue du Général Cassagnou, 68300, Saint-Louis, France,
johanna.boutillier@isl.eu
[2] Institut de Recherche Biomédicale des Armées (IRBA), Brétigny sur Orge, France

Abstract. Blast pulmonary trauma are common consequences of modern war and terrorism actions. To better protect soldiers from that threat, the injury risk level when protected and unprotected must be assessed. Knowing that the lung injury risk level when unprotected is correlated with the maximum of incident impulse for shock wave duration below 6 ms, the objective is to correlate parameters related to large animal chest response under blast loading with a pulmonary risk level. It implies to determine parameters which are themselves correlated with the incident impulse for short-duration waves. Twelve post-mortem swine (PMS), lying on the ground, were exposed to shock wave of increasing intensity. Two groups of 6 PMS were set-up and exposed to shock waves of constant positive phase duration. Group 1 and 2 were respectively exposed to shock waves of duration 1.0 ms and 1.8 ms. Their thorax were instrumented with a piezo-resistive pressure sensor, an hydrophone placed into the esophagus and an accelerometer, screwed onto a mid-torso rib, directly exposed to the shock wave. Combining those scenarios with others performed at a duration 1.4 ms allowed to show that two parameters were correlated with incident impulse ranging from 35 kPa ms to 160 kPa ms: the maximum chest wall velocity (V_{MAX}) and intra-thoracic pressure impulse ($\Delta Ieso$). Lung injury tolerance limits from no injury to severe ones (hemorrhage involving up to 60% of the lung) were redefined with those parameters. The lung injury threshold for near wall scenarios in terms of incident impulse is 58.3 kPa ms, corresponding to a V_{MAX} of 2.78 m/s and a $\Delta Ieso$ of 137.8 kPa ms. This study allowed the definition of injury criteria for the evaluation of lung injury risk when unprotected, which is a first step toward the proposition of tolerance limits to evaluate thoracic protective system regarding injury outcomes.

1. INTRODUCTION

Until now the design of a new thoracic protective equipment for soldier and law enforcement did not considered the risk of blast injury. Indeed, focus is made on ballistic, knives and fragments protection, leaving the uncertainty for protection against explosive devices. However, studies have proven that some thoracic protections can induce an amplification of the blast threat behind the protection, which increase the risk of chest and abdominal injuries (lungs and gastrointestinal injuries) [1-5]. Understanding the reason of such amplification of the injury risk (or reduction) is important, but being able to predict the risk level when a person is unprotected or protected is also a challenge for the development of future thoracic protective systems.

In order to evaluate a thoracic protection as regard to injury risk level, an adapted injury criterion is needed and should be measurable on a dummy to reduce the use of animal model. Different injury criteria were defined regarding blast wave characteristics [6-9], maximum chest wall velocity [10-11] or the irreversible work [12]. However, those models are not directly usable to estimate the lung injury risk under a thoracic protection [13]. Nevertheless, they can be used as a basis for the development of adapted criteria able to estimate lung injury risk level when protected/unprotected. Indeed, Boutillier *et al.* [9] demonstrated that the lung injury risk level (from no injury to severe ones) is correlated with shock wave incident impulse (ΔI_I) of duration below 6 ms. A good injury criterion for thoracic protective system evaluation regarding lung injury risk could then have these properties:

- Be correlated with the incident impulse;
- Be able to discriminate thoracic protective systems and estimate the correct injury risk with a thoracic protection.

Data measured on instrumented anthropomorphic manikins or on animal models can be examined to check those two conditions. Magnan *et al.* [14], Bass *et al.* [15] and Bouamoul *et al.* [16] respectively exposed the "U"-shape membrane, the Hybrid III and the MABIL (Mannequin for Blast Incapacitation and Lethality) to blast with and without thoracic protective equipment. Regarding animal models, thoracic response of post-mortem swine (PMS) exposed to blast threats of increasing intensity at a constant positive phase duration have been analyzed by Boutillier *et al.* [17]. Moreover, thoracic

 https://doi.org/10.52202/078352-0029

response and injury outcomes on living swine exposed to a specific blast threat for three levels of protection (none, soft ballistic pack and hard ballistic pack) have been shown by Prat *et al.* [5].

The aim of this study is to propose relevant parameters for the definition of a good lung injury criterion related to large animals' chest response. Three datasets were used, corresponding to post-mortem swine exposed near a wall to shock waves of different short positive phase durations. Data obtained from new experiments from shock wave of duration 1.0 ms and 1.8 ms were compared to data previously obtained with shock waves of duration 1.4 ms [17]. This strategy allowed the evaluation of parameters related to large animal chest response that correlated with ΔI_I for short-duration blast waves. After selecting such parameters, lung injury tolerance limits related to these latter were defined. For that purpose, lung injury thresholds related to the maximum of incident impulse from Boutillier *et al.* [9] were used. In this latter study, lung injury thresholds were defined for 50 kg large animals exposed near a wall to short-duration Friedlander blast waves. The threshold from no injury to trace/slight injury (superficial petechial or ecchymotic hemorrhages involving less than 10% of the lung surface) was defined at a ΔI_I of 58.3 kPa ms. A ΔI_I of 119.1 kPa ms led to moderate injuries (subpleural ecchymotic hemorrhage with superficial involvement of 11–30% of the lung surface) and above 232.8 kPa ms, severe injuries (diffuse ecchymotic hemorrhage extending into parenchyma involving 31–60% of the lung) were observed. The hypothesis underlying this study is that the chest response of a post-mortem or a living animal model can reasonably be considered identical in the field of high-speed loadings [18] (even if this statement needs to be verified).

2. METHODS

The animals used in our study were purchased from a recognized source producing high quality animals for experimental testing. They were sacrificed for the specific purpose of testing under the European directive (2010/63/EU), adopted in February 2013 in France.

2.1 Animals

Experiments were carried out on twelve post-mortem swine (53.1 ± 5.3 kg) at a rate of one per day. Due to the absence of gastrointestinal tract, the abdominal cavity, before being sutured, was filled with natural sponges placed on a plastic bag to ensure the morphology necessary for the tests.

A uniaxial accelerometer (PCB 3501A12, 60 kG) was screwed with a small rigid target (on top of it for displacement tracking purpose) on the 8th-9th rib, counting from top of the thorax. The sensor was fixed at the highest point of the rib (compared to ground level) when the post-mortem swine is lying on its left flank. In addition, two pressure sensors were used:

- A piezo-resistive pressure sensor (Kulite XCQ 093, 35 bar) was placed near the accelerometer, sutured on the skin. It measures the pressure experienced by the PMS thorax;
- A hydrophone (RESON TC4013) was placed inside the esophagus to measure the intra thoracic pressure. Its sensitive part was located underneath the accelerometer (8th-9th rib).

The instrumentation cables were tunneled under the skin for protection and to avoid interferences related to their shake.

2.1 Near wall blast experiments

Figure 1 illustrates the experimental setup. Animals were placed on the ground, lying on the left flank. Spherical charges of Composition-4 (C-4) were hanging over the animal at two different heights of burst (HoB). In addition to the animal instrumentation, two Free-Field ICP® pencil probes (type 137B22, PCB piezotronics) were used to measure the threat characteristics at the same distance from the explosive charge that the animal.

Two high-speed camera were installed on the proving ground at 30 m from the charge to delay the video instability during the passage of the shock wave:

- A Photron SA-Z (color, 40,000 fps) filmed the whole setup. Sphericity and homogeneity of the fireball can then be checked after the detonation;
- A Phantom V1610 (black and white, 40,000 fps) provided a zoomed view of the animal thorax (instrumented part) in order to record the chest displacement during the interaction with the shock wave. The optical center of the camera was at the same height from ground than the target fixed on the accelerometer.

https://doi.org/10.52202/078352-0029

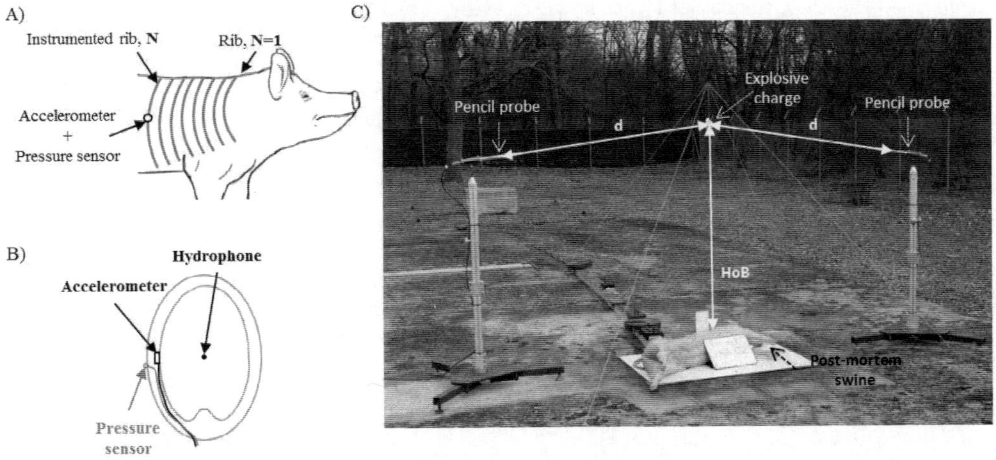

Figure 1. A) and B) Animal instrumentation setup; C) Experimental setup. d: distance explosive charge / pencil probe sensitive part. Illustration of the configuration with HoB = 140 cm

Post-mortem swine were exposed to Friedlander shock waves of increasing intensity with two different positive phase durations. Two animal groups were built:

- Group 1: N=6. Exposed to 1.0 ms shock waves duration;
- Group 2: N=6. Exposed to 1.8 ms shock waves duration.

Table 1 summarized the test matrix performed. Five scenarios were performed per group and distributed to check the reproducibility of the measurement on different animals. Around forty experiments per group were performed.

Table 1: Test matrix

Group 1 (T+=1.0 ms)			Group 2 (T+=1.8 ms)		
C-4 (kg)	HoB and d (cm)	Nb test	C-4 (kg)	HoB and d (cm)	Nb test
0.13	140	12	0.40	270	10
0.35	140	7	0.75	270	7
0.50	140	6	1.80	270	7
0.80	140	8	2.30	270	7
1.00	140	5	3.50	270	6

3. RESULTS

The acquisition was made by a high-speed range data acquisition system (MF instruments), with a sampling rate of 1 MHz. Data were then filtered with a 6th order Bessel at 100 kHz for the pencil probes and at 60 kHz for the other sensors.

Due to slight differences on animals' weight, scaling laws of Baker *et al.* [19] were used. The scaled factor was calculated to scale measured data to a 50 kg animal. Several parameters were analyzed:

- The incident/reflected pressures and corresponding impulses (which are defined as the time integration of pressure profiles);
- The intra thoracic (esophageal) pressure and corresponding impulse;
- The swine chest wall kinematic through acceleration, velocity and displacement of the instrumented rib. Chest wall acceleration was obtained from the accelerometer while both displacement and velocity were obtained through video tracking of the small rigid target fixed on the accelerometer.

3.1 Pressure measurements

https://doi.org/10.52202/078352-0029

Figure 2-A and 2-B summarized the average pressure measurement characteristics in terms of overpressure (ΔP) and maximum of impulse (ΔI) for all scenarios tested for group 1 and 2, respectively. Group 1 were exposed to Friedlander waveform of ΔP_I ranging from 101 kPa to 475 kPa, with an average positive phase duration of 1.02 ± 0.06 ms. It induces ΔI_I from 36 kPa ms to 117 kPa ms. Group 2 faced similar threat, with higher positive phase duration. Indeed, they faced threats of ΔP_I from 60 to 301 kPa with an average positive phase duration of 1.78 ± 0.11 ms, corresponding to ΔI_I from 39 kPa ms to 145 kPa ms.

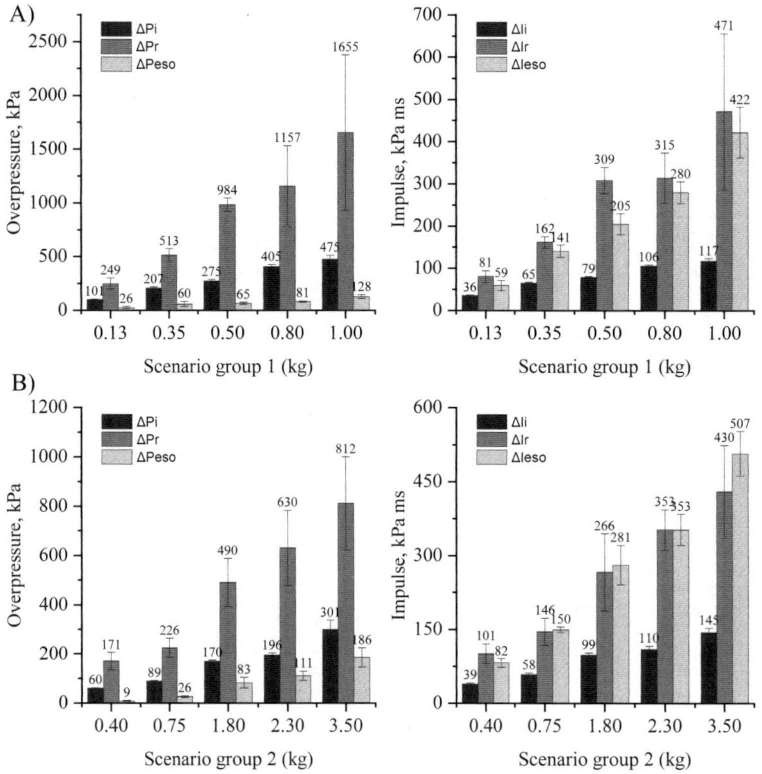

Figure 2. Characteristics of the pressure measurements for tested scenarios in (A) group 1 and (B) group 2. The subscripts 'I', 'R' and 'eso' correspond to the incident, reflected and esophageal pressure measurement characteristics.

Incident pressure and impulse profiles from the detonation of 0.50 kg of C-4 (group 1) and 1.80 kg of C-4 (group 2) are illustrated in Figure 3-A and 3-B respectively (graphs on the left side). Reflected pressure measurements, illustrated in Figure 3-A and 3-B for same scenarios (graphs in the middle) are more disturbed than the incident pressure. This is noticeable on Figure 2, where the pressure characteristics are plotted. This is partially due to the difficulty to correctly place the sensor on the post-mortem swine skin regarding the explosive charge but bad weather conditions during some tests (rain) could have affect measurements because of the sensor used. Two intra thoracic (esophageal) pressure measurements are also illustrated on Figure 3-A and 3-B (on the right side). Due to filtering effect of the chest wall, the rise time is much longer than the two previous pressure measurements. Moreover, as illustrated in Figure 2, the esophageal overpressure ($\Delta Peso$) is 1.6 to 5 times lower than the incident overpressure while the reflected overpressure (ΔI_R) is 2.5 to 3.5 times higher than this latter. Indeed, for Group 1, the overpressure ranged from 101 to 475 kPa, from 249 to 1655 kPa and from 26 to 128 kPa, respectively for ΔP_I, ΔP_R and $\Delta Peso$. This is different for the maximum impulse, where both the reflected (ΔI_R) and the esophageal impulses ($\Delta Ieso$) are in the same order of magnitude and are higher than ΔI_I.

https://doi.org/10.52202/078352-0029

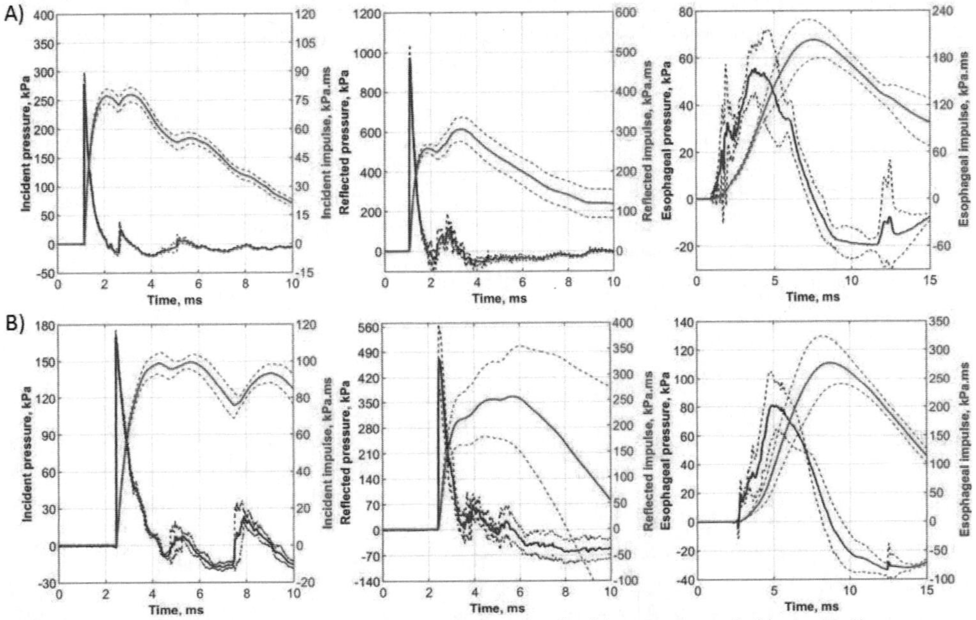

Figure 3. Pressure time histories from the pencil probe (incident pressure, left), the Kulite sensor (pressure on the chest skin, middle) and the hydrophone (intra thoracic pressure, right). Data from scenarios 0.50 kg at 140 cm ((A), group 1) and 1.80 kg at 270 cm ((B), group 2) are plotted

3.2 Chest wall motion

Figure 4-A and 4-B illustrate measured and calculated chest acceleration, velocity and displacement for the detonation of 0.50 kg of C-4 (group 1) and 1.80 kg of C-4 (group 2), respectively.

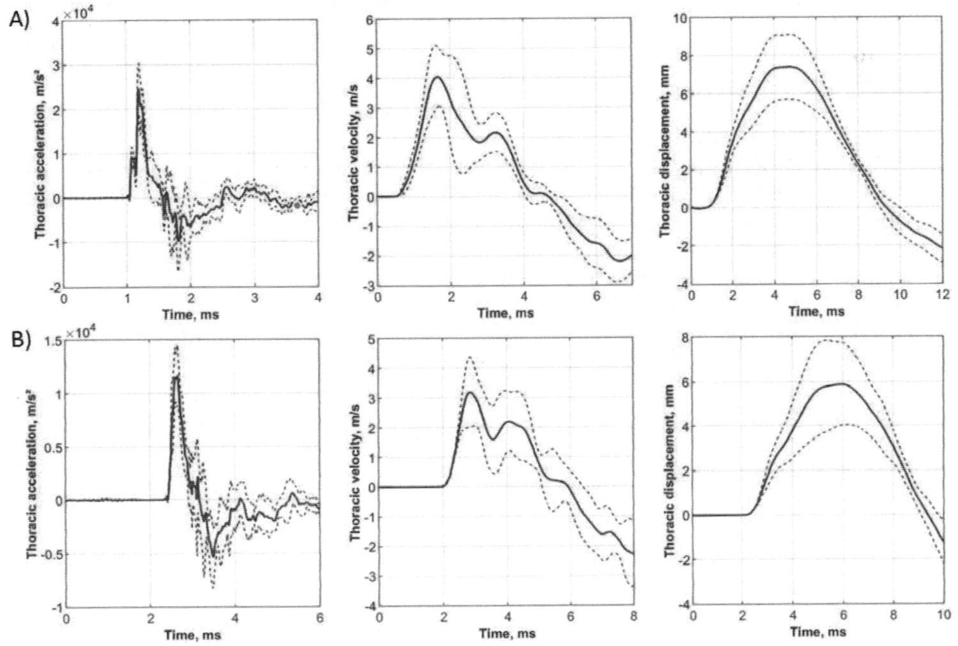

Figure 4. Chest wall motion response from scenarios 0.50 kg at 140 cm ((A), group 1) and 1.80 kg at 270 cm ((B), group 2). Chest wall acceleration (left side), velocity (middle) and displacement (right side) time histories are illustrated

https://doi.org/10.52202/078352-0029

Characteristics of these profiles are summarized on Figure 5, where standard deviations are around 20% for the maximum chest wall acceleration (Γmax), velocity (Vmax) and displacement (Dmax). For Group 1, with ΔI_I ranging from 36 kPa ms to 117 kPa ms:

- Γmax from 6413 ± 1346 m/s² to 45610 ± 12767 m/s² were measured;
- Vmax from 1.6 to 6.6 m/s were calculated (tracking was not possible for 1.00 kg of C-4);
- Dmax from 2.6 mm to 11.1 mm were calculated (tracking was not possible for 1.00 kg of C-4).

For Group 2, with ΔI_I ranging from 39 kPa ms to 145 kPa ms:

- Γmax from 4063 ± 607 m/s² to 30932 ± 5887 m/s² were measured;
- Vmax from 1.5 to 6.9 m/s were calculated;
- Dmax from 2.3 mm to 11.6 mm were calculated.

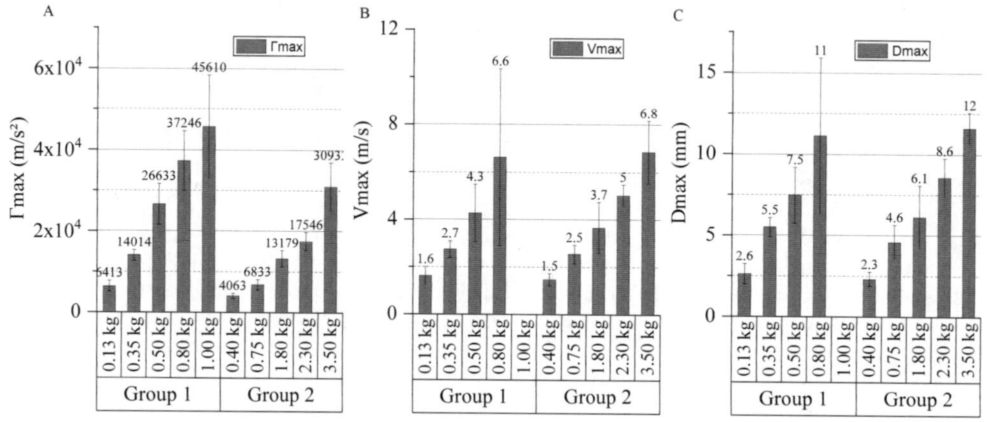

Figure 5. Characteristics of the chest wall motion for all tested scenarios. Maximum of: A) chest wall acceleration (Γmax), B) velocity (Vmax) and C) displacement (Dmax) are plotted.

4. DISCUSSION

Lung injury risk level was found to be correlated with shock waves incident impulse of duration below 6 ms [9]. For near wall exposition, the threshold from no injury to trace/slight injury was defined at a ΔI_I of 58.3 kPa ms. A ΔI_I of 119.1 kPa ms leads to moderate injuries and above 232.8 kPa ms, severe injuries are observed. However, this parameter is not sufficient to evaluate the injury risk variation when wearing a thoracic protective system. The parameter that could be used as a lung injury criterion for thoracic protective system evaluation could then be correlated with incident impulse and be able to discriminate thoracic protective system by a correct estimation of the injury risk under the protection.

Figure 6 illustrates the chest wall motion parameter and the intra thoracic pressure and impulse against the maximum incident impulse for PMS expositions to three different shock wave durations: 1.0 ms and 1.8 ms from current study and 1.4 ms from Boutillier *et al.* [17]. It can be noticed that only the maximum chest wall velocity and intra thoracic impulse are well correlated with the incident impulse for short duration waves. Vmax and ΔIeso are then good candidate parameters for injury criteria definition for short duration shock waves, at least for unprotected scenario.

https://doi.org/10.52202/078352-0029

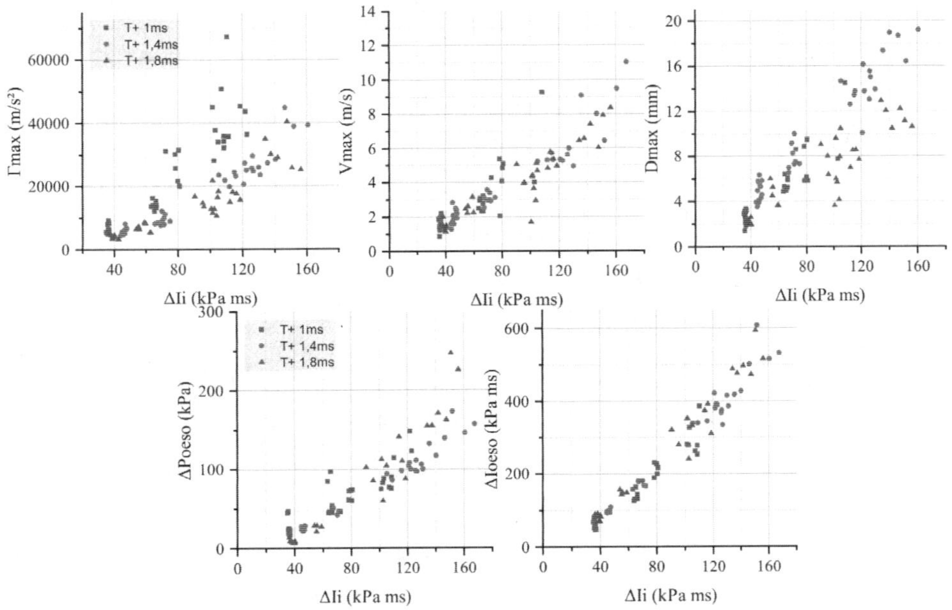

Figure 6. Comparison of the measured parameters on PMS exposed to shock waves of three different positive phase durations: 1.0 ms (black square), 1.4 ms (red circle) and 1.8 ms (blue triangle). Chest motion parameters and intra thoracic pressure and impulse are plotted against ΔI_I.

Figure 7 shows fitted curves for ΔIeso and Vmax plotted against ΔI_I in addition to 95% confidence intervals (CI). The obtained correlations are the following, with R^2 values of 0.99 (2nr order polynomial fit) and 0.95 (linear fit), respectively (p-value<0.001):

$$\Delta I_{eso} = 1.68906\Delta I_I + 0.01156\Delta I_I^2 \quad (1)$$

$$V_{MAX} = 0.04776\Delta I_I \quad (2)$$

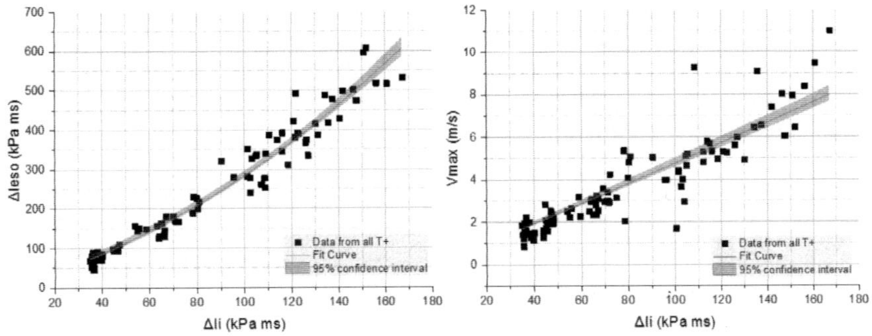

Figure 7. Fitted curves for ΔIeso (left) and Vmax (right) plotted against ΔI_I, in addition to 95% confidence intervals. Data from all positive phase durations are plotted.

Using lung injury risk level thresholds defined by Boutillier *et al.* [9], for near wall experiments, threshold values in terms of maximum chest wall velocity (Vmax) and intra thoracic pressure impulse (ΔIeso) have been defined. Table 2 summarized the defined thresholds, valid for an unprotected 50 kg animal exposed to short-duration Friedlander waveform against a wall. Thresholds in terms of Vmax for near wall scenarios are of the same order of magnitude then the ones proposed by Axelsson *et al.* [10]. Those latter were defined using a one degree of freedom mathematical model whose aimed was to reproduce the human thorax response exposed to shock wave (Friedlander or complex), while lung injury risk were obtained from sheep experiments. The trace/slight+, moderate+ and severe+ lung injury thresholds defined by Axelsson *et al.* are respectively 3.6-7.5 m/s, 4.3-9.8 m/s and 7.5-16.9 m/s, while the proposed 50% risk values are respectively 2.7-2.9 m/s, 5.4-6.0 m/s and 10.6-11.6 m/s.

https://doi.org/10.52202/078352-0029

Both types of measurement (velocity and internal pressure impulse) could be measured on an adapted dummy and under a thoracic protection. In order to be used as injury criteria on such dummy (unprotected), correlations between dummy chest response and swine chest response must be evaluated (Vmax or ΔIeso). A validated finite element model (FEM) of the dummy could be a good alternative to experimentation. Because of ethical reasons and costs, a swine FEM could also be validated regarding the large database available on swine chest response to blast. Such model could allowed the definition of tolerance limits regarding biomechanical parameters to be defined.

To check if Vmax and ΔIeso can discriminate thoracic protective system and estimate the correct lung injury risk with thoracic protection, data on living animals are needed. Such data were presented by Prat *et al.* [5], where anesthetized swine (unprotected and protected) were exposed to a high-intensity blast in free-field. Both parameters seemed able to discriminate thoracic protective system, with higher value when lung injury were more severe, and inversely. Nevertheless, data with several shock wave characteristics, in free-field and in near wall configuration, are needed.

Table 2. 50% risk of a given lung injury level for an unprotected 50 kg animal model exposed to Friedlander waveform against a wall. Threshold values in terms of ΔI_I, Vmax and ΔIeso are presented

Lung injury risk level	ΔI_I, kPa ms	Vmax, m/s	ΔIeso, kPa ms
None	< 58.3	< 2.78 (95% CI [2.66 2.91])	< 137.8 (95% CI [128.4 147.1])
Trace/Slight+	> 58.3	> 2.78 (95% CI [2.66 2.91])	> 137.8 (95% CI [128.4 147.1])
Moderate+	> 119.1	> 5.69 (95% CI [5.43 5.95])	> 365.1 (95% CI [355.9 374.4])
Severe+	> 232.8	> 11.12 (95% CI [10.62 11.62])	> 1019.7 (95% CI [952.6 1086.7])

5. CONCLUSION

In order to evaluate the efficiency of thoracic protection against the blast threat, adapted injury criteria must be developed. Two angles are considered: the chosen criterion must be correlated with shock wave incident impulse and be able to predict injury level for both unprotected and protected thorax. This study investigated the first angle.

Combining published [17] and new experimental data obtained on post-mortem swine exposed to short-duration blast waves against a wall pointed out two physical parameters that respect the first angle: the maximum chest wall velocity (Vmax) and the maximum esophageal impulse (ΔIeso). Those two parameters can be considered as adapted criteria for the estimation of lung injury risk when unprotected and exposed to Friedlander wave (near wall). Data on living animal response to blast wave with different thoracic protection are needed to verify the second angle.

Acknowledgments

This work was partially supported by the French ANR program ASTRID (ANR- 16-ASTR-0025-DA), led by the French Ministry of Defense, Direction Générale de l'Armement (DGA), under the contract name ''REELTHOR''.

References

[1] Cooper G.J., Townend D.J., Cater S.R., Pearce B.P., The role of stress waves in thoracic visceral injury from blast loading: modification of stress transmission by foams and high-density materials. J. Biomech., 1991; 24 (5); 273–285.

[2] Jetté F.X., Dionne J.P., Williams K., Anctil B., Makris A., Development of a mannequin for assessment of blast injuries and lethality – assessment of protective systems. Proceedings of the MABS 18th Conference. Schneizlreuth, Germany, 2004.

[3] Phillips Y.Y., Mundie T.G., Yelverton J.T., Richmond D.R., Cloth ballistics vest alters response to blast. J. Trauma, 1988; 28 (1); S149–S152.

[4] Thom C., Cronin D., Ouellet S., Makris A., Numerical simulation of primary blast injury amplification by fabrics. J. Biomech., 2007; 40 (S2); 56–57.

https://doi.org/10.52202/078352-0029

[5] Prat N.J., Boutillier J., Cardona V. *et al.*, Injury and biomarkers from pigs following isolated open-field blast exposure depend on ballistic protection level. Poster from the Military Health System Reseach Symposium, Kissimmee, Florida, August 2019.
[6] Bowen I.G., Fletcher E.R., Richmond D.R., Estimate of Man's Tolerance to the Direct Effects of Air Blast. Technical Progress Report, DASA-2113. Defense Atomic Support Agency, Department of Defense, Washington, DC, 1968
[7] Richmond D.R., Cooper P.W., Evaluation of Bowen's curves. Unpublished manuscript. 2002
[8] Bass C.R., Rafaels K.A., Salzar R.S., Pulmonary injury risk assessment for short-duration blasts. J Trauma, 2008; 65; 604–615.
[9] Boutillier J., Deck C., De Mezzo S., Magnan P., Prat N. *et al.*, Lung injury risk assessment during blast exposure. J. Biomech., 2019; 86; 210-217.
[10] Axelsson H., Yelverton J.T., Chest wall velocity as a predictor of nonauditory blast injury in a complex wave environment. J Trauma, 1996; 40 (3); 31–37.
[11] Teland J.A., Review of blast injury prediction models. FFI-rapport 2012/00539. 2012
[12] Stuhmiller J.H., Ho K.-H.H., Vander Vorst M.J., Dodd K.T., Fitzpatrick T., Mayorga, M., A model of blast overpressure injury to the lung. J. Biomech., 1996; 29 (2); 227–234.
[13] Boutillier J., Deck C., Magnan P., Naz P., Willinger R., A critical review on primary blast thorax injury and their outcomes. J. Trauma Acute Care Surg., 2016; 81 (2); 371–379.
[14] Magnan P., De Mezzo S., Heck S., Boehrer Y., Approche métrologique du chargement par le blast de cibles anthropomorphiques instrumentées – Essais ISL/DGA Tt. Report S-R 108/2012 of the French-German Research Institute of Saint Louis, France. 2012.
[15] Bass C.R., Davis M., Rafaels K., A methodology for assessing blast protection in explosive ordnance disposal bomb suits. Int. J. Occupat. Safety Ergonom., 2005; 11 (4); 347–361.
[16] Bouamoul A., Numerical Calculation of Blast Effect on Human and on the Mannequin for the Assessment of Blast Incapacitation and Lethality. Technical memorandum. DRDC Valcartier TM 2008-285, Defense R&D Canada. 2008.
[17] Boutillier J., De Mezzo S., Deck C., Magnan P., Naz P., Willinger, R., Chest response assessment of post-mortem swine under blast loadings. J. Biomech., 2017; 65; 169–175.
[18] Bir C., Viano D., King A., Development of biomechanical response corridors of the thorax to blunt ballistic impacts. J. Biomech., 2004; 37; 73–79.
[19] Baker W.E., Westine P.S., Dodge F.T., Similarity Methods in Engineering Dynamics: Theory and Practice of Scale Modelling. Elsevier, Amsterdam, New York, 1991.
[20] Boutillier J., Contribution à la compréhension et à la modélisation des effets lésionnels sur le thorax des ondes de choc aériennes. Ph.D. Thesis, Strasbourg University, France, 2017.

 https://doi.org/10.52202/078352-0029

Role of Army Combat Boot in influencing calcaneus and distal tibia injuries and risk curves from underbody blast loading

N. Yoganandan[1], X. Yayun[2], A. Banerjee[2], M. Schlick[1], S. Chirvi[1], Frank Pintar[1], David Barnes[3], and Kathryn Loftis[4]

[1]*Department of Neurosurgery, Medical College of Wisconsin, 8701 Watertown Plank Road, Milwaukee, WI 53226, USA*
yoga@mcw.edu
[2]*Division of Biostatistics, Medical College of Wisconsin, 8701 Watertown Plank Road, Milwaukee, WI 53226, USA*
[3]*SURVICE Engineering Co., Contractor to U.S. Army CCDC-DAC, Building 4501, Room 124, Aberdeen Proving Ground, MD, USA*
[4] *U.S. Army CCDC-DAC, Aberdeen Proving Ground, MD, USA*

Abstract. Calcaneus and distal tibia fractures are the most common injuries from underbody loading events. Human cadaver foot-ankle-tibia complexes have been subjected to underbody blast impacts, and tests have been done with and without boot use. While peak forces have been determined for both boot conditions, the role of boots in influencing calcaneus and distal tibia injuries and risk curves has not been determined. The objectives of the study were to analyze our previously conducted tests and delineate the role of the boot for two common types of fractures. Forty-five foot-ankle-lower specimens were subjected to vertical impacts using custom vertical accelerator. For the statistical analysis, the peak force data were used as the primary response variable, and the presence or absence of the boot was treated as a covariate for the two most conservative human injury probability curves (HIPCs). Twenty-seven sustained calcaneus fractures and ten specimens sustained distal tibia fractures. Calcaneus fractures occurred in ten and tibia fractures occurred in nine specimens with the booted condition. The HIPCs based on parametric survival analysis are provided in the paper for the presence and absence of boots for the two most conservative (lower bound) estimates. The boot modulated the forces by approximately ten percent for both conservative estimates of the HIPCs. The plus and minus 95% percent normalized confidence intervals and quality of HIPCs at discrete probability levels are given the body of the paper along with the various HIPCs. The greater occurrence of calcaneus than tibia fractures with the booted condition is in line with the field studies associated with underbody blast environments. Additional tests posture-based studies are needed to delineate the role of posture on the greater occurrence of calcaneus injuries.

1. INTRODUCTION

Combat-related activities have shown that they can result in lower leg injuries in the form of fractures of the calcaneus and or distal tibia complex to military personnel [1-4]. Studies have shown that they can occur from vertical impacts during events such as underbody blast loading from improvised explosive devices [5-8]. This vertical loading mode to the human foot is also prevalent in other scenarios [9, 10]. To investigate the biomechanics of these injuries, describe injury criteria, and develop manikin or anthropomorphic test devices, it is important to conduct impact tests under the vertical loading mode. Unembalmed human cadaver specimens are routinely used for such purposes. These types of tests can be conducted using whole body surrogates or subsystem/component models [11-14]. In the former model, human cadavers have been exposed to blast loading in a field-type environment, and the resulting injuries have been identified to the calcaneus, tibia, and other body regions [11]. In a more controlled setting whole body human cadavers have been exposed to simulated impact loads, and similar injuries have been found. In the latter experimental model, isolated lower leg specimens including the foot-ankle-tibia complex have been subjected to simulated vertical impact loading, and this model produced injuries to the calcaneus, tibia and other bones or joints [13]. The cited literature is not all inclusive. This has formed a primary dataset in the understanding of these injuries from vertical loading to the plantar surface of the human foot, and as applied to military environments.

Whole body human cadaver experimental models described above have included the use of the boot to paralleled field conditions in military scenarios [11]. In contrast, more controlled and focused laboratory studies with the subsystem or component experimental model have evaluated the biomechanics of injury with and without the presence/use of the boot [12, 14-17]. Different researchers have used different types of boots, based on the needs of the study design and intended application [15, 17-20]. The introduction of an energy absorbing material, in this case, the boot, influences the load path

https://doi.org/10.52202/078352-0030

within the foot-ankle-distal tibia fibula complex, i.e., the multiple bones and surrounding joints. The level of influence depends on numerous factors. Some studies have developed injury risk curves for the entire dataset; however, the role of the boot in influencing the impact biomechanics of internal load transfer, injury outcomes and HIPCs has not been investigated. The purpose of the present study is therefore, to determine the role of the boot in modulating the foot-ankle-distal tibia fibula complex injuries and associated risk curves from vertical impacts. The role of the boot in producing the two most common fractures are delineated.

2. METHODS

2.1 Experimental methodology and biomechanical data

Tests were conducted using unembalmed human cadaver foot-ankle-distal tibia fibula complexes. While detailed test information is provided elsewhere, the following descriptions are pertinent to meet the objectives of the present analysis-based study [13]. The specimens were screened for pre-existing fractures, and x-rays and computed tomography (CT) scans were obtained. The foot-ankle-distal tibia fibula complex was at fixed at the proximal end, a six-axis load cell was attached, and a simulated femur was fixed to the load cell in some preparations, based on the device used to apply the impact loading. A military combat boot (Style 3187; McRae or Belleville Footware Inc., Baltimore,MD) was donned in some specimens. Its size was based on the anthropometry of the foot of the specimen. A new boot was used for each specimen. Vertical impact loading was directed along the longitudinal axis of the tibia using one of the two devices: custom vertical accelerator or pendulum (Figure 1).

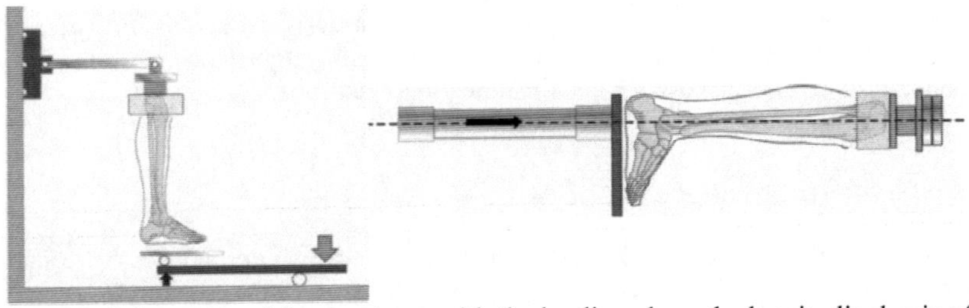

Figures 1 and 2. Schematics of the specimen with the loading along the longitudinal axis of the tibia in the vertical accelerator (Figure 1a, left) and pendulum (Figure 1b, right) devices.

For tests with the vertical accelerator, the preparations were aligned such that the plantar surface of the preparation, longitudinal tibial axis, and longitudinal axis of the femur were orthogonal to each other before impact, termed as 90-90-90 posture in literature. For tests with the pendulum device, they were aligned in a similar posture in the pendulum device for applying the load that was aligned along the tibia axis with the exception that the simulated femur was replaced by equivalent ballast mass, mounted superiorly to the load cell. Thus, the preparation sustained the intended vertical impact loading vector in both devices, i.e., inferior-to-superior accelerative forces via the plantar surface to unbooted and booted foot-ankle-distal tibia fibula complex preparations. Following the final impact test, the specimens underwent radiography, followed by CT and gross dissection to identify the injuries, and orthopedic surgeons of our team assisted in the assessments of injuries. Fractures to the calcaneus and tibia identified in these experiments were used in the analysis of data, described below. The peak forces recorded by the load cell was identified using the force-time histories, and the fracture outcomes from the pre- and posttest images, described above. Parametric survival analysis techniques were used to construct the human injury probability curves, HIPCs, for the following five groups.

2.3 Grouping for human injury probability curves

Group A consisted of HIPCs for calcaneus fractures without the presence of tibia fractures, i.e., specimens with tibia-only injuries were removed. Group B consisted of HIPCs for tibia fractures without the presence of calcaneus fractures, i.e., specimens with calcaneus-only injuries were removed. Group C consisted of HIPCs associated with the presence of either outcomes, tibia or calcaneus. Group D consisted of HIPCs for calcaneus fractures without the presence of tibia fractures, and specimens with tibia only injuries were considered as no injuries. Group E consisted HIPCs for tibia fractures without

 https://doi.org/10.52202/078352-0030

the presence of calcaneus fractures, and specimens with calcaneus-only injuries were considered as no injuries. All these selection groups have statistical relevance, and they are discussed later.

2.4 Survival analysis

In all cases, specimen outputs considered as non-injury and injury data were treated as right and exact censored observations in the survival analysis. Reasons for selecting this censoring scheme are given in the discussion section. The survival analysis modeling was performed using the R-software (version 3.6.4). The lowest Brier Score Metric, BSM, and its associated distribution were used to calculate the final injury risk curves [21]. The cumulative density functions evaluated in the analysis were the Weibull, lognormal, and log-logistic distributions. The Normalized Confidence Interval Size, NCIS, was defined as the ratio of confidence interval width to the magnitude of the peak force estimate. The NCIS magnitudes of <0.5, between 0.5 and 1, >1 to 1.5, and >1.5 were assigned the adjectival ratings of good, fair, marginal, and unacceptable, respectively [22]. They were reported as tabulated data at 5%, 10%, 25%, 50%, 75%, 90%, and 95% probability levels. First, from the group of five HIPCs, the two most conservative HIPCs serving as the lower bound estimates for calcaneus or tibia injuries were identified. For these two datasets, the effect of the boot was determined by treating its presence or absence in the survival analysis. The boots were treated as a covariate. This process yielded four HIPCs: with and without boots for the two most conservative cases.

3. RESULTS

The mean age and stature of these human cadavers were 56 ± 12 years and 1.79 ± 0.06 m. All specimens were from male human cadavers. The dataset used in the analysis consisted of 45 specimens with calcaneus injuries to 27 and tibia injuries to ten, while the remaining eight specimens did not sustain any injury. Calcaneus fractures occurred in ten and tibia fractures occurred in nine specimens with the booted condition.

3.1 HIPCs from five groups

A comparison of the HIPCs for the five groups of specimens are shown in Figure 3. The quality indices ranged from fair to good at the 5%, 10%, 25%, 50%, 75%, 90%, and 95% probability levels for all risk curves in all groups. Groups A and C were considered as the two most conservative HIPCs. Figure 5 shows the HIPCs and NCIS magnitudes at different probability levels for the group A dataset without the use of the boot. Forces associated with the 10%, 25%, and 90% injury probability levels were 4.4 kN, 7.4 kN, and 10.3 kN, respectively. The quality indices were in the fair, good, and good categories at these probability levels, respectively.

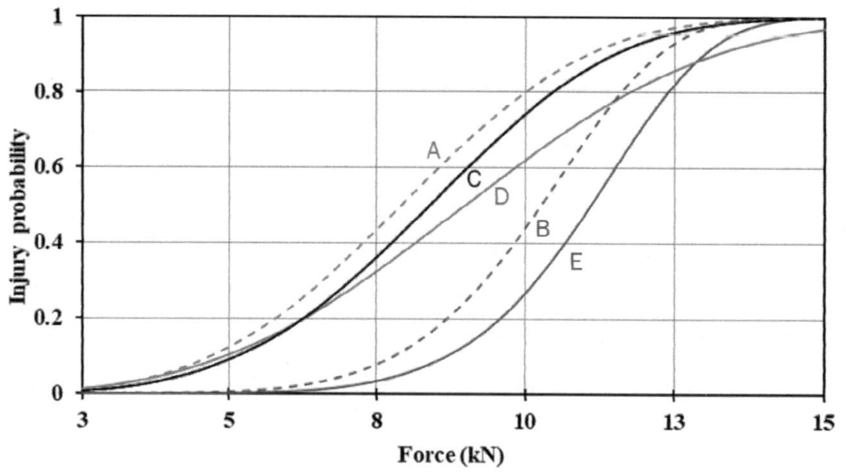

Figure 3. HIPCs for the five groups. See text for grouping details.

https://doi.org/10.52202/078352-0030

3.2 Boot-based injury probability curves

Group A and group C data were used to determine the role of the boot in this analysis. Figure 4 shows the HIPCs and NCIS magnitudes at different probability levels for the group A dataset without any consideration regarding the use of the boot. Table 1 includes the NCIS and other data.

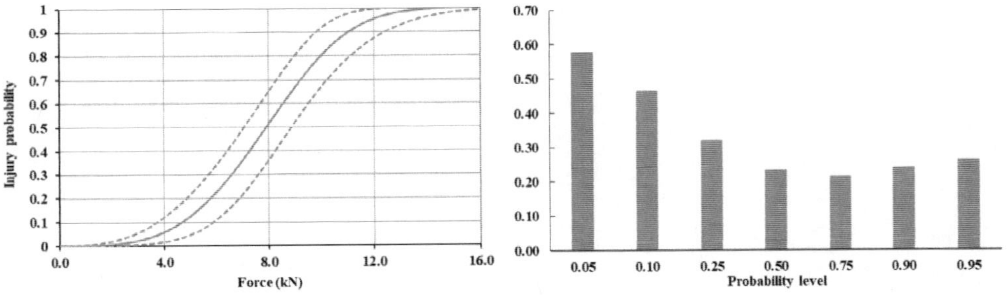

Figure 4. HIPCs for group A without considering the use of the boot as a covariate. Dashed lines show the 95% confidence intervals. Bar chart shows the NCIS at different probability levels.

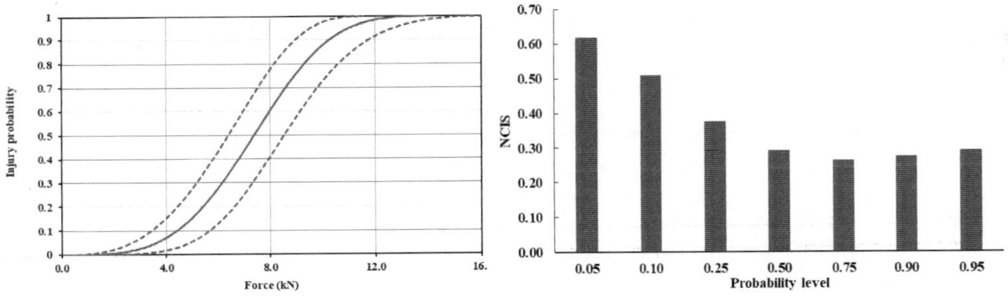

Figure 5. HIPCs for group A without the use of the boot as a covariate. Dashed lines show the 95% confidence intervals. Bar chart shows the NCIS at different probability levels.

Table 1: HIPC data for the group A dataset without the use of a boot

Risk Level	Mean force (kN)	95% Confidence interval		NCIS	Quality index
		Lower bound	Upper bound		
		Group A data			
0.05	3.62	2.67	4.91	0.62	Fair
0.10	4.41	3.43	5.67	0.51	Fair
0.25	5.81	4.82	7.00	0.37	Good
0.50	7.39	6.40	8.53	0.29	Good
0.75	8.94	7.84	10.18	0.26	Good
0.90	10.27	8.97	11.76	0.27	Good
0.95	11.04	9.56	12.74	0.29	Good

Figure 6 shows the HIPCs and NCIS magnitudes at different probability levels for the group A dataset with the use of the boot. Forces associated with the 10%, 25%, and 90% injury probability levels were 5.2 kN, 8.7 kN, and 12.4 kN, respectively. The quality indices were in the good category at these probability levels. Data at other levels are shown in Table 2. Table 3 shows the change in the forces at different probability levels with and without the use of the boot, with respect to the peak forces independent of its use. The following equations were used to determine the percentage changes.

https://doi.org/10.52202/078352-0030

Table 2: HIPC data for the group A dataset with boot use

Risk Level	Mean force (kN)	95% Confidence interval		NCIS	Quality index
		Lower bound	Upper bound		
		Group A data			
0.05	4.25	3.17	5.69	0.59	Fair
0.10	5.17	4.05	6.62	0.50	Good
0.25	6.81	5.61	8.28	0.39	Good
0.50	8.67	7.30	10.29	0.34	Good
0.75	10.48	8.82	12.46	0.35	Good
0.90	12.04	10.00	14.51	0.37	Good
0.95	12.95	10.64	15.74	0.39	Good

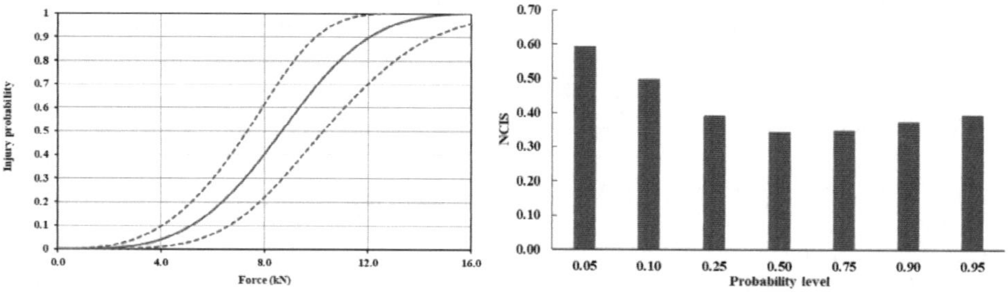

Figure 6. HIPCs for group A with the use of the boot as a covariate. Dashed lines show the 95% confidence intervals. Bar chart shows the NCIS at different probability levels.

$$\% \; change \; with \; boot \; use = 100 \, X \, \frac{\text{Force with boot use} - \text{Force independent of boot use}}{\text{Force independent of boot use}} \quad (1a)$$

$$\% \; change \; without \; boot \; use = 100 \, X \, \frac{\text{Force without boot use} - \text{Force independent of boot use}}{\text{Force independent of boot use}} \quad (1b)$$

The mean difference across all risk levels for the mean value was 6.5% without boot use and 9.6% with boot use. In other words, at any risk level, the absence or presence of boot modulated (decreased or increased) the magnitudes of forces by approximately 7% and 10% compared to the case wherein the boot effects were ignored.

Table 3: Effect of boots on HIPC magnitudes for group A dataset

Risk level	Mean difference (kN)	% difference	Mean difference (kN)	% difference
	Without Boot		With Boot	
0.05	0.23	-5.9%	0.40	10.4%
0.10	0.29	-6.1%	0.48	10.2%
0.25	0.40	-6.4%	0.61	9.8%
0.50	0.53	-6.6%	0.75	9.5%
0.75	0.66	-6.8%	0.89	9.3%
0.90	0.77	-7.0%	1.01	9.1%
0.95	0.84	-7.1%	1.07	9.0%
Average		-6.5%		9.6%

https://doi.org/10.52202/078352-0030

Figures 7 shows the HIPCs and NCIS magnitudes at different probability levels for the group C dataset without any consideration regarding the use of the boot. Figure 8 shows the HIPCs and NCIS magnitudes at different probability levels for the group C dataset without the use of the boot. Forces associated with the 10%, 25%, and 90% injury probability levels were 4.7 kN, 7.5 kN, and 10.1 kN, respectively. The quality indices were in the good category at these probability levels. Data at other levels are shown in Table 4. Figure 9 shows the HIPCs and NCIS magnitudes at different probability levels for the group C dataset with the use of the boot. Forces associated with the 10%, 25%, and 90% injury probability levels were 5.8 kN, 9.2 kN, and 12.3 kN, respectively. The quality indices were in the good category at these probability levels. Data at other levels are shown in Table 5. Table 6 shows the change in the forces at different probability levels with and without the use of the boot, with respect to the force magnitudes independent of its use. The mean difference across all risk levels for the mean value was 10.5% without boot use and 9.6% with boot use. In other words, at any risk level, the absence or presence of boot modulated (decreased or increased) the magnitudes of forces by approximately 10% compared to the case wherein the boot effects were ignored.

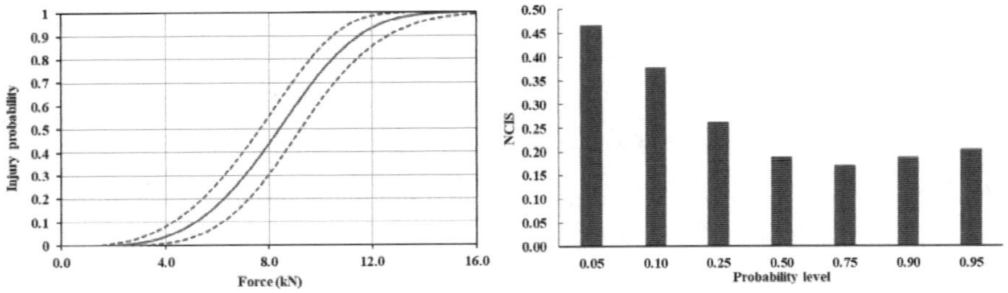

Figure 7. HIPCs for group C without considering the use of the boot as a covariate. Dashed lines show the 95% confidence intervals. Bar chart shows the NCIS at different probability levels.

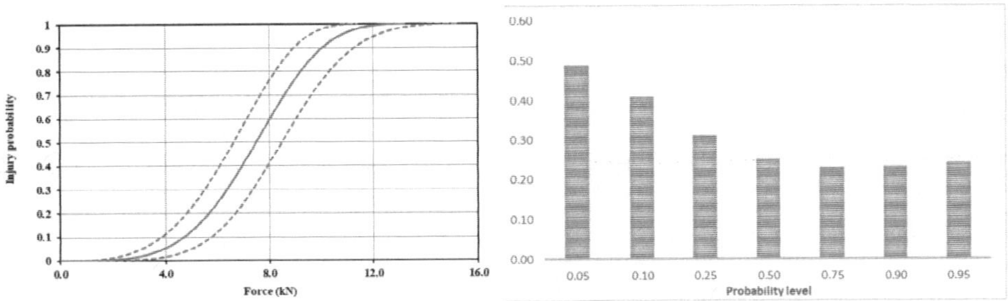

Figure 8. HIPCs for group C without the use of the boot as a covariate. Dashed lines show the 95% confidence intervals. Bar chart shows the NCIS at different probability levels.

4. DISCUSSION

As stated in the introduction, the purpose of the study was to determine the role of the boot in modulating injuries and risk curves to the tibia and calcaneus from vertical impacts simulating underbody blast environments and develop HIPCs. The role was determined using a human cadaver model that consisted of unembalmed human cadaver foot-ankle-distal tibia fibula complexes that were fixed at the proximal end with a load cell to record the peak forces. Fracture outcomes and force data were obtained from injury and non-injury tests and from specimens that were tested with and without the presence of the boot. These form the basic steps for achieving the objectives of the present analysis. The process of using parametric survival analysis has become a de facto norm in impact-injury biomechanics to determine the human tolerance in the form of risk curves [9, 17, 20, 23-27]. Peak forces were used in the development of HIPCs. They were treated as right censored observations for the non-injury and exact censored observations for the injury data points. The reason for selecting the right

 https://doi.org/10.52202/078352-0030

censoring option was that the survival analysis, unlike traditional binary regression, allows this flexibility, and adds statistical content to the ensuing HIPCs. The uncensored option for injury data points was used based on the following reasons. First, the peak force represents the greatest force sustained by the specimen due to the application of the impact load. Second, the injuries observed following the experimentation was associated with this force level. Previous studies using human cadaver foot-ankle complexes have shown that the peak force is associated with injury [17, 19, 28].

Table 4: HIPC data for the group C dataset without boot use

Risk Level	Mean force (kN)	95% Confidence interval		NCIS	Quality index
		Lower bound	Upper bound		
		Group A data			
0.05	3.94	3.10	5.02	0.49	Good
0.10	4.71	3.84	5.77	0.41	Good
0.25	6.03	5.16	7.04	0.31	Good
0.50	7.49	6.61	8.48	0.25	Good
0.75	8.88	7.93	9.96	0.23	Good
0.90	10.07	8.96	11.31	0.23	Good
0.95	10.74	9.52	12.12	0.24	Good

Table 5: HIPC data for the group C dataset with boot use

Risk Level	Mean force (kN)	95% Confidence interval		NCIS	Quality index
		Lower bound	Upper bound		
		Group A data			
0.05	4.82	3.87	6.02	0.45	Good
0.10	5.76	4.79	6.92	0.37	Good
0.25	7.38	6.42	8.48	0.28	Good
0.50	9.16	8.17	10.28	0.23	Good
0.75	10.87	9.72	12.16	0.22	Good
0.90	12.32	10.93	13.89	0.24	Good
0.95	13.15	11.58	14.93	0.25	Good

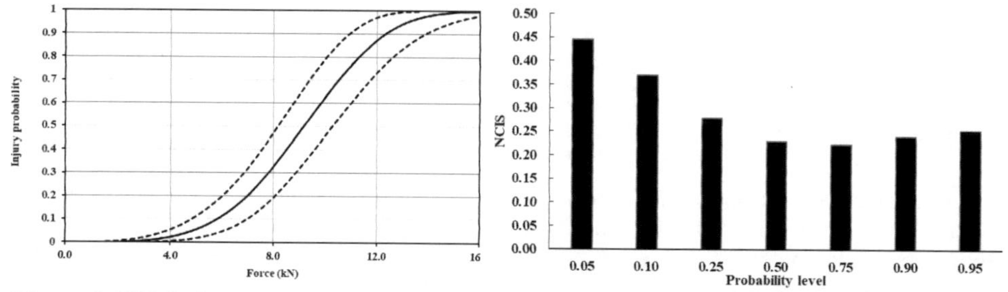

Figure 9. HIPCs for group C with the use of the boot as a covariate. Dashed lines show the 95% confidence intervals. Bar chart shows the NCIS at different probability levels.

Multiple datasets were considered to analyze the biomechanical data and develop the HIPCs (Figure 3). Grouping by the presence of only calcaneus injuries represents the cases wherein the curve is associated and applicable to no other injuries (group A), and this is also true for the grouping of cases with tibia-only injuries. These HIPCs focus on the fracture of the specific bone from vertical loading.

https://doi.org/10.52202/078352-0030

Field data show that this type of fractures occurs in the military environments to Service Members [29]. The grouping of injuries by the tibia-only cases (group B) fall along these lines of discussion. From this perspective, the present study has provided HIPCs specific to these two injuries, treated separately. The cases where the presence of either injury, calcaneus or tibia (group C), represent the HIPCs in which one of these two injuries can be expected and need protection. Because the proximity of the calcaneus to the impact vector and the relatively stronger long bone, HIPCs for tibia-only fractures are right shifted compared to the calcaneus only HIPCs, even when the dataset considered calcaneus-only injuries as no fractures (group E). Likewise, the HIPCs the calcaneus-only fractures are left shifted compared to the tibia-only HIPCs, even when the dataset considered tibia-only injuries as no fractures (group D). These are also in line with the expected outcomes for the choice of the datapoint in the HIPC analysis.

Table 6: Effect of boots on HIPC magnitudes for group C dataset

Risk level	**Mean difference (kN)**	**% difference**	**Mean difference (kN)**	**% difference**
	Without Boot		With Boot	
0.05	0.32	-7.5%	0.56	13.2%
0.10	0.43	-8.5%	0.62	12.0%
0.25	0.65	-9.7%	0.70	10.5%
0.50	0.91	-10.8%	0.76	9.1%
0.75	1.18	-11.7%	0.81	8.0%
0.90	1.42	-12.3%	0.84	7.3%
0.95	1.56	-12.7%	0.85	6.9%
Average		-10.5%		9.6%

The present study focused on determining the two most conservative HIPCs (lower bound estimates) to determine the role of the boot in modulating the injuries and resulting injury risk curves. This type of analysis led to the identification of group A and group C HIPCs for which the boot was used as a covariate and boot dependent HIPCs were derived. As expected, the HIPCs without boots resulted in lower forces than the HIPCs with boots at the same injury risk level. The differences in the use or nonuse of the boot was however, limited to approximately ten percent when compared to the combined (regardless of the boot presence) datasets, true for both groups A and C (Tables 3 and 5). The present analysis quantifies the modulating role of the boot in shifting the HIPC from left to right (absence to presence of the boot), and it should be noted that the underlying biomechanical tolerance of the bone does not change. That is, bone fractures when it is loaded above its threshold. and any intervening end condition modulates the transmitted force to the bone. In other words, the true fracture force limit does not change and is independent of the end condition (boot use in this case). The changing HIPCs from left to right with the boot shows the greater magnitudes of the impact loading that can be applied to the plantar surface of foot from the no boot to with boot condition, (demonstrating the protective effect of the boot use, an intended feature in the military environment), and in this case, as discussed, it is approximately ten percent across the entire probability curve for the two most conservative HIPCs.

The presence of the boot resulted in more calcaneus fractures than tibia fractures in this experimental study. This suggests that while the boot acts as a medium for transmitting the impact loads and protects the military personnel, calcaneus fractures occur more than tibia fractures. In an analysis of injuries to Warfighters in underbody blast environments, Danelson et al., found that calcaneus fractures were the most prevalent followed by forefoot and distal tibia fractures [29]. The differences between the outcomes of the present series of experiments and field data may be due to the controlled, nominal posture in the PMHS tests compared to varied postures of the Soldiers in the field environment. However, the greater occurrence of calcaneus fractures, in this small sample study, is in line with field outcomes. Additional tests are needed, however, in other postures to delineate the role of posture on the greater occurrence of calcaneus injuries with the use of the boots.

As stated in the methods section, a new boot was used for each leg. While not presented in the manuscript, a separate series of tests was done to determine the attenuation characteristics of the boot. Variations in the attenuation forces were 1.3%, 2.5%, and 3.8% for 3 different boots, all compressively tested 10 times in a material testing device without the presence of the biological surrogate. Thus, the use of the same boot for each specimen when it was repeatedly loaded (interval censoring design of the experiment wherein noninjury and injury tests were done) induced minimal changes to the responses of

https://doi.org/10.52202/078352-0030

the human cadaver leg. It should be noted, however, that the responses may differ if different types of boots are used. This is a future investigation topic.

Introduction of a boot to the specimen influences the original load path, and the load transmitted to the leg depends on the properties of the boot (expected reduction in amplitude and increase in time). Mass recruitment effects come into play depending on the relative contributions of the amplitude and duration of the applied pulse to the specimen, and in addition, specimen characteristics play a role. Previous whole-body studies that applied inferior to superior acceleration to the pelvis have shown that a lower magnitude and a longer duration pulse tends to injure the lumbar spine while a larger amplitude and a shorter time duration pulse tends to fracture the pelvis [30]. A similar phenomenon is expected although the clear differentiation of the mass recruitment effects, and pulse profiles were not investigated in the present study. This is also a future research topic.

Acknowledgments

This work was supported under contract #N00024-13-D-6400, sponsored by the U.S. Army Research Lab in support of the WIAMan Program, W81XWH-16-01-0010, Department of Veterans Affairs Medical Research, and the Department of Neurosurgery at the Medical College of Wisconsin. This material is the result of work supported with resources and use of facilities at the Zablocki VA Medical Center, Milwaukee, Wisconsin. The views expressed are those of the authors and do not necessarily represent the official position or policy of the U. S. Government, the Department of Defense (or its branches), or the Department of the Army.

References

[1] Holcomb, J. B., 2012, "Associated injuries in casualties with traumatic lower extremity amputations caused by improvised explosive devices," Br J Surg, 99(3), pp. 362-366.

[2] Morrison, J. J., Hunt, N., Midwinter, M., and Jansen, J., 2012, "Associated injuries in casualties with traumatic lower extremity amputations caused by improvised explosive devices," Br J Surg, 99(3), pp. 362-366.

[3] Ramasamy, A., Masouros, S. D., Newell, N., Hill, A. M., Proud, W. G., Brown, K. A., Bull, A. M., and Clasper, J. C., 2011, "In-vehicle extremity injuries from improvised explosive devices: current and future foci," Philos Trans R Soc Lond B Biol Sci, 366(1562), pp. 160-170.

[4] Ramasamy, A., Hill, A. M., Phillip, R., Gibb, I., Bull, A. M., and Clasper, J. C., 2011, "The modern "deck-slap" injury--calcaneal blast fractures from vehicle explosions," J Trauma, 71(6), pp. 1694-1698.

[5] Yoganandan, N., Nahum, A. M., and Melvin, J. W., 2015, "Accidental Injury: Biomechanics and Prevention," Springer, NY, p. 851.

[6] Danelson, K. A., Frounfelker, P., Pizzolato-Heine, K., Valentine, R., Watkins, L. C., Tegtmeyer, M., Bolte, J. H., Hardy, W. N., and Loftis, K. L., 2019, "A Military Case Review Method to Determine and Record the Mechanism of Injury (BioTab) from In-Theater Attacks," Mil Med, 184(Suppl 1), pp. 374-378.

[7] Loftis, K. L., Mazuchowski, E. L., Clouser, M. C., and Gillich, P. J., 2019, "Prominent Injury Types in Vehicle Underbody Blast," Mil Med, 184(Suppl 1), pp. 261-264.

[8] Ramasamy, A., Newell, N., and Masouros, S., 2014, "From the battlefield to the laboratory: the use of clinical data analysis in developing models of lower limb blast injury," J R Army Med Corps, 160(2), pp. 117-120.

[9] Yoganandan, N., Arun, M. W., Pintar, F. A., and Szabo, A., 2014, "Optimized lower leg injury probability curves from postmortem human subject tests under axial impacts," Traffic Inj Prev, 15 Suppl 1, pp. S151-156.

[10] Yoganandan, N., Pintar, F. A., Boynton, M., Begeman, P., Prasad, P., Kuppa, S. M., Morgan, R. M., and Eppinger, R. H., 1996, "Dynamic axial tolerance of the human foot-ankle complex, Stapp Car Crash Conference. Albuquerque, New Mexico, United States."

[11] Danelson, K. A., Kemper, A. R., Mason, M. J., Tegtmeyer, M., Swiatkowski, S. A., Bolte, J. H. t., and Hardy, W. N., 2015, "Comparison of ATD to PMHS Response in the Under-Body Blast Environment," Stapp Car Crash J, 59, pp. 445-520.

[12] Gallenberger, K., Yoganandan, N., and Pintar, F., 2013, "Biomechanics of foot/ankle trauma with variable energy impacts," Ann Adv Automot Med, 57, pp. 123-132.

[13] Yoganandan, N., Chirvi, S., Pintar, F. A., Uppal, H., Schlick, M., Banerjee, A., Voo, L., Merkle, A., and Kleinberger, M., 2016, "Foot-Ankle Fractures and Injury Probability Curves from Post-mortem Human Surrogate Tests," Ann Biomed Eng, 44(10), pp. 2937-2947.

[14] Newell, N., Masouros, S. D., Ramasamy, A., Bonner, T. J., Hill, A. M., Clasper, J. C., and Bull, A. M., 2013, "Use of cadavers and anthropometric test devices (ATDs) for assessing lower limb injury

https://doi.org/10.52202/078352-0030

outcome from under- vehicle explosions," IRCOBI Conference Proceedings, Dublin, Ireland, pp. 296–303.
[15] Chirvi, S., Pintar, F. A., and Yoganandan, N., 2016, "Human surrogate leg response with and without military boot," PASS conference.
[16] Yoganandan, N., Pintar, F., Banerjee, A., Schlick, M., Chirvi, S., Uppal, H., Merkle, A., Voo, L., and Kleinberg, M., 2015, "Hybrid III Lower Leg Injury Assessment Reference Curves Under Axial Impacts Using Matched-Pair Tests," Biomed Sci Instrum, 51, pp. 230-237.
[17] McKay, B. J., and Bir, C. A., 2009, "Lower extremity injury criteria for evaluating military vehicle occupant injury in underbelly blast events," Stapp Car Crash J, 53, pp. 229-249.
[18] Pandelani, T., Sono, T. J., Reinecke, D., and Nurick, G. N., 2015, "Impact loading response of the MiLLx leg fitted with combat boots," Int. J. Impact Eng.
[19] Chirvi, S., Pintar, F. A., Yoganandan, N., Banerjee, A., Schlick, M. S., Curry, W. C., and Voo, L. M., 2017, "Human foot-ankle injuries and associated risk curves from under body blast loading conditions," Stapp Car Crash J, 61, pp. 189-210.
[20] Mildon, P. J., White, D., Sedman, A. J., Dorn, M., and Masouros, S. D., 2018, "Injury Risk of the Human Leg Under High Rate Axial Loading," Human Factors and Mechanical Engineering for Defense and Safety, 2(1), p. 5.
[21] Yoganandan, N., and Banerjee, A., 2018, "Survival Analysis-Based Human Head Injury Risk Curves: Focus on Skull Fracture," J Neurotrauma, 35(11), pp. 1272-1279.
[22] Petitjean, A., Torsseille, X., Yoganandan, N., and Pintar, F. A., 2015, "Normalization and scaling for human response corridors and development of risk curves " Accidental Injury: Biomechanics and Prevention, N. Yoganandan, A. M. Nahum, and J. W. Melvin, eds., Springer,, NY, pp. 769-792.
[23] ISO, 2013, "TS18506: Road Vehicles - Procedure to construct injury risk curves for the evaluation of road users protection in crash tests. International Standards Organization," American National Standards Institute, New York.
[24] Petitjean, A., Trosseille, X., Praxl, N., Hynd, D., and Irwin, A., 2012, "Injury risk curves for the WorldSID 50th male dummy," Stapp Car Crash J, 56, pp. 323-347.
[25] Yoganandan, N., Banerjee, A., Hsu, F. C., Bass, C. R., Voo, L., Pintar, F. A., and Gayzik, F. S., 2016, "Deriving injury risk curves using survival analysis from biomechanical experiments," J Biomech, 49(14), pp. 3260-3267.
[26] Yoganandan, N., Chirvi, S., Pintar, F. A., Baisden, J. L., and Banerjee, A., 2018, "Preliminary female cervical spine injury risk curves from PMHS tests," J Mech Behav Biomed Mater, 83, pp. 143-147.
[27] Yoganandan, N., Moore, J., Pintar, F. A., Banerjee, A., DeVogel, N., and Zhang, J., 2018, "Role of disc area and trabecular bone density on lumbar spinal column fracture risk curves under vertical impact," J Biomech, 72, pp. 90-98.
[28] Funk, J. R., Crandall, J. R., Tourret, L. J., MacMahon, C. B., Bass, C. R., Patrie, J. T., Khaewpong, N., and Eppinger, R. H., 2002, "The axial injury tolerance of the human foot/ankle complex and the effect of Achilles tension," J Biomech Eng, 124(6), pp. 750-757.
[29] Danelson, K., Watkins, L., Hendricks, J., Frounfelker, P., Team, W. I. C. R., Pizzolato-Heine, K., Valentine, R., and Loftis, K., 2018, "Analysis of the Frequency and Mechanism of Injury to Warfighters in the Under-body Blast Environment," Stapp Car Crash J, 62, pp. 489-513.
[30] Yoganandan, N., Moore, J., Arun, M. W., and Pintar, F. A., 2014, "Dynamic Responses of Intact Post Mortem Human Surrogates from Inferior-to-Superior Loading at the Pelvis," Stapp Car Crash J, 58, pp. 123-143.

 https://doi.org/10.52202/078352-0030

Dynamic Mechanical Properties of Human Skin

K. Kong[1], M. DeWitt[2], M. Danilich[2], B. Gillich[3], **M. B. Panzer**[1]

[1]*University of Virginia, 4040 Lewis and Clark Dr. Charlottesville, 22911 VA., USA*
panzer@virginia.edu
[2]*Luna Innovations, 706 Forest St., Charlottesville, 22901 VA., USA,*
[3]*U.S. Army Aberdeen Test Center, 6943 Colleran Rd B400, APG, 21005-5059 MD., USA*

Abstract. Skin is the largest human organ which has many functions, including the initial protection of the human body from injury. Therefore, understanding of mechanical properties of the skin is crucial for various applications, including non-lethal ballistics development where the goal is effective deterrence while avoiding penetrating or irreversible damaging of the skin during impact. The objective of this study is to quantify the tensile mechanical properties and failure response of human skin through dynamic tests. A total of 108 skin samples were obtained from the posterior torsos of six post-mortem human subjects for experimental testing. The skin samples were tested in comprehensive battery of tests, including quasi-static and dynamic tension (uniaxial) to failure, tensile stress relaxation (uniaxial), compression, and dynamic indentation (biaxial) to failure. The skin samples were marked in the *in-situ* reference state before removal. On one side of the torso, dog-bone test samples for uniaxial tension tests were prepared by orienting the axis of loading at 0° and 90° respectively with respect to identified Langer lines. On the other side, square patches (approximately 75 x 75 mm) were prepared for the dynamic indentation tests. All samples were stored in a dampened environment of saline to prevent moisture loss. During the testing, all samples were preloaded to match the *in situ* stretch state before any mechanical loading. Destructive tensile tests, compression tests, and stress relaxation tests of human skin were performed across different strain-rates (up to 180 1/s) and *in situ* orientations. The results of this study will be used for informing the development of constitutive models of skin for human body modeling, and skin simulants used in the physical test and evaluation environment for assessing the safety of protective body armor and/or blunt non-lethal weapons.

1. INTRODUCTION

Human skin is a delicate organ that can easily result in injury when subjected to loading conditions such as scraping, tearing, and penetrating. Open skin wounds can potentially allow severe blood loss and lead to infection or chronic conditions, which affects 4.5 million people in the United States [1]. Understanding the biomechanics of the skin, and its threshold for failure, is vital for improving the efficacy of safety equipment and protective gear in preventing skin injury in many applications including transportation safety, law enforcement, and in the military. Knowing the biomechanics of skin is equally important to the development of non-lethal blunt projectiles, which are not intended to cause penetration.

To study skin injury mechanics, animal skin tissue collected from mice and goats have been studied but were found to not represent the human skin respectively in terms of ultimate stress [2,3]. Porcine skins have been reported to be like the human skin when looking at elastic modulus and ultimate stress but was found to rupture at lower strain levels [4]. Chamois skin was also used as part of skin surrogate's development for skin penetration assessment showed acceptable biofidelic results, however, its thickness inconsistency and reproducibility reduce its practicality [5,6]. Efforts have been made to develop affordable synthetic skin simulant (silicone and urethane) with a stable shelf life to exhibit biomechanical failure equivalence to human skin also show little success [7].

Skin is reported to be non-linear, anisotropic, and viscoelastic [8]. Generally, skin shows stiffer response when loaded along its collagen fiber direction than across fiber direction under tension. The fiber recruitment contributes to the anisotropic behavior and therefore provides full strength of the skin when loaded under tension compared to compression and shear. Skin rupture data is necessary to study open skin injuries and therefore in-vitro skin testing is desired compared to in-vivo skin testing. Several destructive tensile tests of in-vitro human skin have been conducted in the past [9-11] looking into either the effect of loading rate or skin orientation with respect to Langer line on skin tensile failure behavior. To date, Ottenio et al. (2015) have provided a first comprehensive dataset regarding mechanical failure properties of human skin by considering anisotropy, strain rate effect (0.06/s-167/s) and skin failure simultaneously despite only one post-mortem human subject (PMHS) was studied [12]. The lack of having inter-subject variation simplifies investigation on the strain rate and orientation effect on skin behavior but this also limits the practicality of the skin data for representing the human population. In addition, neither of these tests provides viscoelastic data of the human skin although it has been shown the human skin response is strain rate dependent [12]. The collection of viscoelastic data is deemed

https://doi.org/10.52202/078352-0031

necessary for skin constitutive model development. Viscoelastic tests of human skin were conducted in-vivo where small stretch/stress levels were applied to living human subjects through suction [13,14]. However, these in-vivo studies do not allow viscoelastic information at high stretch levels to be determined and effects from surrounding tissues cannot be eliminated.

This study provides new viscoelastic and destructive tensile stress-stretch response of human skin by considering inter-subject variation with strain rate applicable in between automotive crash and ballistic impact event in the field of injury biomechanics. The in-vitro viscoelastic dataset is expected to inform the selection of an appropriate constitutive model that can be used to model human skin tensile response. The destructive tensile test series will provide a relationship between strain rate and sub-failure initiation for different skin orientations. Identification of the sub-failure initiation will also help to bound the stress-stretch region that contains the skin undamaged response in which the pristine constitutive model can be applied to. Ultimately, this data will be used to develop constitutive models of skin for use in simulating blunt impact events with human body models.

2. METHODS

Skin samples were excised from the back of six male PMHS at the Center for Applied Biomechanics, University of Virginia (UVA), USA. All test procedures were approved by the UVA Institutional Review Board prior to any testing and the PMHS were screened for pre-existing pathologies to avoid skin diseases that may affect skin quality. The PMHS represent an average of 57±11 years old, 178.6±3.8 cm in height and weighed 88.4±19.8 kg adult male (Table 1).

Table 1. Post-mortem human subject information

Specimen ID	Age (years)	Height (cm)	Weight(kg)
795	60	175.3	83.9
757	49	185.4	122
702	42	178	86
733	74	180.3	78.9
919	59	177.8	96.6
680	58	175	63

2.1 Sample Preparation

All the PMHS were thawed for three days in a room temperature environment before skin excision took place. The orientation of the skin sample is determined with reference to anatomy illustration of Langer line [15], and both parallel (0°) and perpendicular (90°) samples with respect to the Langer line were included for testing (Figure 1, left). Custom printed stencils and sharpie markers were used to mark the contour and in-vivo dimension of the skin samples prior to excision (Figure 1, center). After marking, a surgical scalpel was used to excise the skin samples and excessive adipose tissues were carefully scraped off (Figure 1, right). This completes the skin excision process and each skin sample was wrapped in saline dampened gauze pad to prevent moisture loss and refrigerated at 4°C until testing which was within 48 hours. For each PMHS, two different sizes of skin samples were excised on the left side of the back for uniaxial tensile (static and dynamic) and stress relaxation tests, the contralateral side of the back was reserved for other skin samples for a different study. An additional cutting step was performed during the day of testing where the skin was cut into a dogbone shape using a custom made hardened steel cutter based on ASTM D412 standard test method for the dynamic tensile (DT) test while a scaled version of the standard was used to cut the dogbone skin samples for the static tensile (ST) and stress relaxation (SR) test. The purpose is to maximize the numbers of skin sample within individual PMHS specimen for different test series and a total of 72 (48 tensile test samples and 24 stress relaxation test samples) skin samples were collected from the six PMHS. Each skin sample thickness was measured using a digital caliper prior to testing and the measured mean thickness was 3.33±0.87 mm.

https://doi.org/10.52202/078352-0031

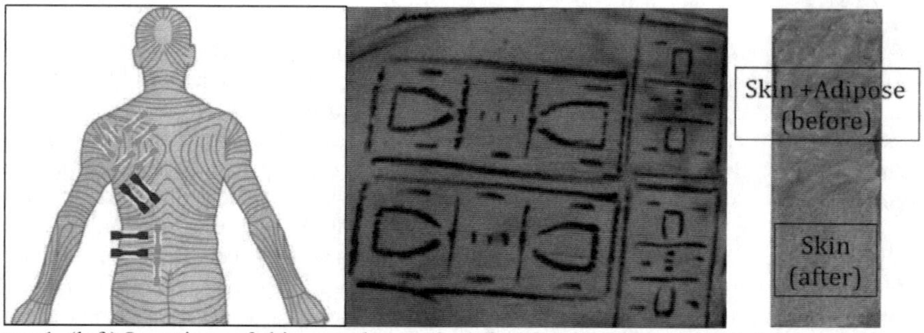

Figure 1. (left) Locations of skin samples on the left side of the human back (blue – static tensile samples (n=4) and stress relaxation samples (n=4), red – dynamic tensile test samples (n=4)). (center) Marked in-vivo state of human skin before excision. (right) Skin samples with underlying adipose (before) versus isolated skin samples without adipose (after).

2.2 Uniaxial Tensile Tests

The ST tests were performed using the Instron Model 8874 servohydraulic actuated test machine (Instron, Canton, MA) and a custom-built gravity-based drop tower was used for the DT tests (Figure 2). Dogbone sample was clamped with additional 80 grit sandpapers at both ends of the test fixture to avoid slippage during testing. After clamping, the skin sample was loaded until reaching its in-vivo length by comparing to an in-vivo reference length rod (Figure 4) before initiating the test. The test was initiated by moving the crosshead of the machine until skin failure occurred. Both parallel and perpendicular samples were tested at static (1/s) and dynamic (75/s and 180/s) engineering strain-rates, similar strain-rate levels conducted by Ottenio et al. (2015). A 1000 lbf (4.4 kN) Honeywell model 31 piezoresistive load cell (Honeywell, Charlotte, NC) was used to measure the force at 1000 and 10,000 sampling rates for ST and DT tests, respectively. Similarly, videos were recorded at 1000 fps for ST tests and 10,000 fps for DT tests using a Memrecam GX-1 high-speed camera (NAC Image Technology, Simi Valley, CA). Sharpie markers which were predetermined on the PMHS before skin excision were used for determining stretch values through video tracking software (Figure 3) (Tracker, ver. 4.11.0). A trigger box was utilized to activate data acquisition of the load cell and video recording simultaneously when the test was initiated.

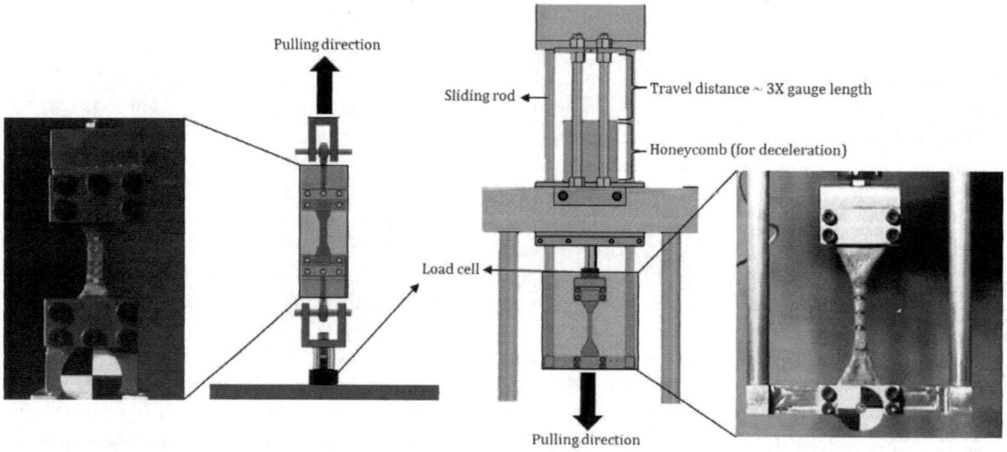

Figure 2. (left) Uniaxial static tensile test setup. A similar setup is also used for stress relaxation test. (right) Uniaxial dynamic tensile test setup. The indicated red and blue circles are video tracked to determine stretch ratio history

2.3 Stress Relaxation Tests

Stress relaxation tests were performed using the Bose ElectroForce Testing Machine (TA Instruments, New Castle, DE). The objective of these tests is to measure viscoelastic response of the tissue for implementation into constitutive models for future use in human body models. While stress relaxation tests are capable of measuring long-term viscoelastic response, the rapid step-hold process that is utilized

https://doi.org/10.52202/078352-0031

in this test type, and the methods used for processing this data, permit the measurement and identification of short-term viscoelastic responses and time-constants that are applicable to blunt impact loading. Like the uniaxial tensile test series, the dogbone samples were loaded to in-vivo length prior to testing. A displacement-controlled step-hold profile was used by subjecting the skin samples to 10%, 15%, 20%, and 30% engineering strain at the fastest loading rate (0.3m/s) achievable by the test machine without overshooting the programmed profile to obtain the instantaneous elastic response, each step was followed by a 60 seconds dwell period to collect the relaxation response. Similarly, both parallel and perpendicular samples were used for testing. The force was measured using a 50lbf (222.41N) Bose model WMC-50-456 load cell (TA Instruments, New Castle, DE) and stretch values were determined through video analysis as described in the tensile test series.

2.4 Measurements and Metrics

Engineering stress-stretch curve was constructed based on the measured force and the video tracked displacement. The engineering stress was calculated by dividing the measured force by the undeformed cross-section area. The stretch ratio was defined as the ratio between the current gauge length and original gauge length. An example of the engineering stress-stretch curve is plotted as reference (Figure 3), showing the 5 skin parameters of interest. The definition of each parameter is explained as follows:

- Ultimate tensile strength (UTS): Maximum force divided by the original cross-section gauge area.
- Strain energy: Area underneath the stress-stretch curve.
- Initial Young's modulus (E_1): Slope of the stress-stretch curve taken at the initial 5% of the UTS.
- Young's modulus (E_2): Slope of the stress-stretch curve taken in between 30% and 70% of the UTS.
- Failure stretch (λ_f): stretch ratio where the skin is completely separated.

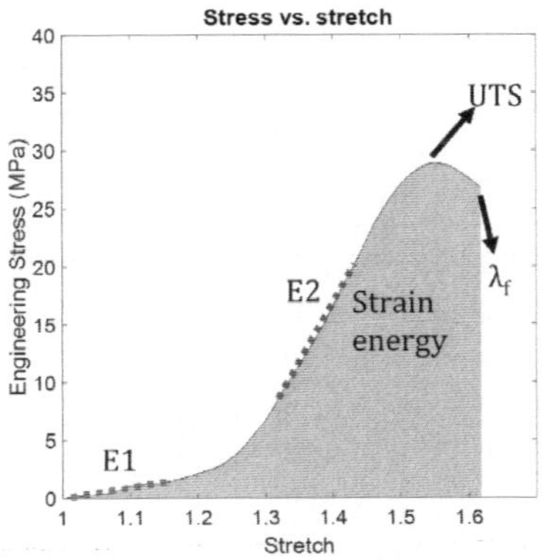

Figure 3. Typical mechanical response of human skin in tension.

Sub-failure initiates when the collagen fibers gradually break, and this was determined by investigating stiffness history (i.e. stress/stretch) of the test sample. A decrement of the stiffness after reaching the maximum indicates the sub-failure initiation and the corresponding stretch refers to the damage initiation stretch (λ^*).

2.5 Data Processing

Pre and post-test data history were recorded and a range of data of interest was determined through video analysis and force measurement data. The initial data point was determined from video analysis by tracking movement initiation of the marker within the sample gauge length. The reason to use the video analysis technique is to make sure the deformation history of the sample is captured (i.e. initial elongation with very little force) as opposed to relying on load cell information. The end data point where the failure occurred was determined when the skin sample was completely separated (i.e. no tension) is detected. The raw data was filtered using the SAE J211 standard with channel frequency class (CFC) of 600 Hz. To calculate the average response of the tensile test series where different failure stretch happened, a procedure was used where each stress-stretch curve was divided into two sections at the UTS point.

https://doi.org/10.52202/078352-0031

2.5 Statistical Analysis

A Wilcoxon Rank Sum Test (a non-parametric equivalent of two-sample t-tests) was performed using the ranksum function in MATLAB R2018b to determine the statistical significance ($p<0.05$) from the effect of strain-rate and skin orientation with respect to the Langer line on the skin response

3. RESULTS

3.1 Uniaxial Tensile Tests

The uniaxial tensile test matrix and average responses of the uniaxial tensile skin samples is summarized in Table 2. Among the 48 skin samples, 5 samples slipped out of the test fixture and were permanently deformed during testing, therefore, these samples were excluded from data analysis.

Table 2. Average uniaxial tensile test results of human skin

Orientation	Strain-rate (/s)	UTS (MPa)	Strain Energy (MPa)	E_1 (MPa)	E_2 (MPa)	λ^*	λ_f
Parallel	1 (n=11)	28.4±6.3	8.5±2.8	8.4±3.1	89.6±24.2	1.48±0.08	1.76±0.14
	75 (n=5)	20.6±7.8	7.6±3.5	10±6.5	74.2±27.3	1.36±0.10	1.75±0.18
	180 (n=6)	25.6±5.4	8.6±2.2	8.3±1.4	82.9±28.2	1.47±0.04	1.75±0.07
Perpendicular	1 (n=10)	22.6±4.6	7.1±1.2	4.2±1.0	49.5±16.5	1.74±0.12	1.97±0.15
	75 (n=5)	16.6±5.8	6±1.8	5.6±1.6	52.7±24.7	1.43±0.06	1.78±0.12
	180 (n=6)	20.7±4.9	8.5±2.4	11.0±3.3	64±21.1	1.41±0.03	1.81±0.10

For the skin samples tested at the same strain-rate, the parallel and perpendicular samples are compared and average responses for each test condition are illustrated (Figure 4). Results from the statistical analysis show the skin response to be significantly different in terms of UTS, failure stretch ($p=0.004$), initial Young's Modulus (E_1) and Young's Modulus (E_2) at the static loading rate, but no significant differences were found in both dynamic loading groups (Figure 6). The strain energy was also insignificantly different between the parallel and perpendicular samples regardless of loading rate.

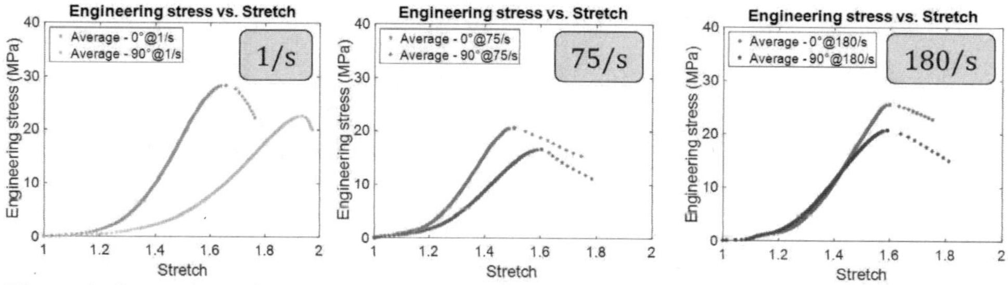

Figure 4. Comparison of skin response between different orientations under three loading rate groups

Average responses are grouped between parallel and perpendicular group to investigate the effect of strain-rate on the skin response (Figure 5). From the statistical analyses of the perpendicular samples, failure stretch ($p=0.013$) and initial Young's modulus (E_1) ($p=0.028$) were significantly different when comparing tests between 1/s and 75/s. Similar results were also observed on failure stretch and initial Young's modulus (E_1) between 1/s and 180/s loading rate group (Figure 6). However, the UTS, Young's modulus (E_2) and strain energy of the perpendicular samples were similar, and no statistical difference was found among the parallel samples between different loading rate situations.

https://doi.org/10.52202/078352-0031

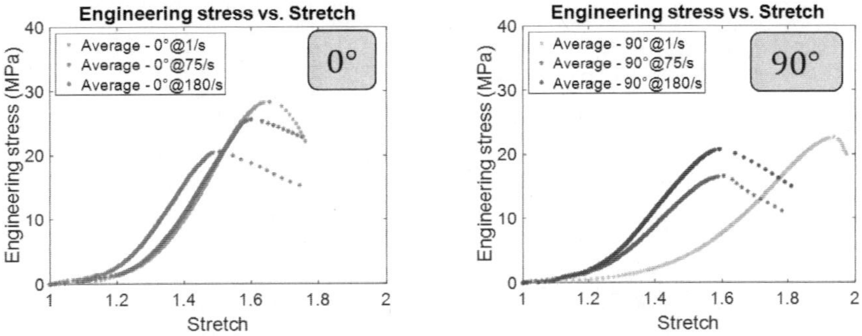

Figure 5. Comparison of skin sample response between different strain-rates at two different skin orientations (left - parallel samples, right - perpendicular samples)

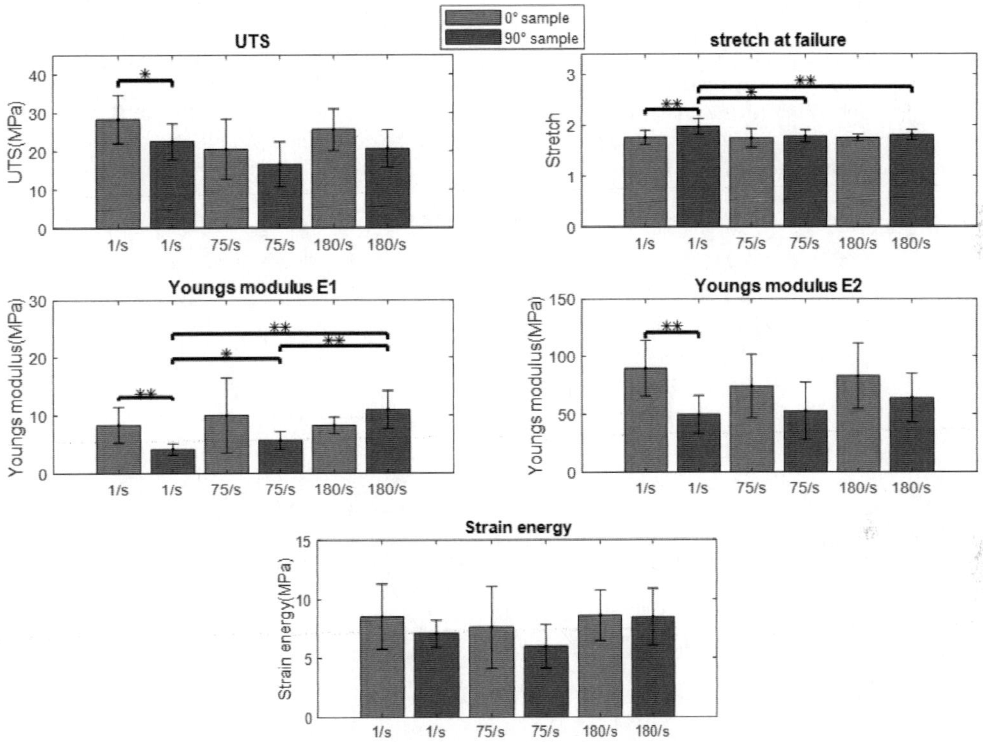

Figure 6. Wilcoxon Rank Sum Test results – 1. Effect of skin orientation with respect to Langer line on skin response under same loading rate. 2. Effect of loading rate on skin response under same skin orientation. (*p<0.05, **p<0.01)

3.2 Stress Relaxation Tests

All skin samples (n=24) were tested successfully without having slippage in the stress relaxation tests. Viscoelasticity was observed in both parallel and perpendicular samples with increasing peak stress at each ramp process followed by decaying of stress during the dwell period (Figure 7, left). Reduced relaxation function was also determined by normalizing stress decay response at each step with its corresponding peak stress. For each step-hold, the average reduced relaxation response (n=24) was calculated and an overall average reduced relaxation function (n=120) taken from all step-hold was also determined along with responses of one standard deviation (Figure 7, right).

 https://doi.org/10.52202/078352-0031

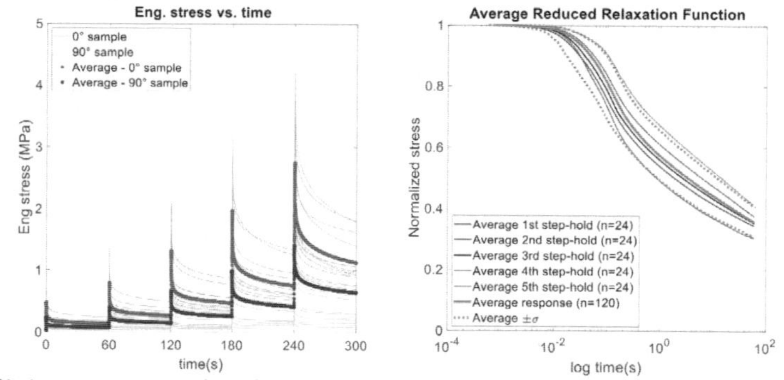

Figure 7. (left) Average stress relaxation response of parallel (thick red) and perpendicular (thick blue) skin. (right) Average reduced relaxation function of individual step-hold profile.

3.3 Sub-failure Initiation

Sub failure initiation was identified by finding maximum stiffness of the skin sample (Figure 8, left) where the stiffness was calculated at each incremental stretch value. The damage initiation stretch (λ^*) of both parallel and perpendicular samples against the strain-rate is illustrated (Figure 8, right). On average, comparing the parallel and perpendicular samples, the damage initiation stretches are 1.48 vs. 1.74 in static loading group, 1.36 vs. 1.43 and 1.47 vs. 1.41 in dynamic loading groups (75/s and 180/s).

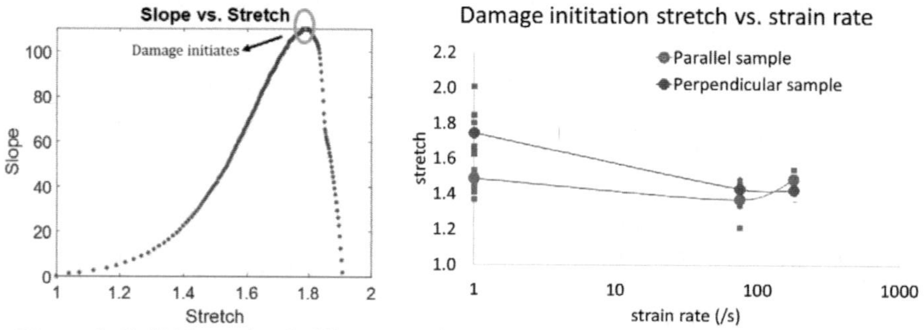

Figure 8. (left) Example of stiffness vs. stretch history of the skin sample. (right) Damage initiation stretch vs. strain-rate. Square dots are from individual samples and circular dots represent the average.

4. DISCUSSION AND CONCLUSION

4.1 Discussion

A comprehensive test matrix has been developed to fulfill the current human skin database by considering anisotropy, viscoelasticity, damage, and inter-subject variability (six human subjects) for constitutive modeling. The current study investigated skin data from a pool of a wider human population group aged between 42 and 74, producing average skin responses under different strain-rates and skin orientations. In this study, stress relaxation tests were performed to characterize skin viscoelasticity while the destructive tensile test series provide mechanical properties and damage response of the skin. To maximize the number of skin samples for the test matrix, the mechanical properties of the human back region are treated to be homogeneous, and effect of location was not considered in this study. Although skin properties have been shown to differ with body region [10], this was mainly believed due to the variation of Langer line distribution across these body regions. By characterizing two categories of skin orientation with respect to perceived Langer line, this has considered the spectrum of the skin properties at different body regions.

In the stress relaxation test, the average of the parallel skin samples exhibits higher peak stress at each step hold process when compared to the perpendicular skin samples. The peak stress on both sample types increases in a non-linear fashion as the stretch level increases. For the relaxation response, both parallel and perpendicular samples show a similar average rate of relaxation. This indicates the fiber component in the parallel sample only contributes to the instantaneous elastic response and the relaxation response is mainly dominated by the ground substance of the skin. Overall, the stress relaxation test shows the skin elastic response was non-linear and stretch independent. The results satisfy the

https://doi.org/10.52202/078352-0031

requirement of Quasi-Linear Viscoelastic (QLV) constitutive model [16] which is capable to model the temporal response of biological tissues, and this can be coupled with established hyperelastic (isotropic or anisotropic) constitutive models. The selection of the QLV model is consistent with previous efforts in the viscoelastic testing of mice skin [17].

Uniaxial tensile test data from this study were similar and within the range when compared to data from previous studies conducted on human back skin under similar loading rate conditions. However, there were differences observed on Young's modulus of the parallel and the perpendicular samples when comparing between our study and Ottenio et al. [12] at the dynamic strain-rate group (167-180/s), where we observed a lower Young's modulus. Despite having similar loading rate condition, the finding of the difference can be due to several reasons. First, the Ottenio study drew conclusions utilizing a single PMHS that was 90 years old. The elderly skins are generally believed to be stiffer and less extensible due to straightening of collagen fibers as part of the aging process [18]. Second, different testing protocol from the previous study where 2N preload was applied may have brought the skin samples to different stress state prior to testing. Unfortunately, this was the only comparable study in terms of applied strain-rate looking into the effect of loading rate and Langer line orientation on human skin. In terms of failure stretch, the definition of failure was not clearly defined across literature, hence, it was difficult to make a fair comparison of the failure stretch data. Only Ottenio et al. defined the failure as superficial tear visible on the epidermis layer. This was different from the current study as the load was still carried by the skin during superficial damage and therefore, the failure was defined as complete separation of the skin. This resulted in higher failure stretch in the current study when compared to literature.

Finally, impact velocities to the body from non-lethal blunt impacts or behind armor blunt trauma (BABT) are often an order of magnitude higher than the loading rates utilized in this study. The choice of loading conditions in this study was made based on the overarching objective of characterizing the material properties of skin for developing constitutive models that can be implemented into human body models for blunt impact simulation. Testing at higher loading rates will typically result in data with many confounding factors related to inertial effects that would limit the ability to identify the appropriate mechanical properties of the tissue. However, the loading rates tested in this study may not be a limiting factor for developing constitutive models of skin for high-rate blunt impact simulation: A) In non-lethal blunt impact and BABT, the impact occurs perpendicular to the plane of the skin and the deformation of the skin is driven by the impactor deforming into the flesh. The tensile loading rates that result from this mechanism will be lower than that of the compressive rates of the subcutaneous flesh; B) Maximum strain-rates during high-rate blunt impact only occur at an instance in time, likely during the initial impact phase where the strains are relatively low. Most of the loading and unloading of the tissue will occur at lower strain-rates, and it is likely that incorporating higher rate mechanical properties than what is currently considered will not substantially alter the overall response of the skin; and C) Our data suggests that differences in the mechanical response and failure thresholds of human skin at our highest strain-rates are relatively small and not statistically significant. This suggest that higher loading rate data will be consistent with the data collected in this study, and extrapolation of the constitutive model to higher loading rates would not result in non-biofidelic material response.

4.2 Conclusion

The current study investigated the effect of strain-rate and Langer line orientation on human skin mechanical response which produced viscoelastic and uniaxial tensile failure average curves of human skin. The following statements highlight the takeaways from this study:

- Human skin exhibits strong anisotropy behavior (UTS, E_2 and failure stretch) at static strain rate and evolves into isotropic material as strain-rate increases.
- Skin response loaded across (perpendicular) the collagen fibers are rate dependent while the response loaded along (parallel) the collagen fibers are rate independent.
- The collagen fibers within the skin are observed to be elastic with minimum viscosity due to similar relaxation response when two different skin orientations were loaded in the stress relaxation tests.
- A power trendline describing the relationship between damage initiation stretch and strain-rate is developed for the perpendicular sample. The damage initiation stretch can help to bound undamaged response of the skin where the pristine constitutive model can be applied.
- The data generated in this study is likely applicable to non-lethal blunt impacts and behind armor blunt trauma (BABT), with the caveat that the strain-rates tested in this study may not encompass the entire loading history of these types of impact events.

https://doi.org/10.52202/078352-0031

With the experimental curves presented in this study, the test curves can be converted into true stress-stretch response before developing a constitutive model to capture dynamic tensile test response of human skin. Ultimately, this data will be used to develop constitutive models of skin for use in simulating blunt impact events with human body models, and will be beneficial for developing a biofidelic, physical skin simulants for the future testing and evaluation of blunt impactor performance.

Acknowledgments

This study was supported by the U.S. Army Aberdeen Test Center under Contract No W91CRB-17-C-0032.

References

[1] Jones, R. E., Foster, D. S., and Longaker, M. T. (2018). Management of Chronic Wounds—2018. JAMA, 320(14), 1481–1482.

[2] Haut, R. C. (1989). The Effects of Orientation and Location on the Strength of Dorsal Rat Skin in High and Low Speed Tensile Failure Experiments. Journal of Biomechanical Engineering, 111(2), 136–140.

[3] Arumugam, V., Naresh, M. D., and Sanjeevi, R. (1994). Effect of strain rate on the fracture behaviour of skin. Journal of Biosciences, 19(3), 307–313

[4] Pramudita, Jonas A., Shimizu, Y., Tanabe, Y., Ito, M. and Watanabe, R. (2014). Tensile Properties of Porcine Skin in Dorsal and Ventral Regions. Journal of JSEM, Vol.14, Special Issue (2014) s245-s250.

[5] Papy, A., Robbe, C., Nsiampa, N., Oukara, A., and Goffin, J. (2012). Definition of a standardized skin penetration surrogate for blunt impacts. IRCOBI Conference 2012.

[6] Bir, C. A., Resslar, M., & Stewart, S. (2012). Skin penetration surrogate for the evaluation of less lethal kinetic energy munitions. Forensic Science International, 220(1–3), 126–129.

[7] Pramudita, Jonas A., Sasaki, M., Ito, M., Watanabe, R. and Tanabe, Y. (2016). Mechanical Characterization of Soft Tissue Simulant Materials. Advanced Experimental Mechanics, Vol.2, (2017), 135-140.

[8] Joodaki, H., and Panzer, M. B. (2018). Skin mechanical properties and modeling: A review. Proceedings of the Institution of Mechanical Engineers, Part H: Journal of Engineering in Medicine, 232(4), 323–343.

[9] Jacquemoud, C., Bruyere-Garnier, K., and Coret, M. (2007). Methodology to determine failure characteristics of planar soft tissues using a dynamic tensile test. Journal of Biomechanics, 40(2), 468–475

[10] Ní Annaidh, A., Bruyère, K., Destrade, M., Gilchrist, M. D., and Otténio, M. (2012). Characterization of the anisotropic mechanical properties of excised human skin. Journal of the Mechanical Behavior of Biomedical Materials, 5(1), 139–148.

[11] Gallagher, A.J., Ní Annaidh, A., Bruyere, K., Otténio, M., Xie, H., and Gilchrist, M.D. (2012). Dynamic Tensile Properties of Human Skin. IRCOBI Conference 2012.

[12] Ottenio, M., Tran, D., Ní Annaidh, A., Gilchrist, M.D., and Bruyère, K. (2015). Strain rate and anisotropy effects on the tensile failure characteristics of human skin. Journal of the Mechanical Behavior of Biomedical Materials, 4(1), 241-250.

[13] Khatyr, F., Imberdis, C., Vescovo, P., Varchon, D., and Lagarde, J.-M. (2004). Model of the viscoelastic behaviour of skin in vivo and study of anisotropy. Skin Research and Technology, 10(2), 96–103.

[14] Piérard, G. E., Piérard, S., Delvenne, P., and Piérard-Franchimont, C. (2013). In Vivo Evaluation of the Skin Tensile Strength by the Suction Method: Pilot Study Coping with Hysteresis and Creep Extension. ISRN Dermatology, 2013, 1–7.

[15] Langer, K., 1978. On the anatomy and physiology of the skin—I. The cleavability of cutis. Br. J. Plast. Surg. 31, 3–8.

[16] Fung, Y.-C. (1993). Biomechanics: Mechanical Properties of Living Tissues. New York: Springer-Verlag. p. 568. ISBN 0-387-97947-6.

[17] Wang, Y., Marshall, K. L., Baba, Y., Lumpkin, E. A., and Gerling, G. J. (2015). Compressive Viscoelasticity of Freshly Excised Mouse Skin Is Dependent on Specimen Thickness, Strain Level and Rate. PLOS ONE, 10(3), e0120897.

[18] Tonge, Theresa K., Atlan, Lorre S., Voo, Liming M., and Nguyen, Thao D. (2019). Full-field bulge test for planar anisotropic tissues: Part I – Experimental methods applied to human skin tissue. Acta Biomaterialia, 9(2013) 5913-5925.

https://doi.org/10.52202/078352-0031

Influence of adhesive geometry and material property on the ballistic protection performance of ceramic composite armour panels

P. Tan

Land Division, Defence Science and Technology Group, 506 Lorimer Street, Fishermans Bend, Melbourne, Victoria 3207, Australia
Ping.tan@dst.defence.gov.au

Abstract. Ceramic composite body armour is generally a system composed of a ceramic strike face plate and fibre reinforced composite backing plate. It absorbs an impact by blunting and shattering hard projectiles, capturing the fragments, distributing and absorbing the kinetic energy across the armour panel. A technique to combine the ceramic and fibre reinforced composite plates is the use of adhesives. However, investigation of the adhesive geometry and material property on the damage/energy-absorbing mechanisms and ballistic protection performances of ceramic composite body armours is scanty. In this study, various finite element models were developed using the commercial finite element software ANSYS/AUTODYN to investigate the effects of adhesive interlayer thickness, number, shape, material type and the size of cohesion between adjacent ceramic tablets on the ballistic protection performances and damages of the selected ceramic composite armour panels, including ballistic limit velocity, maximum backface deformation, damage area and pattern. Two types of ceramic strike face plates were considered in this investigation; one was made of monolithic ceramic plate(s) and the other composed of different size ceramic tablets. The ceramic materials used in this preliminary study were chosen to be silicon carbide and alumina. The composite backing material was selected to be Kevlar fibre reinforced composite. Seven different types of resin materials were used as adhesives to bond the ceramic composite armour panels respectively. The ceramic strike face and fibre reinforced composite backing plates considered were assumed to be perfectly bonded by the adhesive at the interface without defects. The corresponding modelling methodology and techniques were validated by comparing the present predicted results with those previously reported. It was found from this numerical study that for the panels considered, the ballistic limit velocity and damage are generally affected by the key parameters of the adhesive, whereas the influence of the key parameters on the predicted value of maximum backface deformation is not significant. This information may offer advantages to meet the requirement for design of future body armours.

1. INTRODUCTION

Ceramic composite body armours are generally made of ceramic strike face plate and fibre reinforced composite backing plate. One of the most common methods for combining these materials is using the adhesive, which are crucial for their further development and improvement. Hence, it is necessary and required to understand the influences of adhesive geometry and material property on the behaviour of ceramic composite body armours under various impacts. In order to investigate the effects of adhesive layer type and thickness on the ballistic behaviour of ceramic/metal armours subjected to low calibre projectile impact, Zaera et al [1] conducted numerical and experimental studies. Both numerical and experimental results showed that the ceramic damage is greater in the armours bonded using polyurethane adhesive than that using epoxy resin. The numerical study also shows that the thicker the adhesive layer the greater the damage to the ceramic, whereas it was observed from a set of full-scale fire tests that fragmentation increases in inverse proportion to the thickness of the adhesive. Lo´pez-Puente [2] conducted experimental and numerical studies to investigate the influence of adhesive layer thickness on the ballistic limit of ceramic/metal armours, in which the toughened epoxy resin was used for the adhesive layer, of different thickness. It was noted that for alumina/aluminium configurations considered, a variation of the adhesive layer thickness affects the efficiency of the armour and a thickness value of 0.3mm was found optimum. Übeyli et al [3] experimentally investigated the effects of mechanical properties of backing material and laminating type as well as the adhesive type on the ballistic performance of Al_2O_3/Al2024 (alumina/aluminium) laminated composites armour against 7.62 × 51 mm armour piercing projectiles. It was noted that composites bonded with polyurethane exhibited more resistance to spalling of ceramic tiles than those bonded with epoxy. However adhesive type had no appreciable effect on the ballistic performance of the composites. Tasdemirci et al [4] experimentally and numerically investigated the effects of rubber, Teflon and aluminium foam interlayer materials on the ballistic performances of ceramic/composite armour targets. It was reported that the presence of interlayer altered the stress wave transmission

 https://doi.org/10.52202/078352-0032

between the ceramic and composite layers. Damage in the ceramic layer was highly localized around the projectile impact zone for without interlayer and rubber interlayer configuration, while aluminium foam and Teflon interlayer spread the damage zone in the radial direction. Grujicic et al [5] performed finite element analyses to investigate the role of adhesive layer in the ballistic/structural performance of ceramic/polymer–matrix composite hybrid armour. It was noted that significant improvements in the ballistic protection performance and durability of hybrid armour can be attained by proper modifications in the adhesive layer mechanical properties. Prakash et al [6] conducted a numerical study to investigate the influence of adhesive thickness on the dynamic responses of Al_2O_3/Al5083 H116 composite targets subjected to ogive nosed projectile impact. They pointed out that the impact responses of the ceramic/metal composite panels were influenced to different degree by the adhesive thickness. Seifert et al [7] carried out experimental study to understand the effect of adhesive stiffness on the failure of ceramic tiles adhered to metallic backings, in which four different types of adhesive were tested. It was found from testing results that the damage behaviour of the ceramic/metal composites can be controlled either by the adhesive thickness or stiffness. Jiusti et al [8] experimentally investigated the influence of filling materials on the ballistic performance of Al_2O_3 mosaic armours. They found that the epoxy-filled mosaics exhibited a significantly superior performance than the filling-free mosaics. Gao et al [9] conducted experimental and numerical investigations on the influence of adhesive layer on the high velocity impact performance of the ceramic/metal composite armour. It was reported that the size of fractured ceramic was decreased with the thickness of adhesive layer. Wang et al. [10] performed a drop weight test to investigate the effects of partitioned tile layer, impact location, stagger mode, tile shape and size, adhesive type, as well as fiberglass mesh on the low velocity impact resistance of a layered and staggered bio-inspired building ceramic composite. It was noted that the elastic adhesive interlayers had higher efficiency than the rigid ones in improving the fracture toughness of the composite.

It is indicated from literature that the ballistic protection performance of armour plates are affected by the geometrical parameters and material property of adhesive interlayers used for bonding ceramic composite panels. Hence, testing and modelling of adhesive and cohesion in ceramic composite armour panels are crucial for identifying their behaviours under ballistic impact and for their further development and improvement. However, the research on the influence of adhesive and cohesion on the ballistic performance of armour panels composed of ceramic plate(s)/tablets as strike face and Kevlar fibre reinforce composite as backing plate is still scanty. This investigation aims at numerical study on the effects of adhesive/cohesion parameters, including adhesive interlayer thickness, number, material properties and shape, ceramic tablet pattern (or ceramic tablet size and location), size of cohesion between the adjacent tablets, on the ballistic protection performances and damages of the selected armour panels, including ballistic limit velocity (V_{bl}), maximum backface deformation (MAXBFD), damage area and pattern. Also, the effect of ceramic type on the sensitivity of ballistic protection performance of armour panels to the adhesive interlayer thickness is discussed. Comparisons of the ballistic performance and dynamic response for the ceramic composite armour panels with and without filling materials between the adjacent tablets are conducted. In this study, strike face plates are made of monolithic ceramic plate(s) or composed of different size of ceramic tablets, and the backing plates are made of Kevlar fibre reinforced composite (KFRC). Their corresponding finite element (FE) models were generated using the commercial FE software ANSYS/Autodyn [11].

2. NUMERICAL MODELLING

Figure 1 shows the FE models for three typical types of ceramic composite panels, which are named as SLMS-SiC for the panel having a strike face plate made of the single-layer monolithic silicon carbide (SiC) plate and a KFRC backing plate; TLMS-SiC for the panel having a strike face made of triple-layer monolithic SiC plates and a KFRC backing plate and TLST-SiC for the panel having a strike face composed of triple-layer SiC tablets/resin and a KFRC backing plate, respectively. The adhesive and cohesion materials used to bond the SiC plate(s)/tablets and KFRC plate are epoxy resin. All panels shown in Fig. 1 have the same panel thicknesses of 9.5 mm, same total thicknesses of 3mm SiC plate(s), 1 mm resin layer(s) and 5.5 mm KFRC plate. For the panel composed of ceramic tablets, the size of cohesion between two adjacent ceramic tablets (i.e., Tc in Fig. 1 (c)) was selected to be 0.5 mm.

The present FE models were developed using the commercial finite element software ANSYS/Autodyn [11], in which the zero x- and y-velocity boundary conditions were applied to the top edges of the panels as shown in Fig. 1(a). The panels are subjected to an impact from a 30 caliber fragment-simulating projectile (FSP). Only half of the panel and the 30 caliber FSP above the central line are shown in Fig. 1 due to the symmetry of the FE models. Gauge 1 in Fig. 1(c) located at the

https://doi.org/10.52202/078352-0032

centre of the FSP for measuring the velocity of FSP was used to predict the value of the required ballistic limit velocity. In this study, the predicted value of V_{bl} was obtained by averaging the initial velocity of the projectile that led to a partial penetration (V_0^P) and the initial velocity that led to a complete penetration (V_0^c). The difference between V_0^P and V_0^c was chosen to be 20 m/s [12]. Gauge 2 in Fig. 1(c) measured the displacement of the selected point is located on the rear surface of the KFRC backing plate with the same vertical coordinate as Gauge 1. This measurement was used to obtain the predicted value of MAXBFD.

The KFRC material used for composite backing plate was considered to be homogeneous and orthotropic. It was modelled using the orthotropic equation of state (EOS), elastic strength model and material stress/strain failure model. The ceramic materials used for strike face plate, including SiC and alumina 99.7% (Al_2O_3-99.7) were considered to be homogeneous and isotropic. They were modelled by the polynomial EOS, Johnson–Holmquist strength model and failure model. The steel 4340 material used for the FSP was modelled using linear EOS, the Johnson–Cook strength and failure models. The adhesive and cohesive materials used to bond ceramic composite panels were considered as fluid resisting high pressure due to its low strength compared to those of other materials [2]. Hence, the adhesive and cohesive materials used in this study was modelled using the Mie Gruneisen EOS, in which the relationship between the shock velocity (U) and particle velocity (u_p) is expressed as

$$U = c_0 + su_p \tag{1}$$

where c_0 and s are parameters which are generally determined by experiments.

All material models mentioned above are available in the Autodyn material library [11]. The mesh sizes for panels and FSP were selected to be less than 0.8 mm based on the sensitivity analysis results.

For validating the modelling capability, a comparison of the ballistic limit velocity between the FE and testing results was conducted for the selected Kevlar fibre reinforced composite panels under the 30 caliber FSP impact. The difference between them is 2.2%. The present modelling methods and techniques were also validated in author's previous paper for the selected armour components including armour hard panels and helmets subjected to FSP impacts [12-13].

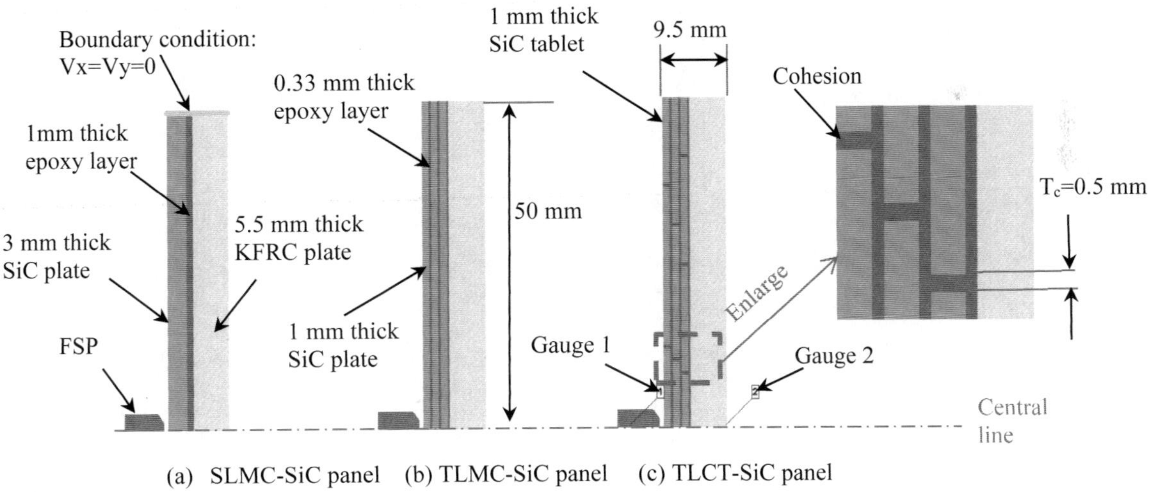

Figure 1. Schematics of the finite element models for the selected ceramic composite panels under a 30 caliber FSP impact

3. NUMERICAL RESULTS AND DISCUSSIONS

In order to understand the correlations between the ballistic protection performances (e.g., V_{bl} and MAXBFD) and the key parameters of the adhesive and cohesion, including adhesive interlayer thickness, number, material property and shape, ceramic tablet pattern, size of cohesion between the adjacent tablets, an extensive numerical study was conducted using the present FE models. The influence of the ceramic type on the sensitivity of ballistic protection performance of armour panels to

https://doi.org/10.52202/078352-0032

the adhesive interlayer thickness was also investigated. The key parameters considered were varied in individual simulations: adhesive layer thicknesses of 0.5, 1 and 2 mm; adhesive layer number of 1 and 3; adhesive types of EPOXY RES1, EPOXY RES2, POLYETHYL, POLYRUBBER, POLYSTYREN and POLYURETH available in Autodyn material library [11] and EPOXY RES3 in [2]. It is worth to mentioning that for the adhesive materials considered in this study, only the EPOXY RES3 is modelled using the Von Mises strength model and Hydro (Pmin) failure model in addition to the Mie Gruneisen EOS; five different adhesive interlayer shapes shown in Fig.2 below, in which the triple-layer SiC panels are bonded with the same type of adhesive but with different adhesive interlayer shapes, and the thickness of ceramic layer bonded with wavy adhesive layer is not constant at 1mm; six different tablet patterns shown in Fig.3 below, in which the triple-layer SiC tablets having different size and location are bonded using the same type of adhesive; size of cohesion between two adjacent ceramic tablets (i.e., Tc in Figure 1 (c)) of 0.5, 1 and 1.5 mm, and ceramic type of SiC and Alumina (Al_2O_3-99.7) available in Autodyn material library [11]. A comparison of the ballistic protection performances and damages between the panels composed of SiC tablets filled with and without epoxy resin is also discussed.

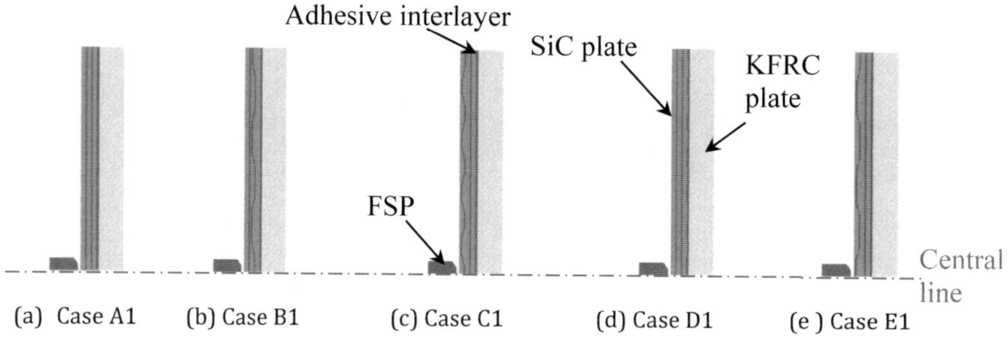

Figure 2. Schematics for the selected panels made of triple-layer monolithic SiC and KFRC plates having different adhesive interlayer shapes

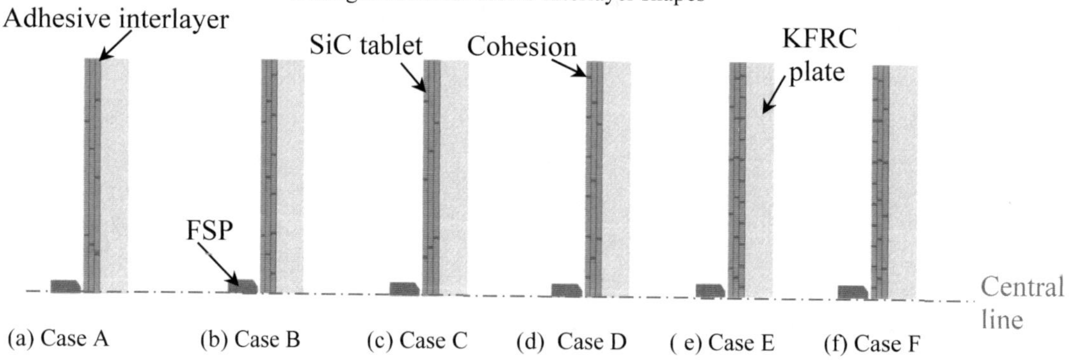

Figure 3. Schematics for the panels composed of KFRC backing plate and triple-layer SiC tablets having different sizes and locations

Figure 4(a) plots the variations of the predicted V_{bl} with the total adhesive layer thickness (T_{ad}) for the armour panels having a strike face made of single-layer monolithic SiC plate (Fig.1(a)), triple-layer monolithic SiC plates (Fig.1(b)), and triple-layer SiC tablets/resin (Fig.1(c)), respectively. It is noted from Fig. 4(a) that for the selected panels, the predicted V_{bl} reduces with an increase in T_{ad}. The predicted V_{bl} for the single-layer monolithic SiC plate is more sensitive to the change in T_{ad} than others. Figure 4(a) also indicates that for the selected panels made of monolithic SiC plate(s) and having the same areal density, an increase in the adhesive layer number from one to three results in a reduction of predicted V_{bl} ranging from 33 to 39%. To investigate the effect of ceramic type on the sensitivity of predicted V_{bl} to T_{ad}, the ceramic material of Al_2O_3 is used to replace the SiC in the panels made of triple-layer monolithic ceramic plates and those composed of triple-layer ceramic tablets/resin, respectively. Their corresponding variations of the predicted V_{bl} with T_{ad} were plotted in

https://doi.org/10.52202/078352-0032

Fig. 4(b). It is noted that for the Al_2O_3 panels considered, an increase in T_{ad} results in a slight reduction or remain unchanged in the predicted V_{bl}. A comparison of the variations between Fig. 4(a) and (b) indicates that for the cases considered, the predicted V_{bl} for the panels made of SiC is more sensitive to the change in T_{ad} compared to those made of Al_2O_3. In addition, it was found in Fig. 4 that for the panels having strike face made of SiC or Al_2O_3, replacement of strike face made of triple-layer monolithic ceramic plates with that composed of triple-layer ceramic tablets/resin leads to a slight decrease or remain unchanged in the predicted V_{bl}. This implies that for the strike face plate considered, replacement of the multi-layer monolithic ceramic plates with multi-layer ceramic tablets could improve flexibility of armour panels without significant reduction in V_{bl}. Under an impact of a 30 caliber FSP having an initial velocity of 200 m/s, the predicted MAXBFD between the panels is slightly affected by the T_{ad} and adhesive interlayer number. For example, the different between the selected panels having T_{ad}=0.5 mm and T_{ad}=2 mm ranges from 0.12 mm to 0.35mm, and those between the selected panels having adhesive interlayer number of 1 and 3 ranges from 0.42 mm to 1.12mm.

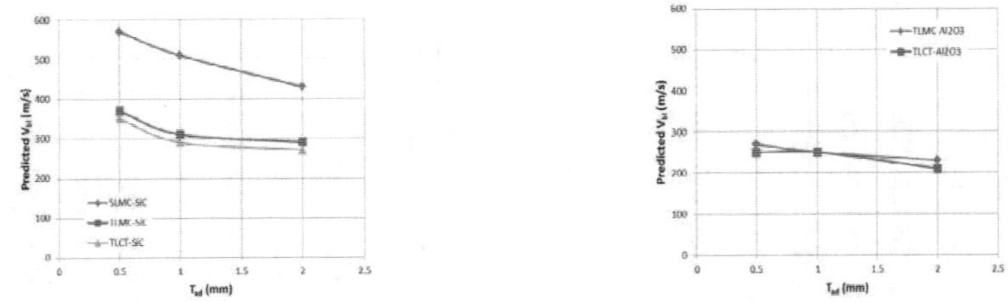

(a) For the case of SiC (b) For the case of Al_2O_3

Figure 4. Variations of predicted V_{bl} vs T_{ad} for the selected panels having strike face plates made of SiC or Al_2O_3

Figure 5 illustrates the damage patterns and areas at the time of 0.2 ms for the selected SLMC-SiC, TLMC-SiC and TLCT-SiC having T_{ad} of 0.5 mm or 2 mm respectively. The initial impact velocity of the FSP was chosen to be 200 m/s. It is noted that for the panels considered, damages in the ceramic strike face plates are localized around the projectile impact zone. This is consistent with the results obtained by Tasdemirci et al [4] (i.e., damage in the ceramic layer was highly localized around the projectile impact zone for without interlayer and rubber interlayer configuration). Figure 5 also indicates that the damages in the ceramic strike plates are greater in the panels having T_{ad} of 2 mm compared to those having T_{ad} of 0.5 mm. This finding is similar to that reported by Seifert et al [7] (i.e., the damage behaviour of the ceramic/metal composites can be controlled by the adhesive thickness.) and the numerical result from Zaera et al [1] (i.e., the thicker the adhesive layer the greater the damage to the ceramic).

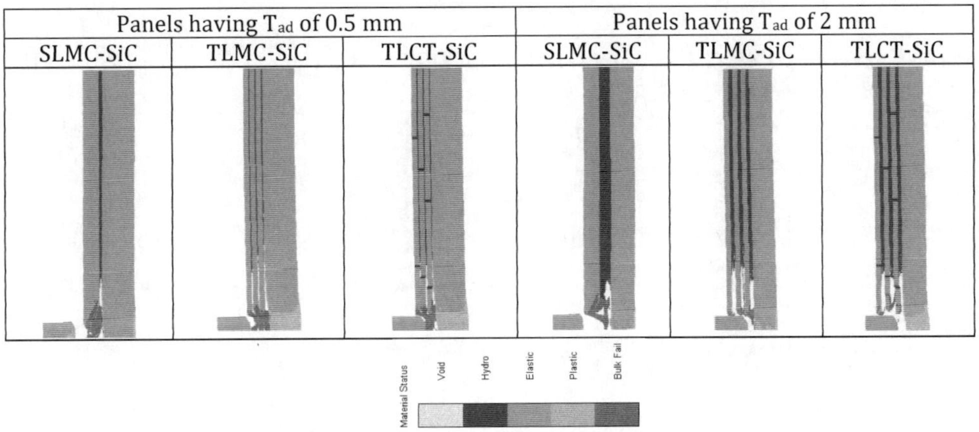

Figure 5. Damage patterns and areas for the selected panels having different adhesive interlayer thickness/number under an impact of 30 caliber having the initial velocity of 200 m/s (at t = 0.2 ms)

https://doi.org/10.52202/078352-0032

Figure 6 illustrates a comparison of the predicted V_{bl} and MAXBFD for the panels, which were made of single-layer monolithic SiC plate and KFRC backing plate as shown in Fig. 1(a), but bonded with seven different types of adhesive materials respectively, including EPOXY RES1, EPOXY RES2, EPOXY RES3, POLYETHYL, POLYRUBBER, POLYSTYREN and POLYURETH. It is noted from Fig.6(a) that for the epoxy resins considered, changes in parameters c_0 and s required using the Mie Gruneisen EOS or changes in the values of shear modulus, Yield Stress and Hydro tensile limit do not affect the predicted V_{bl}. The predicted values of V_{bl} for panels bonded with POLYETHYL, POLYRUBBER, POLYSTYREN and POLYURETH are the same. However, the predicted V_{bl} for the panels bonded with epoxy resin is 8.51% higher than those bonded with POLYETHYL, POLYRUBBER, POLYSTYREN and POLYURETH, respectively. The predicted values of MAXBFD for the selected panels, which are subjected to an impact from a FSP having initial velocity of 300 m/s, are shown in Figure 6(b). It is interesting to note that the predicted MAXBFD for the panel bonded with EPOXY RES3 is higher than others. However, for the panels considered, the difference of the predicted MAXBFD between them ranges from 0.003 mm to 0.37 mm, which are not significant and can be ignored.

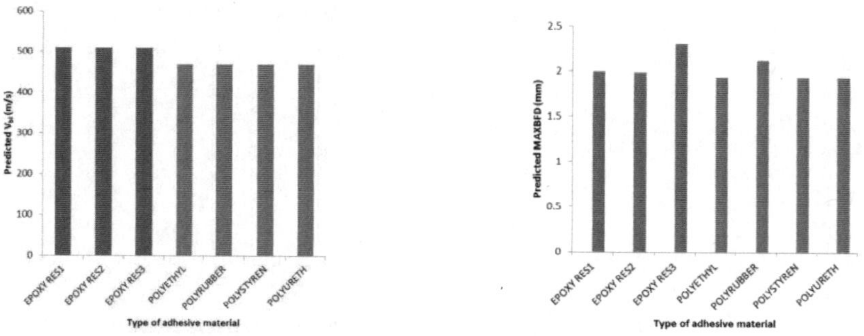

(a) Predicted V_{bl} (b) Predicted MAXBFD

Figure 6. Predicted values of V_{bl} and MAXBFD for the selected SLMC-SiC bonded with different types of adhesive materials

Figure 7 demonstrates the damage patterns and areas for the selected SLMC-SiC bonded with different types of adhesive materials and subjected to an impact of a FSP having initial velocity of 300 m/s. It is found that the debonding damage between the ceramic strike face plate and KFRC backing plate, which could result in the degradation in overall strength of the panel, is significant for the panels except for that bonded with EPOXY RES3. This implies that changes in the values of shear modulus, Yield Stress and Hydro tensile limit may significantly affect the debonding damage between the ceramic strike face plate and Kevlar composite backing plate.

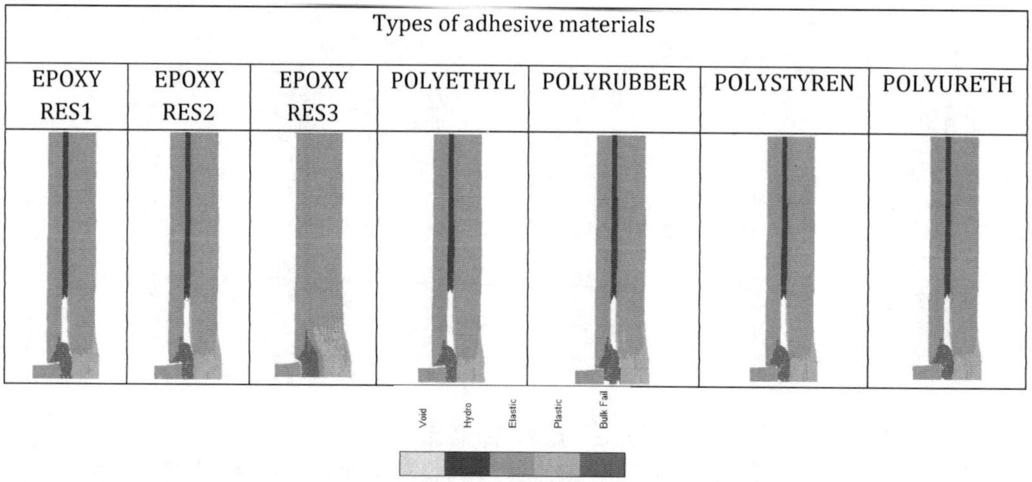

Figure 7. Damage patterns for the panels bonded with different types of adhesive materials under an impact of a 30 caliber FSP having initial velocity of 300 m/s (at t=0.2 ms)

https://doi.org/10.52202/078352-0032

For the panels shown in Fig.2, it is noted from the numerical study that the predicted V_{bl} is not affected by the adhesive layer shapes, whereas the corresponding predicted values of MAXBFD, which are shown in Fig. 8, are slightly sensitive to the change in the adhesive interlayer shapes. The difference of the predicted MAXBFD between the panels ranges from 0.07 mm to 0.86 mm. A comparison of the damage patterns and areas for the panels is illustrated in Fig. 9. It is interesting to note that for the cases considered, the debonding damages generally occur in flat adhesive interlayers. There is almost not debonding damage in the waving adhesive interlayers. This implies that replacement of flat adhesive layer with waving layer in body armour may lead to improve the overall strength of the armour panels subjected to FSP impact.

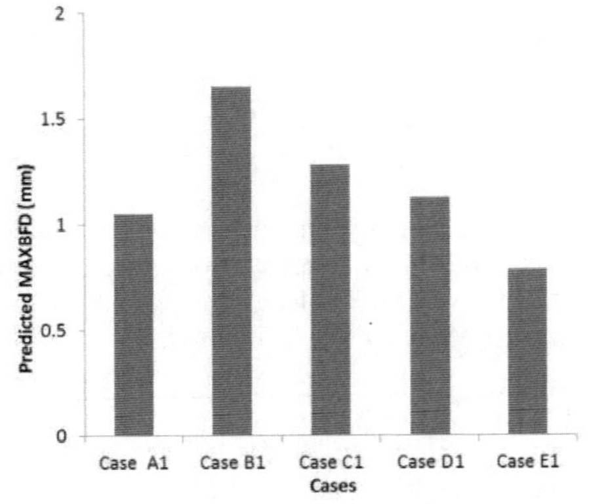

Figure 8. Predicted values of MAXBFD for the panels having different adhesive interlayer shapes under an impact of a 30 caliber FSP having initial velocity of 200 m/s. (at t = 0.2 ms)

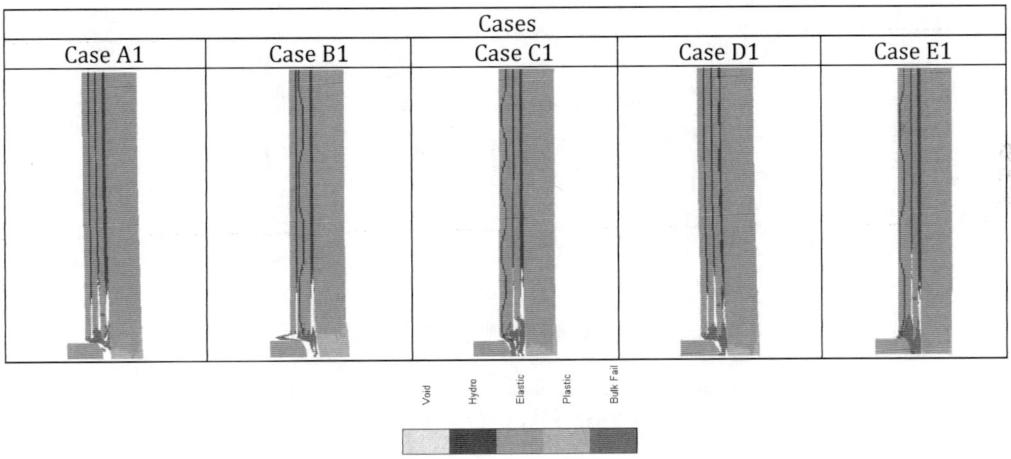

Figure 9. Damage patterns and areas for the selected TLMC-SiC having different adhesive interlayer shapes and under an impact of a 30 caliber FSP having initial velocity of 200 m/s (at t = 0.2 ms).

The predicted values of V_{bl} and MAXBFD for the selected panels, which have different SiC tablet patterns (i.e., different sizes and locations of SiC tablets) as shown in Fig. 3, are illustrated in Fig. 10. It is noted from Fig. 10(a) that for the panels considered, the percentage difference of the predicted V_{bl} between the panels ranges from 0 to 12.12%. This implies that optimal selection of the SiC tablet size and location may improve the ballistic performance of body armour composed of ceramic tablets. Figure 10(b) shows that the predicted values of MAXBFD for the panels considered, which are subjected to an impact of the FSP having initial velocity of 200 m/s, are slightly affected by the tablet pattern. The difference of the predicted MAXBFD between the selected panels ranges from 0.02 mm to 0.54 mm, which are not significant and can be ignored.

https://doi.org/10.52202/078352-0032

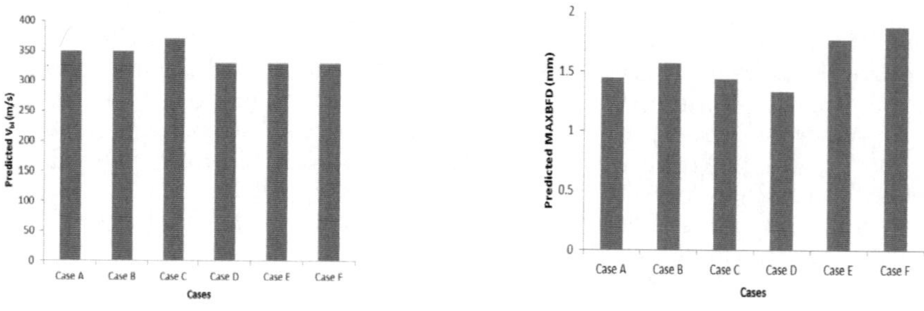

(a) Predicted V_{bl} (b) Predicted MAXBFD

Figure 10. Predicted values of V_{bl} and MAXBFD for the panels having different tablet patterns

Figure 11 illustrates the damage patterns and areas for the panels with different SiC tablet patterns shown in Fig.3. It is noted that for the cases considered, the damage patterns and areas are affected by the tablet pattern. Hence, optimal selected tablet size and location could improve the overall strength of the armour panels subjected to a FSP impact.

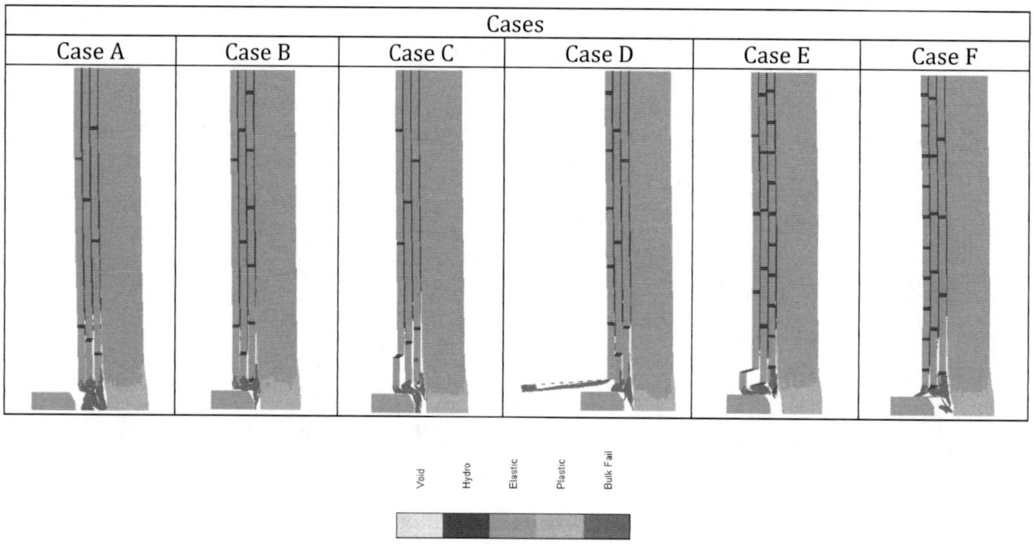

Figure 11. Damage patterns and areas for the panels having different tablet patterns and under an impact of a 30 caliber FSP having initial velocity of 200 m/s (at t=0.2 ms)

Figure 12 plots the variations of the predicted V_{bl} with the size of cohesion between the adjacent tables (i.e., T_c in Fig. 1(c)). It is noted that for the panels having T_{ad} of 0.5 mm, the predicted V_{bl} remains unchanged when T_c increases, whereas for those having T_{ad} of 2 mm, the predicted V_{bl} decreases slightly with an increase in T_c. This indicates that the predicted V_{bl} is more sensitive to T_c for the panels having higher values of T_{ad} compared to those having lower values of T_{ad}. For the panels impacted by a FSP having initial velocity of 200 m/s, the effect of T_c on the predicted MAXBFD is not significant. For example, for the panels having T_{ad} of 0.5 mm, the predicted MAXBFD is almost remain unchanged, while for those having T_{ad} of 2 mm, the predicted MAXBFD slightly increases as T_c increases. The difference of the predicted MAXBFD between panels is less than 0.45 mm, which is insignificant and can be ignored. A comparison of the damages between the panels having T_c=0.5 mm and T_c=1.5 mm for the case of T_{ad} = 0.5 mm or T_{ad} = 2 mm is illustrated in Fig. 13 for the epoxy-filled panels. It is noted that for the cases considered, the effect of T_c on the damage is not significant.

https://doi.org/10.52202/078352-0032

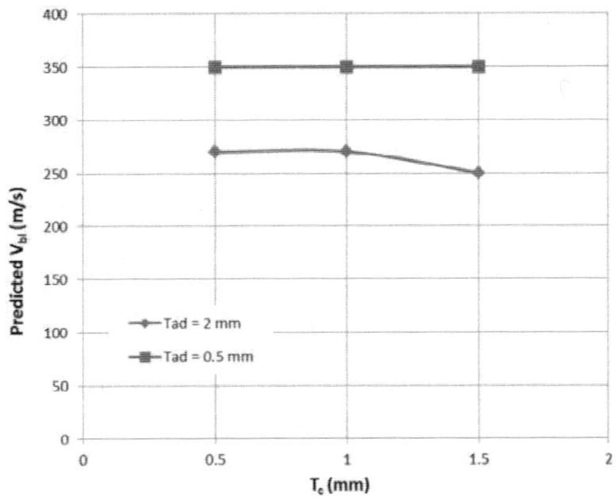

Figure 12. Variations of predicted V_{bl} vs T_c for the selected panels

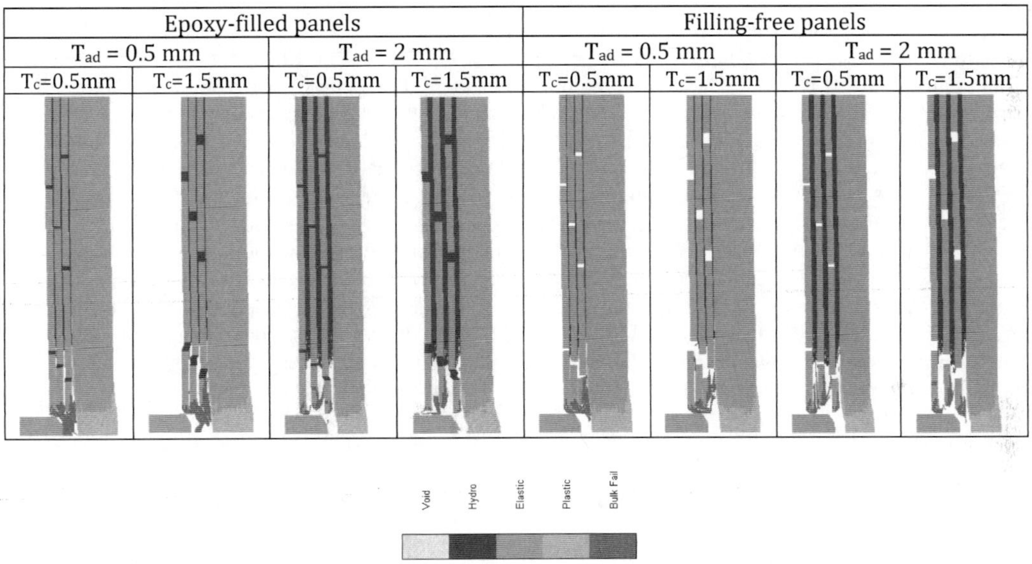

Figure 13. Damage patterns and areas for the selected panels filled with and without epoxy resin between the adjacent tablets under an impact of a FSP having initial velocity of 200 m/s (at t=0.2 ms).

In order to compare the ballistic protection performances and damages of ceramic composite armour panels with and without filling materials between the adjacent tablets, four typical TLST-SiC panels were considered and shown in Fig. 14 for the cases filled with or without epoxy resin, respectively. It is noted from the numerical results that for the cases considered, the difference of the predicted V_{bl} and MAXBFD between the panels filled with or without epoxy resin is not significant. For example, for the selected panels having T_{ad} of 0.5 mm, the differences of the predicted V_{bl} between the epoxy-filled and filling-free panels are zero, and those of the predicted MAXBFD between them range from 0 to 0.2 mm, which are not significant and can be ignored. However, the damage pattern and area in the panels are sensitive to the existence of cohesion, which are illustrated in Fig. 13. It is noted that the damage area in the filling-free panel is generally greater than that in the corresponding epoxy-filled panel.

Epoxy-filled panels	Filling-free panels

https://doi.org/10.52202/078352-0032

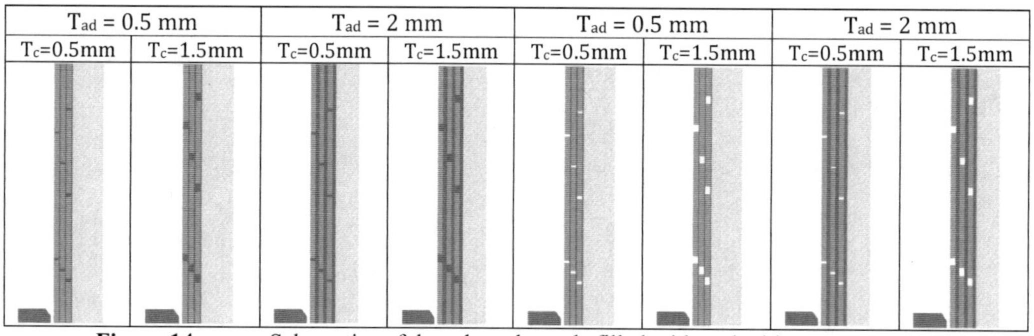

Figure 14. Schematics of the selected panels filled with and without epoxy resin

4. CONCLUSIONS

In this paper, the influences of adhesive geometry and material property on the ballistic protection performances and damages of the selected ceramic composite armour panels were studied by using the present finite element models, which were generated using the commercial finite element software ANSYS/AUTODYN. It was found from the corresponding numerical results that for the selected panels, an increase in total adhesive interlayer thickness (T_{ad}) or adhesive interlayer number results in a decrease in the predicted ballistic limit velocity (V_{bl}). The sensitivity of the predicted V_{bl} to T_{ad} is affected by the type of ceramic used for the strike face plate. The predicted V_{bl} for the panels bonded using epoxy resin is higher than those boned using POLYETHYL, POLYRUBBER, POLYSTYREN and POLYURETH. Also, the predicted V_{bl} is slightly affected by the change in the tablet pattern, but it is not highly sensitive to the change in the adhesive interlayer shape and the size of the cohesion between adjacent tablets. For the panels considered, the predicted MAXBFD is not significantly affected by the adhesive interlayer thickness, number, shape, material type and the size of cohesion between adjacent tablets. The numerical study also shows that for the cases considered, the difference of the predicted V_{bl} and MAXBFD between the panels filled with or without epoxy resin is not significant. However, the damage patterns and areas are generally affected by the adhesive interlayer thickness, number, shape, material type and the size of cohesion between adjacent tablets. The damage area in the filling-free panel is generally larger than that in the epoxy-filled panel. Also, it is interesting to note that replacement of flat adhesive layer with waving layer in body armour may lead to improve the overall strength of the armour panels subjected to FSP impact.

References

[1] Zaera, R., Sa´nchez-Sa´ez, S., Pe´rez-Castellanos, J.L. and Navarro, C., COMPOS PART A-APPL S., 2000; Volume 31; 823–833.

[2] Lo´pez-Puente, J., Arias, A., Zaera, R. and Navarro, C., INT J IMPACT ENG., 2005; Volume 32; 321-336.

[3] Ubeyli, M., Yildirim, OR, Bilgehan, O., J. Mater. Process. Technol., 2008; Volume 196; 356-364.

[4] Tasdemirci, A., Tunusoglu, G. and Güden, M., INT J IMPACT ENG., 2012; Volume 44; 1-9.

[5] Grujicic, M., Pandurangan, B. and d'Entremont, B., Mater. Des., 2012; Volume 41; 380–393.

[6] Prakash, A., Rajasankar, J., Anandavalli, N., Verma, M. and Iyer, N.R., INT J ADHES ADHES.; 2013; Volume 41; 186–197.

[7] Seifert, W., Strassburger, E., Grefen, S. and Schaare, S., Defence Technology, 2016; Volume 12; 188–200.

[8] Jiusti, J., Kammer, E.H., Neckel, L., Lóh, N.J., Trindade, W., Silva, A.O., Montedo, O.R.K. and De Noni Jr., A., CERAM INT.; 2017; Volume 43; 2697–2704.

[9] Gao, Y., Zhang, W., Xua, P. Caia, X. and Fana, Z., INT J IMPACT ENG., 2018; Volume 122; 60–72.

[10] Wang, Z., Sun, Y., Wu, H. and Zhang, C., CONSTR BUILD MATER.; 2018; Volume 169; 851–858.

[11] ANSYS Autodyn, ANSYS Workbench Release 17.2, ANSYS, Inc., 2016.

[12] Tan, P., Finite element simulation of anti-ballistic/fragment performance of military helmets, Australian Simulation Congress, Sydney, Australia, 28th-31th, August, 2017.

[13] Tan P., COMPOS PART B-ENG; 2014; Volume 59; 50-59.

https://doi.org/10.52202/078352-0032

Assessing the Service Life of Aged Hard Armour Composite Material Products

R. Ratrout[1] and H. Al-Ta'amneh[2] and M. Al Afifi[3] and S. Al Majali[4] and I. Rawashdeh[5] and A. Al Sardyah[6] and A. Aldaradkeh[7] and S. Obeidat[8] and A. Hijazi[9] and A. Al Khateeb[10]

King Abdullah II Design and Development Bureau (KADDB), Amman 11190, Jordan
dr.riyad@kaddb.mil.jo

Abstract. World-leading companies have been racing in the past few decades to design and develop ultra-light and durable protective hard armour materials and inserts. Storage of UHMWPE personal armour inserts for long times or exposure to UV light tend to cause physical and chemical degradation and hence reduction in the mechanical properties and ballistic protection capabilities. Similarly, extensive use tends to degrade the protection capabilities of these inserts (Wear). NIJ test standards assess the ballistic performance of hard armour inserts at the time of manufacturing and guaranty their performance for a period of time (Warranty period). However, there are only a few studies on the effect wear/ageing has on the ballistic performance of hard armour inserts. Therefore, this study was conducted to generally investigate the effect ageing has on the ballistic resistance capabilities of hard armour (inserts) by investigating the ballistic protection capabilities of five excessively used (worn-out)/aged inserts (older than 5 years; exceeding the warranty period) as well as five Non-Used (Aged)/ Stored Inserts also exceeded their warranty period. The main finding was that excessively used (worn-out)/aged inserts have constituted failure when tested for their perforation resistance according to NIJ standards at fair impact velocities unlike the Non-Used (Aged)/ Stored inserts, indicating that they could retain their ballistic protection capabilities. Furthermore, some of the excessively used (worn-out)/aged inserts from the same batch number were re-built by "re-pressing" using temperatures and pressures similar to the ones used when the inserts were originally made, as a cost-effective solution to utilize the excessively used (worn-out)/aged hard armour inserts. It was found that "re-pressing" excessively used (worn-out)/aged hard armour inserts increased the ballistic protection capabilities and could hence be a potential cost-effective solution.

1. INTRODUCTION

Wear and ageing of composites in personal hard armour inserts are important topics to understand as many technologies in lightweight hard armour inserts are made of UHMWPE. The use of UHMWPE personal armour inserts for long periods of time over many years and their exposure to different environmental conditions during service tend to cause physical and chemical degradation and hence reduction in their mechanical properties and ballistic protection capabilities [1].

Test standards such as NIJ 0101.04 and NIJ 0108.01 assess the protective armour's resistance to penetration and back face signature P-BFS at the time of manufacturing or use [2 and 3]. However, they do not provide any guidance or warranty regarding the validity of test results (protection capabilities) after the test inserts have aged, this results in an increased cost on the customer because of the necessity to replace the personal armour inserts purchased after the warranty period is over. Therefore, studying the effect ageing and wear have on the ballistic resistance of personal armour is important in determining their protection capabilities after the warranty period is over. In the meantime, there is not enough investigation on the effect ageing has on the ballistic resistance capabilities of personal armour. Because of this lack of knowledge, customers world-wide do not know the suitable commercial and technical warranty period that must be agreed upon with the manufacturer before purchasing personal armour inserts.

In 2007, the former Canadian police research center CPRC have carried out a program to develop an aged armour replacement protocol. This protocol is very detailed and technical providing guidance on "investigating the issue of life expectancy of personal hard armour with respect to issues including the manufacturer's warranty period and replacement time" [4]. This study aims to simplify the topic studied previously by the CPRC to demonstrate the effect ageing/wear has on the ballistic protection capabilities of hard armour inserts older than 5 years. The main objectives of this study were to demonstrate the effect ageing/wear has on the ballistic perforation resistance of vests and inserts and to investigate the validity of re-build solutions "re-pressing" using temperatures and pressures similar to the ones used when the inserts were originally made on the excessively used (worn-out)/aged hard armour inserts as a cost-effective solution allowing the extension of the warranty preiod of those inserts.

 https://doi.org/10.52202/078352-0033

2. LITERATURE REVIEW

2.1 Degradation of hard armour inserts due to wear

The main cause of degradation on personal hard armour inserts is abrasive wear due to continuous use resulting in loss of mechanical strength and hence ballistic protection capabilities. There are three main components to abrasive wear: (a) friction, (b) surface cutting and (c) fiber plucking. Abrasive wear is seen on the affected personal hard armour as surface damage to individual fibers; cracks and fiber failure [5].

2.2 Previous work on warranty periods and replacement time [4], [6].

The most suitable ways to assess future aged hard armour performance is perforation testing at a fair range of speeds rather than V50. Assessing the ballistic protection capabilities of excessively used (worn-out)/aged inserts must be conducted by selecting inserts from the same lot at 5 years of age and then repeated annually until the inserts show no resistance to perforation at fair speeds according to the level of protection they provide according to NIJ 0101.04.

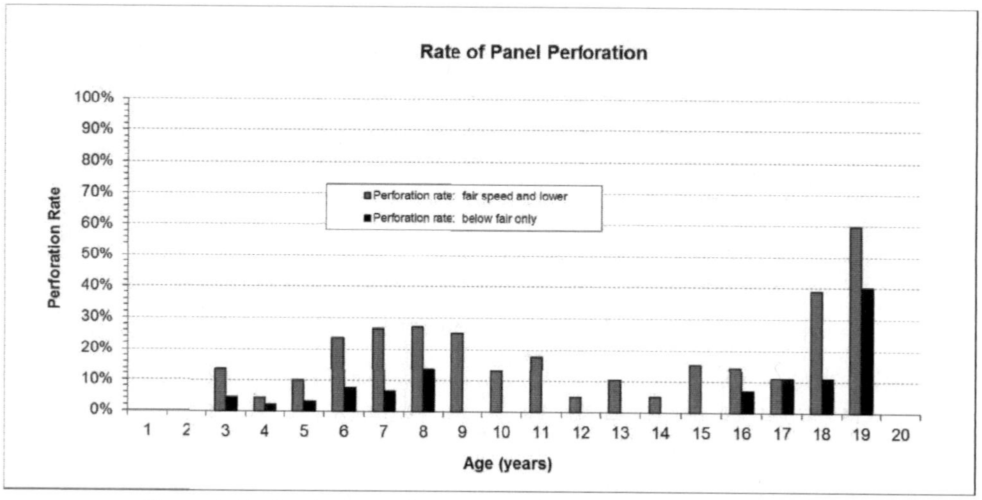

Figure 1. Perforation rate of armour panels as age increases in years [5].

Figure 1 shows that there is no clear correlation between age and perforation rate or (ballistic protection). However, it shows an increase in perforation rate after 5 years until 8 years. This implies that there are other factors affecting the perforation rate other than wear or ageing. Therefore, this study will be conducted on hard armour inserts older than 5 years in age.

This paper was prepared following previous publication by the same authors of this paper. The last publication investigated the ballistic penetration resistance capabilities of worn-out "Used" inserts and the effectivity of "re-pressing" as a potential cost-effective solution. However, this paper aims to further ensure the findings from the previous study by investigating the ballistic penetration resistance of three times the insert size of worn-out "Used" inserts as well as un-used "aged" stored inserts and the effectivity of "re-pressing" worn-out inserts as a potential cost-effective solution.

https://doi.org/10.52202/078352-0033

3. METHODOLOGY

Step 1: A random selection of fifteen hard armour inserts manufactured in 2011, 2012, 2013, 2014 and 2015 were selected form the same batch. Three inserts from each year of manufacture were selected to bring the total number of inserts in this study to fifteen.

Step 2: Five excessively used (worn-out)/aged inserts manufactured in 2011, 2012, 2013, 2014 and 2015 were tested ICW 3A 9mm Vests (2015 Production) according to NIJ 0108.01 test standard [2] by firing five (7.62x39mm PS) Special Type shots at each impact side placed at 15 m from the muzzle to test ballistic penetration only.

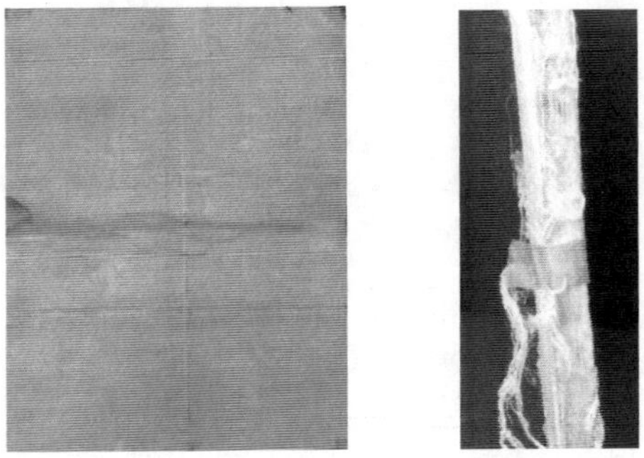

Photo No. 1 Excessively Used (Worn-out)/Aged Hard Armour Insert (2013 Production).

Step 3: Five Non-Used (Aged)/ Stored Inserts manufactured in 2011, 2012, 2013, 2014 and 2015 were tested ICW 3A 9mm Vests (2015 Production) according to NIJ 0108.01 by firing five (7.62x39mm PS) Special Type shots at each impact side placed at 15 m from muzzle to test ballistic penetration only.

Photo No. 2 Non-Used (Aged)/ Stored Hard Armour Insert (2013 Production).

Step 4: Five Re-built Excessively Used (Worn-out)/Aged hard armour inserts (re-built by re-pressing polyethylene hard layers with the temperature and humidity used when they were first manufactured) were tested ICW 3A 9mm Vests (2015 Production) according to NIJ 0101.04 test standard [3] by firing six (7.62x39mm PS) Special Type shots at each impact side placed at 15 m from the muzzle to test ballistic penetration and back face signature P-BFS.

Photo No. 3 Re-built Excessively Used (Worn-out)/Aged Hard Armour Insert (2013 Production).

https://doi.org/10.52202/078352-0033

Step 1
Random Selection of:

3 Body Armour Insert (2011 Production)
3 Body Armour Insert (2012 Production)
3 Body Armour Insert (2013 Production)
3 Body Armour Insert (2014 Production)
3 Body Armour Insert (2015 Production)

↓

Step 2
Excessively Used (Worn-out)/Aged Inserts - Ballistic Resistance Testing (NIJ 0108.01) Special (7.62x39mm PS)

1 Body Armour Insert (2011 Production) ICW 3A 9mm Vest (2015 Production)
1 Body Armour Insert (2012 Production) ICW 3A 9mm Vest (2015 Production)
1 Body Armour Insert (2013 Production) ICW 3A 9mm Vest (2015 Production)
1 Body Armour Insert (2014 Production) ICW 3A 9mm Vest (2015 Production)
1 Body Armour Insert (2015 Production) ICW 3A 9mm Vest (2015 Production)

↓

Step 3
Non-Used (Aged)/ Stored Inserts - Ballistic Resistance Testing (NIJ 0108.01) Special (7.62x39mm PS)

1 Body Armour Insert (2011 Production) ICW 3A 9mm Vest (2015 Production)
1 Body Armour Insert (2012 Production) ICW 3A 9mm Vest (2015 Production)
1 Body Armour Insert (2013 Production) ICW 3A 9mm Vest (2015 Production)
1 Body Armour Insert (2014 Production) ICW 3A 9mm Vest (2015 Production)
1 Body Armour Insert (2015 Production) ICW 3A 9mm Vest (2015 Production)

↓

Step 4
Re-built Exsessively Used (Worn-out)/Aged Inserts - Penetration and Backface Signature Testing (NIJ 0101.04) Special (7.62x39mm PS)

1 Body Armour Insert (2011 Production) ICW 3A 9mm Vest (2015 Production)
1 Body Armour Insert (2012 Production) ICW 3A 9mm Vest (2015 Production)
1 Body Armour Insert (2013 Production) ICW 3A 9mm Vest (2015 Production)
1 Body Armour Insert (2014 Production) ICW 3A 9mm Vest (2015 Production)
1 Body Armour Insert (2015 Production) ICW 3A 9mm Vest (2015 Production)

Figure 2. Methodology of work including sample selection criteria, number of samples and testing sequence.

https://doi.org/10.52202/078352-0033

4. RESULTS

4.1 Excessively Used (Worn-out)/Aged Inserts - Ballistic Resistance Testing (NIJ 0108.01)

Table 1. Ballistic Penetration Test Results - Hard Armour Inserts (2011 and 2012 Production)

Shot No @15m 7.62x39mm PS	Hard Armour Insert (2011 Production) ICW 3A 9mm Vest (2015 Production)		Hard Armour Insert (2012 Production) ICW 3A 9mm Vest (2015 Production)	
	Bullet Velocity m/s	Status CP / PP	Bullet Velocity m/s	Status CP / PP
1st	741	PP	736	PP
2nd	747	PP	741	PP
3rd	743	CP	733	PP
4th	-	-	746	CP
5th	-	-	-	-

Table 2. Ballistic Penetration Test Results - Hard Armour Inserts (2013 and 2014 Production)

Shot No @15m 7.62x39mm PS	Hard Armour Insert (2013 Production) ICW 3A 9mm Vest (2015 Production)		Hard Armour Insert (2014 Production) ICW 3A 9mm Vest (2015 Production)	
	Bullet Velocity m/s	Status CP / PP	Bullet Velocity m/s	Status CP / PP
1st	728	PP	728	PP
2nd	728	PP	744	PP
3rd	734	PP	731	PP
4th	726	CP	736	PP
5th	-	-	740	CP

Table 3. Ballistic Penetration Test Results - Hard Armour Inserts (2015 Production)

Shot No @15m 7.62x39mm PS	Hard Armour Insert (2015 Production) ICW 3A 9mm Vest (2015 Production)	
	Bullet Velocity m/s	Status CP / PP
1st	736	PP
2nd	740	PP
3rd	739	PP
4th	728	PP
5th	741	CP

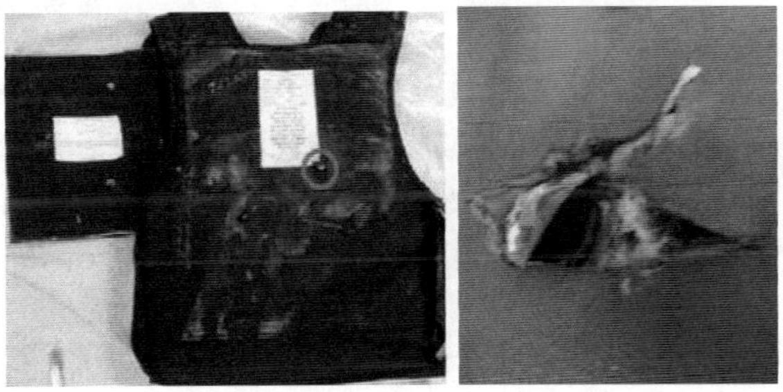

Photo No. 4 Status of Excessively Used (Worn-out)/Aged Inserts (2013 Production) ICW 3A 9mm Vest, Complete Penetration

https://doi.org/10.52202/078352-0033

4.2 Non-Used (Aged)/ Stored Inserts - Ballistic Resistance Testing (NIJ 0108.01)

Table 4. Ballistic Penetration Test Results - Hard Armour Inserts (2011 and 2012 Production)

Shot No @15m 7.62x39mm PS	Hard Armour Insert (2011 Production) ICW 3A 9mm Vest (2015 Production)		Hard Armour Insert (2012 Production) ICW 3A 9mm Vest (2015 Production)	
	Bullet Velocity m/s	Status CP / PP	Bullet Velocity m/s	Status CP / PP
1st	728	PP	730	PP
2nd	723	PP	734	PP
3rd	727	PP	734	PP
4th	728	PP	727	PP
5th	734	PP	729	PP

Table 5. Ballistic Penetration Test Results - Hard Armour Inserts (2013 and 2014 Production)

Shot No @15m 7.62x39mm PS	Hard Armour Insert (2013 Production) ICW 3A 9mm Vest (2015 Production)		Hard Armour Insert (2014 Production) ICW 3A 9mm Vest (2015 Production)	
	Bullet Velocity m/s	Status CP / PP	Bullet Velocity m/s	Status CP / PP
1st	749	PP	738	PP
2nd	746	PP	726	PP
3rd	745	PP	732	PP
4th	753	PP	729	PP
5th	746	PP	740	PP

Table 6. Ballistic Penetration Test Results - Hard Armour Inserts (2015 Production)

Shot No @15m 7.62x39mm PS	Hard Armour Insert (2015 Production) ICW 3A 9mm Vest (2015 Production)	
	Bullet Velocity m/s	Status CP / PP
1st	729	PP
2nd	733	PP
3rd	731	PP
4th	740	PP
5th	732	PP

Photo No. 5 Status of Non-Used (Aged)/ Stored Inserts (2013 Production) ICW 3A 9mm Vest No Complete Penetration

https://doi.org/10.52202/078352-0033

4.3 Re-built Excessively Used (Worn-out)/Aged Inserts - Penetration and Backface Signature Test (P-BFS) (NIJ 0101.04)

Table 7. Ballistic Penetration Test Results - Hard Armour Inserts (2011 and 2012 Production)

Shot No @15m 7.62x39mm PS	Hard Armour Insert (2011 Production) ICW 3A 9mm Vest (2015 Production)			Hard Armour Insert (2012 Production) ICW 3A 9mm Vest (2015 Production)		
	Bullet Velocity m/s	BFS Depth (mm)	Status CP / PP	Bullet Velocity m/s	BFS Depth (mm)	Status CP / PP
1st	739	14	PP	735	12	PP
2nd	725	11	PP	738	13	PP
3rd	733	11	PP	744	10	PP
4th	749	6	PP	753	11	PP
5th	738	6	PP	741	6	PP
6th	738	7	PP	746	7	PP

Table 8. Ballistic Penetration Test Results - Hard Armour Inserts (2013 and 2014 Production)

Shot No @15m 7.62x39mm PS	Hard Armour Insert (2013 Production) ICW 3A 9mm Vest (2015 Production)			Hard Armour Insert (2014 Production) ICW 3A 9mm Vest (2015 Production)		
	Bullet Velocity m/s	BFS Depth (mm)	Status CP / PP	Bullet Velocity m/s	BFS Depth (mm)	Status CP / PP
1st	755	28	PP	736	10	PP
2nd	743	12	PP	743	17	PP
3rd	738	22	PP	732	17	PP
4th	725	10	PP	740	10	PP
5th	741	9	PP	732	15	PP
6th	736	9	PP	724	23	PP

Table 9. Ballistic Penetration Test Results - Hard Armour Inserts (2015 Production)

Shot No @15m 7.62x39mm PS	Hard Armour Insert (2015 Production) ICW 3A 9mm Vest (2015 Production)		
	Bullet Velocity m/s	BFS Depth (mm)	Status CP / PP
1st	738	20	PP
2nd	730	16	PP
3rd	735	17	PP
4th	743	16	PP
5th	749	13	PP
6th	738	14	PP

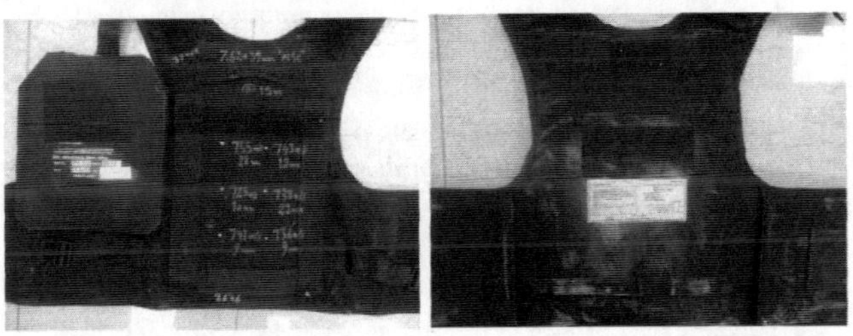

Photo No. 6 Status of Re-built Excessively Used (Worn-out)/Aged Inserts (2013 Production) ICW 3A 9mm Vest, No complete Penetration and BFS Depths are less than 44mm

https://doi.org/10.52202/078352-0033

5. DISCUSSION

Personal hard armour inserts tend to degrade with time due to many factors. The major cause of wear according to literature in personal hard armour including, vests and inserts is abrasive corrosion. This reduction in protection is very common in excessively used (worn-out)/aged hard armour inserts. Therefore, a re-build or replacement protocol must be adopted and followed to ensure the functionality of such inserts after exceeding their warranty period.

Re-building hard armour inserts could be a cost-effective solution instead of total replacement. There are special techniques used in re-building hard armour inserts. Some of those techniques add more protective fabric layers to the impact and rear face of the excessively used (worn-out)/aged inserts to compensate for their reduction in protection. However, in this study, the excessively used (worn-out)/aged inserts were re-pressed using the same temperature and pressure during manufacturing of the hard inserts. The re-building techniques and the finished inserts need to be validated, tested and inspected annually to ensure the delivery of the intended protection level at fair impact velocities.

Table 2 and Photo 4 represent a clear example of an excessively used (worn-out)/aged hard armour insert manufactured in 2013. The 2013 (worn-out hard armour) showed a complete penetration at 726 m/s. Comparing the penetration resistance of the 2013 worn-out insert with the same after rebuild, the rebuilt insert showed partial penetration against the same threat at a slightly higher velocity as could be seen in Table 8 and Photo 6. For further assurance, a number of rebuilt hybrid hard armour inserts which passed the penetration resistance test according to NIJ 0108.01, were further tested according to NIJ 0101.04 to evaluate their back face signature values which they passed resulting in values less than 44mm in P-BFS depth and therefore validating the rebuild process adopted to be an effective solution in contrast to complete replacement.

Tables 4, 5 and 6 clearly show the results of the penetration test using stored inserts manufactured at the same years as the worn-out and re-built inserts for fair comparison and analysis. The tables clearly show no complete penetration in any insert leading to the conclusion that non-used (aged)/ stored inserts stored at ambient temperatures away from direct UV light or excessive humidity and moisture tend to retain their penetration resistance capabilities even after exceeding their designated warranty period (5 years).

Those findings were a living proof that personal hard armour inserts tend to degrade with extensive use and age. Those inserts could be re-built as a cost-effective solution instead of replacements specially if bought in large quantities. Hard armour inserts need to be checked and tested annually after they exceed their warranty period as there is a high probability that they fail at fair velocities.

6. CONCLUSIONS

- The main cause of degradation on personal hard armour is abrasive corrosion due to continuous use and environmental conditions.
- There is no clear correlation between age and perforation rate (ballistic protection capabilities). However, there is a very high probability that a worn-out product fails after exceeding the warranty period at fair impact velocities.
- Inserts stored at ambient temperatures away from direct UV light or excessive humidity and moisture tend to retain their penetration resistance capabilities even after exceeding their designated warranty period (5 years).
- Excessively used (worn-out)/aged personal hard armour inserts could be re-built as a cost-effective solution instead of replacement.
- Re-build or replacement protocols must be adopted and followed to ensure functionality and protection of all excessively used (worn-out)/aged personal hard armour inserts.
- Annual inspection must be carried out on hard armour inserts after exceeding their warranty or service life to ensure their functionality.

https://doi.org/10.52202/078352-0033

REFERENCES

[1] Grant, H., Phd. (n.d.). Hard Armour Use, Care, and Performance (USA).
[2] NIJ Standard 0108.01 for Ballistic Resistant Protective Materials, September 1985.
[3] NIJ Standard 0101.04 for Ballistic Resistance of Personal Hard Armour, June 2001.
[4] (n.d.). Update of the hard armour purchase and replacment protocol (Rep.).
[5] R. (2012). Degradation of military hard armour due to wear: Laboratory testing. Textile Research.
[6] R. RATROUT, H. AL-TA'AMNEH, S. AL MAJALI, I. RAWASHDEH, A. AL SARDYAH, A. ALDARADKEH and S. OBEIDAT. Assessing the Service Life of Aged Hard Armour Inserts (Helmet-Vest-Insert), 31st International Symposium on Ballistics, Hyderabad, India, November 4-8, 2019, PROCEEDINGS, VOLUME 2 (31st International Symposium on Ballistics, Hyderabad, India, November 4-8, 2019); pp. 1384-1392.

https://doi.org/10.52202/078352-0033

Small Arms Ammunition and Personal Armour – Standards versus the Real World

PL Gotts

Phil Gotts Consulting Ltd, 23 Thorney Rd, Capel St Mary, Ipswich, Suffolk, IP9 2HL, UK
phil@philgotts.com

Abstract. Much personal armour is designed to protect the wearer from specified types of small arms ammunition. The definition of these small arms ammunition threats may be included within a user requirement, or may be referenced in a ballistic test standard or method. Either way there will be a definition of an ammunition type, combined with an impact velocity, a shot pattern and perhaps some other associated pieces of information, all of which theoretically allow the armour testing to be conducted. The easiest option for the specifier of personal armour is to use a test standard which includes specified levels of ammunition. However, the available ammunition found in the test standards may not reflect the required protection levels from an operational perspective. There is also the possibility that ammunition required for armour testing from an operational perspective may not be suitable as test ammunition for some reason. This paper explores the relationship between ammunition specified in armour standards and the ammunition most relevant for real life scenarios. Both military and law enforcement environments are considered. As a common ammunition calibre used for soft body armour testing, 9 mm ammunition of different configurations is discussed. As perhaps the most common real world threat to hard armour upgrade plates, 7.62 x 39 mm ammunition of different configurations is discussed. As one example, the latest version of the UK Home Office Body Armour test standard is discussed, with respect to the choice of test ammunition, both what has been included and what has been omitted or made optional. These ammunition types are compared to their terminal effects related to armour and the reality of use.

1. INTRODUCTION

One of the major threats for which personal armour is designed, is that of small arms ammunition. The armour may include soft armour components for the defeat of low velocity handgun and sub-machine gun ammunition and/or upgrade plates for the defeat of high velocity rifle and machine gun ammunition. When the user of personal armour, or their representative, needs to specify the small arms ammunition threat regime for their armour, they need to make decisions as to which ammunition types to include. The immediately obvious choice would be for ammunition which the user of the personal armour is likely to encounter during the conduct of their duties. However, this may not always be the best option.

One ammunition type which is included within many Western body armour standards is the 9 mm FMJ (full metal jacket) Luger and the history and use of this ammunition has issues associated with it.

An interesting ammunition type with respect to the trade-off between realism and repeatability is the 7.62 x 39 mm Russian.

At the end of July 2017 the UK HO CAST (Home Office Centre for Applied Science and Technology) published its first new body armour standard for 10 years [1], and this includes a different approach to the ammunition specified, which in part addresses some of the aspects of realism versus repeatability, as well as the blue-on-blue possibility.

2. TEST STANDARD AMMUNITION VERSUS REAL LIFE

In an ideal world, the ammunition chosen to be used for the testing of personal armour should be both realistic and repeatable. In order to be realistic, it should represent a ballistic threat which the particular user community would be expected to encounter during the conduct of their duties. In order to be repeatable the chosen ammunition should be manufactured within close tolerances, such that one shot interacts with the personal armour in the exact same manner as all other shots for the same impact scenario. Unfortunately in many cases the requirements for repeatability and realism may not be achievable with the same ammunition type.

In those situations where the requirements for repeatability and realism cannot both be met, a decision needs to be made as to which of the two requirements takes priority. In most cases the need for repeatability takes priority, and this determines the ammunition types usually included in the test standards. This is one reason that many test standards are criticised as being unrealistic with regards to test ammunition.

https://doi.org/10.52202/078352-0034

One source of ammunition which is renowned for causing repeatability issues is anything manufactured within the former Eastern Bloc countries. It is for this reason that former-Soviet ammunition is quite rare in Western body armour test standards (it is however more common in vehicle armour standards). Some former-Soviet ammunition types do appear in VPAM (Vereinigung der Prüfstellen für angriffs-hemmende Materialien und Konstruktionen) APR 2006 [2] levels 6, 8 and 10 and have also been included within levels B4, C4 and C6 of AEP-2920 [3]. In both of these cases, there are also Western and/or NATO ammunition types included within the standard. The Russian GOST 50744-95 Armour Clothing standard [4] uses only ex-Soviet ammunition types.

3. 9 mm TEST AMMUNITION

The 9 x 19 mm FMJ ammunition is often used as a test round for soft body armour. Over the years certain specific designations of this 9 mm ammunition have been used. The choice of the ammunition has partly been dictated by its availability, and partly by its consistency and/or variability.

3.1 Background of 9 mm Parabellum / Luger Ammunition

The 9 mm FMJ ammunition used in body armour testing mainly complies with the generic description of 9 x 19 mm Parabellum or Luger.

The cartridge was developed in 1902 by Georg Luger in Germany, in order to improve the stopping power of his 9 mm pistol, and was adopted by the German Armed Forces before World War One. Initially the bullet was cylindro-conoidal with a flat tip. However, this was found not to always feed smoothly and thus led to weapon jams. Therefore the modification to an ogival shaped bullet was made and this entered service in 1917. It is this later ogival design which is normally used for the testing of soft body armour today.

Figure 1: Typical 9 mm FMJ Bullets

3.2 Body Armour Test Standards and Bespoke Requirements

This section will focus predominantly upon the 9 mm requirements of the UK, US and European Police body armour test standards, although there will also be reference to the historic UK MOD (Ministry of Defence) approach.

3.2.1 PSDB, HOSDB, HO CAST Test Standards

For the UK Police the relevant body armour test standards are those produced by the Home Office, Centre for Applied Science and Technology (HO CAST) and its predecessors; PSDB (Police Scientific Development Branch) and HOSDB (Home Office Scientific Development Branch). Since the first documented standard of 1993, there have been a number of versions of this standard. The 9 mm ammunition appears in more than one test level in every version of the standard.

3.2.2 PSDB Publication No: 12/93 (1993) [5]

Two levels in this standard use 9 mm FMJ Dynamit Nobel ammunition:

- HG1 - 9 mm DM11A1B2 at 330 ± 10 m/s
- HG2 - 9 mm DM11A1B2 at 425 ± 10 m/s

3.2.3 PSDB Publication No: 2/96 (1995) [6]

Two levels in this standard use 9 mm FMJ Dynamit Nobel ammunition:

- HG1 - 9 mm DM11A1B2 at 360 ± 10 m/s
- HG2 - 9 mm DM11A1B2 at 425 ± 10 m/s

Therefore the only change has been an increase in velocity of the HG1 level.

https://doi.org/10.52202/078352-0034

3.2.4 PSDB Publication No: 7/03/B (2003) [7]

Two levels in this standard use 9 mm FMJ Dynamit Nobel ammunition:

- HG1 - 9 mm DM11A1B2 at 360 ± 10 m/s
- HG2 - 9 mm DM11A1B2 at 425 ± 10 m/s

The use of 9 mm ammunition in this version of the standard is exactly the same as the previous version.

3.2.5 HOSDB Publication No: 39/07B (2007) [8]

Three levels in this standard use 9 mm FMJ Dynamit Nobel ammunition:

- HG1 - 9 mm DM11A1B2 at 365 ± 10 m/s
- HG1A - 9 mm DM11A1B2 at 365 ± 10 m/s
- HG2 - 9 mm DM11A1B2 at 430 ± 10 m/s

The 2007 version includes the addition of the HG1A level which is identical for the 9 mm ammunition and velocity, but allows for 44 mm back-face signature instead of 25 mm.

3.2.6 Summary of UK Home Office Body Armour Test Standards to 2017

In summary, 9 mm ammunition has been included in all versions of the standard and has always been to the DM11A1B2 specification. DM11A1B2 is a generic German specification for the ammunition and does not specify a manufacturer. Initially the ammunition was manufactured by the German company Dynamit Nobel Ammotec GmbH, but this company was acquired in 2002 by the Swiss company RUAG and became RUAG Ammotec AG. RUAG Ammotec AG continued to manufacture the DM11A1B2 for a few years. The Dynamit Nobel ammunition is in theory the same as that manufactured by RUAG. As can be seen from the different versions of the standard, the choice of the Dynamit Nobel / RUAG Ammotec DM11A1B2 has remained constant, but with slightly varying impact velocities. There have also been slightly different levels with the notable addition of Level HG1A for the 2007 version.

3.2.7 HO CAST 2017 [1]

This latest version of the HO standard incorporates the following levels and ammunition:

- HO1 - 9 mm DM11A1B2 (MEN) at 365 ± 10 m/s
  - 9 mm Federal Premium JHP P9HST1 at 365 ± 10 m/s
- HO2 - 9 mm DM11A1B2 (MEN) at 430 ± 10 m/s
  - 9 mm Federal Premium JHP P9HST1 at 430 ± 10 m/s

The addition of the 9 mm JHP will be discussed later.

3.3 MEN DM11A1B2

The DM11A1B2 specification ammunition is also manufactured by MEN (Metallwerk Elisenhütte GmbH), and this is now specified in the 2017 HO CAST Body Armour standard. It cannot however, be guaranteed that the MEN version performs against soft body armour in exactly the same way as the RUAG version. The MEN DM11A1B2 has previously been shown to be more penetrative versus some armour designs than the RUAG ammunition. Jones and Barnes-Warden [9] presented results of a recent study at PASS 2018, which emphasises and explains the differences between the RUAG and MEN versions of the DM11A1B2.

The author has been involved with a series of ballistic testing for a European military organisation, in which the ammunition was suspected of producing a rogue result. The ammunition in this case was identified as an NSN (NATO Stock Number)-marked version of the DM11A1B2 manufactured by MEN, thus further indicating variability in the MEN version of the DM11A1B2.

3.4. NIJ (National Institute of Justice) Test Standards

The US Law Enforcement community are supported by the NIJ (National Institute of Justice) and its predecessor, NILECJ (National Institute of Law Enforcement and Criminal Justice). As with the HO

https://doi.org/10.52202/078352-0034

CAST standards, the 9 mm FMJ ammunition is used in different levels of the standards. The main difference with the HO CAST standard is that for much of its life the NIJ series of standards, did not provide a specification for the ammunition used, except for it being a 124 grain 9 mm FMJ. This changed only with the NIJ-0101.06 version, for which a specific supplier was suggested.

3.4.1 NILECJ-0101.00 (Mar 1972) [10]

The first version of the NIJ-0101.0n series of body armour test standards did not include any 9 mm ammunition within its test levels.

3.4.2 NILECJ-0101.01 (Dec 1978) [11]

This test standard includes the following levels which incorporate 9 mm 124 grain (8.0 g) FMJ ammunition:

- Level IIA – 9 mm FMJ 124 grain (8.0 g) at 1,090 fps (332 m/s)
- Level II – 9 mm FMJ 124 grain (8.0 g) at 1,175 fps (358 m/s)

3.4.3 NIJ-0101.02 (Mar 1985) [12]

This test standard includes the following levels which incorporate 9 mm 124 grain (8.0 g) FMJ ammunition:

- Level IIA – 9 mm FMJ 124 grain (8.0 g) at 1,090 fps (332 m/s)
- Level II – 9 mm FMJ 124 grain (8.0 g) at 1,175 fps (358 m/s)
- Level IIIA – 9 mm FMJ 124 grain (8.0 g) at 1,400 fps (427 m/s)

3.4.4 NIJ-0101.03 (Apr 1987) [13]

This test standard includes the following levels which incorporate 9 mm 124 grain (8.0 g) FMJ ammunition:

- Level IIA – 9 mm FMJ 124 grain (8.0 g) at 1,090 fps (332 m/s)
- Level II – 9 mm FMJ 124 grain (8.0 g) at 1,175 fps (358 m/s)
- Level IIIA – 9 mm FMJ 124 grain (8.0 g) at 1,400 fps (427 m/s)

3.4.5 NIJ-0101.04 (Sep 2000) [14]

This test standard includes the following levels which incorporate 9 mm 124 grain (8.0 g) FMJ ammunition:

- Level IIA – 9 mm FMJ RN 124 grain (8.0 g) at 1,120 fps (341 m/s)
- Level II – 9 mm FMJ RN 124 grain (8.0 g) at 1,205 fps (367 m/s)
- Level IIIA – 9 mm FMJ RN 124 grain (8.0 g) at 1,430 fps (436 m/s)

Shortly after this revision was accepted NIJ instructed test laboratories only to use the Remington bullet, although this was never formally adopted in the 04 version of the standard.

3.4.6 NIJ 2005 Interim Requirements for Bullet-Resistant Body Armour

The test level requirements for the 2005 Interim are the same as those of NIJ-0101.04.

3.4.7 NIJ-0101.06 (Jul 2008) [15]

This test standard includes the following levels which incorporate 9 mm 124 grain (8.0 g) FMJ ammunition:

- Level IIA – 9 mm FMJ RN 124 grain (8.0 g) at 355 m/s for conditioned armour and 373 m/s for new armour
- Level II – 9 mm FMJ RN 124 grain (8.0 g) at 379 m/s for conditioned armour and 398 m/s for new armour
- Level IIIA – there is no longer any 9 mm ammunition in this level.

Appendix A of NIJ-0101.06 states that the 9 mm ammunition to be used is 9 mm Luger FMJ RN (round nose) Remington 23558.

https://doi.org/10.52202/078352-0034

3.4.8 NIJ-0101.07 – Draft January 2018[16]

This draft test standard includes the following levels which incorporate 9 mm 124 grain (8.0 g) FMJ ammunition:

- Level HG1 – 9 mm FMJ RN 124 grain (8.0 g) Remington 23558 at 398 m/s
- Level HG2 – 9 mm FMJ RN 124 grain (8.0 g) Remington 23558 at 448 m/s

3.5 UK MOD / Military Test Standards

The UK MOD / Military do not have a usual requirement for protection against low velocity bullets. However, there have been exceptions.

3.5.1 Mk 2z Luger

During the 1990s, there were some limited requirements for specialised body armour with this level of protection. Initially the UK MOD specified a test programme using the 9 mm Mk 2z Luger bullet. The test results obtained using the Mk 2z showed to be very inconsistent and a detailed investigation into the possible causes was conducted. The eventual conclusions were that the cause of the inconsistencies related to a variable jacket thickness between different bullets, particularly around the nose. This variation was apparent even between bullets from the same box. It was therefore determined that the Mk 2z was not a suitable ammunition type to be used for testing body armour. Therefore the UK MOD also moved over to using the DM11A1B2 as its standard 9 mm FMJ body armour test ammunition.

3.6 DAG 9 mm Luger

DAG 9 mm Luger has been previously specified as an alternative to the Dynamit Nobel / RUAG DM11A1B2. The DAG 9 mm Luger is also a 124 grain (8.0 g) FMJ bullet, with a copper jacket. In theory this may also be a 9 mm GECO as it is now manufactured under RUAG in Switzerland. The head-stamp may be DAG 9 mm Luger or GECO.

3.7 9 mm FMJ DM41

There have been a number of occasions when the future manufacture of the DM11A1B2 has been uncertain, and an alternative has been suggested. This alternative has usually been the DM41. The DM41 specification is predominantly used by the German police. Like the DM11A1B2, it is a 124 grain (8.0 g) FMJ. The jacket is described as being made of steel and copper and tin plated. The DM41 is also manufactured by RUAG Ammotec AG. The DM41 can be found in some European body armour test standards, such as the Technische Richtlinie Ballistische Schutzwesten (TR 03/2008; Rev 10/2008) [17] and the VPAM APR 2006 [2] levels 2 and 3. Ballistic testing with this ammunition has shown some inconsistency, particularly with angled shots. For this reason it has not been adopted as a test round in the UK or the USA.

3.8 Swedish Norma / Bofors M39B

This 9 mm ammunition has been specified in specific body armour tenders. It is of a slightly different construction to the DM11A1B2, as the rough sectioning of a bullet shows (figure 2). It consists of a lead core and a copper full metal jacket, but with a steel under-jacket of significant thickness. This means that it has a higher penetration performance than a DM11A1B2.

This ammunition has been included within Scandinavian Police Body Armour tenders, for example.

Figure 2. Sectioned M39B

3.9 Summary of 9 mm Ammunition in Body Armour Standards

The same type of 8.0 g FMJ 9 mm ammunition has been used in body armour standards for 40 years. Specified velocities have changed over the years. A few variations on the theme have occurred in some standards, such as the DM41 and the M39B.

https://doi.org/10.52202/078352-0034

4. 7.62 x 39 mm AMMUNITION

The 7.62 x 39 mm is probably the most prolific high velocity rifle ammunition currently in existence. As such it is produced in a number of configurations and by many different factories, in many countries. It would perhaps be naïve to expect all of the ammunition from all of the factories, to be identical and to perform in exactly the same way, and indeed they do not.

4.1 Ammunition Development [18]

Soviet development of an intermediate rifle cartridge began in the 1930s, but it was not until 1943 that a design was approved. On July 15, 1943, the Technical Council of the People's Commissariat for Armaments met to discuss the introduction of a Soviet intermediate cartridge. The job of designing the Soviet intermediate cartridge was assigned to a committee led by chief designer NM Elizarov. Elizarov collaborated closely with some leading weapons designers, including Fedorov, Tokarev, Simonov, and Shpagin.

A first variant of the new cartridge was officially adopted for service after completing range trials in December 1943; it was given the GRAU index 57-N-231. This cartridge actually had a case length of 41 mm, so it is sometimes referred to as the 7.62 x 41 mm. The bullet it contained was 22.8 mm long and had a core made entirely of lead. This bullet has a somewhat stubbier appearance than later 7.62 x 39 mm bullets and it was lacking a boat tail.

After more detailed testing results became available, starting in 1944 the cartridge was modified in order to improve its accuracy and penetration. Initially, the boat tail had been omitted because the Soviet designers had assumed (incorrectly) that it would only make a difference at long ranges when the bullet became subsonic, and the accuracy of the intermediate cartridge at these ranges was considered inconsequential. However, further testing showed that the boat tail improved accuracy even at shorter ranges, where the bullet was still supersonic. In order to maintain the overall mass of the bullet, after adding the boat tail, the ogival head section of the bullet was lengthened as well, making the bullet more streamlined overall. Additionally, the new bullet had a core made of low-carbon steel wrapped in lead and a jacket of Gilding Metal Clad Steel (GMCS). This bullet was given the acronym '7.62 PS' (7.62 ПС). The 'S' initially stood for 'surrogate', but later the letter was taken to refer to the steel component of the core, which accounted for about 50% of the core volume. The 7.62 x 39 mm cartridge equipped with the PS bullet finally overcame all objections of the GRAU in mid-1947, when it was ordered into series production, and given the index 57-N-231S. In an effort to simplify terminology, sometime thereafter the 57-N-231 designation was recycled to denote all steel-core 7.62 x 39 mm Soviet ammunition, irrespective of case material.

4.2 Increased Core Hardness [18]

After 1989, the regular (PS) Russian bullets started to be manufactured with a steel core with a higher carbon concentration and subjected to heat treatment. This change improved their penetration by 1.5 to 2 times. It is not possible to externally distinguish these bullets from the earlier, softer PS ones except by year of fabrication. At about the same time, tool steel was adopted for a normal velocity 7.62 x 39 mm bullet. Called BP, this bullet was developed in the 1980s and 1990s. It was officially adopted for Russian service in 2002 under the service name '7.62 BP', and with the GRAU designation 7-N-23. The BP bullet is claimed to achieve over three times the penetration of the PS bullet. The BP cartridge has the tip of its bullet painted black. The BP bullet itself is slightly longer (27.4 mm) compared to the PS bullet, but has the same mass of 7.9 g.

When armour manufacturers state that their armour defeats the 7.62 x 39 mm mild-steel core PS ball, it is critical that they have completed their testing on ammunition later than 1989, and preferably later than 2002. This information is critical for armour testing and design.

4.3 Improvements [18]

The original Soviet M43 bullets are 123 grain (7.97 g) boat-tail bullets with a copper-plated steel jacket, a large steel core, and some lead between the core and the jacket. The cartridge itself consisted of a Berdan-primed, highly tapered (usually steel) case which seats the bullet and contains the propellant charge.

The complete solidity of the M43 projectile causes its only drawback - it is very stable, even while traversing tissue. It begins to yaw only after traversing nearly 26 cm of tissue. This greatly reduces the potential wounding effectiveness of the projectile against humans.

https://doi.org/10.52202/078352-0034

In the 1960s, Yugoslavia experimented with new bullet designs to produce a round with a superior wounding profile, velocity, and accuracy to that of the M43. The M67 projectile is shorter and flatter-based than the M43. This is mainly due to the deletion of the mild steel insert. This has the side effect of shifting the centre of gravity rearward in comparison to the M43. This allows the projectile to destabilize nearly 17 cm earlier in tissue. This causes a pair of large temporary cavities at a depth likely to cause effective wound trauma. When the temporary cavity intersects with the skin at the exit area, a larger exit wound will result. Additionally, when the temporary cavity intersects a dense organ such as the liver, it will cause damage to that organ.

Many people believe that the lead cored ammunition was the initial design, which was then phased out and hence less likely to be encountered. However, in reality the lead-cored ammunition was a later development to increase wounding potential. Therefore it may well still be available in significant numbers.

4.4 Types of 7.62 x 39 mm Ammunition [18]

The 7.62 x 39 mm round is also referred to as 7.62 mm M43, 7.62 mm M1943, 7.62 x 39 mm Soviet, 7.62 mm short, 7.62 mm Kalashnikov and 7.62 mm Obr 43 g.

The 7.62 x 39 mm round can deliver bullets of different configurations for different purposes. These include: FMJ PS ball, AP (armour-piercing), AP – hard core, API (armour-piercing incendiary), subsonic, soft point and frangible types

4.5 Implications of Ammunition Type and Differences for Armour

The design and the variability of 7.62 x 39 mm ammunition have implications for body armour upgrade plates.

The first modern body armour upgrade plates consisted of a ceramic strike-face with some level of composite backing. This composite backing may be just a couple of layers, or may be a substantial thickness. The backing textile-based composite material may be glass-fibre reinforced plastic (GFRP), para-aramid (Kevlar®, Twaron® etc), or now more commonly ultra-high molecular weight polyethylene (UHMWPE, such as Dyneema®, Spectra® etc).

The ceramic strike-face breaks up or erodes the steel core of the bullet, as its hardness greatly exceeds that of the core. The kinetic energy of a typical 7.62 x 39 mm bullet as it leaves the muzzle of a typical weapon is approximately 2 kJ, whereas the muzzle kinetic energy of a typical 7.62 x 51 mm bullet is approximately 3.3 kJ. This means that armour that is capable of defeating the higher energy 7.62 x 51 mm ball round, should be capable of defeating the lower energy 7.62 x 39 mm ball round, thus negating the testing against the 7.62 x 39 mm round. For ceramic-faced armour this is, in fact, the case.

More recently however armour has been produced of monolithic ultra-high molecular weight polyethylene (UHMWPE) composite, for which the above rule no longer holds. If armour is designed to defeat the 3.3 kJ of the 7.62 x 51 mm NATO ball, it will not defeat the lower energy 7.62 x 39 mm PS ball, as it contains a mild steel, rather than lead, core.

For armour considerations the main two 7.62 x 39 mm ammunition types of interest are the PS ball and the API-BZ.

4.5.1 PS Ball

The 7.62 x 39 mm PS ball (figure 3) has been extensively described earlier, during which three main types have been identified:

1. Original Steel-Cored – This will be easily defeated by ceramic-faced armour designed to defeat 7.62 x 51 mm NATO ball (e.g. to HOSDB 2007 Level RF1 [2]), but will defeat monolithic UHMWPE armour of an areal density designed for the 7.62 mm NATO ball.
2. Later Steel-Cored (post-1989) – This should be easily defeated by a ceramic-faced plate designed to defeat 7.62 x 51 mm NATO ball (e.g. to HOSDB 2007 Level RF1), but will defeat monolithic UHMWPE armour of an areal density designed for the 7.62 mm NATO ball.
3. Lead-Cored – This should be easily defeated by either ceramic-faced or monolithic UHMWPE armour designed to defeat 7.62 x 51 mm NATO ball (e.g. to HOSDB 2007 Level RF1).

Monolithic UHMWPE armour can defeat the mild steel-cored ammunition if the thickness, and hence the areal density, of the armour is increased above that required for the 7.62 x 51 mm NATO ball.

https://doi.org/10.52202/078352-0034

However, if specifying UHMWPE it is critical to know exactly which factory and year of manufacture of PS ball the armour is being tested with.

The UK HO CAST has acknowledged both that the 7.62 x 39 mm PS ball is a realistic threat and that it also experiences significant variability as a test projectile. Therefore in their 2017 Body Armour standard they have included a surrogate PS ball projectile. This surrogate is discussed further during the discussion of this standard.

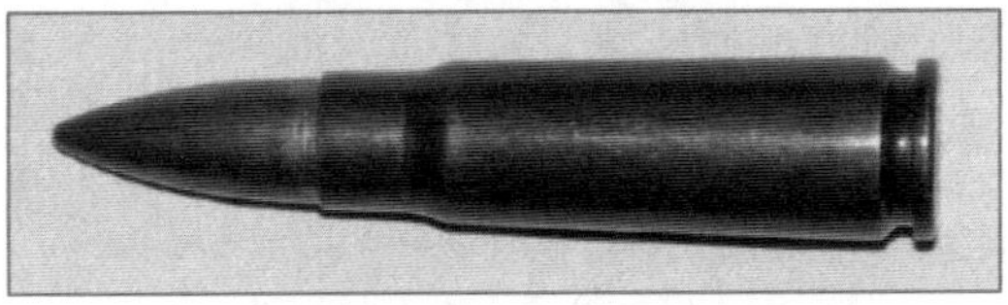

Figure 3. 7.62 x 39 mm PS Ball

4.5.2 API BZ [18]

The 7.62 x 39 mm API BZ contains a hardened steel core and also an incendiary component in the tip, and possibly in the tail, of the bullet. This incendiary ignites upon impact with a hard target. The incendiary ammunition is designed for the engagement of lightly armoured / soft-skinned vehicles, and particularly with the aim of igniting fuel tanks and systems. The API BZ is to the same dimensions as the PS ball and is ballistically matched. The API BZ is significantly more penetrative than a 7.62 x 39 mm PS ball, and also a 7.62 x 51 mm NATO ball. However it is not as penetrative as a 7.62 x 51 mm AP or the 30.06 AP M2 as found in the NIJ-0101.06 Level IV.

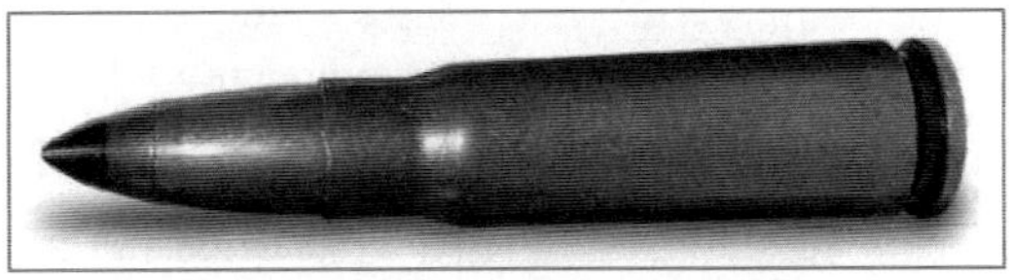

Figure 4. 7.62 x 39 mm API BZ

It should be noted that even 16 years ago, this API BZ ammunition was shown to be defeated by a simple ceramic-faced armour with an areal density of 34 kg/m^2. The armour tested was of a 1980s design and consisted of an alumina strike-face bonded to a para-aramid composite backing. This particular armour was designed to defeat the 7.62 x 51 mm NATO ball and the 5.56 x 45 mm SS109, but was known to have a significant safety margin.

For identification purposes the bullet tip of the API BZ has markings which comply with the standard Russian system. The tip is painted black with a red band just below the tip, as shown in figure 4. This paint is known to wear off with handling over time.

A recent test programme used 7.62 x 39 mm BZ API of a specific Russian factory and year. They were procured in factory-sealed tins. During the test process it became obvious that the projectiles within the sealed tin were of two different constructions. The bullets were sectioned to reveal two different constructions of steel AP core (figure 5), each of the same length (22.7 mm) and diameter (6.1 mm). The mass of the boat-tailed core was an average of 4.04 g and the one without the boat tail was 4.18 g. The two different core types were hardness tested (Vickers diamond indenter with 30 kg load), the results of which showed that the hardness of the two designs was consistent within experimental error at 788 to 811 VHN for the flat base and 811 to 812 VHN for the boat-tail.

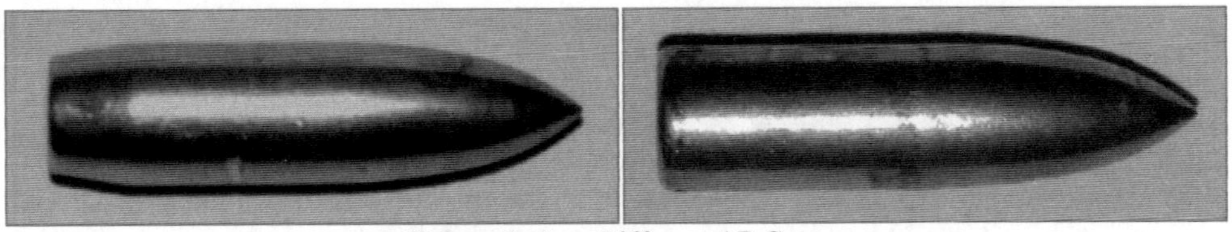

Figure 5. Two Different AP Cores

Testing has also been conducted using a surrogate for the BZ API, and this has been shown to significantly over-perform compared to typical worst case real threats, such as the Factory 539. The construction of the surrogate included an exposed core nose and it is suspected that this significantly modified the terminal ballistic effects versus certain armour constructions. More work would be required to confirm this. Any surrogate should only enter service as a test projectile, once it has undergone extensive validation trials versus a wide range of armour materials and constructions and in all typical test scenarios.

https://doi.org/10.52202/078352-0034

5. HO CAST 2017 BODY ARMOUR STANDARD [1]

July 2017 saw the arrival of the long-awaited replacement for the 2007 UK Home Office Body Armour Standard. This 2017 standard [1] is not a minor update, but a significant change in many aspects. One of the major changes is the ammunition types now included in the new series of test levels.

5.1 Standard Ammunition Categories

The standard ammunition categories are listed as HO1 through to HO4 and SG1.

Levels HO1 and HO2 are low velocity 9 mm ammunition levels. Level HO1 includes 9 mm FMJ and 9 mm JHP at 'handgun' velocities (365 ± 10 m/s) and level HO2 includes the same ammunition types at 'sub-machine gun' velocities (430 ± 10 m/s). Levels HO1 and HO2 allow for 44 mm BFS in the clay-based backing material. Levels HO1 and HO2 are met by the use of soft body armour.

Level HO3 includes the 7.62 x 51 mm NATO ball (L2A2 or L44A1) at 830 ± 15 m/s, along with the 7.62 x 39 mm surrogate projectile at 705 ± 15 m/s. Level HO3 allows 25 mm BFS. Level HO3 may be met by either ceramic-faced or monolithic UHMWPE upgrade plates. The 7.62 x 39 mm surrogate based upon a typical and higher performing Russian PS ball is referenced as PT11009 (manufactured by EPA Manufacturing Ltd). The surrogate use is an attempt to address the issues of both repeatability and realism.

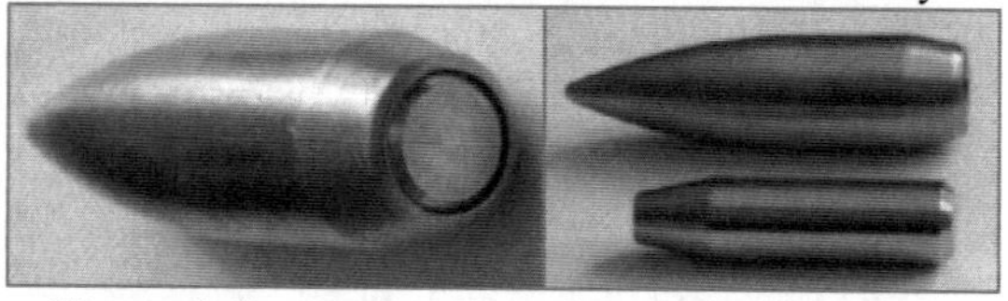

Figure 6. Surrogate – Complete and Components

Level HO4 includes two options of .308 ammunition, which are the SAKO .308 Win 480A Powerhead or the Barnes .308 TSX BT both at a velocity of 820 ± 15 m/s. Level HO4 allows 25 mm BFS. Level HO4 is expected to require a ceramic-faced upgrade plate.

Level SG1 includes the Winchester 1 oz rifled 12 RSE.

5.2 Optional Ammunition Categories

There are three optional ammunition types listed as follows:

- .357" Magnum Soft Point Flat Nose Remington R357M3 at either 390 or 455 ± 10 m/s
- 5.56 x 45 mm SS109 (L17A1 or L15A1) at 920 ±15 m/s
- 5.56 x 45 mm Federal Tactical Bonded (LE223T3) at 750 ± 15 m/s

5.3 Implication of Ammunition Choices

5.3.1 Levels HO1 and HO2

These levels include 9 mm ammunition of both FMJ and JHP designs. Both projectile types are of the same mass, and projected at the same two velocities. It may be expected that the most aggressive of these two ammunition types would be the FMJ, as the JHP is designed to expand on impact with the target. However, with some types of armour material, and particularly with the higher test velocity the JHP could be the most aggressive, as it punches a hole before it expands.

5.3.2 Level HO3

The HO3 level includes two ammunition types which are very different. The first is the 7.62 x 51 mm NATO ball, which equates to the RF1 level of the 2007 HOSDB standard. The second ammunition type is a surrogate of a 7.62 x 39 mm PS ball. This is the first use of a small arms bullet surrogate in a test standard. The surrogate was developed by Dstl and the Defence Academy of the UK.

It should be noted that a 7.62 x 39 mm surrogate projectile is also intended for inclusion in NIJ-0101.07 [17] which is due to be published during 2020. The indicated surrogate is included in both levels RF1 and RF2 and is a 7.81 g (120.5 grain) projectile with a specified impact velocity of 725 m/s.

5.3.4 Level HO4

Level HO4 includes two types of ammunition which are new to the body armour standard. These are the SAKO .308 480A Powerhead and Barnes TSX BT, both .308 calibre with 10.7 g (165 grain) solid copper expanding bullets, predominantly used for large deer hunting.

https://doi.org/10.52202/078352-0034

7. SUMMARY

There is often a trade-off between realism and repeatability with the types of small arms ammunition used for the testing of personal armour. 9 mm FMJ ammunition has been used for many years for testing of soft body armour within the UK and USA including in many law enforcement test standards. 7.62 x 39 mm ammunition is often a very realistic ammunition type, but also one which has introduced issues regarding repeatability. The use of a repeatable surrogate bullet in the 2017 HO CAST body armour standard is the first step towards addressing this trade-off. The 2017 HO CAST body armour standard has introduced a different focus on the ammunition being specified for testing body armour for UK police, specifically including UK Police ammunition to allow for blue-on-blue incidents.

References

[1] Payne, T; O'Rourke, S; Malbon, C; 'Body Armour Standard (2017)'; CAST Publication No. 012/17, July 2017

[2] VPAM APR 2006 (English); 'General basis for ballistic material, construction and product testing'; May 2009

[3] AEP-2920; 'Procedures for the Evaluation and Classification of Personal Armour; Bullet and Fragmentation Threats' Edition A Version 2; Sep 2016

[4] GOST R 50744-95; 'Armour Clothing'; July 1995

[5] Pettit, MJ; 'PSDB Ballistic Body Armour Standard (1993)'; PSDB Publication 12/93

[6] Pettit, MJ; 'PSDB Ballistic Body Armour Standard (1995)'; PSDB Publication 2/96

[7] Croft, J; 'PSDB Body Armour Standards for UK Police (2003), Pt 2 Ballistic Resistance'; PSDB Publication No 7/03/B

[8] Croft, J; Longhurst, D; 'HOSDB Body Armour Standards for UK Police (2007), Part 2: Ballistic Resistance' Publication No 39/07/B; July 2007

[9] Jones, AH; Barnes-Warden, J; 'Determination of the Cause of the Differing Ballistic Performance of 9 mm DM11 Bullets from Two Manufacturers' Proc PASS 2018; Sep 2018

[10] NILECJ-STD-0101.00; 'The Ballistic Resistance of Police Body Armor NIJ 0101.00; NILECJ; Mar 1972

[11] NILECJ–STD–0101.01; 'Ballistic Resistance of Police Body Armor, NILECJ–STD–0101.01'; Washington, D.C.: U.S. Department of Justice; December 1978

[12] NIJ Standard-0101.02; 'Ballistic Resistance of Police Body Armor'; US Dept of Justice; March 1985

[13] NIJ Standard-0101.03; 'Ballistic Resistance of Police Body Armor'; US Dept of Justice; February 1987

[14] NIJ Standard-0101.04; 'Ballistic Resistance of Body Armor NIJ Standard-0101.04'; US Dept of Justice; June 2001

[15] NIJ Standard-0101.06; 'Ballistic Resistance of Body Armor NIJ Standard-0101.06'; US Dept of Justice; Jul 2008

[16] Greene, ME et al; 'The Next Revision of the NIJ Performance Standard for Ballistic Resistance of Body Armor, NIJ Standard 0101.07: Changes to Test Methods and Test Threats'; PASS 2018; Sep 2018

[17] TR 03/2008; 'Technische Richtlinie Ballistische Schutzwesten'; Rev 10/2008

[18] Edited by Cutshaw, CQ and Ness, L; 'Jane's Ammunition Handbook 2003 -2004'

https://doi.org/10.52202/078352-0034

Techniques to assess the quality of armor performance measurement

D. Bourget, M. Bolduc

Defence Research and Development Canada – Valcartier Research Center, 2459 Pie-XI Blvd North
Québec (Québec) G3J 1X5, Canada, Daniel.Bourget@drdc-rddc.gc.ca

Abstract: Ballistic limit and proofing velocity tests are used for decades to assess the performance of armors in general. Many papers discussing the subject have been presented in the preceding PASS conferences concerning both: methods to efficiently execute testing and methods to calculate the ballistic limit and proofing velocity. This paper discusses about the later. Recent important changes in this field include the introduction of fitting techniques to calculate the ballistic limit using either Logit (NIJ 0101.06) or Probit (NATO STANAG 2920) link functions and the calculation of the uncertainties of the ballistic limit values. Different link functions and techniques to assess their Goodness of Fit towards the experimental data are discussed in this paper. In particular, the Wald test, the Likelihood Ratio test, the Anderson-Darling test, the Sensitivity/Specificity Analysis and the Information Criteria are discussed.

1. INTRODUCTION

Armour performance, against specific threats can be evaluated using ballistic limit (BL or V_{50}) and proofing velocity (V_{proof}) tests. While the V_{proof} test is used to assess if the armour actually protects against a threat and does not consider the residual capacity of the armour, the BL test enables the assessment of the absolute resistance of an armour against a threat and therefore, enables the ranking of armour performance. For this reason BL tests are specified in most ballistic standard along with V_{proof} tests. Up until recently, BL value was calculated based on the simple average of the impact velocity of 3 (or 5) partial and 3 (or 5) complete perforations of the armour within a specific velocity spread. Discussions within the different Standard committees showed that this simple way of calculating the BL value does not provide the full picture of the armour capability and that more information can be drawn out of the raw data. Amongst them, the estimation of the impact velocity corresponding to low probability of perforation (e.g. V_{05} for 5% probability of penetration and V_{01} for 1% probability of penetration) is of interest as these values are highly related to the protection capability of the armour and the V_{proof} test results. In addition, the uncertainty related to the V_{50}, V_{05} and V_{01} estimations are needed for further armour performance comparison purposes. To assess the parameters mentioned above, techniques consisting of fitting probabilistic curves through the pass/fail BL test data point were put forward and consequently, made their way to ballistic standards. Not a simple calculation anymore, BL test data exploitation necessitates techniques that need to be programmed to enable a robust and trustworthy evaluation. In close relation with the estimate of the V_{50}, V_{05}, V_{01} values and their uncertainties, the statistical significance and the goodness of fit of the model to estimate these values need to be assessed. This enables the analyst to determine how well the fitted model used in the fitting process corresponds to the reality of the experimental data.

To fulfill these needs, the BLC (Ballistic Limit Calculator), a MS Excel VBA program was created to:

a. Calculate the BL and V_{proof} related values,
b. Calculate the errors on these values, and
c. Assess the goodness of fit of the model, be it Logit, Probit or others, to the experimental data.

This paper describes the statistical aspect of the BLC and the different probabilistic models that are used within the software. It also provides examples of how to use the software for ballistic data analysis.

2. BALLISTIC AND STATISTICAL BACKGROUND

2.1 General

The Ballistic Limit is the velocity at which the probability of perforation of an armour is 50% and it is defined for a specific armour against a specific threat. It is calculated based on a series of shots fired at an armour following a pre-defined impact pattern for a range of impact velocity above and below the estimated BL value. The sequence of impact velocity follows the up and down procedure. It is well described in NATO STANAG 2920 (Edition 3) [1] and [2] and in the NIJ 0101.06 [3] documents. The

https://doi.org/10.52202/078352-0035

up and down procedure is used to minimize the number of shots necessary to determine the 50% probability value. For each firing, the dichotomous impact status (either a perforation (1) or a non-perforation (0)) is recorded.

Using the impact velocity as independent variable and the impact status as the dependent variable, the probability of perforation versus impact velocity is modelled using specific probabilistic "link functions" for dichotomous data. Amongst the large number of possible link function, the Probit and Logit cumulative distributions are the most often specified, [1] to [3]. The model parameters are defined using a curve fitting technique that uses Maximum Likelihood estimation (ML). Both, NIJ and NATO STANAG suggest using ML estimation and have described the general procedure ([1] and [3]). For research purposes the current version of BLC includes 3 other link functions: the Gompit distribution, the Scobit distribution and the Weibull distribution. As part of the BL and its variance estimation, it is necessary to assess how well the data are modelled by the different model functions. This assessment can be done using typical goodness-of-fit tests available for dichotomous dependent variables.

2.2 Link functions

The Probit link function, also called the cumulative normal distribution (Φ), has a mean μ and a variance σ^2. The Logit link function also has two parameters, α and β. Both link function probability of perforation distribution (Pp) equation for an impact velocity (V_S) can be found in Table **1**, 2nd column. Those link functions are symmetrical around their means (i.e their skewness is null) and they have very similar shape as can be seen in Figure 1. The Probit link density probability function is steeper in the middle and more quickly approaches zero on the left and on the right compared to the Logit link function [4].

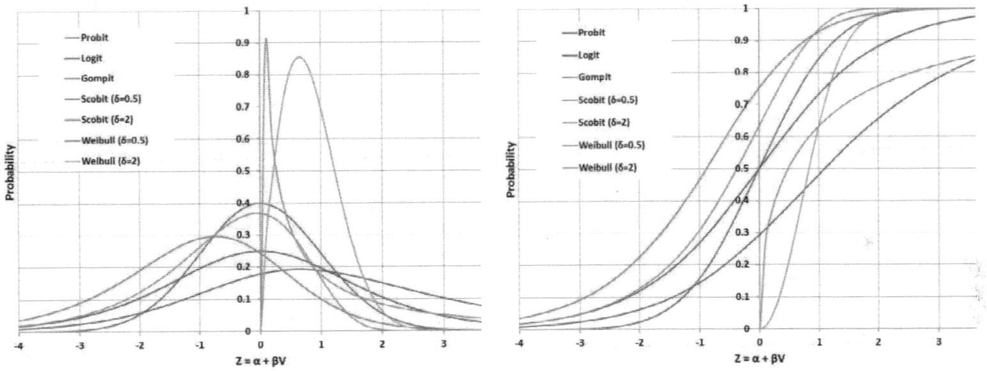

Figure 1 – Density probability (left) and cumulative probability (right) distribution curves for the Probit, Logit, Gompit, Scobit and Weibull distributions. For the Scobit and the Weibull distributions, curves for 2 different shape parameter values are shown (δ = 0.5 and δ = 2.0). Notice the difference in the slope around Z = 0 for the different cumulative probability distributions.

The Gompit link function is also called the cumulative complementary log-log (CLogLog) distribution. It has two parameters: α is the location parameter and β is the scale parameter (Table **1**). That link function is not symmetrical and has a fixed negative skewness of -1.1396 (skewed to the left). This means that its cumulative probability function is S-shaped and that it approaches 0 slowly but approaches 1 much more sharply. The Scobit link function [4] is also called the Burr-10 distribution or a Type I skewed-logit function. It has tree parameters: α is the location parameter, β is the scale parameter and δ is the shape parameter (Table **1**). The Logit distribution is a special case of the Scobit distribution: for $\delta = 1$, the Scobit and Logit distributions are equivalent. The Scobit link function is not symmetrical, except when $\delta = 1$. Contrary to the Probit, Logit and Gompit distributions, it does not have limitations in the amount of skewness it can model. The Weibull link function has tree parameters: α is the location parameter, β is the scale parameter and δ is the shape parameter (Table **1**). Like for the Scobit distribution, the Weibull distribution is not symmetric and does not have a fixed skewness value. The typical curve shapes for the different link functions presented above are shown

https://doi.org/10.52202/078352-0035

in Figure 1. The Scobit and Weibull distributions are presenting a variety of shapes and skewness depending on the shape parameter (δ) value. As concluded by Mauchant in [6] relative to the analysis of ballistic data using the Probit, Logit and Gompit link functions: "*This work shows that the choice of link function, between the Logit, the Probit and the complementary log-log link functions, is not the most important issue in V_{50} ballistic limit performance estimation, since the different GLMs (Generalized Linear Model) examined all gave similar results*". This might have been the case because the Probit, Logit and Gompit link functions have a constant shape parameter and a fixed skewness, which is not the case for the Weibull and the Scobit link functions.

Table **1** presents a summary of the information related to the different link functions presented above. For all link functions, the standard deviation σ of the distribution is equal to $1/\hat{\beta}$. The V_{50} value can be calculated directly from the estimated α, β and δ values following the formulas shown in the last column.

Table 1 – Summary of link function characteristics and V_{50} formula

Link function	Pp (at velocity V_S)	Pp scaling $(\hat{\alpha} + \hat{\beta}V_S) =$	V_S (for probability Pp)	V_{50} value
Probit	$\frac{1}{\sqrt{2\pi}/\hat{\beta}} \int_{-\infty}^{V_S} e^{-\frac{1}{2}(\hat{\alpha}+\hat{\beta}x)^2} dx$	$\Phi^{-1}(Pp)$	$(\Phi^{-1}(Pp)-\hat{\alpha})/\hat{\beta}$	$-\hat{\alpha}/\hat{\beta}$
Logit	$\left[1+e^{-(\hat{\alpha}+\hat{\beta}V_S)}\right]^{-1}$	$-\ln\left(\frac{1-Pp}{Pp}\right)$	$\left[-\ln\left(\frac{1-Pp}{Pp}\right)-\hat{\alpha}\right]/\hat{\beta}$	$-\hat{\alpha}/\hat{\beta}$
Gompit	$1-e^{-e^{(\hat{\alpha}+\hat{\beta}V_S)}}$	$\ln(-\ln(1-Pp))$	$[\ln(-\ln(1-Pp))-\hat{\alpha}]/\hat{\beta}$	$(-0.3665-\hat{\alpha})/\hat{\beta}$
Scobit	$\left[1+e^{-(\hat{\alpha}+\hat{\beta}V_S)}\right]^{-\hat{\delta}}$	$-\ln\left[\left(\frac{1}{Pp}\right)^{1/\hat{\delta}}-1\right]$	$-\left(\ln\left[\left(\frac{1}{Pp}\right)^{1/\hat{\delta}}-1\right]+\hat{\alpha}\right)/$	$-\left(\ln\left[(2)^{1/\hat{\delta}}-1\right]+\hat{\alpha}\right)$
Weibull	$1-e^{-(\lvert\hat{\alpha}+\hat{\beta}V_S\rvert)^{\hat{\delta}}}$	$[-\ln(1-Pp)]^{1/\hat{\delta}}$	$\left([-\ln(1-Pp)]^{1/\hat{\delta}}-\hat{\alpha}\right)/\hat{\beta}$	$\left([0.6931]^{1/\hat{\delta}}-\hat{\alpha}\right)/\hat{\beta}$

2.3 Curve fitting process

The fitting process consists of finding the estimates of the different distribution parameters (α, β and δ) that maximise the likelihood that the fitted curve corresponds to the experimental data (Maximum Likelihood Estimation, ML). For curve fitting purpose, the ML techniques should be equivalent to the Minimum Least Square (LS) method if the error distribution between the regressed curve and the experimental data points follows a normal distribution. For dichotomous dependent data, the error distribution is not normal and therefore both techniques are not equivalent, ML technique being optimal. The search for the optimal distribution parameters is done using an optimisation algorithms: the Downhill Simplex algorithm. It was published by Nelder and Mead [7] and uses a single-objective optimization approach (in our case, maximising likelihood) for searching the space of p-dimensional real vectors. The p dimensions correspond to the number of parameters in the link function. It only uses the values of the objective functions without any derivative information and therefore, it falls into the general class of direct search methods. It is therefore suitable for problems with non-smooth functions and discontinuous functions which occur frequently in statistics and experimental mathematics.

2.3.1 Maximum likelihood estimation

Briefly, the log-likelihood equation (LL) is the log of the likelihood equation (L) for M Bernoulli trials (dichotomous, i.e. 0 or 1 response, value 1 representing a success) with success probability $P_P(V_S)$. It can be written as follows:

https://doi.org/10.52202/078352-0035

$$LL\,(X|\alpha,\beta,\delta) = Ln\big(L\,(X|\alpha,\beta,\delta)\big) = \sum_{i=1}^{M} Resp_i\, Ln\big[P_P(V_{S_i})\big] + (1 - Resp_i)\, Ln\big[1 - P_P(V_{S_i})\big] \quad (1)$$

Where X are the experimental data points composed of $Resp_i$ (the perforation status, either 0 or 1) and the impact velocity V_{S_i}. Note that $P_P(V_s)$ in equation (1) is any of the link functions presented above.

The desirable properties of consistency, normality and efficiency of the ML estimators are asymptotic, i.e. these properties have been proven to hold as the sample size approaches infinity. This has consequences when considering small samples [8]:

i. The small sample behavior of ML estimators is largely unknown. There are no hard and fast rules for selecting sample size. It is risky to use ML with samples smaller than 100, while samples over 500 seem adequate.
ii. While the standard advice is that with small samples larger p-values should be accepted as evidence against the null hypothesis, given that the degree to which ML estimates are normally distributed in small samples is unknown, it is more reasonable to require smaller p-values in small samples.

3. TESTS TO ASSESS GOODNESS-OF-FIT OF MODELS

To assess the Goodness-of-fit of the models is at least as important as fitting the model itself as it provides an indication of how well the model describes the data. Some of the test presented below require an estimation of the Standard Error (SE) of the model parameters. Estimation of the SE of the parameters is also important for the determination of the error bounds of the different parameters. The SE of the parameters are never mentioned when BL values are published, although they should be, at least for quick Goodness-of-fit assessment purpose and error estimation purposes.

3.1 Standard error of the model parameters

Standard errors (SE) of the model parameters are calculated based on the estimated variance (Var, with $SE = \sqrt{Var}$) and covariance of the fitted parameters. The variance of a model with N parameters (β_i, i = 1 to N, in our case N = 3, α, β and δ) is given by equation (2). In BLC, once the best fit parameters for a link functions are defined using the ML technique, a numerical double derivative of the Log-Likelihood function (LL) is calculated for each parameters to define the variance/covariance matrix. For the V_{50} and σ values, which are function of α, β and δ as shown in
Table 1, the SE is calculated based on the arithmetic of error propagation described in [9], Section 1.4 as well as [10], [11] and [12].

$$\begin{bmatrix} Var_{11} & Var_{21} & Var_{31} \\ \dots & Var_{22} & Var_{32} \\ \dots & \dots & Var_{33} \end{bmatrix} = \begin{bmatrix} -\dfrac{\partial LL^2}{\partial\beta_1\,\partial\beta_1} & -\dfrac{\partial LL^2}{\partial\beta_2\,\partial\beta_1} & -\dfrac{\partial LL^2}{\partial\beta_3\,\partial\beta_1} \\ \dots & -\dfrac{\partial LL^2}{\partial\beta_2\,\partial\beta_2} & -\dfrac{\partial LL^2}{\partial\beta_3\,\partial\beta_2} \\ \dots & \dots & -\dfrac{\partial LL^2}{\partial\beta_3\,\partial\beta_3} \end{bmatrix}^{-1} \quad (2)$$

3.2 The Wald test

The Wald test is obtained by comparing the ML estimate of the parameter (say $\hat{\beta}$) to its standard error ($SE(\hat{\beta})$). The resulting ratio (W) follows a normal distribution under the assumption that $\beta = 0$ [13].

$$W = \frac{\hat{\beta}}{SE(\hat{\beta})},\ p - value = P(|z| > W). \quad (3)$$

If p-value $\leq \alpha_{lim}$ then we can reject the null hypothesis that $\hat{\beta} = 0$. The Wald test was found to behave in an aberrant manner often failing to reject the null hypothesis when the test was significant. It is therefore a conservative test to assess if the parameters are significant or not.

3.3 Likelihood Ratio test

https://doi.org/10.52202/078352-0035

This test consist of comparing the model likelihood with all its parameters ($\widehat{\alpha}, \widehat{\beta}$ and $\widehat{\delta}$) to the model without one of its parameters ($\widehat{\beta}$). For the purpose of comparing the models with and without the independent variable, the values of the Likelihood functions are compared using equation (4).

$$G = -2\ln\left[\frac{L\left(X|\widehat{\alpha},\widehat{\delta}\right)}{L\left(X|\widehat{\alpha},\widehat{\beta},\widehat{\delta}\right)}\right] \quad (4)$$

The test statistic G is called the Likelihood Ratio. Under the hypothesis that $\widehat{\beta} = 0$, the G statistic asymptotically follows a χ^2 distribution with 1 degree of freedom. When the sample size is small, the χ^2 distribution has to be corrected. Many different corrections exists, and reference [14] examined 5 of them in terms for Type I and Type II errors. The authors recommends the Bartlett correction. It is implemented in BLC and its details can be found in [15].

3.4 Anderson-Darling Test

The Anderson-Darling (AD) Test [16] enables the comparison between an assumed cumulative distribution function (a link function with its fitted parameters, $F(x,\theta)$) and the empirical distribution function (EDF, $F_n(x)$) based on the experimental data points. θ is a vector of the parameters ($\widehat{\alpha}, \widehat{\beta}$ and $\widehat{\delta}$). $F(x,\theta)$ is compared to $F_n(x)$ using the AD statistic [16]. The AD test can be used to detect the goodness-of-fit for the Probit, Logit, Weibull and exponential distributions. Due to its weight function, the AD test usually makes a more powerful test statistic by emphasizing the tail differences between the empirical distribution function and the assumed cumulative distribution function.

References [17] and [18] compared the power of different tests to the AD tests for small and large samples. They found that: a) the AD test results in excellent power levels compared to the other tests or it is marginally close to the best test, b) sample sizes above 50 can reach power levels above 80% and c) the increase of the allowable Type I error level results in higher power level for the same sample size. It is therefore advisable to accept higher Type I error (15 to 20%) in order to reach acceptable Type II error rates for the typical types of sample size used in ballistic studies.

3.5 Sensitivity, Specificity and Area Under the Curve (AUC)

The use of Sensitivity, Specificity and AUC tries to answer the following question: How good a job the model does of predicting outcomes? Or said another way: What percent of the observations the model correctly predicts? Can my model discriminates between positive and negative outcomes (between complete and partial perforations)?

These questions can be answered by measuring how good the model is to measure true positive and true negative answers while minimizing the number of false positive and false negative. All of these are calculated by comparing the model outcome to the sample measured outcome. First, here are some definitions:

a) **Sensitivity** (or true positive rate) is the proportion of positives (complete perforation) correctly identified by the model. This value has to be has high as possible.
b) **Specificity** (or true negative rate) is the proportion of negative (partial perforation) correctly identified by the model. This value has to be has high as possible. On contrary the value of 1-Specificity is the false negative rate, and it has to be as small as possible
c) **ROC**: Receivers Operating Characteristics. It is a plot of the model sensitivity versus the model false negative rate (1-specificity)
d) The **AUC** is a measure of how well a variable (in our case, impact velocity) can distinguish between two diagnostic groups (complete perforation/partial perforation). AUC is the area under the ROC curve.

Without getting into the details of how the sensitivity, specificity and AUC are calculated we can say that the AUC is at minimum 0.50 (model equivalent to chance) and at maximum 1.0. The higher the AUC between those two limits, the better the model is in predicting the outcome. Theoretically, the AUC is the probability that a randomly selected complete perforation velocity has a higher value (higher impact velocity) than a randomly selected partial perforation velocity.

3.6 The Information Criterion

https://doi.org/10.52202/078352-0035

The Log-Likelihood Ratio cannot be used to compare non nested models. In that case, the Information Criterion (IC) has to be used [19]. It is based on the information theory using the Kullback-Leibler information equation that measures the 'information' lost when approximating reality using a function. The IC accounts for how well the model fits the data and also accounts for the complexity of the model. Model complexity is its ability to fit any data set and can be approximated as the number of parameters in the model [19]. It is well known that some models can be so complex that they can fit any data set [19]. Details on the development of the IC can be found in [20], [21], [22] and [23]. The IC, called the AIC (Akaike Information Criterion [23]) can be calculated using the following: $\mathrm{AIC} = -2\mathrm{LL}(\mathrm{x}|\hat{\theta}) + 2\mathrm{p}$.

Variable p is the number of parameters in the model and $\mathrm{LL}(\mathrm{x}|\hat{\theta})$ is the Log-Likelihood value of the model (equation (1)) for the p estimated parameters ($\hat{\theta}$). For comparison between models, the AIC value as to be as small as possible. AIC is true asymptotically, i.e. when the number of data point is large. The rule of thumb from [19] is that for cases where sample sizes are small ($n < 100$) or where the number of free parameters is large ($p > 5$), the AIC value is biased and therefore it has to be corrected. For the for Logit and Probit link functions, a small sample empirical estimation is used [24]. For the Gompit, Scobit and Weibull link functions the Bootstrap method is used (section 8.3 of [21]).

4. HOW TO USE ALL THESE STATISTICS? (SO WHAT?)

The various outputs of BLC can be used to calculate the following:

1. Statistics for a sample (mean, standard deviation, median, 1st quartile, 3rd quartile, zone of mixed results, skewness, kurtosis, lowest complete and highest partial perforation)
2. Values and uncertainties of the V_{50} and the standard deviation (σ) of an armour for a specific confidence level
3. Based on the calculated uncertainty on the V_{50} and σ values, should more shots be done?
4. Extreme values of V_{50} and σ related to the NATO STANAG 2920 and the NIJ 0101.06 procedures as well as other error distributions
5. Values and uncertainties related to any probability of perforation (e.g. V_{01}, V_{05}, V_{95}, V_{99}) as well as for the V_{proof} value based on normal error and binomial error distributions
6. Identification of the model that most accurately represents the experimental data or: Is the selected model valid?

4.1 Sample statistics

Sample statistics are found under the heading “Sample Statistics” in BLC (Figure 2, part A) and in the “Box Plot” page (Figure 3).

https://doi.org/10.52202/078352-0035

Part		Logit		Probit		Scobit	
				SAMPLE STATISTICS			
A	Sample median (m/s) =			513.53			
A	Sample mean (m/s) =			513.53			
A	Sample σ (m/s) =			17.13			
A	Spread (m/s) =			84.00			
A	ZMR (m/s) =			16.60			
A	LC (m/s) =			505.30			
A	HP (m/s) =			521.90			
A	Skewness =			-0.32			
A	Kurtosis =			2.86			
A	Number of data points =			20			
	MONOBIT test p-value for sample randomness =			Cannot reject that data sample is random (p = 0.824)			
	RUN test p-value for sample randomness =			Cannot reject that data sample is random (p = 1.185)			
	Cumulative Sum test p-value for sample randomness =			Cannot reject that data sample is random (p = 0.918)			
				CURVE FIT STATISTICS for CI = 95%			
		Logit (NIJ 0101.06)		Probit (STANAG 2920)		Scobit	
B	V_{50} ± SE (m/s) =	515.25	± 3.8825	515.20	± 3.8295	517.24	± 0.7923
B	σ ± SE (m/s) =	6.6496	± 3.3908	10.9791	± 5.3795	0.0841	± 0.0001
B	α ± SE (m/s) =	-77.4866	± 24.5426	-46.9257	± 19.3767	-6260.7103	± 9.4536
B	β ± SE (m/s) =	0.1504	± 0.0478	0.0911	± 0.0378	11.8953	± 0.0179
B	δ ± SE (m/s) =					0.0064	± 0.0022
	Standard Error of Estimate at V_{50} =	0.4205		0.4202		0.4309	
E	Wald test p-value on α =	Reject that α is null (p = 0.001)		Reject that α is null (p = 0.008)		Reject that α is null (p = 0.000)	
E	Wald test p-value on β =	Reject that β is null (p = 0.001)		Reject that β is null (p = 0.008)		Reject that β is null (p = 0.000)	
E	Wald test p-value on δ =					Reject that δ is null (p = 0.002)	
E	Likelihood Ratio test p-value =	Reject that α and β are null (p = 0.005)		Reject that α and β are null (p = 0.005)		Reject that α, β and δ are null (p = 0.004)	
E	Osius-Rojek Goodness-of-Fit test p-value =						
E	Anderson-Darling Goodness-of-Fit test p-value =	Cannot reject that distribution is Logit (p = 0.382)		Cannot reject that distribution is Probit (p = 0.222)		Reject that distribution is Scobit (p = 0.000)	
E	Stukel test for fit of distribution tails p-value =	Logit model tails do not fit the data (p = 0.001)					
	ROC AUC =	0.8586		0.8586		0.8586	
	Model Sensitivity (Proportion of complete perforation correctly identified by the model)	0.7778		0.7778		0.7778	
	Model Specificity (Proportion of partial perforation correctly identified by the model)	0.9091		0.9091		0.9091	
	Threshold probability =	0.5100		0.5100		0.4400	
	Deviance =	18.7279		18.6390		18.2416	
	Pearson X^2 =	16.3956		16.2809		15.8247	
	Nagelkerke pseudo R^2 =	0.4761		0.4799		0.4968	
	Tjur pseudo R^2 =	0.3636		0.3669		0.3830	
	Log Likelihood =	-9.364		-9.320		-9.121	
	Least square =	1.7683E-01		1.7656E-01		1.8566E-01	
	Small Sample Corrected Information Criterion Bias =	2.409		2.223		2.932	
	Small Sample Corrected IC ± SD (m/s) =	23.5451	± 0.0000	23.0840	± 0.0000	24.1046	± 2.6762
C	V_{50} (m/s) CI length from STANAG 2920 method =	15.06		14.46		3.74	
C	V_{50} max (m/s) from STANAG 2920 method =	523.04		522.54		517.20	
C	V_{50} min (m/s) from STANAG 2920 method =	507.98		508.09		513.45	
C	Sigma max (m/s) from STANAG 2920 method =	50.97		83.44		0.20	
C	Sigma min (m/s) from STANAG 2920 method =	2.07		3.99		0.08	
C	V_{50} (m/s) CI length from Normal CI curves =	19.29		26.92		7.78	
C	V_{50} max (m/s) from Normal CI curves =	526.13		532.95		534.10	
C	V_{50} min (m/s) from Normal CI curves =	506.83		506.03		-	
				VPROOF STATISTICS			
				$V_{5\%}$ and CI = 95% Normal error distribution			
		Logit		Probit		Scobit	
D	Vproof (m/s) =	495.67		497.14		487.07	
D	Max prob of perforation (%) =	27.85		38.24		37.49	
D	Min prob of perforation (%) =	0.72		0.14		0.67	
D	Vproof max value (m/s) =	504.76		506.21		2199.77	
D	Vproof min value (m/s) =	-		-		502.90	

Figure 2 – Example of BLC output

4.2 Values and uncertainties of V_{50} and σ (are more shots necessary?)

Those statistics can be found for each model under the heading "Curve Fit Statistics", Figure 2, part B (Weibull and Gompit models are omitted for clarity). Clearly, the 3 models show similar V_{50} values while the value of σ varies significantly and the standard error also varies significantly. This difference is due to the varying shape and slope of the link functions as seen above (Figure 1). If unsatisfied with the V_{50} standard error obtained, another shot can be done and statistics recalculated to provide another estimate of V_{50} and σ with their SE.

4.3 Extreme values of V_{50} and σ

V_{50} and σ values as calculated using the NATO STANAG 2920 procedure can be found under the heading "Curve Fit Statistics" (Figure 2 Part C, data highlighted in green) for a specific confidence level (e.g. 95%). Maximum and minimum V_{50} values assuming a normal and a binomial error distribution can also be found. The difference between the STANAG 2920 and the normal distribution error as well as the description of other variables are presented graphically in Figure 4. It shows the fitted curve (e.g. Logit) in red along with its upper (light green) and lower (magenta) 95% percentile error curves based on normal distribution of the error. The confidence interval lengths, as defined in the STANAG 2920 are shown in blue, based on STANAG 2920 procedure and in yellow, based on the normal error curves.

https://doi.org/10.52202/078352-0035

As they are calculated in different ways using different distributions, the confidence interval lengths (CI), the maximum V_{50} values and the minimum V_{50} values are all different. Although not explicitly written, the V_{50} and σ values and their SE calculated using the NIJ 0101.06 procedure are highlighted in yellow in Figure 2 Part B.

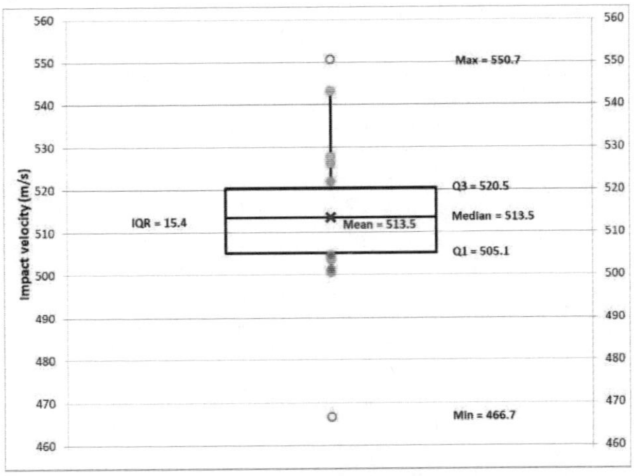

Figure 3 – Box plot page

4.4 Values and uncertainties related to any probability of perforation, including V_{proof}

Descriptions of the different values calculated by BLC are shown graphically in Figure 5. For a given probability of perforation (e.g. 10%), the associated impact velocity (V_{proof}), the minimum and maximum possible probability of perforation are displayed for either a normal error or a binomial error distribution. Similarly, the maximum expected V_{proof} value, based on the lower confidence limit curve is also displayed. These data can be found under the heading "V_{PROOF} Statistics", Figure 2 Part D. Notice that values are varying depending on the model used. Of particular interest in Figure 2 Part D is the error on the actual perforation probability. Although 5% probability is expected, the error of the model can result in probabilities of perforation as high as 38% for the Probit model with a 95% confidence level. This emphasises the use of V_{proof} values that corresponds to very low probability of perforation so that the maximum probability of perforation expected can be acceptable.

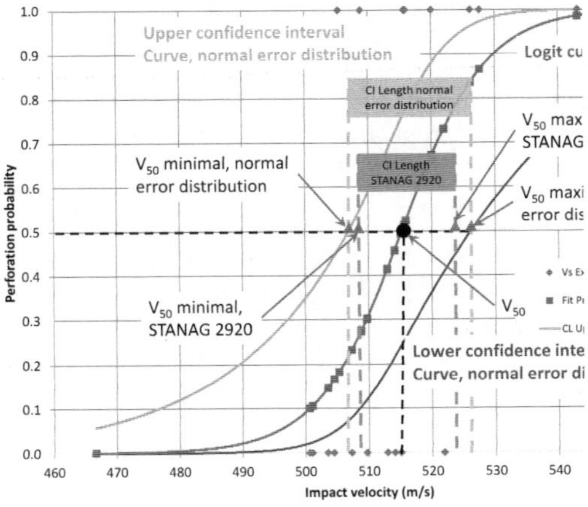

Figure 4 – Illustration of V_{50} statistics, extreme values and CI length

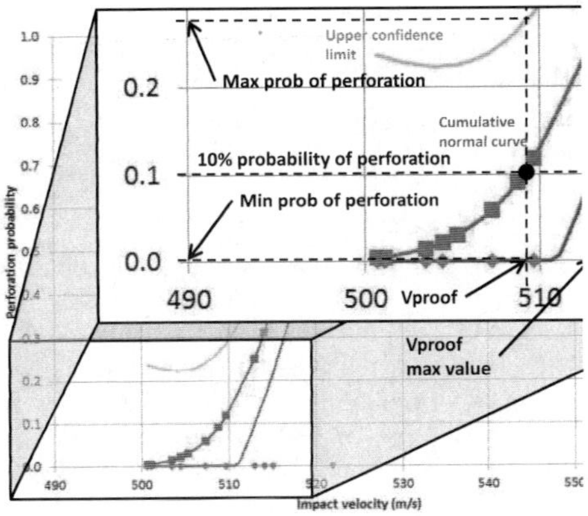

Figure 5 –Illustration of V_{proof} statistics with an example for 10% probability of perforation.

4.5 Identification of the best model or: Is the selected model valid?

https://doi.org/10.52202/078352-0035

This last assessment is a rather complex issue and necessitates the use of many statistics and tests described above. A flow chart of the process is presented in Figure 6.

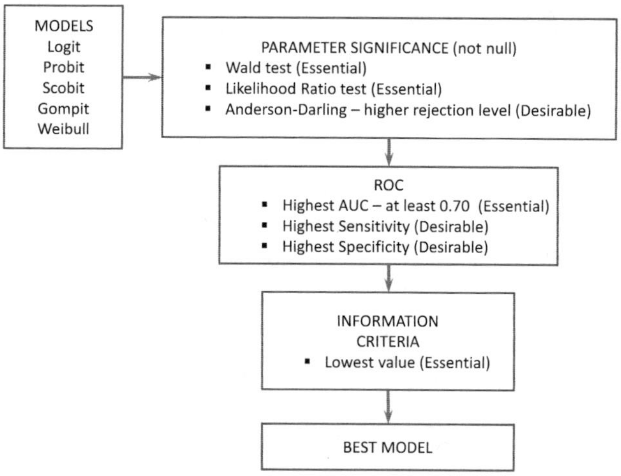

Figure 6 - Decision process to determine best model

To assess model validity, the parameter significance and the ROC criteria must be achieved. To assess which model best represents the experimental data, all 3 criteria (parameter significance, ROC and IC) must be achieved.

4.5.1 Parameter significance

The first thing is to verify if the calculated parameters ($\widehat{\alpha}, \widehat{\beta}$ and $\widehat{\delta}$) are significant. If one of them is null, the model is worthless. To verify that point, the Wald test and Likelihood Ratio results can be analysed. Each of the 2 or 3 parameters have to be significant using the Wald Test. This enables to determine if each parameter is significant (not null). The Likelihood Ratio test compares the full model to the constant only ($\widehat{\alpha}$ parameter only) model. The comparison enables to determine if the $\widehat{\beta}$ parameter is significant (not null). The use of the AD test to assess if the experimental data sample is a good fit to the ML fitted link function can be done keeping in mind the conclusion relative to the acceptable level in section 3.4.

4.5.2 Goodness-of-fit

Goodness-of-fit of the model can be verified using the ROC. The ROC AUC has to be as close as possible to 1.0. A qualitative scale for the ROC AUC value is presented in [9]. Also, [9] recommends that the ROC AUC value be above 0.70, which corresponds to at least an acceptable level of discrimination. Because they represent the proportion of complete and partial perforation correctly identified by the model, sensitivity and specificity of the model has to be as close as possible to 1.

4.5.3 Information Criteria

Finally, the lowest the IC value, the best the model is. Of all the criteria presented in this section, this is the only one that can be used directly to compare between the different models. The variance on the IC value can be quite large. Therefore, when compared to each other's, the difference between the values of IC can be not significant. Reduction of the variance on the bias estimate can be achieved by increasing the number of experimental data point or by providing a more accurate evaluation of the bias and its variance by increasing the total number of Bootstrap samples used to assess variance. This can only be done at the cost of longer execution duration.

4.5.4 Example

In the present case, the 3 models should be considered valid (Figure 2 Part E). All the parameters are non-null as shown by the Wald and Likelihood ratio tests. The ROC AUC is 0.86, which means that each model presents excellent discrimination [9]. Sensitivity and specificity are also high, which means that the number of partial and complete perforation correctly identified by each model is high.

Assessment of the best model is more difficult. The IC values for the Probit is the lowest of the 3 models. On the other hand, the Scobit model have a large standard deviation value and as a result, the observed difference in the mean IC values for each model is not significant.

https://doi.org/10.52202/078352-0035

5. CONCLUSIONS

Capabilities of the BLC for the assessment of validity and goodness of fit of different models for V_{50} and V_{proof} assessment is explored. A variety of tests are discussed, accounting for the effects of small sample size.

6. REFERENCES

[1] AEP 2920 – Procedures for the Evaluation and Classification of Personal Armour, Bullet and Fragmentation Threats, Edition A, Version 1, 22 June 2015, NATO Standardization Office, Brussels.

[2] STANAG 2920 – Classification of Personal Armour (EDITION 3), 22 June 2015, NATO Standardization Office, Brussels.

[3] NIJ Standard-0101.06 – Ballistic Resistance of Body Armour, July 2008, U.S. Department of Justice, Office of Justice Programs, National Institute of Justice.

[4] Nagler, J., Scobit: An Alternative Estimator to Logit and Probit, American Journal of Political Science, Vol. 38, No. 1, February 1994, pp 230-255.

[5] Brathwaite, T. and Walker, J.L., Asymmetric, closed-form, finite-parameter models of multinomial choice, Journal of Choice Modelling (2017), http://dx.doi.org/10.1016/j.jocm.2018.01.002 .

[6] Mauchant, D., Rice, K.D., Riley, M.A., Leber, D., Samarov, D., and Forster, A.L., Analysis of Three Different Regression Models to Estimate the Ballistic Performance of New and Environmentally Conditioned Body Armor, U.S. Department of Commerce, National Institute of Standards and Technology, NISTIR 7760, February 2011.

[7] Nelder, J.A. and Mead, R.A., A simplex method for function minimization. Computer Journal, Vol. 7, 308–313, 1965.

[8] Williams, R., Maximum Likelihood Estimation, University of Notre Dame, http://www3.nd.edu/~rwilliam/, Last revised January 14, 2016, Accessed 13 July 2016.

[9] Hosmer, D.W. and Lemeshow, S., Applied Logistic Regression, Second Edition, Wiley Series on Probability and Statistics, 2000.

[10] Ku, H.H., Notes on the Use of Propagation of Error Formulas, Journal of Research of the National Bureau of Standards - C. Engineering and Instrumentation, Vol. 70C, No.4, October- December 1966.

[11] Lorenz, V., Yang, L., Grosse Perdekamp, M., Hertzog, D., Clegg, R., Error Analysis, https://courses.physics.illinois.edu/phys403/sp2016/lectures/ErrorAnalysis_Lorenz2016.pdf . PHYS403 Course lecture, University of Illinois at Urbana, College of Engineering, Spring 2016.

[12] Heo, J.H., Salas, J.D, Kim, K.D., Estimation of Confidence Intervals of Quantiles for the Weibull Distribution, Stochastic Environmental Research and Risk Assessment, Vol. 15, 2001, pp 284-309.

[13] Watkins, J., Topic 15 – Maximum Likelihood Estimation, University of Arizona, http://math.arizona.edu/~jwatkins/o-mle.pdf , November 1 and 3 2011. Accessed 13 July 2016.

[14] Das, U., Dhar, S.S. and Pradhan, V., Corrected likelihood-ratio tests in logistic regression using small-sample data, Communications in Statistics - Theory and Methods, 2017.

[15] Tanizaki, H., Power Comparison of Empirical Likelihood Ratio Tests - Small Sample Properties Through Monte Carlo Studies, Kobe University Economic Review, Vol. 50, 2004.

[16] Stephens, M.A., The Anderson-Darling Statistic, Prepared under Grant DAAG29-77-G-0O31 For the U.S. Army Research Office, Technical Report No. 39, October 31, 1979

[17] Razali, N.M. and Wah, Y.B., Power comparisons of Shapiro-Wilk, Kolmogorov-Smirnov, Lilliefors and Anderson-Darling tests, Journal of Statistical Modeling and Analytics Vol.2 No.1, 21-33, 2011.

[18] Bispo, R., Marques. T.A. and Pestana, D., Statistical power of goodness-of-fit test based on empirical distribution function for Type-I right censored data, Journal of Statistical Computation and Simulation, 2011, pp 1-9.

[19] Cousineau, D. and Allen, T.A., Likelihood and its use in Parameter Estimation and Model Comparison, Mesure et évaluation en éducation, Vol 37, No. 3, pp 63–98, 2015.

https://doi.org/10.52202/078352-0035

[20] Anderson, D.R and Burnham, K.P., Understanding information criteria for selection among capture-recapture or ring recovery models, Bird Study, 46 (suppl.), 1999, pp S14-21.

[21] Konishi, S. and Kitagawa, G., Information Criteria and Statistical Modeling, Springer Series in Statistics, 2008, ISBN: 978-0-387-71886-6.

[22] Takeuchi, K., Distribution of informational statistics and a criterion of model fitting, Suri Kagaku (Mathematic Sciences), 153, 1976, pp.12-18 (in Japanese).

[23] Akaike, H., A new look at the statistical model identification. IEEE Transactions on Automatic Control, Vol. 19, No. 6, pp 716–723, 1974.

[24] Imori, S., Yanagihara, H., and Wakaki, H., General Formula of Bias-Corrected AIC in Generalized Linear Models, Scandinavian Journal in Statistics, Vol. 41,No. 2, pp 535–555, 2014.

https://doi.org/10.52202/078352-0035

The Development of the Chisel Nosed Fragment Simulating Projectiles for Personal Armour Testing

PL Gotts[1], M Helliker[2]

[1]Phil Gotts Consulting Ltd, 23 Thorney Rd, Capel St Mary, Ipswich, Suffolk, IP9 2HL, UK
phil@philgotts.com
[2]Policing and Security Group, Defence Science and Technology Laboratory, Porton Down, Salisbury, SP4 0JQ

DSTL/CP118995

Abstract. The testing of personal armour for fragmentation protection typically uses the chisel-nosed fragment simulating projectile (C-N FSP) of different masses. The relevance of these projectiles to modern conventional munition and improvised explosive device (IED) fragmentation has become of interest to the personal armour community in recent years. This is partly because the origin of the design of the C-N FSP has been all but forgotten, to most in the armour community. This paper describes the original development of the FSP at Watertown Arsenal, USA in 1943 and its evolution to its current forms, as specified in MIL-P-46593B and SCRDE A3/6723. From the 1970s onwards studies have been conducted to evaluate the C-N FSP when compared to both conventional munition and IED-formed fragmentation. The paper describes studies conducted within defence research laboratories in both the USA and the UK.

1. INTRODUCTION

In order to test personal armour against the fragmentation threat from exploding munitions, the obvious solution would be to detonate those munitions in an arena surrounded by personal armour worn by mannequins. However, this approach is not simple for a number of reasons:

- Detonating munitions in test arenas will not produce the same result from one detonation to the next, i.e. this test lacks consistency or repeatability.
- The results will only give a pass / fail result without any numerical assessment of the performance of the armour. This means that the test will not discriminate between armours which are borderline performance and those which are over-engineered.
- Of perhaps more relevance in the 21st century, the conduct of arena tests is an expensive process and hence not suited to the development of personal armour.

Therefore an alternative method of testing the performance of personal armour against fragmentation is required. This need for an easy, cheap, objective and repeatable test method led ultimately to the development of the fragment simulating projectile (FSP). There are now few people left in the armour community who are aware of the history of the FSP, and hence this paper aims to describe the history of the FSP as we know it today.

Sullivan [8] in 1945 summarised the limitations which led to the development of the FSP thus, 'Such a test ideally would consist of actual fragmentation of service projectiles, and such tests were promptly suggested by this laboratory, but, because of the inherent variability of fragmentation, the number of rounds, necessary to give results from which any valid conclusions could be drawn, discouraged the application of such methods, and shifted emphasis in the direction of a test that would be simple, reproducible and measurable.' *JF Sullivan, Watertown Arsenal, Oct 1945*

2. TYPES OF FRAGMENTS

The first question to address is what is the fragmentation that needs to be replicated by any FSP. Fragmentation caused during the detonation of a munition may consist of both what is known as primary and secondary fragmentation. There are a number of differences in how primary and secondary fragmentation can be described, tested against and protected against.

https://doi.org/10.52202/078352-0036

2.1 Primary Fragmentation

Primary fragmentation is that which is produced by the casing or components of the munition itself. Primary fragmentation from munitions, such as artillery shells, mortars and grenades can be considered to be one of two categories:

i) explosively formed 'natural' fragments or
ii) pre-formed fragments [1]

2.1.1 Natural Fragments

Figure 1. Selection of natural fragments from artillery shell (Photo: PGC)

Natural fragments, or explosively formed fragments, are formed when the shell casing of a munition fractures during an explosive detonation. For artillery shells, the overriding consideration in the design of the shells is the need to withstand the forces during launch. Often these requirements detract from the desire for the munition to produce as many fragments as possible (which is desirable to improve the probability of a hit on the target), whilst maintaining sufficient kinetic energy to cause incapacitation[1]. These integrity requirements reduce the efficacy of pre-formed fragments or shell scoring, both of which encourage more regular fragment shapes and distributions, but result in less structural strength during launch. During the detonation of the shell, the pressures within the case increase rapidly forcing the case to expand. Stress fractures in the case will form and the case will ultimately fragment, but in a fairly random pattern. These stress fracture patterns will vary due to the manufacturing of the shell casing. As a result, natural fragments vary in size, shape, mass and velocity. Figure 1 shows a few fragments from a 155 mm high explosive (HE) artillery shell. In Figure 2 the fragment distribution from a typical mortar is presented (by the region of the shell from which they originated). It can be seen that the fragments vary in mass and distribution due to the geometry and material properties of the shell.

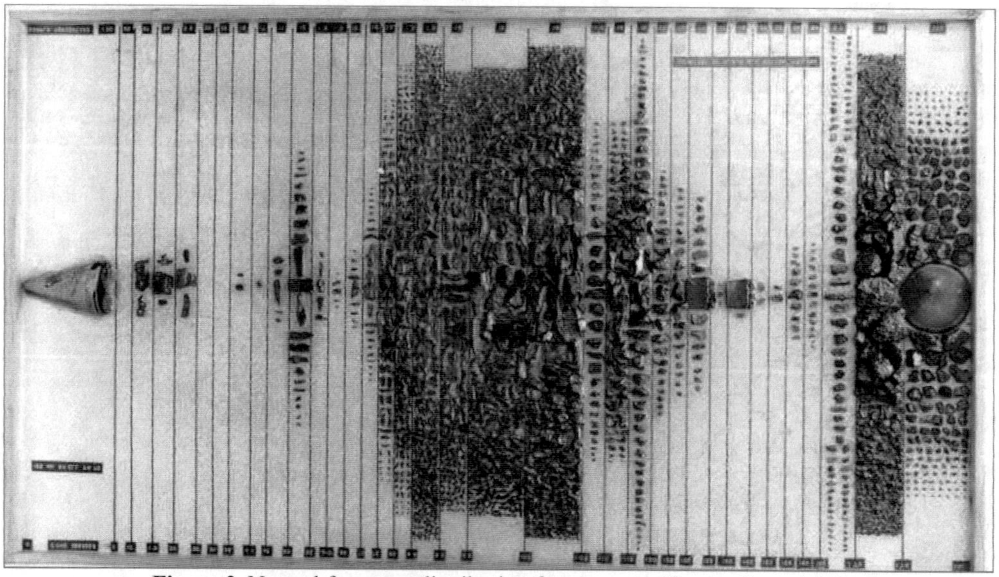

Figure 2. Natural fragment distribution from mortar (Crown Copyright)

[1] An energy of 80 J is often cited as being required 'to stop a man', but is more closely linked with the time in which incapacitation is required [1]

https://doi.org/10.52202/078352-0036

2.1.2 *Pre-Formed Fragments*

Weapons such as grenades, mortars, rockets and mines, do not need to withstand extreme forces during launch, and usually contain scored cases, internal components or pre-formed fragments. The first examples of pre-formed fragments were the development, by Lieutenant Henry Shrapnel, of steel balls within the munition (Figure 3), and these were referred to as 'shrapnel' [2]. Often also referred to as "proof shot", these steel spheres were used in the assessment of fragment protective equipment during World War I (WWI) and through to the end of World War II (WWII) [3].

Pre-formed fragments can take the form of notched wire, cubes, metal spheres or flechettes, although varying shapes and sizes have been used for specialist applications (such as anti-aircraft missiles).

Pre-formed fragments are common in many modern military munitions, but may also be found in many types of improvised explosive devices (IEDs).

Pre-formed fragments are designed to provide an optimum of fragment distribution and kinetic energy, in order to optimise hit probability and terminal effect.

It should be noted that even pre-formed fragments do not always reach the target in the designed geometry. For example, steel balls may fracture under detonation and notched-wire fragments may break into strips of 2, 3 or 4 fragments rather than always as individual fragments.

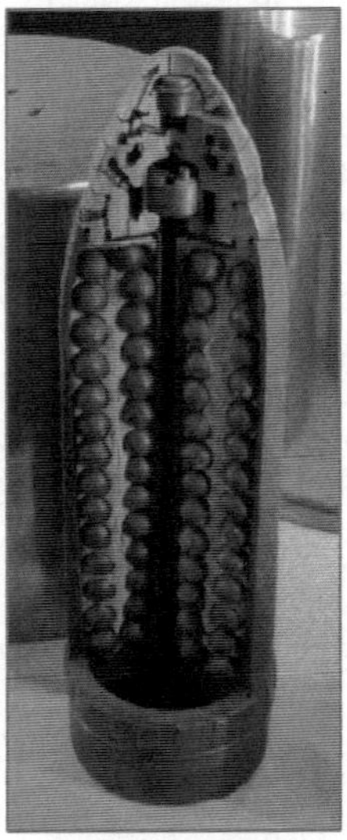

Figure 3. Sectioned WWI artillery shell showing shrapnel (Photo: PGC)

2.2 Secondary Fragmentation

Secondary fragments are those produced, not by the structure of the munition, but by the environment in which the explosive device detonates. The fragments tend to be part of the environment, which are accelerated to high velocity by the explosion. An example would be the sand and gravel in which an anti-personnel blast mine is buried (figure 4). There are some scenarios for which the definition of whether the fragments produced are primary or secondary fragmentation is not always clear, and may depend upon the focus of the interested party. One example of this is a vehicle-borne improvised explosive device (VBIED). In this case, most of the fragmentation is likely to be metallic components of the vehicle in which the device is contained. This fragmentation can be considered in two possible ways: the vehicle is the environment in which the detonation occurs (i.e. secondary fragmentation), or the vehicle is the casing of the device (i.e. primary fragmentation).

Figure 4. Secondary fragmentation from a buried anti-personnel mine (Courtesy: OTS Ltd)

3. TYPES OF FRAGMENT SIMULATING PROJECTILES (FSP)

When designing FSPs to test fragment protective materials and armours, testing agencies are posed several problems:

i) the need for a representative projectile which provides consistent results and
ii) an understanding of what velocity or energy level to test at.

In theory the first problem can be addressed by the design of a Fragment Simulating Projectile (FSP), which represents the threat to be countered by the armour. The velocity of fragments from any

https://doi.org/10.52202/078352-0036

given device, at a given range, can be calculated with a reasonable level of confidence from data obtained from arena trials.

For armour testing, a number of FSP types and sizes have been developed and used. These include chisel-nosed fragment simulating projectiles (CN-FSP), right circular cylinders (RCC), parallelepipeds, cubes, spheres and darts [2]. Through extensive analysis it has been shown that no one design or mass of FSP truly reflects the full range of fragments from all explosively formed fragmenting munitions [2, 4, 5, 6] and this is discussed later in this paper.

During WWI the first fragment protective equipment for general service was introduced in the form of the Steel Combat Helmet [3]. The effectiveness of candidate armours was assessed using steel shrapnel balls. As fragmenting artillery shells typically used these steel balls as pre-formed fragments, they were an obvious and simple choice for Fragment Simulating Projectiles (FSP). Other methods to assess the suitability of materials for providing protection from fragmentation included low velocity bullets, typically of .45 cal [7].

Through WWII and the subsequent Korean War, testing of materials and armours to protect both aircrew and later ground troops used a variety of FSPs, including parallelepiped, cubes, ball bearings, right circular cylinders and chisel-nosed FSP [2]. Whilst cubes, ball bearings and parallelepiped FSP were selected due to their similarity to pre-formed fragments, chisel-nosed FSPs, were developed to better represent natural fragments [8].

The original development of the chisel-nosed FSP was conducted at Watertown Arsenal in 1943, in order to aid the development of body armour for US aircrew. The initial requirement for FSPs to replicate natural fragments from German anti-aircraft flak and 20 mm HE shells started in September 1943, and three types of FSP were developed [7]. The chisel-nosed FSP (originally designated the G2 – see Figure 5) was selected for future development work, after initial trials on materials were reported in December of that year [7]. The original G2 FSP of 16 grains (1.04 g) was skirted to engage with a rifled barrel and had a Rockwell Hardness C (RHC) of 20 – 25, with some variation noted in manufacture (see Figure 6). This was later changed to 17 grains (1.10 g) with a RHC of 30 ± 2, to better reflect fragments from US munitions after WWII [4].

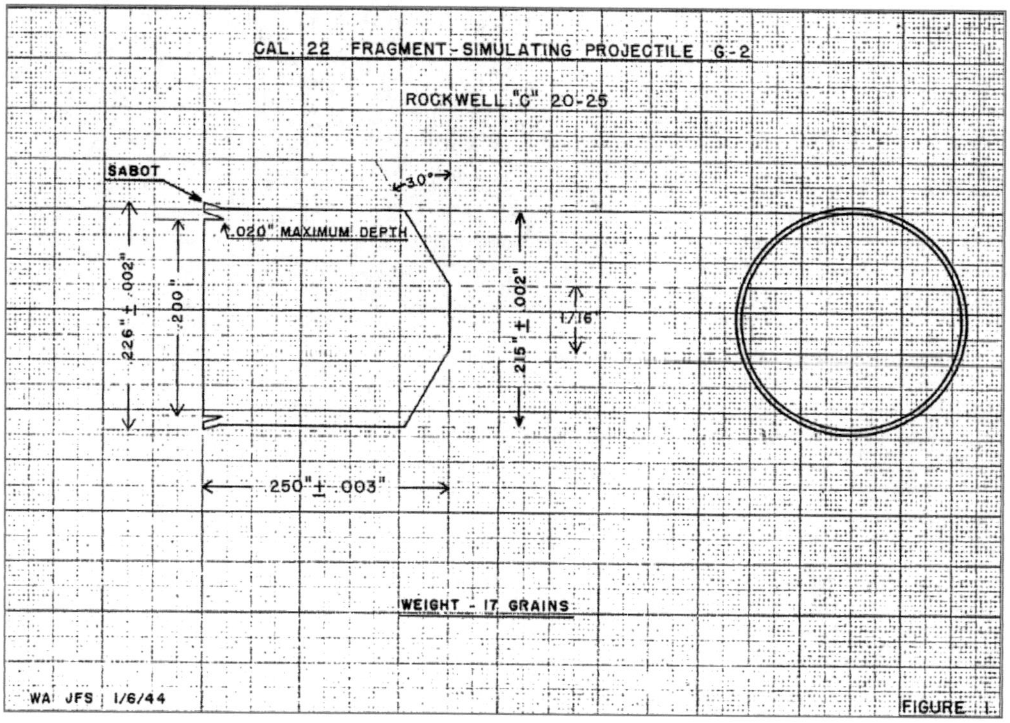

Figure 5. Original drawing of G2 FSP [9]

https://doi.org/10.52202/078352-0036

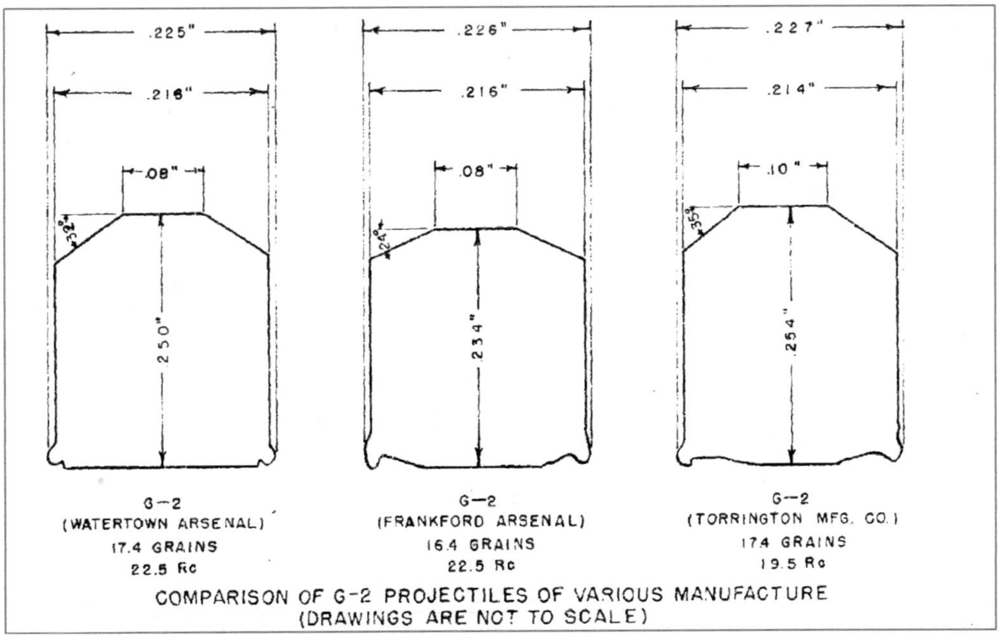

Figure 6. Comparison of G2 FSP of different manufacture [10]

The designation of the FSP changed to the T37, the design of which has not changed since 1948 and this forms the basis of the range of FSPs, which are scaled up versions of the original 17 grain projectile specified in MIL-P-46593B [11] (see Figure 7). The T37 FSPs are fired from an appropriate calibre barrel, which allows the skirt to engage in the rifling of the barrel. The 17 grain (1.10 g) T37, for example would be fired from a 5.56 mm barrel.

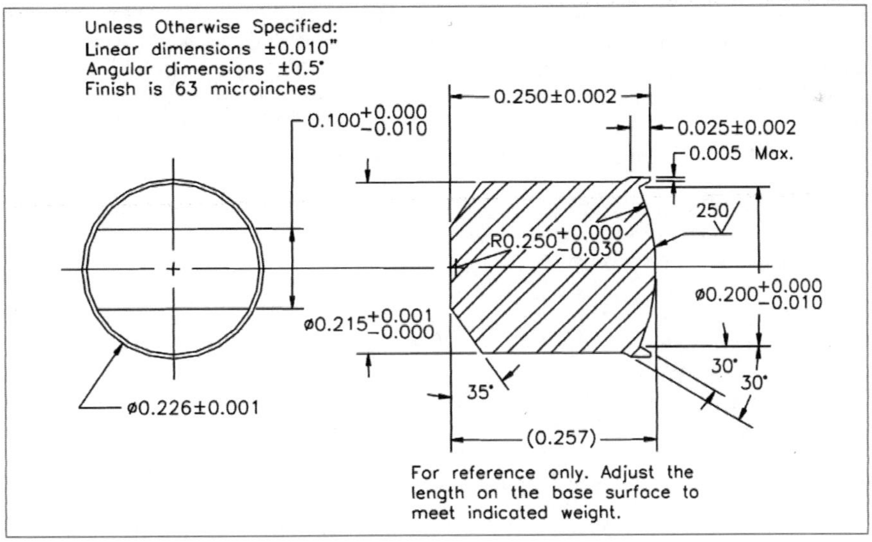

Figure 7. T37 FSP Types 1 and 2 [11]

The UK later adopted the same general projectile, but removed the skirt and scaled the FSP across a mass range from 0.16 g to 2.8 g, as specified in drawing SCRDE/A3/6723 [12] (see Figures 8 and 9). As the UK versions no longer had the skirt, they were fired from a polymer sabot which would engage in the rifling of the barrel. This meant that the FSPs were fired from a larger calibre barrel than the FSP. For

Figure 8. Typical chisel-nosed FSPs (Photo: PGC)

https://doi.org/10.52202/078352-0036

example, the 1.10 g FSP is fired from a sabot in a 7.62 x 51 mm barrel.

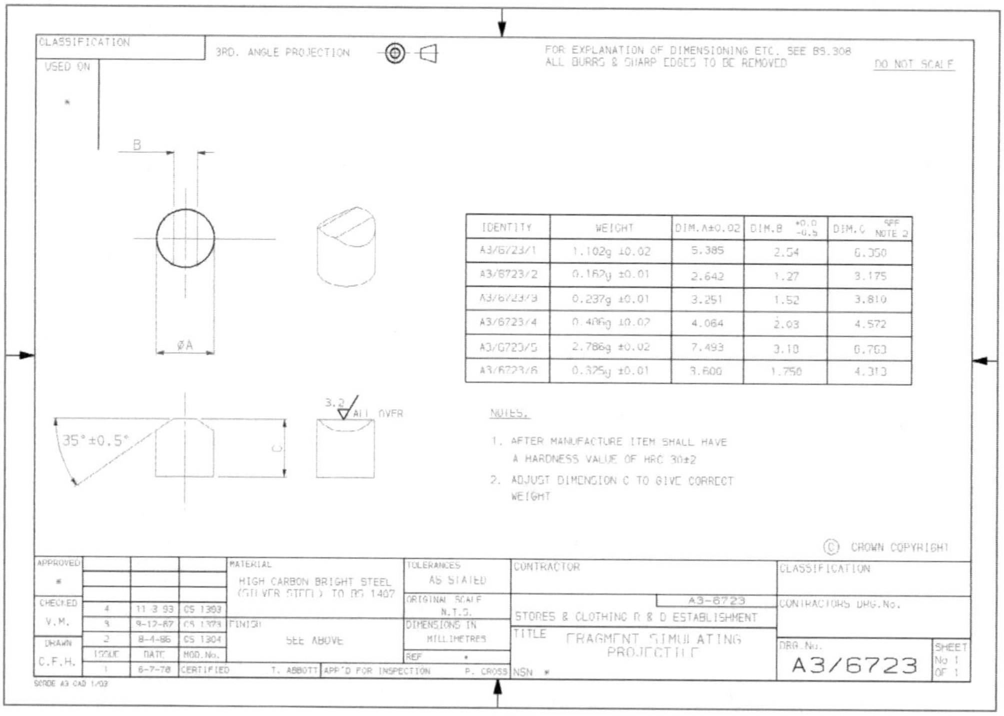

Figure 9. SCRDE drawing A3/6723 for Fragment Simulating Projectile (Crown Copyright)

The drawing in figure 9 above contains the phrase 'all burrs and sharp edges to be removed'. This phrase has been interpreted differently by different manufacturers of the FSPs. Many manufacturers tumble the finished FSPs together, while others do the same, but with addition of a grinding paste. The two types are easy to distinguish apart as those tumbled without the paste still retain a shiny metallic appearance whereas those tumbled with the paste will have a matt grey appearance. For certain textile materials it is known that the different sources of FSP can affect the test results. In these cases the textile material will exhibit a higher performance when tested with the FSPs tumbled in the grinding paste.

The current NATO test standard AEP-2920 [13] includes the options for a number of different FSPs within Annex C. these include: chisel-nosed FSPs to SCRDE drawing A3/6723, T37 type chisel-nosed FSPs to MIL-P-46593, and RCCs.

4. COMPARISON OF FSPs AND REAL FRAGMENTS

The original comparison of the performance of the G2 FSP against the fragments it was designed to replicate, showed a lack of correlation. Sullivan wrote in his report:

"While a disappointing lack of correlation has resulted between the relative resistance of materials to perforation by this projectile and to perforation by actual fragments of statically detonated 20 mm high explosive shells, due to the uniformly efficient manner in which this projectile perforates material and the variable inefficiency with which an actual fragment, because of its random behavior in flight, perforates material, it is believed that the results of firings with this projectile may at least be used to advantage as a basis for evaluation of the control of quality being exercised by the supplier of armor material." (Sullivan 1945 [8])

What Sullivan means by this very long and complicated sentence, is that although the FSP does not replicate the performance of the real fragments, it can be used to grade or rank materials, particularly for quality assurance purposes. The use of the G2 FSP in Quality Assurance testing ensured its continued use [8].

https://doi.org/10.52202/078352-0036

Few studies presented in the open literature compare the performance of FSP to natural fragments. Aside from the limited comparison work conducted during the development of the G2 chisel-nosed FSP, which did not present data comparing natural fragments to the FSP under development, most comparison work has remained under limited circulation.

By the early 1970s, a number of research laboratories within the US were involved in ballistic trials with various designs of FSP. In 1974 a number of the establishments conducted a coordinated research programme to compare and quantify the performance of various FSPs with real fragments of the type which posed a threat to dismounted troops [4]. In this study, fragments from a range of fragmenting munitions were analysed. A range of fragments of masses commensurate with a range of FSP masses were collected and fired at five different target types, representative of a range of materials considered suitable at that time for body armour applications. The natural fragments were collected using an arena trial and the fragments recovered from the strawboard witness packs. It should be noted that there is anecdotal evidence to suggest this method of "hard" recovery may result in blunting of the fragments, when compared to "soft capture" methods [14].

In the work by De Luca, the performance of the recovered natural fragments compared with RCC, chisel-nosed, cubes, spheres and parallelepiped FSP designs was conducted. The comparison included V_{50} and residual velocity assessments against titanium, nylon 728 fabric, XP polymer film and glass re-enforced plastic (GRP) targets [4].

The comparison between natural fragments (approximate mass of 1.04 g) and the chisel-nosed FSP (mass of 1.10 g) showed a significant difference in both V_{50} performance and residual velocity measurements. The reduction in V_{50} for the chisel-nosed FSP for the nylon target was approximately 5 – 10 % and for the XP polymer film approximately 15 – 20 %. The 16 grain (1.04 g) RCC with a 5 – 10 % decrease against the nylon and between a 5 - 10% increase against the XP film, proved a closer match. The study concluded that all fragment simulators (chisel-nosed FSP, RCC cubes and spheres) of aspect ratio 1:1 were more penetrative than real fragments of comparable mass. The report goes on to state that the family of RCCs with the same aspect ratio as the natural fragments provided the closest residual velocity data of any of the FSPs [4]. It is not possible to determine from these results whether the natural fragments were blunted in the capture process, or whether the disparity between the masses of the natural fragments and the chisel-nosed FSP would account for the difference in performance.

The results of the work conducted by De Luca were used as the basis for the recommendation that the RCC should be adopted for future testing of fragment protective body armour [15][2].

A later study, conducted at Cranfield University [14], compared the performance of natural fragments recovered from UK mortar shells to the CN FSP, and confirmed some of the findings identified by De Luca [4]. Three masses of CN FSP were compared to equivalent masses of natural fragments (0.24 g, 0.49 g and 1.10 g). This study also showed that the chisel-nosed FSP proved more penetrative than natural fragments of equal mass. This study compared the performance of natural fragments and chisel-nosed FSPs against textile armour typical of modern combat body armours.

Other research in the open literature has compared the performance of FSPs of differing geometries and materials. Work by Prosser compared the performance of chisel-nosed FSPs with RCC FSPs against woven fabric panels [16, 17]. In this study it was determined that the chisel-nosed FSPs were more penetrative than RCC FSPs of the same aspect ratio and mass. The cause of this difference in performance was due to the increased occurrence of fibre slip observed with chisel-nosed FSPs. This was ascribed to the chamfered edges of the chisel-nosed FSP enabling the pushing aside of the primary yarns instead of completely engaging them in tension. This resulted in less energy transfer between primary and secondary yarns and fewer broken yarns encountered with the chisel-nosed FSP. These are mechanisms that have been identified as contributing to the ability of a textile to dissipate energy from a projectile [18, 19]. Other studies have investigated the effect of projectile geometry on their ability to perforate armour systems (for example, Abbott [20], Gibbon [21], Ipson [5], Tan [19], Montgomery [22]).

A study on the effect of FSP material was conducted at Cranfield University, to determine if the material of the projectile played a significant role [23]. In this study, the spherical FSP of differing materials was investigated against textile armour panels of varying layers. It was noted in this study that the softer material (aluminium in this case) deformed upon impact and had a reduced residual velocity as a result. It was also observed that the deformed projectile on average broke more yarns than a spherical FSP of greater hardness. It was also noted that a glass FSP broke upon impact with the textile, although it was not possible to determine whether this affected its ability to perforate the textile target.

---

[2] A reference to the letter describing the recommendation has been identified by the author, but not the content. The reference is included as much as is possible for completeness.

https://doi.org/10.52202/078352-0036

5. DESIGN PARAMETERS RELEVANT TO THE CHOICE OF FSP

The choice of FSP for armour testing should be carefully considered, in order to represent the anticipated threat. The mechanisms by which the projectile interacts with the target should be replicated wherever possible. It can be argued that the choice of a projectile that has a greater ability to slip through a textile rather than fully engage with the fibres may lead to the development of textiles to prevent such slippage. There is then a risk that these textiles will possess other properties that are disadvantageous to the development of a body armour system. For example, to prevent fibre slip, a tighter weave may be employed to reduce the movement of the yarns around a projectile. This textile is likely to be stiffer with less flexibility which may adversely impact the comfort and movement of the wearer.

Of all of the designs of FSP, the RCC, CN FSP and the spherical FSP are the most commonly used in armour testing. The spherical FSP is often chosen due to its inherent stability of flight and low cost. The RCC and CN FSP can present more challenges in launching in a stable flight, particularly low mass variants or at low velocity. The RCC and CN FSP are manufactured specifically for armour testing, whereas spherical FSP are easily available in many materials and with differing properties.

From the summary of the research into the effect of FSP design on performance presented here, different simulators exhibit significantly different properties due to their geometries. The RCC presents a uniform edge in contact with the armour that minimises the amount of fibre slip, in comparison with CN and spherical FSP. There is evidence that this is a better representation of natural fragments of irregular shape where fibre slip would be minimal.

The CN FSP presents two chamfered edges, which allow more yarns to slip past the FSP, reducing the ability of the fabric to transfer energy away from the projectile. With the chamfered edges, the high kinetic energy density of the impact face due to its length to diameter ratio, make the CN FSP more severe than natural fragments of the same mass.

Spherical FSP have a much greater ability to slip through the weave of a fabric, potentially increasing the performance of the FSP against some armour materials. The ability of the spheres to slip through the fibres is affected by its surface finish and hardness. If a spherical FSP is chosen to replicate a preformed fragment (such as a sphere), it is important to consider the hardness and surface finish in the selection of the FSP, as these can have an impact upon the results. In comparison to natural fragments, the reduced kinetic energy density of the spherical FSP can offset the increased fibre slip from its curved surface and reduces the performance of the FSP.

Natural fragments are characterised as being irregular with sharp edges and are less able to push yarns apart than pre-formed fragments which may be rounded or have flat edges. Natural fragments from artillery shells will have a high hardness in comparison with some pre-formed fragments. Secondary fragments (such as stone and sand, or components of IEDs) may present other challenges such as being easily broken up upon impact or be of a low density. As fragments vary greatly in the range of properties discussed in this paper, it would not be possible to develop a single FSP representative of all fragments. Selecting unrepresentative FSPs in these scenarios presents a risk of developing inappropriate protection levels when designing protective systems.

Testing with a wide range of different FSPs can take a significant amount of time and hence be very expensive. With knowledge of the specific armour type to be tested and the realistic threat regimes, and consideration of the information discussed above, it is possible to narrow down the choices of FSPs to conduct the testing with.

6. SUMMARY

There has been little published work in the open literature comparing natural fragments with fragment simulators. Work to date suggests that the RCC FSP may present a better simulator of natural fragments, though there are limitations with this work. Careful consideration should be given to the design and material of any FSP for armour testing to ensure that accurate representation of the threat is made. The inappropriate selection of FSP may result in the introduction of unintended properties, that may have a negative impact upon aspects of an armour's design.

7. ACKNOWLEDGEMENTS

The authors would like to acknowledge Troy Shaver (US DoD) for sourcing some of the harder to obtain historical US technical reports.

https://doi.org/10.52202/078352-0036

References

[1] Courtney-Green PR. Ammunition for the land battle, volume 4. Exeter, UK: Brassey's; 1991.

[2] Laible RC. Ballistic materials and penetration mechanics: Elsevier Scientific Pub. Co.; 1980.

[3] Dean B. Helmets and body armour in modern warfare: Yale University Press; 1920.

[4] De Luca E, Frost RH, Prifti JJ. Terminal Ballistic Simulation of Munition Fragments. Watertown, Massachusetts 02172: Army Materials And Mechanics Research Center, Division PD; 1974 AMMRC TR 74-32.

[5] Ipson TW, Recht RF. Ballistic Perforation By Fragments Of Arbitrary Shape. Denver, Colorado, USA: Denver Research Institute, 1977 NWC TP 5927.

[6] Ipson TW, Recht RF, Schmeling WA. The Effect of Projectile Nose Shape Upon Ballistic Limit Velocity, Residual Velocity, and Ricochet Obliquity. Denver, Colorado, USA: Denver Research Institute, 1973 NWC TP 5607.

[7] Sullivan JF. Development of Projectiles to Be Used in Testing Body Armor, to Simulate Flak and 20 mm H.E. Fragments. Watertown, USA: 1943 Memorandum Report WAL 762/247.

[8] Sullivan JF. Resume of Programs on Development of Body Armor Undertaken at Watertown Arsenal during World War II. Watertown, USA: Watertown Arsenal, 1945 WAL Experimental Report 710/747.

[9] Sullivan JF. Development of Projectiles to Be Used in Testing Body Armor, to Simulate Flak and 20 mm H.E. Fragments. Watertown, USA: 1944 Memorandum Report WAL 762/253.

[10] Sullivan JF. Comparison of G-2 Projectiles of Various Manufacture. Watertown, USA: 1945 Memorandum Report WAL 762/314.

[11] Department of Defence. MIL-P-46593A - Projectile,Calibers .22, .30, .50 and 20 mm Fragment-Simulatlng. 1962. MIL-P-46593A.

[12] Ministry of Defence. Drawing A3/6723 - Fragment Simulating Projectile. 1993. A3/6723/3.

[13] North Atlantic Treaty Organisation. AEP 2920 Procedures for the Evaluation and Classification of Personal Armour - Bullet and Fragmentation Threats. 2015. AEP 2920 Edition A Version 1.

[14] Cant, D. S., Ashmore, A., Dray, J., et al. 2016. A comparison of fragment simulating projectiles and real fragments with respect to soft body armour performance. 29th International Symposium on Ballistics. Edinburgh, UK.

[15] Ballistic Research Laboratory. 1973. Letter subject: Ballistic Evaluation and Acceptance Testing of Fragmentation Protective Systems. Ballistics Research Laboratory. AMXBR-VL

[16] Prosser RA. Penetration of Nylon Ballistic Panels by Fragment-Simulating Projectiles1: Part I: A Linear Approximation to the Relationship between the Square of the V50 or Vc Striking Velocity and the Number of Layers of Cloth in the Ballistic Panel. Textile Research Journal. 1988;58(2):61-85

[17] Prosser, R. A. 1988. Penetration of Nylon Ballistic Panels by Fragment-Simulating Projectiles: Part II: Mechanism of Penetration. Textile Research Journal, 58(3):161-165.

[18] Roylance, D., Chammas, P., Ting, J., et al. 1995. Numerical Modeling Of Fabric Impact. Procedings of the National Meeting of the American Society of Mechanical Engineers. San Francisco, US.

[19] Tan, V. B. C., Lim, C. T. & Cheong, C. H. 2003. Perforation of high-strength fabric by projectiles of different geometry. International Journal of Impact Engineering, 28:201-222.

[20] Abbott, T. A. 1990. The variation of the geometry of fragment simulators. In: TOBIN, L. (ed.) The Ballistic Testing of Personal Armours. Colchester, UK: Stores and Clothing Research and Development Establishment.

[21] Gibbon, R. A. 2012. Effect of fragment morphology on the perforation of body armour materials. MSc., Cranfield University.

[22] Montgomery, T. G., Grady, P. L. & Tomasino, C. 1982. The Effects of Projectile Geometry on the Performance of Ballistic Fabrics. Textile Research Journal, 52(7):442-450.

[23] Steier, V., Carr, D. J., Crawford, C., et al. 2014. Effect of FSP material on the perforation of a typical body armor fabric. 28th International Symposium on Ballistics, 2:880-885.

https://doi.org/10.52202/078352-0036

Effect of Backing on Residual Armour Deformation

M. Bevan[1], C. Peitsch[1], J. Clark[1], D. Rose[1], D. Drewry[1], Q. Luong[1] and M. Maffeo[2]

[1]*The Johns Hopkins University Applied Physics Laboratory, Laurel MD, 20723 USA, Matthew.Bevan@jhuapl.edu*
[2]*US Army Combat Capabilities Development Command Soldier Center, Natick, MA 01760 USA*

Abstract. A key challenge in transitioning from a clay to force-based measurement of armour backface forces is identifying a suitable pad material to place between the armour and rigid force plate that allows for meaningful force measurements and for armour to deform as it would when coupled with a clay backing. This study focuses on comparing the residual deformation of ceramic armour when tested with a clay versus pad backing material, and on the effects of pad material and thickness on spatial distribution of forces. APM2 (.30 calibre) projectiles impacted both flat and curved ceramic armour plates backed with soft armour. The armour was tested on clay, as a baseline, and on a custom force plate fixture that measured the impact forces in two zones: at the centre (< 30 mm diameter) and outer (approximately 30-150 mm ring) regions. A polymeric pad was placed between the armour and force plate to dampen the force and allow the armour to deform. A wide variety of pad materials and thicknesses were evaluated. The mechanical properties of the pads were characterized using Dynamic Mechanical Analysis, allowing estimation of the strain rate effects on the material's storage modulus using time-temperature superposition. The depth, diameter and volume of residual armour deformation was measured for each impact. Test results showed that pad material and thickness has a significant effect on the residual armour deformation and peak forces measured. The pad material and thickness also affected the force distribution between the centre and outer regions. The flat tiles and curved ceramic plates produced different force and back face deformation measurements.

1. INTRODUCTION/BACKGROUND

Body armour ballistic test standards have been using clay to measure backface deformation as way to estimate the potential for Behind Armour Blunt Trauma (BABT) injury [1]. First established for testing soft armour against lower-velocity, handgun projectiles, the test method and acceptance criteria have been applied to thicker woven fibre, composite and ceramic armour impacted with heavier, higher-velocity projectiles, as well as armour piercing projectiles [2]. Variation of clay properties over the decades and the inherent instability of clay's mechanical properties [3] are a couple of factors motivating a consideration of transition from a clay standard to one that measures forces.

The focus of this study is identifying a suitable pad material to place in between ceramic armour system (including multilayer fabric backing, i.e. shootpack) and the force plate that allows armour to deform as it does on clay while also allowing for meaningful force measurements. For this series of tests, armour deformation was characterized by the maximum residual deformation depth, diameter and volume. Testing armour with different backing materials, including pads and clay, can produce a range of armour deformations (Figure 1). The goal of this project was to identify a pad material mounted on a force plate that permits the armour deformation, i.e. the depth, diameter and general shape, to match that for a ceramic tile/shootpack armour backed with clay when impacted with a .30 calibre APM2 at the rated velocity. If a similar armour deformation is achieved when backed with either clay or a pad, the force transmission is likely to be more similar than if the armour deformation is different.

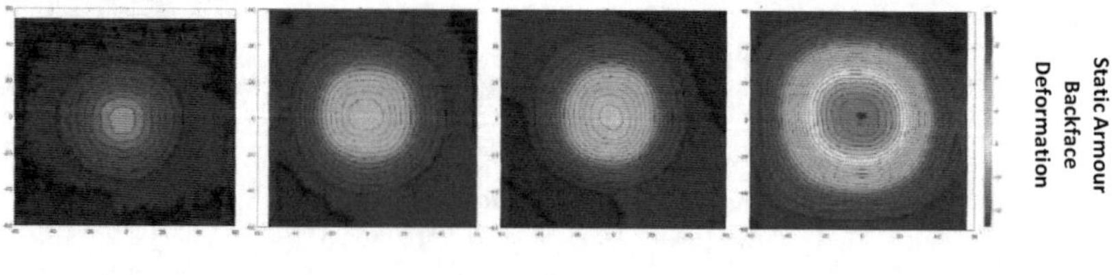

(a) (b) (c) (d)

Figure 1. Contour plot of armour deformation after ballistic impact with different backing materials: (a) high durometer neoprene, (b) Roma Plastilina Clay No. 1, (c) 30A durometer neoprene, (d) soft HD60 foam

https://doi.org/10.52202/078352-0037

2. EXPERIMENTAL METHOD

Ballistic testing was conducted on flat tile plates and curved, torso plates. Test series 1-3 assessed the behaviour of flat tile plates, while test series 4-6 evaluated the curved plates. For each series of tests, armour was first tested with a clay backing in accordance with NIJ test protocols to provide reference armour curvature to compare with the pad results [2]. The armour was then tested with various pad backing materials using a custom force plate test system, described below. The armour was impacted with .30 calibre APM2 projectiles at 853 (-0, +15) m/s in the centre of the flat tile and on the plate crown for the curved armour. Lessons learned from prior test series informed the materials selection and test matrix for each subsequent test series. In general, earlier test series had a lower sample size as the goal was to screen a large number of material combinations in order to choose the best performing pads for later test series. Details of the armour, impact measurement and pad materials follow.

2.1 Armour Tested

The flat tile plates were fabricated using St. Gobain SiC tiles (8.9 mm x 102 mm x 102 mm) mounted on 152 mm x 152 mm Dyneema HB80 ultrahigh molecular weight polyethylene (UHMWPE) panels. Flat tile plate armour areal density (without cover) was 41.5 kg/m^2. The shootpack tested was Kevlar KM2, 600 denier plain weave, 86 cm x 86 cm (end/picks), 28 plies with an areal density of 5.3 kg/m^2. The curved torso plates were previously fielded ceramic-faced armour provided by the US Army.

2.2 Pad Materials

Many combinations of materials and thicknesses were tested. The pad materials tested were selected for their mechanical properties and availability. For some tests, different thicknesses and layered combinations of pads were evaluated. Twenty-nine combinations were tested on flat plates, 11 combinations for curved plates. Each pad was impacted only once to ensure that the potential effect of damage was eliminated. Table 1 lists the armour and pad materials and thickness combinations tested in each test series.

2.3 Dynamic Mechanical Analyser Characterization

Since the pad mechanical properties were critical, the pads were tested using a dynamic mechanical analyser (DMA), TA instruments RSA G2, to measure the storage modulus, a measure of the elastic modulus. By making these measurements across a range of test frequencies and temperatures, the elastic properties at high strain rates, similar to those seen in ballistic testing, can be determined using time-temperature superposition. This measurement is important because the elastic properties of polymers measured at low strain rates may be considerably different if measured at high strain rates [4]. DMA measurements are made using small displacements and do not reflect the large amplitude behaviour seen in ballistic testing.

2.4 Clay Ballistic Testing

Clay ballistic testing was conducted following current NIJ-0101.06 methods to provide reference armour curvature to compare with the pad results [2]. The clay was preconditioned and drop-tested to insure its properties were within the range and mean limits. The armour was impacted with .30 calibre APM2 projectiles at 853 (-0, +15) m/s in the centre of the flat tile and on the plate crown for the curved armour. Each armour plate was impacted only once. For the flat tiles, the tiles and shootpack backing were placed against the flat surface. For the curved armour, a clay preform was used to assure contact of the clay and armour during testing. The clay was laser scanned before and after impact so that the depth, diameter and volume of the displaced clay could be measured.

https://doi.org/10.52202/078352-0037

Pad Material and Thickness	Durometer (Shore #)	Part Number and Supplier	Number of Flat Plate Tests			Number of Curved Plate Tests		
			Series 1	Series 2	Series 3	Series 4	Series 5	Series 6
Evazote: 25 mm	50 (OO)	Evazote VA-35		2				
Evazote: 51 mm	50 (OO)	Evazote VA-35		2				1
Sil. Rub./Neop. layers: 6.3 mm/38 mm	60(A)/30 (A)	Home Depot/Warco Biltrite	1					
Sil. Rub./Neop. layers: 13 mm/38 mm	30 (A)	Home Depot/Warco Biltrite	1					
HD30: 51 mm	72 (OO)	Plastazote HD-30		1				
HD60: 32 mm	78 (OO)	Plastazote HD-60		1				
HD60: 48 mm	78 (OO)	Plastazote HD-60		2				
Neoprene (Hard): 13 mm	70 (A)	Warco Biltrite Black 70A	1					
Neoprene (Hard): 25 mm	70 (A)	Warco Biltrite Black 70A	1					
Neoprene (Hard): 38 mm	70 (A)	Warco Biltrite Black 70A	1					
Neoprene (Hard): 51 mm	70 (A)	Warco Biltrite Black 70A	1					
Neoprene (Medium): 13 mm	50 (A)	Warco Biltrite Black 50A	1					
Neoprene (Medium): 25 mm	50 (A)	Warco Biltrite Black 50A	2					
Neoprene (Medium): 38 mm	50 (A)	Warco Biltrite Black 50A	2					
Neoprene (Medium): 51 mm	50 (A)	Warco Biltrite Black 50A	1					
Neoprene (Soft): 25 mm	30 (A)	Warco Biltrite Black 30A	1	2	5	6		
Neoprene (Soft): 38 mm	30 (A)	Warco Biltrite Black 30A	1		6	6	7	
Neoprene (Soft): 44 mm	30 (A)	Warco Biltrite Black 30A	1					
Neoprene (Soft): 51 mm	30 (A)	Warco Biltrite Black 30A	1	2	6		9	5
Neoprene (Soft): 76 mm	30 (A)	Warco Biltrite Black 30A	1					
Neoprene (Ultra-Soft): 51 mm	20 (A)	Warco Biltrite						4
Plastazote: 25 mm	61 (OO)	Pastazote LD-45		2				
Plastazote: 51 mm	61 (OO)	Pastazote LD-45		2				
Poron (XRD-20): 51 mm	N/A	Rogers Corp.						6
Poron (XRD-25): 51 mm	N/A	Rogers Corp.						2
Roma Plastilina No. 1: 140 mm	N/A	Roma Plastilina	5	4	5	6	7	4
Silicone: 25 mm	40 (A)	Sponsor Provided		2				
Silicone: 51 mm	40 (A)	Sponsor Provided		2			7	
Silicone – Soft: 51 mm	10 (A)	Stockwell SSP4749-10D						3
Soft/Firm Neoprene layers: 13 mm ea.	30 (A)/70 (A)	Warco Biltrite		2	5	6		
Soft/Firm Neo. Comp.: 25 mm ea.	30 (A)/70 (A)	Warco Biltrite		2	5	6	7	
Sorbothane: 25 mm	70 (OO)	Sorbothane 0266100-70-10		2				
Sorbothane: 51 mm	70 (OO)	Sorbothane 0266100-70-10		2	6			
VN600: 41 mm	N/A	Der-Tex VN600		2				

N/A: Not Available or Not Applicable

2.5 Force Plate Test Systems

The pad materials were tested on two test fixtures (Figure 2), one flat and the other curved. Both systems shared a common impact plate geometry of a centre 30 mm diameter impact cap, surrounded by a 152 mm diameter force ring whose shape matched that of the flat or curved armour. These two test fixtures were used to assess the effect of the pad on force distribution. The centre plate force measurement was collected by a single load cell while the outer ring was supported by four load cells whose measurements were summed. Data from the five (5) load cells (PCB Model 200C20) was collected at 1 $x10^6$ samples/s during impact and filtered to 50-kHz using a digital, 20-pole low pass filter.

2.6 Armour Deformation Analysis Methods

After ballistic impact testing, the armour was scanned using a 3D laser scanner (ROMER Absolute Laser Scanning Arm) to create a point cloud of the back surface of the armour. The point cloud was manually analysed using Geomagic Wrap & Geomagic Control software where the point clouds from an undamaged armour plate and impacted armour plate were compared. Measurements were made of the maximum residual deformation of the armour, average diameter of the deformation (an average of 3 diameter measurements) and volume displaced from undamaged armour.

https://doi.org/10.52202/078352-0037

(a)

(b)

Figure 2. Views of (a) flat and (b) curved plate test systems

3. RESULTS

3.1 DMA Results

DMA testing was conducted to characterize the strain rate sensitivity of the elastic modulus of the pads (Figure 3). In comparing the curves, the storage modulus of the different durometer neoprene materials at low strain rates were different, but becomes similar above $1x10^3$ strain/s. For reference, a 50 mm thick pad that is compressed at a velocity of 100 m/s is 2×10^3 strain/s. While Sorbothane and neoprene both become significantly stiffer above 1 strain/s, the Evazote and Plastazote foams did not exhibit the similar changes. Silicones dramatically change stiffness with strain rate and over a narrower range of strain rates than seen in neoprene and Sorbothane. Poron becomes stiffer at much lower strain rates than neoprene and Sorbothane.

3.2 Clay Backface Deformation Results

Testing was conducted with a clay (Roma Plastilina No. 1) backing to collect reference armour deformation measurements for comparison. Table 3 summarizes the measurements of flat tile and curved plate armour against clay. The flat tile clay BFD data for Series 2 testing were statistically significantly different ($p<0.017$) than data collected for Series 1 and 3. Despite the difference in clay BFD, the residual armour deformation for the three flat tile test series showed no statistically significant differences. This suggests that the difference in clay BFD had no effect on the armour deformation.

The clay deformation results are summarized in Table 4. There was no statistically significant difference in the clay BFD behind the curved plates and the flat tile test series 1 and 3. However, there was a statistically significant difference between the flat tile and curved plate armour deformation values ($p<0.0001$ for all armour values) for the three armour deformation parameters measured, with the difference in volume being the greatest, an increase of 180%.

https://doi.org/10.52202/078352-0037

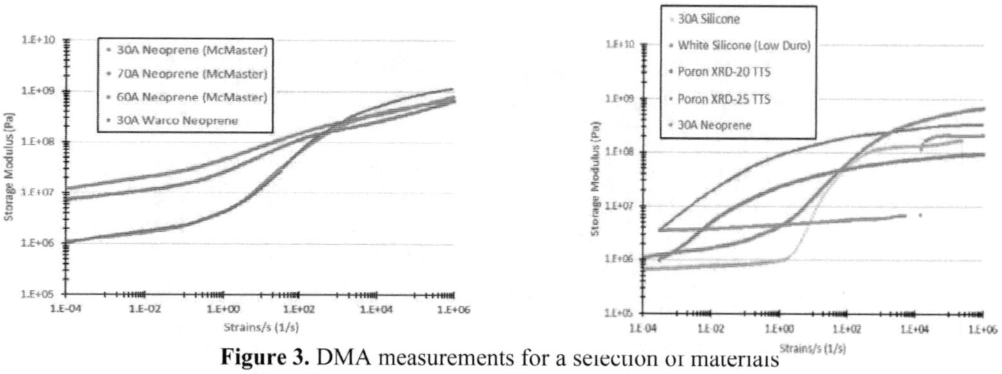

Figure 3. DMA measurements for a selection of materials

Table 3. Results of testing flat tiles and curved plate armour against clay

Characteristic	Flat Tile			Curved Plates		
	Series 1	Series 2	Series 3	Series 4	Series 5	Series 6
Avg. Clay BFD (mm)	31.8	26.7	33.6	32.1	33.0	31.4
Avg. Residual Armour Deformation (mm)	5.2	6.0	6.1	9.6	10.0	8.5
Armour Avg. Dia. (mm)	88.8	87.0	88.1	120.1	117.5	121.3
Avg. Armour Volume (mm^3)	10628	12416	12664	33396	31069	36711

Table 4. Summary of flat tile and curved plate armour results on clay

Measurement	Flat Avg.	Curved Avg.	Increase	p-Value
Clay BFD (mm)	31.0	32.3	4%	0.26
Residual Armour Deformation (mm)	5.73	9.49	66%	<0.0001
Average Diameter (mm)	88	119	35%	<0.0001
Volume (mm^3)	11866	33207	180%	<0.0001

3.3 Residual Armour Deformation Results

3.3.1 Flat Tile Armour - Clay and Pad Results

Figure 4 compares the residual armour deformation collected from flat plate testing on clay and force plate. A number of pad combinations showed results similar to those measured against clay. The similarity between clay and the different pad materials and thicknesses was calculated using Dunnett's method [5] to quantify the difference in armour deformation between the clay and pads. A value of 1 shows the two test conditions are similar and a value of 0 shows very poor similarity. Table 5 summarizes the average values measured and the similarity between clay and pad measurements.

The pads that best matched the flat tile armour deformation seen in clay were:

- Soft (30A) neoprene at 25 mm, 38 mm, 44 mm, and 76 mm thicknesses
- Soft/Firm layered neoprene, 25 mm thick
- Sorbothane, 51 mm thick

https://doi.org/10.52202/078352-0037

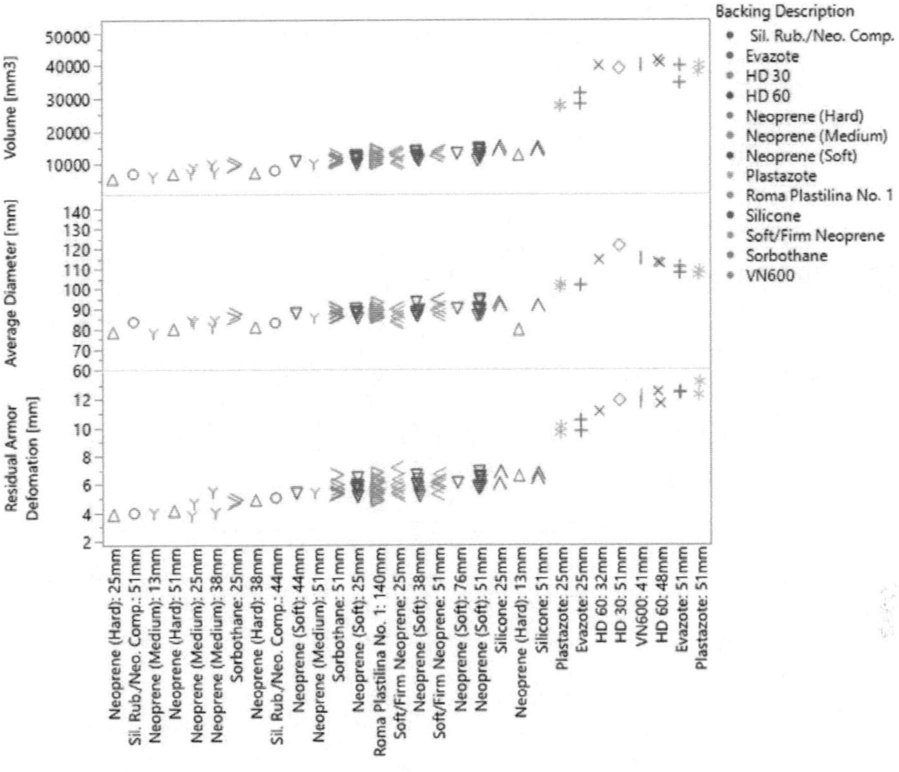

Figure 4. Residual depth of deformation, average deformation diameter and deformed volume results for flat tile armour, rank ordered by residual armour deformation

3.3.1 Curved Plate Armour - Clay and Pad Results

In Test Series 4-6, with curved plate armour, 86 tests were conducted against clay and 12 pad material and thickness combinations. The best pads identified in Series 1-3 were tested in Series 4 to determine how well they repeated the armour deformation behaviour seen over clay. The results differed significantly between Series 1-3 and 4. In Series 5 and 6, additional pads not tested in Series 1-3 were evaluated to improve the similarity in armour deformation between clay and pads.

Overall, the armour deformation seen on the curved plates over pads was greater than that seen on flat plates for clay and similar pads. However, Series 4 results showed that the pads that best matched armour deformation on the flat tiles produced much less deformation on the curved plate. The materials that produced too much deformation in the flat tile tested produced results that better matched clay results on the curved plate (Table 6). The results of the curved plate armour testing are summarized in Table 6 and Figure 5.

The armour deformation for the 51 mm Poron XRD-25 pad was the best match for the curved plate armour deformation seen when tested over clay. However, the conclusion is based on 2 tests, and additional tests may show that the Poron XRD-20 may perform similarly.

3.4 Force Plate Measurement Results

Force measurements were made during each ballistic impact. These measurements showed a wide range of force-time behaviours, examples of which are shown in Figure 6. From these measurements, peak centre force and peak total forces were extracted for analysis (Figure 7).

In examination of the plots, a number of observations may be made:

- The peak centre force was always greater for the curved armour plate than the flat tile.
- The peak total force was generally, but not always, less for the curved armour plate than the flat tile.

 https://doi.org/10.52202/078352-0037

- Foam material generally had smaller centre and total forces.
- The peak total and centre force generally decreased with increasing pad thickness. There were a couple of exceptions that were probably due to test variability and the limited number of tests conducted at a given thickness.
- The ratio of centre to total force was greater for the curved plate.

Table 5. Comparison of armour deformation of flat tiles (Series 1-3) backed by clay and pads. Value of 1 is very similar; value of 0 is very poor similarity. Highlighted lines were the best match with clay results

Backing & Thickness	Deformation Depth (mm)		Avg. Diameter (mm)		Volume (mm^3)		Mean Similarity
	Mean	Similarity	Mean	Similarity	Mean	Similarity	
Roma Plastilina No. 1: 140 mm	5.7	1.00	88	1.00	11866	1.00	1.00
Neoprene (Soft): 25 mm	5.7	1.00	88	1.00	11433	1.00	1.00
Neoprene (Soft): 38 mm	5.9	1.00	89	1.00	12410	1.00	1.00
Neoprene (Soft): 44 mm	5.4	1.00	88	1.00	10857	1.00	1.00
Soft/Firm Neoprene: 25 mm	5.9	1.00	87	1.00	11862	1.00	1.00
Neoprene (Soft): 76 mm	6.1	1.00	90	1.00	13112	1.00	1.00
Sorbothane: 51 mm	5.6	1.00	88	1.00	11282	1.00	1.00
Neoprene (Medium): 51 mm	5.4	1.00	86	1.00	10000	0.975	0.99
Soft/Firm Neoprene: 51 mm	6.1	0.989	91	0.354	12992	0.716	0.69
Neoprene (Hard): 13 mm	6.6	0.962	79	0.007	12450	1.000	0.66
Sorbothane: 25 mm	4.8	0.441	86	1.000	9606	0.355	0.60
Sil. Rub. / Neo. Comp.: 44 mm	5.1	0.998	83	0.522	7928	0.073	0.53
Neoprene (Soft): 51 mm	6.2	0.543	90	0.484	13212	0.272	0.43
Silicone: 51 mm	6.6	0.566	92	0.572	14526	0.135	0.42
Silicone: 25 mm	6.4	0.848	92	0.283	14723	0.077	0.40
Neoprene (Hard): 38 mm	4.9	0.944	81	0.045	7012	0.008	0.33
Sil. Rub. / Neo. Comp.: 51 mm	4.0	0.056	84	0.703	6925	0.007	0.26
Neoprene (Medium): 38 mm	4.8	0.362	83	0.044	8571	0.019	0.14
Neoprene (Medium): 25 mm	4.2	0.011	84	0.296	7838	0.001	0.10
Neoprene (Hard): 51 mm	4.1	0.116	80	0.012	6706	0.004	0.04
Neoprene (Medium): 13 mm	4.0	0.064	78	0.001	6108	0.001	0.02
Neoprene (Hard): 25 mm	3.9	0.031	78	0.001	5282	<.0001	0.01
Evazote: 25 mm	10.1	<.0001	102	<.0001	29927	<.0001	0.00
Evazote: 51 mm	12.5	<.0001	110	<.0001	37356	<.0001	0.00
HD30: 51 mm	11.9	<.0001	121	<.0001	38963	<.0001	0.00
HD60: 1.25 in	11.1	<.0001	114	<.0001	39835	<.0001	0.00
HD60: 1.875 in	12.1	<.0001	113	<.0001	41102	<.0001	0.00
Plastazote: 25 mm	9.8	<.0001	102	<.0001	27632	<.0001	0.00
Plastazote: 51 mm	12.7	<.0001	108	<.0001	38928	<.0001	0.00
VN600: 1.626 in	12.0	<.0001	115	<.0001	40191	<.0001	0.00

Table 6. Results of Series 4-6 testing on curved plates. Poron XRD-25 was the best match with clay results. Boxes highlight how poorly the curved plate results matched the flat plate results

Backing & Thickness	Deformation Depth (mm)		Average Diameter (mm)		Volume (mm^3)		Series 1-3 Similarity Mean	Series 4-6 Similarity Mean
	Mean	Similarity	Mean	Similarity	Mean	Similarity		
Roma Plastilina No. 1: 140mm	9.5	1.00	119	1.00	33207	1.00	1.00	1.00
Poron (XRD-25): 51 mm	9.1	1.00	123	0.986	28158	1.00	Not Tested	1.00
Poron (XRD-20): 51 mm	10.2	0.952	113	0.435	39951	0.97	Not Tested	0.79
Silicone (White): 51 mm	8.3	0.851	108	0.238	22824	0.79	Not Tested	0.63
Evazote: 51 mm	12.2	0.468	120	0.115	52694	1.00	0	0.53
Neoprene (Ultra-Soft): 51 mm	7.4	0.101	106	0.114	22633	0.48	Not Tested	0.23
Soft/Firm Neoprene: 51 mm	7.6	0.009	110	0.007	23400	0.44	0.69	0.15
Neoprene (Soft): 51 mm	8.2	0.195	101	0.017	23999	0.003	1.00	0.072
Neoprene (Soft): 38 mm	7.9	0.060	105	0.005	22681	0.065	1.00	0.043
Neoprene (Soft): 25 mm	7.2	0.014	97	0.001	18581	0.005	1.00	0.006
Silicone: 51 mm	7.0	0.005	97	0.0002	17219	0.007	0.42	0.004
Soft/Firm Neoprene: 25 mm	6.2	0.0003	93	0.0001	13578	0.005	1.00	0.002

https://doi.org/10.52202/078352-0037

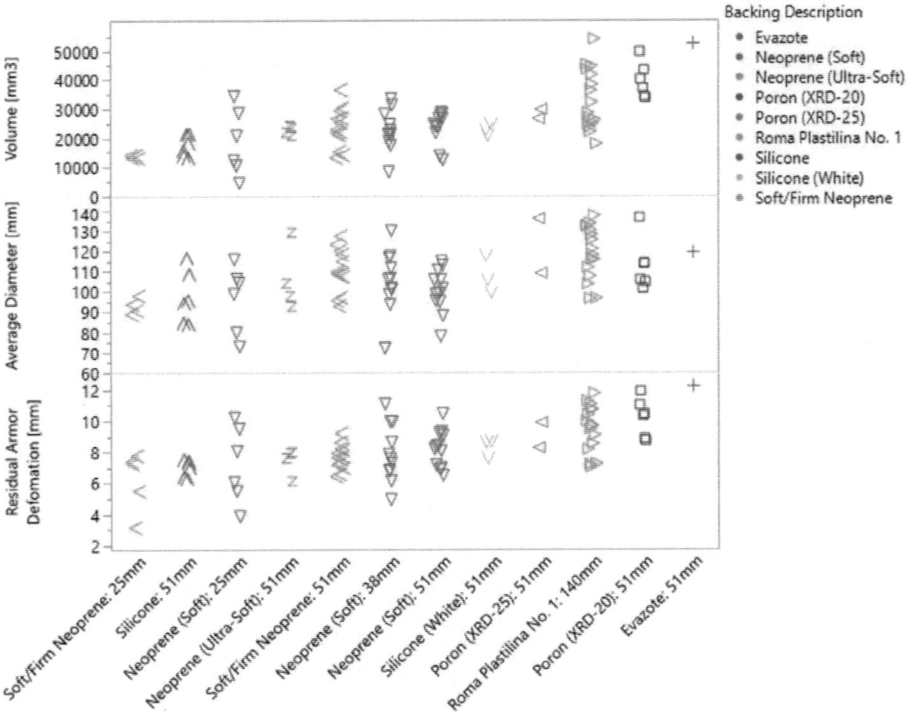

Figure 5. Comparison of curved plate armour deformation on clay and pads

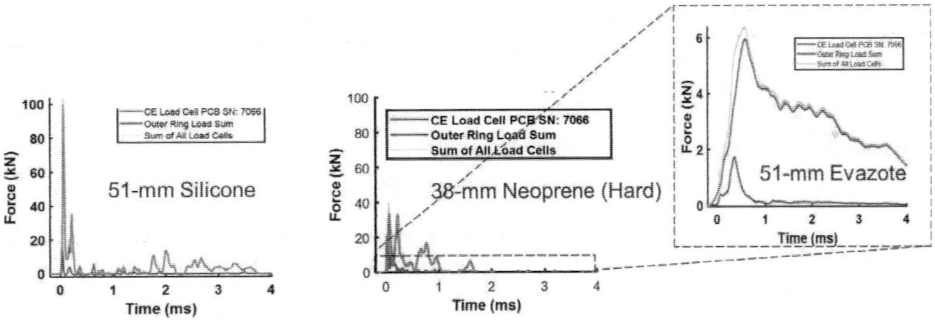

Figure 6. Examples of force-time plots measured behind different pad materials

https://doi.org/10.52202/078352-0037

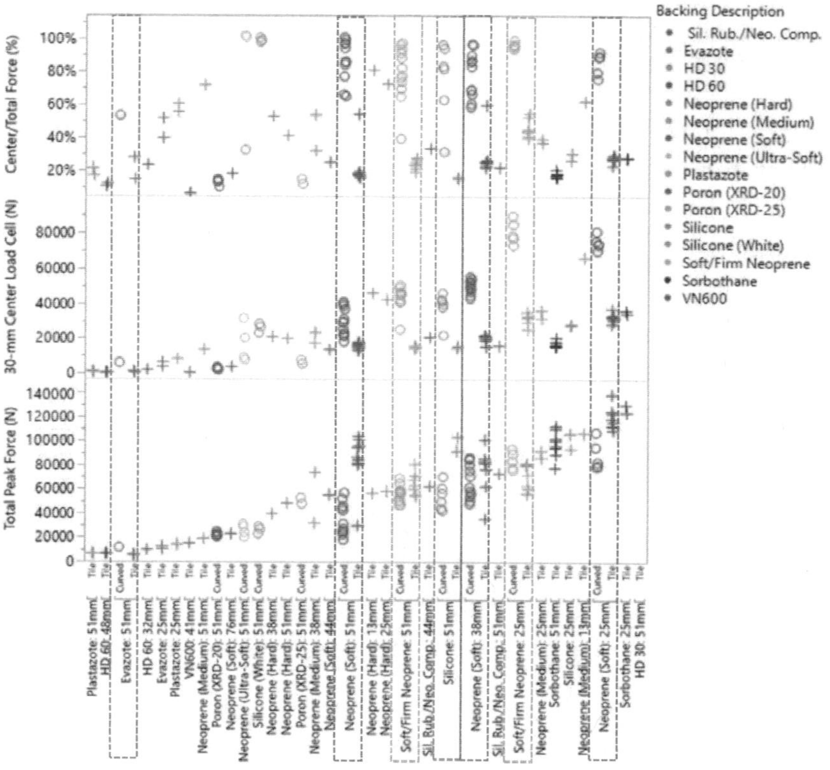

Figure 7. Centre, total and ratio of centre to total force measurements behind flat tiles and curved plates. Pad material/thickness combinations tested under both flat and curved armour are enclosed in a box

4. DISCUSSION AND FUTURE WORK

Comparison of the clay results in Table 4 shows that the clay BFD values measured were similar between flat and curved armour plates. However the data showed the magnitude and scatter of the residual armour deformation, deformation diameter and deformation volume was greater for the curved plates than the flat tiles. This suggests that there were fundamental differences between plates and/or test methods. Given the lack of cross-over tests, the differences are difficult to attribute to one source. The statistically significant differences in clay BFD results between Series 2 versus Series 1 and 3 may indicate poor reproducibility in the clay test protocol despite all clay blocks passing the clay calibration drop test.

Differences between flat tiles and curve armour plates performance were also observed when testing was conducted on force plates. The pads needed to match the deformation over clay for the flat tiles were much firmer than pads needed behind curved plates. There were also differences in the distribution of forces behind the flat tiles and curved armour plates. The curved armour plates had much greater centre loads than flat tiles across the range of pads tested in both conditions.

There are a number of potential explanations for the differences in pad and clay performance against the flat tiles and curved armour plates. They include:

- Differences in the mechanical response of the ceramic and UHMWPE properties and material thicknesses between the flat tiles and curved ceramic plates.
- Effect of plate curvature on the material performance.
- Effect of moulded clay curvature on the clay test. Mechanical properties of the clay may be different due to moulding into a curved insert and geometric differences in flow.
- Effect of curvature on the pad properties. When the flat pad is curved, one side is in tension and the other is in compression.
- Effect of curved impact caps, which presents a convex shape facing the impact
- Differences in measuring force behind flat and curved plates.

https://doi.org/10.52202/078352-0037

Additional ballistic and air cannon testing, as well as analytical modelling, is needed to determine the source of the differences observed. Additional testing can be conducted on the current curved plates to provide greater statistical confidence in the results, particularly for the Poron pad material where limited testing was conducted.

The current testing using ceramic plates and .30 calibre APM2 projectiles at one velocity shows that the pads had a range of effects on the forces and their distribution. The differences between the flat tiles and curved armour plates was sufficient that different patterns of behaviour were observed. Testing with different armour plates, armour plate materials and projectiles may produce a similar diversity in results. The current testing focused on armour and test conditions for the 44 mm clay BFD criteria; additional work can be done to assess armour pad suitability for a 58 mm clay BFD criteria.

5. SUMMARY

The goal of the project was to identify pad materials that can be placed in between ceramic armour and a rigid load cell impact plate that allows armour to deform similarly as when tested on clay. The project developed flat and curved test fixtures and analytical methods to assess armour response for different backing materials and established a database of impact pad configurations and armour response. It measured the peak centre and peak total forces associated with the pad material and thickness. This effort lays a groundwork of initial test results that can be leveraged by future efforts to develop force- based ballistic test devices.

The findings of this project include: discovering that pads and plates respond as a system; and the best pad materials for flat tiles was different than curved plates. For flat tiles, the pad materials most similar to clay were: soft (30A durometer) neoprene at 25 mm, 38 mm, 44 mm and 76 mm thicknesses, soft/firm layered neoprene at 25 mm thick, and Sorbothane at 51 mm thick. For curved plates, only one pad produced similar residual armour deformation to clay, Poron XRD-25. Additional testing is required to confirm this result since only two Poron samples were tested.

Differences in the clay test block response was observed for the test series, but it did not affect the armour residual deformation depth, diameter and volume deformation measurements of the flat tiles. These results reflect testing with .30 calibre APM2 bullets against ceramic plates back with a shootpack. Other armour and bullet combinations may result different results.

Acknowledgements

This research was funded by the US Army Combat Capabilities Development Command Soldier Center under Naval Sea Systems Command (NAVSEA) Contract N00024-13-D-6400, Task Order #VKW03. Any opinions, findings and conclusions or recommendations expressed in this material are those of the author(s) and do not necessarily reflect the views ofNAVSEA.

References

[1] Hanlon E and Gillich M., Origin of the 44 mm Behind-Armor Blunt Trauma Standard Military Medicine, 177, 3:333, 2012, pp 333-339.

[2] Rice K, Riley M. A., and Forster A. Ballistic Resistance of Body Armor, NIJ-Standard-0101.06, National Institutes of Justice, July 2008.

[3] National Research Council 2010, *Testing of Body Armor Materials for Use by the U.S. Army Phase II: Letter Report*. Washington, DC: The National Academies. p.10

[4] Roland C. M., Mechanical Behavior of Rubber at High Strain Rates. Rubber Chemistry and Technology: July 2006, Vol. 79, No. 3, pp. 429-459.

[5] Dunnett C. W., A multiple comparison procedure for comparing several treatments with a control. Journal of the American Statistical Association. (1955) 50: 1096–1121

https://doi.org/10.52202/078352-0037

Modelling dynamic deformation of clay backing in ballistic impact of armour

Timothy G Zhang, Sikhanda S Satapathy

US Army Research Laboratory, Aberdeen Proving Ground, MD 21005, U.S.A.

timothy.g.zhang.civ@mail.mil

Abstract. In this study, a numerical model was developed to model the ballistic impact on an armour composed of ceramics, an Ultra-high-molecular-weight-polyethylene (UHMWPE) plate, and soft shoot-pack with clay backing. A material model, calibrated with experimental data from impact and drop-tests in clay blocks that accounted for the deformation rebound was used to model the clay. The numerical model captured the back face deformation of the armour system consisting of ceramics, composite, fabric, and the interactions with the clay backing. Ballistic impact tests were conducted in two sizes of clay blocks: 0.15 m cylindrical clay block contained in a polyvinyl chloride (PVC) pipe, and a 0.61 m by 0.61 m rectangular clay block contained in a metal frame. Repeated tests were conducted for three different armour thicknesses. The model results were found to be in very good agreement with the experimentally measured final clay deformation for both clay blocks and various armour thicknesses. The quantification of the peak dynamic clay deformation from the simulation and correlating it to residual clay deformation provides more insight into energy and momentum transmission behind armour.

1. INTRODUCTION

Current personal protective equipment (PPE) have drastically reduced battlefield fatalities by using high performance materials, such as Ultra-high-molecular-weight-polyethylene (UHMWPE) and ceramics, to defeat threats. Even when an armour defeats the projectile, impact load is still transferred to the body through the deforming back face of the armour. Roma Plastilina No.1 clay is typically used to characterize the load transfer from the deforming back face of the armour to the human body. Measurement of the residual clay deformation is used to represent the peak armour deformation.

However, the residual clay deformation can often be less than the peak dynamic clay deformation. In [1], ballistic impact tests were conducted to measure the indent depth in the clay backing. X-ray results showed that the clay rebounded after reaching the peak deformation. However, the qualities of X-ray images drop as the clay size increases and introduce errors for the measured dynamic clay indents. To correlate the dynamic to residual clay indents, a numerical approach would be desired.

A clay model was previously calibrated with experimental data from low velocity impact and drop tests [2]. However, there was no rebound in the clay under those impact conditions, or, only residual clay indents were measured; therefore the calibrated model did not account for the rebound. The clay rebound was also observed in the ballistic impact tests at various velocities and drop tests in [3].

In this study, ballistic impact tests were conducted in hard-armour and shoot-pack with clay backing. Three different thicknesses of hard-armour were used to investigate the effect of armour thickness on the clay indent. Two sizes of clay block were used as the backing to study the effect of clay size on the clay deformation response, especially the dynamic deformation vs the residual deformation. The Finite Element models were developed to simulate the ballistic response of the ceramics, hard-armour, shoot-pack, clay backing and their interaction. The clay model developed in [2] was re-calibrated to account for rebound using the test data in [3]. The residual clay indents from the ballistic impact tests of various armour thicknesses and clay sizes were used to validate the model.

2. EXPERIMENTS

Ballistic impact tests were performed on ceramic/composite targets with a clay backing to characterize the back-face response of the armour material. The 0.10 m (4 inch) × 0.10 m ceramic and 0.20 m (8 inch) × 0.20 m composite layer were bonded together. A 0.20 m (8 inch) × 0.20 m shoot-pack layer was placed behind the composite layer and was backed by a clay-block. The test configurations are shown in Figure 1. Two configurations of the clay backing were used, as shown in Figure 2: 1) 0.15 m (6 inch) diameter × 0.28 m (11 inch) thick cylindrical clay block contained in a polyvinyl chloride (PVC) pipe with a 0.03 m (1 inch) thick plywood backing at the distal end of impact, and 2) 0.61 m (2 ft) × 0.61 m × 0.14 m (5.5 inch) thick rectangular clay block contained in a metal container with 0.02 m (3/4 inch) thick plywood in the back. The clay–pipe configuration was used to capture the dynamic deformation of the clay. Due

https://doi.org/10.52202/078352-0038

to the smaller size of clay and the confinement of PVC, the clay response is likely to be affected by the boundary conditions at earlier time. The larger rectangular clay block configuration was used to obtain the relationship between the areal density of armour and the clay residual indents.

In the experiments, a steel core projectile impacted the centres of the ceramic plates at constant velocity. Three thicknesses of hard-armour were used in the tests. The areal densities of the hard-armour were 31.2 kg/m^2 (6.4 psf), 35.1 kg/m^2 (7.2 psf), and 39.0 kg/m^2 (8.0 psf), the first and second were 80% and 90% of the third one. The thickness ratio of the ceramic plate to composite plate was fixed for the three different thickness armours. In order for the composite plates to glide along rods, holes, which were slighter larger than the rod size, were drilled at the four corners of the composite plates and rods were inserted in to the holes.

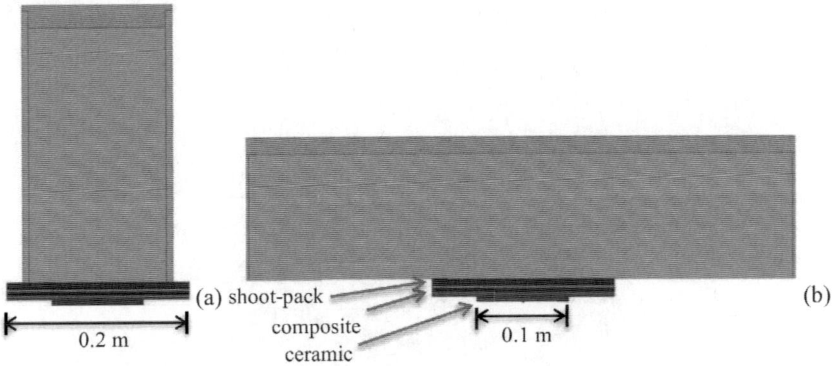

Figure 1 Impact of a hard armour and shoot-pack, with (a) 0.15 m diameter cylindrical clay backing, and (b) 0.61 m × 0.61 m rectangular clay backing (half model)

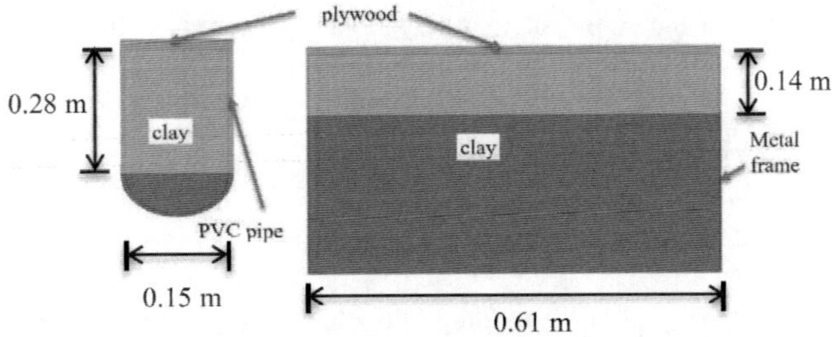

Figure 2 (a) 0.15 m diameter clay block, and (b) 0.61 m × 0.61 m clay block (half model)

Figure 3 shows the normalized residual clay indents measured after the impact tests for both 0.15 m and 0.61 m clay block. The clay indent was normalized by the maximum indent of all the test data. The test data was scattered, especially for the case of 0.15 m clay block. The response of smaller clay was more likely to be affected by the boundary conditions from the PVC pipe. As expected, the clay indents decreased with increase in the armour areal density. For thicker hard-armour, less energy of the projectile was transferred to the shoot-pack and subsequently to the clay block. The residual clay indents were higher in the larger clay blocks since the clay can continue to deform for longer period of time before the boundary conditions affected the deformation.

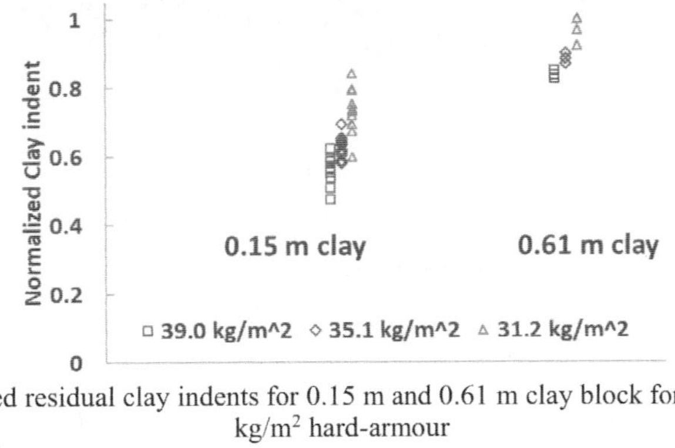

Figure 3 Normalized residual clay indents for 0.15 m and 0.61 m clay block for 39.0, 35.1 and 31.2 kg/m^2 hard-armour

https://doi.org/10.52202/078352-0038

3. FINITE ELEMENT MODELS

Figure 4 shows a schematic of the computational model of 0.61 m × 0.61 m clay block used in the study. The ceramic plate was bonded to a composite plate backed by the shoot-pack, which rested on Roma Plastilina No.1 clay backing. The clay was contained in a metal container with 0.02 m thick plywood in the back. The commercial software, LS-DYNA was used to compute the ballistic interaction between the bullet, armour and the clay block. The geometry was symmetric with respect to orthogonal mid-planes; hence a quarter symmetry model could be used for efficiency. However, as soon as the ceramic material cracks, the random fracture destroy the symmetry, and hence a full model should be used. We used a half symmetry model with symmetric boundary conditions as a compromise between efficiency and accuracy. Holes were created at the corners of the composite plate and the nodes at the hole edges were only allowed to move along impact direction to represent the gliding motion along the rods. A rigid wall boundary condition was used at the plywood location.

Figure 4 Computational model for the hard-armour, shoot-pack and 0.61 m clay block (half model)

The material models for the composite, shoot-pack and clay were calibrated with impact test data, while the material models for the bullet, ceramics, metal frame, PVC pipe and plywood were obtained from the literature or material library from EPIC code [7]. The final clay indent data were used to validate the computational model.

3.1 UHMWPE panels

The UHMWPE composite was made of 0/90° plies of unidirectional laminate sheets. To accurately model the failure in vicinity of the contact areas, a finer mesh was used near the point of impact, whereas a coarser mesh was used elsewhere. An example of the mesh for one layer is shown in Figure 5 (only a quarter model is shown).

Figure 5 Mesh for one layer of UHMWPE composite (quarter model)

A numerical model was developed and characterized for UHMWPE [4], which was used in this analysis. In that model *MAT62, only fibre tension-shear failure and fibre crush failure were included. In addition, the delamination failure was explicitly modelled using tie-break contact between adjacent composite layers. The material model was characterized with a series of ballistic experiments, including V_{50} data, back face deformation data, and delamination failure. The model captured the test data very well. The details for the model can be found in [4].

https://doi.org/10.52202/078352-0038

In the finite element model, a few laminates were fused into a single computational layer for computational efficiency. Ten "fused" layers were used for the 39.0 kg/m^2 (areal density) armour and the number of "fused" layers was proportional to the areal density. Two elements per layer were used along the thickness direction to account for bending effects. Eroding contact algorithm was used to delete failed material near the projectile impact site. Eroding contact algorithm is computationally expensive, and hence was used only in the zone near impact where material failure was expected.

3.2 Clay

A material model *MAT_PLASTICITY_COMPRESSION_TENSION [2] for the clay was developed based on quasi-static compression tests, drop tests and impact tests at medium rates. In the drop tests a 1 kg hemispherical nose cylindrical projectile was dropped from a height of 2 m on the clay block. In the impact tests, a 0.2 kg hemispherical nose projectile with a long tail impacted the clay block at impact velocity up to 55 m/s. The projectiles used in the drop and impact tests were shown in Figure 6. In the impact tests and drop tests, rectangular clay blocks of size 0.30 m × 0.30 m × 0.28 m thick was used, which doubled in thickness compared to the clay block used in this work. The rectangular clay block was placed on a table during the tests, which was modelled by using a rigid surface constraint boundary condition in the numerical model. The other sides were kept stress free. In the clay material model, compression and tension responses were treated differently. The strain rate effect was included in the model.

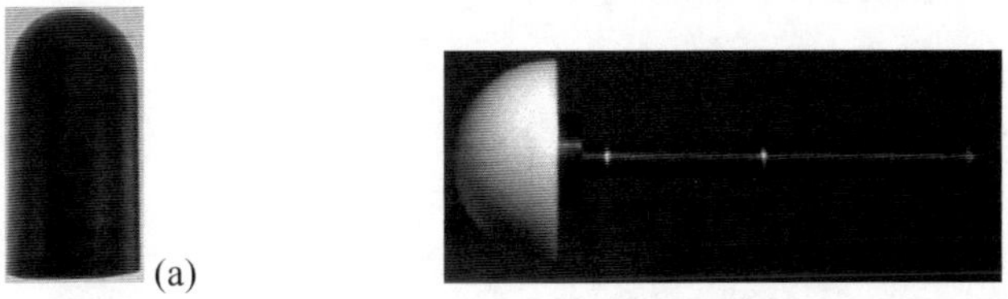
(a) (b)

Figure 6 The projectile used in (a) drop test, and (b) impact test

However, there was no rebound in the clay of 55 m/s impact tests, and only residual clay indents were measured in the drop tests, therefore the calibrated model did not account for deformation rebound [2]. Additional tests were conducted at different impact velocities. Also the drop tests were repeated, and the time history of clay indents were recorded [3]. The deformation rebound was observed from the new test data and the clay model was therefore re-calibrated. Below the results of the new model are discussed.

The comparison between simulation and experimental results are shown in Figure 7 and Figure 8. The clay model captures the response including the rebound observed in the experiments. However, even though the calculated rebound is within the experimental error for the drop tests, the calculated rebound was slightly outside the experimental bounds for the impact tests. The peak clay indent was captured reasonably well in the calculation for various impact velocities.

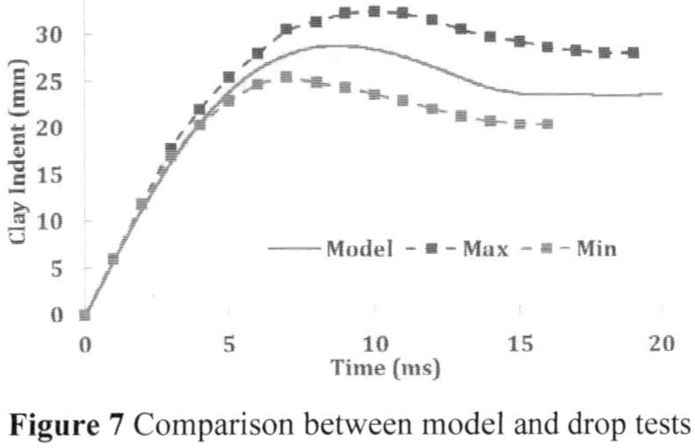

Figure 7 Comparison between model and drop tests

https://doi.org/10.52202/078352-0038

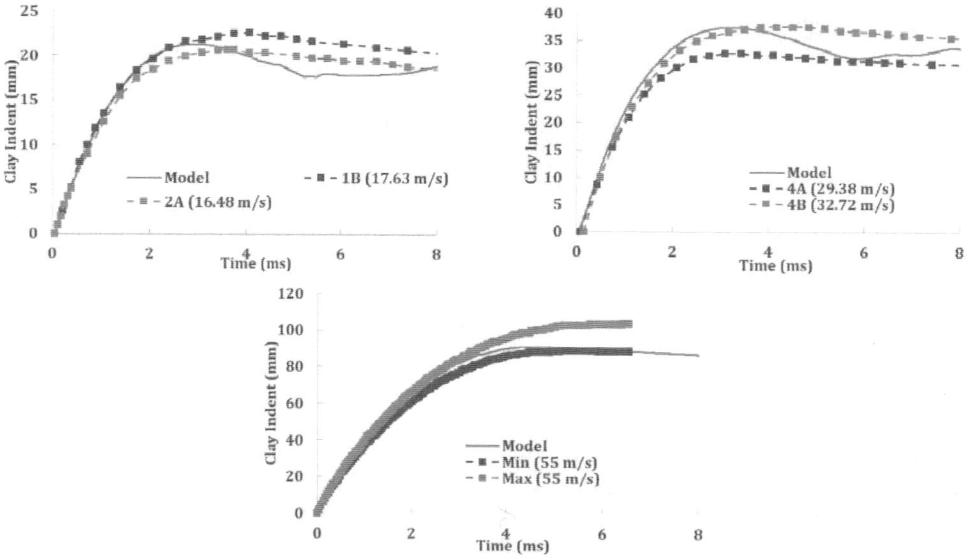

Figure 8 Comparison between model and impact tests

3.3 Shoot-pack

The shoot-pack model was developed previously [5]. The model is briefly described here. The woven structures of yarns were modelled. Due to interaction between yarns, the model was not computationally efficient. Figure 9 (a) shows the FE meshes for yarn model for one layer. Non-uniform meshes were used with finer meshes in the impact zone. Figure 9 (b) shows the meshes in the impact zone. Only a quarter of model was shown here.

The predicted time history of BFD from the yarn model had reasonable agreements with the test data [5]. The model had better accuracy since woven structures was captured in the model.

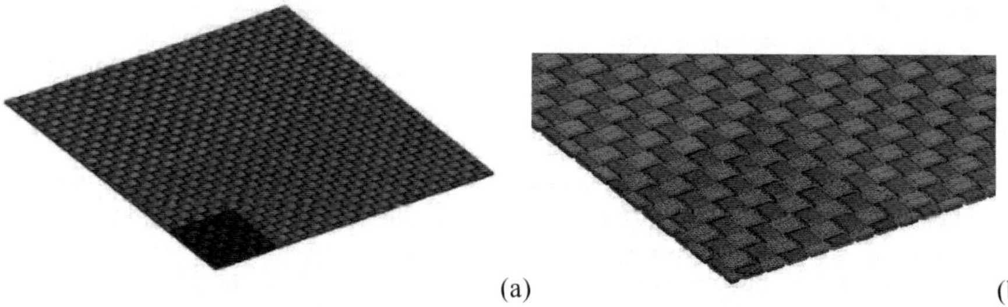

(a) (b)

Figure 9 FE meshes for (a) one layer of weft and warp yarns, and (b) yarn meshes in the impact zone (only a quarter model was shown)

3.4 Ceramics model

The JH-2 ceramics material model available in LS-DYNA was used for the SiC plate. This model tracks damage evolution and adjusts the material strength based on the amount of damage accumulation. The pressure is calculated using a prescribed equation of state. The material parameters for this model can be found in [6].

https://doi.org/10.52202/078352-0038

4. RESULTS

Figure 10 and Figure 11 show the comparisons between the simulation result and residual clay indents measured from the ballistic test on the armour backed by clay. The residual clay indent data are plotted next to end of the simulation time, which was ~2 ms for 0.15 m clay case and ~3.5 ms for 0.61 m clay case. The simulation end time was selected so that the clay deformations reach equilibrium (clay velocity was small). Overall, the calculations showed good agreement with the residual clay indents measured in the experiments. The 0.15 m diameter clay started to rebound at ~1 ms and the total rebound can be up to about 15% of the peak deformation. However, the 0.61 m × 0.61 m clay started to rebound at a later time ~1.5 ms and the rebound was much less pronounced compared to 0.15 m diameter clay case. The 0.61 m × 0.61 m clay continued to deform after the small rebound. The boundary conditions seem to affect the clay response. The final clay indent is less than the peak indent for 0.15 m diameter clay case, whereas the final clay indent is almost the same as the peak indent for 0.61 m × 0.61 m clay case.

The time history of the clay indents were similar for hard-armours of different thicknesses, including the time when the slope change at ~0.1 ms, time when the rebound occurs, and then subsequently the indent depth increases. The major differences are the magnitudes for different thickness armours.

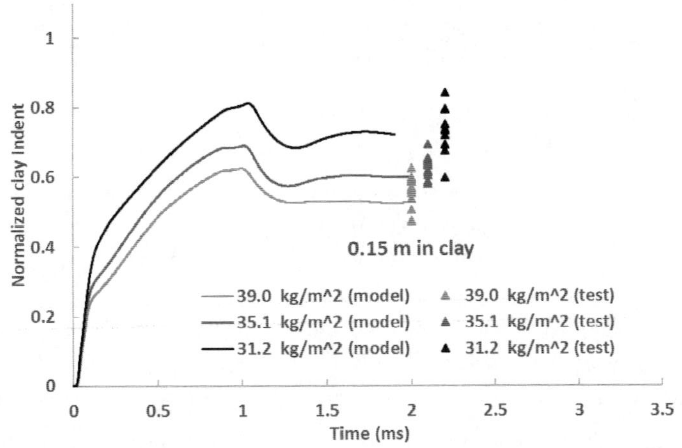

Figure 10 The time history of clay indent for 0.15 m diameter clay (test data is the residual clay indent measured after the impact tests)

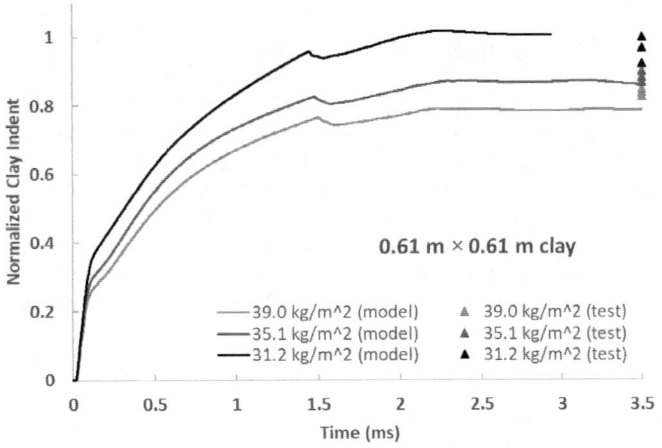

Figure 11 The time history of clay indent for 0.61 m × 0.61 m clay block (test data is the residual clay indent measured after the impact tests)

https://doi.org/10.52202/078352-0038

5. DISCUSSIONS

Figure 12 combines Figure 10 and Figure 11 to understand the effect of clay size on the clay deformation. The solid lines are the time histories of centre indents in 0.61 m clay blocks, and the dashed lines are for 0.15 m clay blocks. The clay deformations are observed to be almost identical until ~0.7 ms when the clay boundary conditions started to affect the response. The clay indent increases rapidly when the shoot-pack starts to deform. Around 0.1 ms the clay deformation slows down and continues to increase to the first peak at around 1.5 ms for 0.61 m clay. After a small rebound, the clay indent continues to increase to its peak after 2 ms when the deformation in the clay slows down to reach a plateau. Due to the confinement from the PVC pipe, the 0.15 m clay starts to rebound at an earlier time, ~1 ms. The residual clay indents in the 0.15 m block were about 10-15% smaller compared to the dynamic peak.

It can be seen from the simulations that the residual clay indents do not always correspond to the peak clay indents, especially for smaller clay block. For 0.61 m clay block, the clay indents at around 3.5 ms are about same value as the corresponding peak values. For the 0.15 m clay block, the clay indents at 2 ms are about 15% less than the respective peak indents.

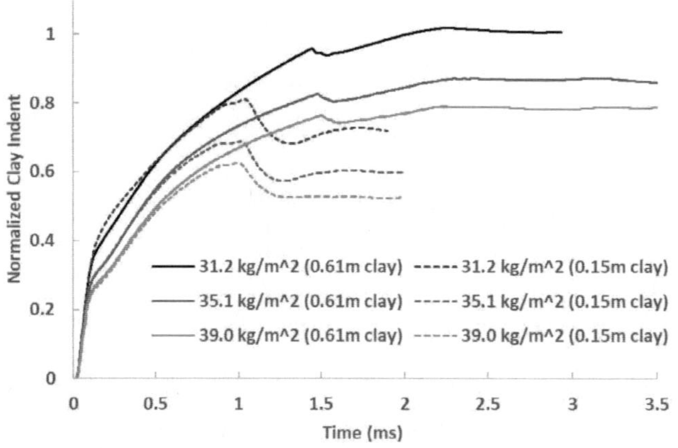

Figure 12 The comparison of clay indent time history between 0.61 m and 0.15 m clay block

Figure 13 shows the penetration at various times for armour areal density of 31.2 kg/m^2, 35.1 kg/m^2, 39.0 kg/m^2, with 0.61 m clay block cases. After impact, the projectile started to penetrate the ceramics. For 31.2 kg/m^2 and 35.1 kg/m^2 hard-armour, the projectile perforated the ceramics at around 50 μs, and 100 μs, respectively. But the projectile was arrested in the ceramics for 39.0 kg/m^2 armour and the composite failed partially right behind the ceramics where they were bonded together.

For all the cases, the projectile rotated either clockwise or counter-clockwise during the impact due to the non-symmetric deformation and failure of the ceramics, which validated the necessity of half model instead of quarter model. The failure of a brittle material like ceramic arises from its order of magnitude lower tensile strength than its compressive strength. As a result, crack appear in the hoop directions causing radial and cone cracks. Under perfectly axially symmetric impact and boundary conditions, a homogeneous target should break into infinite number of radial cracks when the hoop stress exceeds the tensile strength. However, this rarely happens in the experiments due to lack of perfect symmetry in the impact and boundary conditions, and also due to lack of homogeneity in the material causing cracks to preferentially appear at points of weak strength (defects, inclusions, voids, pre-existing micro-cracks, etc.) Cracks nucleating at such sites relieve the surrounding material of stress, leading to finite number of cracks in the target, and hence a non-symmetric crack pattern. While numerical simulations do not model such material heterogeneities explicitly, they do possess geometric asymmetry due to finite element mesh discretization and numerical approximations. While a full symmetry model is preferable to account for full penetrator-target interaction, it is computationally expensive. Therefore, a half symmetry model was employed to improve the numerical accuracy compared to a quarter-symmetry model while enabling a reasonable computation time for the calculations.

The clay initially deformed together with the shoot-pack. When the clay deformed faster than the shoot-pack, a gap was generated between them, at around 100 μs. Due to the gap, no further energy was transferred to the clay and the clay deformation slowed down, as shown in Figure 12. The slope of

https://doi.org/10.52202/078352-0038

the time history of clay indents changed suddenly around 100 μs due to the separation of clay and shoot-pack.

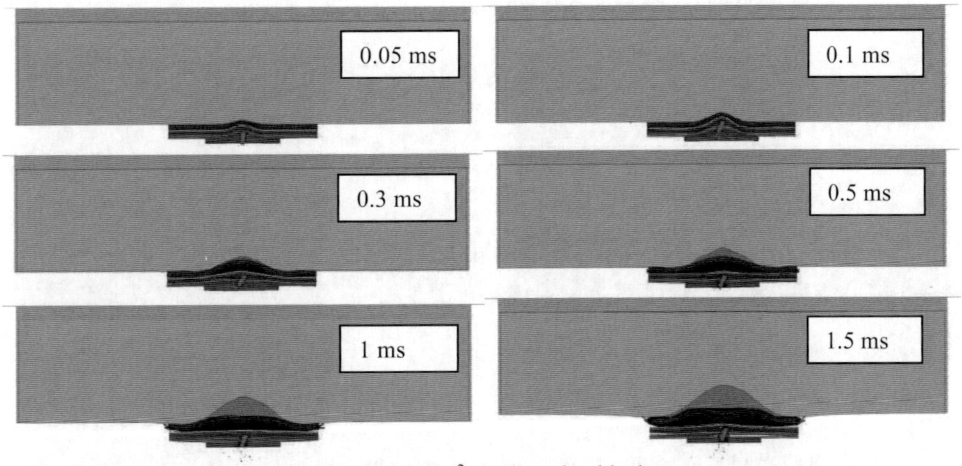

(a) 31.2 kg/m^2, 0.61 m clay block

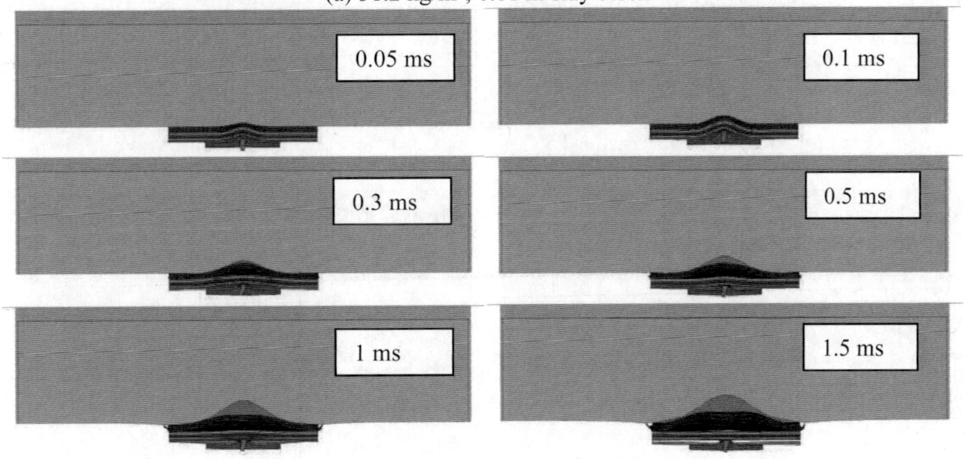

(b) 35.1 kg/m^2, 0.61 m clay block

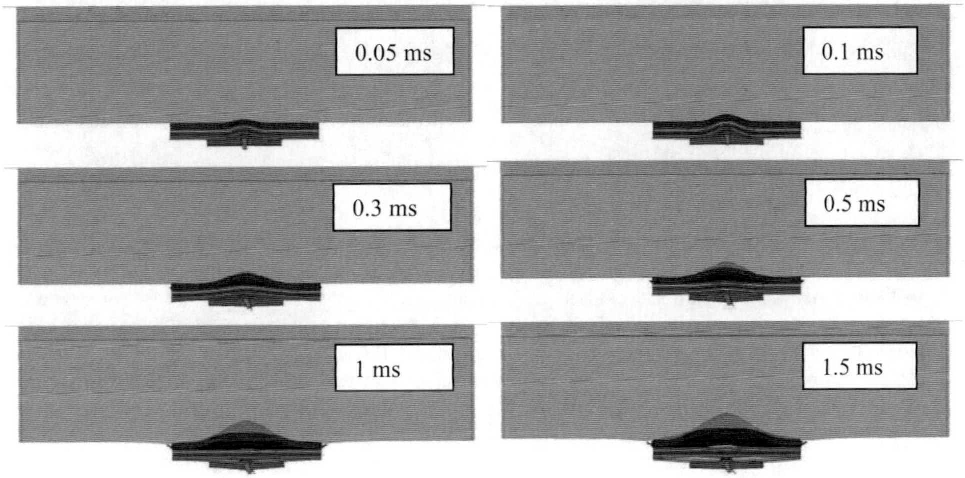

(c) 39.0 kg/m^2, 0.61 m clay block

Figure 13 The penetration at various time for armour areal densities of (a) 31.2 kg/m^2, (b) 35.1 kg/m^2, and (c) 39.0 kg/m^2, and 0.61 m wide clay block

 https://doi.org/10.52202/078352-0038

Figure 14 shows the penetration at various time for armour areal density of 39.0 kg/m^2 with 0.15 m diameter clay block. Compared to Figure 13(c), due to the smaller size in radius, the clay deformation was confined by the PVC pipe, resulting in a smaller clay indent.

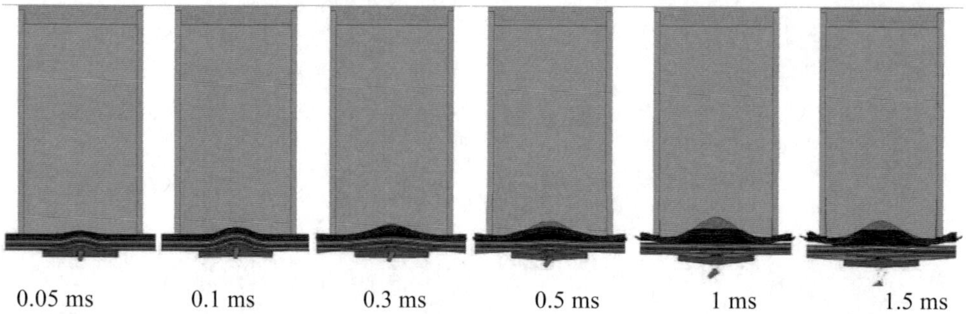

Figure 14 The penetration at various time for 39.0 kg/m^2, 0.15 m diameter clay block

Figure 15 shows the pressure contours in both 0.61 m and 0.15 m clay blocks for various times. The pressure wave propagated in a "spherical" direction towards the edge of the clay. As the pressure wave propagates, its amplitude decreases. The pressure wave, reflected from the end of the clay confined by the plywood, arrived at the impact region at around 1.5 ms, when the clay indent started to rebound, as shown in Figure 12. Similarly in the smaller clay block, the pressure wave arrived at the side earlier and the reflected pressure wave returned to the impact region at around 1 ms. The rebound in the 0.15 m clay was larger probably due to the smaller PVC pipe and stronger reflected pressure wave.

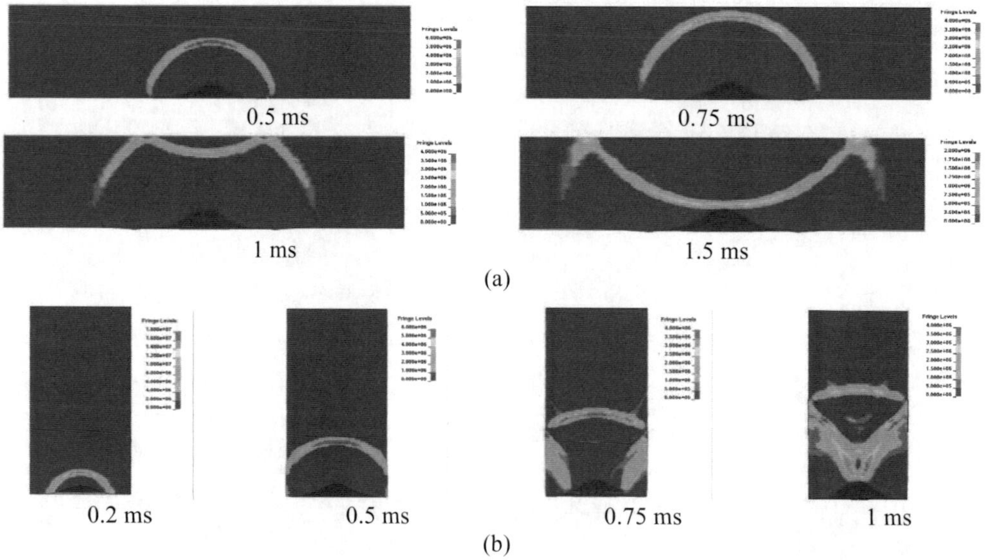

Figure 15 The pressure contour in the (a) 0.61 m, and (b) 0.15 m clay block

Table 1 lists the normalized residual clay indents and standard deviations of test data, predicted peak clay indents and clay indents at computation termination time. The measured residual clay indents were about 30% less in the 0.15 m clay block than in the 0.61 m clay block. The clay indents at computation termination time of the model agree very well with the residual clay indents measured in the experiments. The rebound was calculated by the indents at termination time and peak indents. The clay rebound was up to 15% in 0.15 m clay.

https://doi.org/10.52202/078352-0038

Table 1 Comparison between test and model

Clay size (m)	Hard armour areal density (kg/m^2)	Normalized clay indents					
		Test		Model			
		Average	Standard deviation	Peak	Indents at termination time	Rebound
0.15	31.2	0.73	0.069	0.81	0.72	11.1%
	35.1	0.63	0.029	0.69	0.60	13.0%
	39.0	0.56	0.044	0.62	0.53	14.5%
0.61	31.2	0.96	0.039	1.02	1.01	1.0%
	35.1	0.88	0.015	0.87	0.86	1.2%
	39.0	0.84	0.011	0.79	0.79	0.0%

6. CONCLUSION

In this work, ballistic impact tests were conducted to measure the residual indents in the clay blocks for three different thicknesses of hard-armour with a shoot-pack between the hard-armour and clay block. To understand the effect of boundary conditions, two different sizes of clay blocks were used in the experiments: a 0.61 m × 0.61 m × 0.14 m rectangular clay block and a 0.15 m diameter × 0.28 m thick cylindrical clay block. The test data showed that the residual clay indents decreased with increased thickness of hard-armour, as expected. The residual clay indents were about 30% less in the 0.15 m cylindrical clay block than in the 0.61 m rectangular clay block, which was mainly due to the smaller clay size and confinement effects of the PVC pipe.

A numerical model was developed to capture the interactions between hard-armour, shoot-pack and clay block. The material models for the composite, shoot-pack and clay were calibrated with impact test data in our earlier studies. The simulations using the calibrated clay model captured the rebound in the clay.

The simulations showed that the clay indents increased rapidly together with the deformed shoot-pack and then slowed down when a gap formed between the shoot-pack and clay, and hence no further energy was transferred to the clay. The clay continued to deform until the reflected wave from the rear surface arrived, when the clay started to rebound. The rebound depended on the clay size. For the smaller 0.15 m clay, the clay rebounded much earlier and the magnitude of rebound was about 15%. The clay had a very small rebound at a later time in the 0.61 m clay block. The predicted clay indents agreed reasonably well with the tested values of residual clay indents.

Since the dynamic clay indent is higher than the residual clay indent, the former appears to be a better metric to use to characterize armour performance and relate it to potential behind armour blunt trauma. However, it is not always feasible to measure the dynamic indentation in experiments using radiographic methods due to opacity of large targets to x-ray. Therefore, a modelling approach outlined in this study can be used to correlate the peak dynamic clay indent to the residual clay indent. Other parameters of interest, such as, stress, strain, energy and momentum transfer from the projectile to the clay can also be quantified from the simulation.

References

[1] M.J. Van der Jagt-Deutekom, Dr. E.P. Carton, M.M.G.M. Philipens, Separation phenomena between armour plate and clay backing during projectile impact, TNO report R10755, 2012.

[2] Timothy G. Zhang, Juliana Ivancik, Randy A Mrozek, Sikhanda S. Satapathy, Material Characterization of Ballistic Roma Plastilina No. 1 Clay, 30th International Symposium On Ballistics, Long Beach, CA, Sept. 11-15, 2017.

[3] Mary J Graham, Timothy Zhang, Experimental Investigation of Geometric Effects on Ballistic Clay Backing Material, ARL-TR-8717, , U.S. Army Research Laboratory, Aberdeen Proving Ground, MD, June 2019.

[4] Zhang TG, Satapathy SS, Vargas-Gonzalez LR, Walsh SM. Modeling ballistic response of ultra-high-molecular-weight polyethylene (UHMWPE), ARL-TR-7723, U.S. Army Research Laboratory, Aberdeen Proving Ground, MD, 2016.

https://doi.org/10.52202/078352-0038

[5] Timothy G Zhang, Phillip A Jannotti, Sikhanda S Satapathy, Jason R Mcdonald, Brian E Schuster, Indent depth and volume in the clay backing for soft and hard armour, Proceedings of Personal Armour Systems Symposium 2018, Washington, DC, Oct. 1-5, 2018.
[6] D.S. Cronin, K. Bui, C. Kaufmann, G. McIntosh, and T. Berstad, "Implementation and validation of the Johnson-Holmquist ceramic material model in LS-Dyna", Proc. 4th Eur. LS-DYNA Users Conf. 1, 47-60 (2003).
[7] G. R. Johnson, S. R. Beissel, C. A. Gerlach, T. J. Holmquist, User instruction for the 2018 version of EPIC, Southwest Research Institute, Sept. 2018.

https://doi.org/10.52202/078352-0038

Estimation of Armour Backface Velocity

K. A. Rafaels[1], K. Choi[2], G. Glasser[2], and C. A. Bir[3]

[1]Army Futures Command, CCDC Army Research Laboratory, Aberdeen Proving Ground, MD, USA, karin.a.rafaels.civ@mail.mil
[2] University of Southern California, Keck School of Medicine, 1333 San Pablo Street, Los Angeles, CA, USA
[3]Department Chair of Biomedical Engineering, Wayne State University, Detroit, MI, USA

Abstract. The velocity of an impact plays a significant role in the response of biological tissue. Current techniques to assess behind-armour blunt trauma (BABT) only measure the static post-test depth in clay and cannot capture the velocity of deformation. Assessments of focal, high-energy impacts that include velocity have been shown to better predict increased injury severity. Previous work has described a technique to estimate the backface velocity (BFV) of armour in clay tests; however, the accuracy of that technique was not reported. To overcome this limitation, experiments have been conducted on ballistic gelatine to measure the BFV of the armour from a subset of the impact conditions previously conducted on clay. The estimates of armour BFV using the conservation of momentum, areal density of the armour, and the area of the clay impression at the impact surface were biased lower than the BFV calculated from high-speed video analysis of ballistic gelatine tests. In other words, the BFV from the gelatine tests tended to be faster than the estimated BFV from clay. Closer inspection of these cases indicated that often many layers of the armour were penetrated, suggesting that the mass of the involved-armour may be overestimated in the calculations. One case exhibited the opposite trend (estimated clay BFV was faster than gelatine BFV). This case involved an impact near the edge of the armour, possibly underestimating the impact area in the clay. This would lead to an underestimation of the mass of the involved-armour, and therefore, an increased BFV estimate. This study begins to identify impact conditions where the BFV of armour can be estimated from clay testing. As velocity may be an important indicator of injury, being able to include a parameter from the backface signature that incorporates it would improve the assessment of injury risk of BABT.

1. INTRODUCTION

Behind-armour blunt trauma (BABT) is the unique blunt injury mechanism resulting from the backface deformation (BFD) of body armour that defeats the bullet or projectile, but still causes injury. That is why before body-armour systems can be purchased or issued, they must be able to pass standards for BFD. However, the current standard [1] used to evaluate BABT does not have a direct relationship to injury and was developed for soft body armour using a low-velocity handgun round [2]. Since the dynamics of BABT loading are not necessarily the same between armour systems or against different threats [3], and the response of human tissue is dependent on strain rate [4], the metric in the standard may not apply to the types of armour and threats currently being evaluated. The current metric is based on a static deformation measurement, but deformation alone, especially for high-rate impacts, does not adequately reflect the viscous properties of the chest and, therefore, injury risk [4]. It then follows that, metrics that include the rate of deformation are better predictors of increased injury severity than ones with just deformation [4-6]. However, current techniques to measure the backface velocity (BFV) requires testing armour with a different methodology than currently used to assess BABT [7-18].

Prather et al. described a procedure to estimate the velocity of the armour deformation from tests similar to the current standard [19]. This method was used to calculate the mass and BFV of the armour in BABT cases re-created on clay to relate the injuries seen to the Blunt Criterion (BC), a metric that considers the rate of deformation and is related to the amount of energy available to cause injury [6]. As BC was better able to predict increased injury severity in those cases than clay depth or clay volume, it might be beneficial to use this metric to evaluate armour systems. However, the accuracy of the Prather method to calculate the BFV of the armour has never been validated. To overcome this limitation, this study will compare the calculated velocity from a subset of the impact conditions previously conducted on clay to the measured BFV from ballistic gelatine experiments at the same impact conditions.

2. METHODOLOGY

This study expanded on previous efforts to re-create cases of real world BABT injuries on Roma Plastilina No. 1 clay [6, 20-21] by re-creating a subset of 10 of these cases on 10% ordnance ballistic gelatine to measure the BFV from the cases. Generally, the gelatine was prepared according to the

https://doi.org/10.52202/078352-0039

guidelines described in [22]. To obtain the 10% concentration of ordnance gelatine, 10 parts by weight (1,000 g) of gelatine (250 Type A Ordnance Gelatine, available from Kind and Knox) was mixed with 90 parts by volume (9,000 ml) of water. The mixture was allowed to stand for approximately one hour before being poured into a 15.2 cm by 15.2 cm by 40.6 cm (6 inch by 6 inch by 16 inch) aluminium pan. The pan was then placed in an environmental conditioning chamber between 3°C and 5°C (37°F and 41°F) for 30 hours prior to use. To calibrate the gelatine, a 0.177 calibre copper-plated sphere BB was fired at 179 +/- 4.5 m/s (590 +/- 15 fps) from 2 m (6.5 ft) into one block from each batch of gelatine that was made. The resting position of the BB was required to be 8.5 ± 1 cm (2.95 ± .39 inch) for 10% in order for the gelatine to pass and be used for testing.

The round identified in each case was fired from an appropriately-sized barrel, housed in a universal receiver. The receiver was remotely fired using a computer-controlled pneumatic firing system. The end of the barrel was positioned at the reported standoff distance from the gelatine for each case. Several shots were performed prior to each test to confirm the sighting and projectile velocity. The velocity of each round was recorded with three light screens (Oehler Research Inc., Model 57, Austin, Texas) attached to an Oehler 35P chronograph or by using the Caldwell Ballistic Precision Chronograph (#721122). All velocity measurements were taken with the front screen measuring 0.9 m (3 ft) from the target. Where point-blank contact shots were required for a re-creation, the velocity of the bullet was determined in a "test run"' shooting at a standoff of 1.5 m (5 ft).

The backface response of the armour was obtained using two high-speed cameras: one overhead and one side view, with a sampling rate of 35,000 frames per second. The overhead camera was a Phantom Miro LC310 (Vision Research, Inc., Wayne, New Jersey) with a resolution of 320 × 240. The side camera was a Vision Research Phantom V1212 with a resolution of 512 × 384. The gelatine block was backlit to increase resolution and reduce glare. To calibrate the camera images for calculating the dynamic gelatine deformation and velocity, images were captured by the cameras prior to testing that included scales oriented along the projectile path and plane and placed within the frame of the cameras. The high-speed cameras were triggered by the universal receiver with sufficient duration to capture the whole impact event.

The gelatine block was placed in a specially constructed 1.27 cm (1/2-inch)-thick acrylic enclosure to provide a mounting surface for the armour, as shown in Figure 1. The dimensions of the enclosure were 60.0 cm by 60.0 cm by 45.7 cm (24 inch by 24 inch by 18 inch). The armour was secured using Velcro straps. The impact location on the armour was placed so that it was positioned over the centre of the front face of the gelatine block. All blocks were tested only once per side.

a. **b.**

Figure 1. Gelatine test armour mounting fixture: a) acrylic fixture with a representative block of similar size to the ordnance gelatine placed inside and b) example of an armour system mounted to the fixture.

The time history of the BFD of the re-created cases was digitised using frames taken from the high-speed video of the top view (Figure 2). The leading edge of the deformation was traced (green line) to capture the deformation profile at discrete times for the entire loading phase of the impact event for each of the re-created cases.

https://doi.org/10.52202/078352-0039

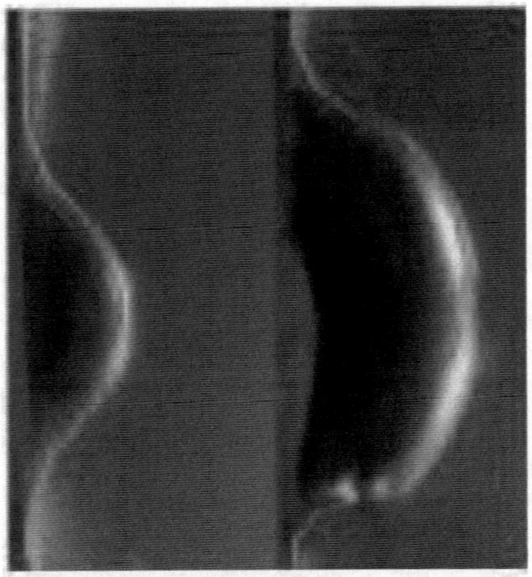

Figure 2. Representative high-speed video images used to calculate the time history of the BFD of the armour. Green line demonstrates the deformation profile determined in each frame.

To estimate the BFV from the re-created BABT cases on clay, the method to estimate the mass and velocity of the armour backface described by Prather et al. was used [19]. In this method, it is assumed that mass and velocity of the projectile at impact is equal to the mass and velocity of the armour and projectile after impact because momentum should be conserved. Standard procedures during the ballistic tests involve weighing the projectile making it easy to obtain the mass of the projectile. The velocity of the projectile in the re-creations was measured using a chronograph as described in the work by Hewins et. al. [20] or by the method described above. To estimate the mass of the armour involved in the BABT impact, the areal density of the armour was multiplied by the area of the armour involved in the impact. The areal density of the armour systems was obtained via manufacturers' and retailers' catalogues, or directly from the manufacturer. To get the amount of the armour involved, it was assumed that the number of penetrated layers of the armour is negligible, and that the area of the crater in the clay from the previously-conducted tests represents the area of the armour involved in the impact. Since the projectile did not fully penetrate the armour, its mass should also be included in the post-impact mass. In summary, the calculation of the effective mass of the armour backface is described by Equation 1,

$$\mathrm{m}_{eff} = \rho_A A_{crater} + m_{bullet} \tag{1}$$

where ρ_A is the areal density of the armour, A_{crater} is the area of the crater along the initial plane of the clay box, and m_{bullet} is the mass of the projectile. Once the effective mass is determined, using the conservation of momentum, the effective velocity of the armour (v_{eff}), or BFV, can be calculated using Equation 2.

$$v_{eff} = \frac{m_{bullet} v_{bullet}}{\mathrm{m}_{eff}} \tag{2}$$

A Bland-Altman plot analysis was performed to assess the agreement between the measurement of BFV in gelatine tests and the calculation of BFV from the clay tests [23]. A Shapiro-Wilk test was used to evaluate the normality of the differences in BFV [24]. As it is expected that the BFV measured in the gelatine tests would be more accurate than the calculation from the clay tests, a percentage similarity analysis was evaluated using Equation 3 [25]. The analysis was carried out using JMP, version 12.0.1, statistical software (SAS Institute, Inc., Cary, NC, 2015).

$$\%_{sim} = \left(\frac{\frac{BFV_{gelatine} + BFV_{clay}}{2}}{BFV_{gelatine}} \right) \times 100 \tag{3}$$

https://doi.org/10.52202/078352-0039

3. RESULTS

A total of 10 case studies were included in the analysis. Table 1 provides a summary of the peak BFVs determined from the measurements of the gelatine tests and from the calculations from the clay tests for each case.

Table 1. Summary of cases

Case	$BFV_{gelatine}$ (m/s)	BFV_{clay} (m/s)
BABT ID 029	64.4	62.1
BABT ID 031	68.1	26.8
BABT ID 019	91.6	85.1
BABT ID 043	101.5	91.1
BABT ID 010	111.4	25.6
BABT ID 036	122.7	94.9
BABT ID 046	123.8	131.1
BABT ID 004	128.7	136.7
BABT ID 017	133.7	202.5
BABT ID 027	202.2	30.3

Figure 3 compares the BFVs determined from clay and gelatine to the unity line, or the line that indicates when the BFVs calculated from each method are the same. The plot indicates that there are a number of cases where the two methods are equivalent and roughly 4 cases where they deviate from the unity line. The regression analysis of these parameters did not indicate a strong correlation between the two BFV methods, with an R^2 of only 0.018.

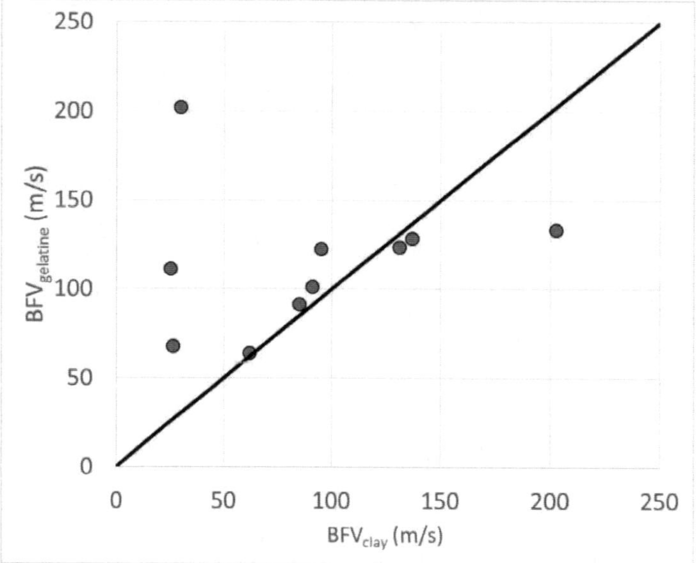

Figure 3. The BFV determined from the gelatine and clay tests compared to the line of equality.

To better identify the differences between the BFV methods, Figure 4 plots the differences between the methods and the average of the peak gelatine and clay BFVs for each case. The distribution of the difference data was determined to be normal. The mean difference between the methods is 26 m/s ± 65 m/s, indicating that the BFV calculated from the clay is generally less than the peak BFV measured from the gelatine tests. The differences are normally distributed, so about 95% of the differences between the two methods will be between the mean difference ± 2 times the standard deviation (SD). In other words, the BFV calculated from the clay could be 155 m/s above or 103 m/s below the peak BFV measured from the gelatine tests.

https://doi.org/10.52202/078352-0039

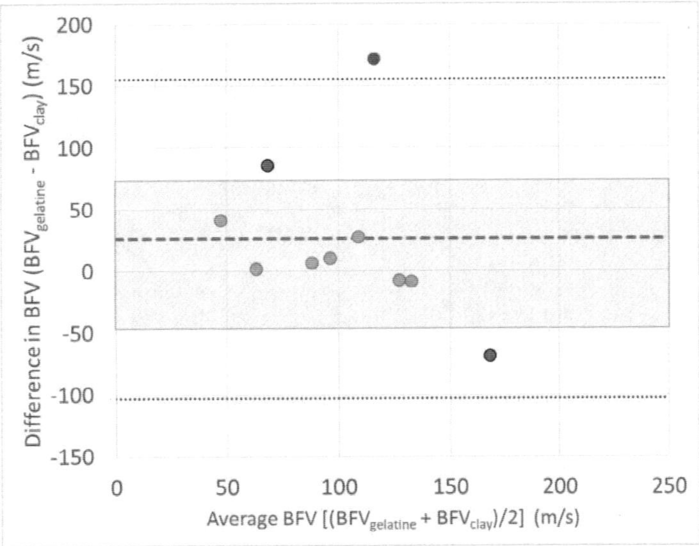

Figure 4. The difference in BFV against the average BFV of the two methods for each case. The bold dashed line indicates the mean difference between the methods. The finer dashed lines demonstrate the mean ± 2SD. The datapoint highlighted in red is considered an outlier. The light grey box indicates the 95% confidence interval (CI) for the mean difference. The 95% CI for the agreement limits are wider than the scale of the plot.

Within Figure 4, the datapoint highlighted in red indicates an outlier in the dataset. Inspection into that test, BABT ID 027, revealed that this case involved an impact near the edge of the armour to represent the real world case. As a result, portions of the armour exhibited a significant amount of extrusion along the edge, as shown in Figure 5. This extrusion caused the clay to have a deformation crater wider than the armour. As a consequence, the estimated amount of armour involvement, i.e., effective mass of the armour, was overestimated for this test, thereby reducing the estimated BFV in the clay test. Since testing procedures for armour do not involve tests so close to the edge, even for the so-called "edge shots", the outlier was removed from further analysis in this study.

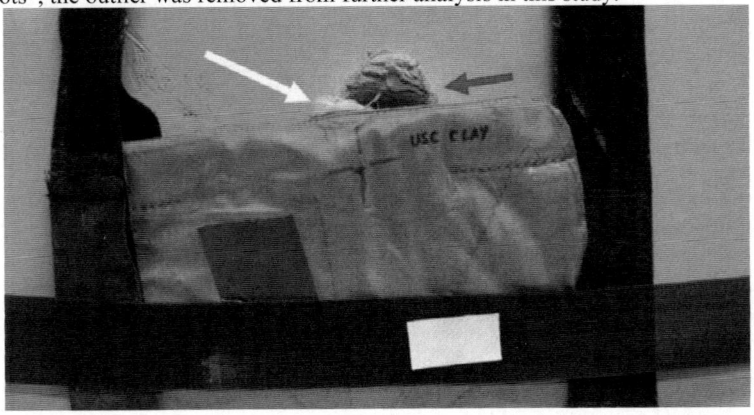

Figure 5. A picture of the armour panel on the clay block post-impact demonstrating the extruded fibres of the armour panel (white arrow) and the crater diameter exceeding the edge of the armour (red arrow).

Another analysis of the differences between the methods and the average of the $BFV_{gelatine}$ and BFV_{clay} for each case was done after removing the outlier, as displayed in Figure 6. The distribution of the difference data was still determined to be normal, even after the outlier was removed. The mean difference dropped from 26 m/s to 10 m/s when the outlier was removed. The mean difference is also a positive value, indicating that the BFV calculated from the clay is generally less than the peak BFV measured from the gelatine tests. With the outlier removed, the BFV calculated from the clay could be 94 m/s above or 74 m/s below the peak BFV measured from the gelatine tests.

https://doi.org/10.52202/078352-0039

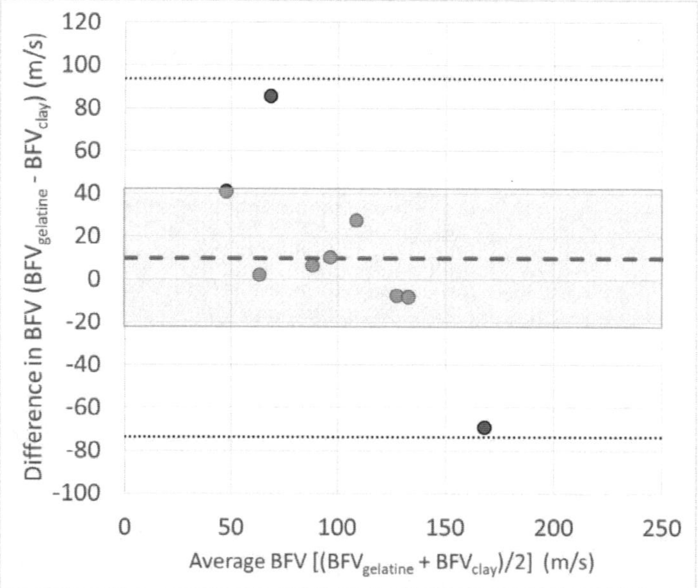

Figure 6. The difference in BFV against the average BFV of the two methods for each case with the outlier removed. The bold dashed line indicates the mean difference between the methods. The finer dashed lines demonstrate the mean ± 2SD. The light grey box indicates the 95% CI for the mean difference. The 95% CI for the agreement limits are wider than the scale of the plot.

Figure 7 displays the distribution of the data of the percentage similarity values for the analysis with the outlier removed. The mean percentage similarity value for this data is 93.5%, which indicates that the BFV calculation from the clay tests has a mean bias of 6.5% less than the BFV measurements from the gelatine tests. The coefficient of variation, a metric that reflects accuracy and precision, is 20.3%.

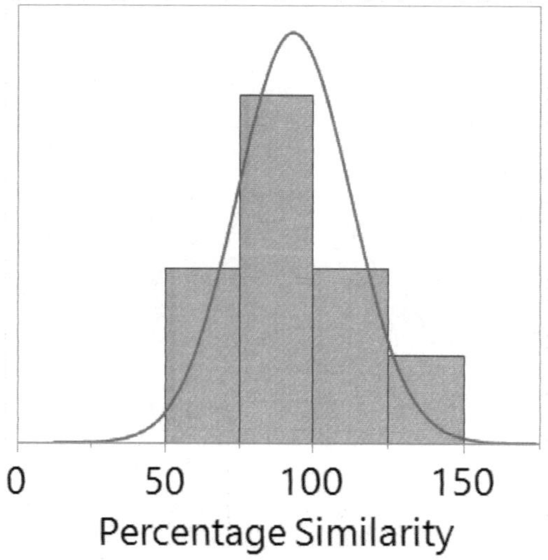

Figure 7. A histogram of the percentage similarity comparing the clay-calculation method for determining BFV with the more accurate gelatine method. A percentage similarity value of 100% indicates that the two methods give the same results, whereas a value of 0% indicates that the results of the two methods are not similar. The red line illustrates the normal distribution.

https://doi.org/10.52202/078352-0039

4. DISCUSSION

This study has measured the peak BFV of the armour for various BABT impact events, a potentially important parameter to consider in the injury risk from BABT impacts. Current methodologies to evaluate body armour for BABT do not currently consider the velocity of the BFD. The method to estimate the BFV from the common clay tests used in the evaluation of BABT is advantageous because it can easily be implemented into the current methodology and would not add a significant cost to testing. However, the accuracy of that technique had never been explored. Therefore, the calculated BFV from the clay tests were compared with the peak BFV measured from tests on gelatine using high-speed cameras to visualize the dynamic BFD. The analysis revealed important differences in the BFVs determined between the two methods.

On average, the BFV calculated from the clay tests underestimated the peak BFV measured from the gelatine tests by 10 m/s. For the BFV to be underestimated in the clay tests, the effective mass for those tests must be overestimated. Investigating the cases more deeply, it was revealed that the cases that have a large deviation above the unity line in Figure 3 (Cases BABT ID 010 and BABT ID 031) have numerous layers of the armour fully penetrated. If fewer layers of the armour system were deforming into the clay, then the areal density used to calculate the effective mass would not represent the actual areal density of the armour involved, leading to an overestimate of the mass. Figure 8 shows how the layers separated within the armour panel in Case BABT ID 010 suggesting the raised layers were not significantly contributing to the mass of the armour during the impact, leading to an underestimate of the BFV.

a. b.

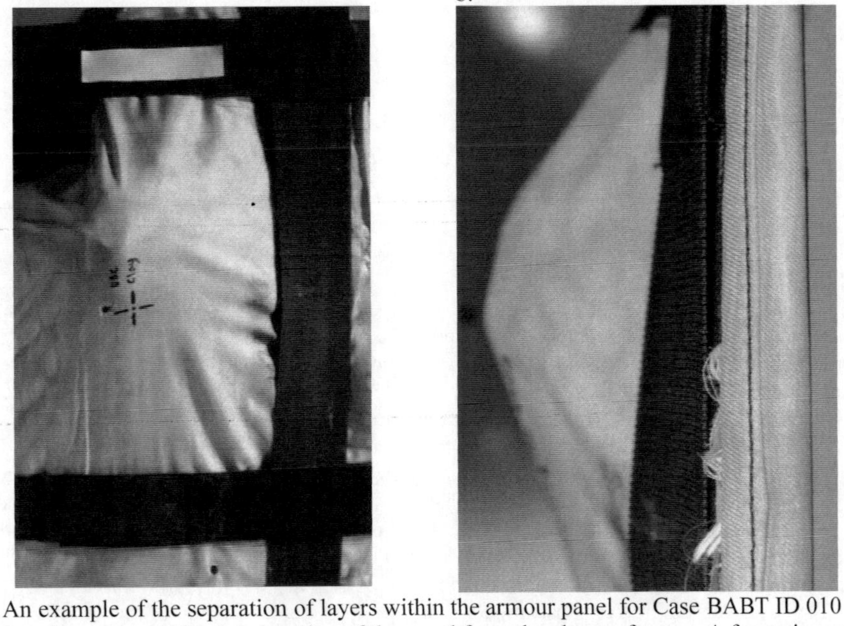

Figure 8. An example of the separation of layers within the armour panel for Case BABT ID 010 after the impact, as demonstrated by the elevation of the panel from the clay surface. a. A front view of the armour on the clay block post-impact. b. A side view of the elevation of the armour panel post-impact.

There was one case, BABT 017, that had a relatively large deviation such that the BFV from the clay test was estimated to be greater than the peak BFV measured on gelatine. This case was also near the edge of the armour, approximately 3.5 cm from the edge. For this case, it was observed that the area of clay residue left behind on the back side of the armour was greater than the area of the crater in the clay. Since the impact location was near the edge of the armour, it is possible that the armour along that edge could collapse, or fold, into the crater since there is limited resistance on the free edge of the armour. If more armour was involved during the impact than estimated from the area of the crater, then the effective mass for this case would be underestimated, thereby calculating a higher BFV. Case BABT ID 004 demonstrated a combination of factors that might affect the estimate of the effective mass of the armour. This case demonstrated many layers of armour penetration, but this impact location was also near the edge. Considering the difference between the two methods was not very large, it is likely that these two effects cancelled each other out with regards to over- and underestimating the effective mass

 https://doi.org/10.52202/078352-0039

of the armour. The rest of the cases demonstrated very little to no penetration of layers. If the cases BABT ID 027, BABT ID 010, BABT ID 031, BABT 017, and BABT ID 004 are removed from the analysis, that leaves a sample size of only 5. With a sample size so small, it is difficult to properly evaluate that the distribution of the differences is normal, an assumption in Bland-Altman analysis, since normality tests have little power to reject the null hypothesis with small sample sizes [26]. Small sample sizes also typically expand the confidence intervals since sample sizes are included in the calculation of those intervals. Yet, in this study, removing the cases associated with indeterminate armour involvement, improved the limits of agreement and the confidence intervals in the Bland-Altman analysis for the remaining cases (Figure 9). Considering the limited sample size, caution should be exercised when applying these results, but the mean difference, now 8 m/s, still indicates a bias towards an underestimate of the BFV in the clay tests.

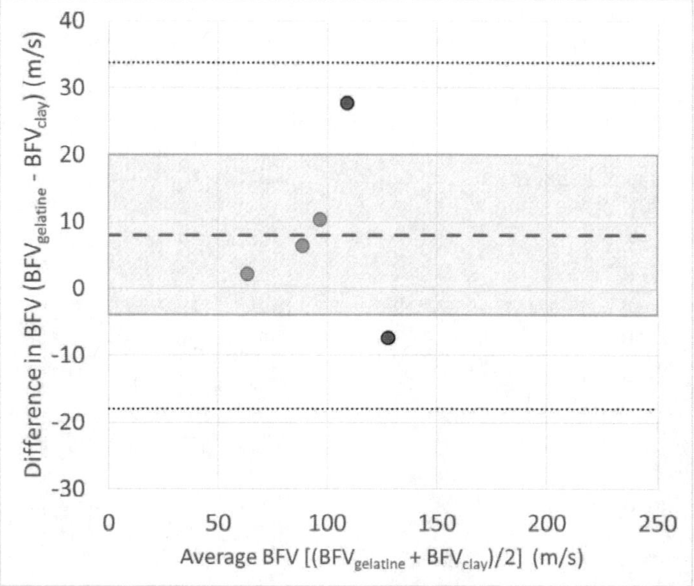

Figure 9. The difference in BFV against the average BFV of the two methods for each case with the cases having suspect effective masses removed. The bold dashed line indicates the mean difference between the methods. The finer dashed lines demonstrate the mean ± 2SD. The light grey box indicates the 95% CI for the mean difference. The 95% CI for the agreement limits are wider than the scale of the plot.

Even with the improvements in the limits of agreement and confidence intervals, the range of the limits of agreement in the velocity are greater than the velocity deviations allowed within armour testing (±9.1 m/s). More comparison testing will likely improve the confidence intervals and the limits of agreement, but increasing the number of samples will not address the unique contributions the specific armour systems played in the calculation of the BFV in the clay tests. Additional analysis to identify the necessary armour parameters post-impact to improve the determination of the effective mass of the armour is required for this method to be able to be incorporated in a testing environment. For example, a potential source of error in the estimation of the mass in the clay tests could be from using the manufacturer-provided areal density. There are several methods to calculate areal density that do not provide equivalent values [27], and it is not clear which method was utilised for each armor system. Future estimates of BFV from clay tests should include experimentally determining the areal density of the armour systems. Furthermore, the added complexity of obtaining those armour parameters during the testing will have to be evaluated for feasibility and financial and time costs. The method to calculate the BFV from the clay tests does show promise for conditions in which the armour witnesses limited penetration and the impact locations are an adequate distance from the edge.

5. CONCLUSION

This study has measured the peak BFV of body armour for 10 re-created BABT impact events, a potentially important parameter to consider in the injury risk from BABT impacts, using a clay-based

https://doi.org/10.52202/078352-0039

method and a gelatine-based method. Armour testing on clay is more common than on gelatine, but can only infer the BFV, whereas the gelatine-based method can directly measure the BFV. Comparing the two methods, the BFV calculated from the clay tests underestimated the peak BFV measured from the gelatine tests by 10 m/s, a difference greater than the velocity deviations allowed in armour testing. Cases where very few layers of the armour system were penetrated and the impact locations were an adequate distance from the edge showed good agreement between the two methods. Because the threat-armour interaction affected the BFV calculations, the clay-based method for estimated BFV should be implemented with care. Additional research and armour mass estimation methods, including an increased number of experiments are recommended before the clay-based method can be implemented in standard qualification testing.

Acknowledgments

This project was supported, in part, by Award No. 2011-IJ-CX-K006, awarded by the National Institute of Justice, Office of Justice Programs, U.S. Department of Justice, and Cooperative Agreement No. W911NF-17-2-0072 with the U.S. Army Research Laboratory. The opinions, findings or conclusions or recommendations expressed in this publication are those of the authors and do not necessarily reflect those of the Department of Justice or the Department of Defense. The authors would like to thank, first and foremost, all of the survivors who consented to participate in the study. We would also like to thank Rodrigo Villalta and the students in the Biomechanics Injury Research Laboratory group at the University of Southern California.

References

[1] National Institute of Justice. Ballistic Resistance of Body Armor, NIJ Standard - 0101.06. (Available at: National Criminal Justice Reference System https://www.ncjrs.gov/pdffiles1/nij/223054.pdf, 2008, accessed 19 August 2011).

[2] Prather R. 2010. "The Lightweight Body Armor Program – A History," presented to the Committee to Review the Testing of Body Armor Materials for Use by the US Army Phase III, August 9-11, 2010, Aberdeen, MD.

[3] Rafaels KA, Loftis KL, Moholkar NM, and Bir CA, Comparing the Backface Deformation Behavior between Soft and Hard Body Armors, 30th International Symposium on Ballistics, Long Beach, CA, Sept 11-15, 2017.

[4] Viano DC and Lau IV, J Biomechanics, Volume 21 (1988), 387-399.

[5] Sturdivan LM, Viano DC, and Champion HR, J Trauma, 2004; Volume 56; 651-663.

[6] Rafaels K.A., Loftis K.L., and Bir C.A., Can Clay Tell Us More Than Deformation?, Proceedings of the Personal Armour Systems Symposium, Washington D.C., USA, Oct. 1-5, 2018.

[7] Nader J., and Dagher H., Exp Tech, 2011; 35(2); 55-60.

[8] O'Masta MR, Compton BG, Gamble EA, Zok FW, Deshpande VS and Wadley HNG, Int. J. Impact Eng., 2015; 86, 131-144.

[9] Bass C.R., Salzar R.S., Lucas S.R., Davis M., Donnellan L., Folk B., Sanderson E., and Waclawik S., Int J Occup Saf Ergon, 2006; 12(4); 429-442.

[10] Amarilio I.B., Benes D., Asaf Z., Ya'akobovich A., Shmulevich I., Mouradjalian A., Wolf A., Grunner S., and Kluger Y., Proceedings of the Personal Armour Systems Symposium, Nuremburg, Germany, Sept. 17-21, 2012.

[11] Stuivinga M., Carton E.P, Verbeek, H.J., and van Bree J.L.M.J., Proceedings of the Personal Armour Systems Symposium, Nuremburg, Germany, Sept. 17-21, 2012.

[12] Broos J.P.F, van der Jagt-Deutekom, M., Halls V.A., and Zheng J.Q., Proceedings of the Personal Armour Systems Symposium, Nuremburg, Germany, Sept. 17-21, 2012.

[13] Metker L.W., Prather R.N., and Johnson E.M., A Method for Determining Backface Signatures of Soft Body Armors, U.S. Army Edgewood Arsenal, Technical Report, TR-75029, 1975.

[14] Mauzac O., Paquier C., Debord E., Barbillon F., Mabire P., and Jacquet J.F., Proceedings of the Personal Armour Systems Symposium, Quebec City, Canada, Sept. 14-17, 2010.

[15] Goode T., Shoemaker G., Schultz S., Peters K., and Pnakow M., Compos. Struct., 2019; 220, 687-698.

[16] Hinsley DE, Tam W, and Evison D, Behind Armour Blunt Trauma to the Thorax – Physical and Biological Models, Proceedings of the Personal Armour Systems Symposium, The Hague, Netherlands, Nov 18-22, 2002.

https://doi.org/10.52202/078352-0039

[17] Bourget, D, B Anctil, D Doman, and D Cronin, Development of a Surrogate Thorax for BABT Studies, Proceedings of the Personal Armour Systems Symposium, The Hague, Netherlands, Nov 18-22, 2002.
[18] Arborelius UP, Tryberg A, Gustavsson J, Malm E, Gryth D, Olsson LG, Skoglund M, Rocksen D, Proceedings of the Personal Armour Systems Symposium, Nuremberg, Germany Sept. 17-21, 2012, pp. 305-314.
[19] Prather RN, Swann CL, Hawkins CE, Backface Signatures of Soft Body Armors and the Associated Trauma Effects, U.S. Army Armament Research and Development Command, Technical Report, TR-77055, 1977.
[20] Hewins K, Anctil B, Stojsih S, and Bir C, Ballistic Blunt Trauma Assessment Methodology Validation, Proceedings of the Personal Armour Systems Symposium, Nuremburg, Germany, Sept 18-21, 2012; pp. 315-323.
[21] Bir C, Lance R, Stojsih-Sherman S, and Cavanaugh J, Behind Armor Blunt Trauma: Recreation of Field Cases for the Assessment of Backface Signature Testing, Proceedings of the 30th International Symposium on Ballistics, Long Beach, CA, USA, Sept. 11-15, 2017.
[22]Minisi, M. and Spickert-Fulton, S., *Guidelines for Gelatine Block Testing*, T.J.S.W.B. IPT, Editor. 2004, Joint Services Wound Ballistic Team.
[23] Bland JM and Altman DG, Lancet, 1986; 8476; 307-310.
[24] Shapiro SS and Wilk MB, Biometrika, 1965; 52; 3–4.
[25] Scott LE, Galpin JS, and Glencross DK, Cytometry, 2003; 54B; 46-53.
[26] Oztuna D, Elhan AH, and Tuccar E, Turk J Med Sci, 2006; 36(3); 171–6.
[27] Cronin J, Kinsler R, and Allen J, Lightweight Ballistic Composites: Military and Law-Enforcement Applications, (Woodhead Publishing, Duxford, UK, 2016); pp. 311-326.

https://doi.org/10.52202/078352-0039

Calibration of velocity light screens

J.P.F. Broos BSc[1)], W.E.T. Spruit BSc[1)]

[1)] *TNO Defence, Security and Safety, P.O. Box 480, NL-2501 Rijswijk, the Netherlands, hans.broos@tno.nl*

Abstract. Correct projectile velocity measurement is essential for shooting ranges, in order to determine the ballistic performance of sample materials. It determines the accuracy of the provided measurement result and therefore requires flawless and precise functioning of the velocity measurement system, including light screens, counters and calculations. A regular, traceable calibration of the velocity measuring instrumentation is mandatory for ISO 17025 accredited laboratories. This could be achieved by returning the complete system to the manufacturer for calibration. At TNO, infrared (IR) light screens are mounted on the wall inside the shooting ranges at a fixed position. Removing and re-installing the light screens could induce small deviations in the distance between the light screens, causing an error in the velocity measurement. For this reason it was chosen to calibrate the light screens in their original position. With this approach potential effects of outdoors calibration, like transportation and re-positioning are prevented.
Because a constant projectile velocity cannot be generated over the measurement range, considering the effect of drag resulting from variance of air pressure, temperature and humidity, a high speed video system is used for calibration. The in-house developed and precise velocity measurement device visualizes the projectile position during its flight. With the known projectile position over time and the resulting displacement between two recordings, the projectile velocity can be accurately measured. It results in an accurate average velocity over the distance between infrared light screens, for comparison with the reading obtained with the to be calibrated device.
The results of different sets of calibrations with two different projectiles show that projectile shape and velocity can influence the velocity measurements of the IR light screen. Also effects potentially resulting from vibrations due to moving of mobile velocity measurement equipment to different locations have been found.

1. INTRODUCTION

At TNO's Laboratory for Ballistic Research, ballistic tests according to standards are performed on daily basis. As an ISO17025 accredited laboratory, regular traceable calibration of measuring equipment is mandatory. Velocity screens are one of the most important devices since they determine the projectile impact velocity and decide together with the end result whether the sample has passed or failed the test. Since there is no such thing as a 'precise projectile velocity generator' that also corrects for drag for the given air pressure, temperature and humidity, TNO has developed a precision velocity measurement instrument to accurately calibrate their light screens.

Although it will be used mainly for light screens a variety of velocity systems can be calibrated, for example:

- Light screens
- Laser screens
- Contact foil/wires
- Doppler radar

Calibrating the velocity measurement instruments is an action which takes place on a regular basis. The outcome is important to monitor, not only to make sure that the readings are ok but also because the yearly gathered information could provide information about the possible causes for deviating readings.

This paper describes the method TNO is using for calibrating their velocity measuring instruments as performed on regular basis. A description of the system together with the outcomes of the measurements and lessons learned are presented.

 https://doi.org/10.52202/078352-0040

2. CALIBRATION PRINCIPLE

The calibration system and its components consists of two instrumented units which are capable to measure the projectile velocity with high accuracy. The units are triggered by a projectile passing a (laser) screen in the units. At the trigger, a shadowgraph picture of the projectile in flight is made by means of a short duration flasher. The position of the projectile imaged by both units is then measured with a calibrated ruler. By placing the units just in front and behind the velocity measuring instruments, the readings of both systems will give the same (average) velocity. In figure 1 a schematic setup of the device is shown. Boxes 'A' represents the velocity measuring instrument to be calibrated and 'B' the calibration system. The red dots C1 and C2 are the positions where the cameras are located. The distance between C1 and C2 is calculated using a calibrated aluminum ruler. The ruler contains several small holes (markers) at calibrated distances over the length of the ruler. With this tool different velocity measuring devices with varying distances between start and stop trigger can be calibrated.

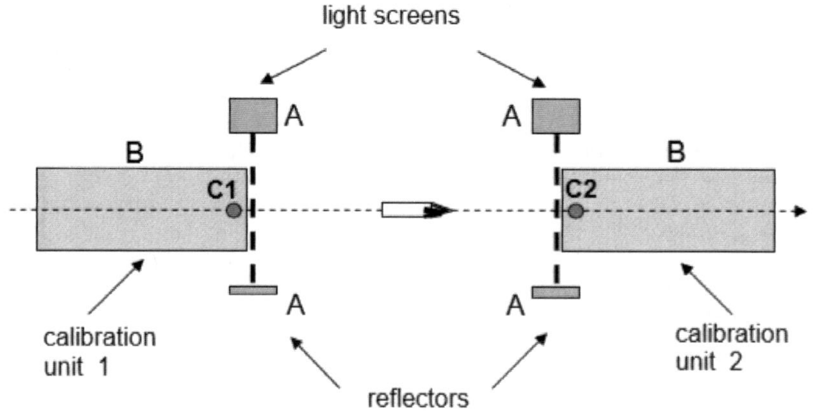

A = device to be calibrated
B = calibration system
C1, C2 = camera position

Figure 1. Schematic top view of the setup for calibrating a velocity instrument (A in this case).

To ensure a sharp image of the projectile in flight, a short duration flash is combined with a parallel collimated optical system. The two lenses that provide the image and background illumination of the projectile have a double function. These imaging lenses used in the calibration units are based on the combination of the 2 lenses mounted in each cone holder in the units (see Figure 2). Thus, each lens in the speed calibrator is a combination of the large spherical lens located in the front of the cone holder and the small correction lens located in the back of the cone. This large lens is a spherical lens and therefore has spherical aberration. I.e. that the focus point of the light rays passing through the edge of the lens does not coincide with that of the light rays passing through the lens near the center. To correct for this, a correction lens is included in the back of the cone. The lens in the speed calibrator is therefore an aspherical lens. With this lens, the focus point of the edge rays does coincide with the focus point of the more central rays.

However, the result is that there is a quality difference in the depicted image. This is corrected in image processing. The dual function of these imaging a-spherical lenses consists on the one hand of creating as smooth a background illumination as possible and on the other hand of creating a good sharp image of the projectile. These are two different functions combined in one lens system. The beam path in this lens system is as follows. The flash emits a diverging beam during the flash. The spectrum of this beam ranges from UV to over 750 nm. This causes problems when focusing on the camera because it is not an a-chromatic lens and this a-spherical lens is therefore not corrected for this. A filter is used to correct for this and is discussed later. 50% of the light is transmitted through the 50R / 50T mirror and falls on the first large lens via the 100

https://doi.org/10.52202/078352-0040

% mirror. The flash electrodes are exactly in the focal point (f = 30cm) of this lens. This creates a parallel beam behind this lens (in the pipe). This very wide spectrum beam is then focused by the second large lens on the side towards the camera (see Figure 2). Due to the wide spectrum and the non-chromaticity of this lens, the location of the focal point depends on the wavelength. This is the reason that a narrowband filter is placed in front of the camera. This filter is composed of two colored glass filters, namely a band filter and a long pass filter. The choice of these filters is determined by the final bandwidth and thus the resulting amount of light (energy) required to create adequately exposed images with current flashes. The bandwidth is preferably chosen as narrow as possible because the better the focus point for the background lighting is determined and the image of the projectile (shadow recording) is sharper the narrower this bandwidth. The focus point of the beam that is now very well determined should exactly coincide with the aperture in the camera lens.

If the beam path is optimally aligned, the aperture hole can be small and there is still sufficient background lighting. Because the aperture hole is small (F≈11 to 16), the image of the projectile is also as sharp as possible after focusing. The latter is therefore the second function of the lens system. Only the second large lens is involved in imaging the projectile. The first large lens ensures that a parallel light beam is present in the pipe. A sharp image of the projectile can now be captured during a flash against a well-lit background.

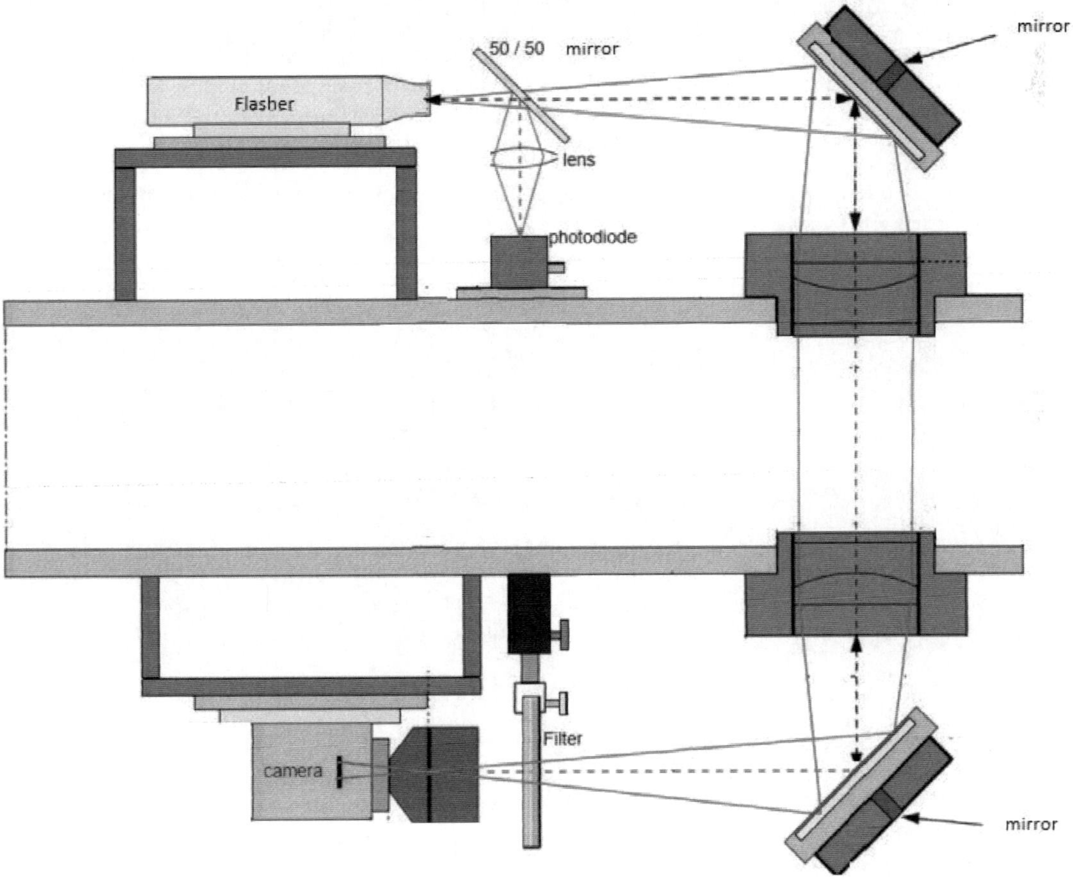

Figure 2. Schematic representation of the optical system of a calibration unit.

Figure 3 shows the calibration set-up in one of TNO's small caliber ranges. The calibration units are placed just in front and behind the light screens that are mounted on the wall.

https://doi.org/10.52202/078352-0040

Figure 3. Overview of the setup in one of the small caliber ranges. The red boxes on the right are wall mounted IR light screens, the black boxes on the left are the reflectors. The calibration units are positioned in the line of fire and just in front of light screen 1 and behind light screen 2.

To resume: the two calibration units are identical and combined create a precise velocity measurement system. The spacing between the units can be accurately varied by using the calibrated ruler. In that way different velocity measurement instruments with varying spacing (up to 2 m) can be calibrated. Figure 4 shows two pictures of both the calibration units seen from two sides.

Figure 4. Side view of the unit with camera (left) and flasher on the opposite side (right)

The calibration system is composed of two identical units containing:

- Optical components
- Lasers
- High speed flash units
- Light detection components
- Cameras
- Optical filters
- Laser light detector system

Used additional equipment consists of:

- Electronic counter (externally calibrated)

https://doi.org/10.52202/078352-0040

- Reference distance ruler (externally calibrated)
- Delay units
- Linux PC
- Image analysis software

Specifications of the system:

- Velocity range: 20 – 3500 m/s
- Distance over which the system can measure the average velocity: 800 – 2000 mm

The measurement accuracy depends on:

- Distance over which the measurement is carried out
- The projectile velocity (a very small contribution)
- The used camera lenses (our 35mm objective increases the accuracy 1.8 times)

The above variables results for a typical projectile velocity in a measurement accuracy of 0.019% at 1 meter distance between units and 0.009% at 2 meter distance between units.

3. CALIBRATION PROCEDURE

The units are accurately positioned in the line of fire and at each side of a velocity measurement instrument. The ruler is then placed in between the calibration units (see figure 5). A picture is taken by the cameras in the units, identifying and storing the ruler marker positions using the image analysis software. After that, the ruler is removed and a projectile is fired through the holes in the units. At the moment the projectile passes a laser screen at point C1 (see figure 1) a shadowgraph picture is taken with a short duration (nanoseconds) flasher. The same procedure holds for point C2.

Figure 5. The ruler with a marker (hole) positioned in one of the calibration units.

The projectile position is determined from the exact center of the projectiles in flight, eliminating the influence of yaw on the measured projectile position. With image analysis software the contour of the projectile is determined and from that also the center of the projectile. The center is digitally marked with a cross. The two pictures of the projectile in flight are combined with the digitally stored marker positions that are now visualized as dots in the photograph (see figure 6). The software is now able to determine the exact position of the projectiles relative to the marker positions.

https://doi.org/10.52202/078352-0040

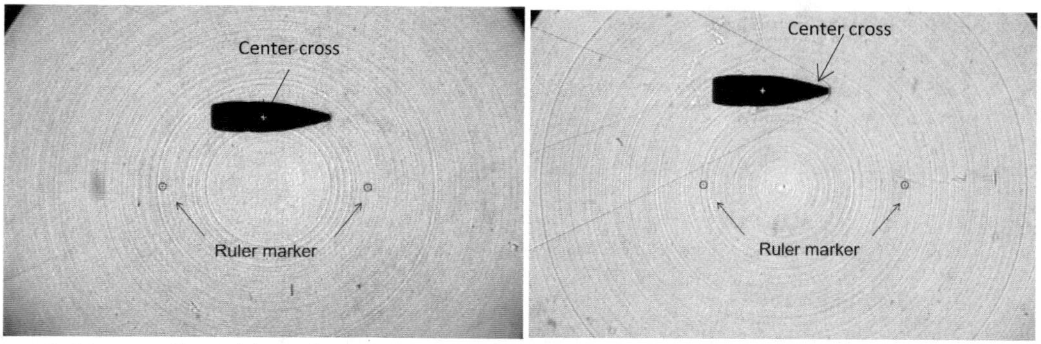

Figure 6. Two recordings of the projectile in flight from unit 1 (left in figure 1) and unit 2 (right). The digitally stored ruler markers are also projected in the pictures.

The time elapsed between taking the two pictures is measured exactly (within 0.2 nanoseconds) with an externally calibrated counter. The counter is started and stopped on C1 and C2 respectively by light sensors that are triggered by the flash. By means of a beam splitter the light is divided 50:50 between the picture and the light sensor. The picture on the left in figure 7 shows the flasher, beam splitter and sensor on one side of the unit. The picture on the right shows the camera on the opposite side looking at the flasher via a lens and mirror.

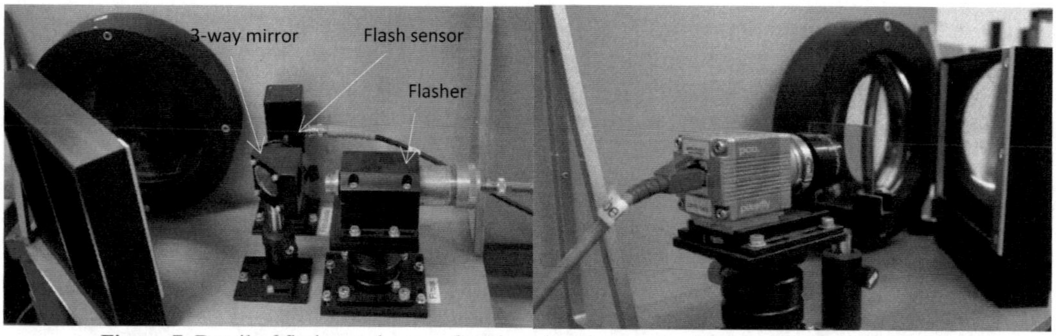

Figure 7. Detail of flasher and sensor for triggering the counter (left) and the camera (right).

The distance between the center of the projectiles on both pictures can be calculated from the distance between C1 and C2 combined with the position of the projectile center relative to each set of ruler markers, see figure 5. These ruler markers are digitally combined on the image of the projectile at C1 and C2.

With the retained information the average projectile velocity between C1 and C2 can now be calculated. Because C1 and C2 are as close as possible to the velocity measuring instrument that is being calibrated, the calculated average speed should be the same for both. When the values deviate, the distance that is used to calculate the velocity for the velocity instrument will be adjusted to a new value.

4. RESULTS

This chapter discusses some of the results found during annual calibrations of velocity light screens. As an ISO 17025 accredited testing institute, TNO must regularly calibrate its velocity measuring equipment. The calibration method described in this article is used for this and has been accepted by the Dutch Accreditation Council [2] (RVA) for application under ISO 17025.

There are mainly two type of velocity measuring systems that are used at TNO: stationary and movable systems. The small caliber shooting ranges have stationary Infrared (IR) light screens. These light screens are mounted on the wall of the shooting range and have a reflector mounted on the opposite wall (see also figure 3).

https://doi.org/10.52202/078352-0040

The movable systems can be separately installed in a test setup. Examples are wheeled laser units and Doppler radar. An example of a wheeled laser screen device can be seen in figure 8.

Figure 8. Example of a wheeled mobile velocity measurement instrument

Below results are shown for calibration measurements performed with both a stationary and a mobile velocity measurement system. The calibrations were performed for two projectile types:

- 9x19 mm Ball @ 400 m/s
- 7.62x51 mm Ball @ 850 m/s

Nose shape and velocity range are chosen differently, representing realistic conditions during a year of ballistic testing.

4.1 Stationary system

Multiple stationary systems are available in the small caliber ranges. Below calibration results are presented for one set IR light screens in small caliber range 1 (KKW1). Each IR light screen has both IR transmitter and receiver in one box and a reflector opposite of that at 3 m distance. The mutual distance between the IR light screens is 1.5 m.

In total 10 experiments are performed with each projectile. The velocity measurements of the IR light screens were subtracted from the calibration measurements resulting in a deviation ΔV. Therefore a negative value means that the IR light screen measurement was lower than the calibration measurement. A maximum deviation of - 0.6 m/s was found for one experiment with a 7.62mm projectile in 2018. In 2019 a maximum deviation of +0.3 m/s was found also with a 7.62 mm projectile. Figure 9 and 10 show the deviation between the velocity measurements with the IR light screens and the calibration device measured in 2018 and 2019 respectively.

https://doi.org/10.52202/078352-0040

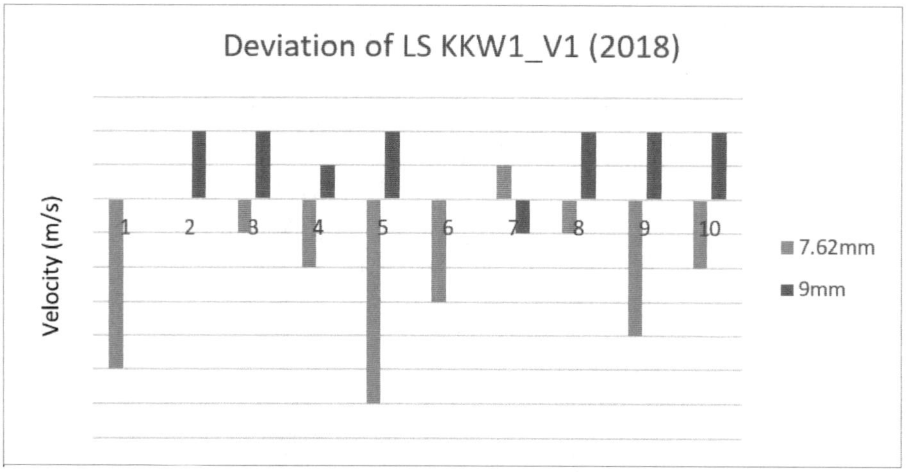

Figure 9. Deviation in velocity between the IR light screens and the calibration device, measured in 2018. A negative value means that the IR light screen measurement was lower than the calibration measurement.

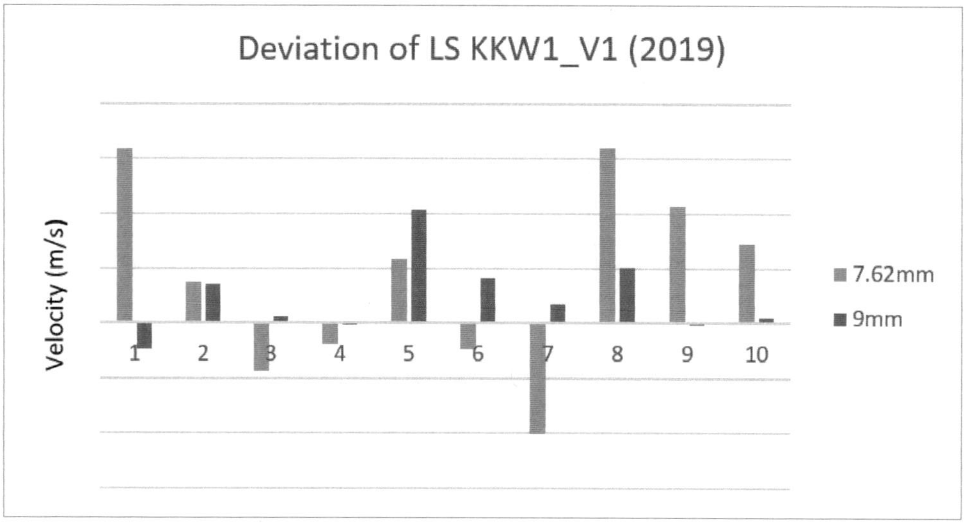

Figure 10. Deviation in velocity between the IR light screens and the calibration device, measured in 2019. A negative value means that the IR light screen measurement was lower than the calibration measurement.

The results in figures 9 and 10 show that the measured deviations between IR light screens and the calibration measurement are well within 1 m/s. It also shows a difference in deviation between the 9mm and the 7.62 mm bullet particularly in 2018. Only one 9 mm bullet measurement turns out to be lower than the calibration value. For the 7.62 mm bullet it's the other way around: only one measurement turns out to be higher than the calibration unit. The different results for 7.62 mm and 9 mm projectile calibration is unknown but may be caused by the operating principle of the light screen. When the projectile passes through a light screen, it obscures the IR receivers. This obscuring of the IR receivers and the underlying electronics in the device determine the trigger moment at a certain level. This trigger moment depends on the shape of the projectile and the speed of the projectile. The low-velocity 9 mm projectile's obscuration is more than the smaller, 7.62 mm, fast-flying bullet. Both start light screen and stop light screen probably have no problem with the 9 mm bullet, while for the 7.62 mm bullet it may be more critical because of its high speed and smaller size and therefore less obscuration. This might cause a slight discrepancy in sensitivity between the start light screen and the stop light screen. The results in 2019 show a less pronounced effect.

https://doi.org/10.52202/078352-0040

4.2 Mobile system

Examples of mobile velocity measurement systems are:

- Doppler radars
- Laser screens (wheeled)
- IR light screens (wheeled)

These systems are used as additional equipment for triggering a yaw measurement system or triggering of High Speed (HS) cameras. Doppler radars are mainly used to measure velocity decrease of small fragments or for more complex test setups.

Below calibration results are obtained for a wheeled laser screen (see figure 7). The system has a laser transmitter in one box and a receiver in the opposite box. The mutual distance between the boxes is 1m.

Also in this case 10 experiments are performed with each projectile type. In figure 11 and 12 the results are shown from the measurements in 2018 and 2019.

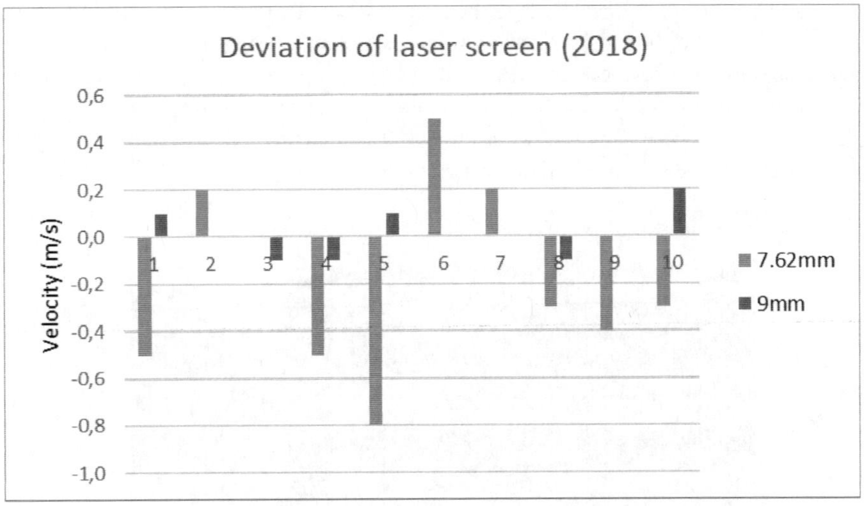

Figure 11. Deviation in velocity between the mobile laser screens and the calibration device, measured in 2018. A negative value means that the IR light screen measurement was lower than the calibration measurement.

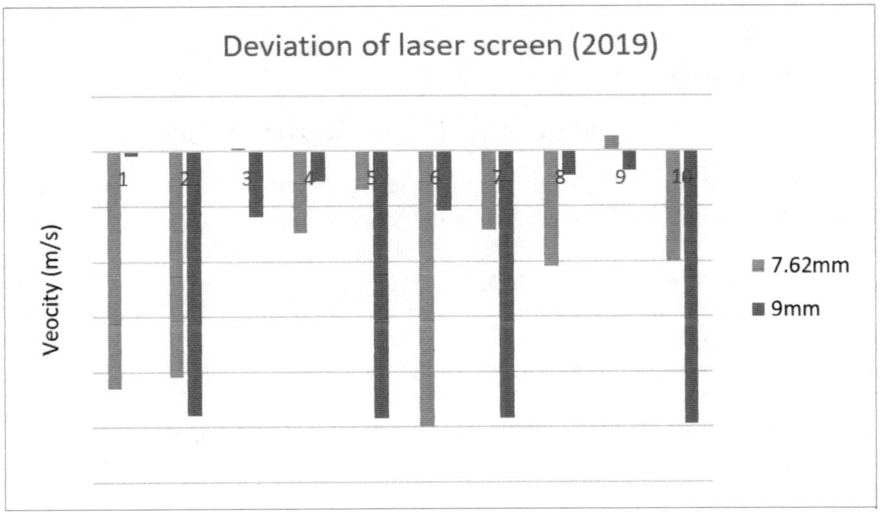

https://doi.org/10.52202/078352-0040

Figure 12. Deviation in velocity between the mobile laser screens and the calibration device, measured in 2019. A negative value means that the IR light screen measurement was lower than the calibration measurement.

The results show an increase in deviation of the mobile laser screens after one year of usage. The deviations measured in 2019 show all negative values except one. The maximum measured deviation is -1 m/s which is at the limit of what is acceptable according to some standards (like NIJ 0101.06 [1]). The increased deviation could possibly be caused by vibrations that occur by regularly moving of the systems. Since almost all values changed to negative, this can be an indication that the vibrations might have changed the mechanical position of detectors and/or lens systems.

4.3 Damaged optics

When ballistically testing armour materials, fragments are ejected from the target and projectile during impact. These fragments will fly also in lateral directions towards the velocity measurement instruments mounted on the wall and can sometimes damage the optics of the IR light screens. In figure 13 an example of a damaged lens in one of the IR light screens that was caused by a fragment impact is shown. When this damaged velocity measurement system was calibrated, it surprisingly turned out that the deviation was still within 1 m/s. Probably the cracked lens was not influencing the optic path of the IR beam at the projectile line of flight in this particular case. However after replacing the lens with a new one, a major correction had to be made. The position of the lens in its frame turned out to be more critical.

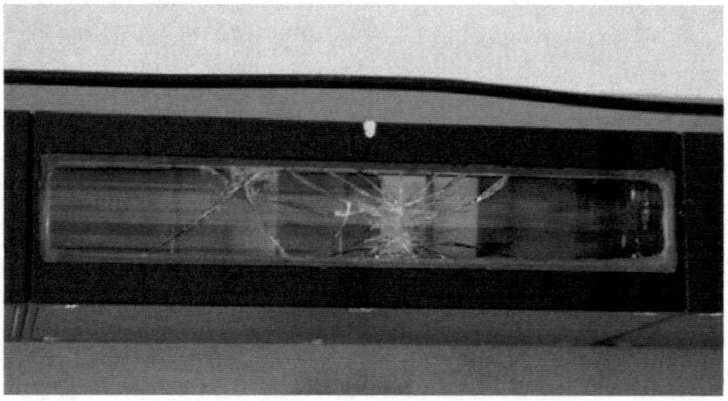

Figure 13. A cracked lens of an IR light screen due to fragment impact.

5. GENERAL EXPERIENCES

Below the main findings of calibrating different types of light screens at TNO are listed:

- No major deviations are witnessed for stationary IR light screens.
- Mobile systems should be aligned accurately in the line of fire since this can influence the velocity reading.
- Doppler radar systems can give fluctuating velocity readings because of software settings.
- A cracked lens from an IR light screen (see figure 8) showed (in this case) to have minor deviation on the velocity reading.
- Replacing a lens can have a major influence on the velocity reading, the deviation can be up to more than 3 m/s.
- Aging of electronics should be monitored and reflectors and lenses should be frequently cleaned to ensure an constant performance of the light screens.
- Calibration of the velocity instruments should be performed at least once a year (but preferably more frequent).

https://doi.org/10.52202/078352-0040

6. CONCLUSION

TNO has developed a calibration system for velocity measurement devices at ballistic ranges. Using this tool, it was found that our stationary IR light screens showed a small deviation over the past 2 years. It seems that mobile laser screens are more sensitive to vibration as results show a higher deviation after one year of usage. This could be caused by regularly moving the wheeled systems. These vibrations could influence the mechanical position of the detector and/or lens system. Externally calibrated light screens can be seen as mobile velocity systems that could also experience vibration during transport. Other deviations can be caused by:

- ageing of electronics
- dirt/dust or damage to components
- accuracy of installing (alignment)
- replacing of components (lenses)
- incorrect use of device and/or software (Doppler radar)
- rough handling of the equipment

Therefore regular maintenance, careful handling and regular calibration of light screens is necessary to keep the velocity readings reliable.

References

[1] Ballistic Resistance of Body Armor NIJ Standard-0101.06, July 2008

[2] Dutch Accreditation Council, https://www.rva.nl/en/scopes/details/L275

https://doi.org/10.52202/078352-0040

NIJ Standard 0101.07, Ballistic Resistance of Body Armor: Changes and Challenges

Mark E. Greene[1], Jeffrey Horlick[2], Daniel A. Longhurst[3], Lance L. Miller[3], Casandra Robinson[2] and Richard A. Sundstrom[3]

[1]*National Institute of Justice, 810 7th Street NW, Washington, DC, 20531, USA, mark.greene2@usdoj.gov*
[2]*National Institute of Standards and Technology, 100 Bureau Drive, Gaithersburg, MD 20899, USA*
[3]*RTI International, 3040 Cornwallis Road, Research Triangle Park, NC 27709, USA*

Abstract. The National Institute of Justice (NIJ) has completed development of its seventh revision to NIJ's standard for ballistic resistance of body armor, NIJ Standard 0101.07. This revision takes into consideration the changing landscape of policing in the U.S. and implements lessons learned over the past decade of testing using NIJ 0101.06. The improvements include more widespread stakeholder engagement, improved test methods and procedures, and updated test threats and protection levels, all of which result in better protection for officers wearing body armor. This standard was developed with the guidance and input of a large group of end users and technical experts, and it incorporates by reference ten ASTM standards that were developed by an expanded group of stakeholders, including manufacturers. The test threats have been updated to reflect the evolving threats faced by U.S. law enforcement end users, including a wider range and more severe ballistics threats. The protection levels have been modified accordingly, with the protection level nomenclature also changed for better clarity and to reduce officer and end user confusion. Several new rifle test threats were added, and one of those is a 7.62 x 39 mm mild steel core round, which is commonly seen by officers but is known to have huge variability in performance. That variability has driven the development of a surrogate test round to ensure consistency in testing. Additionally, the laboratory accreditation process has been addressed in order to inform technical assessors of potential issues to be aware of and specific processes to observe during laboratory assessments. This paper will describe the improvements and changes to the NIJ standard and related conformity assessment activities and how those changes ultimately lead to enhanced officer safety and protection.

1. U.S. LAW ENFORCEMENT: A DANGEROUS PROFESSION

Law enforcement is a dangerous profession. The United States (U.S.) Bureau of Labor Statistics reports that in 2018 police and sheriffs' patrol officers in the U.S. experienced a fatality rate on the job of 13.7 in 100,000 officers—four times higher than the overall fatality rate on the job of 3.5 in 100,000 workers across all industries in the U.S. that year [1]. A majority of the accidental fatalities each year are traffic-related, while a majority of the felonious fatalities are due to assaults with firearms. The Federal Bureau of Investigation's (FBI) Law Enforcement Officers Killed and Assaulted (LEOKA) statistics show that from 1987 through 2018, over 78,000 officers were assaulted with firearms. Furthermore, of the 1,874 officers feloniously killed in the line of duty by all means over that time span, 1,728 were feloniously killed by firearms. That equates to *over 92% of all felonious deaths* in the line of duty being due to firearms. Handguns alone accounted for at least 1,284 of those firearms fatalities [2].

Research funded by the National Institute of Justice (NIJ) examined detailed information on 1,789 officers who were killed or severely injured between 2002 and 2011 due to assaults with firearms, knives, or other cutting instruments. This information, taken from the FBI's LEOKA dataset, was analyzed for cases where officers were shot in the torso. Analysis revealed that among the 637 officers who were shot by a firearm in the torso, those who wore body armor were 76% less likely to be killed than those who did not wear armor, controlling for an array of individual and incident characteristics [3]. This research is consistent with previous reporting on the benefits of wearing body armor to save lives [4].

2. NIJ STANDARD 0101.07: BALLISTIC RESISTANCE OF BODY ARMOR

NIJ published its first performance standard for ballistic-resistant police body armor in 1972 [5]. The current revision is the seventh, NIJ Standard 0101.07, *Ballistic Resistance of Body Armor*, which includes improved test methods for female body armor and updated body armor protection levels that incorporate additional rifle threats faced by U.S. law enforcement, as discussed previously [6]. This revision takes into consideration the changing landscape of policing in the U.S. and implements lessons learned over the past decade of testing using NIJ Standard 0101.06, published in 2008 [7]. The

https://doi.org/10.52202/078352-0041

improvements include more widespread stakeholder engagement, improved test methods and procedures, and updated test threats and protection levels, all of which result in better protection for officers wearing body armor.

NIJ Standard 0101.07 was developed with the guidance and input of a large group of end users and technical experts, and it incorporates by reference ten ASTM standards that were developed by an expanded group of stakeholders, including manufacturers. NIJ established a Special Technical Committee to aid the development of NIJ Standard 0101.06 in the 2004-2008 timeframe and has continued this approach for developing NIJ Standard 0101.07. NIJ also hosted a workshop in 2018 for manufacturers of body armor and protective materials to provide an update on the progress of the standard's development and field questions from the manufacturing community [8]. Furthermore, ASTM provides a forum and infrastructure that enables all interested stakeholders to collaborate, discuss differences and commonalities, and come to consensus on standards, test methods, and practices [9].

NIJ published a draft of the new standard as well as the new threat specification in 2018 through the *Federal Register* to request comments and input from the public, a practice commonly used by U.S. Government agencies to seek input on important guidance, policy, or regulations they plan to publish [10,11]. The threat specification document is published as a companion to NIJ Standard 0101.07 and, as a standalone specification, may also enable testing of a variety of ballistic-resistant equipment, not just ballistic-resistant body armor, against contemporary U.S. law enforcement threats. However, the threat specification itself does not define any test methods like the body armor standard does.

The test threats have been updated to reflect the evolving threats faced by U.S. law enforcement end users, including a wider range and more severe ballistics threats. Several new rifle test threats were added, and one of those is a 7.62 x 39 mm mild steel core (MSC) round, which is commonly seen by officers but is known to have huge variability in performance. That variability has driven the development of a surrogate test round to ensure consistency in testing, however NIJ Standard 0101.07 makes use of a factory round until the surrogate test round development activities are completed. The protection levels have been modified accordingly, with the protection level nomenclature also changed for better clarity and to reduce officer and end user confusion.

3. ASTM TEST METHODS: THE NEW BUILDING BLOCKS OF BODY ARMOR STANDARDS

Unlike previous revisions of the NIJ body armor standard which have been comprehensive, standalone documents, NIJ Standard 0101.07 incorporates ten standard test methods and practices published by ASTM. In 2013, the U.S. Army, NIJ, and the National Institute of Standards and Technology (NIST) began a partnership to harmonize the standards and their implementation for ballistic-resistant vests. The federal agencies chose to work through ASTM's E54 Committee on Homeland Security Applications to develop standard test methods and practices for the purpose of improving and validating methods, increasing consistency between test laboratories, and ultimately increasing confidence in ballistic-resistant equipment.

Incorporation of relevant ASTM standards into NIJ standards and U.S. Army requirements and testing documents affords the opportunity to harmonize laboratory test procedures and practices for both law enforcement and military ballistic-resistant armor and other ballistic-resistant equipment while allowing those end user communities ultimate control over product specifications, such as the specific threats against which their equipment must protect. Table 1 below lists the standardized test methods, practices, and specifications published by ASTM that are incorporated into NIJ Standard 0101.07. These include standardized methods for laboratory measurements, female armor testing, and data collection, among others.

 https://doi.org/10.52202/078352-0041

Table 1. ASTM standards that are incorporated into NIJ Standard 0101.07.

ASTM Standard	Description
ASTM E3004, *Standard Specification for Preparation and Verification of Clay Blocks Used in Ballistic-resistance Testing of Torso Body Armor*	This standardizes a method for preparation and verification of backing assemblies used in ballistic-resistant torso body armor testing and is intended to reduce variability within and between laboratories. The backing assembly is a clay block that contains ROMA Plastilina No. 1® clay as the backing material, which serves as a witness material and provides a measurable indication of test item performance.
ASTM E3005, *Standard Terminology for Body Armor and Related Items*	This standardizes terminology for body armor providing protection against ballistic threats, stabbing, fragmentation, blunt impact, or a combination of threats. The intent of this terminology is to have terms and definitions that are applicable across federal agencies, law enforcement and corrections agencies, testing and certification bodies, and manufacturers.
ASTM E3062/E3062M, *Standard Specification for Ballistic Test Range Configuration for Small Arms and Fragmentation Testing of Ballistic-resistant Items*	This specifies indoor ballistic test range requirements for small arms and fragmentation testing of the following ballistic-resistant items: soft body armor, hard armor plates, body armor accessories, shields, and helmets. The specification includes requirements for range geometry, range conditions, test equipment, instrumentation, and measurement procedures. The purpose is to specify critical test range parameters in order to achieve consistency and repeatability among test ranges.
ASTM E3068, *Standard Test Method for Contact Measurement of Backface Deformation in Clay Backing During Body Armor Testing*	This describes test methods for the contact measurement of backface deformation (BFD) in clay backing caused by a test threat that produces a partial penetration in a body armor test item. The purpose is to achieve consistent measurements between laboratories and reduce differences that could result from using different measurement techniques. This test method is applicable to testing of planar and nonplanar soft armor and hard armor.
ASTM E3078/E3078M, *Standard Practice for Conditioning of Hard Armor Test Items*	This applies to the conditioning of hard armor test items prior to ballistic resistance testing. This practice describes requirements for subjecting test items to conditions intended to assess the capability of the test item to withstand environmental conditions that induce responses similar to those that might result from conditions encountered during use.
ASTM E3086, *Standard Practice for Creating Appliques for Use in Testing of Nonplanar Soft Body Armor Designed for Females*	This specifies a procedure for creating appliques (e.g., build-up of clay) for use behind nonplanar, soft body armor test items and affixing the appliques to the clay block. The purpose is to specify critical parameters for creating appliques in order to improve consistency of the test setup between laboratories. This practice describes a single applique shape applicable only to nonplanar, soft body armor designed for females.
ASTM E3107/E3107M, *Standard Test Method for Resistance to Penetration and Backface Deformation for Ballistic-resistant Torso Body Armor and Shoot Packs*	This standardizes testing the resistance to ballistic penetration and to BFD for ballistic-resistant torso body armor and shoot packs. This test method is intended for testing of soft body armor, hard armor plates, in conjunction with armor, and shoot packs mounted on a clay block as the backing assembly.
ASTM E3110/E3110M, *Standard Test Method for Collection of V_x Ballistic Limit Data for Ballistic-resistant Torso Body Armor and Shoot Packs*	This describes procedures for collecting ballistic limit data for ballistic-resistant torso body armor and shoot packs. Several commonly used test item mounting procedures and ballistic limit procedures are included in this test method; however, this test method does not specify computation or performance criteria for the ballistic limit.
ASTM E3112/E3112M, *Standard Test Method for Ballistic-resistant Products and Shoot Packs*	This standardizes assessing the resistance to complete penetration by projectiles and applies to ballistic-resistant products or shoot packs, either flexible or rigid, that are intended

https://doi.org/10.52202/078352-0041

	to provide protection against projectiles from small arms. This test method is applicable to products or shoot packs that are constructed with identical layups of ballistic materials over the entire product or shoot pack. This test method does not assess blunt impact resistance.
ASTM E3192/E3192M, *Standard Practice for Soft Body Armor Conditioning By Tumbling*	This describes subjecting test items to conditions intended to provide some indication of the test item ability to withstand conditions of heat, moisture, and mechanical wear (folding and abrasion) that might be encountered during wear. This practice is intended for soft armor for law enforcement applications and applies only to the conditioning of soft body armor via temperature, humidity, and tumbling exposure prior to ballistic resistance testing.

4. NEW NIJ THREAT SPECIFICATION DEFINES BALLISTIC THREATS TO U.S. LAW ENFORCEMENT

NIJ opted to develop a standalone specification of ballistic threat levels and associated test ammunition rather than specify the information directly in NIJ Standard 0101.07, as had been done in NIJ 0101.06 and prior revisions. Its primary purpose will be to specify the test threats—including projectiles and reference velocities—identified by U.S. law enforcement as representative of prevalent threats in the United States which will be used to test ballistic-resistant equipment for U.S. law enforcement applications. It is incorporated into NIJ Standard 0101.07 for body armor and may be incorporated into future NIJ standards for ballistic-resistant helmets and ballistic-resistant shields.

Threat level nomenclature has been revised from previous NIJ standards to be more descriptive of threats and to reduce possible confusion among law enforcement end users of body armor. Level II and Level IIIA have been replaced with "NIJ HG1" and "NIJ HG2," respectively, to represent handgun (HG) threats, as listed in Table 2 below. The lowest threat Level IIA that appears in NIJ Standard 0101.06 has been dropped with no equivalent threat level in the new NIJ specification.

Table 2. Proposed NIJ handgun (HG) threat levels and associated test ammunition.

NIJ HG Threat Level	Test Threat Ammunition	Manufacturer & Model ID	Reference Velocity
NIJ HG1	9mm Luger full metal jacketed (FMJ) round nose (RN) 124 grain	Remington #23558	1305 ft/s (398 m/s)
	.357 Mag jacketed soft point (JSP) 158 grain	Remington #22847	1430 ft/s (436 m/s)
NIJ HG2	9mm Luger FMJ RN 124 grain	Remington #23558	1470 ft/s (448 m/s)
	.44 MAG jacketed hollow point (JHP) 240 grain	Speer #4453 or #4736	1430 ft/s (436 m/s)

Level III and Level IV have been revised to three levels representing rifle (RF) threats as "NIJ RF1," "NIJ RF2," and "NIJ RF3." These rifle threats are listed in Table 3 below. Noteworthy about the specification is that several new rifle test threats were added to the ones specified in NIJ Standard 0101.06. One of those threats is a 7.62 x 39 mm mild steel core (MSC) round, which is commonly seen by law enforcement officers but is known to have huge variability in performance. That variability has driven the development of a surrogate test round to ensure consistency in testing; however, NIJ Standard 0101.07 makes use of sealed lots of Type 56 factory rounds of Chinese origin until the surrogate test round development activities are completed.

NIJ engaged ammunition experts to investigate the possible solutions to define a 7.62x39mm MSC test round for the purposes of NIJ certification testing and Follow-up Inspection Testing (FIT) testing. Physical characteristics of the ammunition as well as factors such as availability of supply were factored into the assessment of the various factory rounds available. An audit procedure, included as an appendix

https://doi.org/10.52202/078352-0041

to the NIJ threat specification document, was used for a multi-laboratory evaluation of the candidate ammunition, which performed consistently across the NIJ-approved laboratories.

Use of a purpose-built surrogate round for testing to 7.62x39mm MSC has many advantages and should prove to be the most sustainable solution in the long run, rather than relying on the finite supply of a factory round. NIJ had previously proposed use of a surrogate under development by the U.K. Government and has also coordinated with the U.S. Department of Defense's Combatting Terrorism Technical Support Office on the development of a U.S.-based surrogate [12]. Given the status of the surrogate round development, NIJ determined that the best solution at the time of publication is to specify a factory round for certification testing.

Table 3. Proposed NIJ rifle (RF) threat levels and associated test ammunition.

NIJ RF Threat Level	Test Threat Ammunition	Manufacturer & Model ID	Reference Velocity
NIJ RF1	7.62x51mm M80 ball NATO FMJ steel jacketed spire point boat tail (BT) 147 +0/-3 grain	U.S. military supply or rounds meeting NATO specifications	2780 ft/s (847 m/s)
	7.62x39mm mild steel core (MSC) ball ammunition Type 56 from Factory 31 123 grain	Factory 31 Ammunition conforming to specifications in sealed lots only.	2400 ft/s (732 m/s)
	5.56mm M193 BT 56 +0/-2 grain	U.S. military supply or rounds meeting NATO specifications	3250 ft/s (990 m/s)
NIJ RF2	7.62x51mm M80 ball NATO FMJ steel jacketed spire point BT 147 +0/-3 grain	US military supply or rounds meeting NATO specifications	2780 ft/s (847 m/s)
	7.62x39mm mild steel core (MSC) ball ammunition Type 56 from Factory 31 123 grain	Factory 31 Ammunition conforming to specifications in sealed lots only.	2400 ft/s (732 m/s)
	5.56mm M193 BT 56 +0/-2 grain	U.S. military supply or rounds meeting NATO specifications	3250 ft/s (990 m/s)
	5.56mm M855 BT 61.8 ± 1.5 grain	U.S. military supply or rounds meeting NATO specifications	3115 ft/s (950 m/s)
NIJ RF3	30.06 M2 Armor Piercing (AP) FMJ spire point AP 165.7 +0/-7 grain	U.S. military supply or rounds meeting NATO specifications	2880 ft/s (878 m/s)

5. BODY ARMOR COMPLIANCE TESTING

NIJ operates a body armor certification program called the NIJ Compliance Testing Program (CTP) to provide U.S. law enforcement confidence that the body armor they purchase and use performs according to minimum performance requirements to protect against common handgun and rifle threats. NIJ established body armor compliance testing in the late 1970s after publishing the first 0101.00 standard, and the NIJ CTP has evolved along with the standard through the decades.

Today, the NIJ CTP is the recognized authority on body armor for law enforcement use—not only in the United States, but across the world. Use of NIJ-certified body armor is ubiquitous among U.S. law enforcement agencies, and many law enforcement agencies outside the U.S. make use of NIJ standards and NIJ-certified armor as well. While participation in the program by manufacturers is voluntary, it is recognized by industry as the standard in body armor quality assurance, with manufacturers representing approximately 100 manufacturing locations and 19 countries participating. Since the CTP began accepting armor submissions to meet the requirements of NIJ Standard 0101.06 in 2009, over 1,400

https://doi.org/10.52202/078352-0041

unique models of ballistic-resistant body armor have been submitted to the CTP for compliance testing through the end of 2019.

With the publication of NIJ Standard 0101.07, the NIJ CTP anticipates accepting armor for certification to the new standard sometime in 2021. The primary purpose of NIJ Standard 0101.07 is for use by the NIJ CTP for testing, evaluation, and certification of ballistic-resistant body armor. The standard will be used by both ballistics laboratories that test body armor and body armor manufacturers participating in the NIJ CTP. The standard will be added by the National Voluntary Laboratory Accreditation Program (NVLAP) to the Personal Body Armor scope of accreditation used to accredit ballistics laboratories that participate in the NIJ CTP.

While NIJ will begin to certify body armor to the new NIJ Standard 0101.07 and will publish a new Compliant Products List (CPL) for those armor models, it will also continue to accept armor for certification to NIJ Standard 0101.06 during a transitional period. NIJ will also continue to maintain its CPL for armor models compliant with NIJ Standard 0101.06 for a period of time and will continue to require FIT on these models. This will allow law enforcement agencies the time needed to transition their equipment as smoothly as possible over a reasonable amount of time.

6. CONCLUSION

NIJ develops body armor performance standards, and the seventh revision, NIJ Standard 0101.07, *Ballistic Resistance of Body Armor*, includes improved test methods for female body armor and updated body armor protection levels that incorporate additional rifle threats faced by U.S. law enforcement as well as more widespread stakeholder engagement, all of which result in better protection for officers wearing body armor. Unlike previous revisions of the NIJ body armor standard which have been comprehensive, standalone documents, NIJ Standard 0101.07 incorporates ten standard test methods and practices published by ASTM. NIJ has also developed a standalone specification of ballistic threat levels and associated test ammunition rather than specify the information directly in NIJ Standard 0101.07, which may also enable more consistent testing of a variety of ballistic-resistant equipment, not just ballistic-resistant body armor, against contemporary U.S. law enforcement threats. Several new rifle test threats were added, including a 7.62 x 39 mm mild steel core (MSC) round, which is commonly seen by law enforcement officers but is known to have huge variability in performance. NIJ has specified a factory round for certification testing but is also pursuing the use of a surrogate round as a long-term solution. With the publication of NIJ Standard 0101.07, the NIJ CTP anticipates accepting armor for certification to the new standard sometime in 2021. However, NIJ will also continue to maintain its list for NIJ Standard 0101.06-compliant armor for a period of time to allow law enforcement agencies the time needed to transition their equipment as smoothly as possible over a reasonable amount of time.

Acknowledgments

The authors would like to acknowledge the ASTM staff and the participants in the ASTM E54 working groups who contributed to the development of the ASTM standards referenced in this paper. The authors would like to acknowledge the participation of the NIJ Special Technical Committee on Body Armor for contributing to the development of the two new NIJ documents discussed in this paper. Individuals who provided public comments for the draft of the new standard as well as the new threat specification are acknowledged for their contributions to the revision process. Finally, the laboratory personnel and subject matter experts who have been consulted on various aspects of ballistic testing and evaluation of body armor and ammunition during the revision process are acknowledged for their insights and perspectives.

References

[1] "Injuries, Illnesses, and Fatalities: Fact Sheet – Police Officers 2018," U.S. Department of Labor, Bureau of Labor Statistics, July 2020, https://www.bls.gov/iif/oshwc/cfoi/police-2018.htm.

[2] U.S. Department of Justice, Federal Bureau of Investigation, Law Enforcement Officers Killed and Assaulted, https://ucr.fbi.gov/leoka.

[3] Weiwei Liu and Bruce Taylor, "The effect of body armor on saving officers' lives: An analysis using LEOKA data," *Journal of Occupational and Environmental Hygiene*, Vol. 14, No. 2, 73-80, 2017, https://doi.org/10.1080/15459624.2016.1214272.

 https://doi.org/10.52202/078352-0041

[4] Tom LaTourrette, "The Life-Saving Effectiveness of Body Armor for Police Officers," *Journal of Occupational and Environmental Hygiene* 7, no. 10, (2010): 557-562. doi:10.1080/15459624.2010.489798.

[5] *NILECJ Standard on the Ballistic Resistance of Police Body Armor*, NILECJ Standard 0101.00, March 1972, https://www.ncjrs.gov/pdffiles1/Digitization/7037NCJRS.pdf.

[6] M. E. Greene, J. Horlick, D. A. Longhurst, L. L. Miller, M. O'Shea, D. Otterson, C. Robinson, D. A. Stoe, and R. A. Sundstrom, "The Next Revision of the NIJ Performance Standard for Ballistic Resistance of Body Armor, NIJ Standard 0101.07: Changes to Test Methods and Test Threats," *15th Personal Armour Systems Symposium*, Washington, DC, October 1-5, 2018.

[7] *Ballistic Resistance of Body Armor*, NIJ Standard-0101.06, July 2008, https://www.ncjrs.gov/pdffiles1/nij/223054.pdf.

[8] "Body Armor Manufacturer Workshop," *Federal Register*, 83 FR 40568, August 15, 2018, https://www.federalregister.gov/d/2018-17466.

[9] C. Robinson, D. Stoe, and J. Zheng, Improving the Testing of Ballistic-Resistant Body Armour through Harmonization of Standards," *14th Personal Armour Systems Symposium*, Amsterdam, Netherlands, September 19-23, 2016.

[10] "Request for Public Comment on Proposed Revision of NIJ Standard 0101.06, Ballistic Resistance of Body Armor," *Federal Register*, 83 FR 7776, February 22, 2018, https://www.federalregister.gov/d/2018-03674.

[11] "Request for Public Comment on Proposed Specification Threat Levels and Associated Ammunition To Test Equipment Intended To Protect U.S. Law Enforcement Against Handguns and Rifles," *Federal Register*, 83 FR 7776, February 22, 2018, https://www.federalregister.gov/d/2018-03672.

[12] Combatting Terrorism Technical Support Office/ Technical Support Working Group (CTTSO/TSWG), Broad Agency Announcement 19S3011, January 7, 2019, https://legacy-bids.cttso.gov/TSWG/bids.nsf/0/35F93ED00B35E3F885258313004F54C9/$FILE/19S3011+2019-01-07.pdf.

https://doi.org/10.52202/078352-0041

Test and Assessment Methods to Evaluate Combat Helmets for Rotation-Induced Injury

T. Plaisted[1] and R. Neice[1,2]

[1]CCDC Army Research Laboratory, Composite and Hybrid Materials Branch, 6300 Rodman Road, Aberdeen Proving Ground, 21005 MD, USA
thomas.a.plaisted.civ@mail.mil
[2]Oak Ridge Associated Universities, Fellowship Programs P.O. Box 117 MS-36, Oak Ridge, 37831 TN, USA

Abstract. Military helmets are typically evaluated for blunt impact protection with a drop tower impact apparatus, where head motion is restricted to a single direction and acceleration is measured with a uniaxial accelerometer. Experimental methods that introduce rotational kinematics have been developed for the evaluation of recreational helmets across a number of sports and activities. This is in response to growing consensus among the medical community that excessive rotational kinematics can cause traumatic brain injury and further reflects the reality that typical head impacts involve both linear and rotational loading. This paper reports on a blunt impact test method of combat helmets that imparts combined linear and rotational loading to a headform. To better relate experimental head kinematic outputs to soldier experience, a recently published brain deformation assessment tool is explored that translates kinematic data into a predicted brain strain.

1. INTRODUCTION

An estimated 1.7 million Traumatic Brain Injuries (TBIs) occur in the United States each year as a result of automotive crashes, recreational activities, sports, falls, or other accidents [1]. TBI is also a significant threat to the United States military and has been described as the signature injury of recent conflicts in Iraq and Afghanistan. The threat of TBI in military populations is magnified due to the unpredictable and physically demanding nature of both training and combat environments. The Defense and Veterans Brain Injury Center (DVBIC) has tracked the incidence TBI from 2000 to 2019, reporting that over 400,000 TBIs were diagnosed in military service members [2]. Over 80% of the diagnosed TBIs in the military are classified as concussions, otherwise referred to as mild TBI[1].

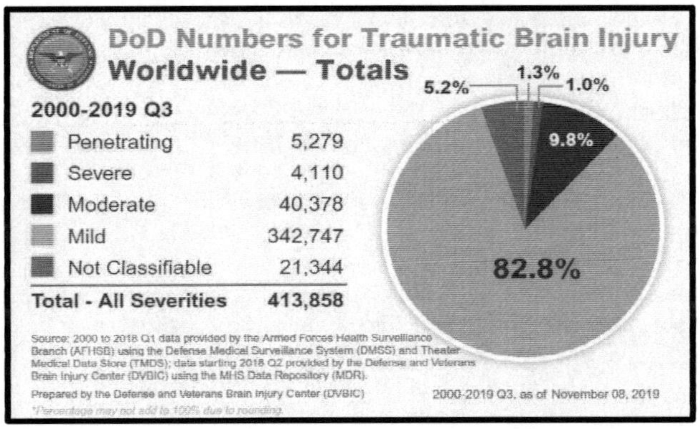

Figure 1. Medically diagnosed TBIs in United States military from 2000 to 2019 (DVBIC) [2].

The military related TBI considered in this research is solely blunt impact induced TBI. Other research groups have produced findings that probe the phenomenon of blast induced TBI [3]. In the soldier population, blunt impacts to the head can occur in a variety of situations, many of which are similar to those experienced by the general population, but may be more severe due to the inherent hazardous conditions of training and combat scenarios. Likewise these impacts rarely, if ever, consist of purely translational or rotational motion, but rather some combination of the two. The principle

[1] Mild TBI is classified by DVBIC as including the following symptoms: confusion or disorientation for less than 24 hours, memory loss lasting less than 24 hours, and loss of consciousness for up to 30 minutes. Brain imaging associated with mild TBI cases as classified by DVBIC do not contain any abnormalities.

https://doi.org/10.52202/078352-0042

protection a soldier will have from all of these blunt impact scenarios is the combat helmet. Improvements to helmets can aid in reducing TBI if there exists better understanding of brain injury causations and methods that evaluate a helmet's ability to mitigate such causations.

Recent as well as historical biomechanics research suggests that mitigating excessive rotational kinematics during head impacts is a critical component in developing effective protection against impact induced TBI. The theory that rotational kinematics are the primary cause of brain deformation was first proposed in 1943 [4]. This theory has grown stronger in recent years, particularly with advancements in biomechanics tools such as human brain finite element models that have been used to demonstrate the severity of brain deformation resulting from rotational motion versus translation alone [5]. This response is due in large part to the inherent tissue level characteristics of the brain, where the bulk modulus is understood to be several orders of magnitude greater than its shear modulus [6]. Brain deformation has been proposed as a metric that can be related to the risk of sustaining a TBI based on available injury data [7]. Multiple kinematic based metrics have been proposed [8]. For example, studies have used angular velocity and acceleration to propose injury risk thresholds related to football helmet collisions to predict the likelihood of a TBI based on head kinematics [9]. The obvious advantage of such metrics is that head kinematics can be more easily measured than brain deformation, particularly on living human subjects. An improved understanding of the link between head kinematics and TBI risk is essential for creating effective performance standards to evaluate the protective capacity of helmets. Further there is a need to assess the performance of military helmets in more relevant impact conditions that impart rotational head kinematics, in conjunction with the current blunt impact test methods that solely evaluate a helmet's ability to protect against linear translation accelerations.

2. MATERIAL AND METHODS

The current Army blunt impact performance standard for the ACH is based on a modified version of the Federal Motor Vehicles Safety Standard 218 (FMVS 218) for testing motorcycle helmets [10]. The test consists of a headform and helmet assembly that is dropped from a height to achieve a specific striking velocity, guided by a monorail, so that impact occurs onto a rigid stationary anvil with a hemispherical shape. The headform specified is a magnesium DOT headform instrumented with a single accelerometer. The current performance standard requires that helmets must limit linear headform acceleration to less than 150 g (g-force, 9.81 m/s^2) for impact at 3.05 m/s (10 ft/s) at each impact site. The seven impact sites include the crown, front, rear, sides, and nape regions. There are currently no performance standards in place to evaluate the effectiveness of military helmets to mitigate rotational kinematics. The purpose of this research study is to determine suitability and feasibility of using an alternative format based on a pneumatic ram impact to evaluate military helmet technologies.

The pneumatic ram impact test is commonly employed in the evaluation and certification of American football helmets. The largest professional football organization, the National Football League (NFL), has adopted the pneumatic ram test method to certify or ban certain helmet designs (NFL Helmet Test Protocol 2019) [11]. Additionally NOCSAE, the National Operating Committee on Standards for Athletic Equipment, which certifies all amateur football helmets in the United States, instituted a similar test as part of its helmet as a certification process (NOCSAE DOC (ND) 081) [12]. The pneumatic ram format was chosen to test helmets in this study.

Within the pneumatic ram format, also known as the linear impactor format, the ram is propelled by compressed air, launching it at a repeatable velocity in a linear motion toward the headform. A head-neck assembly, initially at rest, is attached to a sliding carriage and is free to translate along the direction of the impact vector once struck. The head and neck may be rotated and posed to achieve desired head orientation with respect to the ram. The manufacturer of the machine used in this study was Biokinetics LLC. The impacting ram mass is 14.3 kg and the mass of the head-neck and sliding table assembly is 17.7 kg. The impactor on the end of the ram consists of a metallic cap with a radius of curvature of 127 mm backed by a compliant elastomer, as used in NOCSAE testing.

The neck presents a critical boundary condition in defining the degree of rotation that the head will experience during an impact. Within this study the Hybrid III head and neck were chosen to measure impact kinematics, where rotational motion is achieved at the head-neck joint, as well as in the flexible neck itself in this setup. The tension in the Hybrid III neck is controlled by cables that run along its primary axis, which in this study were torqued to 1.4 N.m (12 in.lb_f) and checked intermittently between testing to maintain consistent response. The Hybrid III neck is the most widely used surrogate in evaluating helmets for blunt impact protection due in part to its availability and compatibility with the Hybrid III and NOCSAE headforms [13,14]. However, the biofidelity of the Hybrid III neck in impact conditions has been a topic of debate, as it has been validated for inertial loadings specific to automotive

https://doi.org/10.52202/078352-0042

crashworthiness testing. A recent head impact study suggests that the Hybrid III neck may be too stiff in lateral impact scenarios, when compared to neck surrogates specifically validated for lateral inertial loading [15]. Investigations into the human neck response during short duration impact events, including the effect of muscle activation and pre-tensioning, are ongoing in the field of biomechanics and can be potentially used to improve upon available neck surrogate designs in the future [16].

The Hybrid III head is instrumented with a nine-accelerometer array package, or NAP, which consists of nine linear accelerometers in a 3-2-2-2 arrangement. The first set of three accelerometers in this package are located at the center of gravity of the Hybrid III head, while the additional three sets of two accelerometers are set a fixed distance away. This accelerometer package allows for the calculation of linear and rotational kinematics of the head as detailed in Padgaonkar 1975 [17]. A NAP check tool is used at the beginning of each test series to verify that the accelerometers of the NAP are giving consistent results. This spreadsheet based tool is made available by Biocore LLC [18].

The example helmet tested in this methods paper is an Advanced Combat Helmet (ACH), a widely fielded combat helmet that consists of a ballistic protective shell, 7-pad suspension system, and a 4-point chin to nape retention strap system. The suspension system (Zorbium Action Pad – ZAP, Team Wendy) employs seven pads made from a polyurethane based foam enclosed in a moisture resistant membrane. The ACH meets the existing U.S. Army blunt impact performance standard as described in the purchase description, where blunt impact testing by traditional monorail format is performed at 3.05 m/s (10 ft/s) at 7 specific impact locations around the helmet [19]. Those impact locations were mirrored in the pneumatic ram testing described herein, where Figure 2 illustrates a single frame during an impact to the rear location on the Hybrid III head-neck fitted with an ACH, along with accompanying kinematic traces.

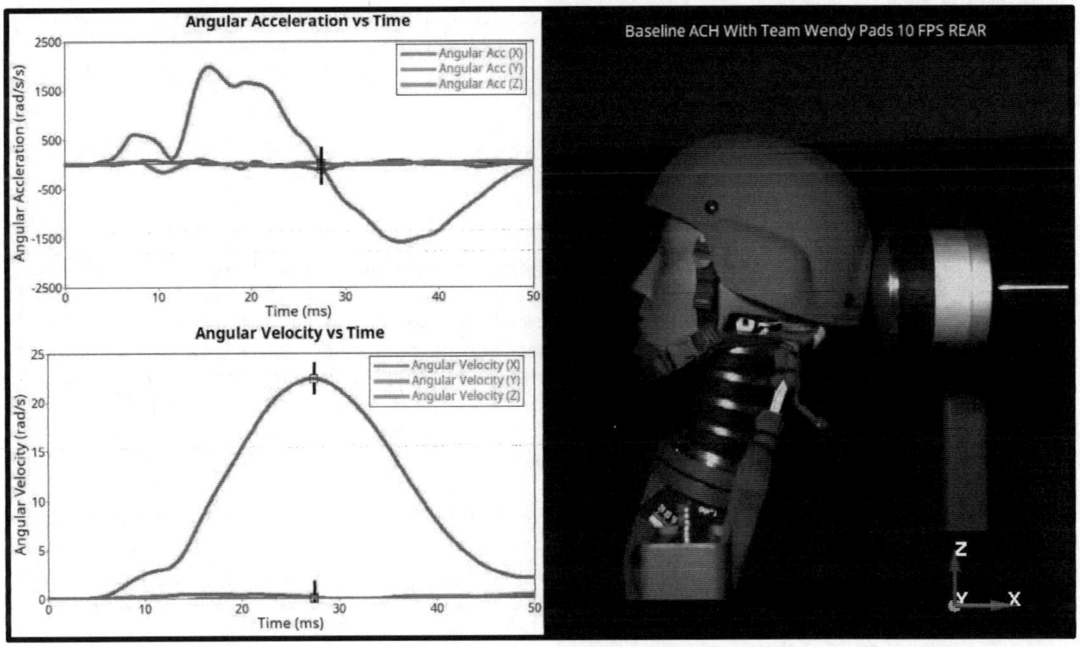

Figure 2. Rotational acceleration and velocity data traces with a synched video frame at the time of peak angular velocity of the Hybrid III head.

3. DISCUSSION

The assessment of combined linear and angular kinematic data and its relation to brain injury is a topic of ongoing research in the biomedical community. While the tissue level damage mechanisms and related injury thresholds are being studied and developed, brain deformation has been identified as one of the metrics often correlated to TBI, and is commonly estimated through computational modeling by calculating maximum principal strain (MPS) using finite element analysis [20,21,22]. On its own this metric can serve as a convenient quantitative assessment for relative comparisons between head protection systems. Towards injury prediction, a relationship between MPS and injury risk at various severity levels has been proposed by National Highway Transportation Safety Administration (NHTSA) based on scaled animal injury data, automotive crash and football impact data using ATDs, and strain

https://doi.org/10.52202/078352-0042

calculations using the SIMon and GHBMC brain models [7]. These relationships serve as a starting point while more extensive testing and human response data can be applied across a number of loading scenarios and durations to develop injury risk curves with wide acceptance.

To alleviate the computational expense and time of modeling every impact scenario, various analytical tools have been developed to offer a quick assessment and prediction of brain strain. The Diffuse Axonal Multi-Axis General Evaluation (DAMAGE) is a multibody kinematic assessment tool that translates linear and rotational kinematic data traces into a predicted maximum brain strain and has demonstrated high correlation to finite element model results across a wide range of impact durations [23]. In contrast to other models that rely solely on the peak linear and/or angular kinematics, the DAMAGE model considers the full duration of the kinematic event.

Currently the U.S. Army does not currently endorse any one particular brain injury risk metric or risk function for evaluation of helmet performance under combined translational and rotational impact testing. For demonstration purposes the DAMAGE score for the example impact in Figure 2 was calculated using a script available from Biocore LLC and related to the probability of AIS2 and AIS4 brain injury based upon the injury risk functions in equations 1 and 2 from NHTSA [7].

$$P(mTBI) = 1 - e^{-(\frac{MPS}{0.505})^{2.84}} \quad (1)$$

$$P(sTBI) = 1 - e^{-(\frac{MPS}{0.823})^{2.84}} \quad (2)$$

Relating back to the DVBIC study, AIS2 injuries correspond to mild TBI (mTBI) whereas AIS4 injuries correspond to severe TBI (sTBI). For the impact to the rear of the helmet at 3.05 m/s (10 ft/s), the DAMAGE score was 0.207, which equates to a predicted maximum principal strain (MPS) of 20.7%. This MPS would estimate a probability of mTBI to be 8% and sTBI 2%. Note that the current Army blunt impact test method does not induce any rotational head kinematics and is unable to generate rotation based DAMAGE scores.

4. CONCLUSION

TBI remains a significant threat to military service members. The combat helmet is the primary piece of head protective equipment to protect a soldier from the severity of blunt impact loading. Current U.S. military blunt impact test and performance standards are based on methods that solely evaluate the helmet's ability to mitigate linear acceleration. The biomedical community has begun to converge on a consensus that rotational kinematics are a significant contributor to blunt impact induced TBI. This research outlines a helmet evaluation method that utilizes an experimental apparatus to impart combined linear and rotational motion in a soldier-relevant head impact scenario. Headform kinematics were measured for impact to the rear location of a helmeted ATD and, using the kinematic assessment tool DAMAGE, were used to predict a maximum principal strain in the brain. The ability to quickly determine predicted brain response from a complex head loading scenario is useful, particularly when paired with injury risk functions to determine severity and probability of TBI. As improvements in the understanding of TBI injury classification and human brain finite element models continue greater confidence can be attributed to determination of injury risk.

https://doi.org/10.52202/078352-0042

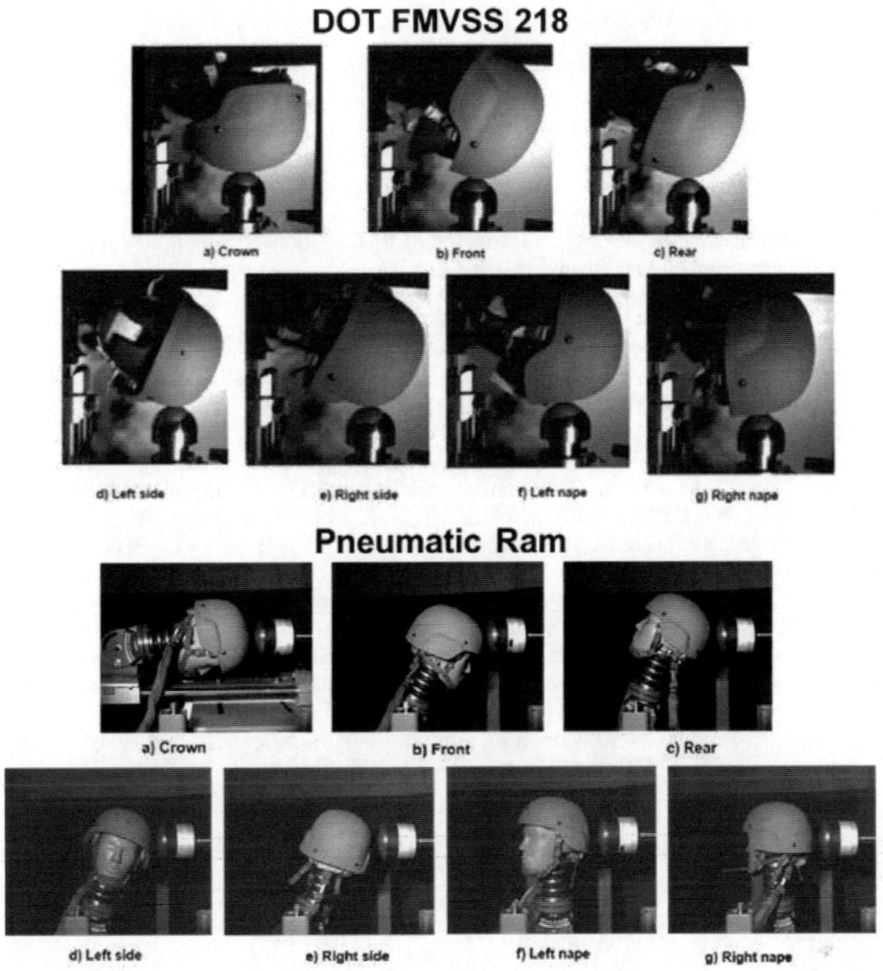

Figure 3. Top) Helmet impact location per current guided monorail U.S. Army Blunt Impact testing format. Bottom) Helmet impact locations for pneumatic ram test configuration, with reference to the existing current impact locations. Reporting on these tests is planned for future release.

Acknowledgments

The authors would like to acknowledge PEO Soldier Technical Management Division for supporting this research. Additionally the authors would like to acknowledge contributions from The University of Southern Mississippi School of Kinesiology: Dr. Mark Jesunathadas, Dr. Scott Piland, Ms. Elizabeth Edwards, and Dr. Trenton Gould as well as contributions from Mr. Alex Lurski and Dr. Eric Wetzel of CCDC Army Research Laboratory. This project was supported in part by an appointment administered by Oak Ridge Associated Universities (ORAU) for ARL through Cooperative Agreement (CA) W911NF-16-2-0008.

References

[1] Faul M, Xu L, Wald MM, Coronado VG. Traumatic Brain Injury in the United States: Emergency Department Visits, Hospitalizations and Deaths 2002 – 2006. Atlanta (GA): Centers for Disease Control and Prevention, National Center for Injury Prevention and Control; 2010.

[2] Defense and Veterans Brain Injury Center (DVBIC) "DoD Numbers for Traumatic Brain Injury Worldwide – Total" provided by the Armed Forces Health Surveillance Center. 2019.

[3] Azar et al. "Protective Headgear Attenuates Forces on the Inner Table and Pressure in the Brain Parenchyma During Blast and Impact: An Experimental Study Using a Simulant-Based Surrogate Model of the Human Head". Journal of Biomechanical Engineering. 2019.

[4] Holbourn "MECHANICS OF HEAD INJURIES". The Lancet. Volume 242. Issue 6267. 1943.

 https://doi.org/10.52202/078352-0042

[5] Kleiven, S. "Why Most Traumatic Brain Injuries are Not Caused by Linear Acceleration but Skull Fractures are." Front Bioeng Biotechnol 1: 15. 2013.
[6] Budday, S., et al. "Fifty Shades of Brain: A Review on the Mechanical Testing and Modeling of Brain Tissue." Archives of Computational Methods in Engineering. 2019.
[7] Takhounts "Development of Brain Injury Criteria (BrIC)". Stapp Car Crash Journal. Vol 57. PP 243-266. 2013.
[8] Sanchez, E. J., et al. "Evaluation of Head and Brain Injury Risk Functions Using Sub-Injurious Human Volunteer Data." J Neurotrauma 34(16): 2410-2424. 2017.
[9] Rowson, S., et al. "Rotational head kinematics in football impacts: an injury risk function for concussion." Ann Biomed Eng 40(1): 1-13. 2012.
[10] US Department of Transportation (DOT). Laboratory test procedure for FMVSS No. 218 motorcycle helmets. Washington (DC): DOT; 2011 May 13. p. 33.
[11] Biocore LLC. "Helmet Test Protocol". September 11. 2019.
[12] NOCSAE DOC 081-18am19a. "Standard Pneumatic Ram Test Method and Equipment used in Evaluating the Performance Characteristics of Protective Headgear and Faceguards". September. 2019. Effective November 1. 2019.
[13] Karton, Clara, Gilchrist, Michael D, Hoshizaki, Thomas "The Influence of Impactor Mass on the Dynamic Response of the Hybrid III Headform and Brain Tissue Deformation" Mechanism of Concussion in Sports. June 2014.
[14] Macalister, Anna "Surrogate Head Forms for the Evaluation of Head Injury Risk" Brain Injuries and Biomechanics. April 2013.
[15] Begonia, M., et al. "Development of a Methodology for Simulating Complex Head Impacts With the Advanced Combat Helmet." Mil Med 184(Suppl 1): 237-244. 2019.
[16] Reynier, K, et al. "Evaluating the effect of muscle activation on head kinematics during non-injurious impact in human volunteers" Injury Biomechanics Symposium, Ohio State University 2019.
[17] Padgaonkar, A. J., et al. "Measurement of Angular Acceleration of a Rigid Body Using Linear Accelerometers." Journal of Applied Mechanics 42(3): 552-556. 1975/
[18] Gabler, Bailey-Good, and Crandall Biocore LLC. "Evaluation Dummy Sensor Consistency using Rigid Body Constrains". May 8. 2018.
[19] Purchase description - helmet, advanced combat (ACH): AR/PD 10-02 rev A. Ft Belvoir (VA): Program Executive Office Soldier; 2012.
[20] Menichetti, Andrea et al. "Local head injury criteria for cebral contusion: a numerical study on their predictive capability." 8th World Congress of Biomechanics. July 2018.
[21] Taylor, Karen et al. "The influence of impact force redistribution and redirection on maximum principal strain for helmeted head impacts" Computer Methods in Biomechanics and Biomedical Engineering 22:13 1047-1060. June 2019.
[22] Knowles, Brooklynn "A Laboratory Study on the Ability of Head Kinematics to Predict Brain Strains in Helmeted Head Impacts" University of Alberta Graduate PhD Thesis. 2018.
[23] Gabler, L. F., et al. "Development of a Second-Order System for Rapid Estimation of Maximum Brain Strain." Ann Biomed Eng 47(9): 2019.

https://doi.org/10.52202/078352-0042

Area of coverage and sizing definitions of personal armour for UK Armed Forces personnel

R. Fryer, J. Breeze[2] and E. Lewis[3]

[1]*Defence Science and Technology Laboratory, Platform Systems Division, Portsdown West, Fareham, Hampshire, UK. PO17 6AD.*
rfryer@dstl.gov.uk
[2]*Royal Centre for Defence Medicine, Universities Hospital Birmingham, Birmingham, UK. B15 2SQ.*
[3]*Defence Equipment and Support, Ministry of Defence Abbey Wood, Bristol, UK. BS34 8JH.*

Abstract. Personal armour is worn by Armed Forces personnel and aims to prevent or mitigate the damage caused by projectiles to structures that are likely to result in death or life-changing long-term morbidity. Such injuries remain the leading cause of potentially survivable deaths on the modern battlefield. Defining anatomical coverage is necessary to enable objective comparisons between body armour designs and ensure Armed Forces personnel are sufficiently protected. Historically, protection has generally been provided to cover the whole population, as insufficient evidence existed to justify the coverage that should be provided for a given individual. This paper aims to summarise recent work that has been undertaken to define anatomical coverage for all areas of the body for hard armour plates and soft armour. Coverage was grouped into distinct areas that require coverage:

- Head and face
- Neck
- Torso (thorax and abdomen)
- Upper arm/axilla
- Thigh/pelvis

A systematic review of the literature was undertaken to identify those anatomical structures that, if damaged were highly likely to result in death or life-changing long-term morbidity. Anthropometric landmarks were identified for each area and Computed Tomography (CT) scans were utilised to determine how the internal anatomical structures corresponded to anthropometric landmarks and to define variation in the population. In addition, the lower borders of coverage for the upper arms and legs were related to the application of tourniquets. This is the first time that the medical area of coverage has been defined for personal armour for UK Armed Forces personnel. This paper also describes how the area of coverage is defined to industry and how coverage is compared between potential suppliers. Finally, plans for future studies using Magnetic Resonance Imagery (MRI) scans to determine 3D positions of structures in supine and upright positions is outlined; which will enable high fidelity coverage and modelling studies.

1. Introduction

Historically, protective equipment was designed to prevent death, but there is an increasing recognition that prevention of those injuries causing significant long-term morbidity is also required [1][2]. For example, ballistic eyewear has been worn for many years, but more recently pelvic protection was introduced due to the long- term morbidity from genital injuries [3]. However, any protective system will be a compromise, between the degree of protection and the encumbrance or 'burden' on the wearer. A programme to procure new body armour for UK Armed Forces personnel is currently underway and part of this programme is to optimise the anatomical coverage of the armour, which can subsequently be modified by factors such as tactical considerations on the ground, weight restrictions and equipment integration. This paper summarises this coverage work in its entirety; it is built upon numerous other papers that have been published for each body region [4-10]. For more insight into the rationale for each area the reader is encouraged to read the specific papers. Protection levels of armour should be selected to correspond to the prevalent threat that will be encountered, within the constraints of acceptable human factors considerations, however, this is outside the scope of this paper.

1.1 Methodology

The exact medical coverage requirements for UK body armour have not been openly published until recently, making objective comparisons between designs more difficult; Breeze *et al.* [4] introduced the terms *essential* and *desirable* structures to enable such comparisons. The essential and desirable medical coverage provided by a particular element of body armour are medical judgments and should be independent of the ballistic protective protection and material used. Essential medical coverage is the

https://doi.org/10.52202/078352-0043

minimum coverage that should be provided to all Armed Forces personnel, although in reality, this will be subject to a degree of modification due to human factors considerations such as equipment integration and interoperability. The boundaries of the soft armour will be eventually determined by the 'trade-offs' in the requirements for mobility and acceptable thermal burden, and for all other areas a suitable level of basic armour should be used.

1.1.1 Essential medical coverage

Those anatomical structures that, if damaged would likely lead to death prior to definitive surgical intervention being available, for example, bleeding from the thorax that cannot be compressed and requires surgical access (thoracotomy) to arrest it. In recent military operations such as Afghanistan, it is recommended that damage control surgery be performed within 60 minutes [*].

1.1.2 Desirable medical coverage

Those anatomical structures potentially responsible for mortality which, if damaged, would cause morbidity necessitating lifelong medical treatment or that result in significant disability. This includes physiological disability as well as psychological disability, for example, damage to the lower parts of the spinal cord (lumbar or sacral parts) may result in significant loss of function of limbs, or damage to the genitalia may result in psychological trauma.

1.2 Vulnerable structures

A review of the medical literature was undertaken in order to ascertain those structures within the thorax and abdomen likely to lead to death or significant long-term morbidity. Using the Preferred Reporting Items for Systematic Reviews and Meta-Analyses (PRISMA) methodology, PubMed, ProQuest, Web of Science and Google Scholar were searched. Four limited-access sources (Ministry of Defence online library, the Barrington digital library at Cranfield University, the Dstl Athena electronic library and the proceedings of the Personal Armour Systems Symposia conferences) were also interrogated.

Anatomical structures were identified that, if damaged were highly likely to result in death within 60 minutes (essential medical coverage) or would cause death after that period or result in significant long-term morbidity (desirable medical coverage). A time period of 60 minutes from time of injury to arrival at either a Role 2 or 3 Medical Treatment Facility (MTF) capable of performing Damage Control Surgery (DCS) was chosen as this is the target of the UK Ministry of Defence [*] and the US Department of Defense [@]. Strictly defining the time to surgery and not just 'time to medical care' as used in the past is important as surgery is the only means of arresting non-compressible haemorrhage; fluid resuscitation, compression and novel haemostatic agents merely buy time. The essential and desirable coverage structures are defined in the following tables.

Table 1. Anatomical structures comprising essential coverage

Head and Face	Neck	Torso/Abdomen	Arms	Pelvis/legs
Brain	Spinal Cord (C1-C5)	Heart	Axillary Arteries	Iliac Arteries
Brain Stem	Carotid Arteries	Aorta	Brachial Arteries	Femoral arteries
Cerebellum	Vertebral Arteries	Vena Cava		
	Larynx	Liver		
	Trachea	Bronchial Arteries		
		Pulmonary Arteries		
		Pulmonary Veins		
		Spleen		
		Subclavian Artery		
		Subclavian Vein		

https://doi.org/10.52202/078352-0043

Table 2. Anatomical structures comprising desirable structures

Head and Face	Neck	Torso/Abdomen	Arms	Pelvis/legs
Eyes	Oesophagus	Oesophagus	Median Nerve	Testis
Optic Nerve	Pharynx	Pharynx	Ulnar Nerve	Anus
Nose	Vagus Nerve	Lungs	Radial Nerve	Rectum
Lips	Brachial Plexus	Trachea		Sacral nerve
Ears	Vocal Cords	Kidneys		Femoral nerve
	Spinal Cord (Below C5)	Intestines		Urethra
		Spinal Cord (Below C5)		Ureters
		Spinal Nerves		
		Pancreas		
		Ovary		

Rationale into these definitions is not included in this paper due to brevity, however full details are available in the relevant papers [4-10].

1.3 Protection Levels

This paper focuses on coverage and not the impact on human factors of wearing protective equipment. In an ideal situation, personnel would be protected from every threat from every angle, however this is not feasible and so we define three general types of protection that can be used for coverage:

1. Hard armour has highest impact on human factors as it is often rigid, heavy and bulky;
2. Soft armour has less impact on human factors as it is flexible, but it still restricts movement and comfort;
3. Basic armour material is similar to regular clothing material, so is designed to have minimum impact on human factors, however it offers a low level of protection.

Therefore some compromises are always necessary. These armours are defined as follows.

1.3.1 Hard Armour

A rigid ballistic protective material designed to protect against high velocity bullets. This is currently fulfilled by a ceramic and composite plates.

1.3.2 Soft Armour

A flexible ballistic protective material designed to protect against low velocity bullets and high energy fragments. This is currently fulfilled in most systems by layers of Para-aramid and/or Ultra High Molecular Weight Polyethylene (UHMWPE) but could be comprised of various types of and/or combinations of materials. This can have water repellent treatment and is normally encased in a water repellent and ultra violet resistant cover. Soft armour material can also be pressed into a rigid material, for example, if used as a helmet.

1.3.3 Basic Armour

A flexible material that provides protection against lower energy fragmentation, such as the knitted silk used in the Tier 1 pelvic protection *(ballistic underwear)* currently utilised by the UK armed forces, and the Ultra-high-molecular-weight polyethylene (UHMWPE) used in the neck collar of the Enhanced Protection Under Body Armour Combat Shirt (EP-UBACS). Basic armour should be used in areas of coverage where a hard armour would severely impair mobility and/or comfort.

https://doi.org/10.52202/078352-0043

2. Coverage definitions by body region

External anthropometric landmarks are identified that define the coverage boundaries of each of the body regions and correspond to the internal anatomical structures. Coverage is further defined as threshold and objective area of coverage

- Threshold coverage are areas that must be covered by a given armour and is associated with a high level of mortality (usually corresponding to the essential structures, subject to human factors considerations);
- Objective coverage are areas where coverage would be advantageous, but may only be achieved by a lower protection level or only used in a scalable system.

2.1 Head and face

The head and face are defined as the area above the base of the skull.

2.1.1 Area of coverage definition

Threshold coverage:
- Helmet coverage to be from the margins of the brain which relate to the nasion, external auditory meatus and superior nuchal line, as shown in Figure 1;
- This coverage only protects against horizontal trajectories (such as ‘a’ in Fig 1). It is unrealistic to protect from lower angle trajectories (such as ‘b’ in Fig 1) using a helmet.

Objective coverage:
- All areas;
- This could be achieved using a mandible guard and nape protection, but these should only be employed in roles considered high risk and when not limited by human factors.

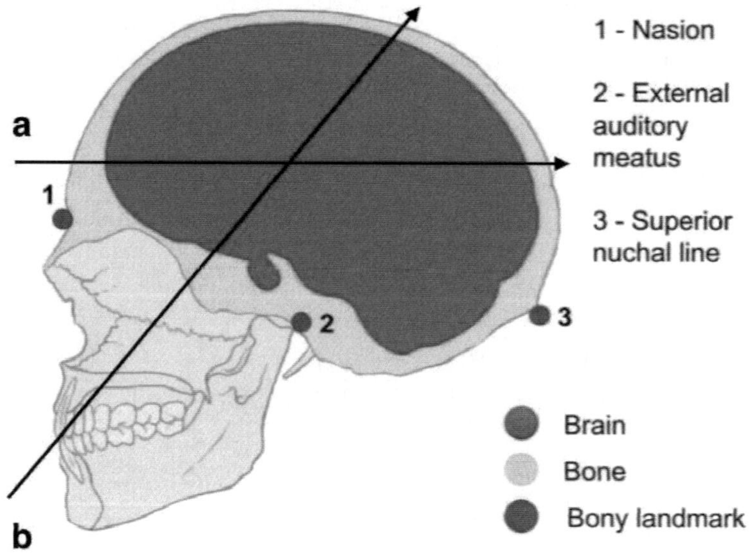

Figure 1. Landmarks for the brain with example trajectories ‘a’ and ‘b’. Reproduced from [6] with permission of BMJ.

2.2 Neck

The neck is defined as the area below the base of the skull and above the suprasternal notch.
2.2.1 Area of coverage definition

Anatomical coverage of the neck according to essential and desired structures alone is not feasible due to human factors considerations. Therefore coverage of the neck is divided into 3 zones as depicted in Figure 2. These are defined as:

- Zone 1: Suprasternal notch to the cricoid cartilage;
- Zone 2: Cricoid cartilage to the lower border of the mandible;
- Zone 3: Lower border of the mandible to base of the skull.

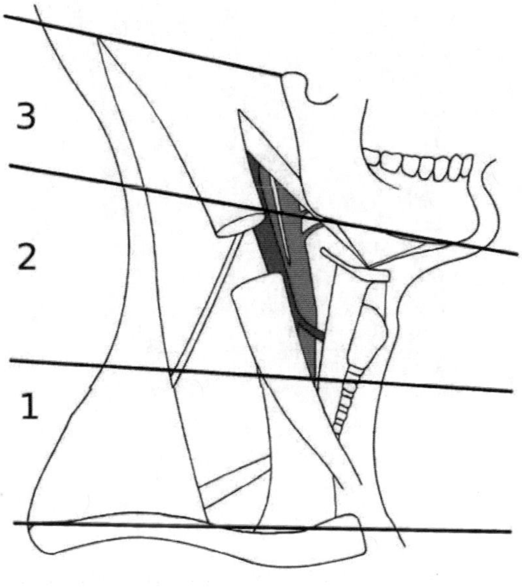

Figure 2. Landmarks of the neck.

Threshold coverage:

- Coverage must be afforded to Zone 1;
- The level of protection should be basic armour at the threshold level.

Objective coverage:

- Coverage to zones 1 and 2.
- Using combination of soft and basic armour.
- It is deemed unrealistic to cover zone 3 using neck protection with current technology.

2.3 Torso

The torso is defined as the area below the suprasternal notch and above the iliac crest; it is bordered laterally by the axillary fold.

2.3.1 Area of coverage definition

Coverage of the torso (thorax and abdomen) is defined from the three landmarks depicted in Figure 3.

1. Suprasternal notch;
2. Lower border of ribcage (10th rib);
3. Iliac crest.

These correspond to the boundaries of the most vulnerable structures that are identified as the aortic arch, heart, liver and spleen, and bifurcation of aorta respectively [5].

 https://doi.org/10.52202/078352-0043

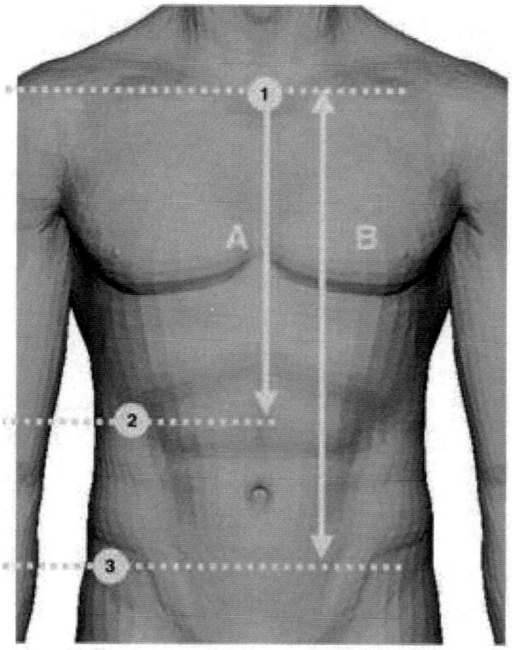

Figure 3. Suprasternal notch (1), lower border of ribcage (2) and iliac crest (3). Threshold height is A, objective height is B.

Threshold coverage:

- Coverage must be afforded from suprasternal notch to the lower border of rib cage.
  - The level of protection in this area should be commensurate to the threat of either small arms or fragmentation (corresponding to hard or soft armour respectively).
  - If hard armour is utilised then a further restriction must be placed upon the width, otherwise no movement would be feasible. This is summarised in Figure 4.
    - The top width must cover the heart
    - The bottom width must cover the liver and spleen.
- Coverage of all areas by basic armour that are not covered by hard or soft armours.

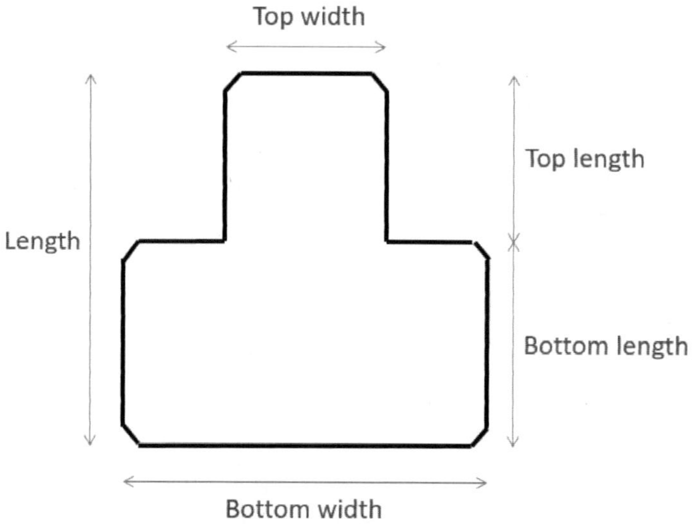

Figure 4. Definition of hard plate widths.

https://doi.org/10.52202/078352-0043

Objective coverage:

- Coverage of all areas by soft armour.
- This should only be achieved up to a level suitable to human factors needs.

2.4 Upper arm/axilla

The Upper arm/axilla is defined as the areas lateral to the axillary fold, extending to the elbow.

2.4.1 Area of coverage definition

Coverage of the arm is defined from the three landmarks depicted in Figure 5.

1. Acromion;
2. Axillary fold;
3. Deltoid insertion.

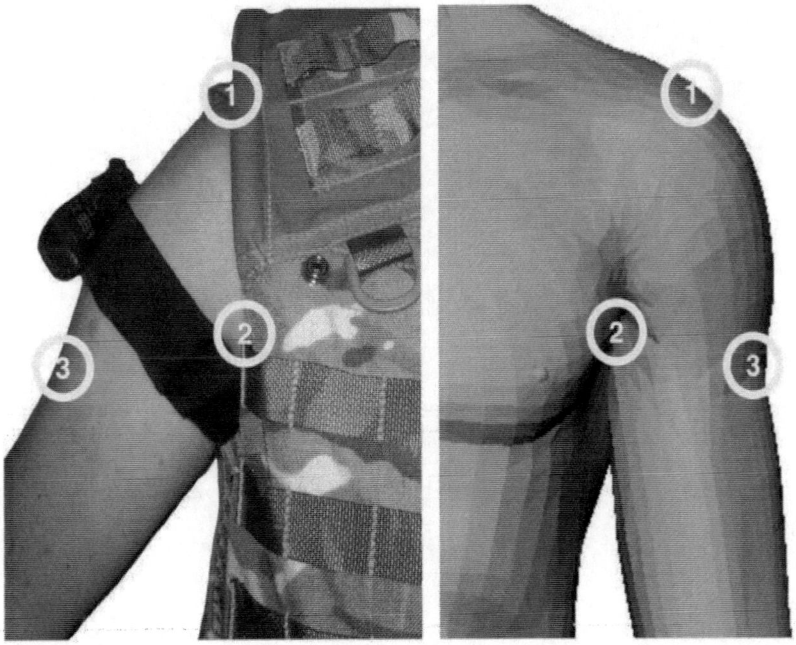

Figure 5. Landmarks of the arm/axilla, with example of application of tourniquet (left).

Threshold coverage:

- Coverage must be afforded to a sufficient distance below the deltoid insertion so that a tourniquet remains in place and is effective.
  - For the UK Armed Forces a distance of 40 mm below the deltoid insertion [9] was selected.
- Basic armour level of protection.

Objective coverage:

- Threshold area of coverage, but with soft armour level of protection.

2.5 Thigh/pelvis

The thigh/pelvis is defined as the area below the iliac crest, extending to the knee.

 https://doi.org/10.52202/078352-0043

2.5.1 Area of coverage definition

Coverage of the thigh/pelvis is defined from the two landmarks in Figure 6.

1. Iliac crest;
2. Ischial tuberosity.

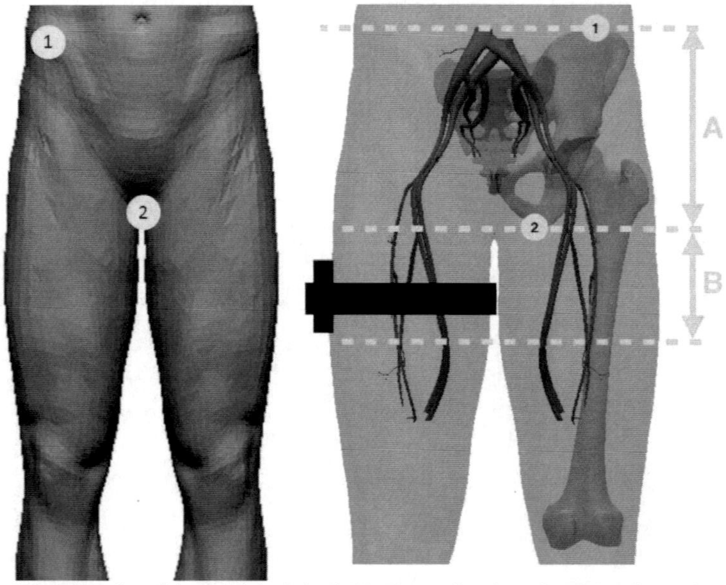

Figure 6. Landmarks of the thigh/pelvis (left). Example of application of tourniquet (right).

Threshold coverage:

- Coverage must be afforded to a sufficient distance below the ischial tuberosity so that a tourniquet (or two adjacent tourniquets) remain in place and are effective.
    - For the UK Armed Forces a distance of 100 mm below the ischial tuberosity was selected [10].
- Coverage at the threshold level is only required for the front, underneath and rear. Protection of the sides of the leg/pelvis area is not required.
    - This level was determined as it is commensurate with the current protection offered to UK Armed Forces personnel, which is deemed to be acceptable from a human factors perspective.
- Basic armour level of protection.

Objective coverage:

- Threshold area of coverage, but with soft armour level of protection.
- Coverage to a sufficient distance below the ischial tuberosity (so that a tourniquet remains in place) and is effective from all directions by basic armour.

3. Comparison of armours

When multiple armour solutions exist that achieve the threshold requirements, there is a requirement to objectively compare the coverage they offer so that decisions can be made on the best armour solution to procure or use. It is vitally important that the human factors performance is also assessed during this process so that an optimum solution is used and not just the one that offers the most coverage. There are two tools available to UK MOD for this purpose [11]:

- Coverage of Armour Tool (COAT)
    - COAT is a simple shotline tool. Geometrical elements that model the armour and body are represented. Vulnerable structures from Tables 1 and 2 are selected, the tool then calculates the percentage of coverage from azimuths and elevations that are selected by the analyst. An example grid is shown in a screen shot of the tool in Figure 7;

https://doi.org/10.52202/078352-0043

- The percentage coverage of different armour solutions for areas of the body can then be objectively compared;
- This method assumes that all vulnerable structures are equally important and that uncovered shotlines pass through the entire body.

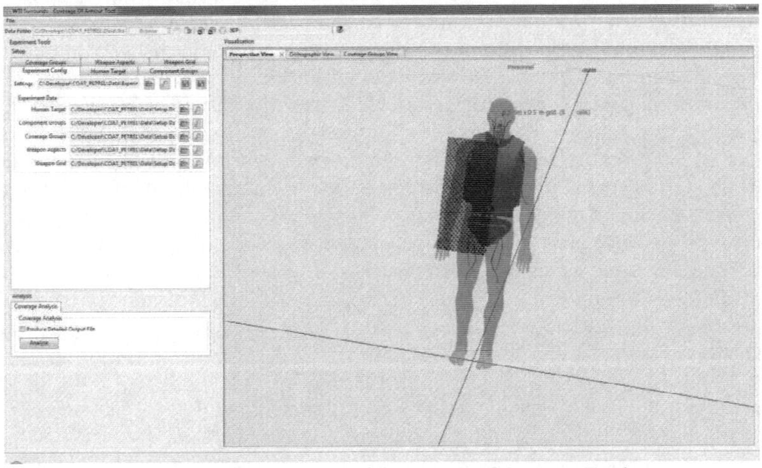

Figure 7. Screen shot of Coverage of Armour Tool.

- Weapon Target Interaction (WTI)
  - WTI is a terminal effects vulnerability model. It simulates the penetration through a human geometry, determines the volume of damaged tissue and outputs injury scores that are commensurate with the Abbreviated Injury Scale [12];
  - The injury scores for grids of shotlines, from user defined azimuths and elevations, can be calculated. The scores can then be weighted for different levels of injury to determine an objective coverage score;
  - Examples of a hard plate of the threshold dimensions from 0° and 20° azimnuth both at 0° elevation is shown in Figure 8.

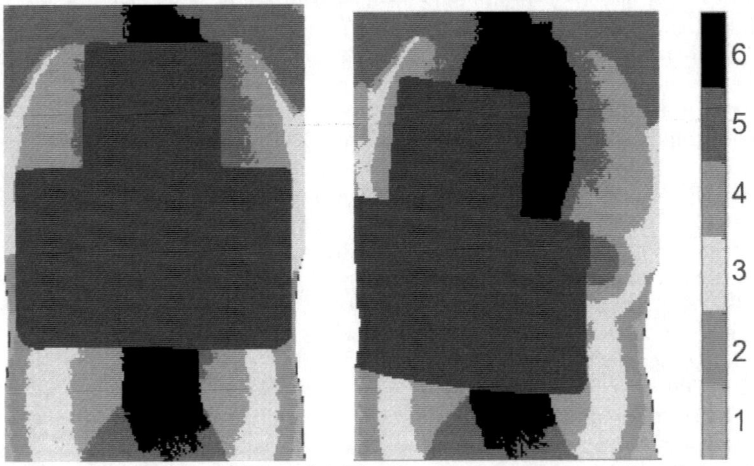

Figure 8. Example of output from WTI model. The blue area represents shotlines being stopped by the hard plates, the other colours correspond to AIS 1-6 as shown in the scale.

4. Future direction

The sizing for UK personal armour coverage is currently based on a range of measurements of the previously described anatomical boundaries from a variety of anthropometric data sources [13] and interrogation of anonymized Computed Tomography (CT) scans of injured military personnel undertaken at the Queen Elizabeth Hospital, Birmingham [5][9][10]. Whilst these represent the best data

https://doi.org/10.52202/078352-0043

set available at the time, there are limitations such as the small amount of measurements of females and the lack of 3D geometry on which to assess ranges of sizes of armour. Currently all assessments using COAT and WTI are undertaken on the Zygote® geometry, which is close to a 50th percentile UK serviceman and STANAG 4512 target so allows for comparison, but there is no account taken on the effect of posture and it has been shown that the Zygote® geometry has many inaccuracies [14].

To address these shortcomings, a study to conduct MRI scans on a range of male and female subjects that represent the UK Armed Forces population is underway in association with the Sir Peter Mansfield Imaging Centre at the University of Nottingham. Subjects will be scanned in supine and upright positions, with major organs segmented to create representations of 3D geometry. From these scans the aim is to create an atlas of geometries that will cover all sizes of males and females of the population in a range of postures. In addition, the subjects will also be scanned wearing the current UK VIRTUS body armour system in the upright position to determine the current level of coverage as worn. When complete, this work will allow assessments of all sizes of body armour in a range of postures.

Finally, a medical area of coverage Defence Standard is in preparation from the recommendations in this paper which will form the basis of future UK body armour procurement.

Acknowledgments

The authors would like to thank all co-authors on the coverage papers for the different body areas.

References

[1] Morrison JJ, Stannard A, Rasmussen TE, et al. Injury pattern and mortality of noncompressible torso hemorrhage in UK combat casualties. J Trauma Acute Care Surg 2013;75(2 Suppl 2):S263–8.
[2] Breeze J, Allanson-Bailey LS, Hunt NC, et al. Mortality and morbidity from combat neck injury. J Trauma Acute Care Surg 2012;72:969–74.
[3] Lewis EA, Pigott MA, Randall A, et al. The development and introduction of ballistic protection of the external genitalia and perineum. J R Army Med Corps 2013;159 (Suppl 1):i15–17.
[4] Breeze J, Lewis EA, Fryer R, Hepper AE, Mahoney PF, Clasper JC. Defining the essential anatomical coverage provided by military body armour against high energy projectiles. J R Army Med Corps. 2016;162(4):284–90.
[5] Breeze J, Lewis EA, Fryer R. Determining the dimensions of essential medical coverage required by military body armour plates utilising Computed Tomography. Injury. 2016;47(9):1932–8.
[6] Breeze J, Baxter D, Carr D, Midwinter MJ. Defining combat helmet coverage for protection against explosively propelled fragments. J R Army Med Corps. 2015;161(1):9–13.
[7] Breeze J, Fryer R, Hare J, Delaney R, Hunt NC, Lewis EA, et al. Clinical and post mortem analysis of combat neck injury used to inform a novel coverage of armour tool. Injury. 2015;46(4):629–33.
[8] Breeze J, Fryer R, Lewis EA, Clasper J. Defining the minimum anatomical coverage required to protect the axilla and arm against penetrating ballistic projectiles. J R Army Med Corps. 2016 Aug;162(4):270–5.
[9] Breeze J, Davis J, Fryer R, Lewis E. Sizing of ballistic arm protection for the VIRTUS body armour and load carriage system. BMJ Military Health: 201910.1136/jramc-2019-001254 on 20 Feb 2020.
[10] Lewis E, Breeze J, Fryer R. Defining the medical coverage of ballistic protection to the pelvis and thigh. Accepted for publication in BMJ Military Health on 07 July 2019.
[*] Hodgetts TJ, Mahoney PF, Kirkman E. Damage control resuscitation. J R Army Med Corps 2007;153:299–300.
[@] Bastian ND, Brown D, Fulton LV, et al. Analyzing the future of army aeromedical evacuation units and equipment: a mixed methods, requirements-based approach. Mil Med 2013;178:321–9.
[11] Fryer R, Breeze J, James G. Development of a personnel vulnerability numerical model. Proceedings of Personal Armour Systems Symposium (PASS) 2016, Amsterdam, The Netherlands.
[12]Gennarelli TA, Wodzin E, editors. Abbreviated injury scale 2005 handbook, update 2008. Barrington, IL: Association for the Advancement of Automotive Medicine; 2008.
[13] Wilson S, Usher D. Dismounted Anthropometric Data Collection. Final Report TIN 3.182. Version 1, dated 29 November 2016. 2016;705.
[14] Laing S, Jaffrey, M. Thoraco-abdominal Organ Locations: Variations due to Breathing and Posture and Implications for Body Armour Coverage Assessments. DST-Group-TR-3636.

Evaluation of test methods on personal protective equipment for blast overpressure

B.A.H. Wilson[1], T.A.T. Westerhof[1], M.M.G.M. Philippens[1]

[1] *TNO, Defence, Safety and Security, Explosions, Ballistics & Protection, Ypenburgse Boslaan 2, 2496 ZA Den Haag, Netherlands*
Ben.wilson@tno.nl

Abstract. In recent conflicts, a major source of injury for military personnel are Improvised Explosive Devices (IEDs), which generate blast overpressure (BOP), fragments and heat. Personal Protective Equipment (PPE) of dismounted soldiers are primarily tested for protection against bullets, fragments, heat and stabbing using accepted formal standards. Until now there are no formalised standards specifying test methods to assess the PPE performance on protection against BOP. This study presents an overview of test methods for PPE BOP protection for the thorax and eyes which are published in open literature. A discussion of the features of the methods will be concluded by a proposal for an optimized method with indications of knowledge gaps or shortcomings to be filled out in the future.

1. INTRODUCTION

Explosions were the dominant mechanism of injury (72%) affecting NATO coalition forces operating in Iraq and Afghanistan during 2001 – 2011 [1]. Explosions include fragments, heat and blast overpressure, which all have their specific injury mechanism. Dismounted soldiers are likely to be protected by PPE (i.e. helmet, soft and/or hard body armour, eyewear, etc.), where ballistic and heat resistance properties are tested and certified using internationally accepted test standards [2-3]. However, accepted standards and test methods for evaluating PPE on BOP are not available. Although, such standards are relevant as the number of overpressure casualties cannot be neglected, as 3.6% of injured soldiers in Iraq/Afghanistan had evidence of thoracic BOP injury [4]. Ocular BOP injury is not specifically documented in epidemiology studies. The Birmingham Eye Trauma Terminology system (BETT) [5] definition of closed globe injuries could give an indication of ocular BOP injuries, since these injuries were found at blast-exposed patients with documented Traumatic Brain Injury (TBI) [6].
The goal of this research is to draft a test method for PPE BOP protection, including indication of knowledge gaps, based on a literature review of published methods for the thoracic and ocular region.

The following elements are considered to be essential for a test method:

- Relevant BOP
- Performance Criteria & Tolerances
- Test setups
    - Threat definition
    - Test device

2. LITERATURE REVIEW RESULTS

2.1 Scope

The selected body regions of interest are thoracic and ocular. The thorax seems to be the most investigated within the field of bodily BOP protection and the ocular area as an emerging topic. Both regions are likely to be protected for (dis)mounted soldier.
The literature review results are split into thoracic (2.2) and ocular (2.3). The assessment of literature was conducted through open access and journal papers, as well as conference proceedings from Personal Armour Systems Symposia (PASS) and Military Health System Research Symposium (MHSRS). Main keywords for identifying papers were: blast, protection, injury, overpressure, test, experiments. These were extended using thorax, thoracic, lung, ocular, eyes for analysing specific body parts.

2.2 Thoracic PPE BOP Test Methods

2.2.1 Relevant BOP

A relevant BOP testing regime is shown by the Bowen curves [7] describing the correlation between lethality and pressure wave characteristics incident overpressure and positive phase duration for an

 https://doi.org/10.52202/078352-0044

unprotected man. More specifically, lung injury risk curves to describe the probability of lung injury are described in [8], discriminating between slight, moderate and severe lung injury.
Investigated literature shows three loading scenarios, namely open field, shock tube and confined testing. Each of these loading environments has a specific impact on the effective blast wave characteristics. No standard blast loading definition could be found for testing thoracic PPE BOP performance [8-9].

2.2.2 Performance Criteria

The thoracic BOP protection is evaluated by measuring specific performance criteria. Specific tolerance values are used to classify the level of protection. Some are injury risk related others are not. Table 1 gives an overview of found criteria, tolerances, injury models and the corresponding references. The criteria are shown in the first column, commonly used names for injury models are given in the second column, references giving thoracic BOP injury tolerances are given in the third column and finally, references are listed in last column.

Table 1: an overview of performance criteria for the evaluation of thoracic PPE w.r.t. lung injury or lethality risk.

Criterion	Injury model	BOP injury tolerances	References
Product of chest velocity (V) and the normalized compression (C) (VC) [m/s]	Viscous Criterion [10]	Lung injury risk [8]	[8-13]. [8],[10-14]
Lung mass [kg]	Fluid in lungs	x	[15]
Chest wall acceleration (a_{cw}) [m/s^2] Sternum acceleration (a_{st}) [m/s^2]	x	Lung injury risk [8] [15]	[8] [15]
		Indirect correlation with lethality risk [16]	[16]
Chest wall velocity (v_{cw}) [m/s]	x	Lung injury risk [8]	[8]
	Axelsson Injury Model (AIM) [21]	Adjusted Severity of Injury Index (ASII) [22]	[17-22]
	Stuhmiller Injury Model (SIM) [23]	Lung injury risk Lethality risk	[23]
Chest wall displacement (d_{cw}) [m]		Lung injury risk [8]	[8]
Incident pressure (p_i)[kPa]	Bowen curves [7]	Lethality [7]	[7],[24-25]
Incident pressure impulse m(i_i) [kPa*ms]		Lung injury risk [8]	[8]

2.2.3 Test setups

The thoracic BOP protection is evaluated through performing simulated BOP tests. The test setups are evaluated by threat definition and test device.

- The threat definition of test setups is described by the loading environment: shock tube, free field and confined space. Also, the characteristics of the blast wave are given.
- The test devices are described and mentioned by their abbreviation.

Table 2 gives an overview of test devices, measured parameters, mention of injury model, threat definition. The tested protective system and effect on the measured parameter are also included.
This table is an update and extension of an overview published earlier [26]

https://doi.org/10.52202/078352-0044

Table-2: an overview of test setups for evaluating thoracic BOP PPE.

TD	Test device	Measured parameters	Injury model	Armour config.	Results & remarks	Ref
S[1]	Animals & Thoracic Rig	Lung mass & a_{cw}	x	(1) Textile (2) Plate	Comparison of performance between armour config (2) and (1): decreased lung injury, lung mass and peak a_{cw}	[15]
S[10]	Animals	Lung mass, Fatality	x	(1) None (2) Vest	Comparison of vest versus no vest: severe lung injury increased and a higher average fatality of test animals	[27]
F[2]	MABIL, (BTD not used for results)	a_{cw} Incident pressure Lethality risk	Bowen curves [7]	(1) Textile (2) Vest (3) Vest+Plate	Config (2) versus (1): similar lethality probability, (2) has increased peak a_{cw} Config (3) versus(1): lower lethality probability and decreased peak a_{cw}	[16]
C[3]	BTD MBTD	Surface pressure, v_{cw}, W_e	SIM	(1) None (2) Materials	Config (2) versus (1): decreased severity of injury, overpressure, v_{cw} and W_e	[28]
F[4]	HSTM	Pressure in organs a_{cw} Compression	X	(1) Textile (2) Vest+Plate	Comparison of performance between armour config (2) and (1): maximum pressure decreased in left lung, increased in left lobe of liver. Pressure impulse increased in stomach and liver.	[29]
C[5]				1) Textile (2) Vest (3) Vest+Plate	Config (2) versus (1):max. pressure increased in liver right lobe, decreased in liver left lobe. Pressure impulse increased in stomach, liver right lobe, decreased in liver left lobe. Config. (3) versus (2): peak pressure increased in left lung, decreased in liver left lobe. Pressure impulse increased in stomach, liver right lobe, decreased in liver left lobe.	
S[6]	Steel pipe	Surface pressure	Bowen curves [7] + other	(1) None (2) Vest (3) Vest+Plate	Config (2) versus (1):peak pressure decreased and pulse duration increased, lung injury probability increased or remained within the same regime Config (3) versus (1): Peak pressure and lung injury probability decreased	[30]
C[7]	Flat plate	Surface pressure	x	(1) None (2) Plate	Config (2) versus. (1): transmitted pressure and impulse decreased.	[31]
	MABIL	a_{cw}			Config (2) versus. (1) a_{cw} decreased	

https://doi.org/10.52202/078352-0044

Table 2 (continued)

C[8]	CS	Surface pressure	Bowen	(1) None (2) Vest (3) Vest+Rigid Material	Comparison of performance between config (2) and (1):pressure and lethality probability increased. Config(3) versus (1): peak pressure and lethality probability decreased. Positive phase duration increased.	[32]
F[9]	Hybrid II	a_{cw} p_{cw}			Comparison of performance between config (2) and (1) maximum a_{cw} and lethality probability increased. Config (3) versus. (1) maximum a_{cw}, overpressure and lethality probability decreased.	
F[11] C[12]	Hybrid III MBTD BTD	Pressure profiles a_{cw}	SIM VC	(1) Vest (2) Plate+Vest	Comparison of config (2) versus (1): Average We and maximum v_{cw} decreased, no VC change	[33]

Threat Definition (TD): Shock tube (S), Free field (F), Confined space (C) loading conditions.
Where a_{cw}, v_{cw} and d_{cw} are chest wall acceleration, velocity and displacement respectively,
1 = unknown; 2 =20 kg C4 at 6.25 m and 10 kg C4 at 4 m; 3 = unknown; 4 = 1.4 kg C4; 1.8 kg C4 and 2.3 kg C4 at 2.3 m on 1 m height; 5 = 0.7 kg; 0.9 kg and 1.2 kg thermobaric charge; 6 = incident overpressure peaks ranging between 83 – 640 kPa; 7 = 0.25 C4 at 1 m; 8 = 0.12 or 0.25 kg C4 at 0.65-0.55-0.45 m distance; 9 = 0.05-0.1-0.2 kg C4 at 0.66-0.68 m distance;10 = 115, 230, 295 and 420 kPa ;11 = 0.7, 0.9, 1.2, 1.4, 1.8 kg; 12 = 0.7, 0.9, 1.2 kg

Three categories in used test devices can be distinguished:

- Non anthropomorphic: measuring behind armour pressure test devices CS, BTD, MBTD;
- Lower complexity anthropomorphic thoracic chest wall motion test device MABIL;
- Higher complexity anthropomorphic test devices HYBRID III and HSTM;

The devices listed in Table 2 are described below.

- Thoracic Rig: A rig replicating the peak acceleration response of the thoracic pig wall [15].
- Blast Test Device (BTD): a rigid vertical instrumented cylinder based on thoracic dimensions of sheep [19]. Pressure transducers are mounted flush with the cylindrical surface.
- Modified Blast Testing Device (MBTD): is very similar to the BTD but modified to handle protective materials [28],[34-35]. The MBTD collects pressure data using many pressure sensors to map out the pressure contour around the entire protective material.
- Mannequin for the Assessment of Blast Incapacitation and Lethality (MABIL) [36] is a anthropomorphic test device. Accelerometer measurements are performed mid-torso, sternum and abdomen.
- Hybrid III originated from the automotive industry and is an anthropomorphic test device with acceleration, force and displacement sensors in the different body parts [37]. Extra sensors are sometimes added for a specific study purposes [38].
- Human Surrogate Torso Model (HSTM) is a high fidelity anthropomorphic test device developed to for armour blunt trauma [29]. High fidelity due to a detailed skeletal structure, major thoracic and abdominal organs, mediastinum, flesh and skin and the number of varying materials used. Instrumentation consists of pressure sensors, within different organs, sternum accelerometers, displacement sensor to measure chest displacement and load cell.
- Chest Simulator (CS) [32]: curved aluminium plate (12.7mm) thick bolted on a flat aluminium base plate. The curved plate is roughly similar to the human torso and pressure sensors are mounted flush at the surface.

2.3 Ocular PPE BOP Test Methods

2.3.1 Relevant BOP

Living rodents have been used to investigate short and long term injuries after single and repetitive BOP [39-47]. The BOP leading to ocular injures provide an estimate for the relevant BOP regime. Using

https://doi.org/10.52202/078352-0044

animals with human size eyes (larger than rodent eyes) better emulates effects for the human eye [48]. Two studies are highlighted here:

- Living rabbits, revealing relevant ocular trauma (corneal thickness, retinal thickness) caused by survivable pressure levels: 120-132 kPa peak overpressure, 96-104 Pa*s impulse [49].
- *In vivo* porcupine eyes also sustain primary blast ocular trauma caused by survivable pressure levels: 113 kPa peak overpressure, positive phase duration of 2.4 ms, impulse 128 Pa*s [50].

2.3.2 Performance Criteria

The ocular PPE protection performance for BOP is evaluated by use of injury criteria and tolerances [51-52] or by making a relative comparison between PPE [53-55]. Table 3 gives an overview of available ocular injury criteria, -models and -tolerances shown in the first, second and third column respectively.

Table-3: ocular injury criteria

Injury criterion	Injury model	Injury Tolerances
Incident peak pressure [kPa]	risk of vision loss calculator [51]	risk function [51]
Reflective peak pressure [kPa]	injury risk prediction [52]	risk function (IOP) [52], [56]
Intra-Ocular Pressure (IOP) [kPa]		

2.3.3 Test Setups

The test setups are characterized by the threat definition and test device.

- The laboratory threat simulation is defined by the blast overpressure parameters: peak overpressure, positive phase, impulse, complex waves and the threat source: shock tube or free field.
- The test devices used in each test setup are described by head form and PPE.

The setups are listed in Table 4.

- Three configurations can be distinguished, namely the bare head form, open eyewear and closed eyewear configurations. Open eyewear implies gaps between the head form facial region and the eyewear and closed eyewear implies that the gaps between head form facial region and eyewear are closed by material.
- Head forms: Most studies used a custom anthropomorphic head form, one study a standardized FOCUS head form and one included a porcine eye surrogate.
- Combat helmet: Most studies include a combat helmet placed upon the head form.
- Important varying parameters are described by Remarks (Rm) in Table 4.

Table 4 gives an overview of threat definition, test devices, measured parameters, mention of injury model, threat definition. Mention of tested protective system, effect on the measured parameter and remarks regarding important variations are also included.

https://doi.org/10.52202/078352-0044

Table-4: an overview of test setups for evaluating ocular BOP PPE.

TD	Test device	Measured parameters	Injury model	PPE	Results (Rs) & Remarks (Rm)	Ref
S[1]	Small and large head form + combat helmet + ocular PPE	Reflective pressure at corneal surface	x	(1) Bare (2) Open eyewear (3) Closed eyewear	Rm: Varying head form size and orientations. Rs: Config (2) versus (1): decreased reflective peak pressure for a few orientations and increased reflective pressure for most orientations. Config (3) versus (1): decreased reflective peak pressure for most orientations and increased reflective pressure for a few orientations.	[53]
S[2]	Head form + combat helmet	Reflective pressure at corneal surface	x	(1) Bare (2) Open eyewear (3) Closed eyewear	Rm: Varying head form orientations & loading. Rs: Config (2)&(3) versus (1) decreased reflective peak pressure for most orientations and increased reflective peak pressure for a few orientations (3) versus (2)&(1) increased reflective pressure impulse for some orientations	[54]
S[3]	Head form + combat helmet	Reflective pressure at corneal surface	risk of vision loss calculator	(1) Bare (2) Open eyewear (3) Closed eyewear	Rm: Varying headform orientations & loading Rs: Config (3)&(2) versus (1) decrease in "risk of vision loss" for an orientation (2) versus. (1) increase in "risk of vision loss" for an orientation	[51]
S[4]	Head form + porcine eye	IOP Facial pressure sensors	(1) eye dissection (2) eye injury risk calculation	(1) Bare (2) Open eyewear (3) Closed eyewear	Rm: Varying loading Rs: Config (2) versus (1) decreased IOP for 2/3 loading scenarios and an increased IOP, IOP impulse and reflective pressure impulse for 1/3 loading scenarios (3) versus (1) decreased IOP, IOP impulse and reflective impulse for all loading scenarios	[52]

https://doi.org/10.52202/078352-0044

Table 4 (continued)

F5	Head form (FOCUS)	Reflective pressure at corneal surface	x	(1) Bare (2) Open eyewear (3) Closed eyewear	Rm: varying loading Rs: (2) w.r.t (1) decreased reflective peak pressure for both loading scenarios (3) w.r.t (2) decreased reflective peak pressure for both loading scenarios	[56] [55]

1 = Using a Friedlander blast wave characterised by; a measured peak overpressure of ~11 kPa, a positive phase duration of ~1.5 ms); 2 = Using Friedlander blast waves characterised by incident overpressure levels of 70, 140 and 210 kPa 70, 140 and 210 kPa; 3 = Incident pressures varying between 78 and 306 kPa with a positive phase duration between 2.97 – 5.32 ms; 4 = 69 kPa, 138 kPa, 207 kPa static overpressure with a positive duration around 2.5 ms;5 = free-field incident overpressures of ~46 and ~190 kPa (0.5 kg C4 at 3m and 5 kg C4 at 3.8m).

3. DISCUSSION

3.1 Threat Simulation

Blast testing using shock tubes, free field or confined spaces loading conditions have their own characteristics. Shock tubes enable blast research in laboratory conditions. The desired pressure at the specimen can be carefully designed using the driver's length, drivers gas and specimen location. However, special attention should be given to for instance, the rarefaction effects of the open end of the tube, the tube cross section-specimen size ratio and the dynamic pressure component because they all have a significant impact on the loading experienced by the target [56]. Free field blast experiments are convenient due to increased fidelity: inclusion of acoustic, thermal, optical and electromagnetic components found in actual blasts and excluding the boundary effects of a shock tube for instance [58].

3.2 Thoracic PPE BOP Test Methods

BOP and PPE body interaction and on top indicating injurious consequences requires multidisciplinary expertise, which indicates the complexity of assessing thoracic PPE BOP protection performance [29]. However, based on thoracic PPE BOP animal studies, a hard armour material combined with a specific backing could decrease lung injury and peak a_{cw} [15], whilst standalone soft armour is able to increase severe lung injury and average fatality [27].

Non animal studies show that protection using only soft material is able to increase a_{cw} [7], [28] and that hard/plate material combinations were able to decrease the lethality probability and peak a_{cw} [16],[31]. Similar effects observed for behind armour surface pressure as performance measurement [28],[30], [32]. However, the behind armour surface pressure shows significant inconsistencies between different test devices. The physics of behind armour pressure is a topic of discussion.

Therefore, measuring a_{cw} seems as a promising performance parameter able to distinguish between different armour configurations.

The MABIL, HSTM, Hybrid III test devices are equipped for measuring a_{cw}. The MABIL test device appears to be less complex than the HSTM and Hybrid III. The HSTM and Hybrid III measure more performance parameters but the added value for a Thoracic PPE BOP test method remains to be determined.

The following results of acceleration based test devices were found:

- The Hybrid III measuring acceleration based parameters: a_{cw}, v_{cw} w_e showed to be repeatable and discernible between charge levels. [33]
- The MABIL showed a correlation between peak a_{cw} and the peak incident overpressure [36].

Other testing devices found in literature but not included in Table-2 are the Swedish torso surrogate amino [59] and South African Torso Surrogate [60].

There are test methods which do not provide consistent results and are therefore questionable to be used for a standard PPE BOP performance assessment:

https://doi.org/10.52202/078352-0044

- The BTD provided inconsistent and contradictory results during the evaluation of different protective materials [38].
- The BTD, MBTD calculated W_e without armour system were inconsistent [33].
- The Hybrid III anthropomorphic test device, obtained an inconsistent measurement between a_{cw}, peak v_{cw} and W_e. [33]
- The Hybrid III with the criterion VC_{max} was not usable [33].
- The HSTM anthropomorphic test device, an increased impulse was measured within the surrogate stomach and liver but not the lungs when PPE was present [29].
- The influence of a complex pressure time history on performance parameters measured by test devices is investigated in [61-62]

The correlation between BOP and injury risk to assess the protective performance needs further elaboration [13], [60], [63-64]. Especially development of severity risk scales for non-lethal or low lethal injurious consequences [61]. Also the influence of armour materials, e.g. mass and cushioning, on the chest kinematics is not fully understood [28],[33].

3.3 Ocular PPE BOP Test Methods

Ocular BOP injury could be expected from Friedlander blast wave characterised by 113-132 kPa incident peak overpressure or 96-128 Pa*s incident pressure impulse. The usage of ballistic eyewear was not related to reducing closed eye injury [6] but is correlated with decreasing eye injuries (26% to 17%) [22] possibly only mitigating secondary fragments. If closed eye injuries are caused by BOP, this could imply that the used eyewear is not protecting against BOP.
The repeatability and reproducibility of measured reflective pressure at the corneal surface, different types of eyewear or the effect of open and closed eyewear for multiple head form orientations is investigated by all studies. One study included measuring the IOP. Further investigation is required to determine the deviation of these parameters.

Ocular PPE BOP Test methods appear to be in a premature phase. However there is serious potential if to merge available information and methods as a basis for a formal standard. E.g. the FOCUS head form is fit to be used for PPE BOP evaluation according to [65].
Further research and elaboration of published research is required . Relevant publications and associated knowledge gaps are listed below for respectively Injury criteria, performance parameters and choice of test configuration:

Injury criteria:

- The injury risk predictions are based on animal test data and their applicability to humans needs to be determined;
- Risk of vision loss calculator [51]. An ocular injury criterion was formulated for the applied pressure. However, specifics to this injury criterion and tolerances were not available.
- Intra Ocular Pressure (IOP) [52]. IOP and reflective pressure were measured at a similar position. The measurements were correlated with injury by assessing the affected eyes for injuries and through using injury risk curves for projectile impacts [56]. The calculated risks are <2% for all eye injuries when protection was worn but are only calculated for the head-on situation;
- Protective eyewear is mainly evaluated using reflective pressure at the corneal surface. However, other parameters such as positive phase duration, corresponding impulse or IOP could also be important [53], [52];
- The reliability of ocular injury curves in relation to post-mortem [52].

Performance parameters:

- Repeatability/reproducibility cannot be assessed based on available publications;.
- Using injury risk as performance parameter;
- The variety of head forms, combat helmets and the influence onto the performance parameter;
- Head-form rotations up- and downward has not been found within the available studies.

https://doi.org/10.52202/078352-0044

4. CONCEPT TEST METHOD

The prementioned performance criteria and test setups for testing thoracic and ocular PPE against BOP help to specify the first elements for a concept test method.

4.1 Thoracic PPE BOP Test Methods

- BOP regime: the lung injury criteria and tolerances in [8], discriminating between none, light, severe lung injury seems promising to determine desired blast loading.
- Performance parameter(s):
  Measuring a_{cw} for evaluating the performance of PPE seems a good starting point, able to distinguish between relevant loading levels, correlated to lung injury and incident overpressure.
- Test device
  A relatively simple test device designed and constructed for evaluating PPE against the effects of BOP and able to measure a_{cw} is preferred. For instance the MABIL or the thoracic rig.

4.2 Ocular PPE BOP Test Methods

- BOP regime: a Friedlander blast wave characterised by 113-132 kPa incident peak overpressure or 96-128 Pa*s incident pressure impulse produces ocular injury at animals and can be used as starting point.
- Performance parameter: measuring reflective pressure at the corneal surface.
- Test device: an anthropomorphic head form equipped with pressure sensors plus eyewear.

5. CONCLUSION

A standard test methodology for evaluating thoracic or ocular PPE performance against the effects of BOP is not available. The available custom test methodologies show variability in both measured parameters and test devices.
BOP regimes for testing thoracic and ocular PPE have been found in literature, some based on animal test data. Some BOP injury criteria and test devices for both body regions were found throughout literature.

- Thoracic hard armour material seems able to decrease lung injury, peak a_{cw} [15] and lethality [16], [31-32]. While soft armour seems able to increase severe lung injury, average fatality [27] and a_{cw} [16], [32] This effect is also shown by the studies using transmissive pressure as performance measurement [28], [30-31].
- Open and closed eyewear seem able to decrease and increase measured reflective pressure at the corneal surface, corresponding impulse and IOP depending on head form orientation.

The results of experimental testing has shown that ocular and thoracic PPE are able to improve and worsen the performance parameter in question.

A basis for concept methods are formulated for both, thoracic and ocular PPE. This needs further elaboration to create preferable international homologation of a performance standard.

6. ACKNOWLEGMENTS

This study was commissioned by the Clothing & Personal Equipment Branch from the Netherlands Ministry of Defence.

https://doi.org/10.52202/078352-0044

6. REFERENCES

[1] Hoencamp, R., "Systematic review of the prevalence and characteristics of battle casualties from NATO coalition forces in Iraq and Afghanistan." Injury 45.7 (2014): 1028-1034.

[2] STANAG 2920 Ed 3 Ballistic Test Method for Personal Armor Materials and Combat Clothing. (2015).

[3] NEN-EN 13087-7:2000, Protective helmets - Test methods - Part 7: Flame resistance. (2000).

[4] Smith, J. E., et al. "The epidemiology of blast lung injury during recent military conflicts: a retrospective database review of cases presenting to deployed military hospitals, 2003–2009." Philosophical Transactions of the Royal Society B: Biological Sciences 366.1562 (2011): 291-294.

[5] Kuhn, F., et al. "The Birmingham eye trauma terminology system (BETT)." Journal francais d'ophtalmologie 27.2 (2004): 206-210.

[6] Cockerham, G. C., et al. "Closed-eye ocular injuries in the Iraq and Afghanistan wars." New England journal of medicine 364.22 (2011): 2172-2173.

[7] Bowen I.G., et al. Estimate of man's tolerance to the direct effects of air blasts, Technical Progress Report, DASA-2113, Defense Atomic Agency, Department of Defense, Washington, D.C., 1968

[8] Boutillier, J., "Lung injury risk assessment during blast exposure. " Journal of biomechanics 86 (2019): 210-217.

[9] White, C. S., et al. The biodynamics of airblast. Lovelace Foundation For Medical Education And Research Albuquerque nm, 1971.

[10] Viano, D. C., "A viscous tolerance criterion for soft tissue injury assessment." Journal of Biomechanics 21.5 (1988): 387-399.

[11] Cooper, G. J., et al. "An experimental investigation of the biokinetic principles governing non-penetrating impact to the chest and the influence of the rate of body wall distortion upon the severity of lung injury." International IRCOBI conference on the biomechanics of impacts. 1986.

[12] Roman H., et al. "Blast injury research: modeling injury effects of landmines, bullets, and bombs." Clinical Orthopaedics and Related Research 422 (2004): 97-108.

[13] Bass D., et al. "A methodology for assessing blast protection in explosive ordnance disposal bomb suits."International journal of occupational safety and ergonomics 11.4 (2005): 347-361.

[14] Van Der Horst, M. J. "Criteria and test methodologies for injury assessment of vehicle occupants threatened by landmines and/or IED; an approach by HFM-148/RTG." (2010).

[15] Cooper, G. J. "Protection of the lung from blast overpressure by thoracic stress wave decouplers." Journal of Trauma and Acute Care Surgery 40.3S (1996): 105S-110S.

[16] Philippens M., et al. "DRDC Mabil Kleding blast bescherming", TNO Rijswijk memo 10DV3-116, 2010.

[17] Li, E., et al. "Weathervane: a single point model for blast injury approximations."20th Symposium on Military Aspects of Blast and Shock. 2008.

[18] Van Doormaal, J. C. A. M., et al. "An approximation of the Axelsson model for quick injury predictions." 21th Symposium on Military Aspects of Blast and Shock. 2010.

[19] Teland, J. A., et al. Single point methods for determining blast wave injury. 2011.

[20] Teland, J. A., et al. A single point pressure approach as input for injury models with respect to complex blast loading conditions. Norwegian Defence Research Establishment Kjeller, 2010.

[21] Axelsson, H., et al. "Chest wall velocity as a predictor of nonauditory blast injury in a complex wave environment. "Journal of Trauma and Acute Care Surgery 40.3S (1996): 31S-37S.

[22] Thomas, Roger, et al. "Ocular injury reduction from ocular protection use in current combat operations." Journal of Trauma and Acute Care Surgery 66.4 (2009): S99-S103.

[23] Stuhmiller, J. H., et al. "A model of blast overpressure injury to the lung." Journal of biomechanics 29.2 (1996): 227-234.

[24] Bass C.R., Rafaels K.A., and Salzar R.S, Pulmonary Injury Risk Assessment for Short-Duration Blasts, The Journal of TRAUMA, Injury, Infection, and Critical Care, Vol 65, No 3, pp. 604-615, September 2008

[25] Rafaels K.A., Bass C.R., Panzer M.B. and Salzar R.S., Pulmonary Injury Risk Assessment for Long-Duration Blasts: A Meta-Analysis, The Journal of TRAUMA, Injury, Infection, and Critical Care, Vol 69, No 2, pp. 368-374, August 2010

[26] Van Der Jagt-Deutekom, M.J. et al. "Status overzicht: Blast borstkas letsel effecten voor gevechtssoldaat met beschermende kleding", TNO Rijswijk rapport TNO 2012 M11179.

[27] Phillips, et al. Cloth ballistic vest alters response to blast. Walter Reed Army Inst of Research Washington DC, 1988.

[28] Chan P., et al., "Study of Material Effects on Blast Lung Injury Using Normalized Work Method", PASS 2010, Quebec, Canada, 2012

https://doi.org/10.52202/078352-0044

[29] Merkle A., et al., Evaluation of the Human Surrogate Torso Model Response to Ideal and Complex Blast Loading Conditions, PASS 2010, Quebec, Canada, 13-17 September 2012
[30] Wood G.W., et al., Attenuation of blast overpressure behind ballistic protective vests, PASS 2010, Quebec, Canada, 13-17 September 2012
[31] Ouellet, Simon., et al. "Parametric study on rigid plates, compressible foams and air gaps combinations for mitigating blast in personal protection applications" Proceedings of personal armour system symposium. 2008
[32] Nerenberg J., et al., The effectiveness of different personal protective ensembles in preventing blast Injury to the Thorax, Fourth international symposium on Technology and the Mine problem, Monterey, California, 2000
[33] Carboni M., et al., Evaluation of a blast overpressure test device for personal protective equipment, PASS 2010, Quebec, Canada, 2012
[34] Department of defense annual report to congress Efforts and Programs of the Department of Defense Relating to the Prevention, Mitigation, and Treatment of Blast Injuries for FY 2008. Deputy Under Secretary Of Defense Washington DC Advance Systems, 2008.
[35] Stuhmiller, J. H. Mathematical Modeling of Physical and Cognitive Performance Decrement from Mechanical and Inhalation Insults. L-3 Communications San Diego ca, 2009.
[36] Ouellet, S., et al. "Characterisation of defence research and development Canada's mannequin for the assessment of blast incapacitation and lethality (DRDC MABIL)." Proceedings of the Personal Armour Systems Symposium (PASS). 2008.
[37] National Highway Traffic Safety Admin., DOT, 49 CFR Ch. V (10- 1-11 Edition), Part 572—Anthropomorphic Test Devices, Subpart E—Hybrid III Test Dummy (1997)
[38] DeCristofano B., Carboni M., Maffeo M. and Segars R., Development of a methology to evaluate blast overpressure protection, PASS 2010, Quebec, Canada, 13-17 September 2012
[39] Hines-Beard, J, et al. "A mouse model of ocular blast injury that induces closed globe anterior and posterior pole damage." Experimental eye research 99, 2012; 63-70.
[40] Mohan, K., et al. "Retinal ganglion cell damage in an experimental rodent model of blast-mediated traumatic brain injury." Investigative ophthalmology & visual science 54.5 (2013): 3440-3450.
[41] Wang, H. C. H., et al. "Pathophysiology of blast-induced ocular trauma with apoptosis in the retina and optic nerve." *Military medicine* 179.suppl_8 (2014): 34-40.
[42] Long, J. B., et al. "Blast overpressure in rats: recreating a battlefield injury in the laboratory." Journal of neurotrauma 26.6 (2009): 827-840.
[43] Zou, Y. Y., et al. "Primary blast injury-induced lesions in the retina of adult rats." Journal of neuroinflammation 10.1 (2013): 849.
[44] Choi, J. H., et al. "Pathophysiology of blast-induced ocular trauma in rats after repeated exposure to low-level blast overpressure." *Clinical & experimental ophthalmology* 43.3 (2015): 239-246.
[45] Shedd, D. F., et al. "Long term temporal changes in structure and function of rat visual system after blast exposure." Investigative ophthalmology & visual science 59.1 (2018): 349-361.
[46] Zhu, Y., et al. "Blast Exposure Induces Ocular Functional Changes with Increasing Blast Over-pressures in a Rat Model." Current eye research 44.7 (2019): 770-780.
[47] Allen, R. S., et al. "Long-term functional and structural consequences of primary blast overpressure to the eye." Journal of neurotrauma 35.17 (2018): 2104-2116.
[48] Panzer, M. B., et al., "Scaling in neurotrauma: how do we apply animal experiments to people?" Experimental neurology 261 (2014): 120-126.
[49] Jones, K., et al. "Low-level primary blast causes acute ocular trauma in rabbits." Journal of neurotrauma 33.13 (2016): 1194-1201.
[50] Sherwood, D., et al. "Anatomical manifestations of primary blast ocular trauma observed in a postmortem porcine model." Investigative ophthalmology & visual science 55.2 (2014): 1124-1132.
[51] Smolek M. K. et al. "Artificially Intelligent Risk Analysis Model for Estimating Primary Blast Vision Loss"
[52] Duma, Stefan M., et al. Biomechanics of Head, Neck, and Chest Injury Prevention for Soldiers: Phase 2 and 3. Virginia Polytechnic Institute and State University Blacksburg United States, 2016.
[53] Williams, Steven T., et al. "Blast wave dynamics at the cornea as a function of eye protection form and fit." Military medicine 182.suppl_1 (2017): 226-229.
[54] Sundaramurthy, A., et al. "Assessment of the effectiveness of combat eyewear protection against blast overpressure." Journal of biomechanical engineering 140.7 (2018).
[55] Bailoor, S., et al. "Effectiveness of eye armor during blast loading." Biomechanics and modeling in mechanobiology 14.6 (2015): 1227-1237.
[56] Duma, S., et al. Biomechanics of Head, Neck, and Chest Injury Prevention for Soldiers. No. CIB-2011-011. Virginia Polytechnic Institute and State University Blacksburg United States, 2011.

 https://doi.org/10.52202/078352-0044

[57] Bentz, V., et al. "Joint live fire (JLF) final report for assessment of ocular pressure as a result of blast for protected and unprotected eyes (Report number JLF-TR-13-01) US Army Aberdeen Test Center." Aberdeen Proving Ground, MD (2013).
[58] Risling, M. et al,. "Experimental animal models for studies on the mechanisms of blast-induced neurotrauma." Frontiers in neurology 3 (2012): 30.
[59] Scott, T., et al. "Modelling primary blast lung injury: current capability and future direction." Journal of the Royal Army Medical Corps volume 163 ed 2, 2017.
[60] Whyte, T. "Adequacy of test standards in evaluating blast overpressure (BOP) protection for the torso.", Cranfield Ceres, 2016.
[61] Boutillier, J., et al. "Shock-wave interaction with reduced-scale simplified torso surrogates." IRCOBI Conference Proceedings. 2016.
[62] Bouamoul, A., K. et al. "Experimental and numerical modelling of a mannequin for the assessment of blast incapacitation and lethality under blast loading." Proceedings of the 23rd International Symposium of Ballistics. 2007.
[63] Teland, J. A. "Review of blast injury prediction models." Norwegian Defence Research Establishment (FFI), 2012.
[64] Boutillier, J., et al. "A critical literature review on primary blast thorax injury and their outcomes." Journal of Trauma and Acute Care Surgery 81.2 (2016): 371-379.
[65] Bull, A. et al, eds. Blast injury science and engineering: a guide for clinicians and researchers. Springer, 2016.

https://doi.org/10.52202/078352-0044

Developing a room temperature replacement for Roma Plastilina #1 as a ballistic backing material

R. Mrozek,[1] E. Bain,[1] E. Beaudoin,[2] S. Cole,[1] J. Cora Cruz,[3] J. Gardner,[3] P. Gillich,[1] E. Napadensky[1]

[1]*CCDC Army Research Laboratory, 6300 Rodman Rd, APG, MD 21005, USA*
randy.a.mrozek.civ@mail.mil
[2]*Program Executive Office – Soldier, U.S. Army, 9102 Iry Road, Fort Belvoir, VA 22060, USA*
[3]*U.S. Army Aberdeen Test Center, 400 Colleran Road, APG, MD 21005, USA*

Abstract. ARL's Reusable, Temperature-Insensitive Clay (ARTIC) is a new backing material for use during ballistic evaluation of protective systems to simulate the deformation resistance of the human body. ARTIC was designed and developed with the sole intent of meeting the needs of the United States Department of Defense and the broader testing community to enhance the evaluation, certification, and development of ballistic protective systems including vests, protective inserts and helmets. ARTIC provides the desired backing material response at room temperature, with a consistent mechanical properties from at least 5 to 38 °C, and is composed of commodity feedstocks with robust supply chains. This paper briefly describes the challenges of designing a material with a controlled response that exhibits dimensional stability while providing minimal elastic recovery. In addition, we demonstrate the ability to tailor the response and compare with heated Roma Plastilina #1 (RP1), an artistic clay that is commonly used as ballistic backing material through a brief summary of initial ballistic and drop impact test results.

1. INTRODUCTION

Quantification of the back face signature (BFS) due to high rate impact is critical for the evaluation of personal protective equipment including helmets and body armour. A witness material is commonly placed behind the protective equipment prior to impacting the front side to serve two functions: 1) to better mimic the deformation resistance of the human body in a worn state rather than no supporting material behind the armour and 2) to quantify the transient deformation behind the armour. In practice, witness materials provide a quantifiable backface signature related to the transient deformation behind the armour that may not be equivalent to the maximum backface deformation. Common witness materials include clays [1], gelatins [2, 3], ballistic soaps [4], and microcrystalline waxes [1]. In 1977, Prather et al. evaluated the back face deformation behaviour of 20 % ordnance gelatin, Roma Plastilina No. 1 (RP1) and Roma Plastilina No. 2 compared to the lethality probability from soft armour impacts [5, 6]. Both RP1 and 20 % ordnance gelatin exhibited a similar suitable response however, gelatin exhibits elastic recovery requiring the use of high speed cameras and image analysis to accurately measure the deformation depth. In contrast, RP1 provides a permanent deformation that can be measured after the impact event. The reduced infrastructure costs and potential for increased throughput has led to RP1's extensive use as a ballistic witness material for different types of hard and soft body armour in the military and civilian sectors [7].

RP1 is a modeling clay used primarily by the artistic community comprised of 20 to 30 proprietary components including petroleum-based oils and waxes, kaolin clay, colouring agents, and sulfur [8]. The key feature of RP1 for use as ballistic witness material is that it is dimensionally stable at rest but will produce permanent, plastic deformation upon impact [9]. However, it is also critical that any ballistic witness material provides a repeatable and reproducible BFS measurement over the broadest range of conditions possible. Since it's initial use in the late 1970s, RP1's formulation and performance have changed due to the needs of the artistic community and changes in component feedstocks. These changes have resulted in significant time and effort by the testing community to maintain body armour test consistency. One of the more significant examples, is that RP1 is no longer used at room temperature. Instead RP1 must now be softened through physical agitation and heating up to 38 °C (100 °F) for it to pass calibration. In addition, RP1 exhibits substantial temperature- and time-dependent performance that results in a small time window before the material will likely no longer pass calibration [10]. In addition, RP1 also suffers from lot-to-lot variability, aging, humidity sensitivity, and a strong odor. In 2010, the United States National Research Council recommended identifying an improved ballistic backing material that can be used at room

 https://doi.org/10.52202/078352-0045

temperature to improve the accuracy and reproduciblity of protective equipment assesment [1]. To our knowledge, no commercial product was identified that met the performance needs to the testing community which required the development of a new backing material.

In this report, we present a new material, ARL's Reusable, Temperature-Insensitive Clay (ARTIC), as a room-temperature replacement candidate for RP1. The material was developed to meet the specific needs of the testing community. ARTIC is composed of three primary materials and one colourant that have robust supply chains to avoid undesired formulation changes in the future. Although tests are ongoing, ARTIC appears to exhibit a shelf life of at least one year with minimal temperature and time sensitivity. ARTIC also exhibits stability to humidity changes, very low toxicity, low flammability, and no detectable odour. The material can be mass manufactured using traditional polymer processing techniques that can enable commercialization of the material at a cost similar to RP1.

2. MATERIAL DESIGN OVERVIEW

It is critical that any RP1 replacement candidate provide dimensional stability at rest but deform readily upon impact while exhibiting minimal elastic recovery. Polymer-based materials were obvious candidates due to similar consistency with RP1 however, the mechanical response of the material is typically described as viscoelastic which is not ideal for a witness material. Too strong viscious character would eliminate the dimensional stability while strong elasticity could allow for the BFS to recover between impact and measurement, reducing accuracy. RP1 accomplishes this balance of properties through 20 or more components whereas a new material would need to accomplish the same performance using a minimal number of commodity components. ARTIC is composed of three primary components: silicone oil, fumed silica, and corn starch. Carbon black is also added to the formulation as a colourant at a loading of 0.004 wt % and has no detectable influence on the performance. Non-crosslinked, linear polydimethylsiloxane (PDMS), within a class of materials commonly referred to as silicones, was chosen due to its long performance lifetimes and uniform performance over a broad range of temperatures. It is important that the PDMS is not cross-linked to limit the potential for elastic recovery. As a single component, the PDMS will readily flow so fumed silica was added as a thickener to produce a dimensionally stable, "grease-like" consistency [11, 12]. Corn starch was added as a third component to reduce the tack adhesion (i.e. "stickiness") of the PDMS-fumed silica material and to promote compatbility with laser-based measurement techniques by virtue of the individual corn starch particles being larger than the laser wavelength (Figure 1). The material is batch mixed using a commercial stand mixer to grossly mix the components and then it is passed through a twin-screw extruder to refine the mixing and enhance the material reproducibility.

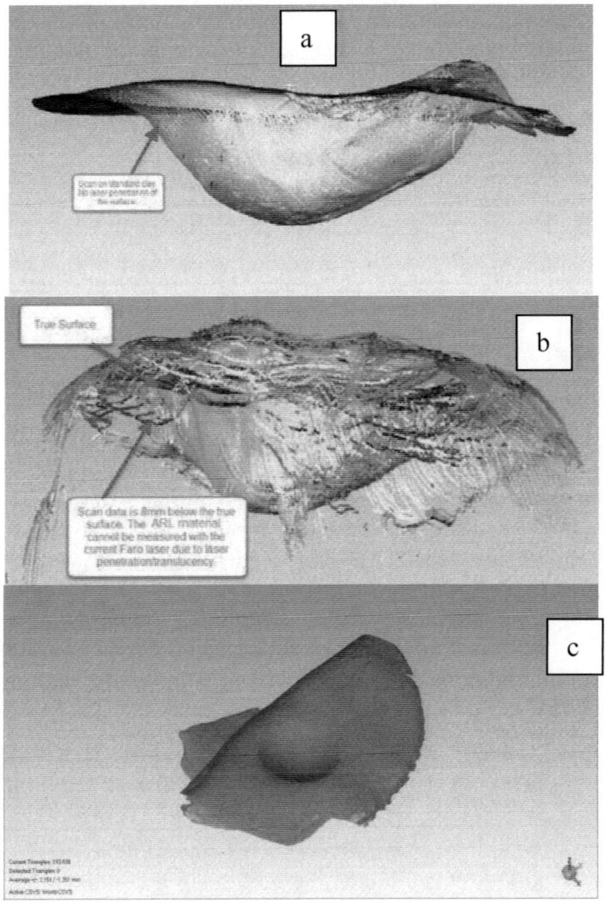

Figure 1. Laser scanning images of a) RP1, b) PDMS and fumed silica, and c) PDMS, fumed silica, and corn starch. Figure 1c was obtained later using a different indenter.

3. TAILORED PERFORMANCE

It was demonstrated in a previous report that ARTIC can provide a uniform performance from at least 5 to 38 °C (41 to 100 °F) and provide a similar response to heated RP1[13]. However, it is important to understand how changes in the material compliance, or resistance to deformation, alters the measured backface deformation. The deformation resistance of ARTIC can be readily controlled through the fumed silica loading. Decreasing the fumed silica loading produced a more compliant soft backing whereas increases in the fumed silica content produced a more rigid material. The materials were produced in 90.9 kg (200 lb) lots, enough to fill a standard ~ 61 cm x 61 cm x 14 cm (24 in. x 24 in. x 5.5 in) test box, with an additional 5 kg of material to repair indents during testing. The quasistatic deformation resistance was measured on every 0.9-1.5 kg of material produced to monitor uniformity of the entire lot (Figure 2). The numbers in parenthesis represent the targeted average value for the penetration force of each formulation and the respective red dotted lines are the manufacturing tolerances for each formulation set at the average ± 0.5 N.

https://doi.org/10.52202/078352-0045

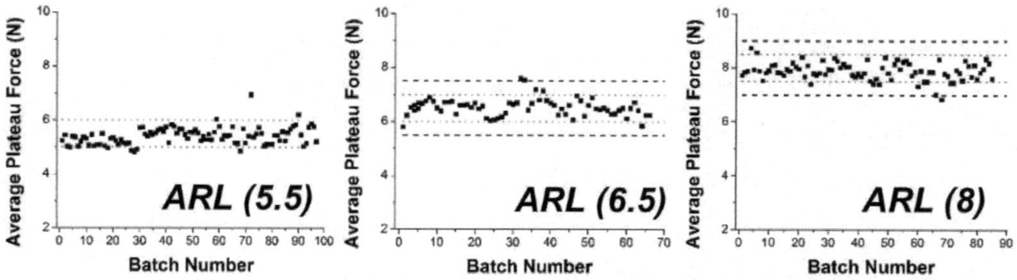

Figure 2. Deformation resistance data on every 0.9 to 1.5 kgs of material to ensure uniformity during manufacturing.

The test boxes were used to back soft armour shoot packs impacted with a 9 mm FMJ projectile at 0° obliquity and matching impact conditions. Figure 3 includes the BFS data for all three formulations side-by-side with heated RP1. From this limited data set, it appears that the softest formulation (ARL 5.5) is much softer than heated RP1 with an average BFS depth 5 mm deeper. However, the medium and relatively rigid formulation were similar to each other and to RP1 exhibiting a 0.7 mm deeper BFS depth and the same depth on average when compared to calibrated RP1, respectively. It also appears that both formulations exhibit a lower standard deviation than heated RP1 however, the time limitations of RP1 required the use of four different RP1 boxes for this study whereas only one box of each ARTIC formulation was used for the testing. This data set indicates that the BFS can be tailored through the formulation and that the ballistic response follows similar trends to those observed in the quasistatic data.

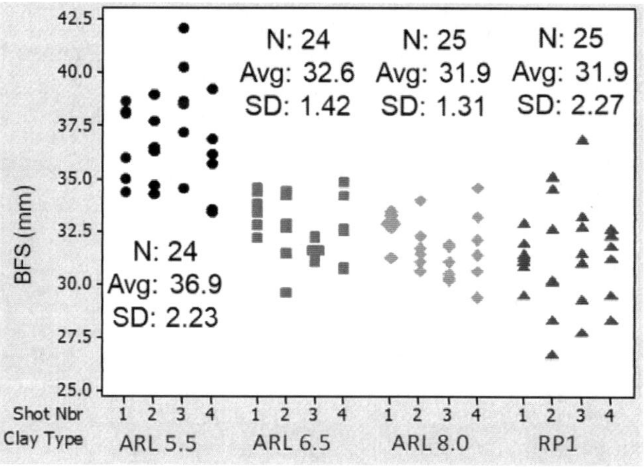

Figure 3. Soft armour BFS data for three different ARTIC formulations compared to heated RP1.

Prior to ballistic testing, heated RP1 is assessed using an impactor drop test (ASTM E3004) that consists of a 1 kg mass with a 45 mm diameter hemispherical end dropped from 2 m. The penetration depth of the impactor needs to fall within 25 ± 3 mm for the heated RP1 to pass calibration. Impactor drop tests were performed under the same conditions on the three different ARTIC formulations prior to testing. The drop depths of all three formulations along with heated RP1 are plotted against the measured BFS in Figure 4. As anticipated, the drop depths follow similar trends as the quasistatic penetration testing and the BFS. What is interesting is that the ARL 5.5 formulation that provided a significantly larger BFS falls within the drop calibration range of heated RP1 whereas ARL 8.0 falls below the calibration range of heated RP1 and provides similar BFSs as heated RP1. The ARL 6.5 falls within the calibration range for heated RP1 but is localized towards the lower end of the range. This data supports that the impactor drop can still be used as a calibration method for the new material however, the range to be within calibration may need to shift lower.

The calibration drops values will continue to be monitored during the material evaluation to better correlate the ARTIC drop values with the measured BFS. More work needs to be done but the differences in depth may be attributed to slightly different shear thinning behavior by RP1 than ARTIC. RP1 exhibits shear thinning behavior at slightly lower strains than ARTIC (0.04 vs 0.4 %) which may reduce the effective friction under low velocity, high surface area impacts.

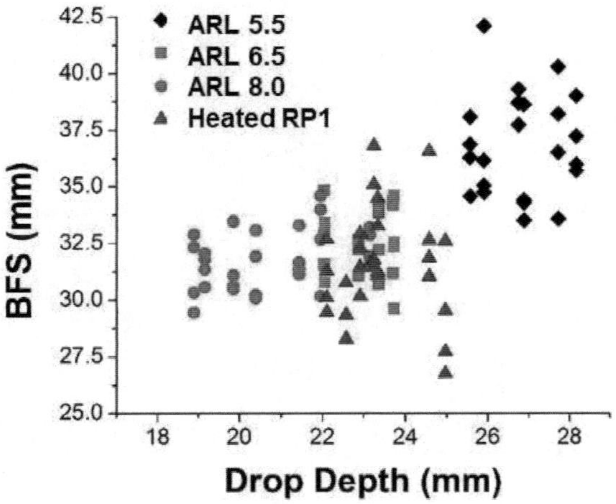

Figure 4. Soft armour BFS data for three different ARTIC formulations compared to heated RP1.

4. DETERMINING FACTORS THAT CONTROL THE DEFORMATION BEHAVIOUR

While the primary goal is to provide a backing material with the desired response at room temperature, a more comprehensive understanding of the deformation behaviour in clay-like materials may benefit the testing community by validating calibration ranges and links between low and high rate testing. Performing these studies is complicated with RP1 due to the time- and temperature-dependent performance. Specifically, it can be a challenge to verify that the RP1 is not changing for the duration of the test. ARTIC's consistent performance over a broad temperature range avoids these issues to increase the confidence of the measurement and the ability to tailor the formulation enables a direct understanding of how backing material compliance influences deformation behaviour.

In addition to ballistic testing, deformation behaviour during impact will be evaluated using a temperature-controlled drop tower. While the drop tower is not capable of reaching ballistic rates, it has the advantage of impacting the backing material at higher rates while eliminating any variability associated with the protection equipment or ballistics. The CEAST 9350 drop tower is spring assisted and enables impact velocities as high as 20 m/s (Figure 4a). It also has a heated chamber that enables temperature-dependent testing and direct comparison of impact behaviour with heated RP1 (Figure 5b). Test boxes with lateral dimensions of 28 cm x 28 cm have been fabricated to fit into the heated chamber at depths of 14, 23, and 28 cm to accommodate the study of clay deformation in excess of 22 cm. Additive manufacturing techniques were used to produce multiple impactor lengths, diameters, and radii of curvature (Figure 5c). Penetration depth and force, as a function of time can provide insight into the behaviour of the backing material when subjected to various impact conditions (Figure 6). The drop tower does not have the ability to monitor depth inherent to the instrument. To measure depth we use a high speed camera to capture the displacement of the cradle as the impactor enters material. During the initial drops, a slight reduction in the depth is observed after it reaches the maximum for some of the impact conditions that would be suggestive of elastic recovery. The extent of that depth change does not follow trends with impactor diameter, mass, or velocity. Current efforts are focusing on determining the cause of this observation however it is suggestive of inherent compliance to the test setup rather than the material itself.

https://doi.org/10.52202/078352-0045

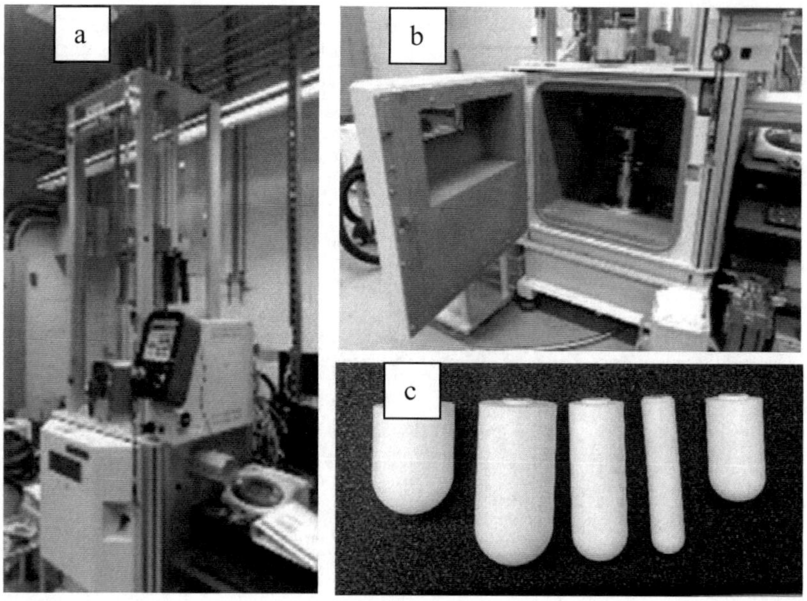

Figure 5. Pictures of the a) CEAST 9350 drop tower, b) temperature controlled chamber, and c) additively manufactured impactors in a range of shapes.

a

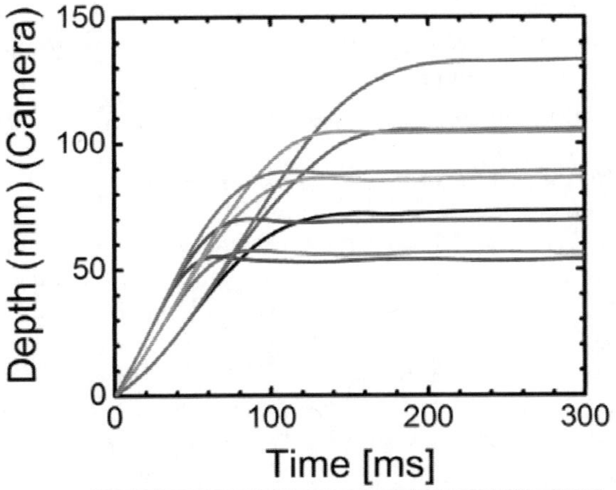

b

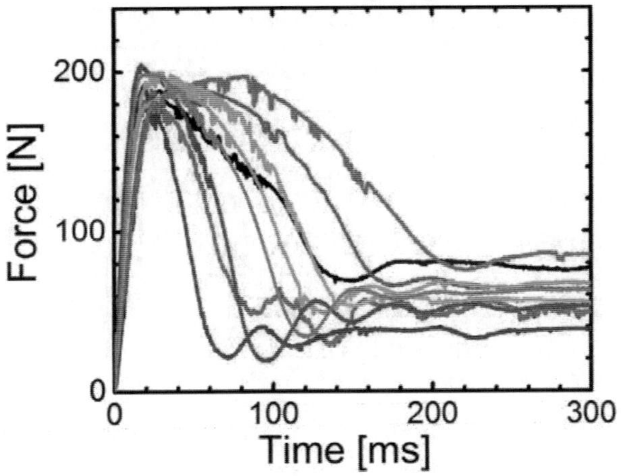

Figure 6. Plots of a) depth as a function of time and b) force as a function of time during an impact event on ARTIC in the drop tower.

5. SUMMARY/CONCLUSIONS

A new clay-like backing material has been developed that does not require heating and exhibits temperature and humidity stability, very low toxicity, low flammability and no detectable odour. Initial ballistic and drop impact tests results indicate that the material can provide a similar performance to heated and RP1 and has the potential to be implemented using existing RP1 infrastructure and calibration techniques. The material has been developed explicitly for the testing community using three commodity materials and a colorant, each with robust supply chains. Therefore it can reasonably be expected that the formulation will not change unless desired by the ballistic testing community.

REFERENCES

[1] L. Lehowicz, M. Denn, W. Fahrenholtz, D.F.J. Ronald, J. McGuffin-Cawley, H. Smith, K. Walker, A. Wilson, in "Testing of Body Armor Materials for Use by the US Army—Phase II: Letter Report", The National Academies Press, 2010.

https://doi.org/10.52202/078352-0045

[2] J. Jussila, Forensic Sci. Int., 141 (2004) 91-98.
[3] R.A. Mrozek, B. Leighliter, C.S. Gold, I.R. Beringer, J.H. Yu, M.R. VanLandingham, P. Moy, M.H. Foster, J.L. Lenhart, J. Mech. Behav. Biomed. Mater., 44 (2015) 109-120.
[4] C.J. Shepherd, G.J. Appleby-Thomas, J.M. Wilgeroth, P.J. Hazell, D.F. Allsop, Int. J. Imp. Eng., 38 (2011) 981-988.
[5] R.N. Prather, C.L. Swann, C.E. Hawkins, in, DTIC Document, 1977.
[6] E. Hanlon, P. Gillich, Military Med., 177 (2012) 333-339.
[7] J. Croft, D. Longhurst, " HOSDB Body Armor Standards for UK Police (2007) Part 2: Ballistic Resistance", Home Office Scientific Development Branch, Sandridge, St. Albans, United Kingdom, 2007.
[8] J.E. Seppala, Y. Heo, P.E. Stutzman, J.R. Sieber, C.R. Snyder, K.D. Rice, G.A. Holmes, J. Mater. Sci., 50 (2015) 7048-7057.
[9] F.A. Andrade, H.A. Al-Qureshi, D. Hotza, Appl. Clay Sci., 51 (2011) 1-7.
[10] H. Ji, E. Robin, T. Rouxel, Mech. Mater., 41 (2009) 199-209.
[11] M.I. Aranguren, C.W. Macosko, B. Thakkar, M. Tirrell, MRS Online Proceedings Library Archive, 170 (1989).
[12] R.S. Ziegelbaur, J.M. Caruthers, J. Non-Newt. Fluid Mech., 17 (1985) 45-68.
[13] T.D. Edwards, E.D. Bain, S.T. Cole, R.M. Freeney, V.A. Halls, J. Ivancik, J.L. Lenhart, E. Napadensky, J.H. Yu, J.Q. Zheng, R.A. Mrozek, Forensic Sci. Int. 285 (2018) 1-12.

Development and validation of knee and elbow guards for combined ballistic and impact protection

M. Keown, J. Levine, C. Hedge, **J.P. Dionne** and A. Makris

Med-Eng Holdings ULC, 2400 St. Laurent Boulevard, Ottawa, ON, Canada, K1G 6C4
jean-philippe.dionne@safariland.com

Abstract. Traditional knee and elbow guards used in the military are worn over garments and only offer impact protection. Often, these guards are secured to the joint using adjustable straps. As such, these guards are bulky and difficult to keep in position over the joint, especially during high-mobility activities like running and jumping. In addition, there can be substantial abrasion, chaffing and perspiration associated with prolonged usage of the protector over the joint. The objective of this work is to illustrate the development and validation of a low profile, multi-threat, knee and elbow guard system that specifically provides protection against impact, projectiles, ballistic threats and blast fragmentation debris, while remaining compatible with existing combat uniforms and equipment. Two distinct solutions have been designed, one meeting more stringent protection requirements, and a lighter and less bulky version meeting lower level protection requirements still deemed of relevance to military activities. Various reference standards were used to justify a minimum level of appropriate protection from fragmentation and from impact, relative to the application. For fragmentation, an analysis of Improvised Explosive Devices threats was used to determine relevant and realistic protection targets. For impact protection, significant efforts were first devoted to identifying appropriate test methodologies to adopt, in view of multiple existing standards from sports and other civilian applications. Drop tower impact attenuation tests were then conducted to optimize the protection provided through the selection of innovative dual-purpose materials. Finally, a limited set of human factors trials was organized, focusing on quantifying the potential benefits of the lighter-weight version in terms of flexibility, comfort and functionality, based on early prototypes built in various sizes. This work resulted in the development of promising fully functional prototypes ready for larger scale human factors evaluation and eventual production.

1. INTRODUCTION

Blast-related injuries from improvised explosive devices (IEDs), land mines and shrapnel predominate battlefield injuries, with approximately 70% these occurring to the upper and lower extremities [1,2,3]. Personal Protective Equipment (PPE) worn over the torso often leaves arms and legs unprotected. Traditional knee and elbow guards used in the military do not protect against blast debris and only offer impact protection or comfort when kneeling in rough terrain. Often, these guards are worn over garments and secured in place using adjustable straps with buckles or Velcro. They are bulky and difficult to keep in position over the joint, especially during high-mobility activities like running and jumping. In addition, there can be substantial abrasion, chaffing and perspiration associated with prolonged usage of the protector over the joint.

The goal of this work is to illustrate the development and validation of a low-profile knee and elbow guard system offering multi-threat protection against impact, ballistic threats and blast fragmentation, while remaining compatible with existing combat uniforms and equipment. Towards this goal, reference standards were first reviewed to justify a minimum level of appropriate protection from impact and from fragmentation. For impact protection, these included standards for knee protection in sports and civilian workplace settings. For fragmentation, an analysis of Improvised Explosive Devices (IED) threats from Defence R&D Canada – Valcartier (DRDC) was used to determine relevant and realistic protection targets. With a test methodology selected, a rigorous set of drop tower impact attention tests was then conducted to optimize the protection provided through the selection and stacking of innovative materials. Finally, limited human factors trials were completed focused on understanding the user requirements for flexibility, comfort and functionality.

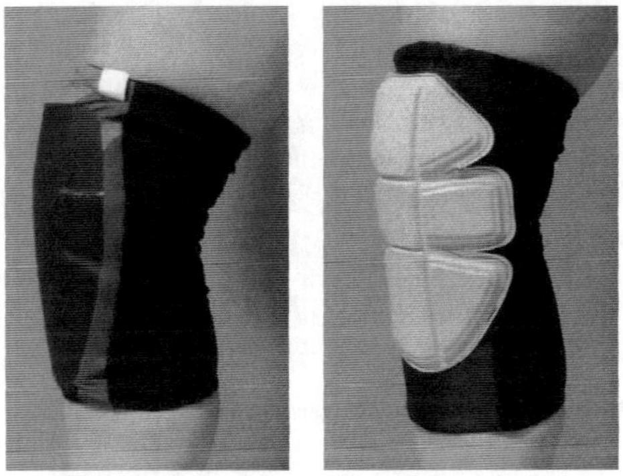

Figure 1: Prototype Knee Protection – Version 1 (Left, higher protection level) and Version 2 (Right, lower protection level)

https://doi.org/10.52202/078352-0046

Two solutions have been designed as shown with the knee version in Figure 1. Both solutions were worn next to skin and used a compression tube to secure the garment over the joint. Version 1 met more stringent protection requirements while Version 2 was lighter and less bulky, meeting lower level protection requirements still deemed of relevance to military activities. The development for both solutions reached a stage where promising fully functional prototypes are ready for larger scale human factors evaluation and eventual production.

2. BACKGROUND

A market review of available protective ballistic and non-ballistic knee and elbow guards identified best practices in design and material selection. Traditional combat knee and elbow guards typically feature rigid outer shells and are cumbersome and difficult to keep in place. They are worn over garments and only offer impact protection. Some tactical trousers now include knee pads integrated into the garment. In sports, like biking, skating, baseball and soccer, the trend is towards soft armour worn next to skin. These compression garments are lighter and more flexible than traditional equipment. Some examples of each are presented in the following sections.

Non-ballistic knee and elbow guards may be issued to soldiers as part of their kit or purchased by soldiers by their own initiative. They offer impact protection but do not protect against ballistic or fragmentation threats. Many major retailers of combat gear for military or police use offer such traditional knee and elbow guard solutions. While the appearance varies, most models use a similar construction of a foam backing with a rigid plastic outer shell, with straps to secure the guard to the knee or elbow joint. In the Canadian Armed Forces, knee protection has been designed into the new Improved Combat Uniform (ICU) combat pant via an exterior pocket positioned over the knee to accommodate a removable, foam knee pad. The uniform does not include integrated elbow pads, but traditional tactical elbow and knee guards are available to wear over the uniform. The UK military kit includes a standard tactical knee pad that is worn over the trouser but does not include elbow pads. The US Army Combat Pant (ACP) has integrated, removable hard-shell knee pads. Traditional knee and elbow pads are also available. The US Army combat shirt features elbow pads, but these are for abrasion resistance only, i.e. they prolong the life of the shirt.

One ballistic knee guard was found during this market research. It is offered by TACARM Tactical Armor (www.tacarm.com) and claims ballistic protection to NIJ Level IIIA. This guard is a traditional tactical construction with the addition of a pocket that can hold one or two ballistic inserts (depending on the desired protection level) constructed from an aramid material.

Recreational guards for sport often mimic the design of tactical guards. For active sports where keeping the guards in place is the priority, traditional straps are replaced by a compression sleeve to secure the guard, but these must be worn directly against the skin. These lightweight, flexible guards use rate-dependent impact attenuating materials sewn directly to the compression sleeve. To maintain comfort and flexibility, a rigid outer shell is avoided. G-Form (g-form.com) has several such products that differ in the coverage and performance of the impact attenuating material. In addition to knee and elbow guards, G-Form also offers shirts, shorts, and shin and ankle guards.

Work knee pads may be used in recreational (e.g. home gardening) or professional (e.g. construction) contexts. Again, the design of these guards is like that of tactical guards but with an emphasis on comfort for working in the kneeling position over extended periods. Work knee pads often use thicker, more compliant foam, forego the rigid outer shell but still use straps to secure the guard. BlakLader (www.blaklader.uk), a manufacturer of work wear, takes an alternate approach to providing knee protection by including a kneepad pocket in their trousers. They offer different grades of kneepads but none that can be worn without the trouser. Elbow pads are not commonly available in this working category.

3. PERFORMANCE STANDARDS

The knee and elbow guards developed for the current application are meant to provide both impact and fragmentation protection. While there is some consensus on which test methodology to use to evaluate fragmentation and ballistic protection performance (e.g. V50 rating [4]), there is a wide range of test methodologies addressing blunt impact mitigation, making the selection of a specific test methodology challenging.

https://doi.org/10.52202/078352-0046

3.1 Description

Performance standards related to impact of the knee and elbow are presented in Table 1. The intended use upon which each standard is developed is listed. Some examples of products discovered during the market research claiming compliance to these standards are also shown. The above listed performance standards involve similar test methodologies in that the test apparatus consists a guided mass falling onto the test specimen placed on an anvil. The impactor and the anvil differ somewhat amongst the standards in material, shape and weight, with some anvils including a complex curvature to better accommodate the knee and elbow guards. Impact velocity and transmitted force are typically measured. The drop height of the impactor is prescribed in order to meet the required impact energy of the given standard. It should be noted that these standards are not limited to impact protection, and often include requirements related to sizing, coverage, ergonomics and restraint. The impact energy and allowable transmitted force for each standard are listed in Table 2. Note that many of these standards feature more than one performance level.

Table 1: Performance Standards for Knee and Elbow Guards

No.	Title	Use
EN 15613	Knee & Elbow Protectors for Indoor Sports [5]	Light Falls
EN 1621-1	Motorcyclists' Protective Clothing against mechanical impact [6]	Hard Falls, Abrasion
EN 14120	Wrist, Palm, Knee & Elbow Protectors for Users of Roller Sports Equipment [7]	Light Falls
EN 14404	Personal Protective Equipment – Knee Protectors for Work in the Kneeling Position [8]	Working Position
EN 7971-4	Protective Clothing in Violent Situations [9]	Blunt Impact
CSA Z617-06	PPE for Blunt Trauma [10]	Blunt Impact

Table 2: Thresholds for Performance Standards

Standard	Level	Impact Mass (kg)	Elbow		Knee	
			Impact Energy (J)	Transmitted Force (kN)	Impact Energy (J)	Transmitted Force (kN)
EN 15613	A	2.5	1	4	2.5	6
	B	2.5	1.5	4	4	6
EN 1621-1	1	5	50	35	50	35
	2	5	50	20	50	20
EN 14120	1	5	6	4	12	6
	2	5	15	4	25	6
EN 14404	N/A	2.5	N/A	N/A	5	3
EN 7971-4	1	5	N/A	N/A	5	10
	2	5	N/A	N/A	15	10
	3	5	N/A	N/A	30	10
CSA Z617-06	N/A	1.46	15	3	20	3

3.4 Comparison Methods

Determining the relative severity of these knee and elbow guard performance standards was not possible by solely plotting input impact energy versus allowable transmitted force. Meeting the performance threshold for a given standard is influenced by the accumulated differences of apparatus, methodology and interpretation of results [11]. Two alternate ranking methods were explored, referred to as "Compression Distance" and "Contact Time", based on conservation of energy and momentum, respectively.

In the first comparison method, it is assumed that the impactor mass will induce a triangular force load on an infinitely thick material specimen (see Figure 2). The height of the triangle is based on the maximum allowable force threshold (F) from the specific standard, and the energy of the impact (E) corresponds to the area under a force-distance curve, as described in Equation 1. This equation can

 https://doi.org/10.52202/078352-0046

then be re-arranged to solve for the distance (d), as shown in Equation 2. The performance standards were then ranked where a greater distance, d, indicated a higher severity impact.

$$E = {}^{1}\!/_{2}\, F d \tag{1}$$

$$d = \frac{2E}{F} \tag{2}$$

The second comparison method was based on the conservation of momentum, using impactor contact time as the comparator. Here, the linear momentum was computed using the velocity (v) and the prescribed mass (m) of the impactor, as shown in Equation 3, and equated to a term involving the maximum allowable force threshold (F). Finally, the contact time (t), i.e. the time required by the material specimen under test to absorb the load, was calculated by dividing the momentum by the performance standard's threshold of transmitted force, shown in Equation 4. The performance standards were then ranked based on a longer contact time indicating a higher severity impact.

$$F\,t = m\,v \tag{3}$$

$$t = \frac{mv}{F} \tag{4}$$

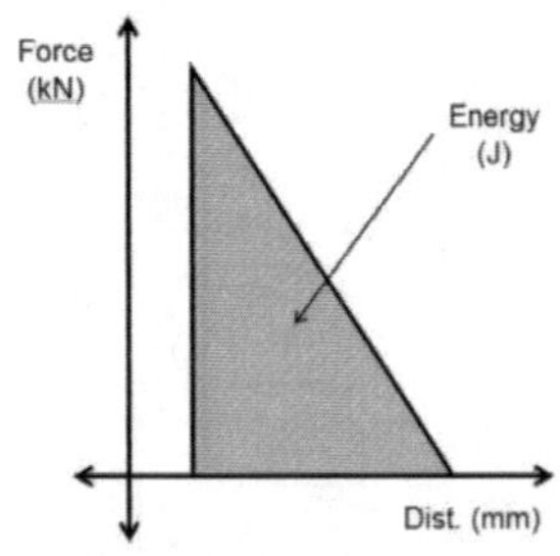

Figure 2: Idealized triangular force (F) versus distance (d) loading, with energy (E) represented as the area under the curve

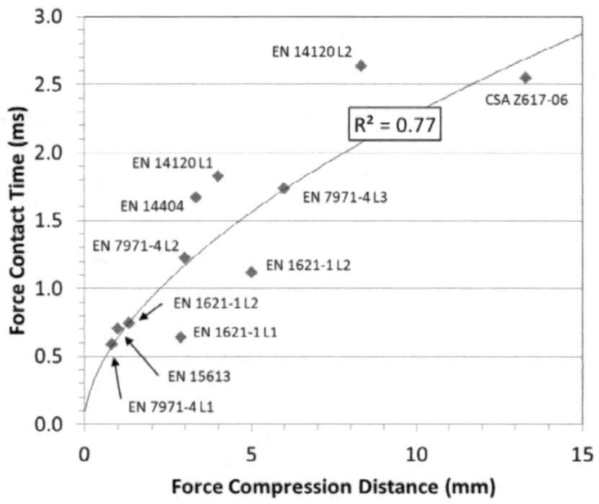

Figure 3: Force Contact Time vs. Force Compression Distance

3.5 Ranking by Test Severity

The estimated force contact time is plotted against the force compression distance, as shown in Figure 3. The R^2 value indicates a strong agreement between the two ranking methods. The computed values were used only to rank the severity levels and not meant to quantify a difference in severity. That is, twice the compression distance does not necessarily equate to twice the severity, or the amount of protection required is doubled, for example.

Given that both ranking methods correlated well to each other, the force compression distance method was arbitrarily selected to list the standards from most to least severe in Table 3. The rank number from the force contact time is also shown and is often near the distance rank value. A method to combine the ranking methods to establish an overall severity ranking was not developed nor applied.

https://doi.org/10.52202/078352-0046

Table 3: Overall Ranking of Performance Standards by Severity (most to least severe)

Standard		Level	Compression Distance Rank	Contact Time Rank
No.	**Title**			
CSA Z617-06	PPE for Blunt Impact	N/A	1	2
EN 14120	Protection for Roller Sports	2	2	1
EN 7971-4	Protective Clothing in Violent Situations	3	3	4
EN 1621-1	Motorcycle Protective Clothing	2	4	7
EN 14120	Protection for Roller Sports	1	5	3
EN 14404	Work in a Kneeling Position	N/A	6	5
EN 7971-4	Protective Clothing in Violent Situations	2	7	6
EN 1621-1	Motorcycle Protective Clothing	1	8	10
EN 15613	Protection in Indoor Sports	B	9	8
EN 7971-4	Protective Clothing in Violent Situations	1	10	9
EN 15613	Protection in Indoor Sports	A	11	11

For a given standard with more than one protection level, the two suggested ranking methods were found to have properly sorted the implied severity of each of those levels. The rankings were also found to align with severity expectations based on the intent of the standard, i.e. equipment for indoor sports is low on the list while riot gear is more severe.

With this information available, the next task consisted in selecting the most appropriate test methodology. Towards that end, former members of the Canadian Forces among Med-Eng staff identified the primary functions of the knee and elbow guards to be comfort and protection from abrasion resulting from normal activities like working in a kneeling position or lying prone on the ground. It was determined that the impact protection levels need not be equivalent to personal protective equipment designed for severe falls or motor vehicle incidents.

As such, the EN 14404 standard for working in a kneeling position was selected, as its intent was deemed representative of the product use, and the impact test severity ranked in 6th place with the distance method and 5th in the time method out of the 11 performance levels. Since it is neither the most nor least severe test, it was deemed suitable for evaluating material options in the current study. Unfortunately, this standard did not address elbow guards so it was decided that the protection level afforded to the knee joint with the novel knee guard would be applied to the elbow area as well.

4. IMPACT TESTING

With the blunt impact test methodology selected, impact attenuating materials of different thickness, grade and type were then combined and subjected to impact testing to guide material selection for the construction of prototype knee and elbow guards.

4.1 Test Method and Set-up

Impact tests were subdivided into two phases. The first phase was conducted using a flat force plate impacted with a large hemispherical impactor. This flips the methodology called for in the second impact test phase in the selected EN 14404 test method, whereby a nearly-flat, shallow curved impactor strikes a hemispherical anvil placed on a force plate. Tests were conducted in this order to accommodate the available flat samples of raw materials. Moreover, test results from the Phase 1 were considered alongside areal density and material availability to select specific samples for testing in Phase 2. Details of each test set-up as well as photos are provided in Table 4 and Figure 4, respectively.

In both test configurations the desired impact energy was 5 J and the drop height was adjusted to compensate for losses in the measured impact velocity due to friction in the guided drop tower. Also measured was the impact force using three force sensors

Figure 4: Impact test set-up for Phase 1 (a) and Phase 2 (b)

 https://doi.org/10.52202/078352-0046

(PCB Piezotronics 208C05) arranged to support the static force plate. Each of the three force sensor signals was post-processed using a CFC600 filter attenuating signals above 1000 Hz and then directly summed to create a single transmitted force, from which a peak value was extracted. The peak value was then normalised with respect to the calculated energy input from the velocity measurement at moment of impact.

Table 4: Impact Test Methods

Set-Up	Method	Force Plate	Impactor	Impactor Mass (kg)	Drop Height (cm)
1	Custom	Flat	Curved	1.04	55
2	CS EN 14404	Hemispherical	Flat	2.56	21

4.2 Test Results

A total of 225 flat sample tests were conducted on 75 variants created by selecting materials from 48 different material options prior to selecting the top performing candidates for the EN 14404 hemispherical testing.

Phase 1 testing had two goals. The first goal was to determine the effect of soft ballistic materials on the impact performance of impact attenuating material, and the second was to characterize materials that were available in different densities. To address the initial goal, two impact materials were selected for impact tests conducted with layers of different ballistic materials with areal densities ranging from 0.64-2.88 kg/m^2. After normalising the transmitted force with respect to trials without any ballistic material, it was found the ballistic material had little to no effect on the transmitted force (see Figure 5), with the vast majority of data points lying within 10% of the average values. This result thus allowed the selection of ballistic material to be based solely on user comfort and known ballistic performance.

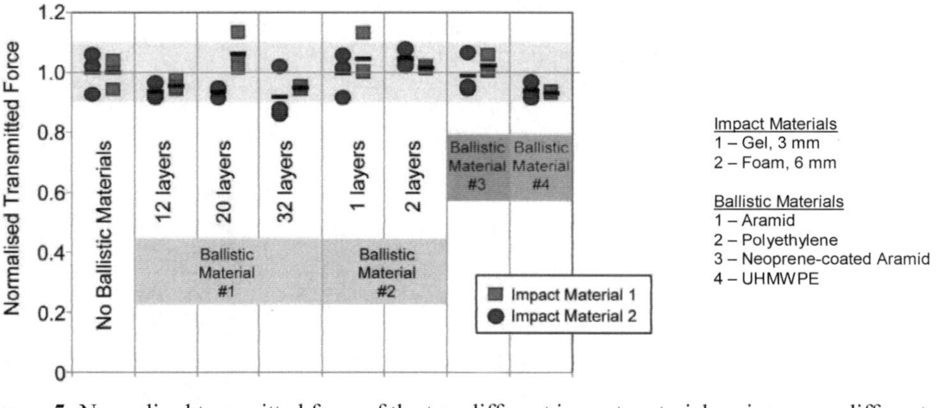

Figure 5: Normalised transmitted force of the two different impact materials using seven different ballistic material options

Towards the second goal of characterizing the performance of potential impact attenuating materials, four different low-density polyethylene and polyurethane foams were used. The low-density polyethylene and polyurethane foams were labelled LDPE-1 to LDPE-4 and PU-1 to PU-4, respectively, representing increasing densities of available foams. Figure 6 illustrates how the two types of impact foams diverged significantly under the chosen loading. The low-density polyethylene foams showed little sensitivity in transmitted force within the chosen density range, as it yielded a strong correlation with the foam thickness. In comparison, the polyurethane foams showed far greater sensitivity to density, whereby the sample thickness played a substantially lesser role.

https://doi.org/10.52202/078352-0046

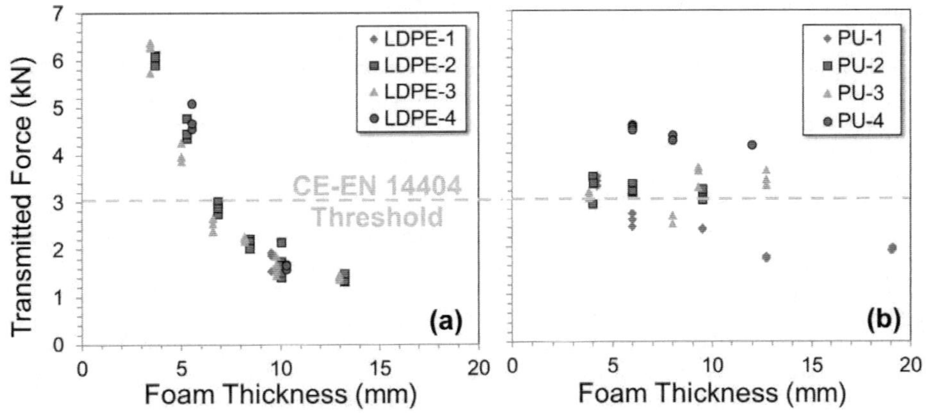

Figure 6: Transmitted force plotted with respect to (a) low-density polyethylene and (b) polyurethane foam thickness for the four different foam densities

After the flat sample Phase 1 testing was completed, 23 novel variants were tested using the Phase 2 EN 14404 hemispherical set-up, alongside 10 commercially available products to establish a baseline level of performance. The results are shown in Figure 7 with only the three best performing novel variants included. Many of the commercially available products reduced the transmitted force to within the threshold of the performance requirement while most of the proposed guards did not. This was due to the semi-rigid outer shell used on the commercially available products to distribute the load across a greater area, whereas the novel guards used only compliant materials aiming to improve comfort and flexibility for the user. Of the novel variants downselected options from Phase 1, only two passed the EN threshold in Phase 2 testing. The top performer was then selected for use in the first prototypes of the knee and elbow guards.

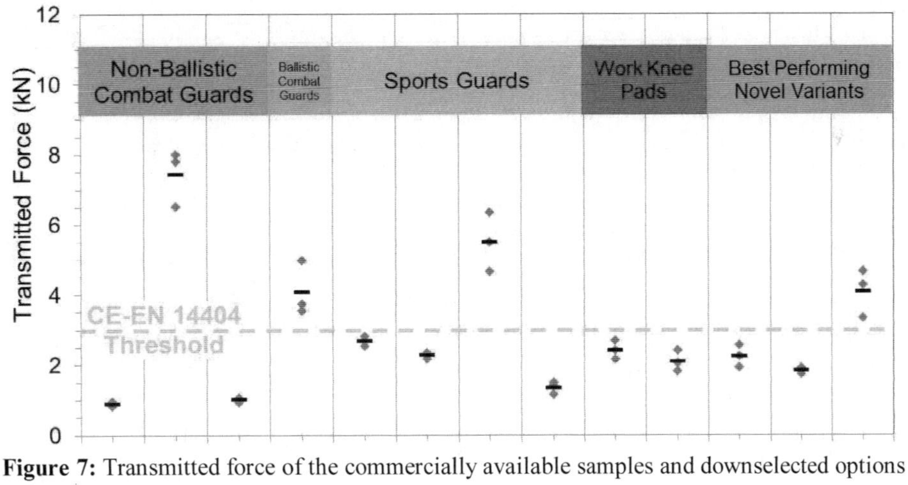

Figure 7: Transmitted force of the commercially available samples and downselected options

5. BALLISTIC PROTECTION

The three downselected options from Phase 2 tests utilised the same felt material to provide ballistic protection, a multi-layer composite fabric of aramid and ultra-high-molecular-weight polyethylene (UHMWPE). This felt was then used to construct the first prototype knee and elbow guards. A neoprene-coated aramid was selected to provide a durable surface to the front, outside of the guards. Finally, a stretchy polyethylene knit fabric was used in the construction of the compression tube that secures the garment over the joint. The V50 ballistic rating [12] with a 17-grain Fragment Simulating Projectile (FSP) was 404 m/s. It was previously shown that the extra ballistic layers would not greatly affect the impact performance. As such, impact testing on the complete ensemble was not deemed necessary.

https://doi.org/10.52202/078352-0046

6. HUMAN FACTORS EVALUATIONS

Prototype knee and elbow guards built from materials selected based on their fragmentation and impact protection performance, were evaluated with human subjects for fit and protective coverage across multiple sizes and for effects on mobility through the completion of a rigorous obstacle course.

6.1 Sizing Study

A sizing study with 22 participants recruited from the Med-Eng workforce was completed to validate the pattern sizing and assess overall fit. This sample size is comparable to the test panel of 25 participants recommended for evaluating first responder protection against chemical, biological, radiological, and nuclear (CBRN) events [13]. The participants in this study did not have military experience and were recruited solely for anthropometric comparison. Body measurements of each participant included circumference of the bicep, elbow and forearm (for the elbow guard) and thigh, knee and calf (for the knee guard). Three females and 19 males participated with body mass and stature ranging from 61-98 kg and 159-197 cm, respectively.

Three sizes (small, medium and large) of the more protective Prototype Version 1 were available for fitting and each participant selected the best fit after trying each size. Participants were also invited to comment on fit issues including bunching (too much material gathering on inside of joint), cuffs (tightness relative to overall guard), grading (difference from one size to the next) and length (coverage over joint). The results of the sizing study drove changes implemented in design, pattern and grading for the Prototype Version 2.

6.2 Mobility Test

A mobility test was conduced to assess how the knee and elbow guards might affect a soldiers' ability to complete typical combat activities. The CAN-LEAP (CANadian Load Effects Assessment Program) combat mobility course provides sequential activities representing common and physically challenging tasks encountered by soldiers [14]. It includes standardized running, climbing, crawling and weapon-related activities to characterize the effect of soldier equipment on the soldiers' mobility. However, the use of CAN-LEAP was beyond the scope of the current study but may be used in future work. Instead, an abbreviated obstacle course was created to replicate the movements necessary by a soldier to complete the CAN-LEAP course, with an emphasis on activities involving the knees and elbows. Since the CAN-LEAP course is often set-up indoors on a concrete floor, two weapons activities were added that placed the participants' knees and elbows into a tray of 6 mm gravel to evaluate comfort over simulated rough terrain.

Participants were recruited from Med-Eng employees but only those with prior police or military training and experience were eligible. The number of participants was based on the only two standards used in this study to assess the impact attenuation that also included ergonomic requirements. BS EN 15613 [5] requires one test subject per guard size to report on mobility and comfort whereas ergonomic testing with EN 14404 [8] allows three to five subjects depending on results, i.e. if 2 out of 3 fail, two additional subjects may be used. The size of these test panels is low in comparison to the sizing study, i.e. 22 participants and a larger sample size would be preferable for aggregating feedback. Only four male volunteers from the subject pool met the service requirements and it was assumed that their opinions reflect that of the target end user, until trials with end users can be accomplished.

These four volunteers all successfully completed the obstacle course wearing one knee pad while leaving the other knee bare. This allowed the participant a direct comparison to evaluate the effect of the knee guard. The same set-up was used for the elbow guard. While on course, comments from the participants were recorded and after finishing the course, each participant completed a survey to rate their level of satisfaction on the knee and elbow guards' weight, bulk, comfort, heat build-up and persistent fit. The survey was a five-point satisfaction rating, as shown in Table 5. Scoring the knee and elbow guards separately, the participants provided a rating on weight, bulk, comfort, heat build-up and secure fit, as well as an overall rating.

Table 5: Satisfaction Rating Scale

1	2	3	4	5
Very Dissatisfied	Dissatisfied	Unsure	Satisfied	Very Satisfied

The average results of the survey responses are shown in Figure 8 for Prototype Version 1 of the knee and elbow guards (higher protection and higher weight). The volunteers used six categories to

https://doi.org/10.52202/078352-0046

rate their ability to successfully complete the activities in the obstacle course. Generally, the elbow guard scored better than the knee guard. This difference, especially with respect to heat and secure fit, may be due to the trouser worn over the knee guard whereas the elbow guard was not covered. Only the weight and bulk average values exceeded the "Satisfied" rating which identified the other categories for improvement. These results, combined with the feedback which prioritized a need for better flexibility, contributed to changes implemented in Prototype Version 2.

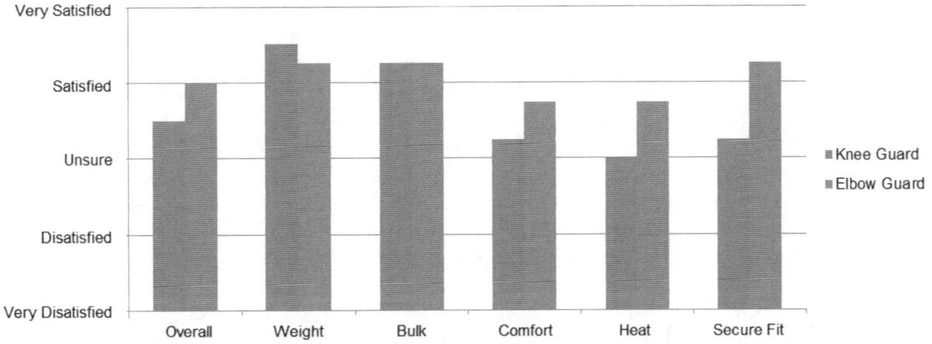

Figure 8: User Satisfaction Rating for Prototype Version 1 (higher protection/higher weight)

7. CONCLUSION AND NEXT STEPS

The project goal was achieved with the development of Prototype Version 1 of a multi-threat protection knee and elbow guard system for impact and ballistic threats. It was usable in the obstacle course and compatible with existing combat uniforms, but user satisfaction ratings were lower than expected. Based on feedback from DRDC Valcartier, the following concessions were made towards the development of a lighter-weight solution. First, protection against blunt impact could be reduced in exchange for better flexibility and mobility. The ballistic protection could also be reduced to focus on soil ejecta (IED) fragmentation instead of blast fragmentation (FSP).

A single set of Prototype Version 2 of the knee and elbow guards was then constructed in the medium size using alternate impact and ballistic materials already tested during this study. Performance indicators are displayed in Figure 9 using the best solution as the baseline. A weight rating was used as a predictor for user satisfaction because lighter weight is expected to reduce bulk and improve flexibility. These performance indicators could be prioritized and weighted accordingly to determine which version of the knee and elbow guard is the preferred solution. Additional performance indicators like comfort and ventilation, from usability trials with a greater number of participants could also be included in this determination and that will be the focus of future work.

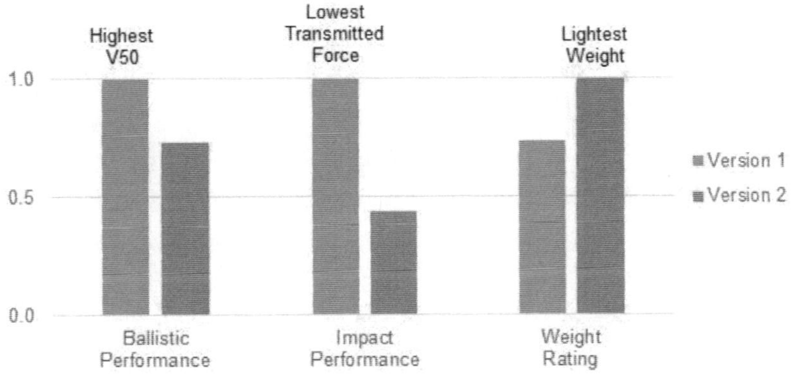

Figure 9: Performance Comparison between Prototype Versions

https://doi.org/10.52202/078352-0046

Acknowledgements

This research was partially supported by the Innovation for Defence Excellence and Security (IDEaS) Program of the Department of National Defence of Canada. We thank our colleagues from DRDC Valcartier for their input throughout the project.

References

[1] Zouris, J., Walker, J., Dye, J., Galarneau, M., "Wounding Patterns for U.S. Marines and Sailors during Operation Iraqi Freedom, Major Combat Phase" Military Medicine, Volume 171 3:246, March 2006.
[2] Ramasamy, A., Harrisson, S., Stewart, M., Midwinter, M., "Penetrating Missile Injuries During the Iraqi Insurgency" Royal College of Surgeons of England, Volume 91, 2009.
[3] Clark, M., Bair, M., Buckenmaier, C., Gironda, R., Walker, R., "Pain and combat injuries in soldiers returning from Operations Enduring Freedom and Iraqi Freedom: Implications for research and practice" Journal of Rehabilitation Research and Development, Volume 44, Number 2, 2007.
[4] NATO STANAG 2920 Ed. 2, Ballistic Test Method for Personal Armour Materials and Combat Clothing, NATO Standardization Agency, 31 July 2003.
[5] Knee and elbow protectors for indoor sports – Safety requirements and test methods, European Standard, European Standard, BS EN 15613, September 2008.
[6] Motorcyclists' protective clothing against mechanical impact, European Standard, BS EN 1621, December 1997.
[7] Protective clothing – Wrist, palm, knee and elbow protectors for users of roller sports equipment – Requirements and test methods, European Standard, BS EN 14120-2003, February 2003.
[8] Personal protective equipment – Knee protectors for work in the kneeling position, European Standard, BS EN 14404:2004, A1:2010.
[9] Protective Clothing in Violent Situations, European Standard, BS EN 7971-4.
[10] Personal Protective Equipment (PPE) for Blunt Trauma, Canadian Standard, CSA Z617-06, March 2006.
[11] Czerwinski, K., Fortuniak K., Obersztyn E., Redlich G., "Comparative Analysis of Impact Strength Tests of the Anti-Blow Multilayer Textile Lower Limb Protectors Devoted for the Police" Fibres & Textiles in Eastern Europe 2015; 23, 2(110): 84-89.
[12] MIL-STD-662F, V50 Ballistic Test for Armor, United States of America Department of Defence, 18 December 1997.
[13] CAN/CGSB/CSA-Z1610-11, Protection of first responders from chemical, biological, radiological, and nuclear (CBRN) events, Canadian Standards Association, March 2011.
[14] Bossi, L., Jones, M., Kelly, A., Tack, D., "A Preliminary Investigation of the Effect of Protective Clothing Weight, Bulk and Stiffness on Combat Mobility Course Performance", DRDC-RDDC-2016-P120, April 2017.

https://doi.org/10.52202/078352-0046

A new biofidelic backing for the evaluation of the ballistic performance of soft armour and lightweight protective fabrics

S. Ouellet[1], G. Pageau[1] and G. James[2]

[1]*Defence R&D Canada – Valcartier, Department of National Defence, 2459, Bravoure Rd, Quebec City, Quebec G3J 1X5, Canada*
simon.ouellet@drdc-rddc.gc.ca
[2]*Defence Science and Technology Laboratory, Porton Down, Salisbury, SP4 0JQ, UK*

Abstract. The Technical Co-operation Panel (TTCP) Land Group (LND) is a multinational defence science and technology collaboration between the governments of Australia, Canada, New Zealand, United Kingdom and United States. Under the auspices of this organisation, a technical panel was convened to focus on personnel protection and vulnerability. This paper presents the development and validation conducted within TTCP of a new backing that supports armour in a more biofidelic way during ballistic testing, providing protective materials with boundary conditions that likely better represent how armour is worn. The measured ballistic performance of soft body armour and lightweight protective fabrics has been shown to depend on the method of support of the samples during testing. Current support methods such as rigid clamping fixtures and deformable clay backing do not provide representative boundary conditions to armour material during testing. The new backing, mimicking the resistance of the human torso, consists of a multi-layer foam pack made of three well-specified materials respectively representing the epidermis (thin neoprene rubber), the dermis (a soft foam), and soft tissue (a stiffer foam). The local dynamic compliance of the pack was successfully adjusted to match abdominal deflection response. The epidermis and dermis material were chosen with physical properties close to the human skin and demonstrated biofidelic skin perforation thresholds, which enables the use of skin perforation as a criterion for V_{50} and V_{proof} testing of lightweight fragmentation protective materials. In addition, the multi-layer construction offers the capability to estimate projectile residual velocity and absorbed kinetic energy through measurement of depth of penetration in the pack, enabling better prediction of injury outcomes when the armour is defeated. Ballistic data obtained with soft body armour and lightweight ballistic combat clothing on the pack is presented and compared to similar data obtained using other retention methods. The pack is shown to be reliable when used as a backing for measuring ballistic performance of materials and provides additional diagnostics and benefits such as ease of use, procurement, shape adjustment, and conditioning with lower cost than clay backing. Its performance, simplicity and ease of use suggest that a final version could be implemented in ballistic laboratories worldwide.

1. INTRODUCTION

Under the auspices of The Technical Cooperation Program, a study assignment involving five cooperating nations was convened to develop a new biofidelic backing for the evaluation of the ballistic performance of soft armour and lightweight protective fabrics.

The process of penetration of a projectile into an armour material is inherently variable. To quantify the performance of body armour against a given ballistic threat with a high level of confidence, controlled and repeatable test methods are required. Body armour test standards such as the AEP-2920 [1] and the NIJ 0101.06 [2] aim at specifying and bounding a multitude of test parameters in order to augment the reliability and reproducibility of test results within a given test laboratory and across test facilities. A critical test parameter is the armour support or mounting method, as it dictates the lateral and transverse boundary conditions that are applied to the armour sample during the ballistic impacts. The two most widely used support methods for soft armour are the rigid frame, which clamps the armour sample around the four edges with a specified level of clamping force and fabric tension, and the clay backing, in which the armour sample is simply strapped over a block of calibrated deformable clay (Roma Plastilina 1 (RP1)) with no edge constraints. These two support methods offer drastically different lateral and transverse boundary conditions to an armour sample. While it is important to note that the main purpose of such methods is to increase consistency and reproducibility in the test results and not to provide realistic "as-worn" test conditions, it appears that they remain prone to errors if not correctly undertaken. For example, clay backings are susceptible to conditioning/calibration issues [3]. The use of clamping fixtures requires the adjustment of the clamping force and fabric tension to prescribed levels. These levels may vary within a test series if they are not verified and adjusted between firings, introducing variations in the boundary conditions applied to the sample.

The ballistic performance of soft body armours, defined here as the V_{50} for a given ballistic threat, may depend on the method used to support the samples during testing due to differences in how

 https://doi.org/10.52202/078352-0047

the armour is allowed to deform during impact. Therefore, the V_{50} of an armour becomes a measurement of its relative performance compared to other armours tested in a similar fashion, or compared to a benchmark performance defined by a standard using the same test method. Measuring the ballistic performance of an armour using realistic boundary conditions would be of little value if the requirements put on the impact velocity, in particular for Fragment Simulating Projectiles (FSP), were not tied to a specific operational scenario. Many defence departments have a growing desire to align the performance requirements put on fragmentation protection with clearly defined operational scenarios. Alternatively, when working towards maximizing the ballistic performance for a defined armour burden, the ability to reduce the physical burden associated with body armour is also contingent on understanding the true protection level offered by a system. As armed forces are moving towards optimising protection systems, partly through modularity and scalability of the protection level [4], there is a need for a test method that can measure true as-worn performance of soft body armour.

There is also a growing requirement for a support method that can provide additional diagnostic capabilities, useful to the evaluation and optimization of protection systems. For example, defence departments often use modelling tools to perform vulnerability analysis on different body armour designs to support procurement programs or research and development efforts [5]. These modelling tools require experimental data on protective materials that include residual velocity data, which is used to predict potential injury outcome following full perforation of an armour. Unfortunately, current support methods are limited in their ability to measure realistic projectile residual velocity. Finally, current support methods may be difficult to deploy in full-scale explosive trials, which have become more important in the context of testing against threats representative of IEDs [6].

In 2015, Technical Panel 5 of the Land group of The Technical Cooperation Program (TTCP LND TP5) launched a study assignment on a biofidelic test backing for soft body armour. The goal of the study was to scope the development of a backing that would provide representative boundary conditions to soft armour during testing, while offering a set of new capabilities compared to current support methods. The work reported herein was carried out in support of this study. The development and validation work was led by Defence Research and Development Canada (DRDC) with support from Defence Science and Technology Laboratory (Dstl), United Kingdom Ministry of Defence during the validation stage.

2. METHOD

First, a statement of requirements defining and quantifying a list of biofidelity, performance, functional and logistic requirements was established in collaboration with members of TTCP LND TP5. Second, a set of biofidelity metrics were selected as performance targets for the three critical features of the backing (compliance, perforation resistance and penetration resistance). Third, an iterative experimental process was followed to progressively refine the backing design in order to match the performance targets. Finally, the response of the backing was validated across a broader range of conditions.

Table 1 was built in consultation with the technical panel team members, which included representatives of Defence Science and Technology Group (AUS), DRDC, the Defence Technology Agency (NZ), Dstl (UK), and US Army Natick Soldier Systems Center.

To be considered biofidelic, the new backing was required to match the compliance of a selected body region as well as the perforation threshold of human skin. Local compliance will differ across body regions, but the region of the upper torso was prioritized since the backing is primarily meant for testing soft body armours. Matching the perforation of the backing with that of human skin was desirable to enable rapid diagnostics of potential injury outcome during laboratory or full-scale fragmentation trials. When gathering experimental data on the residual velocity of a projectile behind a given armour to feed penetrating injury models, the retardation effect of the skin is important to consider.

The new backing had to be usable for the testing of any protective fabrics or soft armour systems relevant to the defence sector, from a single layer (e.g. low areal density combat clothing, Tier 1 Pelvic Protection (PP) systems) to a multi-layer pack (e.g. fragmentation vest). All the relevant fragmentation threats commonly used in armour testing had to be supported in addition to very light weight projectiles which may be used to represent natural ejecta created in buried IED scenarios (e.g. small spheres made of glass). The estimated relevant threat velocity range spanned from 50 m/s to 1000 m/s because the backing would support testing of very light weight protective fabrics such as combat clothing up to heavy fragmentation vests. One of the additional desirable features of the new backing was the ability to be deployable in full scale explosive tests, which may occur in a range of temperatures, to either collect fragment data (act as a witness pack) or evaluate the protective performance of materials in a real environment combining the effects of blast and fragments. Maintaining the integrity of the pack during such explosive tests is important to recover valid fragment data.

Table 1. Performance and logistic requirement for the new biofidelic backing

Biofidelity	Body Region	Thoracic region and/or abdominal region
	Biofidelic characteristics	Local and global compliance; penetration threshold
Protection systems	Type of protection system	Any type of soft armour / lightweight protective fabrics including woven, knitted, felt and uni-directional fabrics
	Protection areal density	0.15 kg/m2 to 4.5 kg/m2
Threat	Threat type	Spheres, RCC, FSP
	Threat material	Metallic (Steel, Aluminum, Magnesium, etc...), Glass, Natural fragments
	Threat mass	2 gr to 130 gr
	Threat velocity	50 m/s to 1000 m/s
Environment	Operating temperature	-10°C to 35°C
Experimental configuration	Type of experiments	Controlled laboratory ballistic experiments and full scale explosive tests
	Sample size	Up to 400 mm by 400 mm flat samples
	Geometry	Possibility to modify the geometry of the backing for fitting PPE in full-scale experiments
	Integrity	Maintains a good level of integrity under blast loadings of relevant severity
	Calibration	Simple and straightforward calibration procedure, if any.
Assessment	Perforation	Rapid and consistent identification of sample and backing perforation/non-perforation
	Striking velocity	Estimation/calculation of projectile striking velocity from DoP measurement for single and multiple simultaneaous impacts
	Residual velocity	Estimation/calculation of projectile residual velocity upon armour perforation
	Projectile recovery	Enable soft recovery of projectiles
Logistics	Procurement	Ease of procurement for Canada, UK, USA, Australia and New Zeland.
	Supply	Ensure security of supply
	Storage	Storage at ambient temperature with no degradation of performance or change in material response
	Cost effectiveness	Reusable or expendable, but remains cost effective compared to other accepted test methods

Based on the shared understanding that the calibration process is often the source of variability in ballistic data generated using RP1 as the backing, it was deemed essential that if there were a calibration procedure for the new backing, it needed to be simple. Likewise, the preparation of the backing needed to be infallible as much as possible. The pack had to enable rapid and consistent assessment of both armour perforation and skin layer perforation, useful to perform V_{50} testing. The criterion on which full perforation of an armour is assessed may be based on armour perforation alone or on pack skin layer perforation, which is analogous to the use of witness materials in the rigid frame method. To support the testing of armour in the overmatch regime or to characterize fragmentation threats, the pack had to enable the estimation of impact velocity from Depth of Penetration (DoP) measurement through DoP vs velocity calibration curves. Such curves may be specific to individual projectile sizes, shapes and densities or be generalized if the pack response allows.

The new pack concept was aimed at providing a long-lasting alternative test method to multiple national departments, so the group agreed on a few logistic requirements. It was deemed important that the base material component for the pack be affordable and easy to procure within the required tolerances. It was important to have a minimum level of resilience and continuity with regards to the supply chain.

2.1 Biofidelity targets

2.1.1 Compliance

Through a review of the open literature, datasets pertaining to the local compliance of the human chest under dynamic loading conditions were identified. Based on a comparison of equivalent mass and impacting velocity [7], two studies in particular used loading conditions comparable to a soft body armour deformation following the ballistic impact of a handgun round. Bir, et al.[8] defined thoracic deflection vs time biofidelity corridors by performing impact tests using long instrumented impactors on Post-Mortem Human Subjects (PMHS). The impactor was 140 g and 38 mm in diameter. Impacts were done at 40 m/s and 60 m/s on the mid-sternum area. The corridors created by Bir were used in

https://doi.org/10.52202/078352-0047

previous work to tailor the response of a Blunt Trauma Thoracic Rig aimed at assessing the risk of Behind Armour Blunt Trauma [9]. Eck [10] followed a similar approach and defined similar deflection-time corridors based on PMHS tests using a 45 g impactor at 65 m/s. Eck defined corridors for both the epigastric and hypogastric region of the abdomen. Unfortunately, PMHS data for loading conditions closer to those of a small fragment impact on soft armour were not available.

For the development of the new backing, it was decided to prioritize matching the compliance of the abdominal region because the reference data was better reported and it represents a region of the torso that is not covered by hard armour. An average corridor built from the hypogastric and epigastric corridors of Eck was used as the objective performance for the pack. During the iterative design process, backing concepts were tested against the Eck target corridor. However, the final solution was evaluated against the corridors from both Eck and Bir.

2.1.2 Skin perforation

Reviews of published experimental work on the ballistic perforation of skin from various projectile sizes and mass are available from Breeze, *et al.* [11], Jussila, *et al.* [12] and Warlow [13]. With the intent of limiting the amount of ballistic testing in the development stages of the new backing, it was deemed desirable to identify only 3 to 4 relevant datasets against which the response of the skin layer of the backing would be adjusted. These datasets were chosen to cover the whole range of projectile mass identified in the requirement table. All the projectiles in these datasets were non-deforming at the velocities used for the skin perforation assessments. The choice of data type to reproduce was also influenced by the availability of the projectiles and capability to launch them at DRDC.

At one end of the mass spectrum, the experiments from DiMaio, *et al* [14] were selected. DiMaio conducted ballistic experiments on whole PMHS lower limb (thigh region) using 9.12 mm bullet (lead round nose, 113 gr) and observed a perforation threshold at approximately 58 m/s. At the lower end of the mass spectrum, some of the experiments from Missliewetz [15] were selected. Missliewetz reported experiments on complete limb (thigh regions) against small brass spheres of 4 mm and 0.31 g (4.28 gr) as well as small silica glass spheres of 4 mm and 0.086 g (1.34 gr). Obtaining PMHS data for the perforation threshold of the 1.1 g (17 gr) chisel nosed FSP was desirable given that it remains today one of the most used projectiles for the evaluation of body armour. Unfortunately, such data could not be found for human PMHS. The data generated by Breeze, *et al.* [11] on porcine skin specimen was used for this purpose. Breeze conducted experiments on fresh pig limbs, over the thigh region, against the 1.1 g chisel nosed FSP. When upper and lower limits of the threshold were identified in the original study, the upper and lower bounds were used to define the performance target range.

2.1.3 Depth of penetration

Breeze, *et al.* [16] obtained data from ballistic experiments using FSPs of different masses against goat skin and muscle from the thigh. The data related projectile velocity to DoP in soft tissue. Although limited, the data was used to benchmark the pack DoP response during development. The performance target was defined as impact velocity minus threshold velocity as a function of DoP normalised by projectile sectional density. The DoP response of the pack was deemed less critical to match closely to the biological datasets. For the purpose of back calculating impact velocity, a transfer function between recorded DoP in the pack and DoP in soft tissue was acceptable.

2.2 Iterative process

The development of the new pack followed an iterative process. The pack design had to enable recovery of projectiles and easy measurement of DoP, therefore a layered construction was the selected option. The initial design of the new backing, provisionally named the 'TP5 pack', was based on the pack construction proposed by James [17], which consisted of a series of 10 mm layers of soft neoprene foams. A skin layer was added to the layered foam concept to fulfil TP5's requirements.

The process started with the ballistic testing of skin materials. Several silicone and neoprene rubber materials were sourced in multiple hardness (50-80 shore A) and multiple thicknesses (1-3 mm). V_{50} tests were conducted with each skin candidate material, placed over 12 layers of neoprene foam. The four selected projectiles were fired from a small calibre gas gun with interchangeable barrels. All projectiles were fired with their corresponding barrel size (i.e. no sabots required). V_{50} estimates were obtained from a Probit regression of a minimum of 12 shots including a minimum of five perforations and five non-perforations within a maximum range of 20 m/s. Shot spacing was maintained at 50 mm. A perforation was defined as the full perforation of the neoprene rubber

https://doi.org/10.52202/078352-0047

material. The best three skin candidates were retained for the next phase.

For the second step of the process, a selection of soft tissue simulant materials was used to construct full packs. Different grades of neoprene foams were sourced, and different layering combinations were created. The local compliance of the various combinations of skin and soft tissue simulants were evaluated by launching a similar 45 g short baton round to the one used in [10], at 60 m/s using a large bore gas gun at DRDC. For each test, the displacement history of the projectile was recorded by tracking the tail of the baton using a high-speed camera placed perpendicularly to the impact direction. Based on the comparison of the compliance response with the target corridor, the best constructions were retained and retested for skin V_{50} and DoP response. DoP tests were conducted with the 1.1 g FSP only, by firing over the range of velocities between the V_{50} and 1.5 x V_{50}, with 25 m/s increments. Once the first round of tests was completed, the process was repeated if further iterations were required.

3 FINAL PACK DESIGN AND VARIANTS

Three iterations of the design process were required to converge to a pack solution that met the biofidelity requirements. There was a significant interaction between the compliance of the first few layers of foam and the skin perforation response. The V_{50} value was influenced by the choice of foam directly underneath the skin layer.

The construction providing the best match to our target values for compliance and skin perforation is shown in Figure 1a. The layering of this first version of the TP5 pack was one layer of a selected neoprene rubber (1.6 mm), one layer of soft neoprene foam (6.35 mm), 12 layers of a harder neoprene foam (6.35 mm each) and 1 layer of isoltop HD sheet (12.7 mm). The first two layers simulate the epidermis and dermis of the skin, while the harder foam is representative of the soft tissues. The pack is 400 mm x 400 mm but can be scaled to any other dimensions to fit within existing target support. Straps may be used to hold the pack together during manipulation but should not affect the pack response. Using straps to hold test samples also enables for a tight fit with minimal constraints, providing more realistic boundary conditions.

a. b. c.

Figure 1, a. TP5 pack b. Pack variants for the conduct of residual velocity tests and c. Pack variant to conduct full-scale explosive tests.

In order to fulfill the requirements with regards to the conduct of residual velocity tests and full scale-explosive tests, two variants of the pack using the same layering scheme were created (Figure 1b and 1c). The first variant was created by adding 15 layers of isoltop HD (12.7 mm), behind the last layer of the original pack to enable characterization of projectile residual velocity over a wide range of impact velocities. Isoltop HD has been used in witness pack applications in the past [18], and a higher density material than the foam was needed to test materials up to a velocity where the energy absorbed by the protective material becomes negligible compared to the residual energy of the projectile. This type of data is important in the context of vulnerability modelling for the generation of empirical models capable of predicting perforation and residual energy.

The second variant of the pack was created to support the characterization of the mass and velocity distribution of natural debris during full-scale explosive testing. It also supports the evaluation of the performance of protective fabrics in the same context. The pack acts as a witness system collecting small natural debris and fragments generated by the detonation of a simulated buried IED. It may not be suitable for capturing metallic fragments directly from ammunitions as the pack materials may not suffice to stop such high energy fragments. The variant in Figure 1c is the same layering scheme, but the dimensions of the pack can be increased to maximize fragment collection and the skin

https://doi.org/10.52202/078352-0047

layer is completely wrapped around the foam layers to ensure that the pack maintains its integrity during an explosion.

3.1 Validation of the TP5 pack response

3.1.1 Local compliance

The deflection time response of the TP5 pack is shown in Figure 2 for both the 45 g short baton at 60 m/s [10] and the 140 g long baton at 40 m/s [8]. Impacts were performed over the central 300 mm x 300 mm area of the 400 mm x 400 mm front surface of the pack. The local compliance of the pack was successfully demonstrated to fall within the Eck target corridor, lying towards the upper bound and following a similar trend to the original PMHS data. Results were deemed reproducible, with a variability on the maximum displacement of approximately ± 1.5 mm.

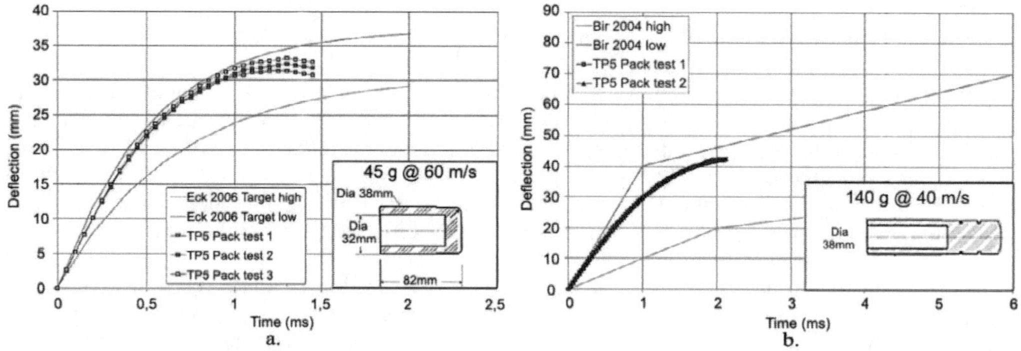

Figure 2, Compliance response of TP5 pack over a. Eck target corridor (n=3) and b. Bir biofidelity corridor (n=2)

Surprisingly, the local compliance also matched the biofidelity corridor from Bir, *et al.* [8] very well. This indicates that the variability in chest compliance between PMHS is probably greater than the difference in compliance between the abdominal and thoracic regions. By targeting an average abdominal corridor, we obtained a system compliance that also falls within the variability of upper thorax compliance. However, in the case of the Bir corridor, maximum deflection of the pack was reached earlier compared with the reference data. This may be due to the method of retention of the PMHS in the original experiments, where some gross motion of the specimen might have occurred, extending the duration of the whole event. During the impact tests on the pack, the pack is rigidly held at the back, preventing any backward gross motion. The purpose of adjusting the local compliance is to provide the appropriate level of resistance to a deforming armour during the ballistic penetration process, which most likely only involves the local dynamics of the abdomen or thorax. Gross motion of the body is most likely occurring beyond 2 ms, when the penetration process of the projectile is over. If indeed the extended duration observed in the PMHS experiments is due to gross motion, then the match obtained with the TP5 pack is appropriate.

3.1.2 Skin perforation

V_{50} estimates were obtained experimentally for the front layer of the pack. The historical data was used to establish the objective performance were perforation threshold values, which should be expected to be slightly lower than the associated V_{50}. After three iteration cycles, it proved to be challenging to match the target values at both ends of the projectile Sectional Density (SD) range. More specifically, the 9 mm projectile and the 4 glass sphere have very different perforation mechanisms in the skin simulant (punch shear vs stretching) which involve competing material properties in the skin simulant. In the end, the skin material that was chosen for the TP5 pack was thought to be the best compromise over the entire SD range.

Multiple V_{50} validation tests were conducted on the TP5 pack using a range of projectile shapes and sectional densities. In addition to the four projectiles used during the iterative development stage, tests using a custom 2 gr. Rock Simulating Projectile (RSP, [19]) as well as a 2 gr Right Circular Cylinder (RCC) were conducted by DRDC. V_{50} tests using a 3 mm glass sphere, a 6 mm glass sphere, a 4.4 mm steel sphere, a 6 mm steel sphere, a 12.7 mm steel sphere and a 9 mm tungsten sphere were also conducted by Dstl. V_{50} estimates were obtained using a Probit regression on all firings of each projectile. Figure 3a shows the 4 original objective performance values and Figure 3b shows all the obtained V_{50} estimates as a function of projectile sectional density. In order to obtain a more rigorous comparison of the perforation behaviour between the TP5 pack skin and human skin, the empirical model of skin perforation developed by James [17] was plotted over the data of Figure 3b. James proposed region specific empirical fits for the V_{50} of human skin as a function of projectile sectional density, based on a broader study of the data in the literature.

It was found that the TP5 pack response approached the estimate of the V_{50} for the thoracic skin, but may underestimate skin perforation, based on the thigh region. This may be satisfactory, however given that the level of confidence in the thoracic skin estimate is lower [17] and that it would be preferable to not underestimate any injury outcome that the pack aims to be able to predict, a thinner skin layer could be preferable. Following the armour penetration model proposed by Cunniff [20], which drew a correlation between armour V_{50}, armour areal density (i.e number of layers or thickness) and projectile sectional density, it is likely that a small change in the skin material thickness would suffice to bring the V_{50} data in Figure 3b at a more conservative level. It is also likely that the V_{50} for the lower sectional density projectiles will be affected more for a given reduction in skin thickness.

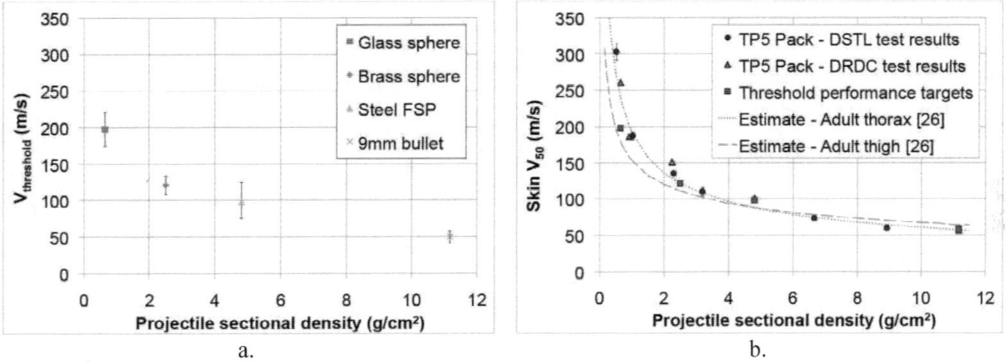

Figure 3, a. Threshold velocity for skin perforation for 4 different projectiles based on [11, 14, 15]
b. TP5 pack skin V_{50} as a function of projectile sectional density

3.1.3 Depth of penetration response

Depth of penetration validation tests were conducted by DRDC and Dstl using the first variant of the pack with the same set of projectiles as for the skin perforation tests. When conducting DoP tests with the pack, it was found that the most efficient way to proceed was to complete a series of firings, starting at the measured skin perforation velocity and moving upward by set increments (25 m/s for the 1.1 g FSP) until the complete pack was perforated. The same method was applied to the armour overmatch test, using the armour V_{50} velocity as the start velocity. Once the firings were completed, the pack was disassembled layer by layer, starting from the top, and the DoPs measured individually. Figure 4 shows the DoP reference data from Breeze [16] as well as all of the DoP results from DRDC and Dstl combined. In order to make the comparison with the data from Breeze easier, results from different projectiles are normalized on both axis. The graphs show impact velocity minus threshold velocity as a function of DoP over projectile sectional density. For the current comparison, it is convenient for removing any differences in perforation threshold values in order to focus on the comparison of the DoP trend alone. It appears that the individual DoP datasets from the TP5 pack can be reasonably represented with a linear regression.

The linear fit of the DoP data for the 1.1 g FSP in the TP5 pack has a higher slope than the biological data from Breeze. Through the iterative development, this was found to be very challenging to adjust further without drastically affecting the compliance of the pack. As mentioned previously, it was deemed acceptable that a transfer function or a simple scale factor between Pack DoP and Soft tissue DoP be required. In the current comparison, that scale factor would be approximately 1.5 for the 1.1 g FSP.

The important aspect of the results presented in Figure 4 is that the pack penetration response follows a linear trend for all the tested projectiles. It is also noticed that the slope of the fit is very similar for a given projectile shape and material (e.g steel spheres). By normalizing DoP by SD, it is seen that the results from spheres of a given material may be collapsed on a single fit. This means that by using this specific normalizing scheme, the impact velocity of projectiles of known or estimated mass, shape and material could be estimated with limited DoP calibration data. This is particularly useful when

https://doi.org/10.52202/078352-0047

conducting explosive testing and using the pack as a witness system. When fragments are recovered or imaged from the pack, one may use the information on fragment shape and material to estimate sequentially the mass and SD of the fragment, the perforation threshold velocity from the skin V_{50} curve and the impact velocity from the fit in Figure 4. Then, if the data is meant to be used for soft tissue injury prediction, the appropriate scale factor or transfer function may be applied to the DoP value.

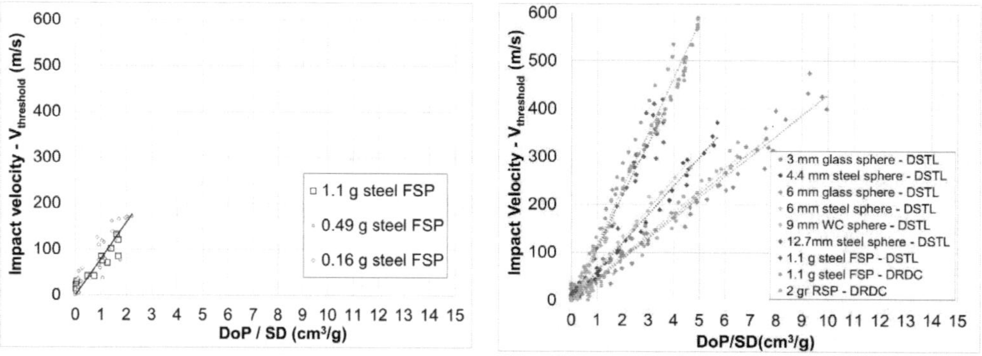

a. b.

Figure 4 a. Depth of penetration data from Breeze, *et al.* [16], b. Depth of penetration response of TP5 pack against various projectiles

4. APPLICATIONS

4.1 Testing of Soft body armour

The TP5 pack was used as a backing material for the V_{50} testing of two generic soft armours against the 1.1 g FSP. The soft armours had an areal density of 3.3 kg/m^2.One was made of Ultra-High Molecular Weight Polyethylene (UHMWPE) Unidirectional fabric and the other, a para-aramid plain weave fabric. Figure 5a shows the resulting V_{50} for the 1.1 g FSP along with additional results from V_{50} tests conducted on the same soft armours, but using different test methods. V_{50} data for a third soft armour made of 27 plies of another woven para-aramid fabric was obtained with these other test methods and is included for comparison. The additional methods considered were; Roma Plastilina 1 (RP1), the Blunt Trauma Thoracic Rig (BTTR) [21], the STANAG frame at two different levels of fabric tension and clamping pressure. All V_{50} estimates were obtained from Probit regression over *n* tests, where *n* varied between 25 and 42. Error bars on the graph indicate the 95% confidence interval, which can be seen here as an indication of the amount of variability observed in the data. When the range of velocities leading to a mixed outcome is large, the 95% confidence interval of the regression on the V_{50} is typically higher.

As expected, V_{50} estimates are significantly affected by the test method, with the highest difference between two methods being on the order of 10%. While this may be deemed acceptable, it is concerning that the performance ranking of the fabrics changes from one method to another. Also, it is highly probable that such variability increases when testing protective fabrics at low areal densities. The TP5 pack yields estimates that fall between those obtained with the RP1 and the STANAG frame. The STANAG frame in particular generated significant variability on the woven material, which is likely due to the tension not being uniform across the fabric surface. The TP5 pack generated narrow confidence intervals on both fabrics tested. While this was also the case for the RP1, the calibration process for RP1 is tedious. Conversely, a short series of DoP tests with the TP5 pack could suffice to validate that the materials of the pack behave as expected.

Figure 5b shows an example of results from an overmatch test where the TP5 pack first variant was used to recover the 1.1g FSP after armour perforation and calculate the residual velocity based on the pack calibration curve for that projectile. The graph shows the energy absorbed by the armour as a function of projectile impact energy. This type of response, where the absorbed energy from a soft armour decreases as the impact energy increases has been reported before [20]. The data was very straightforward to obtain with the TP5 pack. The pack has the considerable advantage of maintaining the coupling of the armour with the backing, as opposed to conducting air backed tests to generate similar residual velocity data.

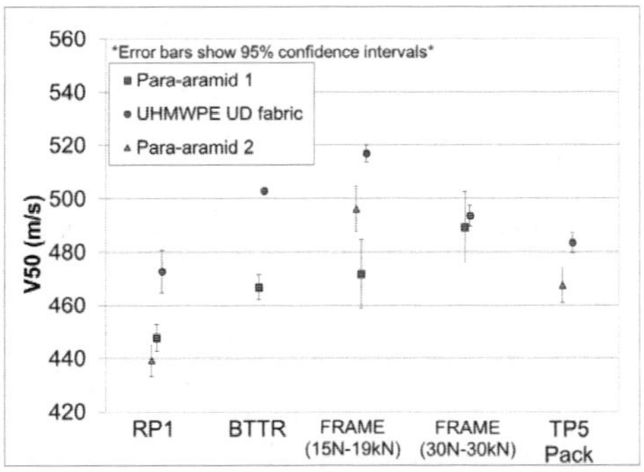

a.

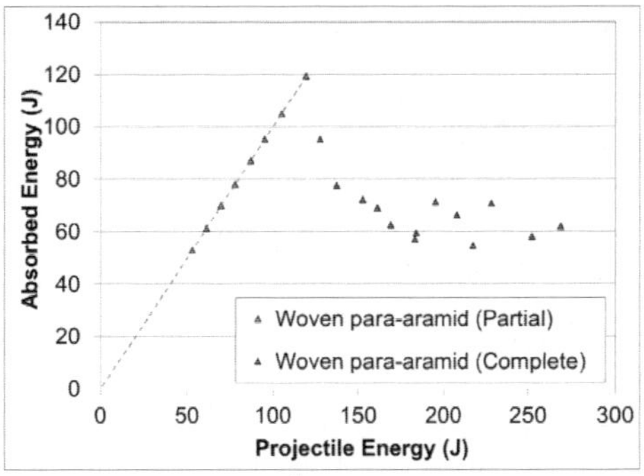

b.

Figure 5 a.V_{50} estimates for 3 soft armour against the 1.1 g FSP obtained with different test methods b. Example results from overmatch tests on a woven para-aramid fabric.

4.2 Full-scale explosive trial

The explosive test variant of the pack was used in a series of experiments simulating the detonation of a buried IED. In one trial, over 30 packs were fabricated and deployed at 1.75 m from a charge made of 5 kg of ammonium nitrate-fuel oil explosive, buried at 0.2 m, in order to characterize the generated ejecta distributions as well as to evaluate the protective performance of several very lightweight fabrics. The TP5 pack was found to be resilient to the blast generated, and aside from a few occasions where a large cluster of fragments tore the skin material, the packs held together and allowed for a complete analysis of the captured fragments post-test. To perform such analysis, a semi-automated method using a CT scanner was developed at DRDC [22]. Figure 6a shows an example of side and top views of a scan of one of the packs used during the explosive trial. While the manual extraction and measurement of fragments is always possible, it can be very labour intensive. The CT scanner method enables the semi-automated extraction of the number of fragments, individual fragment location, DoP, size, shape factor and mass (assuming the material of the fragment is known). The characterization of the mass and velocity distribution of an ejecta cloud associated with a threat is very relevant to vulnerability analysis.

Very lightweight protective fabric with relevance for PPE items, such as enhanced combat clothing and Tier 1 Pelvic Protection systems, were tested over the TP5 packs during the same trial. In addition to the qualitative assessment of the resistance of the fabric, the analysis of the pack enabled comparison of the fabrics using quantitative performance metrics. For example, Figure 6b shows a plot of the reduction of total mass of fragments perforating the skin for various fabrics of different areal density. This plot is obtained by comparing the total mass of fragments embedded in a pack from a protected scenario to a benchmark scenario (bare skin or other material covering).

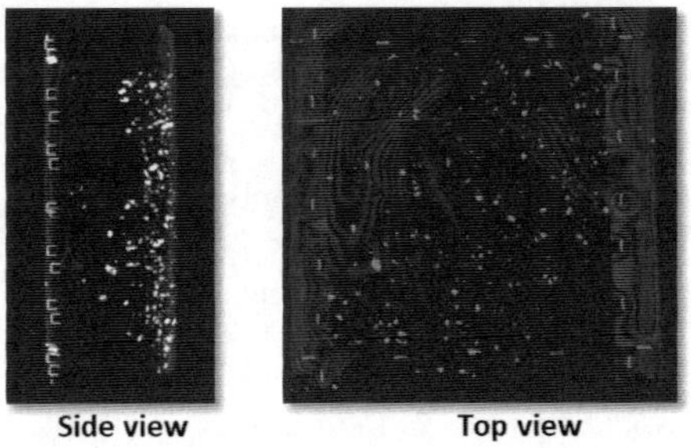

a.

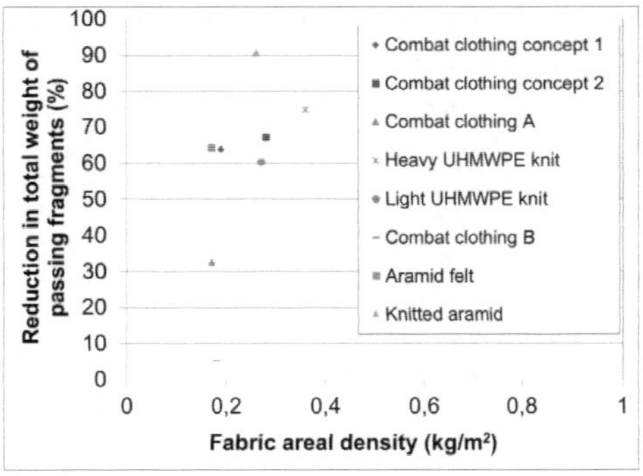

b.

Figure 6, a. CT scans of a pack after exposing it to the detonation of a buried explosive charge b. Example of results from full-scale explosive testing of lightweight fabrics using the TP5 pack.

https://doi.org/10.52202/078352-0047

5. CONCLUSION

Following the launch of a study assignment under The Technical Cooperation Program, a biofidelic test backing, aiming to provide representative boundary conditions for soft body armour during testing, was successfully developed at DRDC. The backing, provisionally called the TP5 pack, exhibits a local dynamic compliance matching abdominal and thoracic deflection responses. The perforation response of the pack skin layer was shown to correlate with skin perforation thresholds over a wide range of projectile SD, which enables the use of skin perforation as a criterion for V50 and Vproof testing. The multi-layer construction was shown to offer the capability to estimate projectile impact velocity, residual velocity or absorbed kinetic energy though measurement of depth of penetration in the pack. While the DoP response of the pack could not directly be tailored to match soft tissue DoP response, the trend of normalized DoP as a function of projectile velocity was similar, which enabled the use of a transfer function for injury prediction. The TP5 pack was shown to be reliable during the testing of soft armours and introduced minimal variability in the testing without necessitating a long calibration process. The pack was also fielded during full scale explosive tests and enabled the ranking of protective materials based on metrics such as reduction of total mass of fragments leading to skin perforation. The pack was deliberately designed using off-the-shelf materials that can be sourced internationally so that it can eventually be readily implemented in any ballistic laboratory worldwide. The final construction is still subject to minor changes.

Acknowledgments

The authors would like to acknowledge the contribution of all TTCP LND TP5 technical panel members to the concept presented herein.

References

[1] NATO-AEP-2920, Procedures for the evaluation and classification of personal armour and fragmentation threats, 2016.
[2] Ballistic resistance of body armor, National Institute of Justice (NIJ) standard-0101.06, US, 2008.
[3] Graham, M. J., & Zhang, T. G. (2019). Experimental Investigation of Geometric Effects on Ballistic Clay Backing Material (No. ARL-TR-8717). CCDC Army Research Laboratory Aberdeen Proving Ground United States.
[4] https://des.mod.uk/what-we-do/army-procurement-support/virtus/, 03-04-2020.
[5] Gillich P, Mermagen W, An Overview of the Operational Requirement-based Casualty Assessment (ORCA) Model and its Military Applications, Personal Armour Systems Symposium 2010, Quebec City, Quebec, PASS, (2010) pp. 328-331.
[6] C. J. Freitas, R. P. Bigger, J. P. Thomas and J. F. Mackiewicz, Threat characterization methodology for IED post-detonation environments to dismounted warfighters, Personal Armour Systems Symposium 2014, Cambridge, UK, PASS, (2014).
[7] Prather R, Swann C, Hawkins C. Backface signatures of soft body armors and the associated trauma effects. Aberdeen Proving Ground (MD): Edgewood Arsenal; 1977. Report No.: ARCSL-TR-7705.
[8] Bir, C., Viano, D., & King, A. (2004). Development of biomechanical response corridors of the thorax to blunt ballistic impacts. Journal of Biomechanics, 37(1), 73-79.
[9] Bourget D., Anctil B., Doman DA, Cronin DS (2002). Development of a surrogate thorax for BABT studies. In: Proceedings from Personal Armour Systems Symposium 2002, The Hague, The Netherlands.
[10] Eck, J. V. (2006). Biomechanical response and abdominal injury due to blunt ballistic impacts. Wayne State University.
[11] Breeze J., Clasper J.C. (2013), Determining the velocity required for skin perforation by fragment simulating projectiles: a systematic review. Journal of the Royal Medical Corps, 159(4), 265-270.
[12] Jussila, J., Leppäniemi, A., Paronen, M., & Kulomäki, E. (2005). Ballistic skin simulant. Forensic Science International, 150(1), 63-71.
[13] Warlow, T. (2016). Firearms, the law, and forensic ballistics. CRC Press.

https://doi.org/10.52202/078352-0047

[14] DiMaio, V. J., Copeland, A. R., Besant-Matthews, P. E., Fletcher, L. A., & Jones, A. (1982). Minimal velocities necessary for perforation of skin by air gun pellets and bullets. Journal of Forensic Science, 27(4), 894-898.

[15] Missliwetz J. Critical velocity in skin: an experimental ballistic study with firearms of 4 mm and 4.5 mm calibers. Beitr Gerichtl Med 1987;45:411–32.

[16] Breeze, J., James, G. R., & Hepper, A. E. (2013). Perforation of fragment simulating projectiles into goat skin and muscle. Journal of the Royal Army Medical Corps, 159(2), 84-89.

[17] James, G. Development of models to assess penetrating injury from ballistic projectiles (PhD). Cranfield University, Cranfield Defence and Security, 2020 (In press).

[18] District, P. O. (1962). The calibration of a collection medium for the determination of particle velocity, Project Thor Technical Report no 50, Ballistic Research Laboratories Aberdeen Proving Ground, Maryland, USA.

[19] Pageau G., Ouellet S., Development of a rock simulating projectile, Personal Armour Systems Symposium 2020, Copenhagen, PASS, (2020).

[20] Cunniff, P. M. (1996). A semiempirical model for the ballistic impact performance of textile-based personnel armor. Textile Research Journal, 66(1), 45-58.

[21] http://www.biokinetics.com/products/blunt-trauma-testing/44-bttr-torso, 03-04-2020.

[22] Gabriel, U., Pageau, G., DesRoches, M. Ouellet, S., Baillargeaon, Y., Francus, P., & Shewchenko, N. Terminal Ballistics Application of X-Ray Computed Tomography (XCT) for the Quantitative In-Situ Analysis of Fragments Embedded in Collection Media. In 31st International Symposium on Ballistics. (2019).

https://doi.org/10.52202/078352-0047

Development and use of an instrumented alternative to the clay box

E. P. Carton and Y.S. Khoe

TNO Group Explosions, ballistics and Protection, Ypenburgse Boslaan 1, The Hague, The Netherlands, erik.carton@tno.nl

Abstract. Roma Clay #1 is frequently used in ballistic testing of body armour systems. It allows the residual depth of the clay indent to be measured, which standards limit for system approval. TNO is developing experimental methods that get more data from these tests. This data can be used to improve the correlation to injury and lethality. An instrumented clay box was developed that allows real time measurement of local pressure, local accelerations, as well as the indent depth using an ultrasound system. The positioning of the sensors in the box was aided by the use of finite element simulation of several bullet to armour systems with a clay backing. Using ultrasound pulses from the rear of the clay box, the formation of the bullet impact induced clay indent could be measured real time as well. This system also provides the maximal dynamic indent in the clay and its elastic recovery.

Experiments with the instrumented clay box on hard and soft body armour systems using various ballistic threats has given insight on the repeatability of the measurements. The use of RP1 required a special repair routine for the clay between each shot to maintain the sensitivity of the ultrasound measurement system. This significantly reduced the cycle time between shots. A second concept was developed where a transparent synthetic gelatin backing material was used which does not require repairs between shots and allows for high speed videos to verify the ultrasound measurements. The signals of the embedded sensors and the indent formation process can be used to compare and validate computer simulation results. The instrumented box allows parameters like pressure, acceleration and velocity of the clay to be measured to assess their possible correlation with injury levels.

1. INTRODUCTION

Measurements of indent depth in Roma plastilina #1 clay boxes have formed the backbone of body armour testing since the 1970s. The 44 mm behind-armour blunt trauma (BABT) standard was selected as threshold as it corresponds to a 6% probability of lethality [1]. TNO believes that the future of body armour lies in scalability of protection to correspond to the level of risk acceptance. This scalability requires that body armour testing should be able to identify different levels of injury consequences (where TNO has chosen to adopt the Abbreviated Injury Scale (AIS)).

The clay box used in ballistic testing of body armour systems only measures the residual backface deformation. TNO is developing experimental methods that get more data from these tests. Such data can be used to improve the correlation with injury and lethality. An instrumented clay box was developed that allows real time measurement of local pressure and accelerations, as well as the indent depth formation using an ultrasound system. The positioning of the sensors in the box was aided by the use of finite element simulation of several bullet to armour interactions with a clay backing. Using ultrasound pulses from the rear of the clay box, the formation of the clay indent could be measured real time as well.

This paper describes the parameter selection, experimental method and future plans for ballistic instrumented backings and TNO's vision on their function in body armour testing.

2. MEASUREMENT PARAMETER SELECTION

An instrumented backing requires selection of measurement parameters. For the instrumented backing, this choice is based on blunt injuries that may arise from non-penetrating impacts. Prediction of blunt injuries can be done using the blunt criterion [2, 3, 4] or the viscous criterion [5, 6].

$$BC = \ln\left(\frac{\frac{1}{2}MV^2}{W^{\frac{1}{3}}TD}\right) \quad (1)$$

$$VC_{max} = \left(\frac{dy(t)}{dt}\frac{y(t)}{D}\right)_{max} \quad (2)$$

The Blunt Criterion (1) correlates impact conditions (projectile mass M, velocity V and diameter D) and the human body (mss W, body wall thickness T) in order to predict injury. The Viscous Criterion

https://doi.org/10.52202/078352-0048

(2) directly correlates torso deformation and velocity ($y(t)$, $\frac{dy(t)}{dt}$, scaled by the torso diameter D) to injury. Sturdivan [2] notes the similarity of both, energy based, approaches. This is reflected in the similarity of the outcomes. Sturdivan also suggests that the VC may be better suited for assessing blunt injury in experimental situations. The added benefit is that the VC is independent of any potentially applied armour. It should be noted that the VC was originally developed for automotive purposes. The application to ballistic impacts is researched and shows chest wall mechanics that are similar [7, 8] for sternum impacts.

However, to apply the VC the torso deformation (or a well correlating parameter) should be measured over time. The current claybox test method prescribes measurement of indent depth [9] and indent velocity [10], but the method does not allow for velocity measurements. This inspired the search of measurement parameters that can be used in future BABT testing equipment and the development of a measurement system with equipment that could measure or correlate to deformation velocity. Additional measurement parameters are also introduced (pressures and accelerations in the backing), so that future correlations to injury could be made based on multiple measured parameters.

3. EXPERIMENTAL METHOD

3.1 Sensor positioning

For the multiple use and repeatability of the instrumented clay box, the sensors should be sufficiently away from the impact and indent zone in order to prevent permanent displacement in the clay as a result of the impact. In order to determine this safe distance both experiments and FEM simulations have been performed. In the experimental set-up the clay consisted of a stack of Roma #1 clay disks with a thickness of about 2 cm and aluminum foil between the layers. Figure 1 shows images of both the FEM model and the stack of clay disks (laminate) after impact of a .44 Magnum bullet (at 400 m/s) with an Aramid surrogate vest as body armour (here protecting the clay stack). From both the FEM and the cross-section of the clay stack, the dent as well as the permanently (plastically) deformed zone can be observed. This allows the safe-distance from the impact point for the sensors in the clay to be determined (150 mm). A cylindrical area in the clay should be free of sensors in order not to damage them for the unwanted situation of a perforation of the armour. Figure 2 shows the position used for the embedded sensors in the clay block. The embedded sensors need a direct line of sight to the impact point of the projectile. Also, their alignment is important as it influences the measurements. For the pressure sensors, the alignment should be such that they are approximately perpendicular to the travel direction of the pressure wave. On the other hand, the acceleration sensors should be approximately aligned with the expected local deformation direction of the clay (which is approximately the travel direction of the pressure wave). The embedded pressure and acceleration sensors measure the pressure wave and local acceleration over time inside the clay.

Figure 1. Simulation (left) and experimental result (right) of an indent (top) in Roma #1 clay (with aluminum foil interlayers) after impact of a .44 Magnum on an Aramid simulant vest

https://doi.org/10.52202/078352-0048

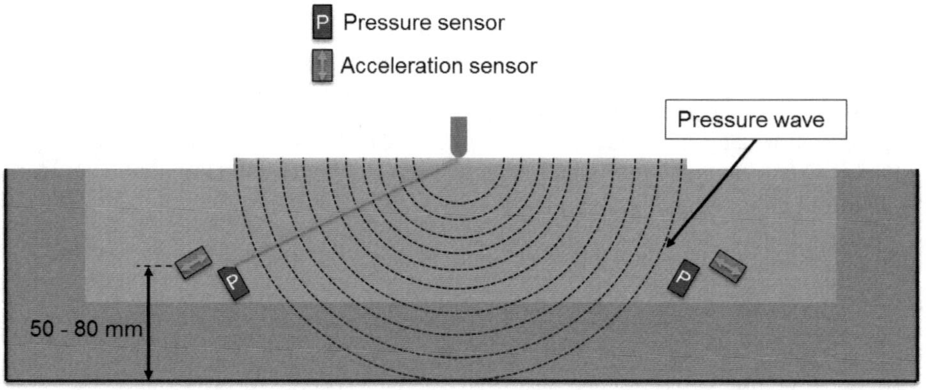

Figure 2. Cross-section of the clay box showing the position of the pressure and acceleration sensors

3.2 Back-face deformation

The back-face deformation is measured dynamically by transmitting ultrasonic pulses rapidly after each other, see figure 3. The two wedges are filled with Roma #1 plastilina clay. The wedge angle is 30 degrees. On each wedge a 250 kHz transducer is mounted. The distance between the two wedges can be varied to change the depth of the beam intersection point (focal point) in the clay. When the clay starts deforming the arrival time between the pulses will change. The time difference can be translated to the indentation depth during the development of the dent.

For the ballistic experiments a pulse repletion time of 100 μs is used. A typical display of a measurement is shown in Figure 4. The horizontal axis is the time scale at which the indentation occurs. The vertical axis is ultrasonic travel time. At 7 ms the pressure wave arrives and the clay starts to deform, a dent is formed which reduces the distance, hence travel time to the reflecting free clay surface. The pressure wave is very strong and overloads the ultrasonic measurement initially for a short time. Just in time the ultrasonic signal is obtained again to determine the maximal indent size and its reduction due to elastic recovery of the clay. The constant value corresponds to the indent depth that is normally measured (many seconds) after an experiment.

The ultrasonic waves reflect from inhomogeneities in a material, for the clay box this means that any voids, foreign objects and residual air pockets resulting from indentation repairs may cause reflections. These unwanted reflections interfere with the indentation echo. A special repair method for the indent of the clay was developed which has to be applied between each shot. Also the temperature sensitivity of the clay provides an issue as its velocity of sound is a function of temperature. In order to keep the clay box at one temperature (39 °C), it was decided to heat the clay box during its operation in the shooting range using two infra-red heaters; one at the back and one in front of the clay box. Figure 5 only shows the infra-red heater in front of the instrumented clay box, which was positioned away from the strike-face in order to allow personnel access for sample manipulation and indent repairs.

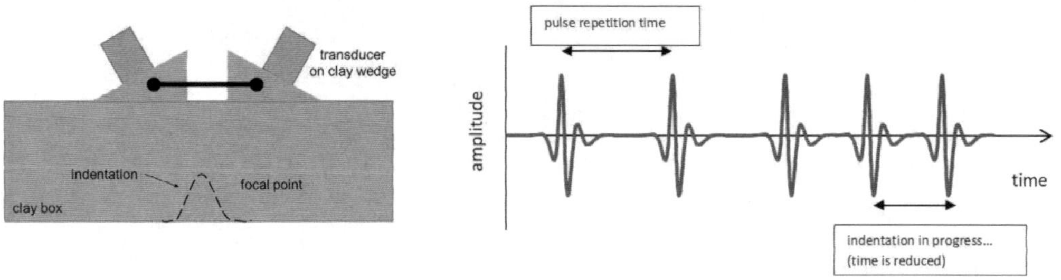

Figure 3. Working principle of the time-resolved ultrasonic clay deformation measurement

https://doi.org/10.52202/078352-0048

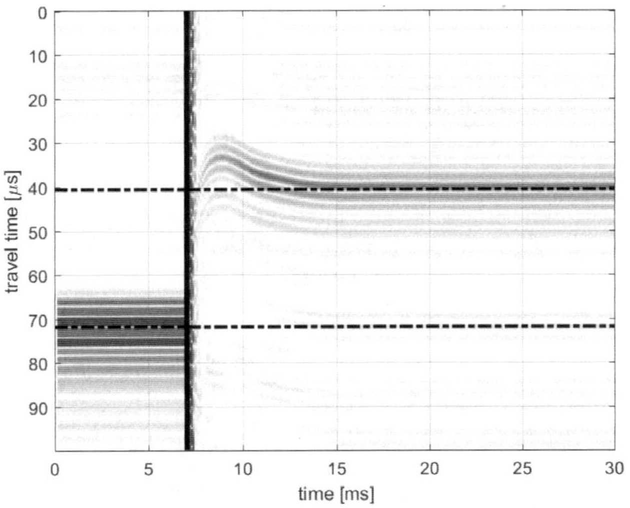

Figure 4. Ultrasonic result of indent formation in clay due to projectile impact. Dashed lines indicate the initial (70 µs) and final position (40 µs) of the dent in the clay

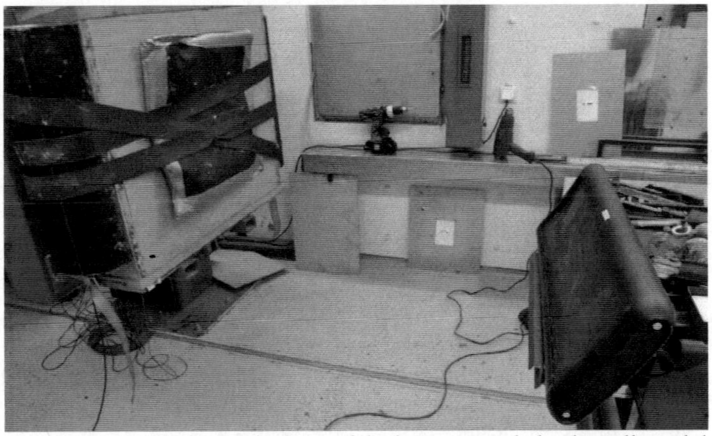

Figure 5. Infra-red heater (lower-right) in front of the instrumented clay box allowed the temperature to be constant and provided access for personnel.

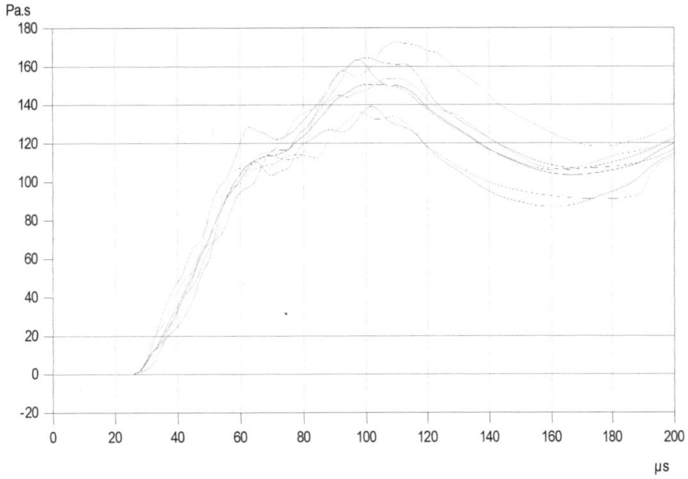

Figure 6. Impulse from pressure signal in clay (P2) of several identical 7.62 Ball shots (at 830 m/s) on a Dyneema® plate in conjunction with a surrogate vest.

https://doi.org/10.52202/078352-0048

3.3 Pressure and acceleration measurements

Figure 2 provides the position of the pressure and acceleration sensors within the clay. The pressure signals were rather noisy, yet after their integration over time, a reproduceable measurement of the impulse is obtained. Figure 6 provides the impulse from one of the embedded pressure sensor signals in clay for several repeated 7.62 Ball shots (at 830 m/s) on a Dyneema® plate in conjunction with a surrogate vest. After about 27 microseconds the pressure wave reaches the sensor and the impulse starts to rise. This behavior is quite repeatable. At about 60 microseconds the rise in impulse is interrupted, yet starts again at about 80 microseconds to reach its maximum of about 150 Pa s around 100 microseconds.

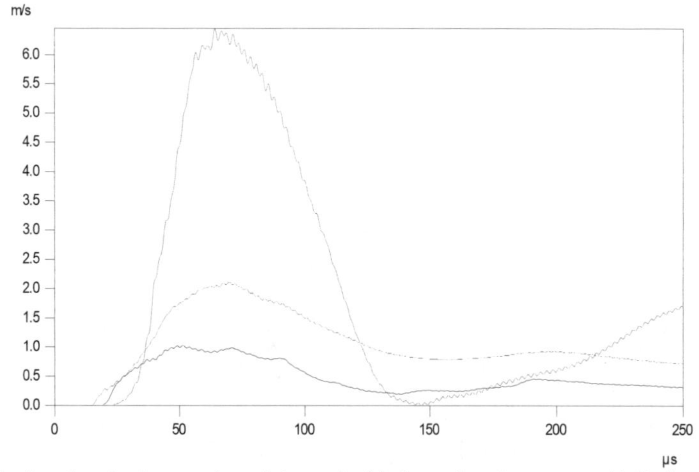

Figure 7. Velocity plots by integration of the embedded acceleration sensor (A1) signal impact of 9 mm FMJ at 400 m/s (black), .44 Magnum at 400 m/s (red) and 7.62 Ball at 830 m/s (blue).

Also the acceleration signals were very noisy, but again after their integration a reasonable smooth velocity history of the embedded sensor in the clay is obtained. Figure 7 compares velocity plots for the three projectiles used in this work: 9 mm FMJ (black), .44 magnum (red) and 7.62 Ball (blue). From figure 7 we see that for each projectile-target combination the maximal (local) clay velocity is reached at about 25 microseconds after the start of the signal. However, the maximal velocity reached is about 1 m/s for the 9 mm FMJ, 2 m/s for the .44 Magnum and 6.5 m/s for the 7.62 Ball. The decline of the (local) clay velocity is slower compared to its initial rise.

4. SYNTHETIC GELATIN BACKING

The instrumented ballistic clay box demonstrated the use of ultrasonic sensors for dynamic deformation measurements in clay. However, the inhomogeneity of clay, especially after multiple tests, causes a deterioration of the signal quality. To overcome this issue a search for a different backing material was initiated.

The second version of the instrumented backing uses Clear Ballistics (a commercial transparent synthetic elastomer). The choice for this material is based on;

- the homogeneity of the material for signal transmission
- the reduced need for repairs between shots due to the hyper elasticity
- the ease of repairability when needed (it can be casted)
- transparency to enable high speed imaging as a means to verify measurements

As this elastomeric material is hyper-elastic it will (self) recover directly after each impact, which prevents the shot-to-shot activity of indent repair and its chance on introduction of air gaps. This also saves a lot of time between shots allowing much more shots per hour.

https://doi.org/10.52202/078352-0048

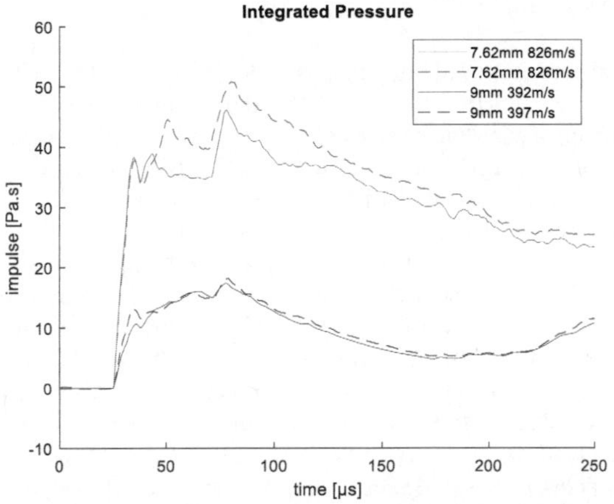

Figure 8. Impulse from pressure signal in clear ballistics (P1) of two 7.62 Ball shots on a Dyneema® plate in conjunction with a surrogate vest and two 9mm shots on a vest.

The homogeneity of the elastomer ensures a much higher signal quality from all of the sensors and improves the repeatability of measurements (see figure 8). If one compares impulse measured for the 7.62 Ball shots on Clear Ballistics (figure 8) and Roma #1 clay (figure 6) there are quite some differences in signal shape and height. This is caused by the difference in both material (properties like density and stiffness) as well as the sensor position.

The transparent clear ballistics also allows a highspeed camera to record the deformation. Holes at the side of the box allow a side view on the impact area. The high speed video was calibrated using static indentations of known dimensions.

Though deformation of the transparent elastomer may result in image distortion due to refraction, the maximum deformation can be reliably extracted. This allows the high speed video to provide verification data of the ultrasonic measurements, see figure 9.

This elastomer based instrumented backing shows the potential benefits of a homogenous, elastic backing material. It provides improved results and reduces practical issues such as the need for repairs. Further investigation is required on the long-term ageing effects and temperature dependency on the performance of this alternative backing material as well as its differences compared to Roma #1 clay.

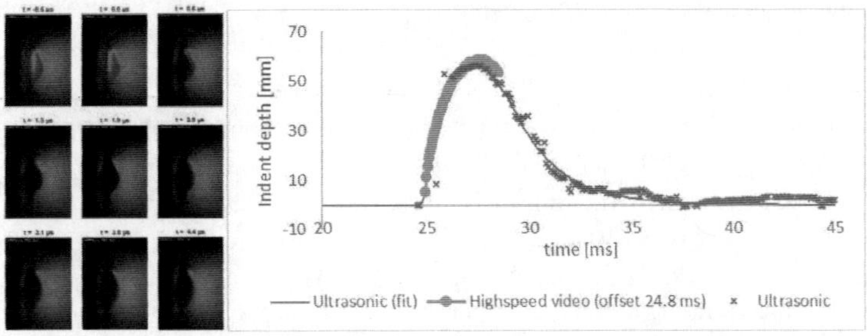

Figure 9. Clay deformation measurements using ultrasonic (blue) and highspeed video (orange) simultaneously.

5. INJURY LEVEL

The Blunt Criterion (1) correlates impact conditions (projectile mass M, velocity V and diameter D) and the those of the human body (mass W, body wall thickness T) to predict an injury level.

The kinetic energy of a 9 mm bullet at 353 m/s equals 500 Joules ($E = \frac{1}{2}\ 0.00804\ 353^2 = 500$ J). The mass of the clay box (body) used is 80 kg. The body (chest) wall thickness is estimated as $0.71*W^{1/3} = 2.3$ cm. The effective diameter D is estimated to be 2 cm (includes diameter of the bullet and thickness of the body armour. Then, using equation 1 the blunt criterion is calculated as $BC = \ln[500/(4.3*2.3*2)] = 3.2$, which can be estimated as $AIS = 0.55 * BC = 1.75$.

As the time-resolved ultrasonic measurement of the indent process allows the displacement (dy) and velocity of the clay to be determined (dy/dt), also the Viscous Criterion (equation 2) can be used for injury assessment.

https://doi.org/10.52202/078352-0048

6 CONCLUSIONS

The clay backing system has its disadvantages, yet its application has helped to significantly reduce BABT related lethality. It remains the most widely used method to assess BABT. That does not prevent the continued development of new measurement systems. More data (besides just the residual indent depth) could provide a more refined appraisal of the PPE performance. Therefore, TNO is continuing research in this field. The presented clay box is a further development of the current setup, extended with pressure, acceleration and indent velocity measurements.

The safe positioning of ultrasonic transducers, as well as pressure and acceleration sensors has been determined both using FEM and experiments. The embedded sensors allow local pressure and acceleration histories to be recorded, while the ultrasonic set-up recorded the formation of the indent in the clay. The latter not only provides the terminal indent depth, but also the maximal dynamic indent depth, as well as the maximal clay velocity upon projectile impact on body armour. The measured parameters can be used to calculate injury levels using the viscous and blunt criteria.

The instrumented clay-box therefore provides a useful research tool for body armour and injury related research. It was shown to provide reproducible results using body armours impacted by 9 mm FMJ, .44 Magnum and 7.62 Ball projectiles.

A second system is assessed that evaluates a synthetic hyper-elastic material as a potential replacement for the clay backing. An elastomeric backing provides a more homogeneous medium, resulting in improved signals from the sensors, as well as a potential backing material that has no plastic deformation and hence does not need repair between tests.

The instrumented clay box with both backing materials are developed to improve current test methodologies. The instrumented system shows that more data can be obtained from a clay backing. It also provides a tool to assess the quality of the state of the clay. Some of the practical drawbacks of clay remain, and alternative backing materials should be explored to improve the quality of testing of body armours.

References

[1] E. Hanlon and P. Gillich, "Origin of the 44-mm Behind-Armor Blunt Trauma Standard," *Military Medicine,* vol. 177, no. 3, pp. 333-339, 2012.

[2] L. Sturdivan, D. Viano and H. Champion, "Analysis of injury criteria to assess chest and abdominal injury risks in blunt ballistic impacts," *J. Trauma,* vol. 56, pp. 651-663, 2004.

[3] D. Cronin, "Behind amour blunt truama evaluation using the blunt criterion," in *23rd International Symposium on Ballistics*, Tarragona, Spain, April 2007.

[4] K. A. Rafaels, K. L. Loftis and C. A. Bir, "Can Clay Tell Us More Than Deformation?," in *Personal Armor Systems Symposium*, Washington D.C., 2018.

[5] D. Viano and I. Lau, "A viscous tolerance criterion for soft tissue Injury assessment," *J. Biomechanics,* vol. 21, no. 5, pp. 387-399, 1988.

[6] National Research Council, Testing of Body Armor Materials: Phase III, Washington, DC: The National Academies Press, 2012.

[7] A. Wickwire, A. Merkle, C. Carneal and J. Paulson, "Thoracic response to high-rate blunt impacts using an advanced testing platform," *Biomed Sci Instrum.,* vol. 48, pp. 485-492, 2012.

[8] C. Bir, "The evaluation of blunt ballistic impacts of the thorax," Wayne State University, Detroit, Michigan, 2000.

[9] National Institute of Justice, Ballistic Resistance of Body Armor NIJ Standard-0101.06, Washington, DC: National Institute of Justice, 2005.

[10] VPAM, BSW 20016 Stand: 16.10.2008 - PRÜFRICHTLINIE "Ballistische Schutzwesten", Münster: Deutsche Hochschule der Polizei, 2008.

https://doi.org/10.52202/078352-0048

The development of knife test blades for use in body armour stab resistant evaluation

Paul Fenne[1], Dirk Landheer[2] and Cameron Hans-Brooker[2]

[1]Physical Protection Group, Metropolitan Police Service, 60 Albany Street, London NW1 4EE, UK, paul.fenne@met.police.uk
[2]SIMPACT, Unit 2, Trojan Business Centre, Tachbrook Park Drive, Warwick CV34 6RS, UK

Abstract. Over the last 30 years there has been a development of the design features of test blades for use in body armour stab test rigs. In the early 90s the test methodology was relatively simple. Since then the design of test rigs and test blades has developed and continues to do so. This paper is a review of that development and the latest efforts in the Metropolitan Police Physical Protection Group to design a new knife to more accurately mimic the performance of real blades typical of those used in crime in the UK. This test blade was developed to investigate the stab resistance of not only body armour related materials but other fabrics used in operational uniform/clothing. The study summarises the work of identifying dimensional details of a large number of crime related knife samples, including measurement of cutting performance of both the tip and the edge, blade metallurgy and the estimation of impact loads derived from forensic analysis of stab incidents. The study is supported by computer modelling to gain further insight into the phenomena occurring in fabrics being stabbed. A new design of test knife has been developed and tested against a range of fabrics, polymers and metallic materials including those used in the construction of body armour. The paper gives details of the manufacturing process of the test blade and the method to sharpen the blade ready for use. One of the criticisms of some of the previous blades is that of cost. Details of how the design of the knife can achieve a cost reduction in the test process is included. This can justify a larger number of test strikes to assess the statistical variability of such a test.

1. INTRODUCTION

The design of body armour for police use is summarised in two main principles, firstly protection from the threat including knife, firearms and blunt impact attacks, and secondly the correct balance between protection and wearability must be achieved. In order to achieve such a balance, the test methodology must be as representative as possible. The design of the test knife is part of that consideration.

1.1 Review of Operational Incidents with Regard to Attack Method and Impact Dynamics.

A number of stabbing incidents involving body armour have demonstrated that the impact dynamics have been less severe than expected. In some cases, the areas of the body armour cover extending beyond the armour panel have been able to stop the knife. This has led us to research the feasibility of applying the best level of knife resistance to these areas, consistent with good wearability and heat dissipation.

1.2 Forensic Analysis of Operational Incidents

A number of forensic analyses have been undertaken in the last few years and it has been necessary to develop tests to assess the point sharpness and also the edge sharpness of knives. This work has led to a number of conclusions, namely the particular influence of both the sharpness of the point and the edge.

The point sharpness test is a quasi-static push of a specimen knife at 90° to a test material at a very low velocity while recording the resistive load. The skin/tissue simulant as used by Department of Engineering, University of Leicester is suitable [1].

Edge sharpness is the ability of the blade to cut whatever material it is designed to cut. So, a sharp razor will be different to a sharp felling axe. Sharpness of a knife can be estimated by measuring the cutting ability of paper. In PPG, a test rig has been developed to measure the load required on the blade under test to cut one layer of a defined paper sample. A simpler test is to use a knife to attempt to cut a sheet of A4 copier paper held by hand vertically. A commonly considered "sharp" edge knife will be able to cut the paper.

 https://doi.org/10.52202/078352-0049

1.3 Analysis of knives associated with crime

During the past 25 years the Metropolitan Police, Physical Protection Group (PPG) has carried out knife analyses of blades associated with crime. These knives include those used in attacks, seized during searches and handed in during amnesties. In all, PPG has undertaken 9 such analyses, the latest being in 2018 [2]. The categories of knives were; domestic knives, fixed blade (not domestic), folding knives, craft knives, butterfly and flick knives, machetes, cleavers/hatchets/axes, and miscellaneous. Of these categories, domestic knives have always been the highest proportion (except for 2008). This proportion has increased from 35% in 1995 to 73% in 2018. See Table 1.

Table 1. Knife analysis results (Percentages of the total number of incidents)

Type	Year								
	1995	1996	1997	2002	2004	2008	2013	2015	2018
Domestic knives	35	43.3	44.2	40.1	42	33.7	66.7	65	73
Fixed blade non-domestic	14	10.5	11	5.2	7.2	4.2	4.1	4.5	7
Folding knives	17	15.9	15.7	32.2	36.6	43.5	15.1	18.8	6
Craft knives	0	0	3.7	3.4	4.4	5.1	2.0	3.1	4
Butterfly & flick knives	15	4	12.5	6.7	2.3	1	0	0	1
Machetes	0	0	0	3.6	0	0	1.6	1.3	2
Cleavers, hatchets, axes	0	0	0	2.5	0.3	0.3	2.5	2.3	1
Miscellaneous	19	26.3	12.9	6.3	7.2	12.2	8.0	5	6

One significant observation from the inspection of domestic knives is that the condition of the point and edge was widely variable and the vast majority of them being dull or damaged to an extent where they were poor in cutting performance. For examples see Figure 1.

Figure 1. Examples of knife points.

1.4 Results of Geometry, Tip and Edge Sharpness Inspection

A second observation from the inspection was the range of typical knife blade shapes. The tip angle of the blade, tangential to any curved surface, was measured and the population distribution calculated. The average angle was found to be 45°. The included flank angle of the blade was measured, the typical angle was found to be 20°. The sharpened edge angle was found to be 30°. See Figure 3.

1.5 Theory of Penetration of Stab Resistant Body Armour

Body armour stab resistant materials are generally thin homogeneous layers (plastics or metals), chain-mail, or woven aramids (plain or with resination).

The theory of penetration of homogeneous materials comes from the study of metals. There is general agreement that the work done in producing a knife type perforation in metal is divided into 3 specific characteristics;

- firstly, the fracture of the material in front of the sharp edge to overcome the molecular binding forces

https://doi.org/10.52202/078352-0049

- secondly, deformation work near the fracture crack and further away from the area which will include elastic and plastic deformation, and
- thirdly, frictional force resisting the perforating knife; a result of material properties and surface finish [3]

For the typical body armour materials, the proportional contribution to the energy dissipation by each of these cutting characteristics will vary. For example, armour designs with metal platelets may typically offer 35% fracture, 25% deformation and 40% friction resistance. Multi-layer aramid panels may offer less frictional resistance. Polymers like polycarbonate in thin platelets will offer a particularly high level of fracture and friction resistance. Chainmail with aramid layers will present a complicated balance of fracture, deformation and friction resistance [3].

1.6 Mechanical Properties of Knives

Knife blades are made from tool steels, carbon steels, and stainless steels. Hardness is required to maintain a sharp edge, but brittleness will lead to blade fracture. However, the blade edge has to be sharpenable; excessive hardness makes this difficult. Toughness, to allow strength without fracture, is very important. Finally, corrosion resistance is essential, particularly for domestic knives, the largest category of blades. A very popular material for knife manufacture is the 400 series stainless steel. Typical mechanical properties for 440C stainless steel heat treated knife blades are: Tensile Strength 2030 MPa; hardness 59 HRC; and impact Charpy 9. It has a high chromium content and therefore a high corrosion resistance, and is able to be easily sharpened.

1.7 The History of Assessment Methods and Test Blades

Table 2. History of test blades since 1990

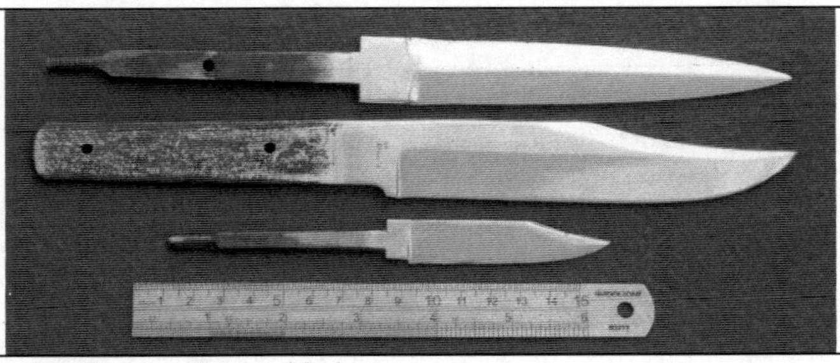	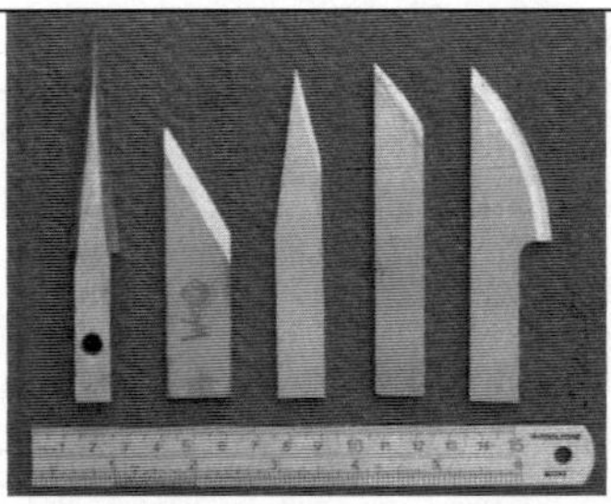	
Swiss-German dagger blade UK PSDB No 5 blade UK PSDB No 1 blade	MPS Triangular blade CEN blade UK HOSDB P1B blade MPS new blades	Shown left to right

1.7.1 Swiss German (1992) [4]

The test rig in the Swiss/German stab test was a vertical drop tower. The double edged test blade was held in a circular sabot, of mass 2.6 kg, to give impact energy of 35 J and a maximum limit of 20 mm perforation was allowed.

1.7.2 Home Office UK (1993) [5]

The original blade research undertaken by UK Police Scientific Branch (PSDB) for the Home Office (HO) stab test selected 2 commercial sheath knife blades (No 1 and No 5) which were fitted into sabots and fired horizontally at 42 J into a vertically mounted target assembly, with the body armour sample backed with a Plastilina block.

 https://doi.org/10.52202/078352-0049

1.7.3 Metropolitan Police (1995)

In the Metropolitan Police (MPS), in 1995, after a series of police officer fatalities from stab attacks, a feasibility study to provide body armour was launched. Body armour samples made to meet the HO 42 J test were not sufficiently wearable for general use, so a development of a new test method was undertaken. Some armour samples were made of platelets (which would pass the 42 J test at 90° impacts but fail at impacts at acute angles). To test at various angles, a swinging arm test rig was introduced. The test blade used was a tapered triangular section engineered blade. A pass/fail criterion of 20 mm maximum perforation at 25 J energy was required.

1.7.4 CEN draft standard (2000)

The CEN Body Armour committee agreed on an engineered blade that would provide a more consistent performance than the test blades which had been in use in the recent years. One difficulty with engineered blades was the problem of weak tips as a result of the need for a grinding process limited to flat surfaces in order to reduce manufacturing costs. The interesting design feature to overcome this was an additional chisel point on the extreme tip to strengthen the point. The design concept was to produce a blade that was one use only and of only 1-2 Euros in price. The CEN methodology involved a vertical drop tower design. The CEN Body Armour Standard was never ratified.

1.7.5 PSDB 2003 Body Armour Standard [6]

The 2003 HO Standard review adopted a vertical drop tower and sabot with an internal damper and a backing material assembly of elastomeric plastic foam layers. The HO Body Armour Standard review in 2003 adopted a new design of engineered blade that was designed to replicate the stabbing performance of the No 1 blade. The product was named the P1B blade.

1.7.6 VPAM Body Armour Standard [7]

In 2004 the VPAM, Test Standard "Stab and Impact Resistance" was introduced and used across mainland Europe. The VPAM used a drop tower, the P1B blade and a backing of Plastilin der Fa. Carl Weible.

1.7.7 Metropolitan Police (2005)

The Metropolitan Police also reviewed their test regime to a drop tower design. However, the damped sabot was removed to improve consistency in the results. The P1B blade was used in these tests.

1.7.8 Metropolitan Police Quality Assurance (QA) test (2008)

In 2008 the Metropolitan Police designed a specific stab test rig to further improve the consistency of results because the previous test methods were not sensitive enough to be used as a QA tool for production. The result was a drop tower design with the blade separated from the sabot, the backing arrangement consists of an anvil to allow the armour to deform, an aperture through which the test blade could perforate and a camera to observe the perforating test blade tip from below. The rig yielded more consistent results and has been used for this purpose since. The P1B blade was used in this test setup.

1.7.9 Home Office, Body Armour Standards HOSDB (2007) [8] and CAST (2017) [9]

The current UK HO Body Armour Standard was issued in 2017. Much of the methodology is similar in principle to the previous issue. The P1B blade is used.

1.8 Lessons from history

From the PPG analysis of knife blades, armour perforation tests, the history of previous test methods and forensic research, we have come to the conclusion that a new design of test knife is necessary. See ref [10].

https://doi.org/10.52202/078352-0049

2. DESIGN FEATURES OF THE NEW BLADE

It is important that the test blade is representative of real blades that may be used in stab related attacks, The geometry of a large number of typical knives was studied and the most common design had a tip angle of 45° at the point and included flank angle of 20° and sharpened edge angle of 30° (see Figure 3). This test blade was developed to investigate the stab resistance of not only body armour panels but other fabrics used in body armour covers and operational uniform/clothing, including:

- A multi-layer resinated anti-stab material panel
- A chainmail and aramid based armour panel
- Thin metal sheet
- Fabrics typically used in body armour covers

2.1 The Manufacturing Process of the Test Blade and Method to Sharpen the Blade Ready for Use.

The material selected for the new blade is AISI type – 01 ground flat stock and is available in lengths of 15 x 2 mm rectangular section. It can easily machined and ground to shape. When hardened and tempered, it can be brought to 62 HRC which is harder than normal knives so the test knife edge will be maintained in use and it can be re-sharpened for future use. It is not as corrosion resistant as ordinary blades, but this is not a disadvantage for a test blade as it is used only in a laboratory.

The shape of the test blade is based on three main dimensions: a tip angle of 45°; a curved sharp edge; a maximum width of 25 mm; and a flank angle of 20° with a sharpened edge angle of 30°.

There are two options for the Metropolitan Police Service (MPS) test blade, one for use in the HO test rig with a 15 x 2 mm shank to fit the sabot, and a 25 mm wide blade (MPS45/80/25). The second option is a parallel sided shank with the same geometry at the impact end but with a reduced width of 15 mm; designed to be used in the MPS QA test rig. This option is necessary because the MPS QA test rig is limited to blades only 15 mm wide (MPS45/80/15). Each has a curved edge radius of 80 mm. They perform identically up to a perforation depth of about 20 mm.

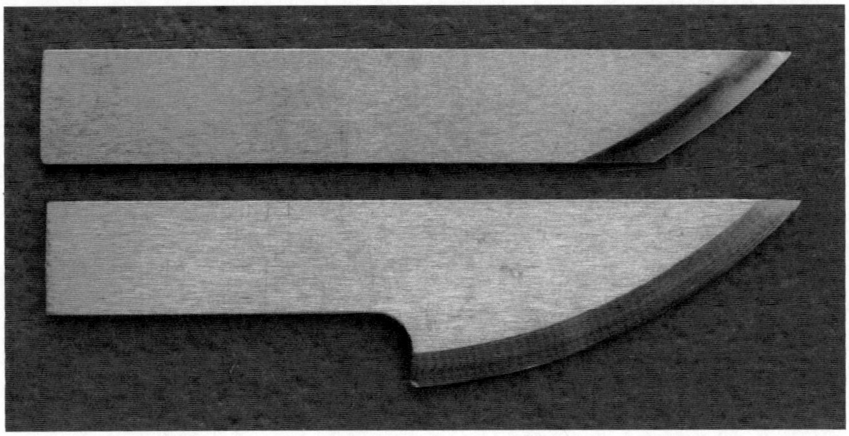

Figure 2. The MPS45/80/15 and MPS45/80/25 test blades (approximately actual size)

One of the criticisms of some of the previous blades is that of cost. The price from suppliers has generally been higher than anticipated, as the practice of single use has made the process of knife testing expensive. The new knife is designed to be sharpened by hand and be able to be used repeatedly. The re-sharpening process involves use of a ceramic whetstone. The sharp edge has an included angle of 30°. It is important to note that any burrs should be removed by stropping and then a simple inspection under magnification to check any burr has been removed. To assess the sharpness of a blade in the test laboratory, a simple test can be performed: the blade is ready for use when it is visually burr free and able to cut one layer of 80 g/m^2 paper held vertically.

 https://doi.org/10.52202/078352-0049

2.2 To Verify the Blade and Compare with the Standard P1B

A comparison trial was undertaken to estimate the perforation with energy characteristics. Using the MPS QA test rig, each blade was used at a range of energy levels up to the point of complete perforation. Owing to the difference in shape between the P1B and the MET blade, for a given perforation depth, the cut width is approximately double with the MET blade. Three different types of armour panel construction were used:

- Resinated aramid: The perforation of the MET blade was similar to the P1B over the whole range of energies up to approximately KR1 level.
- Chainmail and soft aramid quilted pack: The perforation of the MET blade was less than the P1B up to about the energy level of the minimum level of stab protection (KR1) required by the UK HOSDB test level, then the MET blade perforated deeper up to complete perforation with higher energy levels.
- Mild steel 0.5 mm: The perforation of the P1B blade was about double up to the point of complete perforation. (This meant that the crack width formed by each blade were similar)

3. KNIFE GEOMETRY COMPARISON ANALYSIS USING FINITE ELEMENT ANALYSIS

With an aim to identify the effect of blade geometry on cutting ability, finite element (FE) models were created of varying test knife designs. These include a standard test knife (knife 1), a sharpened test knife (knife 3), and a sharpened knife (knife 2) with blunt second edge. Finite element modelling is particularly suited to this analysis style as it is fully repeatable, in addition to providing in-depth results, including local stress/strain levels, without the material itself being deformed during sample extraction from the rig for inspection. Furthermore, material movement can be studied in depth using software, thus not requiring modification of the physical test scenario to enable high speed photographic video recording.

Finite element knife geometry and protective fabric:

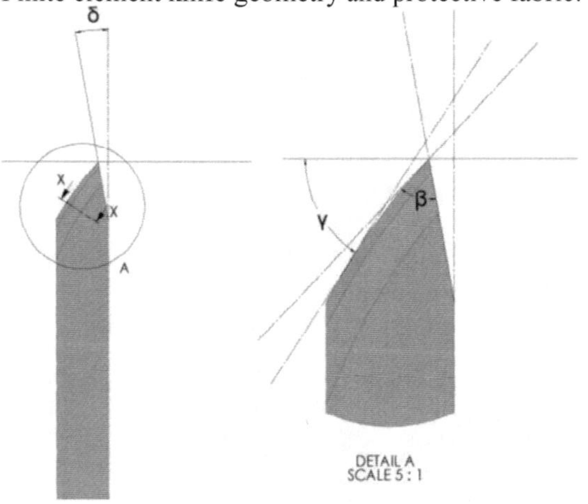

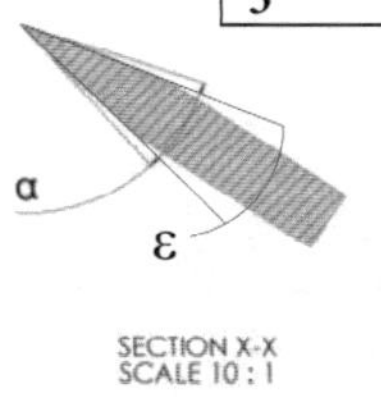

Knife	**Cutting face Angle [°] (α)**	**Blunt Face Angle [°] (δ)**
1	2 x θ	0
2	θ	10.56
3	θ	0

Angle	**Definition**
α	Sharpened Edge Angle
β	Tip Angle
γ	Cutting Edge Angle
δ	Blunt Face Angle
ε	Flank Angle

Figure 3. Geometry of blades and angle identification

The material consists of 2 layers of plain weave aramid, with a tow width of 0.83 mm, a tow thickness of 0.155 mm, and a tow gap of 0.10 mm.

3.1 Modelling Strategy

The model was created to simulate the physical test scenario, carried out with the MPS QA test rig. In order to allow timely completion of the simulation, the model was simplified in terms of: the knife was modelled as rigid (undeformable); material proximal to the tip impact area was modelled as having 5 fibre bundles per tow; material distal to the tip impact area was modelled as 2D (shell) elements; there is no

https://doi.org/10.52202/078352-0049

cross-fibre connectivity at the area proximal to the tip impact; the knife impacting the fabric was loaded with a mass of 0.05 kg, travelling at 3.706 m/s.

The material was first simulated draping over the target disc. The output showing the draped material was then utilised as the input geometry for FE knife impact testing, removing any artificial gap between the material and the disc. The material was trimmed post-drape so that the sample did not overhang the outer perimeter of the target disc. The simulations were run until the impacting knife reached a vertical velocity of 0 m/s.

3.2 Results Overview

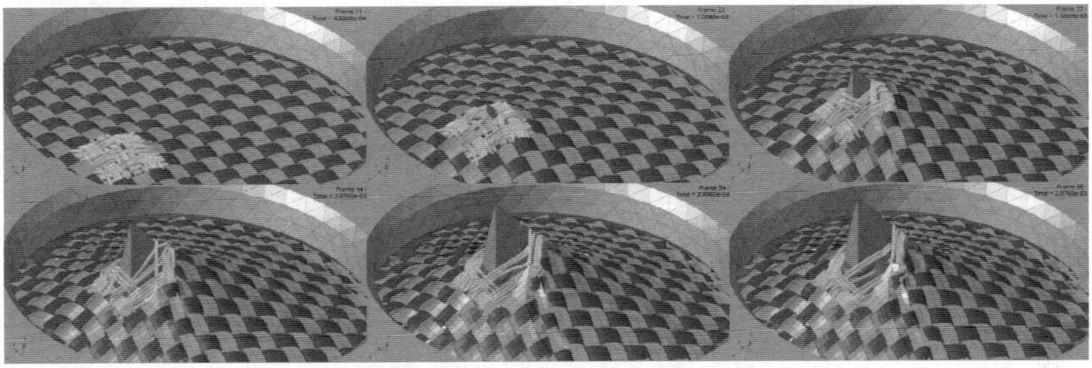

Figure 4. Images of simulated material deformation during knife tip impact.

When observing the material behaviour, three main stages can be observed: 1 – initial puncturing of the material and gathering around blade; 2 – cutting and slippage of fibres causing material relaxation around puncture location; 3 – tightening of material around blade. The differences in these three areas will determine the penetration of the blade into the material. During stage 1 of the impact the material tow is penetrated by the blade, causing fibres to pass over both sides of the blade, in addition to separating to span both the sharpened and blunt faces. This can be seen in Figure 5.

Figure 5. Comparison of physical and finite element model deformation.

Comparing the finite element results to comparable drop tests conducted – the material is shown to behave in a similar pattern, with 4 tows being severely deformed. Fibres travelling parallel to the knife body (horizontally in Figure 5) were deformed less than those travelling perpendicular to the knife face (vertically in Figure 5).

https://doi.org/10.52202/078352-0049

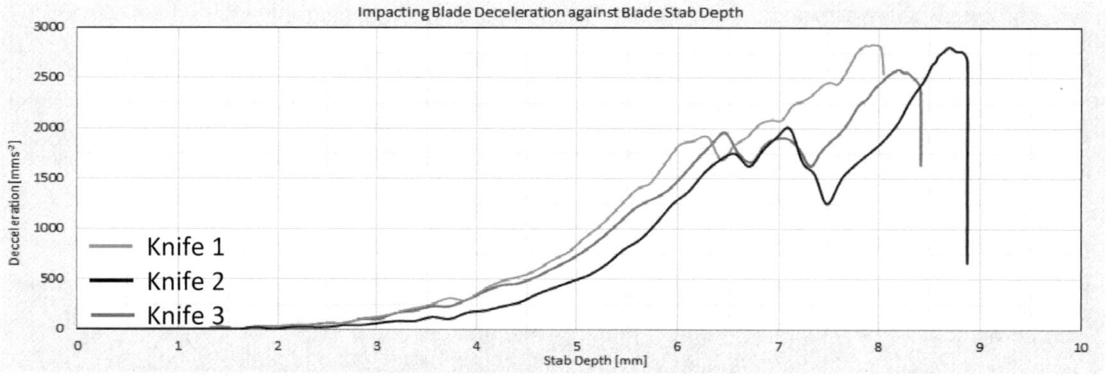

Figure 6. Simulated blade deceleration against blade stab depth.

3.3 Analysis of Results

Figure 7. Comparison of deformation for knife 1 (grey) and knife 3 (purple), created by overlaying results from two tests.

It can be seen from Figure 6 that blade 2 penetrated the material by the furthest distance (8.88 mm). Blade 1 penetrated the material by the shortest distance (8.05 mm), travelling a shorter distance than knife 2 and 3 (8.42mm). The behaviour of the blade can be seen to follow the three stage process described previously, with the first section, puncturing and gathering of material occurring before 6mm. The second stage, slippage and cutting of fibres, occurs between 6mm and 7-7.5 mm. The final stage, material tightening, occurs after 7-7.5 mm.

Knife 1, due to its blunt cutting face, does not penetrate the material as far as knife 3 (Figure 6). It can be seen that the vertical fibres in Figure 7 are required to undergo further deformation in order to accommodate the thicker cross section of knife 1. This tightens the fibres travelling horizontally against the cutting and blunt edges of the knife, thus requiring more force for the knife to pass into the material.

https://doi.org/10.52202/078352-0049

3.4 Observations and Conclusion

Knife 2 penetrates the material further than knife 3. This was found to be caused by the gradient of the cutting edge at the blade's proximal tip; due to the top section of the knife being removed to accommodate the angled blunt face, the gradient of the cutting edge to vertical is reduced from 45° to 35°. The total tip, including blunt face angle, remained at an approximately constant angle of 45°.

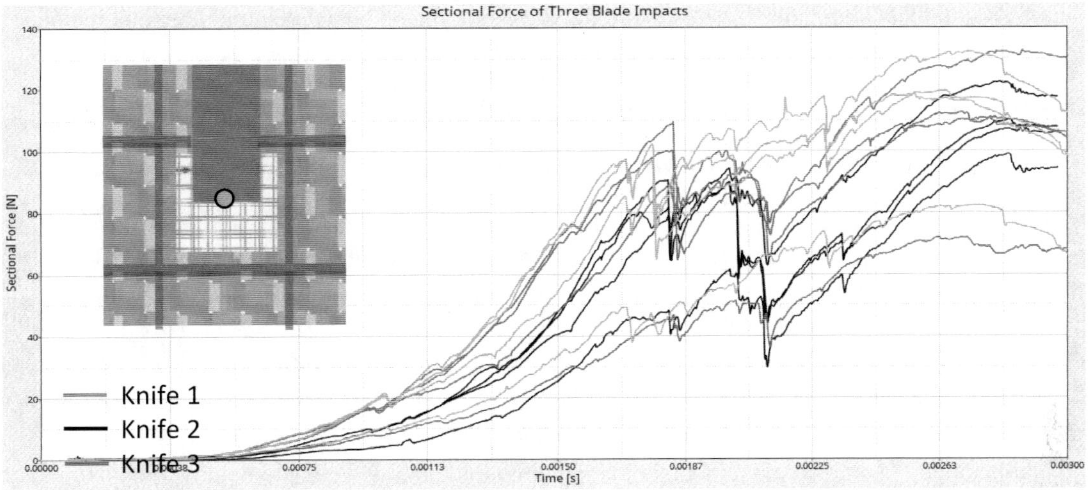

Figure 8. Sectional force (total axial forces measured in a single cross-sectional cut across all fibres) for blade tests; pictorial representation of sections (red) in relation to knife impact area (orange)

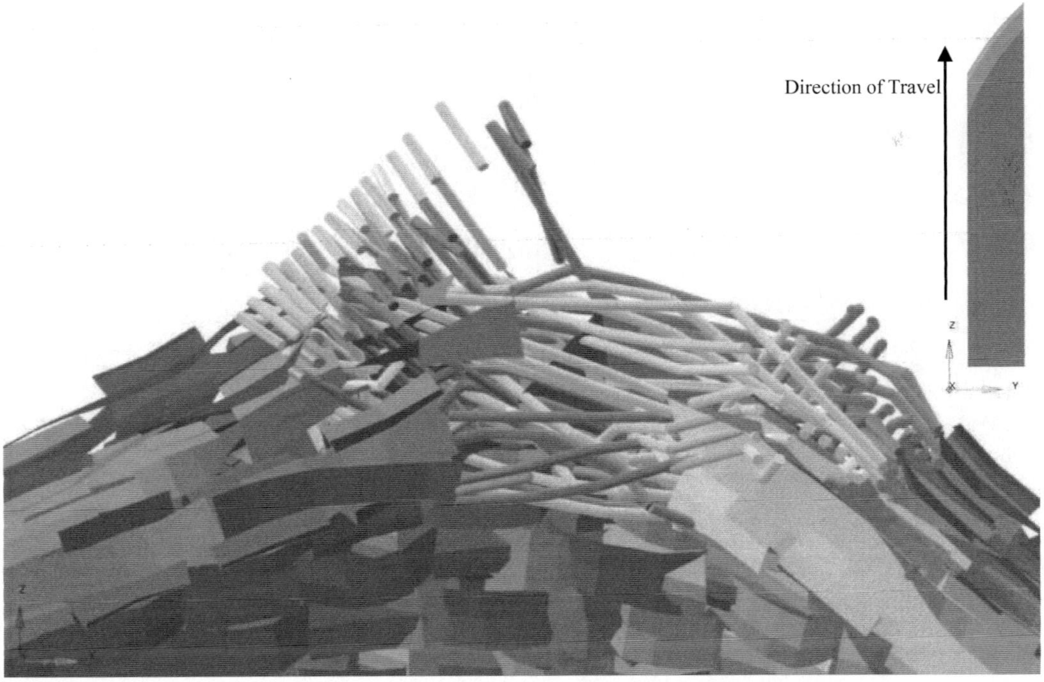

Figure 9. Comparison of material cross-section deformation for knife 2 (purple) and knife 3 (pink), knives omitted for clarity

https://doi.org/10.52202/078352-0049

The difference in cutting edge angle is clearly displayed in Figure 9. The cutting edge is steeper for fibres impacted by blade 2, causing them to slip further down the cutting edge, thus undergoing lower levels of tension, as they are draped over the cutting edge more loosely. This is reflected in the sectional forces within the material (Figure 8). It is clear to see that blade 2 has consistently lower sectional forces than blade 3, with the exception of the lowest trace, produced by fibres positioned horizontally across the vertical blunt face. The sum of the sectional forces is lowest for blade 2.

The total number of 1D elements ruptured (indicating failure of several fibres) is 1 for both knife 1 and 3, compared with 2 for knife 2. This is likely due to the increased tension in the fibres on the blunt edge of the blade, causing more force acting perpendicular to the tow.

3.5 Analysis Conclusion and Connection to Knife Impact Testing

The FEA modelling conducted in this section highlights the difference in stab depth when blade geometry is altered. Decreasing the cutting face angle allows the blade to penetrate further into the woven material, when loaded with a set energy, this suggests that the smaller included angle of the MET knife allows better cutting of the material. Increasing the cutting edge angle, as seen in blade 2, increases the penetration observed in the material, this is reflective of the P1B knife performance. The increased cutting depth gained by including an angled blunt face outweigh the disadvantage of having a larger cross sectional area for the tip region.

It has been observed in the finite element models that the gathering of material, and the distance of fibre spread around the puncture location, demonstrate the complexity of material failure; thus suggesting further work is required on the subject of material behaviour during blade impacts.

4. CONCLUSIONS

Over a long period of time, the Physical Protection Group has believed that an improved blade is necessary. A large amount of body armour testing and research into armour scheme designs and a number of forensic analyses of stabbing incidents have been the motivation behind this project. The blade presented here will be used by the PPG in future. The MET blade has a greater cutting performance than the P1B forming a wider cut for a given perforation depth. The P1B blade has a greater piercing action than the MET blade. The MET blade is a more representative blade and so will be adopted in MPS body armour testing alongside the P1B.

References

[1] Gary Nolan, Forces Required for a Knife to Penetrate a Variety of Clothing Types, J Forensic Sci, March 2013, Vol. 58, No. 2

[2] G A Mapp, MPS – 2018 Knife Blade Analysis, PPG-RPT-18/029, 18/07/18

[3] Horsfall I, Stab Resistant Body Armour PhD, 2000, Page 20.

[4 Technical Guidelines for lightweight personal armour, Schweizerische Polizeitechnische Kommission, (1992).

[5] G. Parker, PSDB Stab resistant body armour test procedure (1993), Police Scientific Development Branch, Publication No 10/93, (1993).

[6] Croft J, PSDB Body Armour Standards for UK Police, Part 3, Publication no 7/03/C, 2003.

[7] VPAM, Test Standard "Stab and Impact Resistance", 2004; Edition 18.05.2011

[8] Croft J, Body Armour Standard (2007), HOSDB Publication no 39/07/C, 2007

[9] T Payne, Body Armour Standard (2017), CAST Publication no 012/17, July 2017

[10] Hainsworth S, Cutting Crimes, Ingenia Issue 37, December 2008.

https://doi.org/10.52202/078352-0049

Review of fragment simulating projectiles definition and associated quality controls needed

F.Barbillon[1], P.Martin[1], P.Mabire[1] and F.Mouhot[2]

[1]DGA Land Systems, Rocade Est, Echangeur de guerry, 18021 Bourges Cedex, France
franck.barbillon@intradef.gouv.fr
[2]DGA Aeronautical Systems, 47 Rue Saint-Jean, 31130 Balma, France

Abstract. STANAG 2920 is a NATO standardization agreement which deals with classification of level of ballistic protection of personal protective equipment. Its associated Allied Engineering Publication (AEP 2920) describes the procedure for classification of personal armour for protection against bullets and fragment threat. The ballistic tests shall be carried out using selected threat ammunition or fragment simulators defined in AEP 2920. Inconsistencies were found in the definition of fragment which can implies misunderstanding for fragment manufacturers, suppliers and test houses. Using inconsistent fragment for assessing level of protection can result in wrong decision in acceptance or qualification process. Analyses of different definition available were conducted and new drawings were built (FSP without skirt, FSP with skirt and RCC). Sizes, tolerances and masses were adjusted to be in compliance with reference standards like MIL-DTL-46593, UK drawings and so-on. Some requirement were added like burrs sizes tolerances. Moreover, STANAG 2920 suggests that the test facility shall take a representative sample from each batch and check the hardness, dimensions and weight. But, guidelines must be given regarding sampling to achieve confidence in new batches. Rockwell hardness measurement is not consistent for small fragments uses for personal protection equipment and hardness measurement methods are inaccurate and must be conducted in respect with available standards like ISO standard. Different results obtained during fragment procurement processes are highlighted and new drawings and requirement proposed for the next STANAG 2920 will be explained and shared with attendees.

1. CURRENT STANAG 2920 FRAGMENTS DEFINITION

Last NATO standardization agreement STANAG 2920 [1] is associated with Allied Engineering Publication AEP 2920 [2]. The purpose of these documents is to establish a standard classification to designate the protection level of Personal Armour (hard armour, soft armour, helmets, face and eye protection) against bullets and fragmentation threat when exchanging tactical information in NATO or to ensure interoperability in multinational missions. Participating nations or test houses agree to use the method of classification described in the standard and to classify on the basis of ballistic tests performed according to the standard. Therefore, fragment-simulating projectiles (FSP[1]) shall meet the specifications detailed in AEP 2920 [2]. Bolduc et al. described [3] which type of fragment are included (or removed) in current edition of STANAG 2920 [1][2]. Three types are defined:

- Chisel nose cylinders launched with sabot (designated by letter "F") or FSP without skirt,
- Chisel nose cylinders launched without a sabot (designated by letter "G") or FSP with skirt,
- Right Circular Cylinders (RCC) (designated by letter "R").

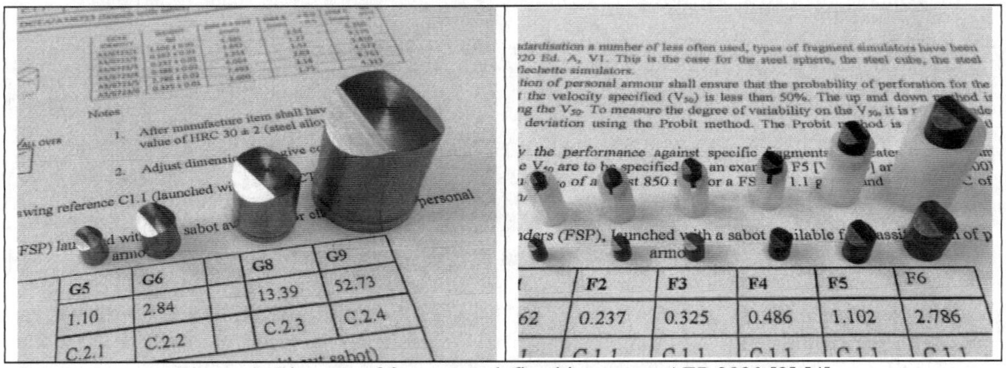

Figure 1. Pictures of fragments defined in current AEP 2920 [2] [4]
- Left: "G" fragments with skirt- Right: "F" fragments without skirt

[1] a projectile of a specific material, shape and size for ballistic test firings so that the effect of typical munition fragments can be simulated

https://doi.org/10.52202/078352-0050

Conducting testing in compliance with AEP 2920 imply (see. § 5.5.2.c in [2]):

- "*….the fragment simulator used to simulate the fragment threat shall conform to the drawings and tables provided in annex…,*
- *Fragment simulators shall be certified by vendors to meet the specifications in annex,*
- *Before use, the test facility shall take a representative sample from each batch and check the hardness (measured on the side), dimensions and weight…*"

Despite these fragment definitions in AEP 2920 [2] perceived as being relevant and rigorous, procurement agencies and test houses were registering irregular results from their testing. It is clear that fragment is only one of few factors of irregular result but variability in ballistic impact performance could be explain by projectile physical properties and dimensions [5]. Nevertheless, using inconsistent fragment for assessing level of protection can result in wrong decision in acceptance or qualification process. Moreover, test houses who are ISO 17025 [6] certified or in an ISO 17025 certification process shall prove they are using fragments which are in compliance with our NATO Standard.
Based on capitalized feedbacks, STANAG 2920 team of experts decided to improve fragment definition and resolve inconsistencies found which can implies misunderstanding for fragment manufacturers, suppliers and test houses. Moreover, next AEP must tightening up some requirements and include guidelines for quality controls and hardness measurement.

The purpose of this article is to share analysis and inconsistencies of different existing FSP definitions and share corrected or selected values for next AEP 2920 (expected in 2021). First paragraphs explain reviewing of weight and dimensions for FSP with or without skirt and associated achievability. In a second part, are highlighted requirements for material to be used for manufacturing fragment, hardness, burrs removal and surface finish. A third part deals with huge difficulties in quality controls mainly regarding hardness. To conclude, this article presents the three new drawings proposed for next AEP 2920 (FSP without skirt, FSP with skirt and RCC).

2. REVIEWING DEFINITION OF FSP WITHOUT SKIRT

The six smallest FSPs without skirt are described in STANAG 2920 edition 1 since 1996 [7] and are based on UK Drawing DCTA/A3/6723 build in the 80's. In the current AEP 2920 [2], they are called F1 to F6; the number defining the weight of the fragment simulating projectile. More recently, FSPs for saboted launch were introduced in MIL-DTL-46593B in 2006 [8,9]. This detail specification also describes a 5.56 mm and 7.62 mm caliber FSPs as in UK Drawing and defines 12.7 mm caliber and 20 mm caliber FSPs without skirt. Therefore, in the way to be comprehensive, these four FSPs coming from MIL-DTL-46593B were included in the AEP 2920 [2]. Table 1 below resume available definitions of FSPs without skirt. Unfortunately, it means existing two definitions of 5.56 mm and 7.62 mm caliber FSPs (called F5 and F6) in AEP 2920. This type of fragment could be launched with both smooth and rifled barrel. For rifled barrel, sabot is used. FSP fragment F5 weight 1.1 g is often called the 17gr FSP.

Table 1. Summary of standards describing FSP without skirt (current standard in bold)

Class [2]	**F1**	**F2**	**F3**	**F4**	**F5**	**F6**		
Caliber (mm)					**5.56**	**7.62**	**12.7**	**20**
2920 Ed 1 [7]	UK drawing DCTA/A3/6723						Not defined	
2920 Ed 2 [10]	As described in UK drawing DCTA/A3/6723[2]						Not defined	
AEP 2920 [2]	UK drawing DCTA/A3/6723[3]							
					"FSP for Saboted Launch" Table C2.1			
MIL-DTL-46593B [9]	Not defined				"FSP for Saboted Launch"			

Moreover, the UK drawing in STANAG 2920 is illegible in current AEP 2920 [2]. Regarding this drawing in AEP, method for expressing the dimension values is not in accordance with international rules [11]. In fact, the numerical value and its tolerance should not be given with an excessive number of digits. It usually suffices to quote value and tolerance to at most two significant digits. It also significantly will simplify manufacturer work and quality controls.

[2] UK drawing DCTA/A3/6723 was not included in STANAG 2920 Ed 2

[3] Same reference and issue number as drawing in 2920 edition 1 [7] but with small differences (date, material…)

2.1 Weight and dimensions correcting

For the smallest fragments F1 to F4, only UK drawing definition exists [2,7,10]. Examples of rewriting UK drawing masses and dimensions in accordance with international rules [11] are given in table 2 for fragment F1 and F2. Same correction are apply to fragment F3 and F4.

Table 2. Corrected weight and dimensions for fragment without skirt F1 and F2

Fragment class	Parameter	Mass	Diameter	Flat size	Length[4]
	Unit	g	mm	mm	mm
F1	Current	0.162± 0.01	2.642± 0.02	$1.27\ ^{0}_{-0,5}$	*3.175*
	Corrected	**0.16± 0.01**	**2.64± 0.02**	$\mathbf{1.27\ ^{0}_{-0,5}}$	***3.17***
F2	Current	0.237± 0.01	3.251± 0.02	$1.52\ ^{0}_{-0,5}$	*3.81*
	Corrected	**0.24± 0.01**	**3.25± 0.02**	$\mathbf{1.52\ ^{0}_{-0,5}}$	***3.81***

For fragment F5 and F6, weight and dimensions must be compare between corrected UK drawing [2] and last MIL-DTL-46593B [9]. Table 3 below gives an overview of weight and dimensions found in standards and proposed values. Table C2.1 in AEP 2920 does not comply with MIL-DTL-46593B [9]. The diameter of F5 fragment is not the same in the two values found in AEP 2920 and seems to come from differences between the UK drawing [2,7,10] and MIL-DTL-46593B [8,9]. For the F6 fragment, differences are found on diameter and flat size. No explanation were found to understand such differences. Corrected value complies with the oldest definition, which come from UK drawing.

Table 3. Selected weight and dimensions for fragment without skirt F5 and F6

Fragment class		Parameter	Mass	Diameter	Flat size	Length[4]
		Unit	g	mm	mm	*mm*
F5	Current	UK drawing [2]	1.102± 0.02	5.385± 0.02	$2.54\ ^{0}_{-0.5}$	*6.35*
		UK drawing corrected	1.10± 0.02	5.39± 0.02	$2.54\ ^{0}_{-0.5}$	*6.35*
		AEP 2920 table C2,1 [2]	1.1± 0.03	5.46± 0.05	2.54± 0.5	*6.35*
		MIL-DTL 46593B[5] [9]	1.10± 0.02	$5.46\ _{0}^{+0.02}$	$2.54\ ^{0}_{-0.25}$	*6.35*
	Selected		**1.10± 0.02**	**5.39± 0.02**	$\mathbf{2.54\ ^{0}_{-0.5}}$	***6.35***
F6	Current	UK drawing [2]	2.786 ± 0.02	7.493 ± 0.02	$3.18\ ^{0}_{-0.5}$	*8.763*
		UK drawing corrected	2.79 ± 0.02	7.49 ± 0.02	$3.18\ ^{0}_{-0.5}$	*8.76*
		AEP 2920 table C2,1 [2]	2.84 ± 0.03	7.52 ± 0.05	3.45 ± 0.5	*8.64*
		MIL-DTL 46593B[5] [9]	2.85 ± 0.03	$7.52\ ^{0}_{-0.02}$	$3.45\ ^{0}_{-0.25}$	*8.64*
	Selected		**2.79 ± 0.02**	**7.49 ± 0.02**	$\mathbf{3.18\ ^{0}_{-0.5}}$	***8.76***

MIL-DTL-46593B standard [8,9] is considered as reference here for definition of 12.7 mm caliber and 20 mm caliber FSPs without skirt. For convenience with other "Fx" FSPs, they are called respectively F7 and F8. Comparisons with table C2.1 included in last AEP 2920 [2] are in table 4.

Table 4. Selected weight and dimensions for fragment without skirt F7 (12.7 mm) and F8 (20 mm)

Fragment	Parameter	Mass	Diameter	Flat size	Length[4]
	Unit	g	mm	mm	*mm*
F7 (12.7 mm)	AEP 2920 table C2.1 [2]	13.39 ± 0.13	12.57 ± 0.05	5.9± 0.07	*14.73*
	MIL-DTL 46593B[5] [9]	13.41 ± 0.13	12.57 ± 0.03	$5.69\ ^{0}_{-0.38}$	*14.73*
	Selected	**13.41 ± 0.13**	**12.57 ± 0.03**	$\mathbf{5.69\ ^{0}_{-0.38}}$	***15***
F8 (20 mm)	AEP 2920 table C2.1 [2]	53.78 ± 0.26	19.92 ± 0.10	9.27 ± 0.6	*22.86*
	MIL-DTL 46593B[5] [9]	53.78 ± 0.26	$19.91\ ^{0}_{-0.05}$	$9.27\ ^{0}_{-0.3}$	*22.86*
	Selected	**53.78 ± 0.26**	**19.91 ± 0.05**	$\mathbf{9.27\ ^{0}_{-0.3}}$	***23.5***

[4] Length must be adjusted to give correct weight (value with no tolerance)
[5] Weight and dimensions are converted from standard (given in grains and dimensions in inches)

 https://doi.org/10.52202/078352-0050

2.2 Achievability

In order to prove achievability, DGA Land Systems decided to analyse weight and dimensions of 200 fragments randomly chosen in a 2200 batch of "F5" fragments received (quantity based on QL 2 in ISO 6509-2 [13]). All measurement were made in an ISO 17025 certified and independent laboratory [14, 15]. Histograms of diameter, flat size and weight are reported in figure 2. Expanded uncertainty on dimension is U=0.005 mm and U=0.0001g on weight. Results obtained on all dimensions, weight but also on chisel nose angles, symmetry and perpendicularity prove that tolerances in new drawing are clearly achievable by (good) manufacturers.

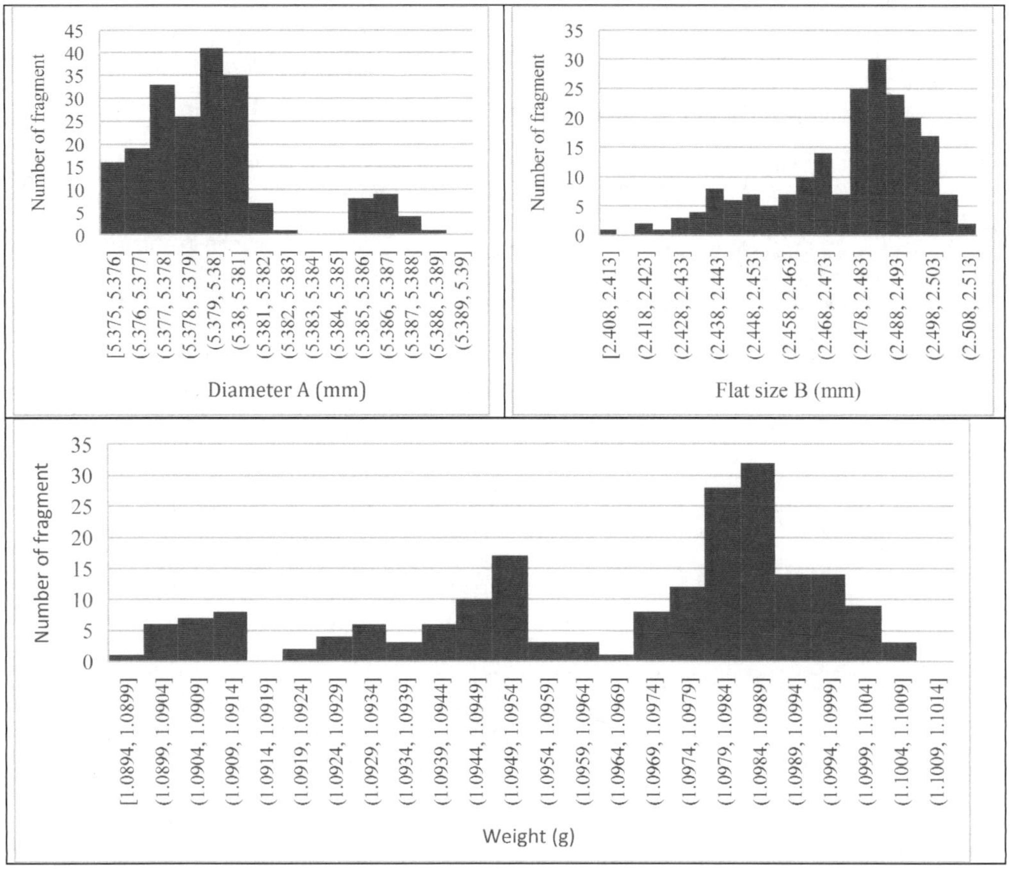

Figure 2. FSP F5 procurement: Histograms of quality control results on 200 fragments

3. REVIEWING DEFINITION OF FSP WITH SKIRT

This type of fragment is launched with rifled barrel. Chisel nose cylinders (FSP) launched without a sabot are designated with character G and an associated number is defining the weight and sizes of the fragment-simulating projectile [2]. Only four fragments exist: caliber 5.56, 7.62, 12.7 and 20 mm. Drawings of FSP with skirt are in MIL-P-46593A [16,17] since 1962 so this standard is assume to be a reference. AEP-55 linked with STANAG 4569 gives a different definition of 20 mm FSP for assessing protection level of armoured vehicles [18]. Table 5 give a summary of values and tolerances given in standards for the 20 mm FSP. Right column gives a proposal for 20 mm FSP weight and sizes based on facts given above. All angle values with their tolerances are similar in all standards analysed and are not reported there. As for FSP without skirt, dimension values are not expressed in accordance with international rules [11]. We can find values and tolerances expressed with a different number of significant digits. Moreover, unit conversion from imperial system to metric system are often wrong[6].

[6] For all analysis here: 1 grain = 0.064799g and 1 inch = 25.4 mm

Values converted must be rounded to the nearest decimal place but tolerances must be rounded to the lower decimal place. For the "*max flat G*" thickness, value is rounded to the lower decimal place because it is a maximum value. Some differences exist between MIL-DTL- 46593B [9] and MIL-P-46593A [16]: tolerances on the length of skirt are reduced for 20 mm and 7.62 mm FSP (from ± 0.005 to ± 0.002") and external diameter tolerance is reduced for 7.62 mm FSP ($^{+0.003}_{-0.001}$ to ± 0.001). All other fragments are analysed in the same way. For convenience with "Fx" FSPs, 12.7 mm caliber G8 is now called G7 and 20 mm caliber G9 is now called G8. (So, F7 and G7 are caliber 12.7mm, F8 and G8 are caliber 20 mm).

Table 5. Reviewing of weight and sizes of 20 mm FSP (current standard in bold) (cf. § 6.1)

Standard	Mil –P-46593A[16]	2920 Ed 2 [10]	**Mil –DTL-46593B [9]**	**AEP 55 Vol 1 [18]**	**AEP 2920 [2]**	Proposal
Year	1962	2003	2008	2014	2016	
Unit of dimension	*inch*	*mm*	*inch*	*mm*	*mm*	*mm*
Unit of weight	*gr*	*g*	*gr*	*g*	*g*	*g*
Weight	830 ± 4	52.73± 0.26 [7]	830± 4	53.8 ± 0.26	53.78 ± 0.26	53,78 ± 0,25
ØA	$0.784^{0}_{-0.002}$	$19.914^{0}_{-0.05}$	$0.784^{0}_{-0.002}$	19.89 ± 0.05	$19.914^{0}_{-0.05}$	$19.91^{0}_{-0.05}$
ØB	$0.823^{0}_{-0.003}$	$20.904^{0}_{-0.076}$	$0.823^{0}_{-0.003}$	$20.83^{+0.08}_{0}$	$20.904^{0}_{-0.076}$	$20.90^{0}_{-0.07}$
ØC	0.740± 0.005	18.796 ± 0.127	0.740± 0.005	18.80 ± 0.12	18.796 ± 0.127	18.80 ± 0.12
Flat size D	$0.365^{0}_{-0.012}$	$9.27^{0}_{-0.3}$	$0.365^{0}_{-0.012}$	$9.27^{0}_{-0.4}$	$9.27^{0}_{-0.3}$	$9.27^{0}_{-0.30}$
E	$0.912^{+0.010}_{0}$	23.165 ± 0.25	$0.912^{+0.010}_{0}$	24	23.165 ± 0.25	$23.16^{+0.25}_{0}$
F	0.091± 0.005	2.31 ± 0.127	0.091± 0.00**2**	2.31 ± 0.05	2.31 ± 0.127	2,31 ± 0.05
Max flat G	0.008	0.2	0.008	0.2	0.2	0.2
Radius R	$0.910^{0}_{-0.030}$	$23.11^{0}_{-0.75}$	$0.910^{0}_{-0.030}$	**No radius**[8]	$23.11^{0}_{-0.76}$	$23.11^{0}_{-0.76}$
H [9]			0.933	not defined		23.7

4. OTHER REQUIREMENTS

4.1 Materials requirements

Since 1962 and based on MIL-P-46593A[16], material composition for the fragment simulating projectile with skirt shall be "*a cold rolled, annealed steel conforming to compositions 4337 H and 4340H*". In current standard [2,9], these compositions are still requested. These steels are nickel-chromium-molybdenum steels as defined in AISI/SAE[10] designation (43xx family). The H suffix denotes hardenability is a major requirement. Since 2008, these steels are also required for heaviest FSP without skirt F5 to F8 defined in MIL-DTL-46593B [9]. For FSP without skirt F1 to F6, material shall conform to UK drawings requirements. In STANAG 2920 edition 1[7], material is an "*alloy steel to BS 970 PT1 817M40 condition T*". Unfortunately, material has changed in current AEP 2920 despite the same reference and issue number of UK drawing [2]. Material shall be now "*high carbon bright steel (Silver Steel) to BS 1407*". Chemical compositions for these different steels are shown in table 6. BS970 steel is equivalent to 4340 steel but is not used anymore in standard. Differences in steel, low availability or price of these specific steels encourage manufacturers to find alternative solutions for material, which do not comply with STANAG 2920. For next AEP 2920, uses of 4337H, 4340H or BS1407 will be encouraged but not mandatory. The fragment-simulating projectile could be manufactured from other steels for quenching and tempering that are capable of hardness uniformity within hardness value specified.

Table 6. Chemical composition in weight (%) for different steels used for fragment

Steel	C	Si	Mn	Cr	Ni	Mo
4337H [19]	0.34-0.41	0.20-0.35	0.55-0.90	0.65-0.95	1.55-2.00	0.20-0.30
4340H [20]	0.37-0.44	0.20-0.35	0.55-0.90	0.65-0.95	1.55-2.00	0.20-0.30
BS970 817M40 [21]	0.40	0.25	0.60	1.20	1.55	0.28
BS 1407 [22]	0.95-1.25	0.40 max	0.25-0.45	0.35-0.45		

[7] Weight was by mistake 52.73 g instead of 53.78 g in STANAG 2920 Ed 2 [12] and in AEP 2920 edition A V1 [13]

[8] Shape of 20 mm FSP is different in AEP-55 (flat rear instead of curved rear) despite referencing MIL-P-46593A [16].

[9] The total length of fragment H is not always indicated but manufacturer shall adjust length on base surface to meet indicated weight. Value given for reference only.

[10] American Iron and Steel Institute/Society of Automotive Engineers

 https://doi.org/10.52202/078352-0050

4.2 Hardness requirements

Hardness for FSP without skirt and RCC is always 30±2 HRC (Rockwell hardness C). For FSP with skirt, all current standards agree on the same hardness 30±2 HRC but it was not always the case. Table 7 resumes variability in hardness requirements.

Table 7. Hardness requirements for FSP with skirt in HRC (current standard in bold)

Class	**Year**	**G5**		**G6**	**G7** [11]	**G8** [12]
Caliber (mm)		**5.56 (type 1)**[13]	**5.56 (type 2)**[14]	**7.62**	**12.7**	**20**
MIL-P- 46593 A [16]	1962	*30±1*	*27±3*	*30±1*	*30±1*	*30±1*
2920 Ed 1 [7]	1996	*Refer to US MIL-P-46593A*				
2920 Ed 2 [10]	2003	*30±1*	*27±3*	*30±1*	*30±1*	*30±1*
MIL-DTL- 46593 B [9]	2008	30±2		30±2	30±2	30±2
AEP-55 vol 1 [18]	2014				Refer to MIL-P-46593A But specified 30±2	
AEP 2920 [2]	2016	30±2		30±2	30±2	30±2

4.3 Burrs removal requirements

Burrs are a residue of machining or of a forming process. Deburring process often conducts to have small deviation inside the ideal geometrical shape of an edge. No requirements regarding burrs removal can be found in MIL-P- 46593 A [16] and in MIL-DTL- 46593 B [9]. STANAG 2920 edition 1[7] mentioned that all burrs and sharp edges are to be removed. STANAG 2920 edition 2 [10] and AEP 2920 [2] mentioned that "*all burrs are to be removed and all surfaces are to be as smooth as possible*". No additional information were given and manufacturers are doing that in their own way. It is assume that different burrs removal can change ballistic results especially on soft ballistic panels. It was decided to add requirement for burrs removal. For tolerancing the geometry (sharpness) of edges a standard ISO 13715 has been developed [23]. Symbol for deburr and break sharp edges according ISO standard 13715 could be included now and is therefore, defined on new drawings. Tolerances for burrs removal are $^{-0.05}_{-0.15}$ and are in respect of values met by mechanical design office of DGA Land Systems.

4.4 Finish

For new drawings, the fragment-simulating projectiles shall have a surface roughness value Ra equal to 1.6 micrometres (or roughness ISO grade N7). This value comply with finish required by MIL-DTL-46593B [9] (63 microinches). UK drawings [2, 7] require only 3.2 micrometers.

5. QUALITY CONTROLS

5.1 Sampling requirements

AEP 2920 [2] defines that the test facility shall take a representative sample from each batch and check the hardness (measured on the side), dimensions and weight, before use. Fragment simulators shall be certified by vendors to meet the specifications. However, a representative sample is not defined and what is expected of vendors too. For MIL-P-46593A [16] hardness of all fragment-simulating projectiles defined shall be tested with an acceptable quality level of 2.5 percent defect per hundred with equipment approved by government [16]. Two years later, an amendment was published [17] and substituted the previous "hardness" paragraph with the following one:

- Caliber .22 type 1 (for armor plate), calibres .30, .50 and 20 mm fragment-simulating projectiles shall be one-hundred percent tested for hardness.

MIL –DTL 46593B [9] therefore requires a one-hundred percent quality control for hardness (for fragments F5 to F8 and G5 to G8). Quality control level or sampling must be enough to build confidence to fragments but with an affordable cost. Achievability of parameter during manufacturing process and

[11] Called G8 in current AEP 2920

[12] Called G9 in current AEP 2920

[13] Type 1 for armor plate and type 2 for body armor

accuracy of measurement means can change the quality control level. Mass and dimensions values are clearly achievable and only few percent controlled can demonstrate compliance with drawings. For hardness, manufacturing process are more complex (machining and heat treatment processes) and inaccuracy in hardness measures is often of the same order than hardness tolerances (see § 5.2). The knowledge and confidence in manufacturer's processes and tools is also very important and can imply quality control level adjustment. Moreover, controls of hardness are destructive testing for small fragments. Nevertheless, certificate of compliance given by vendors or by test houses themselves are not enough and often wrong (very poor sampling, measurement mean not traceable/calibrated to national standard and so on…). These guidelines are proposed to be in the next AEP:

- Test houses shall verify quality of fragment independently from manufacturers/suppliers.
- National Authority[14] shall define level of confidence of quality level. Use ISO2859 [13] for sampling or quality control level choosing is encouraged and 2920 committee will suggest quality control level.
- Hardness, diameters, masses shall be controlled. National authority must define others parameters to be controlled. The use of ISO17025 laboratory for hardness measurement is preferred. Nevertheless, all measurement means used for quality control shall be traceable/calibrated to national standard.

5.2 Quality control of hardness

As define in AEP 2920[2] the fragment-simulating projectile shall be fully quenched and tempered to a Rockwell hardness value as 30±2 HRC. It is proven that (hard) machining can change hardness values and heat treatments are necessary to comply with requirements (See. variability on hardness in figure 3). For small fragment, it is impossible to evaluate the hardness (HRC) because sample is too thin or footprint's penetrator is too large and Vickers measurement methods must be preferred. After measurement conversion from Vickers to Rockwell value must be done and introduce another inaccuracy in results. ISO standards exist to explain hardness measurement (methods, tools, hardness value corrections when testing on cylindrical surfaces, verification and calibration of testing machines and so on…) and how to convert hardness values from different units. Therefore, these guidelines are proposed to be in the next AEP:

- Rockwell hardness measurement shall be done in compliance of ISO 6508 [24],
- Rockwell hardness measures are not always possible and Vickers hardness measures can be used. In this case measurement shall be done in compliance of ISO 6507 [25],
- Conversion from Vickers hardness to Rockwell hardness must be conducted in compliance with ISO 18265 [26] - Small fragments with a mass lower than 2.9 g are to have a Vickers hardness value of HV 295^{+17}_{-18} (based on tables B.2 and C.2 in [26]).
- The position and correction of hardness measurement shall comply with ISO standards (6507 or 6508) [24, 25]. Inaccuracy of hardness measurement shall be evaluate and given within certificate of compliance.

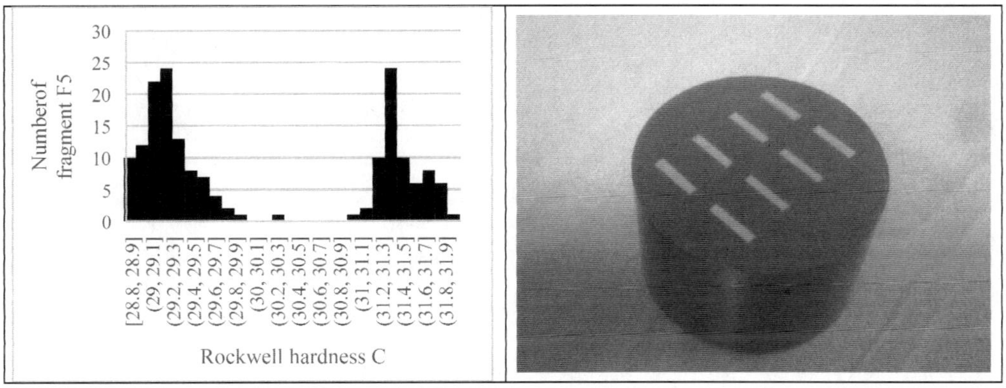

Figure 3. Histogram of hardness quality control on 200 F5 fragments and samples preparation [14,15]

6. NEW DRAWINGS

[14] National Authority (NA): the Authority responsible for the specification of Personal Armour Systems within a Nation or Armed Service [2].

https://doi.org/10.52202/078352-0050

6.1 Proposed drawings for chisel nose Fragment Simulating Projectile

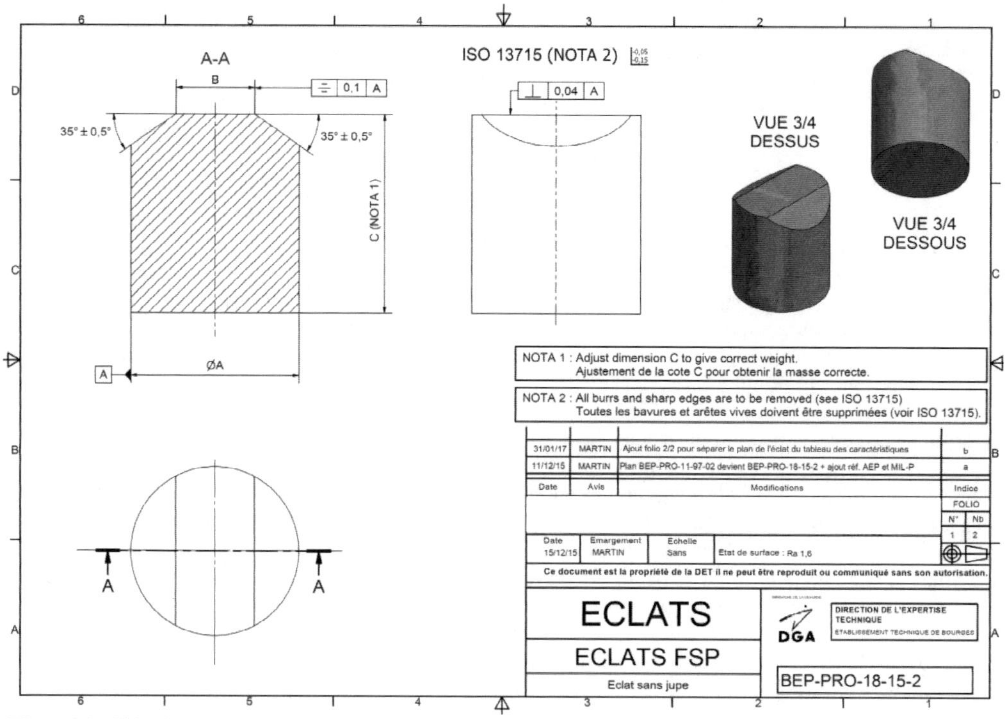

Figure 4. Chisel Nose Fragment Simulating Projectile launched with sabot: F1 (0.16 g) to F8 (20 mm)

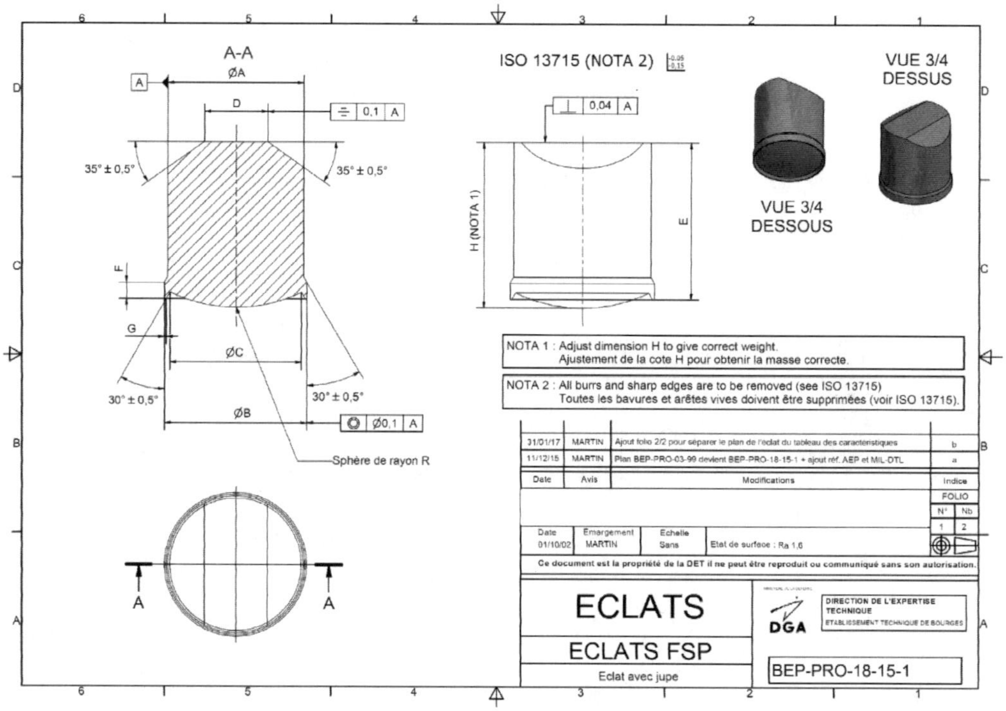

Figure 5. Chisel Nose Fragment Simulating Projectile: G5 (5.56 mm) to G8[15] (20 mm)

6.2 Weight and dimensions for chisel nose Fragment Simulating Projectiles and RCCs

[15] Called G9 in current AEP 2920

https://doi.org/10.52202/078352-0050

Hardness requirement is 30±2 HRC for all fragments (or 295^{+17}_{-18} HV). Tables 8, 9 and 10 give respectively weight and dimensions for FSP without skirt, FSP with skirt and RCCs. Refer to drawing in § 6.1.

Table 8. FSP without skirt characteristics "F"

Class	F1	F2	F3	F4	F5	F6	F7	F8
Mass [g]	0.16 ± 0.01	0.24 ± 0.01	0.33 ± 0.01	0.49 ± 0.02	1.10 ± 0.02	2.79 ± 0.02	13.41 ± 0.13	53.78 ± 0.26
Diameter A (mm)	2.64 ± 0.02	3.25 ± 0.02	3.60 ± 0.02	4.06 ± 0.02	5.39 ± 0.02	7.49 ± 0.02	12.57 ± 0.02	$19.91^{0}_{-0.05}$
B (mm)	$1.27^{0}_{-0.5}$	$1.52^{0}_{-0.5}$	$1.75^{0}_{-0.5}$	$2.03^{0}_{-0.5}$	$2.54^{0}_{-0.5}$	$3.18^{0}_{-0.5}$	$5.69^{0}_{-0.38}$	$9.27^{0}_{-0.3}$
C(mm)[16]	3.17	3.81	4.31	4.57	6.35	8.76	14.8	22.9

Table 9. FSP with skirt launched without sabot characteristics "G"

Class	G5	G6	G7 [17]	G8 [18]
Caliber (mm)	**5.56**	**7.62**	**12.7**	**20**
Mass [g]	1.10 ± 0.03	2.85 ± 0.03	13.41 ± 0.13	53,78 ± 0,25
Diameter A (mm)	$5.46^{+0.02}_{0}$	$7.52^{0}_{-0.02}$	12.57 ± 0.02	$19.91^{0}_{-0.05}$
Diameter B (mm)	5.74 ± 0.02	7.85 ± 0.02	12.95 ± 0.02	$20.90^{0}_{-0.07}$
Diameter C (mm)	$5.08^{0}_{-0.25}$	$6.93^{0}_{-0.25}$	$11.43^{+0,07}_{-0.02}$	18.80 ± 0.12
D (mm)	$2.54^{0}_{-0.25}$	$3.45^{0}_{-0.50}$	$5.69^{0}_{-0.38}$	$9.27^{0}_{-0.30}$
E (mm)	6.35 ± 0.05	8.84 ± 0.05	14.78 ± 0.05	$23.16^{+0.25}_{0}$
F (mm)	0.64 ± 0.05	0.86 ± 0.05	1.47 ± 0.05	2,31 ± 0,05
G (mm) max	0.12	0.12	0.15	0.20
R (radius) (mm)	$6.35^{0}_{-0.76}$	$8.64^{0}_{-0.76}$	$14.73^{0}_{-0.76}$	$23.11^{0}_{-0.76}$
H (mm) [17]	6.53	8.99	14.89	23.7

Table 10. Right Circular Cylinders (RCC) characteristics (R1 to R7)

Class	R1	R2	R3	R4	R5	R6	R7
Mass [g]	0.16 ± 0.03	0.24 ± 0.03	0.33 ± 0.03	0.49 ± 0.03	1.10 ± 0.03	2.84 ± 0.03	4.15 ± 0.03
Diameter A (mm)	2.64 ± 0.27	3.25 ± 0.22	3.60 ± 0.19	4.06 ± 0.14	5.39 ± 0.06	7.49 ± 0.04	8.74 ± 0.03
B (mm) [17]	3.77	3.64	4.07	4.78	6.17	8.19	8.82

[16] Dimension must be adjusted to give correct weight

[17] Called G8 in current AEP 2920

[18] Called G9 in current AEP 2920

https://doi.org/10.52202/078352-0050

In 1996, annex A to STANAG 2920 edition 2 [12] defined steel cylinders or RCCs. Mass, diameter and length values were put in a table for 7 RCC fragments. Masses range from 0.16 g to 4.15 g. In the current AEP 2920 [2] we can find same table. Only mass of 1.10 g RCC change in AEP 2920 by mistake but with same diameter and length (it could be mass of the US 16gr RCC?). It is important to notice that STANAG 2920 RCCs are different from US RCCs. Mass will be corrected in the next AEP and will be back to 1.10 g. Mechanical design office of DGA Land Systems built a drawing although it is not particularly difficult to understand steel cylinders but we can thus add some requirements for manufacturer. Table 10 presents values and tolerances for RCCs.

3. CONCLUSION

Reviewing fragments definition is clearly necessary to correct inconsistencies and mistakes and reduce irregular results from testing. Guidelines for quality controls will also improve acceptance and qualification processes for ballistic protection. Drawings and guidelines detailed here are proposed for the next AEP 2920 expected in 2021.

Acknowledgments

Authors want to acknowledge STANAG 2920 ToE members and manufacturers or test houses who gave (**or will give)** feedbacks on these new requirements.

References

[1] NATO Standard – Standard Agreement STANAG 2920 Edition 3 - Classification of Personal Armour – 06/2015.
[2] NATO Standard - Allied Engineering Publication AEP 2920 Edition A V2 - Procedures for the Evaluation and Classification of Personal Armour - Bullet and Fragmentation Threats- September 2016.
[3] Bolduc et al., Summary of newly ratified NATO standard AEP 2902 Ed A V1 , Personal Armour Systems symposium PASS 2016, Amsterdam NL, Proceeding Book pp 25-40.
[4] https://www.linkedin.com/pulse/new-stanag-2920-standard-davromatic-precision-ltd
[5] Cunniff et al., Variability in ballistic impact performance due to projectile physical properties and dimensions, Proceedings of 24th International Symposium on Ballistics, New Orleans USA - 2008.
[6] International Standard. ISO/IEC 17025; General requirements for the competence of testing and calibration laboratories; Third edition 2017-11.
[7] NATO Standard - STANAG 2920 PPS Edition 1 – Ballistic Test Method for Personal Armours – 12/1996
[8] MIL-DTL-46593B (MR) – Military Detail Specification - Projectile, Calibers .22, .30, .50, AND 20 mm Fragment Simulating – July 2006
[9] MIL-DTL-46593B (MR) w/amendment 1 – Military Detail Specification - Projectile, Calibers .22, .30, .50, AND 20 mm Fragment Simulating – August 2008
[10] NATO Standard, STANAG 2920 Edition 2 – Ballistic Personal Armour materials and combat clothing- 07/2003
[11] JCGM 100:2008, Evaluation of measurement data - Guide to expression of uncertainty in measurement (known as GUM), JCGM Joint committee for guides in Metrology - September 2008 – www.bipm.org
[12] NATO Standard - Allied Engineering Publication AEP 2920 Edition A V1 - Procedures for the Evaluation and Classification of Personal Armour - Bullet and Fragmentation Threats- June 2015
[13] ISO 2859 Sampling procedures for inspection by attributes - Part 2: Sampling plans indexed by limiting quality (LQ1 for isolated lot inspection) – September 1985
[14] Statement of conformity DC 2929-17-18-16 MRT DGA TT – F5 FSP Sans jupe 1,1 g QL2, AMERIS France SARL, Tourlaville - France - 04/04/2017 [*internal report*]
[15] Rapport d'essai COFRAC n°DEM0046413-3, Contrôle de dureté, pesée et contrôle dimensionnel, Laboratoire d'essais CETIM-CTDEC accrédité n°1-0601, Cluses – France, 30/03/2017 [*internal report*]
[16] MIL-P-46593A (MR) Military Specification - Projectile, Calibers .22, .30, .50, AND 20 mm Fragment Simulating – October 1962
[17] MIL-P-46593A (MU) Amendment 1- Military Specification - Projectile, Calibers .22, .30, .50, AND 20 mm Fragment Simulating – October 1964
[18] NATO Standard - Allied Engineering Publication AEP 55 Volume 1 Edition A V1 - Procedures for evaluating the protection level of armoured vehicles – Kinetic energy and artillery threat – April 2014
[19] www.steeldata.info/hard/demo/data/1814.html
[20] www.efunda.com/materials/alloys/alloys_steels/alloys.cfm
[21] www.abbeyforgedproducts.co.uk/images/downloads/PDF/817M40.pdf
[22] www.silver-steel.co.uk/about-bs1407-silver-steel
[23] ISO 13715:2017 Technical product documentation - Edges of undefined shape - Indication and dimensioning
[24] ISO 6508:2016, Metallic materials - Rockwell hardness test, August 2016
[25] ISO 6507-1:2005 "Metallic materials - Vickers hardness test" – March 2006
[26] ISO 18265:2013 (Metallic materials — Conversion of hardness values), October 2013

Intracranial strain and displacements from impacts to a helmeted deformable headform

S. Xu[1], S. Dutrisac[1], S. Ouellet[2], and **O.E. Petel**[1]

[1]*Carleton University, 1125 Colonel By Drive, Ottawa, ON, Canada*
oren.petel@carleton.ca
[2]*DRDC Valcartier, 2459 Pie-XI Blvd North, Quebec, QC, Canada*

Abstract. Novel techniques and tools are needed in helmet performance and injury evaluation. Advances in headforms have been made in recent years, with a shift towards the use of deformable skull and brain surrogates. The development of these headforms has also brought about the integration of new diagnostic capabilities that reach beyond the classic rigid body mechanics approach to injury evaluation. The majority of these diagnostic systems provide only discrete measurements within the headform, that are dependent on the choice of gauge placement. In the present work, we demonstrate advances that have been made in making temporally-resolved full-field intracranial strain and displacement measurements within the surrogate brain of a deformable headform during an impact event. The approach involves the use a custom-built *in situ* high-speed X-ray system. Radiopaque contrast media were integrated into the surrogate to enable tracking of the displacement fields that can then be used to calculate intracranial strain. The measurements presented in this work were taken along a parasagittal plane within the headform. The interaction between the helmet and headform as well as the deformation fields measured within the brain surrogate show tremendous potential for future use in equipment performance evaluation, returning tissue-level metrics that are related to injury mechanisms of traumatic brain injury.

1. INTRODUCTION

Mild-Traumatic Brain Injury (mTBI), whether caused exposure to blast or a direct impact event, remains a primary non-lethal health concern. A recent study of the epidemiology of mTBI among military veterans from the Iraq and Afghanistan war era found that approximately 15.1% had sustained an mTBI relating to their service, of which one third were due to blast exposure, while nearly half were due to blunt impact events (e.g., object hitting head or falls) [1]. Reducing the incidence of these injuries will require improvements to personal protective headgear. Aside from their ballistic performance, military helmets are also subjected to a variety of performance characterization tests for their blunt impact performance [2]. The approaches used to qualify their response are similar to testing methodologies for athletic headgear [3-5]. Those testing procedures, typically involving drop-testing of a helmeted headform onto an anvil in different orientations and measuring the resulting acceleration profiles, focused on reducing the risk of focal injuries (e.g., skull fracture or severe brain trauma) from an impact [3,4]. Given the known link between rotational acceleration of the head and mTBI risk [6,7], linear impactor test protocols have recently been used to set additional performance criteria on athletic headgear [6]. These test protocols have yet to be mirrored in military helmet impact testing. Despite the continued modernization of testing standards, they are not yet able to directly address tissue-level metrics in their testing platforms, focusing exclusively on the rigid body mechanics of the head. Assessing injury risk from these tests has relied on either impact severity criteria [8-11] or coupling the testing protocols to computational models validated against limited cadaveric datasets [12,13]. In the present work, we present the first efforts to establish a deformation-based measurement within a helmeted headform to describe the performance of helmet.

Although current testing procedures use rigid headforms, there have been a considerable effort to design headforms with deformable structures that provide for various modal and inertial responses [14-17]. Several advanced tissue-simulating headforms with elastomeric brain surrogates have been designed in recent years to investigate blast-induced traumatic brain injury (bTBI). These headforms have all been optimized and calibrated for bTBI events. The brain surrogates of these headforms providing researchers with new opportunities to extract local measurements at the brain level. As these headforms improve in their biofidelity, the diagnostic capabilities must be adapted to take advantage of the opportunities for tissue-level metrics to be captured. The headforms have thus far been primarily limited to measurements of global head kinematics and discrete pressure measurements [17]. As such, the full potential of these tissue-simulating headforms has yet to be achieved. Thus, there is a need for new testing methodologies that directly investigate full-field tissue-level metrics of relevance to mTBI, such as brain displacement and deformation.

https://doi.org/10.52202/078352-0051

Monitoring full-field tissue-level metrics associated with blunt trauma required a new diagnostic approach to investigating the headform. The approach chosen involves the use of a custom-built *in situ* high-speed X-ray (HSXR) cinematography system, capable of high capture rates through the headform, to monitor radiopaque marker motion within the headform to determine intracranial displacement and deformation [18]. In the present study, we use this approach to determine the response of the surrogate brain to a frontal impact on a tactical military helmet. The focus was to determine the response of the brain surrogate. This new helmet evaluation technique, once adequately calibrated, represents an opportunity to use tissue-level deformation metrics to evaluate helmet performance and head trauma risks associated with head impacts. At this time, this headform is undergoing a calibration in relation to cadaveric datasets using the same physical setup, however this phase of the work is ongoing and the purpose of the present study is to show the early development of this capability.

2. EXPERIMENTAL CONFIGURATION

The BIPED headform, a thorough description of which can be found in [16,17], was used in the present study. The headform features several intracranial anatomical structures, such as a cerebellum, artificial cerebrospinal fluid (CSF), and falx and tentorium membrane simulants. For the present work, the brain surrogate, which was cast from Sylgard 527, was altered by the insertion of radiopaque markers during the casting process. The markers were placed along a parasagittal plane, 15 mm from the midsagittal plane. Marker composition included a mixture of a thermoplastic gel (Humimic gel #4) with barium sulphate powder (60% by mass). The markers were cylindrical with a 1.5-mm diameter and 2-mm length. Once the brain was sealed within the headform, water was flushed through the intracranial spaces to act as CSF and ensure that the headform was fully perfused prior to testing. During testing, a static pressure of 1 m of H_20 was maintained on the fluid to replicate the vasculature and intracranial repressurization used in prior cadaver studies [12].

For the impact tests, the headform was placed on a linear impactor platform (Cadex) and subjected to a series frontal impacts from a 13.13 kg ram to test the response of the Viper helmet (Galvion). The end cap of the impacting ram was a steel front plate backed by a rubber to simulate the anvil used in military helmet drop testing. The headform was attached to a neck-form variation on the Hybrid III neck that does not have a directional bias [19]. The neck was pre-tensioned to 1ft-lb. The data presented in this paper was generated by two impacts on the same helmeted headform that was subjected to impacts at speeds of 3.15 m/s and 3.54 m/s. These impact conditions correspond to impact energies of approximately 65J and 82J, respectively. The headform was inclined at an angle of 15° for both of these impacts. The headform acceleration history was recorded on a rigidly attached neck extension with an attached DTS 6DX Pro accelerometer and then translated to the centre of gravity of the headform. A photograph of the headform, mounted to a linear impactor table, prior to a test is shown in Figure 1.

Figure 1. Photograph of the helmeted headform with the end cap of the linear impactor visible.

The images captured by the HSXR were processed using an open-source Matlab-based Digital Image Correlation (DIC) script, NCorr [20] to determine the displacement and deformation fields within

https://doi.org/10.52202/078352-0051

the headform. For the DIC analysis, a speckle radius range of 7-10 pixels, a subset radius of 25 pixels, a strain radius of 15 pixels and a subset spacing of 2 pixels were used to generate the results.

3. RESULTS

The linear and rotational acceleration at the centre of gravity of the headform for the impact are shown in Figure 2a and 2b, respectively. The headform was subjected to peak linear and rotational accelerations of 50 g and 2,280 rad/s^2 and 65 g and 3,300 rad/s^2 for the 65J and 82J impacts, respectively. These two impact conditions resulted in peak resultant rotational rates of the headform of 15.8 rad/s and 17.8 rad/s, respectively. X-ray images of this impact were captured with our custom HSXR at a frame rate of 5,000 fps. A series of images captured over the first 70 ms of the 82J impact event are shown at 10 ms intervals in Figure 3. The corresponding shear strain fields in the head fixed reference frame at the associated time intervals are also shown in Figure 3. Note that the strains are shown in a reference frame fixed to the initial skull position in the initial frame of the sequence. The displacement paths of the brain surrogate from this impact, in the skull-fixed reference frame, are shown in Figure 3. These paths are not tracking the displacement of a single marker within the surrogate, but rather are related to the motion paths of pixel subsets. Figure 5 shows the resultant displacement histories of the points from Figure 4 over the first 70 ms following this impact. These results can also be plotted out in the same fashion to demonstrate the shear strain fields within the headform, as shown in Figure 6.

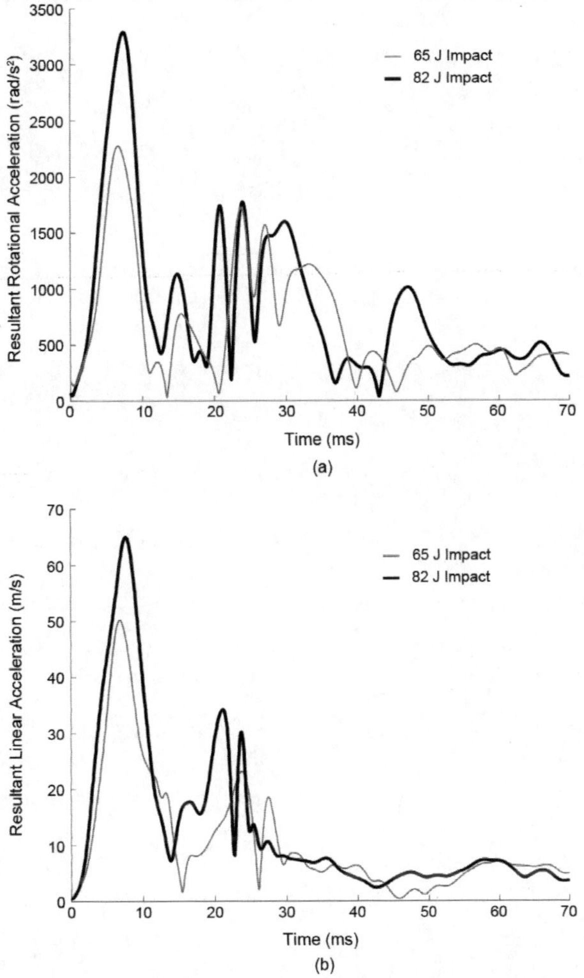

Figure 2. (a) Linear and (b) Rotational acceleration profiles of the headform at its centre of gravity.

https://doi.org/10.52202/078352-0051

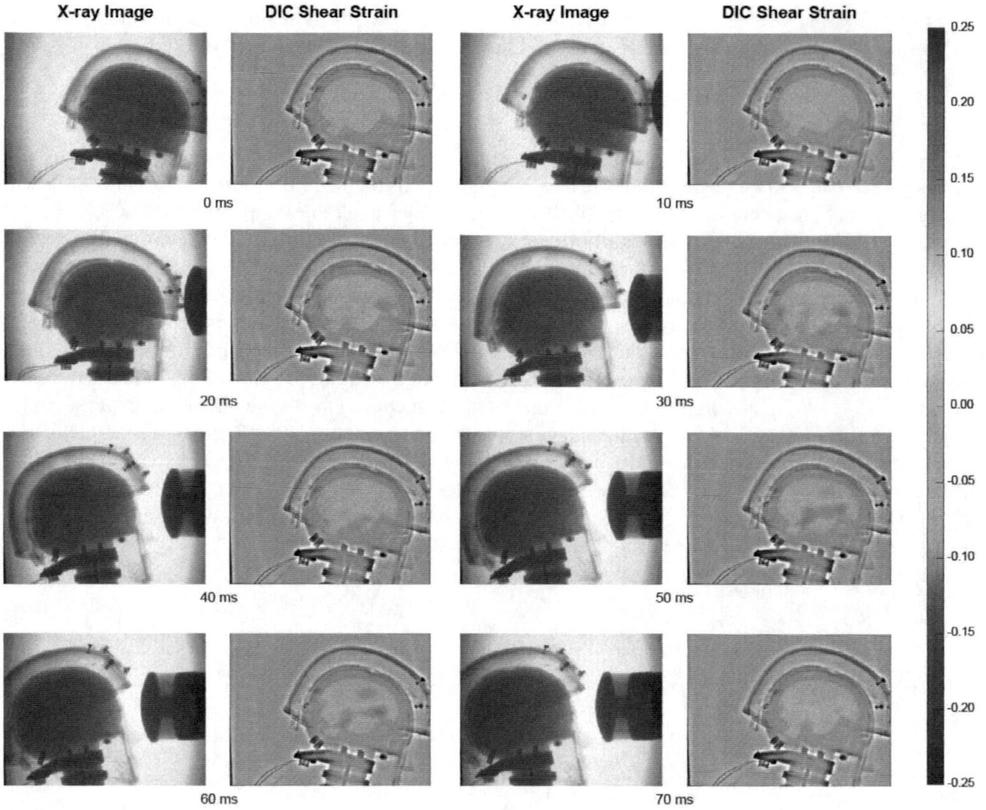

Figure 3. The original X-ray images of the headform impact beside the corresponding X-ray DIC images of the shear strain fields calculated from the image.

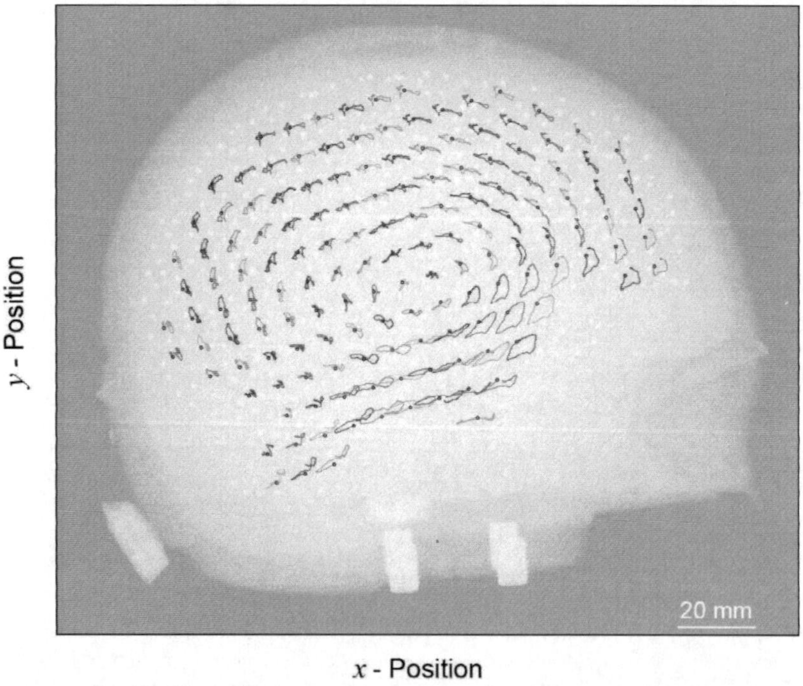

Figure 4. Complete displacement paths of the brain surrogate over a period of 70 ms following impact. The circles mark the start of the paths, and each path has a different colour for clarity.

https://doi.org/10.52202/078352-0051

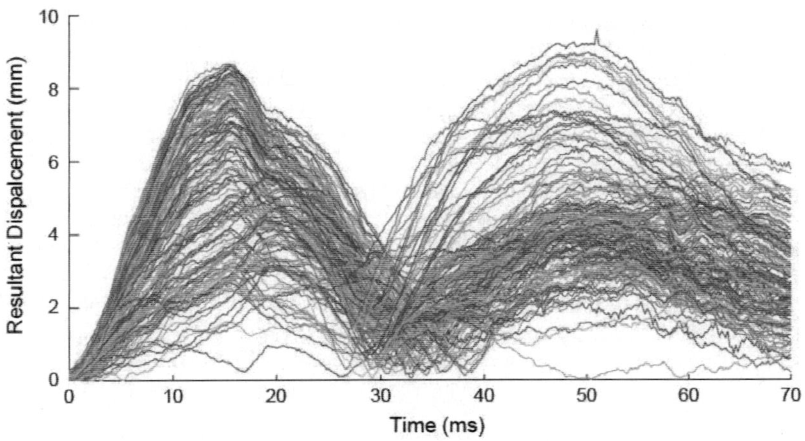

Figure 5. The resultant displacement field within the brain.

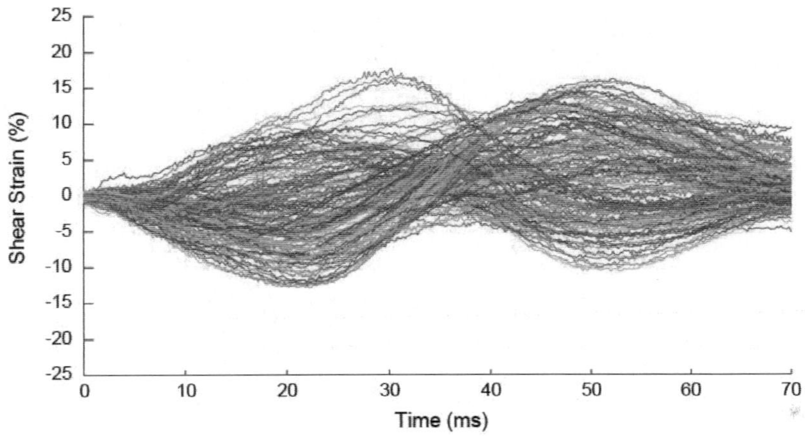

Figure 6. Shear strain histories of individual points within the brain surrogate.

4. DISCUSSION

The results shown in Figures 3 through 6 demonstrate the amount of information relating to tissue-level metrics associated with injury outcomes that can be gleaned from this experimental technique. There are several features in the results that are worthy of note. In Figure 4, there are distinct motion patterns that develop as a result of the impact. The overall motion of the brain surrogate results in a clear rotation of the brain about its centre of gravity. The temporal region is subjected to broader lateral motion than other regions of the brain. When we examine these features in the resultant displacement plots of Figure 5, which are not directional, we see similar features. The displacement magnitude of the brain is notably region dependent. For instance, the peak displacement in the parietal and occipital regions had peak displacements of approximately 4 mm from their initial positions, while the motion in the frontal, temporal, cerebellum regions was notably larger, on the order of 6-7 mm. The displacement histories in Figure 5 also shows a double peak in the displacement field. This is related to the initial motion, change of direction and slowing of the brain, with the second displacement peak coinciding with the full extent of motion of the head. While the primary phase of head acceleration occurs in the first 15 ms of the impact, brain motion continued to occur over a considerably longer timescale. Furthermore, the second displacement peak was slightly larger than the primary peak in certain regions of the brain, such as the temporal region. While the displacement histories represent the resultant motion and are thus non-directional, comparing these paths to Figure 4 shows that the total deviation of a given marker was

 https://doi.org/10.52202/078352-0051

approximately double the peaks seen in Figure 5, as the markers first moved in one direction, then reversed their course.

This motion reversal had a notable effect on the shear strain field calculated within the head, as seen in Figure 3. The shear strain histories show a shear reversal, which is essentially a reversal in the distortion angle of the tissues during the impact event. The inversion of the strains was seen in every region of the brain. The peak shear strain for the impact was on the order of 18% and occurred in the temporal region. This is consistent with the strain fields shown in Figure 3. The parietal, occipital, and cerebellum regions saw the lowest strains. Of particular interest was the response in the frontal and temporal regions, where the strain rate associated with the second strain peak was significantly higher than the initial strain rate associated with the impact. The timing of this inversion is unrelated to the global kinematic response of the head and demonstrates the effect of the decoupled motion of the brain and skull.

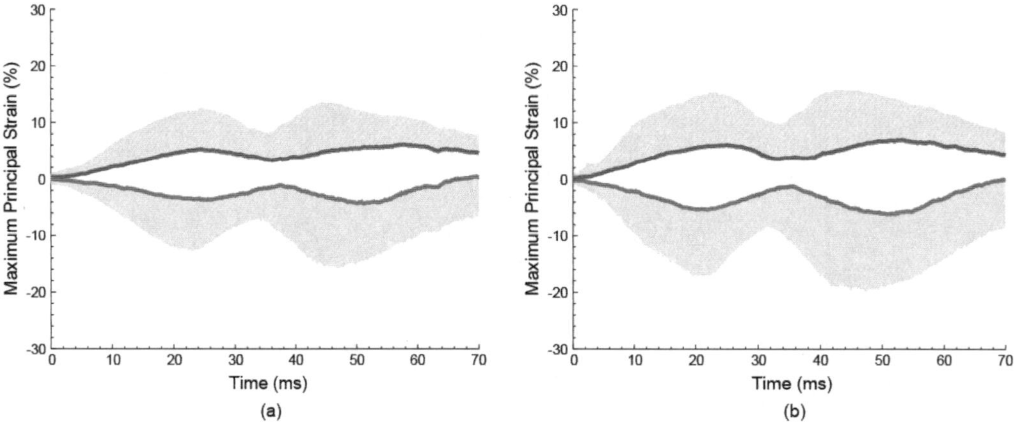

Figure 7. Maximum Principal Strain (MPS) profiles within the brain surrogate for the (a) 65J impact and the (b) 82J impact. The thick lines represent the mean MPS within the brain surrogate, while the shaded bounds represent the 95th percentile MPS values.

The results of the tests were plotted to show the Maximum Principal Strain (MPS) values associated within the brain surrogate for the two impact conditions in Figure 7. The mean and 95th percentile values are shown for tensile and compressive MPS. While the trends in the MPS profiles were the similar under both impact conditions, this comparison showed that there were notable differences in the peak strains within the surrogate, demonstrating that the headform is sensitive enough to measure strain differences associated with increasing impact severity. Thus, this measurement capability has potential as a future helmet performance evaluation tool, returning intracranial strain response measurements.

While there is still work to be done to ensure the biofidelic response of this headform lies within an acceptable human performance corridor, particularly in its impact response, the methods developed in this work are broadly applicable to the field of injury biomechanics. The tissue-level responses that were measured in the present work present a new avenue for research in injury biomechanics, returning metrics that can be directly related to suspected injury mechanisms, such as brain tissue strain. While the datasets currently generated by this headform cannot be directly linked to clinical outcomes, the long-term objective will be such a validation to allow its use as a helmet testing and evaluation platform.

5. CONCLUSIONS

A new testing approach for helmet evaluation has been presented, demonstrating its ability to generate high temporal and spatial resolution tissue-level displacement and strain fields across the entire headform model. A tactical military helmet was tested under a linear impact load and the resulting strain and displacement fields were measured. Using this headform and DIC technique, we have demonstrated some interesting dynamics within the brain surrogate relating to the rotation and inversion of the tissue

https://doi.org/10.52202/078352-0051

distortion field at high strain rate. Further calibration and validation of this headform is warranted and will help it develop into a useful tool for helmet performance evaluation.

Acknowledgments

The authors acknowledge funding support from CIHR-NSERC under grant CHRPJ 508414-17 and financial support for X-ray infrastructure from the Canadian Foundation for Innovation and the Ontario Research Fund under project 39233. The authors thank William Bedard for his assistance in preparing the headform.

References

[1] Lindquist, L.K., Love H.C., and Elbogen, E.B., J Neuropsychiatry Clin Neurosci, 2017; 29(3); 254-259.
[2] Hopping, J. A., Bass, C., Shridharani, J., Perritt, C., Haynes, N., & Masters, K. (2010). Methodologies for blunt trauma assessment in military helmets. Proceedings of the Personal Armour System Symposium; 2010; Quebec City, Canada.
[3] National Operating Committee on the Standards for Athletic Equipment (NOCSAE). Standard Test Method and Equipment Used in Evaluating the Performance Characteristics of Protective Headgear/Equipment, 2012.
[4] ASTM, "Standard Test Methods for Equipment and Procedures Used in Evaluating the Performance Characteristics of Protective Headgear," American Society for Testing and Materials, pp. Standard F1446-13, 2013.
[5] National Operating Committee on the Standards for Athletic Equipment (NOCSAE). Standard Pneumatic Ram Test Method and Equipment Used in Evaluating the Performance Characteristics of Protective Headgear and Face Guards, 2018.
[6] Pellman E.J., Viano D.C., Tuck A.M., Casson I.R., and Waeckerle J.F., Neurosurg, 2003; 53(4); 799-814.
[7] Rowson S., Duma S.M., Beckwith J.G., Chu J.J., Greenwald R.M., Crsico J.J., Brolinson P.G., Duhaime A.-C., McAllister T.W., and Maerlender A.C., Ann Biomed Eng, 2012; 40(1); 1–13.
[8] Gadd C.W., Use of a Weighted-Impulse Criterion for Estimating Injury Hazard. SAE Technical Paper, 1966.
[9] Newman J., Shewchenko N., and Welbourn E., Stapp Car Crash J, 2000, 44:215-247.
[10] Takhounts E.G., Craig M.J., Moorhouse K., McFadden J., and Hasija V., Stapp Car Crash J, 2013; 57;243–266.
[11] Gabler L.F., Crandall J.R., and Panzer M.B, Ann Biomed Eng, 2018; 46(7);972-985.
[12] Hardy W.N., Foster C.D., Mason M.J., Yang K.H., King A.I., and Tashman S., Stapp Car Crash J., 2001; 45; 337–368.
[13] Hardy W.N., Mason M.J., Foster C.D., Shah C.S., Kopacz J.M., Yang K.H., King A.I, Bishop J., Bey M., Anderst W., et al, Stapp Car Crash J., 2007; 51;17-80.
[14] Ganpule S., Mechanics of blast loading on PMHS and surrogate heads in the study of traumatic brain injury using experimental and computational approaches. Ph.D. Dissertation, University of Lincoln Nebraska, 2013.
[15] Merkle A., Wing I.D., Armiger R.A., Carkhuff B.G., and Roberts J.C., Proceedings of ASME International Mechanical Engineering Congress, Lake Buena Vista, Florida, USA, 2009.
[16] Ouellet S., Bir C., and Bouamoul A., Proceedings of the Personal Armour System Symposium; 2014; Cambridge, UK.
[17] Ouellet S and Philippens M., Shock Waves, 2018; 28; 19-36.
[18] Petel O.E., Diagnostic for in situ deformation and strain measurements applicable to traumatic internal injury investigation and prevention. USPTO Patent 9,826,954, 2015.
[19] Walsh E.S., Kendall M., Post A., Meehan A., and Hoshizaki T.B., Sports Eng., 2018; 21;479–485.
[20] Blaber J., Adair B., and Antoniou A., Exp. Mech., 2015; 55;1105-1122.

 https://doi.org/10.52202/078352-0051

Different ballistic performances for reference ammunitions of varied origins

B. Cordeau[1], F. Barbillon[1], F. Miachon[1], M. Essoloh[1] and F. Mouhot[2].
[1]*DGA Land Systems, Rocade Est, Echangeur de Guerry, 18021 Bourges Cedex, France*
benoit.cordeau@intradef.gouv.fr
[2]*DGA Aeronautical Systems, 47 rue Saint-Jean, 31130 Balma, France*

Abstract. As part of an evaluation of a ballistic protection, it is essential to control the parameters influencing the test in order to respect the conditions of a successful test: reproducibility, representability and reliability. We have initiated a certification process for our ballistic evaluation laboratory according to the ISO 17025 standard. This requires prove the laboratory's skills and its ability to produce valid results. One of the influencing parameters on the test results is the ammunition used. Ballistic assessment standards, such as STANG2920, is based on reference assessment ammunition. Those ammunitions are recommended by name, and sometimes with a hardness for the core. For the same designation, we can find different suppliers and different batches of manufacture. In order to better know the ammunitions used in our laboratory, we have carried out protective limit speed determination tests on BLD (NF A36-800) and BLAL aluminum (NF A50-800) reference targets for different ammunition. These tests made it possible to determine a V_{50} and a V_{90} according to a PROBIT method. In parallel, we carried out tests of metallurgical characterization of these ammunitions in order to compare their composition and their hardness. The ammunition concerned are 5.56mm×45 SS109 (5 references), 7.62mm×51 NATO Ball (4 references), and 5.56mm×45 M193 (3 references). Some significant differences in ballistic performance for the same type of ammunition are observed. Those findings lead us to adopt a method of validation of ammunition batch chosen for our assessments. This paper presents the results, ballistic and metallurgical, and the methodology adopted in our laboratory to validate an ammunition.

1. INTRODUCTION

For the evaluation of dismounted soldier protective equipment, DGA Land Systems must master all aspects of its tests. To guaranty its activities, DGA Land Systems have been certifying ISO 9001 [2] already. To go further and be able to guaranty its assessments, the centre is get involved in the creation of a laboratory that, based on ISO 17025:2015 [3], aims to implement tests relating to the evaluation and classification of bulletproof equipment. In addition to a permanent search for progress, this approach makes it possible to give an additional guarantee and greater confidence in the quality of the evaluations conducted by the centre, in particular those contributing to the qualification of the definition of protection for dismounted soldiers. These ballistic protection assessments are mainly carried out using the STANAG2920 [1] standard. The evaluation threats implemented correspond to Annex B proposing four categories of ammunitions: lead core, Mild Steel Core, Hard steel core, and Tungsten Cobalt (WC) Core projectiles. Each threat is defined by the mass of the projectile, the mass of the core and the minimum hardness of the core, excepted for the lead core.

The accreditation process imposed by ISO17025 [3] requires mastering all the parameters of a test. For a test referring to STANAG 2920 [1], one can cite among the parameters most influencing the results of the test: temperature of the samples, the speed of the projectiles, the implementation of the simulant, or the batch of ammunition. Among these parameters, the ammunition used represents one of these influencing parameters and can lead to a significant bias in the evaluation of protection.

Therefore, we decided to qualify the batches of ammunition that we use in our assessments. The interest is to guarantee a continuity of results for evaluations conducted with different batches, and a continuity in the severity of the tests. This article presents the results of comparison obtained between different batches of ammunition, which we have in our laboratory. We compared two category A munitions and one category B ammunition from STANAG2920 [1]. More specifically, class A3 5.56mm×45 M193, and class A5 7.62mm×51 ammunition, as well as class B3 5.56mmx45 SS109. Two comparisons were made: one from the mechanical hardness characteristics, and one from the ballistic characteristics. All the ballistic results are presented and discussed below for 11 batches from various sources. These results should allow us to validate a method of selecting our batches of ammunition in order to have proof of the mastery of our tests to obtain ISO17025 [3] accreditation for tests according to STANAG2920 [1].

2. ISO 17025:2015 AND APPROACH TO QUALIFY AMMUNITION

ISO17025 [3] has been developed with the objective of promoting confidence in the operation of laboratories. It contains requirements for laboratories to enable them to demonstrate they operate

https://doi.org/10.52202/078352-0052

competently, and are able to generate valid results. This norm specifies the general requirements for the competence, impartiality and consistent operation of laboratories. This norm contains objectives for impartiality, confidentiality, and requirements for structure, resource, process and management. It is applicable to all kind of industrials sectors or activities practicing test, and so for ballistic laboratory.

One of the requirement is that the laboratory shall ensure that only suitable externally provided products that affect laboratory activities are used. The ammunitions used in a ballistic laboratory are concerned by this requirement. Also we have to master this element having a first order impact on the performance of the protection solutions evaluated. The approach chosen in our laboratory for controlled the influence of ammunition batch is to have a procedure to justify the change of a batch of ammunition. This procedure should allow us to have continuity in the assessments. Therefore, we have chosen to define a receipt for our batches of ammunition from V_{α} tests (see paragraph 5) to ensure that these munitions have characteristics similar to the previous batches. In order to have a conservative approach to the validation of our protections, we choose to verify that the batches of ammunition that we use have a V_{50} and V_{90} , respectively 50% and 90% of probability of protection, within an acceptable range. The tests described below allow us to define these acceptance ranges for a batch of ammunition specifically for an ammunition type.

3. AMMUNITIONS ASSESSED

The different ammunition compared are listed in the Table 1 and were only drawn from lots of different suppliers, we were looking to draw up a large inventory. In a second step, we plan to refine our results with different batches from same manufacturers with high reproducibility behaviour.

Table 1. Ammunition tested

Munition	Designation	Batch
5.56 mm × 45 M193	M193	LC 10H105 013
		4 ALM 79
5.56 mm × 45 NATO – SS109	DM11	MEN 97
	M855	HK 89
	SS109	FNB 83
		2 MI 01
		20 RG 10
7.62 mm × 51 NATO ball	DM41	2 MEN 03
	C21	-
	M80	9 SFM 86
		LC YZ 65 505

The DGA-Land Systems laboratory also evaluates ballistic protection for vehicles in accordance with STANAG 4569 [4] Vol. 1. The ammunition chosen is also common for KE1 level of this standard to have a common acceptation procedure for our management quality system.

4. AMMUNITIONS CARACTERISTICS

STANAG2920 [1] specifies a minimum hardness for ammunition with a steel core, but nothing for lead cores. In addition, although not specified by STANAG2920 [1], the chemical composition or the mass of the constituent elements of the ammunition can explain differences in mechanical or ballistic characteristics. Therefore, we carried out metallurgical analyses and hardness analyses for all the ammunition.

4.1 Core analysis

Inductively Coupled Plasma Mass spectrometry (ICP) and elemental analysis for the elements Carbon and Sulfur determined the chemical composition of the steel core. We analysed with Energy Dispersive X-ray Spectroscopy (EDS) and Scanning Electron Microscopy (SEM) the soft core. Figure 1 gives details of analysis for one batch. Table 2 gives the results for the soft core of the 7.62 Ball in mass percentage, with the associated calculated error. Table 3 gives the same kind of results for the steel core of SS109 ammunition.

 https://doi.org/10.52202/078352-0052

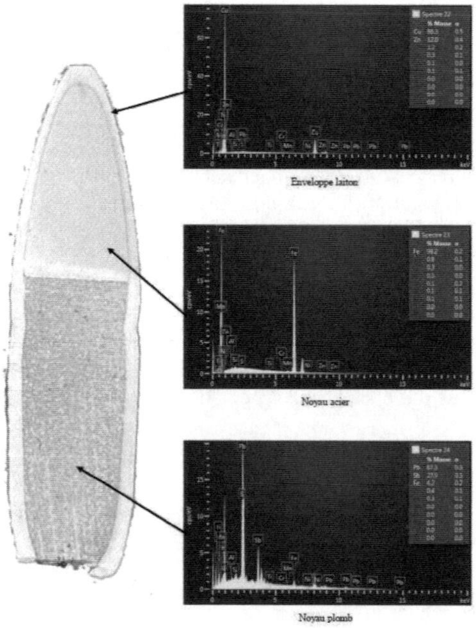

Figure 1. Chemical composition of 5.56mm×45 SS109 – 2 MI 01

For these SS109 batches, it is interesting to note that the steels are different. FNB83's steel core is compatible with C35E steel according NF EN 10083-2 norm. 02IMI01's steel core is compatible with C45E steel according NF EN 10083-2 norm. 20RG10's steel core is compatible with A508 Grade 1 steel according ASTM A 508/A508M-04a norm. The same kind of heterogeneity is observable between the several components (jacket or core) of the several batches for each kind of ammunition.

Table 2. Mass percentage and associated measurement error - Soft core - 7.62mm×51 – M80

	Sb	Pb	Cu	Fe	Al	P	Si
9 SFM 86	**3,55**	**87,45**	**0,23**	**0,43**	**3,48**	**1,13**	**3,28**
	0,14	*1,08*	*0,04*	*0,10*	*0,91*	*0,09*	*0,22*
LC YZ 65 505	**9,37**	**78,20**	**0,24**	**0,94**	**6,58**	**1,02**	**3,00**
	0,22	*1,01*	*0,04*	*0,12*	*0,91*	*0,08*	*0,20*

Table 3. Mass percentage and associated measurement error - Steel core – 5.56mm×45 SS109

	Fe	Mn	Cr	Si	Al	Ti	Zn	Ni
FNB 83	**98,9**	**0,5**			**0,4**	**0,1**		
	0,2	*0,1*			*0,1*	*0,1*		
2 MI 01	**98,2**	**0,9**		**0,3**	**0,3**	**0,1**	**0,1**	**0,1**
	0,2	*0,1*		*0,0*	*0,0*	*0,1*	*0,2*	*0,1*
20 RG 10	**98,3**	**0,9**	**0,2**	**0,2**	**0,3**			
	0,2	*0,1*	*0,1*	*0,1*	*0,1*			

4.2 Ammunition hardness

Hardness measurements, Vickers (HV2, 2kg load) for steel core, or Brinell (HB1 / 2.5, 1 mm diameter ball and 2.5 kg load) for soft core, were made at different points on the jacket and on the projectile core. As example, Figure 2 presents the measurements for the SS109 ammunitions. At last five measurements were made to get the hardness at different points for each core.

https://doi.org/10.52202/078352-0052

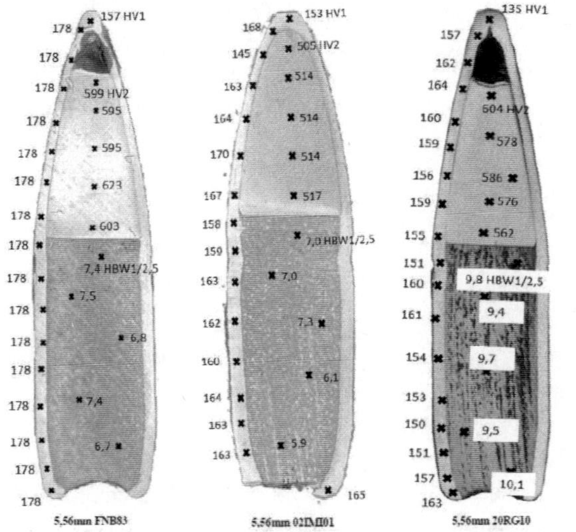

Figure 2. Hardness of core and jackets for 5.56mm×45 SS109

Table 4 summarizes the hardness average for all the batches. Between two batches of the same kind of ammunition, we can observe until 15% of difference for the hardness of the steel core, or between 26% and 40% for hardness of the soft core.

For the SS109 with a mild steel core, STANAG 2920 [1] recommends a hardness greater than 40HRC, all our batches respect this prescription.

Table 4. Hardness measurement for all ammunitions

Ammunition	Designation	Batch	Hardness	
			HB1/2.5	HV2
			Soft core	steel core
5.56 mm x 45 M193	M193	*LC 10H105 013*	8,84	
		4 ALM 79	6,52	
5.56 mm x 45 NATO – SS109	DM11	*MEN 97*		
	M855	*HK 89*		
	SS109	*FNB 83*	7,16	603
		2 MI 01	6,66	512,8
		20 RG 10	9,7	581,2
7.62 mm x 51 NATO ball	DM41	*2 MEN 03*	5,84	
	C21	-		
	M80	*9 SFM 86*	6,4	
		LC YZ 65 505	9,8	

5. AMMUNITION BALLISTIC CARACTERITICS

The ballistic performances of an ammunition can be evaluated from different tests: Depth Of Penetration, Vi/Vr tests, VLP tests, V_α , … [8]. In order to have a relevant comparison, the targets used must be composed of a reference target: armoured steel, reference targets, etc. We have chosen to determine the characteristics of our ammunition by a V_α test on a metal target. This test gives a protection velocity for a probability of protection α. DGA Land Systems implemented two test campaigns with the means and teams of the unit. The processing of shooting data was carried out by the PROBIT method and an evolution of this method. These two treatments are presented below.

The 5.56mm×45 M193 ammunition was evaluated only against 6mm of BLD armour steel. The other munitions, 5.56mm×45 SS109 and 7.62mm×51 O, were evaluated on both 8 mm BLD armour steel and 25 mm BLAL1 armour aluminium.

5.1 Test apparatus

Ammunitions were fired from a pyrotechnical launch system. The firing distance was about 10m. All the munitions were charged specifically to get the attempted velocity on target. The ammunition speeds

https://doi.org/10.52202/078352-0052

were determined by a set of 6 optical screen. The attitude (yaw) of the projectile was measured for each shot by the Projectile Obliquity Measurement system (POM), an orthogonal photographic system. The position of the target was perpendicular to the line of fire and was controlled for each set up. The target was moved horizontally or vertically at each shot to get a new point impact to keep an orthogonality between the target and the line of fire.

5.2 Targets

We have chosen metallic targets. Two types of alloy were chosen: a BLD type armour steel complying with the NFA36-800 standard [5], it is equivalent to an Rolled Homogeneous Armour, .and a BLAL1 type armour aluminium (7020) according to the NFA50-800 standard [6]. They have a minimum ballistic limit imposed by these standards against a 7.62mm×51AP for the thicknesses that we have used, those limits are validated by specific ballistic tests after casting. We used thicknesses of 6 mm and 8 mm for BLD armour steel and 25 mm for BLAL1 armour aluminium.

5.3 Test protocol

The ammunition was fired to frame the expected V_{50} velocity. We carried out series of 15 to 20 shots for each batch of ammunition. In order to distinguish a protection from a perforation, we used a witness control system corresponding to STANAG2920 [1]. This witness consisted of a sheet of aluminium alloy 0.5 mm thick (AlCuMg compliant with ISO / R209 standard and having at least a tensile strength limit of 440 N/mm²). The shots selected complied with the recommendations of the STANAG2920 [1] on influencing parameters: distances from the edge, speed range, yaw, temperature, etc....

5.4 Balistic limit determination

The definition of a ballistic limit for ammunition differs depending on the standard used. This ballistic limit designates a threshold to simplify the transition phenomenon between "protection" and "non-protection". Indeed, the perforation of a protection by an ammunition is a physical phenomenon comprising a random nature that can be quantified by a dispersion. Several statistical methods exist to determine this mixed zone. We determined this transition zone by a series of shots framing the average value. Different statistical methods make it possible to process this sampling of shots to determine the protection limit speeds. We chose to determine this transition zone by two methods: the PROBIT method of the STANAG 2920 [1], and a modified PROBIT method evaluating an uncertainty for each speed of the transition zone.

5.4.1 PROBIT approach

The PROBIT approach [7], described in STANAG 2920 [1], proposes to model this phenomenon by a normal probability density *f* in order to estimate the mean, denoted V_{50}, and the dispersion, noted σ :

$$f(v) = \frac{1}{\sigma\sqrt{2\pi}} e^{-\frac{(v-v_{50})^2}{2\sigma^2}} \tag{1}$$

The probability obtained with regard to this law corresponds to the probability of perforation of the ammunition according to the velocity of impact v of the ammunition on the protection. The mean V_{50} therefore represents the speed at which the probability of perforation, $P_{perforation}$, (or respectively of protection, $P_{protection}$) is equal to 0.5. As a reminder, the probabilities of perforation and protection are dual, formally:

$$P_{perforation} = 1 - P_{protection} \tag{2}$$

The unknown parameters V_{50} and σ are estimated on the basis of N perforation results at different speeds of impact of the munition on the protection studied. The PROBIT method proposes the maximization of a likelihood law. By generalizing the concept of V_{50}, the values V_α, representing the velocity at which a perforation is caused with a probability α, can be deduced from the previous estimates (V_{50}*, σ*) and normal law fractiles, formally:

$$v_\alpha = V_{50}^* - u_\alpha \cdot \sigma^* \qquad \text{for} \qquad \alpha = [0,1] \tag{3}$$

https://doi.org/10.52202/078352-0052

In this equation, u_α represents the normal law fractile associated with the probability α. An example of the result of implementation of the PROBIT method with deduction of the risk curve V_α is represented Figure 3 with the probability of protection for a 5.56mm×45 SS109.

N°	Speed target m/s	Protection (y/n)
1	931,9	n
2	893,9	n
3	864,8	y
4	865,2	n
5	873,2	n
6	863,4	n
7	889,1	n
8	832,1	n
9	844	y
10	851,4	y
11	848,5	n
12	801,6	y
13	819,9	y
14	829,8	y
15	842,3	n
16	843,9	y

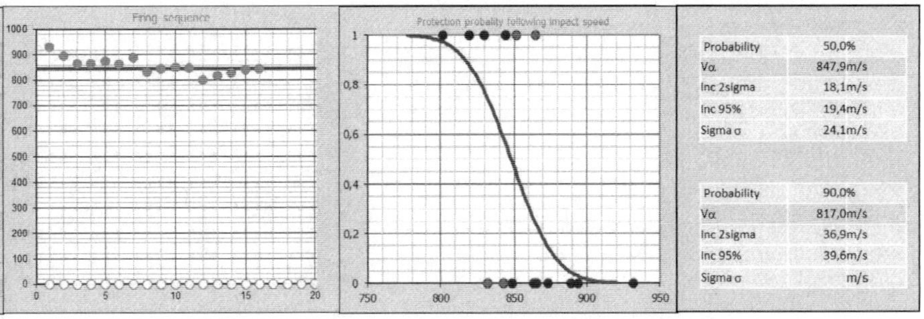

Figure 3. Test round for 5.56mm×45 SS109 - 02IMI01 / BLD 6 mm – Protection probability

5.4.2 PROBIT modified approach

The STANAG2920 [1] proposes the calculation of confidence intervals for the parameters V_{50} and σ. However, nothing about the V_α confidence interval calculation is discussed, while uncertainty on the risk curve is an important element when assessing protection. This uncertainty is a function of the number of shots fired and the intrinsic dispersion of the physical phenomenon of perforation. The proposed approach is to use the two-dimensional likelihood law L (V_{50}, σ) determined using the test data as the Monte-Carlo generation law of the parameters V_{50} and σ. Thanks to the use of normal law fractiles, an empirical distribution of V_α is thus obtained on the basis of the two empirical distributions of V_{50} and σ from Monte-Carlo simulations. The lower and upper bounds of the bilateral confidence intervals on V_α are deduced by identifying the 2.5th and 97.5th percentile of the empirical distribution obtained. This process must be reproduced for any α varying from 0 to 1 with a sufficient resolution step in order to obtain good resolution of the confidence envelope associated with the risk curve. The number n of Monte-Carlo simulations must be reasonably high in order to ensure the convergence of the approach (for example n = 105). An example of the result of calculating the risk curve and the associated 95% confidence envelope is shown Figure 4 for the perforation probability for the 7.62mm×51.

We therefore systematically determined the V_{50}, 50% of protection probability, and the V_{90}, 90% of protection probability, for each munition from a series of shots composed of 15 to 20 shots at different speeds.

Shot number	Target speed	Yaw	Protection	Fair
-	m/s	°	y/n	y/n
65	833,7	0,734	n	y
66	764	0,814	y	y
67	780,3	0,4	n	y
68	780,7	0,708	n	y
69	748,8	2,195	y	y
70	767	0,839	y	y
71	752,1	2,525	y	y
72	772,8	1,938	y	y
73	769,3	0,456	n	y
74	757,1	1,472	y	y
75	771	0,995	n	y
76	756,6	1,365	y	y
77	795,3	0,224	n	y

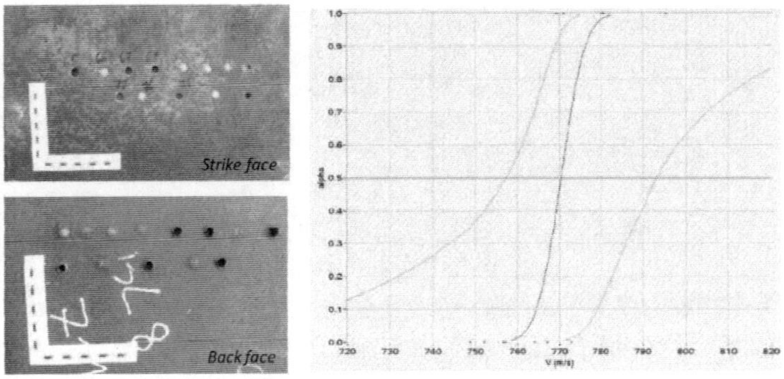

Figure 4. Test round for 7.62mm×51 C21 / 6 mm BLD – Perforation probability

5.4.3 Comparison for the two approaches

These two methods gave us the same V_{50} and V_{90} for draws with shots giving a mixed zone of results around the average value. The differences noted between these two methods during this test campaign are less than 10m/s on few shooting sequences having little shooting with speeds giving both protection

 https://doi.org/10.52202/078352-0052

and non-protection. We present Table 5 here the results of the two methods for the 7.62mm×51 ammunitions face to BLD.

Table 5. Comparison of both approaches for 7.62mm×51 bullet on 8 mm BLD

	7.62 x 51 NATO Ball											
	DM41			C21			M80			M80		
	7.62 DM41 - 02MEN03			7.62 C21			7.62 M80 - 9SFM86			7.62 M80 - LCY762505		
	Campaign 1			Campaign 2			Campaign 1			Campaign 1		
	BLD 8mm			BLD 8mm			BLD 8mm			BLD 8mm		
VLP PROBIT 2920	V50	774,74	m/s	V50	770,37	m/s	V50	781,41	m/s	V50	795,41	m/s
	Inc 2sigma	6,32	m/s	Inc 2sigma	5,49	m/s	Inc 2sigma	9,65	m/s	Inc 2sigma	-	m/s
	Inc 95%	6,78	m/s	Inc 95%	6,05	m/s	Inc 95%	10,14	m/s	Inc 95%	-	m/s
	Sigma	6,42	m/s	Sigma	4,66	m/s	Sigma	9,25	m/s	Sigma	11,18	m/s
	V90	766,51	m/s	V90	764,40	m/s	V90	769,56	m/s	V90	781,08	m/s
	Inc 2sigma	10,90	m/s	Inc 2sigma	9,16	m/s	Inc 2sigma	12,81	m/s	Inc 2sigma	-	m/s
	Inc 95%	11,69	m/s	Inc 95%	10,08	m/s	Inc 95%	13,46	m/s	Inc 95%	-	m/s
VLP PROBIT modified	V50min (95%)	750,51	m/s	V50min (95%)	757,08	m/s	V50min (95%)	767,00	m/s	V50min (95%)	770,52	m/s
	V50	774,73	m/s	V50	770,37	m/s	V50	781,39	m/s	V50	791,97	m/s
	V50max (95%)	792,01	m/s	V50max (95%)	793,08	m/s	V50max (95%)	805,00	m/s	V50max (95%)	808,02	m/s
	V90min	695,92	m/s	V90min	714,37	m/s	V90min	695,45	m/s	V90min	698,85	m/s
	V90	766,54	m/s	V90	764,41	m/s	V90	769,61	m/s	V90	779,34	m/s
	V90max	777,42	m/s	V90max	779,57	m/s	V90max	787,91	m/s	V90max	792,09	m/s

5.5 Ballistic results

The results obtained for 5,56mm ammunition are presented in the Figure 5 to 7 below.

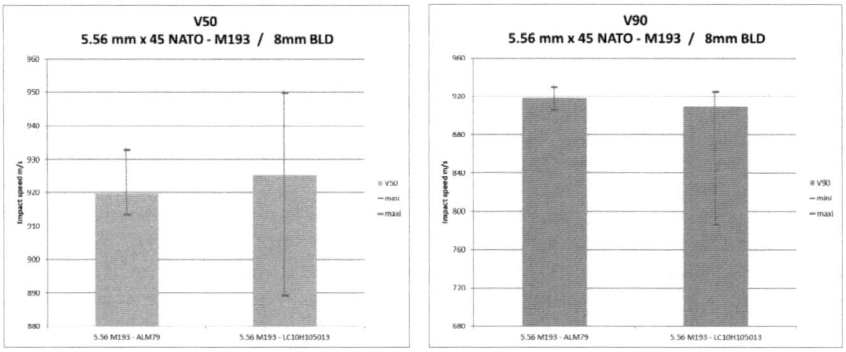

Figure 5. Results for 5.56mm×45 M193 on 8 mm BLD with uncertainties.

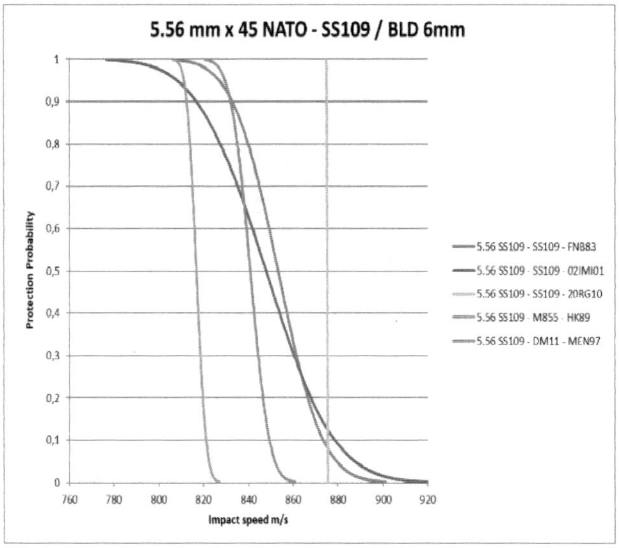

Figure 6. Results for 5.56mm×45 SS109 on 6 mm BLD

https://doi.org/10.52202/078352-0052

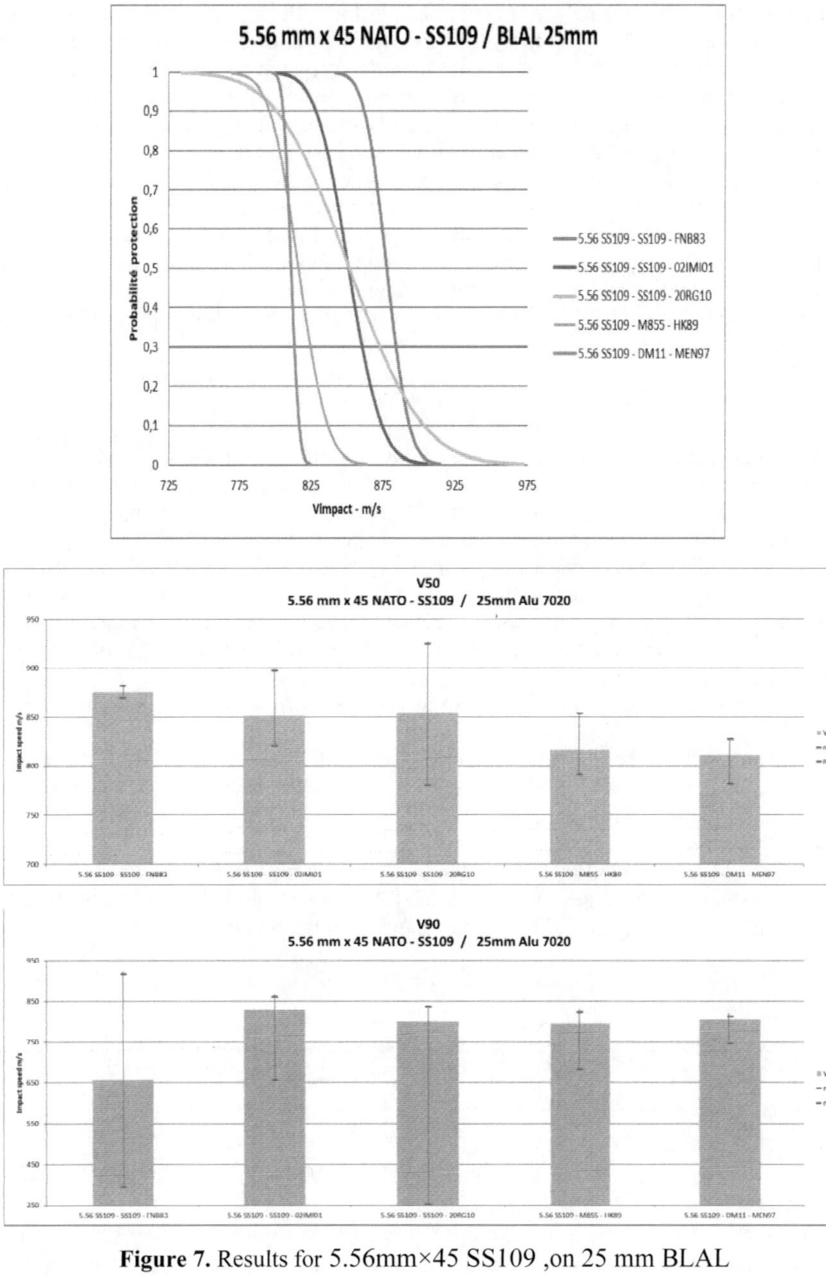

Figure 7. Results for 5.56mm×45 SS109 ,on 25 mm BLAL

5.6 Discussion of ballistic results

The first observation is the "crossing" of the different transition curves. The order of the V_{50} for the different munitions is not the same as the order of the V_{90}. Thus, performance approaches based on the V_{50} or V_{90} separately would not define the same ammunition as the best performing or the least vulnerable. Locally, for the 5.56mm×45 SS109 for example, differences depending on the batches appear:

- Faced with aluminium, batch FNB83 has a higher V_{50} (> 50 m/s compared to the "weak" batches), however, this result is mitigated by a V_{90} at lower than all the other batches.
- Facing steel, the batch HK89 has the weakest V_{50}. However, for the V_{90}, this batch is as efficient as the others are.

These results come from only one series of batch shots. It is a very small sample. A second sequence of shots would probably give different results under the same conditions.

 https://doi.org/10.52202/078352-0052

However, by trying to compare the uncertainties for the ballistic limits obtained from our unique draw, we can get an idea of the high and low limits for each munition: they all seem in the same interval. The uncertainty intervals for our V_{50} and V_{90} values overlap respectively. We can deduce that, without additional firing campaigns to refine our results, macroscopically the batches of ammunition have substantially the same performance in V_{50} and V_{90} on steel armour and aluminium armour.

Depending on the types of targets (steel or aluminium), the bullets do not have the same order of severity. The nature of the target influences the "severity order" of our batches. We believe that the order would be different for another target technology: ceramic materials, composite materials, multi-layers … In addition, protection assessments are carried out on new concepts that by nature have never been assessed; and so we cannot preselect the most sever ammunition for this kind of material. Also, it seems to us more important to have continuity of evaluations than to isolate the most conservative ammunition for metallic targets, this ammunition could become a weak selection ammunition for a target made of another material.

6. VALIDATION OF A BATCH OF AN AMMUNITION

The metallurgic analysis gave us different steel for the same kind of ammunition, and the hardness can be very scattered between two batches ; also those two methods don't seem us adapted to validate a batch in continuity with our reference batch. In addition, those analysis are not easy to realize in our ballistic laboratory, and need to be externalise to a competent and accredited supplier.

The hardness analyses can only be indicative between two batches from different suppliers. The core hardness is nevertheless important because of the STANAG2920 [1] recommendation for the bullet from category B, C, and D giving a minimum. We have to respect this minimum hardness core specified for the ISO17025 [3] accreditation. It is necessary to get those results, but of the ballistic point of view, it is just an indication.

With those different considerations, we plan to validate a change of batch of ammunition with a ballistic assessment.

This validation will consist of a series of 20 shots on a BLD steel plate 6 mm or 8 mm thick. This series will allow us to determine a V_{50} and a V_{90}. We wish to validate the batches for V_{50} and V_{90} intervals listed in the Table 6 below. The tolerances envisaged are ± 20m / s. The batch will have to respect these intervals for both the V_{50} and V_{90}.

More tests are planned to confirm those values, particularly the number of time necessary to repeat the protocol of 20 shots. For example, the standards NFA36-800 [5] and NFA50-800 [6] recommend 7 repetitions of their VLP protocol for validate a new batch of ammunition.

Table 6. Required V_{50} and V_{90}

Munition	Target	V_{90}	V_{50}
5.56 mm × 45 M193	BLD 8 mm	910 m/s	930 m/s
5.56 mm × 45 NATO – SS109	BLD 6 mm	830 m/s	850 m/s
7.62 mm × 51 NATO ball	BLD 8 mm	750 m/s	780 m/s

7. CONCLUSION

The results presented above allow us to propose a ballistic limit, V_{50} and V_{90} values for batches of ammunition used for the ballistic protection assessment in accordance with STANAG2920 [1]. The proposed values will be validated by new tests: same batches on different sheets (other casting), and different batches on the same sheets (same casting).

This method has the advantage of giving access to a ballistic laboratory the possibility of validating a batch using shots, and not from a hardness measurement often outsourced and very scattered between different ammunition suppliers. In addition, the validation of the final performance of an ammunition by the characteristics of the target is a practice already recommended in the STANAG 2310 [9] on the validation of small calibre ammunition.

Nota:
The results presented here are not representative of the terminal ballistics performance of batches of ammunition in operational situations. These batches have their own characteristics depending on the operational conditions of implementation: type of weapon, engagement distances, environmental conditions, storage conditions, nature of the target, etc. These results therefore do not allow ammunition

https://doi.org/10.52202/078352-0052

from different suppliers to be compared for a supply assessment; they are only laboratory results with a test mount and specific pyrotechnic charges on a series of restricted shots.

Acknowledgments

Thank to “Banc National d’Epreuve” from Saint-Etienne in France for their support in the furnishing of ammunition.

References

[1] STANAG 2920 - AEP2920 : Procedures for the evaluation and classification of personal armour – Bullet and fragmentation threat - Ed. 3 - . NATO Publication - June 2015.
[2] ISO9001 – Quality management systems – 2015.
[3] ISO17025 - General requirements for the competence of testing and calibration laboratories – 2017.
[4] STANAG 4569 - AEP55 - Ed. C - Volume 1 : Procedures for the evaluating the protection level of armoured vehicles. NATO Publication – 2013.
[5] NFA36-800 - Tôles en acier soudable laminées à chaud pour blindage – AFNOR – 2015.
[6] NFA50-600 Tôles en alliage d'aluminium soudable laminées à chaud pour blindage – AFNOR – 2014.
[7] D. Finney, Probit analysis: A statistical treatment of the sigmoid curve, Cambridge University Press, Cambridge, United Kingdom (1964).
[8] Jonas A. Zukas, Impact dynamics (1982).
[9] STANAG 2310 – AEP97 : Multi Calibre Manual of Proof and Inspection for 5.56 mm, 7.62 mm, 9 mm and 12.7 mm. NATO Publication.

 https://doi.org/10.52202/078352-0052

The Development of the f-BTTR and its use for Hard Armour Testing

N. Shewchenko[1], E. Fournier[1], T. Bayne[1], S. Magnan[1], and D. Bourget[2]

[1]*Biokinetics and Associates Ltd., 2470 Don Reid Drive, Ottawa, Ontario, Canada K1H 1E1, Shewchenko@biokinetics.com*
[2]*Defence Research and Development Canada (DRDC) – Valcartier Research Center, 2459, Route de la Bravoure, Québec, Québec, Canada, G3J 1X5, Daniel.bourget@drdc-rddc.gc.ca*

Abstract. The development and use of the BTTR (Blunt Trauma Thoracic Rig) has been presented during previous PASS conferences. There were two key issues with its use: a) the membrane deformation was measured only at one point which limits the capacity of the system to measure the transient peak behind armour deformation, and; b) the cylindrical shape of the BTTR prevented the correct support of rigid armour. In order to solve these two issues, the f-BTTR (Flat - Blunt Trauma Thoracic Rig) was developed. The f-BTTR system now enables 3D transient deformation measurement of the backface deformation. Because it is flat, behind armour boundary conditions are easier to control. This paper describes the design and features of the f-BTTR along with its response to generic impacts. Its 3D LDT (Laser Displacement Transducer) based instrumentation is also described and insights into its spatial and temporal resolutions are provided. The development of a test method for the assessment of BABT for hard armour is also presented and includes an analysis of the f-BTTR membrane response for different boundary conditions. The f-BTTR membrane response is finally compared to ballistic clay deformation data for the same armour/projectile/boundary condition combinations. Advantages of the f-BTTR having 3D transient deformation measurement capability include the ability to assess additional response metrics that may be indicative of injury risk including the loading area, volume of deformation and shape of the deformation profile. Initial results of these capabilities are discussed in relation to their accuracy and implications for injury assessment.

1. INTRODUCTION

Early assessments of thoracic behind armour blunt trauma (BABT) for non-penetrating armour impacts have been based on the measurement of the residual deformation in ballistic clay used to support the armour system. While this method is still widely used for assessing armour penetration resistance and blunt impact performance, the limited biofidelity and ability to assess injury risk has been questioned over the past few decades [1]. More recent efforts to further employ ballistic clay for armour performance studies have been made associating the deformation characteristics of clay, e.g. depth, width, volume) with energy/momentum transfer as an indicator of blunt trauma potential [2, 3]. A further study of clay deformation under ballistic re-enactments involving law enforcement survivor cases indicated poorer correlation of clay deformation and volume with injury classifications in comparison to energy-based methods [4]. It is recognized that while static measurements of clay deformation may be indicative on injury potential, they are not necessarily complete nor do they represent the dynamic loading mechanisms that have been related to blunt injury outcome such as the Viscous Criterion (VC) [5]. Alternative approaches using deformable backing systems having biofidelic responses and the ability to measure the dynamic response are perhaps better suited to assessing thoracic blunt trauma such as the Dstl/DLO BABT rig [6] and the 3-RCS [7]. These surrogates are capable of measuring some of the dynamic deformation characteristics to better represent the biomechanics of injury.

The Blunt Trauma Torso Rig (BTTR) was developed to predict the risk of blunt trauma for defeated ballistic impacts onto armour systems and for direct impacts from kinetic energy non-lethal weapons (KENLW) [8. 9. 10]. The design was based on the average human male chest anthropometry with the ability to measure the inner dynamic chest wall deformation at the mid-sternal level to match post-mortem human subject (PMHS) and animal studies conducted at that time. Its biofidelity originally conformed to the PMHS force-deflection data for behind armour impact conditions approximated by a rigid baton impacts [11], and to the animal surrogate (porcine) responses from baton impacts [12, 13]. Injury assessments with the BTTR were based on a review of multiple dynamic metrics for the anticipated loading regime with the Viscous Criterion (VC) being proposed as the most applicable [14].

While the BTTR demonstrated excellent repeatability (SD <7%), better biofidelity and required less testing time than ballistic clay [15], improvements were desired to better characterize the dynamic deformation (e.g. depth, area, volume) for use with energy based injury criteria and to further improve its applicability in assessing armour performance. Specifically, improvements were desired to; a) overcome the single point for dynamic deformation measurements which limited the ability of the system to assess behind armour blunt trauma as deformation shape and volume were not available, and; b) to simplify the cylindrical shape of the BTTR to better support rigid armour plates and provide more realistic and repeatable performance evaluations. The BTTR test setup in Figure 1 depicts the armour system supported by the curved membrane, and when struck by a defeated projectile, the back-face

https://doi.org/10.52202/078352-0053

deformation is measured at the interior surface of the membrane by a single point laser deflection sensor that is coaxial with the projectile point of impact. Membrane impact locations are not constrained to the centre but must account for edge constraints that may affect response. Rotation of the membrane is also possible to achieve obliquity or for changing the armour impact location. Rotation also permits quick repositioning of the armour on the membrane in case of perforation or damage.

Figure 1. The first generation BTTR.

The cylindrical shape of the BTTR was based on averaged chest breadth and approximated curvature to allow for correct armour fit. However, the use of hard armour plates on the BTTR introduced an air gap between the armour back-face and membrane surface in contrast to the fully supported condition seen with typical ballistic clay infills. This motivated further investigation of the BTTR's response and injury assessment capabilities.

2. STUDY OF ARMOUR SUPPORT

A numerical assessment of hard armour fit on six human subjects was conducted for a variety of configurations including the curved Canadian Forces BRP (Ballistic Resistant Plate) and a flat plate of equal size. Three degrees of plate penetration into the thorax (2 mm, 5 mm, 10 mm) were investigated to approximate the plate support conditions with underlying soft armour systems. It was observed that there was greater contact area for the flat plate, regardless of penetration, with supported areas of 26 cm^2 to 319 cm^2 for compressed penetrations of 2 mm to 10 mm respectively across subjects. The plates were typically supported at three contact points, as depicted in Figure 2. The greater contact area for the flat plate suggested that a better fit would be obtained with a flatter supporting surface.

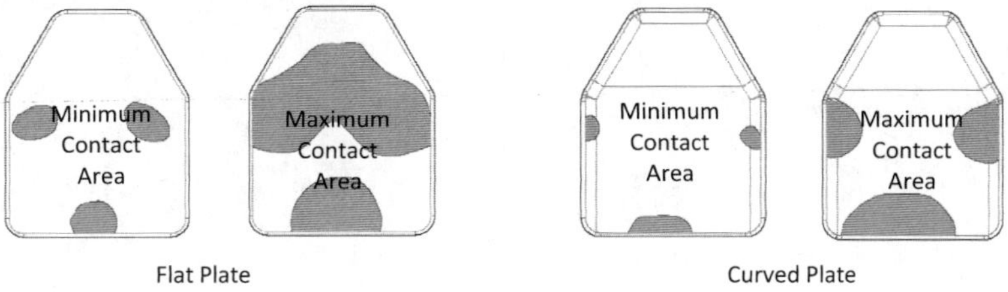

Figure 2: Typical ballistic plate support areas (shaded) for a single subject.

3. F-BTTR DEVELOPMENT

The design of a new flat torso surrogate, the flat-BTTR (f-BTTR), would overcome limitations of single point deflection measurements while providing more realistic ballistic plate support conditions and increased ease of testing. The flat design would also overcome sensitivity of the computed injury metrics due to armour and/or shot misplacement relative to target. For example, the cylindrical shape of the BTTR was noted to require a positioning accuracy better than 3 mm horizontally and 1.5 mm vertically to keep VC errors below 5% for direct impacts thereby requiring close muzzle-to-target distances [16].

The f-BTTR design consists of a flat membrane having inner test surface dimensions of 686 mm wide by 686 mm tall (27" x 27"), comparable to the standard ballistic clay backing configuration of armour performance standards, Figure 3(a). Rear-face support conditions of the mounted ballistic plates could encompass air backed, partially and fully infilled conditions with membrane add-ons. An integrated backfill was also implemented by moulding the shape of the armour plate into the membrane to further facilitate testing. Two horizontal rails were installed in front of the f-BTTR to facilitate

 https://doi.org/10.52202/078352-0053

positioning and support of the backing filler and armour system. Support clamps on the rails allowed for straps to be used to suspend the infill and soft armour as well as providing adjustable feet to support the hard armour plate along the bottom edge. Finally, two horizontal straps straddle the hard armour at the test site and foam blocks were sandwiched between the strap and armour plate to keep it secured.

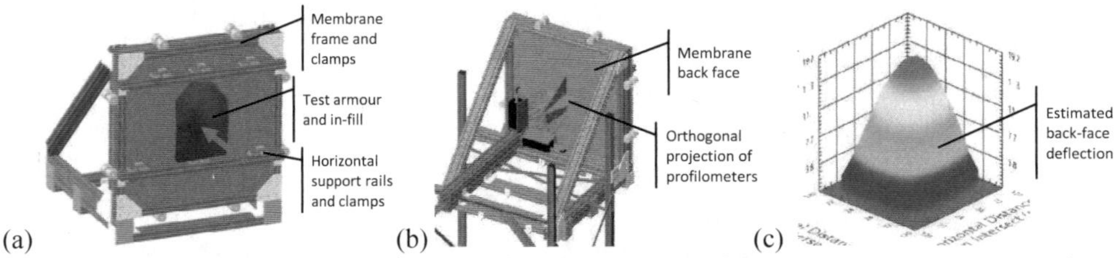

Figure 3: The f-BTTR system; (a) membrane setup, (b) backface measurement system, and; (c) estimated deformed 3D surface.

Measurement of the membrane backface deformation for BABT assessment was approximated in three dimensions (3D) with a Laser Displacement Transducer (LDT) comprising two laser profilometers (Keyence LJ-V7300, ±0.3 mm at 8 kHz, 290 mm depth range, 110-240 mm measurement width) placed orthogonally to each other and offset from the line-of-fire in case of perforation, shown in Figure 3(b). The deformation profiles are captured synchronously for each timestep and are used to create 3D surfaces from the control profiles through scaling of the first profile when propagated along the second profile, Figure 3(c). Custom software was developed to compute metrics associated with injury assessment methods including peak bulge deflection, velocity, volume, and width in two directions. All profile data was transformed to the membrane's coordinate system prior to computation of the metrics. The displacements were smoothed with a cubic spline to remove background noise and velocities were computed employing a two-point central method to remove noise from the differentiation process. Depiction of a typical f-BTTR ballistic impact on a hard and soft armour system can be seen in Figure 4: f-BTTR time histories for; a) peak deflection, b) computed velocity, c) bulge width, and; d) bulge volume..

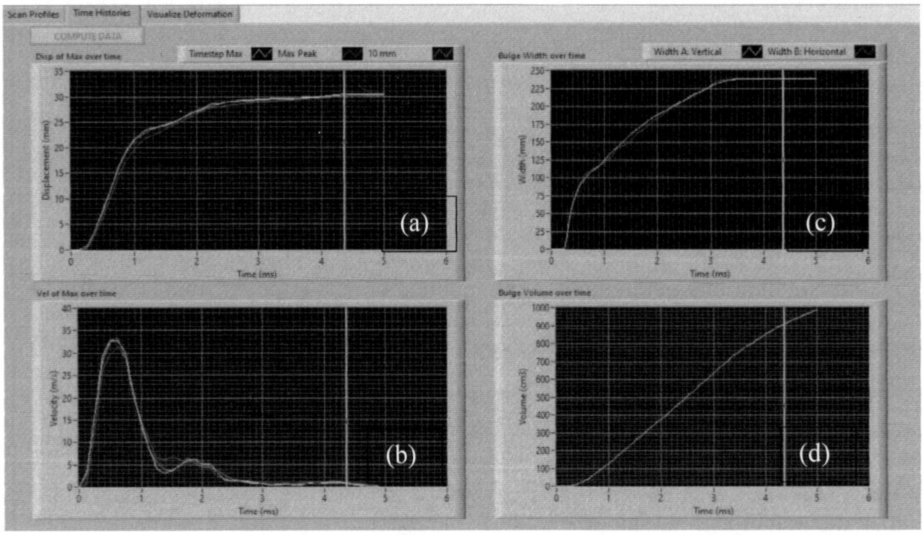

Figure 4: f-BTTR time histories for; a) peak deflection, b) computed velocity, c) bulge width, and; d) bulge volume.

4. BIOFIDELITY TARGETS

The biomechanical response of the f-BTTR was based on PMHS studies [17, 18, 19] to establish the biofidelity of the membrane under impact conditions representative of behind armour loading and direct

https://doi.org/10.52202/078352-0053

impacts from kinetic non-lethal projectiles. A range of target response deflections was defined for each case as detailed in Table 1.

Table 1: Biofidelity response targets for the f-BTTR.

Body Region [Ref.]	Impactor	Mass (g)	Vel (±2 m/s)	Target Deflection (mm)
Thorax [17]	Baton 37 mm dia.	140	40	45-65
Abdomen [18]	Baton 37 mm dia.	45	60	26-34
Thorax [19]	Lacrosse Ball 65 mm dia.	215	27	30-42

A study of different f-BTTR membrane materials and thicknesses was carried out under the various impact conditions to determine the best configuration that would satisfy the biofidelity targets. Furthermore, the peak deflection response variation across the test surface from centre to periphery (100 mm or 4 in from the edge) was assessed with observed differences of up to 17% relative to the centre location when tested. Typical deflection time histories across impact conditions are presented in Figure 5. The average peak membrane deflections were below the lower bound target for the 140 g - 40 m/s baton impacts by 11% and generally met the remaining deflection targets for the 45 g – 60 m/s baton and the 215 g – 27 m/s Lacrosse ball impacts (a 48 g baton was used in place of the 45 g specification for the abdominal corridor). Figure 5 also shows that the membrane's deformation velocity during the first few milliseconds generally meet the requirements for the 3 response targets.

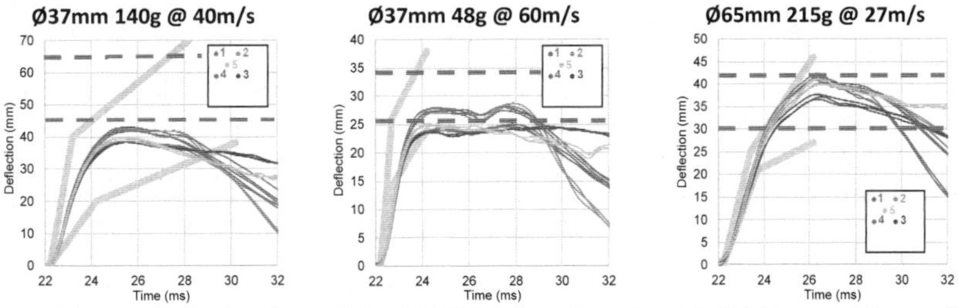

Figure 5: Deflections of the f-BTTR plotted against biofidelity corridors.

5. ACCURACY ASSESSMENT

Initial evaluation of the f-BTTR's accuracy was assessed under a variety of conditions listed in Table 2 with three repeats each. A single high-speed video camera was use for comparative purposes and was skewed from the line-of-fire due to obstructions from the membrane support frame with introduced errors of <1%. Data was collected at 8 kHz for the f-BTTR profilometers and at 3 kHz for the video without any smoothing operations applied. All impacts with the batons implemented a single layer of Kevlar® to prevent surface abrasions.

Table 2: Test conditions for evaluating the accuracy of the f-BTTR.

Loading Condition	Strike Velocity (m/s)	Deflection Error
Ø37 mm, 140 g rigid baton	20	5%
Ø37 mm, 140 g rigid baton	40	-2%
9 mm FMJ bullet, NIJ Level IIA soft armour	373	7%
7.62 mm C21 bullet, NIJ Level III hard armour	847	19%

The ballistic tests with the 9 mm FMJ bullet were carried out with the soft armour supported on the f-BTTR membrane, whereas the combined soft and hard armour were tested without infill for the 7.62 mm C21 bullet. Bullet strike was centred with the membrane and ballistic plate. The error estimates

https://doi.org/10.52202/078352-0053

for the averaged peak data is presented in Table 2 where the results for the 9 mm bullet compared well but less so for the 7.62 mm bullet due to the low sampling rate of the video. Further difficulty in assessing the bulge deflection from video persisted due to the low spatial resolution compared to the profilometers, 1.4 mm vs. 0.6 mm.

The surfacing algorithm which combined the two profiles into a continuous surface was also identified as a possible source of error for off-target impacts. The algorithm's susceptibility to error was assessed by numerically shifting the measured profiles obtained from baton impacts laterally up to 48 mm (1.9 in). Errors of 1%-3% were observed from the peak reference deflection. Differences in backface shape could result in different observations, such as in the case of pencilling with soft body armour, but this affect has not been investigated.

Additionally, the sampling rate of the profilometers was investigated to determine its effect on sampling resolution, accuracy and measurement width. The Keyence LJ-V7300 profilometers are capable of sampling at 8 kHz, 16 kHz and 32 kHz but the higher sampling rates are only made possible by downgrading the measurement width (240 mm to 120 mm) and deactivating the integrated noise reduction features. As a result, the higher rates experienced interference from the intersecting profilometer not seen at the lower sampling rate. The final recommendation was to use an 8 kHz sampling rate to provide reliable scan data while preserving the maximum measurement width.

6. HARD ARMOUR BOUNDARY CONDITIONS

An initial study of armour support conditions was carried out with the f-BTTR to determine the effects of having a soft and hard armour system supported by; a) an air gap between the hard and soft armour, b) an add-on polyurethane (PU) infill to fully support the soft/hard armour system, and; c) an add-on polyurethane infill partially supporting the soft/hard armour system. A typical setup is presented in Figure 7. The partial support was provided at three points with the uppermost corners supported by 50 mm diameter pucks and the lower centre portion by a 120 mm diameter puck similar to that shown in Figure 2 for the curved plate. Impacts were conducted with 7.62 mm NATO Ball rounds (147 grain) striking the target at a nominal velocity of 847±9.1 m/s at the centre of the armour plate. The deformations were sampled at 8 kHz with the profilometers and the peak values extracted without smoothing. The results of the limited test series are presented in Table 3: Test results of hard and soft body armour systems on the f-BTTR and ballistic clay for different support conditions. where larger peak deformations, lower peak velocity and greater deformed volume are observed for the polyurethane infill conditions compared to the air gap condition. The response of the membrane for the fully supported versus partially supported ballistic plate is similar.

Comparison of the air gap and fully supported conditions of the f-BTTR was also conducted with ROMA Plastilina® clay backing and infill. The same 7.62 mm NATO Ball round ballistic impact conditions were used and the clay met the pass-fail criteria for the ball (steel, 50.8 mm diameter, 1.043 kg) drop indentation depth of 19 ±2 mm with no individual value greater than 21 mm or less than 17 mm. The maximum clay indentation depth, volume and surface area of the indentation was obtained from a 3D surface scan acquired with a structured white light scanner (DAVID Pro 3D, ±0.3 mm) and post processed using PTC Creo software for surfacing and volume estimates. The deformed surface was compared to a baseline surface scan of the clay taken prior to each test. Scanned images of the backface deformation from the clay block and extracted volume are shown in Figure 6.

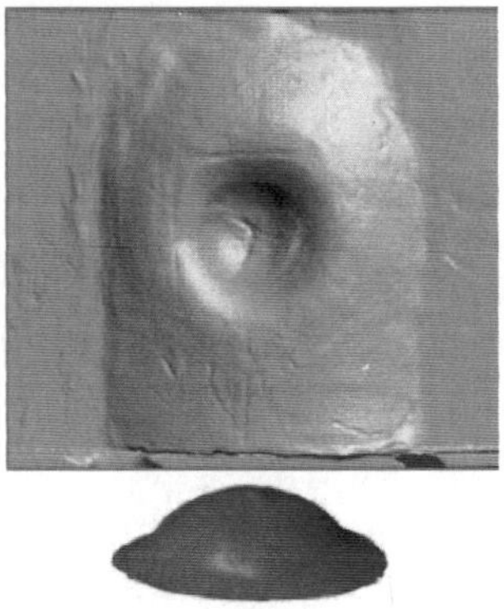

Figure 6: Scanned images of the behind armour indentation in situ and extracted.

Results of the ballistic clay tests are presented in Table 3: Test results of hard and soft body armour systems on the f-BTTR and ballistic clay for different support conditions. which shows that the peak deformation and volume of the indentation are larger for the tests with the clay infill versus the edge supported condition with air gap. The results also indicate that the indentation is less in clay than for the f-BTTR with air gap but greater for the fully supported clay infill condition. It can also be noted that volumes between the clay backing and f-BTTR are very different due to translation of the f-BTTR membrane that occurs for impacts to the hard armour plates. This contrasts with the clay backing which has a higher stiffness and is fully supported by the containment frame. Furthermore, the f-BTTR measures the interior membrane wall deflection and not the back-face of the armour when using clay so through-thickness compression of the membrane is not accounted for.

https://doi.org/10.52202/078352-0053

Table 3: Test results of hard and soft body armour systems on the f-BTTR and ballistic clay for different support conditions.

Test Device	Armour Support Condition	Test #	Test Sample	Impact Velocity m/s	Max. Indentation mm	Max. Velocity m/s	Max. Volume cm^3
f-BTTR	Air Gap / Edge Supported	1	Soft + Hard Armour	846	18.6	23.0	396
f-BTTR	Fully Supported PU Infill	2	Soft + Hard Armour	851	24.1	20.0	610
f-BTTR	Partially Supported PU Infill	3	Soft + Hard Armour	849	22.8	17.6	542
		4	Soft + Hard Armour	841	23.3	18.7	598
		5	Soft + Hard Armour	839	23.9	20.2	672
Clay Block	Air Gap / Edge Supported	6	Soft + Hard Armour	841	15.1	N/A	40
Clay Block	Fully Supported Clay Infill	7	Soft + Hard Armour	848	31.3	N/A	113

A second investigation of the f-BTTR membrane response with different armour infill conditions was carried out and compared to ROMA Plastilina® clay, Figure 7. The tests described in Table 4 were conducted with 7.62 mm NATO Ball rounds (147 grain) striking the target at a nominal velocity of 847±9.1 m/s with minimum shot distances of 60 mm to the edge or 120 mm from adjacent shots. As in the previous study, all tests were conducted at the recommended sampling rate of 8 kHz for the profilometers but enhanced with the data smoothing techniques for deflection and velocity noted earlier. The use of a light block-out curtain also eliminated any external influence from ambient illumination that could contribute to data fluctuations and drop-out by the profilometers. Furthermore, similar controls on ballistic clay consistency were met between shots for meaningful comparisons with the f-BTTR.

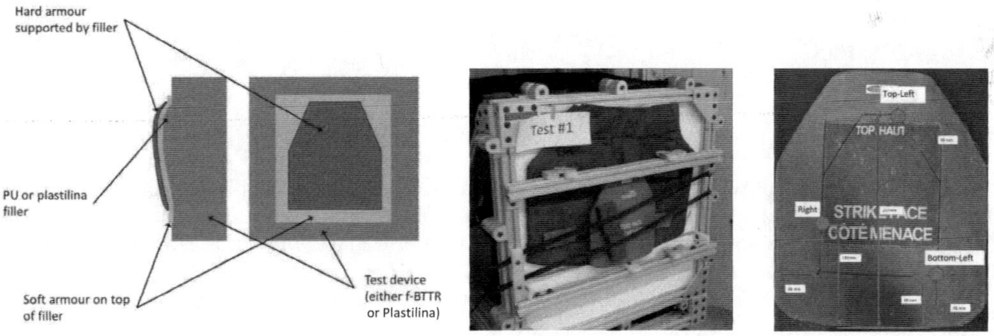

Figure 7: Hard and soft armour layup on top of the backing filler and shot locations.

Table 4: f-BTTR test matrix for different armour support conditions.

Test Device	Soft and Hard Armour Support Condition	No. Samples	Shot Locations
f-BTTR	separate polyurethane infill	5	upper left, right side, bottom left
f-BTTR	integrated polyurethane infill	5	upper left, right side, bottom left
Ballistic Clay	ROMA Plastilina® infill	5	upper left, right side, bottom left

https://doi.org/10.52202/078352-0053

The maximum clay indentation depth, volume and surface area of the indentation was again obtained with a structured white light scanner (DAVID Pro 3D) and post processed using PTC Creo software. The deformed surface was compared to a baseline surface scan of the clay taken prior to each test.

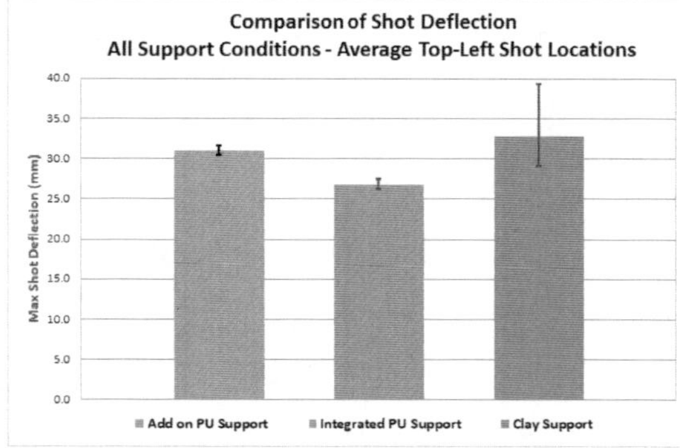

Figure 8: Comparison of f-BTTR deflections.

A comparison of the add-on and integrated polyurethane back-fill elements showed lower deflections and peak velocities for the integrated back-fill condition due to the higher flexural stiffness. Comparisons were not possible across shot locations due to varying armour geometry and differences in response, but similar trends were noted. Additionally, permanent back-face deformation of the armour plates prevented intimate contact with the f-BTTR membrane for subsequent shots, unlike ballistic clay which maintains better contact. Deflections for the top left shot location of the armour plate are shown in Figure 8 with error bars representing the minimum and maximum values. The standard deviations for all test locations across the five hard-soft armour samples were 4% for the f-BTTR with the add-on infill, 6% with the integrated infill and 13% for the ballistic clay infill. An ANOVA on the membrane response with add-on polyurethane back-fill showed that except for the top-left (first) and the right (second) shot deflection and volume, all other values are statistically different. In contrast, similar analysis for clay showed no statistical difference between the 3 shot responses. This is due to the very large standard deviation values associated with the clay results. This also indicates that the f-BTTR is more sensitive to variations in armour support.

Comparison of the shot deformation volume and area could not be made at this time due to bulge width exceeding the measurement width capacity of the current profilometers. While strikes onto soft armour alone would normally be within the capacity of the f-BTTR's measurement system, the hard armour plates resulted in larger width deformations due to a combination of local deformation and plate translation/rotation into the supporting membrane.

7. DISCUSSION

7.1 Biofidelity

The f-BTTR was shown to generally comply with the deflection targets and response corridors with some improvement required to meet the 140 g - 40 m/s baton data [Bir 2000] having peak deflections 11% below target. While this can be realized with a thinner membrane or material change, the current f-BTTR possesses the correct trends to match the biomechanical responses without significant discontinuities and, as a result, it should be possible to develop suitable transfer functions for proper assessment of injury risks across BABT and KENLW loading conditions. It is recommended to limit impacts to the central area for better consistency relative to edge or peripheral impacts. Increasing the lateral extents of the membrane may reduce the edge constraint effects.

It will be important to validate the biofidelity of the f-BTTR against a larger scope of biomechanical studies with PMHS and animal studies as identified by Bourget [14] to increase the relevance and sensitivity of the injury assessments to varying loading conditions. Additionally, compliance of the f-BTTR with the surrogate validation requirements for biofidelity and Viscous Criterion (VC) injury assessment targets presented in NATO STANREC 4744 AEP 99 [20] would be required for use with non-lethal projectiles.

7.2 Accuracy

A preliminary study on the f-BTTR sensor accuracy was conducted for baton and bullet strikes. General agreement for peak deflections was obtained for the batons and 9 mm bullet but not for the 7.62 mm NATO Ball rounds. This is not necessarily a limitation of the f-BTTR sensing system but more a result of the low sampling rate and low spatial resolution of the high-speed video system used. A more complete assessment could be carried out with a Digital Image Correlation system employing two higher

https://doi.org/10.52202/078352-0053

speed cameras with better resolution. Such systems would provide independent validation of the peak deflections, bulge widths and volumes along with computed membrane wall velocities. Any comparisons would have to be conducted through separate tests due to light interference and will require several tests to achieve statistical significance.

The two profile scanners used in the f-BTTR were configured to operate at their upper limits for performance. For achieving sufficient temporal resolution when capturing ballistic events, compromises between spatial and temporal resolution were required. Sampling at rates higher than the recommended 8 kHz would result in half of the available measurement width and increased ambient light sensitivity which reduces measurement reliability and trigger certainty. A black-out curtain was also shown to improve reliability in all cases but prohibits the simultaneous use of external illumination sources such as those used with high-speed cameras. Further limits of the f-BTTR profilometers restricts data width and volume analysis to moderate deformation widths (< 240 mm) which may be exceeded for some cases. As new profilometer technologies emerge, these limitations will likely be overcome.

7.3 Surface Deformation Approximation

The use of two laser profilometer sensors in combination with the custom surfacing algorithm was shown to accurately assess the peak dynamic deformations. Sensitivity of peak dynamic deformation measurements to off-target impacts was small (<3%) for baton impacts with lateral deviations of 48 mm (1.9 in). The errors may increase with greater lateral deviations or decrease in curvature of the deformation profile but further investigation is required.

7.4 F-BTTR Comparison with Ballistic Clay

Greater consistency in the peak deflections measurements was observed for the f-BTTR across the five armour systems and three strike locations in comparison to clay with standard deviations of 6% and 13%, respectively. Similar observations were made in prior studies with soft armour (Level II - NIJ 0101.04, 9 mm 124 gr FMJ bullets 350±9 m/s) with a standard deviations of 3.0% (N=27) vs. 5.7% (N=26) for the BTTR and clay, respectively [15].

7.5 Surrogate Operation

The flat surface of the f-BTTR was selected to provide more representative support of hard armour plates while simplifying back-face deformation measurements. The use of horizontal support bars offset from the front surface worked well to suspend soft armour with straps as well as supporting hard armour plates.

Initial issues with membrane slippage and resulting sagging were overcome with the use of a robust edge lip and clamp. No residual deformation or degradation of the membrane was observed after 29 successive impacts with the baton (140 g, 40 m/s) on the bare membrane or after strikes with a 147 grain, 7.62 mm NATO Ball round at a nominal velocity of 847±9.1 m/s on a soft and hard armour combination, with either the add-on or integrated polyurethane back-fill. In the absence of bullet perforation, membrane durability is expected to be good and matches that of the BTTR which can be used for many years.

Operationally, pre/post test requirements for the f-BTTR are needed and are proposed to be similar to the built-in 2.2 kg 100 mm dia. pendulum impactor employed in the BTTR. Each pre and post test series consists of five consecutive impacts and are intended to be practical and reproducible but not necessarily representative of ballistic impact conditions. The tests are to confirm that the response of the surrogate is within tolerance before and after testing. Large discrepancies between the pre and post tests would indicate a deficiency with the equipment such as a damaged membrane or instrumentation error.

The operational efficiency of the f-BTTR is expected to be similar to the BTTR where the testing time was shown previously with soft body armour tests to be reduced by 44%, including the pre and post verification tests. This was primarily due to the lack of repairs, calibration tests and measurement time needed when using ballistic clay. With use of the integrated polyurethane infill option, further time reductions are expected due to the lack of effort required to create the infill in contrast to that for ballistic clay.

https://doi.org/10.52202/078352-0053

8. SUMMARY

Development of the f-BTTR was intended to provide enhancements over the BTTR by increasing the behind armour blunt trauma measurement capabilities and to provide more realistic armour support conditions known to affect performance.

Introduction of the f-BTTR 3D transient deformation measurement capability allows for the assessment of additional response metrics that may be indicative of injury risk including the loading velocity, area, volume of deformation and shape of the deformation profile. This should result in a more robust and representative system for assessing armour performance and injury risk. Furthermore, it is noteworthy that the initial results have leveraged the operational benefits of the BTTR in terms of repeatability and ease of use making for a more practical ballistic armour performance test methodology.

Future efforts will be focused on improvements and characterization of the f-BTTR response under varying ballistic and KENLW loading for validating injury risk across a wide range of conditions. The f-BTTR in combination with the operational procedures should result in the development of more relevant armour designs and protection.

ACKNOWLEDGEMENTS

The authors would like to acknowledge the support and funding provided by the Defence Research and Development Canada - Valcartier Research Centre.

REFERENCES

[1] Tobin, L., "Feeling Depressed", Proceedings of the Personal Armour System Symposium (PASS), The Hague, The Netherlands, September 2004.

[2] Bevan, M., Luong, Q., "Correlation between Projectile Kinetic Energy and Displaced Clay Volume for Three Classes of Armour", Proceedings of the Personal Armour System Symposium (PASS), Washington D.C., 01-05 October, 2018, pp 71-79.

[3] Zhang, T.G., Jannotti, P.A., Satapathy, S.S., "Indent depth and volume in the clay backing for soft and hard armour", Proceedings of the Personal Armour System Symposium (PASS), Washington D.C., 01-05 October, 2018, pp 292-302.

[4] Rafaels, K.A., Loftis, K.L., Bir, C.A., "Can Clay Tell Us More Than Deformation?", Proceedings of the Personal Armour System Symposium (PASS), Washington D.C., 01-05 October, 2018, pp 303-311.

[5] Bir, C., Viano, D.C., "Design and Injury Assessment Criteria for Blunt Ballistic Impacts", The Journal of Trauma Injury Infection and Critical Care, Vol. 57, pp 1218-1224, 2004.

[6] Hinsley DE, Tam W, Evison D, "Behind Armour Blunt Trauma to the Thorax – Physical and Biological Models", Proceedings of the Personal Armour System Symposium (PASS), The Hague, The Netherlands, November 2002.

[7] Lyon DH, Bir CA, "Injury Evaluation Techniques for Non-Lethal Kinetic Energy Munitions", U.S. Army Research Laboratory, ARL-TR-1868, January 1999.

[8] Anctil, B., Bayne, T., Bourget, D., Pageau, G., Binette, J-S, Rice, K., Toman, A., "An Alternative to Plastilina for Evaluating the Performance of Body Armours", Proceedings of the Personal Armour System Symposium (PASS), Brussels, Belgium, October, 2008.

[9] Bolduc, M., Anctil, B., "Improved test Methods for Better Protection, a BABT Protocol Proposal for STANAG 2920", Proceedings of the Personal Armour System Symposium (PASS), Quebec City, QC, 13-17 September, 2010.

[10] Bourget, D., Anctil, B., "Test Methods to Evaluate Terminal Effects of Kinetic Energy Non-Lethal Weapon Systems", Proceedings of the International Research Council on the Biomechanics of Injury (IRCOBI), KraKow, Poland, 14-16 September, 2011.

https://doi.org/10.52202/078352-0053

[11] Bir. C., Viano DC, King A, “Development of Biomechanical Response Corridors of The Thorax to Blunt Ballistic Impacts”, Journal of Biomechanics, Vol. 37, pp 73-79, 2004.

[12] Cooper GJ, Pearce BP, “The Biomechanical Response of The Thorax to Nonpenetrating Impact with Particular Reference to Cardiac Injuries”, The Journal of Trauma, Vol.22 pp 994-1008, 1982.

[13] Cooper GJ, Maynard RL, “An Experimental Investigation of the Biokinetic Principles Governing Nonpenetrating Impact to the Chest and the Influence of the Rate of Body Wall Distortion Upon the Severity of the Lung Injury”, Presented at the International IRCOBI conference on Biomechanics of Impacts, 1986.

[14] Bourget, D., Anctil, B., “Overview of Thoracic and Abdominal Blunt Impact Injury Models”, Proceedings of the Personal Armour System Symposium (PASS), Brussels, Belgium, 6-10 October, 2008.

[15] Anctil, B., “Are the Sensor Based BABT Test Methods Suitable for Helmet and Body Armour Certification?”, Proceedings of the Personal Armour System Symposium (PASS), The Netherlands, 19-23 September, 2016.

[16] Robbe, C., Nsiampa, N., Papy, A., “Sensitivity Studies of the BTTR Surrogate and Comparison Between NLW and BABT Applications”, Proceedings of the Personal Armour System Symposium (PASS), Washington D.C., 01-05 October, 2018, pp 504-515.

[17] Bir, C., “The Evaluation of Blunt Ballistic Impact of the Thorax”, Wayne State University, Doctorate Dissertation, Wayne State University, 2000.

[18] Bir, C. and Eck, J., "Preliminary Analysis of Blunt Ballistic Impacts to the Abdomen," in IUTAM Symposium on Impact Biomechanics: From Fundamental Insights to Applications, 2005.

[19] Nathan, D., “Development of a Biomechanical Surrogate for the Evaluation of Commotio Cordis Protection”, Wayne State University Dissertation, 2012.

[20] NATO, STANREC 4744 Ed. 3 AEP-99, “Thorax Injury Risk Assessment of Non-Lethal Projectiles”, Edition A, Version 1, Feb. 2017.

 https://doi.org/10.52202/078352-0053

Fragment Characterisation & Threat Modelling - A Multinational Study to Re-Define & Represent the Fragment Threat

J. Weir[1], B. Shaw[2], K. Pizzolato-Heine[3], S. Ouellet[4], L. Martineau[4], R. McGuire[1], M. Mahoney[5], G. James[1], A. Hepper[1], P. Gillich[6], L. Gant[1], N. Eberius[3]

[1]*Dstl Porton Down, Salisbury, Wiltshire, SP4 0JQ, UK, jeweir@dstl.gov.uk*
[2]*Defence Technology Agency, Devonport Naval Base, Devonport 0624, Auckland, New Zealand*
[3]*Data and Analysis Center, Combat Capabilities Development Command, US Army Futures Command, Aberdeen Proving Ground, MD 21005, USA*
[4]*Defence Research and Development Canada – Valcartier Research Center, 2459 de la Bravoure Road, Québec G3J 1X5, Canada*
[5]*Defence Science and Technology Group, 506 Lorimer St, Fishermans Bend, VIC 3207, Australia*
[6]*Army Research Laboratory, Combat Capabilities Development Command, US Army Futures Command, Aberdeen Proving Ground, MD 21005, USA*

Abstract. The Technical Co-operation Panel (TTCP) Land Group (LND) is a multinational defence science and technology collaboration between the governments of Australia, Canada, New Zealand, United Kingdom and United States. Under the auspices of this organisation, a study group was convened to focus on personnel protection and vulnerability. During recent conflicts in Iraq and Afghanistan, this group highlighted the need to gather relevant evidence of the nature of fragmenting threats injuring military personnel and document the relevant test methods. Use of evidence to define protection standards based upon the wrong threat risked providing the wrong level of protection performance: insufficient protection (underestimation of the 'threat') could result in injury but too much protection (overestimation of the 'threat') could result in burden. In the worst case, overestimation of the threat could result in no mitigation being developed if it were thought no practical solution achievable. Experience generated under TTCP LND Technical Panel 5 (TP-5) brought together a community that could challenge legacy beliefs of the fragment threat and protection requirements using updated evidence. Fragment collection and analysis methods were documented for each nation: fragment origin, preparation, examination, classification and attribution methods. A comparative analysis of the fragment collections from some of the nations was conducted. This provided a summary of the content of the fragment collections for use in setting equipment protection or mitigation standards. A record and assessment of fragment threat simulation methods and equipment was made. Some of the models used to assess and communicate the injurious potential of various threats were reported. The work of the panel led to improved understanding of the fragment threat and evidence gathering capabilities for characterising future fragment threats. This provides some commonality between TTCP nations in the analysis, testing and assessment of fragment threats, as well as ensuring that lessons learnt from recent operations are not lost.

1. INTRODUCTION

1.1 Data Content Caveat

In conducting this study, the team has accessed sensitive personal information. The study has complied with all TTCP and national policies in the handling and use of this information in order to preserve anonymity of personnel undergoing treatment or monitoring. In complying with national policies, some information has been removed from the paper in order to allow for its public release.

1.2 Overview

Under the auspices of The Technical Cooperation Program (TTCP) [1], a study assignment involving five cooperating nations was convened to collate evidence and test methods relevant to fragment injury. The project was undertaken to record the lessons identified, the procedures used, the information gathered and the state-of-the-art for fragment threat characterisation for TTCP nations who had been involved in combat operations in Iraq and Afghanistan.

https://doi.org/10.52202/078352-0054

1.3 Requirement

The study team highlighted the need to gather relevant and up to date evidence of the nature of the fragmenting threats injuring military personnel. This was to ensure that protection strategies are based on current and appropriate assessment of the fragmenting threat, reducing the risk of over or under protecting personnel. The expertise within the panel made this possible and there was an ethical imperative to secure this knowledge for the future.

Whilst the study was conducted with the aim of providing a common understanding and some degree of commonality between the TTCP nations in the analysis, testing and assessment of fragment threats, the aim was not to produce a prescriptive standard. Many instances of national differences were reported.

2. FRAGMENT COLLECTION AND ANALYSIS METHODS

2.1 General

Methods used to collect and characterise fragments from various military operational threats were documented, allowing anyone in the future to reproduce the techniques if the national capabilities become dormant or lost.

2.2 Fragment Collection Methods

Opportunities for collecting fragments were identified as being:

- From the site of the incident (e.g. vehicle, surrounding environment) - by post incident investigators;
- Retained within Personal Protective Equipment (PPE) - extracted by trained PPE analysis teams, predominantly within home-nation laboratories;
- Retained within body tissues - commonly removed by medical personnel during treatment; or
- Removed during service police investigations – removed by the investigators.

Collected fragments were placed in a suitable container and labelled, to ensure each could be linked to the incident from which it was recovered.

There were a number of challenges associated with fragment collection, not least the medical, legal and ethical imperatives related to handling *sub-judice* information and preserving anonymity of personnel undergoing treatment or monitoring.

Immediately after an incident, the focus of personnel varied depending on their role. Those responsible for providing casualty treatment or recovering equipment were often not aware of the potential benefit of collecting and retaining any fragments. Additionally, those who survived a bullet or fragment impact were keen to keep the item as a souvenir. Therefore, one of the key elements to the success of the fragment gathering was the ongoing process of educating the stream of deployed people who may be involved in any element of the process, so that they understood the importance of data gathering in shaping future protection requirements.

2.3 Fragment Characterisation Methods

The key fragment characterisation methods were determination of mass, material composition, size, shape, velocity and estimation of source. The fidelity and complexity of measurement varied between nations, due to resources and capability.

Fragment mass was a key parameter due to the ease of measurement and ability to produce a profile of the fragments collected by mass (Figure 1). Mass was also related to existing Fragment Simulating Projectiles (FSPs) to enable determination of the relevance of surrogates. Due to variations in fragment materials, size and shape, mass was the only consistent parameter between the FSPs and collected fragments.

https://doi.org/10.52202/078352-0054

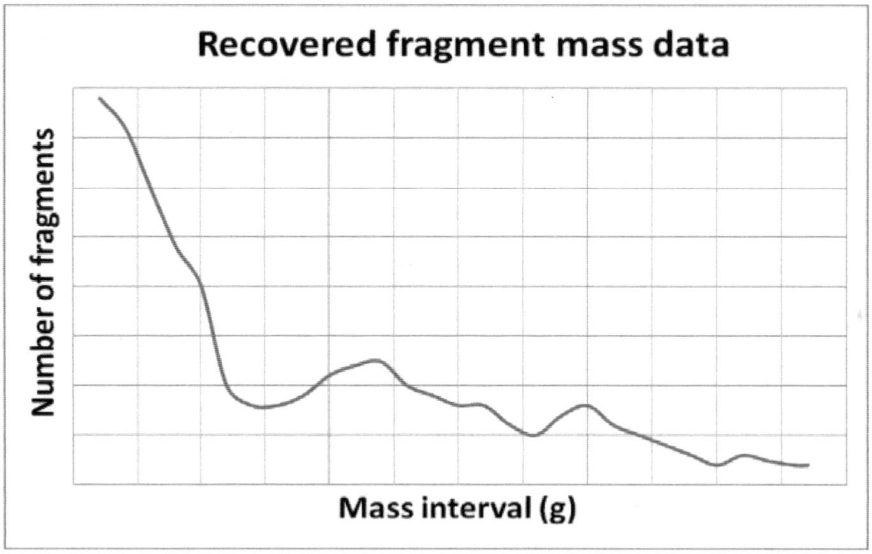

Figure 1. Example of a fragment mass profile based on fragment collections associated with a specific threat. Image © Crown Copyright Dstl

Simple determination of fragment material composition was achieved by visual investigation and by using a magnet. In some nations, qualitative elemental composition was determined using systems such as Scanning Electron Microscopy (SEM) with Energy Dispersive X-ray Spectroscopy (EDS). Detailed metallurgical analysis was achieved for some nations using inductively coupled plasma atomic emission spectrometry (ICP-AES).

The medical community was interested in fragment composition from the perspective of the long-term toxicological implications of any implanted material, which may remain in the tissues of the casualty. Material composition also supported identification of the origin of the fragment and understanding of the threat. Information was also used to determine whether existing FSPs accurately represented the operational threats.

The source of the fragment was determined by subject matter expertise; if the fragment was determined as a part of a component of the threat itself it was classified as a primary fragment. Secondary fragmentation was defined as those fragments that had been energised from the surrounding environment such as from soil, buildings, etc.. It was important to identify the source of the fragment to understand the nature of the threat and its employment. Size was documented for all fragments due to the ease of measurement. This was used in conjunction with weight to understand the distribution of fragments produced by specific threats or to correlate to a known threat type.

A basic level of shape assessment was obtained from size measurements and photographs; detailed investigation of fragment shapes was achieved using 3D scanning, particularly in Canada and the US. This preserved the geometry of a fragment when evidence needed to be returned back to the originating organisation; additionally, shape factor is useful for follow-on vulnerability analysis.

In some cases and nations, fragment velocity was estimated by reverse engineering the outcome: either by using experimental data combined into a look up table to correlate fragment masses, impact velocities and Depths Of Penetration (DOP), or by using numerical injury models, e.g. the Operational Requirement-based Casualty Assessment (ORCA) model developed and maintained by the US Army Research Laboratory. Estimation of fragment velocity provided a further insight into the fragmentation produced by a specific threat and was used to inform Personal Protective Equipment (PPE) testing and protection requirements. In this case, numerous simulations were run, recording depth of penetration for different fragment masses at different velocities, in order to predict the velocity required to achieve a given injury or DOP. An example of this is the prediction of the fragment velocity needed to penetrate different depths into a forearm, shown in Figure 2.

https://doi.org/10.52202/078352-0054

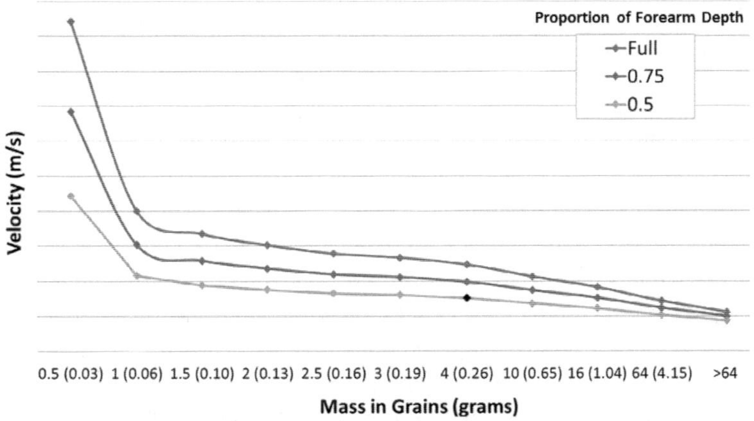

Figure 2. Velocity estimates for fragments recovered from the upper extremity. Outputs from the ORCA model relating to the velocities estimated for different fragment masses to cause different levels of penetration of the forearm
Image © Army Research Laboratory

2.4 Summary

Fragment analysis methods undertaken by TTCP nations varied, due to national capabilities and resources. Fragment threat, mass, size and PPE or tissue location are the four fields of information collected across all of the nations. These are the four easiest measurements to conduct and require minimal resources. The key aspect to success of these methods is documentation of the collection of the fragment, including linkages to the incident to aid determination of the threat.

3. NATIONAL FRAGMENT COLLECTION RESULTS

3.1 Analysis Results

The analysis included the following:

- Number of fragments recovered by each country for each year of conflict;
- Evidence distribution between event types (mounted, dismounted) and threat origins, .e.g. Small Arms Fire (SAF), Improvised Explosive Device (IED), mortar, Rocket-Propelled Grenade (RPG), artillery, grenade and rocket;
- Composition of non-metallic evidence for mounted and dismounted events;
- Primary composition of metallic evidence;
- Mass distribution of metallic evidence – plotted as a cumulative sum of the log fragment mass, to enable the reader to easily extract mass percentiles – also by recovery location & composition;
- Comparison of recovery locations (on the body) of metallic evidence.

3.2 Conclusions and recommendations

Differences in the practices for collection of combat evidence between countries were a limiting factor in some aspects of the analysis, in-particular, that of mass and recovery location.

Recommendations were made as to data fields that should be included in any future analysis. Mandatory fields included:

- Date (dd/mm/yyyy);
- Unique identifier,
- Mass (g);
- Metallic (Y/N);
- Principal material;
- Other materials;
- Recovery location (body/armour/clothing/vehicle/helmet/other);
- Threat class;

 https://doi.org/10.52202/078352-0054

- Status (mounted/dismounted);
- Abbreviated Injury Scale (AIS) body region (1/2/3/4/5/6/7/8).

Optional fields included:

- Source (device/Behind Armour Debris/environment/soldier's equipment),
- Whether armour was worn (yes–missed armour/yes–armour defeated/yes–fragment stopped/no);
- Size/shape of fragment;
- Threat direction (from above / below / front / back / side).

4. FRAGMENT TEST/REPLICATION METHODS

4.1 Overview

A record was made of the various methods that have been used to simulate fragment projection. A selection of key apparatus and methods are described here and their pros and cons summarised. Categories covered are:

- Buried charge assessments;
- Propelled fragments (metallic and non-metallic FSPs);
- Sand Cannon usage (multiple non-metallic fragments).

4.2 Buried Charge Assessments

Three of the TTCP nations have conducted buried charge assessments in the context of PPE testing:

- Starting in 2014, Canada developed a suite of operationally-relevant test methodologies for specific personnel protective equipment. Explosive trials were carried out to characterise the severity of buried charge threats and evaluating the performance of PPE and fabrics;
- The UK conducted a trial to assess dismounted survivability and protective material / PPE performance against Afghanistan type buried IEDs;
- US performed controlled blast trials with conventional explosives and Home-Made Explosives (HMEs) to understand the threat to dismounted personnel from victim operated IEDs.

Trials typically examined the influence of explosive composition, charge mass, standoff, setup, soil test pad dimensions, soil type, granulometry, moisture content, density and burial depths. They also investigated a range of models for assessing fragment injury (see section 5).

A typical trial arena setup is shown in the schematic within Figure 3 below.

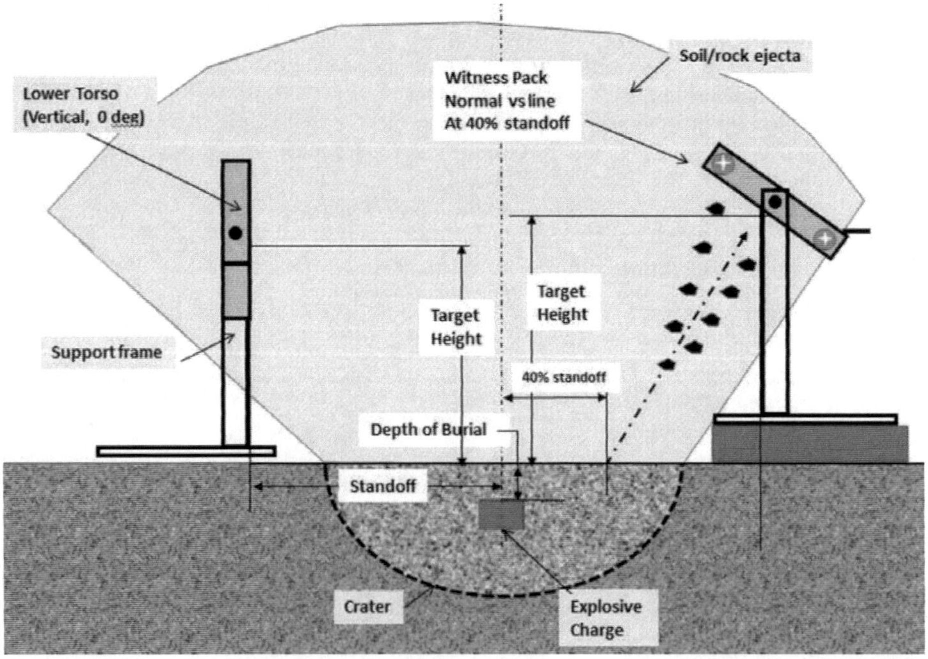

https://doi.org/10.52202/078352-0054

Figure 3. Schematic of a test set-up prior to detonation.
Image © Defence Research and Development Canada

These trials improved TTCP nations' understanding of the key factors to control to achieve repeatable testing, and how to tailor testing to the type of PPE being assessed. For example, soil conditions must be well characterised and controlled in terms of composition, saturation and compaction. Severity of the insult should be repeatable within a test (i.e. it should be radially symmetrical) as well as between tests in the same soil pads. Some specific outcomes were the ability to rank different blast conditions (moisture, burial depth, charge mass, charge composition); knowledge that blast conditions in such tests may be more severe than those survived in theatre and an understanding of which fragment collection models should be applied for which purposes.

All nations noted the resource-intensive nature of this type of trial; they are cumbersome to run and generate large quantities of data, which slows down the reporting of results. This led to Canada developing a more automated analysis method using Computed Tomography (CT) scanning technique to replace manual dissection of targets, post-test [3].

4.3 Propelled Single Fragments (metallic and non-metallic FSPs)

All TTCP nations have some capability to conduct testing with single propelled fragment surrogates in a laboratory environment using pressure housings driven by compress gas or pyrotechnic propellent.

Typical fragments used for testing by TTCP nations are the 0.16 g and 1.1 g Chisel Nosed (CN) cylinder FSPs from Allied Engineering Publication (AEP) Standardisation Agreement (STANAG) 2920 [2] and 1 to 20 mm diameter spheres of steel, ceramic and glass. Less frequently used are the other FSP masses in AEP 2920, cubes of 2 to 10 mm side length (aluminium, steel and tungsten carbide) and spheres of other materials (aluminium and tungsten carbide). The test capability has evolved over the course of recent conflicts, in response to the type of analysis described in previous sections, to include a greater range of velocities and material types. Other 'non-standard' projectiles include stones, soil and irregular fragments (recovered fragments and glass shards).

Methods for launching fragments have evolved too, with a move from pyrotechnically driven weapons, through use of compressed gas cartridges towards custom-designed systems using laboratory gases. This has enabled a greater range of fragment calibres and velocities to be accommodated simply by being able to substitute barrels and adjust pressure; it has also enabled a reduction in associated legislation (no ammunition) and infrastructure (housed in small enclosures, as opposed to ballistic ranges), which has led to faster turnaround times for testing.

Examples of such systems are the Defence Research and Development Canada (DRDC) single-stage gas gun system, which uses nitrogen and helium gas at pressures up to 414 bar and the UK Sabre ballistics gas gun, pictured (Figure 4). These systems have been used to consistently launch projectiles with masses between 0.004-54 grams at velocities from 0-1600 m/s (not across all fragments). DRDC has also developed a set of custom muzzle extension rods to guide the projectile when exiting the barrel in order to ensure stable flight with minimal yaw up to the target.

Figure 4. An example of a modern gas-gun: the UK Sabre gas gun weapon system fitted with an 800 mm barrel
Image © Crown Copyright Dstl

An innovation to bridge the gap between laboratory consistency and blast arena trials is the new gas gun device currently being commissioned by DRDC. This will be capable of simultaneously

 https://doi.org/10.52202/078352-0054

launching up to seven 0.22 calibre FSPs with shot dispersion depending on the barrel length selected (see Figure 5).

Figure 5. DRDC Simultaneous gas gun.
Image © Defence Research and Development Canada

4.4 Sand Cannon Usage

The purpose of a 'sand cannon' is to allow a laboratory based test to bridge the gap between the simple, repeatable, inexpensive but potentially unrealistic single fragment V_{50} test and the complex, expensive, more realistic buried charge assessment.

Some of the TTCP nations have developed methods for firing multiple fragments in a controlled laboratory environment. An example is the UK sand cannon [4]. It is a 20 mm smooth bore barrel fitted with a reduced diameter aperture just after the muzzle. A Low Density Polythene (LDPE) one-piece sabot (20 mm depth and ~1 mm wall thickness) is used to contain multiple fragments. When fired, the sabot walls hit the reduced diameter aperture and the contents continue on at their original velocity.

The UK sand cannon with the 20 mm LDPE sabots can be used to fire approximately seven grams of soil or soil surrogate, which was seen to replicate local aspects of the damage seen from actual buried charges and that from buried charge assessments [4]. The maximum velocity achieved to date is 700 m/s with seven grams of sabot contents, although in excess of 1000 m/s should be achievable using 7.62x51 mm blanks for this mass of projectiles.

The main benefit of this design compared to conventional fragment propulsion equipment is that the muzzle to target distance can be very short, as space for sabot separation is not required as normal for a 2 or 3 piece separating sabot. Muzzle to target distances of 0.5 m to 2.0 m are typical and varying this distance controls the dispersion of the fragments at the target.

Although fully operational, there has been minimal appetite from UK Ministry of Defence (MOD) customers to move away from requesting test fragments other than those in AEP 2920 [2]. There remain some challenges to resolve with the design, should it be adopted for use by test houses, for example, design changes to improve ease of use, separation of sabot and contents and minimise wear on weapon system components.

5. FRAGMENT INJURY MODELS

5.1 Introduction

A variety of methods have been used by TTCP nations during recent conflicts to assess the severity of injury likely to be sustained by personnel for a given threat environment. These methods include legacy models such as the TTCP spaced metal fragment witness pack [5] and strawboard packs [6]. In addition to these, the TP-5 community has generated new methods for their own use (such as witness packs for assessing the effects of non-metallic fragments) and adapted others (such as the use of gelatin in arena trials and mannequins for coverage assessments).

5.2 TTCP Metal Witness Packs

A spaced metal witness pack has previously been developed by a TTCP team for use within physical trials to enable an assessment of the injury to people inside armoured vehicles from a threat that overmatches the vehicle protection. It is often referred to as a Behind Armour Debris (BAD) pack or the 'TTCP metal witness pack'. It is comprised of thin metal sheets, interlaid with 25 mm of expanded

https://doi.org/10.52202/078352-0054

polystyrene. A review of this witness pack was conducted in 2016, which identified new materials to replace those that had become obsolete in the UK. The anti-personnel lethality criterion was also updated.

5.3 Strawboard

Strawboard fragment packs may be deployed in order to characterise both the distribution and typical velocity of sources of metallic fragments. The distribution of fragments is given by the X-Y coordinates of each fragment in the strawboard layer where it comes to rest. The velocity of each fragment is estimated, based on the depth of penetration into the pack. This may then be used to predict a probability of incapacitation using a shot line model.

5.4 Witness Packs for Non-Metallic Fragments

5.4.1 General

Ideally, a witness pack will be able to capture the full range of size and velocity of fragments from a device and store these at a range of depths, without itself being destroyed during the event. Strawboard packs have been shown to be unsuitable for use in buried charge assessments, where many of the fragments are 'secondary' and non-metallic – originating from the soil under which the charge is buried.

TP-5 participants have collaborated to develop models applicable to the characterisation of non-metallic fragment threats, for which other witness packs are not suitable. These models are for use within TTCP nations' national laboratories. Key stages in this process are described below.

5.4.2 UK Multiple Discrete Fragment Physical Injury Model (MDFPIM)

The development of this model was initiated in 2011 following attempts to use other materials and existing models to investigate the injury risk from the penetration of soil and stones projected during testing of buried explosive devices were unsuccessful. The model has been through several design iterations since its initial development and can be applied to scenarios other than just buried charge assessments. The current/most recent documented version is MDFPIM V2.0 (and V2.1).

The MDFPIM V2.0 is based on layers of neoprene foam interleaved with a polythene sheet. The MDFPIM V2.0 allows assessment of:

- The potential for corneal abrasions (a low severity eye injury) by any particles that adhere to the sticky front surface of model;
- The risk of penetrating eye injury, by any perforations to layer 1 of the model;
- The risk of skin perforation, by any perforations to layer 2 of the model;
- The 'severity' of injuries from penetrations deeper into the model, based on estimated impact velocity of the projectile(s);
- Overmatch compared to unprotected cases.

The pack has been deployed extensively within UK and has been very successful in allowing the assessment of potentially injurious projectiles that would not penetrate 'legacy' models. It is calibrated to provide velocity estimates for wide range of projectiles (geometries, diameters, masses and densities) and multiple packs can be tiled to cover large areas or complex shapes.

5.4.3 The 'TP-5 Pack'

Due to material sourcing issues, in 2016, Canada initiated the development of an alternative to the UK MDFPIM in support of a TP-5 study assignment on backing materials for personnel protection. The proposed backing system, provisionally named the 'TP-5 pack', is currently under review by the other TTCP nations and has already been deployed in several trials/studies in Canada. The pack has multiple capabilities and is designed in three versions: the standard version for testing soft armour, the overmatch version for conducting V_s-V_r testing and the witness pack version. The witness pack version (Figure 6) comprises a skin layer (neoprene rubber), several layers of foam and a plywood backing.

 https://doi.org/10.52202/078352-0054

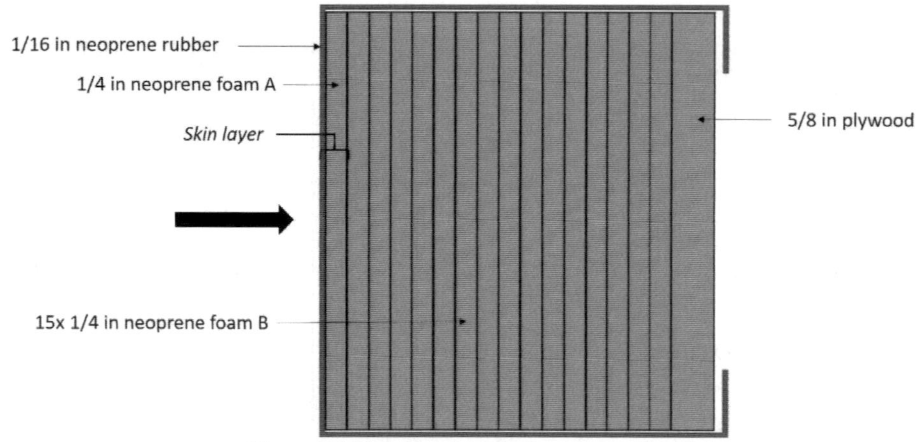

Figure 6. TP-5 Pack in witness pack configuration.
Image © Defence Research and Development Canada

Based on laboratory characterisation data, estimates of the mass, size and velocity distributions of the ejecta cloud can be obtained. The pack has also successfully been shaped in the form of a lower body (pelvis and upper legs) to test pelvic protection systems.

5.5 Other models

TTCP nations have variously experimented with deploying other models, such as tissue simulants and collection media in arena trials using buried explosive charges. Models have included gelatin, clear ballistic gel, mannequins and plywood witness packs [6,7].

5.6 Analysis methods

In order to predict or display injury outcome, a number of analysis methods may be applied to data generated from fragment packs. Two examples of these are the UK-generated Interactive Mapping Analysis Platform (IMAP) and the US Visual Anatomical Injury Descriptor (VAID) tool.

The VAID tool can be used to visualise severity and location of injuries for single or multiple cases, to give a representation of impact frequency distribution for different body regions.

IMAP contains a fixed body model, based on an anthropometrically accurate 50th percentile male and can be used to graphically represent surface impact locations. The tool also enables the geometries of any type of PPE to be imported and overlaid onto the body. This has enabled the effect of PPE coverage to be illustrated. Examples of these are shown in Figure 7.

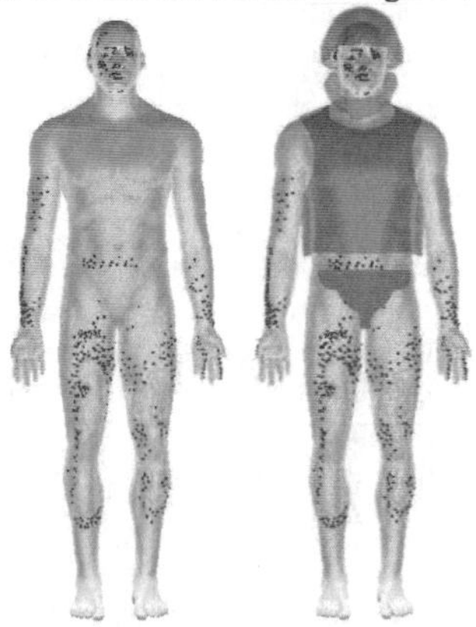

Figure 7. Left: mapped model in IMAP with no PPE. Right mapped model in IMAP with PPE (helmet, soft armour collar, soft armour). Blue dots represent fragment impact locations. Light blue lines indicate a continuously damaged area
Image © Crown Copyright Dstl

https://doi.org/10.52202/078352-0054

6. CONCLUSIONS

The study demonstrates methods for assessing the fragment threat by examining evidence from conflicts. It documents methods for replicating the threat in the laboratory using surrogates, in order that personal protective equipment can be tested against a representative fragment threat. It also documents tools that may be used to assess the effect of the fragmenting threat on personnel, in order for effective protective strategies to be developed.

This work has led to an improved collective understanding among TTCP nations of the fragment threat in recent conflicts. It has also provided methods for assessing fragment threat in any future conflict, ensuring that lessons of the past are not forgotten.

Acknowledgments

The authors would like to acknowledge those who have facilitated access to the evidence base as well as those who have built and maintained the nations' fragment databases over the previous decades. These people include:

- M Clayton, (US Public Health Service) US Army Medical Research and Development Command, Director Joint Trauma Analysis and Prevention of Injury in Combat
- D Jones, Combat Capabilities Development Command Data and Analysis Center, US
- D Lewis, Institute of Naval Medicine, UK
- E Mazuchowski (Lt Col US Air Force), Chief Forensic Services at the Joint Trauma System (JTS) and Medical Examiner for the Armed Forces Medical Examiner System (AFMES)

References

[1] https://en.wikipedia.org/wiki/The_Technical_Cooperation_Program, accessed 26 Feb 2020

[2] NORTH ATLANTIC TREATY ORGANIZATION (NATO) ALLIED ENGINEERING PUBLICATION (AEP) - Procedures for the evaluation and classification of personal armour. Bullet and fragmentation threats, Allied Engineering Publication AEP-2920, Edition A, Version 1, June 2015.

[3] Gabriel, U., Pageau, G., DesRoches, M. Ouellet, S., Baillargeaon, Y., Francus, P., & Shewchenko, N. Terminal Ballistics Application of X-Ray Computed Tomography (XCT) for the Quantitative In-Situ Analysis of Fragments Embedded in Collection Media. In 31st International Symposium on Ballistics, 2019.

[4] James G., and Hepper A., Ballistic simulation of fragmentation from buried improvised explosive devices. Personal Armour Systems Symposium, Robinson College, Cambridge, UK, 2014.

[5] Baillargeon, Y., Lacoursière, P., Sirois, A., Comparative incapacitation study between aluminium-steel and all-steel witness packs used to record behind-armour debris from a shaped charge impact, DRDC-RDDC-2014-R147, 2014, available at: https://cradpdf.drdc-rddc.gc.ca/PDFS/unc198/p802153_A1b.pdf

[6] DEFENCE STANDARD 93-59, Chipboard (For Use at Proof and Experimental Establishments), DEF STAN 93-59 Issue 3, 2005.

[7] Gillich, P., Fragment Penetration Modelling of Anthropometric Ballistic Mannequins, 24th International Symposium on Ballistics, 22 -26 September 2008.

https://doi.org/10.52202/078352-0054

Characterising "Street Knives": A Study of the Tip Sharpness and Penetration Forces for Common Bladed Weapons

A, H Jones[1], I. Elomari[1] and J. Barnes-Warden[2]

[1]*Materials and Engineering Research Institute, Sheffield Hallam University, Howard St, Sheffield, S1 1WB, UK. a.h.jones@shu.ac.uk*
[2]*The Metropolitan Police Service, Physical Protection Group, 60 Albany St, London, NW1 4EE*

Abstract. Data published by the United Kingdom's Home Office (UKHO) details the annual number of UK offences with a knife or sharp implement rose by 44% between 2011 and 2019. Knife crime is prevalent in London, and the knife threat to frontline police officers is significant. It is therefore essential that the threat from "real" blades is defined, and the range of threats present on the street is understood. In order to understand the wide range of real knives typically carried on the streets of London, this paper evaluates a large quantity of knives (n=66) confiscated by the Metropolitan Police. These confiscated knives were compared with new, shop bought, kitchen knives, and the engineered blade (P1B) from the UKHO body amour standard 2017. The penetration capability of the weapons has been measured using quasi-static testing by measuring the forces required to penetrate skin simulants (silicone rubber and foam) formerly used and characterised in forensic studies. These results have been evaluated for correlations between physical parameters, such as tip angle and tip radius, measured using macro-photography and electron microscopy. The lowest force to perforate the skin simulant was recorded for the P1B engineered blade (0.18 kgf). Four new (33% of sample) and four confiscated (6% of sample) knives perforated the skin simulant with less than twice the force required by the P1B. However, the majority of confiscated knives (66% of sample) required more than five times the force of the P1B to perforate the skin simulant. From this data, we have determined a moderately good correlation between the tip geometries of the knives and the loads required to penetrate the skin simulant. The penetration load, L_p, of the skin simulant is proposed as an easily determined and quantifiable measure of knife tip sharpness. Knowledge of the range of sharpness of street knives can be used to assist in future armour design, especially for coverage of hard to protect areas of the torso, head and neck.

1. INTRODUCTION

Data published by the UK Home Office shows that the annual number of UK offences with a knife or sharp implement rose by 44% from 33,669 incidents in 2011 to 47,136 in 2019 [1]. Knife crime is particularly prevalent in London with 179 police recorded crimes per 100,000 population, double the national average for England [2]. In 2019 eighty six fatal stabbings were recorded, many of which are linked to gang violence [3]. Hence, the knife threat to frontline police officers is significant, and therefore it is essential that the threat from "real blades" is defined, and the perforation characteristics are understood. There is also a constant drive to increase the coverage of protection while not overburdening the wearer or making them less effective.

The parameters governing the penetration force of a knife would seem to be related to the knife tip geometry but to date no published study has shown a good correlation between the force required for a knife tip to penetrate in a stabbing action and the geometry of the knife tip. There is also no clear agreement on which parameters relating to the knife tip are most important and how they should be measured. Typically, the "tip angle" the "tip radius", "blade thickness" and the blade's "cutting angle" or "edge angle" are used but these are not universally recognized or defined.

A number of papers have used a limited range of typical bladed weapons such as kitchen knives and determined the forces required to penetrate human cadaveric tissue, simulated human tissue or porcine tissues. A good summary is given by Annaidh [4] and is summarized below in Table 1[1]. There were differences noted between quasi-static tests (QS in table, speed = 0.1 $m.s^{-1}$) and dynamic tests (D in table, speed = 1 $m.s^{-1}$ to 9.2 $m.s^{-1}$) and Annaidh shows that at higher dynamic test speeds of 9.2 $m.s^{-1}$ the penetration force was between 13 % and 20 % lower than the penetration force in quasi-static testing of porcine skin and polyurethane respectively. The penetration forces also varied with the knife type (cook's, carving and utility) with the bluntest knife requiring 85% more force (2.4 kgf) to penetrate skin simulant than the sharpest (1.3 kgf). Much larger differences were observed for "blunt" weapons such as screw drivers and scissors with around 2 to 3 times the force required (2.5 kgf to 3.2 kgf) to penetrate polyurethane or porcine skin compared with the kitchen knives.

Nolan [5] measured the forces required for human participants to stab and penetrate skin simulants with 3 different knives. The knife geometry and the resulting forces for penetrating skin simulant are

[1] Originally report as force in N they have been converted to equivalent loads in kgf for easier comparison with results given later in this paper and for understanding of said forces by non-scientists.

https://doi.org/10.52202/078352-0055

shown in Table 2. Blunt weapons (screw drivers) showed around 4 to 5 times higher forces (63 N to 74 N). The data in the supplementary information for individual knives shows no meaningful correlation between knife geometry and penetration force. The large tip angles and the large standard deviations of +/- 20° from 5 individual knives of the same design is notable.

Hainsworth [6] showed a loose correlation between the "blunt edge radius" of several knives with varying tip geometries and the penetration depth of the knife when stabbing open cell foam with a fixed level of impact force. Below a blunt edge radius of 0.1 mm the penetration was up to100 % greater than knives with a radius > 0.2 mm. For radius from 0.2 mm to 0.7 mm there was no systematic variation in the penetration.

Nolan [7] also studied the ability of 3 types of kitchen knife to penetrate a skin simulant and several types of combinations of clothing (e.g. t-shirts, jeans, jackets). The three kitchen knives had tip angles of 64.8°, 74.1° and 103.5°, but the forces required to penetrate a skin simulant showed little difference (1.1 kgf, 0.97 kgf and 0.99 kgf respectively). However, when penetrating normal clothing the tip angle of 103.5° had a 20 – 30 % higher penetration force and tip angles of 64.8° and 74.1° tip angles showed similar forces.

Table 1: Summary of previous work on forces required to penetrate human and porcine tissues and synthetic materials, adapted from Annaidh (2013).

Test Method	Material Type	Penetration Forces (kgf)
QS, D	Cadaveric	<0.5 (sharp), 3 – 5 (blunt)
QS, D	Cadaveric	<1 (skin), 7 – 10 (clothing)
QS, D	Cadaveric	3.5 – 5.5
QS	Porcine	1.0 – 1.5
QS	Polyurethane (PU)	1.3 – 2.0
QS	Cadaveric, Porcine, PU	1.5 – 1.7 (QS), 1.0 – 1.2 (D)

Table 2: Knife type, Tip Angle and Skin Simulant Penetration Forces adapted from Nolan (2013).

Knife Type	Average Tip Angle, ° (SD)	Average Penetration Force, kgf (Standard Deviation)
All Purpose	120 (20)	1.2 (0.2)
Steak	120 (20)	1.9 (0.2)
Instrumented (Sabatier)	108 (-)	1.3 (0.1)

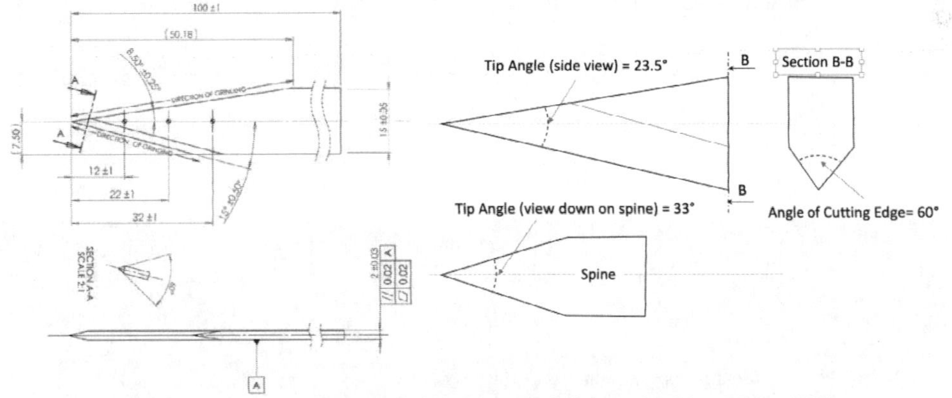

Steel hardness - Rockwell C = 52 – 55 (Vickers Hv30 ~ 550 – 650)

Figure 1: Extract from the engineering drawing of the UK Home Office P1B engineered blade and schematic close-up showing clearer view of tip angles

Assessment of knife resistant amour systems to UK Home Office (UKHO) standards [8] make use of the P1B engineered knife which is machine made blade with high tolerances on its dimensions as shown in Figure 1. It has a highly pointed tip with a tip angle in side view of 23.5°, in a view down on the spine of 33°, a cutting edge angle of 60° and is hardened to a high degree (H_{V30} ~ 600). It has clearly been designed to be a very high threat to the armour. However, data on how it compares in its penetration

https://doi.org/10.52202/078352-0055

ability with the most common threats encountered "on the street", which generally have much larger tip angles, and/or rounded tips, but sharper cutting edges, has, to our knowledge, not been published. A test exists, developed by CATRA[2] in association with UKHO which uses a square section of silicone rubber (Shore A = 40) to determine the force required to push the point of a blade in to the rubber by a fixed distance. The P1B is deemed sharp enough if the force to do this does not exceed 4.5 N (0.46 kgf). However, this test is pass/fail test and does not quantify tip sharpness. Thus, there is a need for test that can accurately determine the "sharpness" of a blade and compare it to that of the standard P1B, while also simulating the materials in to which knifes typically are thrust.

Furthermore, it is of interest to have a statistical measure of the threat presented by the wide variety of "street knives" in the UK and how they compare to the P1B. Data on how the penetration ability of knives differs in terms of cost, quality, design and brand would be of significant interest, as would how such knives that have been in circulation for some time compare with new examples of such knives.

2. EXPERIMENTAL METHODS

2.1 Street Knives

In order to obtain a representative sample of bladed weapons of different types the UK Metropolitan Police Service (MPS) provided access to a large quantity (~2000) of confiscated weapons from which a selection was made that represented the typical range of such confiscated weapons. The choices made were guided by a number of MPS staff who had previously carried out statistical analysis of the types of confiscated weapons. In total 66 confiscated weapons were selected, a sample of which are shown in Figure 2. Additionally, 12 new kitchen knives of various designs and covering a range of costs and quality were purchased (Figure 3). P1B blades were obtained from the MPS or from the manufacturer[3]. A list of the knife categories with a brief description is given below in Table 3.

Table 3: List of Confiscated (C) and New (N) Knives by Type and Blade Length.

N or C	Knife Type	Blade Length (cm)	Description	Number
C	Kitchen - Chef	21 - 23	Wide blade	28
C	Kitchen - Carving	20 - 24	Narrow blade	11
C	Kitchen - Utility	10 - 14	Short narrow blade	7
C	Lock	9	Folding lock knife	7
C	Rambo	23	Large Combat Style	6
C	Combat/Survival	10 - 23	Narrow blade, double edge	3
C	Sword	45 - 47	Straight and "zombie" style	2
C	Filleting	15 - 21	Narrow and tapered	2
N	Kitchen - Chef	21 - 23	Wide blade	4
N	Kitchen - Carving	20 - 24	Narrow blade	4
N	Kitchen - Utility	10 - 14	Short narrow blade	4

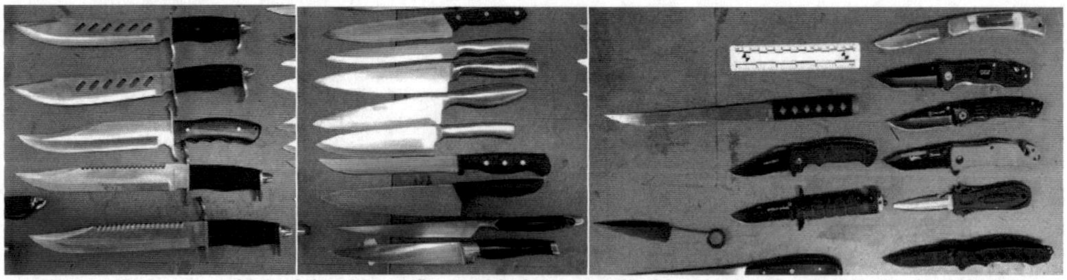

Figure 2: Image of 24 of the 66 confiscated knives supplied by the MPS; Rambo (left), Kitchen (middle), Lock/Combat (right).

[2] Cutlery and Allied Trades Association, Henry St, Sheffield, S3 7EQ
[3] High Speed and Carbide Ltd, Clough Bank, Off Edmund Road, Sheffield, S2 4EL

https://doi.org/10.52202/078352-0055

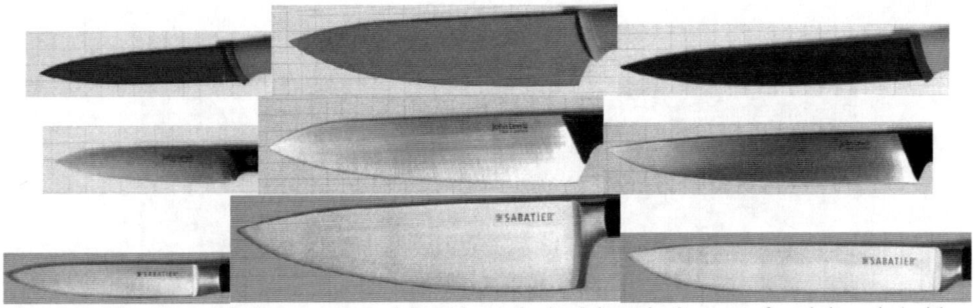

Figure 3: The 12 new knives purchased for this analysis, 3 brands, 3 types of each brand. (grid on paper is 1cm x 1cm)

2.2 Knife Tip Measurement – Optical Methods

A photographic record of all knives was made and macroscopic images of the knife tips were taken for comparative purposes and to measure the side view tip angle (TA). Calibrated digital imaging software was used to measure knife tip geometries. Some examples are shown below in Figure 4. The tip angles (side view tip angle, TA) were measured for all knives at the scale shown in the examples in Figure 4. Where there was an obvious radius to the tip in this view, this was measured and recorded as the tip radius (R_T). Where no obvious flat, radius or damage existed at the tip its condition was recorded as "sharp" meaning it was essentially triangular on the scale of the optical magnification used (estimated $R_T < 0.1$ mm). Where there was an obvious round tip it was recorded as "rounded" and the approximate radius (R_T) measured in mm. Tip thickness was measured with a micrometer at approximately 1mm from the tip.

2.3 Knife Tip Imaging – Electron Microscopy

For safety and practical reasons the end 50 mm of blade was removed from the body of the knife. To observe the detail of the knife tips at higher magnification the ~ 50 mm knife tips were mounted for observation by scanning electron microscopy (SEM) and were observed in two orientations, (i) corresponding to the optical side view but with a slight tilt of the sample to observe the tip fully and (ii) looking down on the knife's spine at the tip to observe the spine tip angle and radius. (SA and R_S). The same magnifications or fields of view were used to capture photographs of the 3-dimensional form of the knife tips and to make measurements of the knife tip geometry using the calibrated SEM software.

2.4 Penetration Force Testing

The material penetrated by the knives was a skin simulant which gives reproducible results that are comparable with those of human skin [Nolan 2013]. A 2 mm thick sheet of silicone rubber with a Shore A hardness of 40 with a self-adhesive backing was adhered to a 100 mm x 100 mm x 100 mm cube of open cell foam[4]. The testing was carried out on a Universal Materials Tester (UMT-2) machine made by CETR (now Bruker) using a load cell with a maximum load of 10 kgf. The load cell and knife tip were located in the upper moving carriage and the skin simulant block was placed below and secured in place with double sided adhesive tape. The penetration speed used was 1 $mm.s^{-1}$ and the travel of the knife (z) was set at 25 mm from the point of initial contact between the silicone and knife tip. Each knife was pushed in to the skin simulant block 5 times, each time in a new location. The load (kgf) and displacement (mm) were recorded during the entire loading cycle. The penetration load (L_p) of the skin simulant was judged as being the first significant event on the force versus displacement curve and was usually characterized by a drop in the force and a change in slope of the line after the event. For some sharp knives (e.g. the P1B) the event was small and hard to observe but nevertheless could be discerned by close inspection of the data. For some of the blunter knives no penetration occurred by a displacement of 25 mm and in these cases the displacement was allowed to continue up to a limit of 40 mm. Some blades did not penetrate even at 40 mm and the load cell reached its maximum value of 10 kgf as the foam was compressed. Average standard deviations from 5 measurements was +/- 7% of the load.

[4] A01 polyether open cell foam, density = 23 to 28 $kgf.m^{-3}$, hardness 125 to 155 N, Acoustafoam Ltd, Shropshire

https://doi.org/10.52202/078352-0055

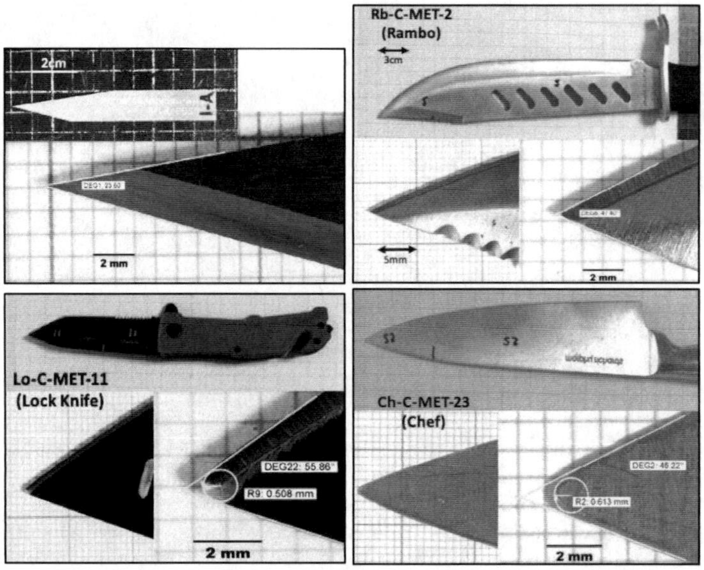

Figure 4: Photographs and macroscopic images of confiscated knives showing the measurement of tip angle (TA) and tip radius (R_T). P1B blade (top left), Rambo with sharp point (top right), Lock knife with rounded tip (bottom left) and Chef's knife with rounded tip (bottom right).

3. RESULTS

3.1. Optical Measurement of Knife Tip Geometry

The data for tip angle (TA), tip radius (R_T) and the blade thickness (t) at 1 mm from the end was measured for all 66 confiscated knives and all 12 new knives. Table 4 shows a comparison of TA and t for a selection of confiscated and new knives.

Of the 66 confiscated knives 23 were judged to be sharp (essentially triangular with R_T <0.1 mm), 27 were judged to be rounded (R_T>0.1 mm) and the remainder (16) were damaged, bent or truncated so that no meaningful measurement could be obtained. Figure 5 shows the distribution of tip angles (TA) for all sharp confiscated knives (23) and the distribution of the tip radius values for the round tipped confiscated knives (27). Of the 12 new knives 3 were judged to have rounded tips.

Table 4: Optically Measured Tip Angle and Tip Thickness for Confiscated (C) and New (N) Knives.

N/C	Knife Type (number)	TA (average), °	TA min/max, °	t (average), mm	t min/max, mm
N	Chef's (4)	55	49/60	0.5	0.5/0.60
N	Carving (4)	55	35/65	0.5	0.4/0.6
N	Utility (4)	52	42/72	0.5	0.4/0.6
C	Chef's (27)	54	33/80	0.7	0.4/1.1
C	Carving (10)	44	37/54	0.9	0.5/1.5
C	Utility (8)	42	30/60	0.7	0.4/1.2
C	Lock (7)	52	29/65	0.9	0.6/1.4
C	Rambo (5)	43	37/47	1.1	1.0/1.3

https://doi.org/10.52202/078352-0055

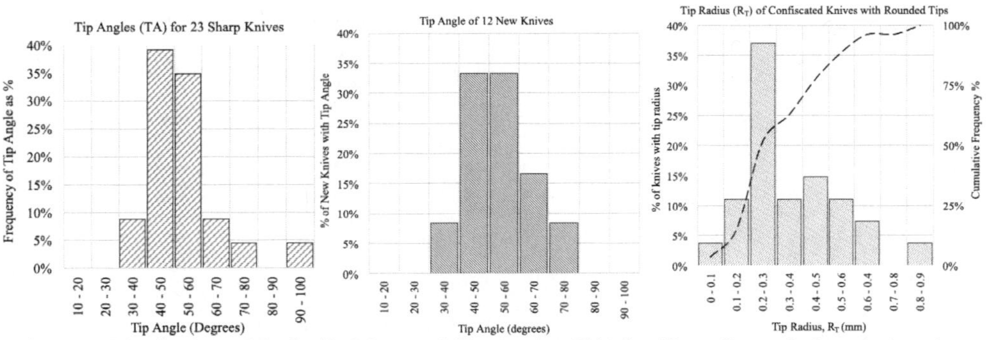

Figure 5: Distribution of Optically Measured Tip Angles (TA) for 23 confiscated, sharp knives (top left) and Tip Angle of 12 New Knives (bottom left). Tip Radius (R_T) of the 27 confiscated knives with rounded tips (bottom right).

3.2 Penetration Loads of Knives using Skin Simulant

The UMT instrument was used with all 66 confiscated knives and the 12 new knives as well as the P1B engineered blade. The distribution of L_p across all knives is shown as histograms in Figure 6. Some examples of the typical load – displacement plots that were used to determine L_p are shown in Figure 7. The average L_p for the P1B was 0.18 kgf +/- 0.014 kgf. Note that 11 of the confiscated knives did not penetrate at the maximum load of 10 kgf or maximum displacement of 40 mm. Across all confiscated knives L_p ranged from 0.31 kgf to 2.5 kgf which is between 1.7 and 14 times that of the P1B. Each knife style (e.g. Cooks, Carving, Rambo, etc.) showed significant variation, for example L_p for the 5 Rambo knives ranged from 0.39 kgf to 1.74 kgf (Figure 7 right). Confiscated kitchen knives had a large range and included some of the lowest (L_p = 0.32 kgf) and highest (L_p = 2.45 kgf) results.

New kitchen knives had a much narrower range of L_p than confiscated kitchen knives (0.3 kgf to 0.8 kgf) as shown in Figure 6 (right). For new kitchen knives, 100% had L_p < 0.9 kgf, with the majority (66%) having L_p < 0.5 kgf and the lowest L_p = 0.32 kgf. For confiscated kitchen knives, only 33% had L_p < 0.9 kgf and only 15% had L_p < 0.5 kgf but the lowest L_p measured was similar to the lowest L_p of the new kitchen knives. Of the 11 (16%) of confiscated knives that did not penetrate (DNP), the majority were the broken, bent or truncated at the tip.

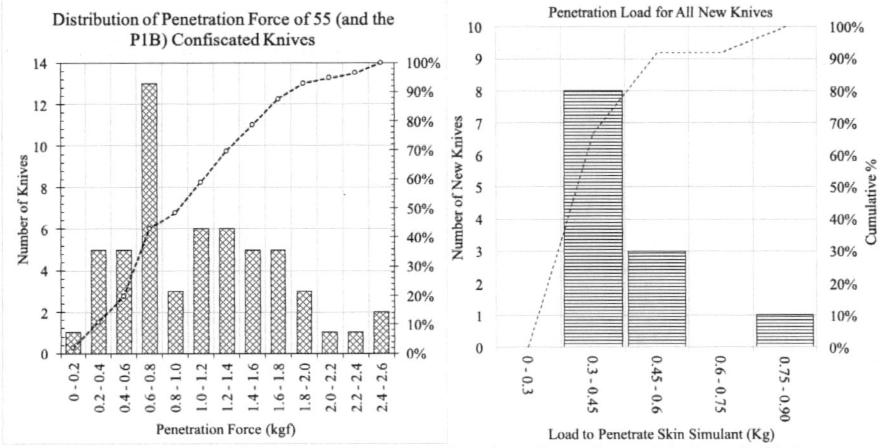

Figure 6: The distribution of the penetration force (L_p) for all confiscated knives (left) that penetrated. Data for new knives (right).

https://doi.org/10.52202/078352-0055

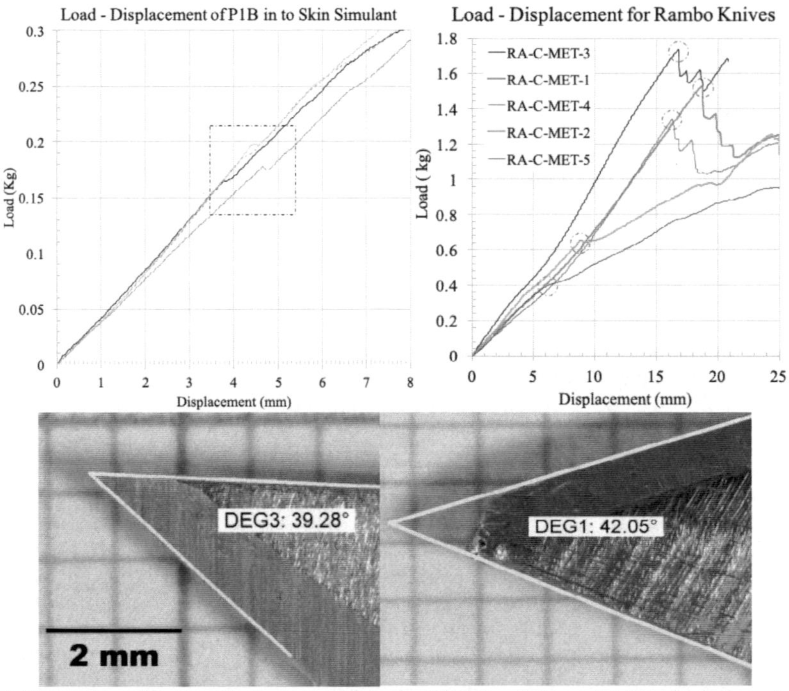

Figure 7: Examples of load – displacement data for P1B (top left) and 5 Rambo knives (top right) using skin simulant. P1B penetration load (L_p) is indicated by features within the dotted square. L_p for Rambo knives is indicated by dotted circles. Rambo knife tips with low (bottom left) and high (bottom right) values of L_p.

3.3SEM Imaging and Measurement of Knife Tips

Examples of the SEM images of the confiscated and new knives are shown in Figure 8. The tip geometries in SEM appear quite different from those observed using macro-photography (Figure 7) with differences in the measured Tip Angles and Tip Radius. In the SEM the Spine Angle (SA) and Spine Tip Radius (R_S) could be observed along with the way the cutting edge intercepted with the tip or did not reach the tip. All tips are different and have complex 3-dimensional geometries which are difficult to describe with 2 or 3 parameters. It was notable how the knife's cutting edge was rarely observed to extend all the way to the tip (top 0.2 mm) and so it plays a negligible role on the initial penetration of the knife tip. In cases where it did extend to the tip, these knives did not appear to have lower values of L_p.

https://doi.org/10.52202/078352-0055

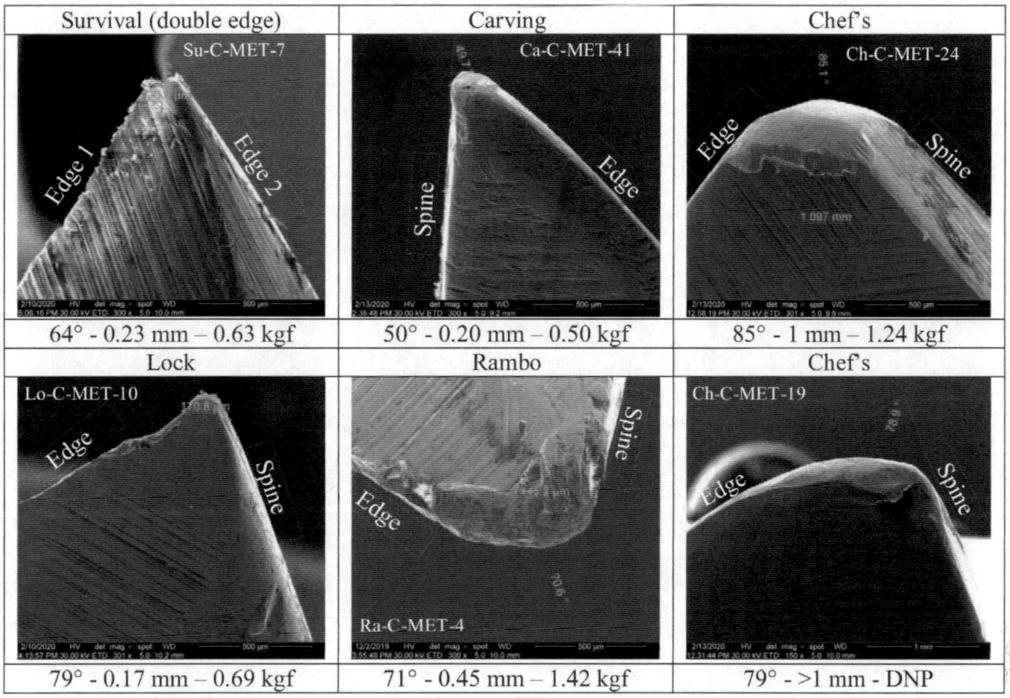

Figure 8: SEM images of some typical knives. SEM measured Tip Angle, Tip Radius and the Penetration Load are shown below each. Note different scale bar for the DNP example

3.4 Correlation between Knife Geometry and Penetration Load

A common view is that the geometry of the knife tip defines its "sharpness" and thus can be used to predict its ability to penetrate materials in a stabbing action. Commons sense would suggest that a knife exhibiting a low penetration load would have a small Tip Angle and a low Tip Radius. However, the tip angle versus penetration force data in Figure 9 (top left) shows high levels of scatter for all the confiscated knives suggesting that Tip Angle alone is not a good predictor of the force required to penetrate.

For 23 of the confiscated classified as "sharp", the tip angles ranged from 37° to 75° and 46% of the these required higher loads to penetrate compared with the new knives with similar tip angles. For confiscated sharp knives L_p increased by 0.43 kgf for every 10° increase in tip angle with a moderately good correlation ($R^2 = 0.43$) as shown in Figure 9 (top left).

For the sub-set of confiscated knives with rounded tips (27/66) there was no meaningful correlation between L_p and the optically measured Tip Radius (R_T), with around 20% of knives lying significantly away from the main group (circled in Figure 9 bottom left). It was possible to identify a "main group" in the data where there was an increase in L_p of 0.28 kgf for every 0.1 mm increase in the radius with a moderately good correlation ($R^2 = 0.41$).

The 9 new knives with sharp tips had tip angles (TA) between 42° and 72° and showed a range of L_p from 0.32 kgf to 0.8 kgf which was noted as being lower than the results of Nolan [5] for similar tip angles. Every 10° increase in tip angle led to an increase in L_p of ~ 0.16 kgf with a moderately good correlation ($R^2 = 0.38$). Note that the gradient (rate of increase of penetration load with tip angle) is substantially different between sharp new and sharp confiscated knives, by a factor of 3.6.

The P1B has the lowest penetration force ($L_p = 0.18$ kgf), while the lowest value for any knife was $Lp = 0.31$ kgf (filleting knife, $TA = 23.5°$, $R_T = 0.33$). Only 6% of the confiscated knives had L_p values less than twice the L_p value of the P1B, where as for new knives this figure was 42%.

A number of additional methods were used to try and improve the correlation between L_p and knife tip geometry. For example, for a rigid spherical indenter (e.g. a rounded tip) indenting an elastic material, the maximum pressure is proportional to $(R_T)^{-2/3}$. For a cone (similar to a sharp knife), it is proportional to tan(180 - TA)/2). However, using these approaches did not yield correlations with lower scatter. Combining the tip parameters in a simple manner, e.g. $x = (TA \times t)/R_T$ and several variations of this, did not improve the correlation or reduce the scatter.

https://doi.org/10.52202/078352-0055

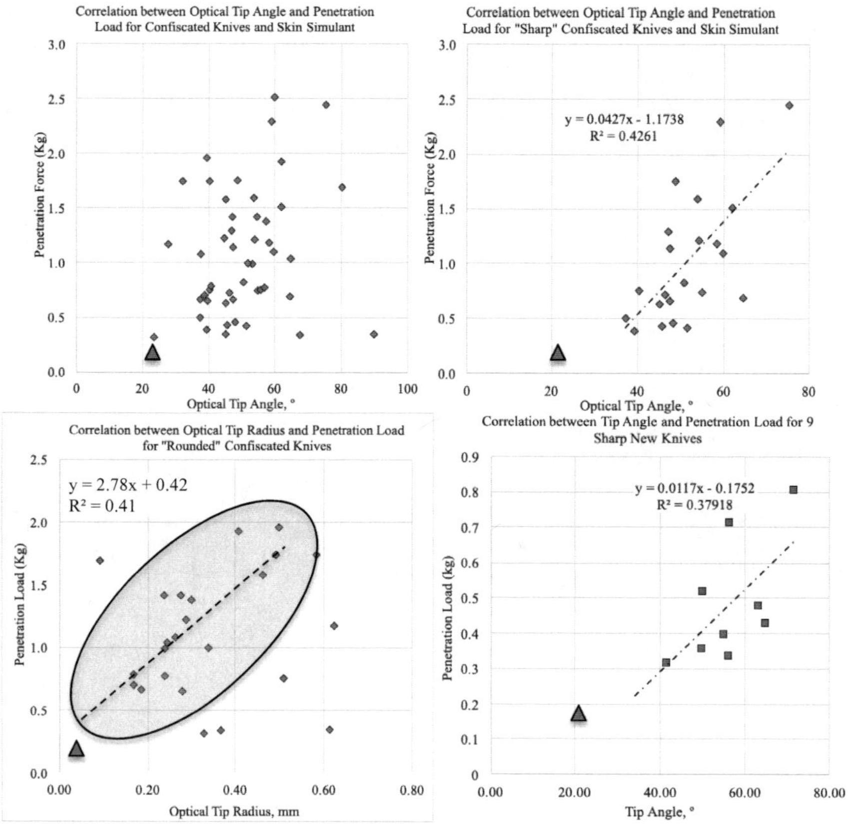

Figure 9: The correlation been between optically measured tip angle (TA) and penetration load (L_p) for all confiscated knives (top left) and only sharp confiscated knives (top right). Correlation between optically measured tip radius and L_p for rounded confiscated knives with "main group" circled (bottom left). Correlation between tip angle and L_p for 9 sharp new knives (bottom right). (triangle indicates result for P1B – TA = 23°, R_T<0.1 mm, L_p = 0.18 kgf)

4. DISCUSSON AND CONCLUSION

This work can be compared with previous published work using very similar skin simulants [5] [6]. In the prior research there has be no, or at best a very weak, correlation between knife tip geometry and penetration ability. As such the sharpness of knives in stabbing action is currently undefined.

This work shows that by categorizing the knives by their tip appearance as "sharp" or "rounded" and then optically measuring the tip angle (in side view) and the tip radius (if rounded) a moderately good correlation can be found between the force for penetration, L_p, (a proxy for sharpness) and the knife tip geometry. The correlation of tip angle with L_p is significantly different for new knives compared with used (confiscated) knives, suggesting a different mechanism is responsible for penetration for the two categories. For confiscated knives the values of L_p for "sharp" and "rounded" knives overlapped, with both having the majority of L_p values in the range 0.5 kgf to 2.0 kgf. This suggests that the observation of a knife tip being "sharp" or "rounded" is not in itself a useful indicator of penetration ability i.e. a low tip radius knife can be as sharp as a low tip angle knife and a high tip angle knife can be as blunt as a large tip radius knife. The actual value of tip angle and tip radius need to be measured carefully and in a consistent manner to determine the penetration ability. Combining the geometrical factors such as tip angle, tip radius and blade thickness in a number of ways so as to account for the 3-dimensional nature of the tip did not produce better correlations with the penetration load. Using classical contact mechanics to estimate the maximum stress imposed by different geometry knife tips did not yield good correlation with L_p.

https://doi.org/10.52202/078352-0055

It is worth noting that the penetration forces measured for a significant number of knives, both new and confiscated, are significantly lower than those reported by previous research [4], suggesting the threat presented by such knives may have been being underestimated. However, no knife measured in this work approached the high sharpness (low penetration force) of the P1B.

If "sharpness" is now defined in terms of the penetration load (L_p), with low values of L_p being sharper, then it can be stated that of all the knives measured (new and confiscated) none were as sharp as the P1B. However, 4 confiscated knives (6%) and 4 new knives (33%) were within a factor of two of the P1B sharpness. For used knives 66% were more than 5 times less sharp than the P1B ($L_p > 0.9$ kgf). Confiscated kitchen knives were some of the sharpest and bluntest knives measured. Confiscated knives such as the Rambo type knife were not sharper than kitchen knives.

The scatter in the data is most likely a result of the 3-dimensional shape of the tip and the large number of possible variations thereof (Figure 8) which are very difficult to mathematically describe using simple measurements. Even accounting for a 3 parameter model (tip angle, tip radius and spine radius) is not sufficient to account for the complex nature of the tip on a scale of < 0.25 mm from the tip. Using SEM to image the tips provided a better insight into the features which give rise to the individual L_p values but measurements made from the SEM images did not in themselves improve the correlation between knife tip geometry and penetration load (sharpness). Capturing 3-dimensional data of the tip on a level comparable to the SEM images would be the most likely method to quantitatively define tip sharpness as it would make measurement of the true surface area of the tip possible.

This work has built on the previous understanding of knife tip sharpness and has shown how weak correlations exist between some aspects of easily measured knife tip geometry and the penetration force. It has also found, however, that the exact relationship between the knife tip geometry and penetration ability is a complex one and that further work is needed to understand it fully.

Statistical analysis of the data from 66 confiscated weapons and 12 new knives shows how the sharpness of these knives (as measured by L_p) varies significantly and ranges from 1.7 to 8 times lower sharpness than the P1B. For confiscated weapons the fact that 66% of confiscated weapons required more than 5 times more force compared to the P1B suggests there is potential for armour designers to use high performance materials to extend armour coverage to currently unprotected areas with materials which are less burdensome than the main body of the armour but still provide high levels of protection against the majority of street knives, alongside comfort and wearability. This may include, for example, under arm protection, forearm/wrist protection and neck and head protection.

5 References

[1] UK Office for National Statistics, Crime in England and Wales: year ending March 2020, Section 6, https://www.ons.gov.uk/peoplepopulationandcommunity/crimeandjustice/bulletins/crimeinenglandandwales/yearendingmarch2019#rise-in-offences-involving-knives-or-sharp-instruments-and-firearms-offences (accessed 13/08/2020)

[2] UK Office for National Statistics, Crime in England and Wales: Police Force Area data tables, 17 Jul 2020, https://www.ons.gov.uk/peoplepopulationandcommunity/crimeandjustice/datasets/policeforceareadatatables, (accessed 13/08/2020)

[3] As for reference 1, Section 5.

[4] Annaidh, A.N., Cassidy M., Curtis, M., Destrade, M., Gilchrist, M.D., Forenci Sci. Int., 2003, 233, 1–3, pp. 7-13

[5] Nolan, G., Hainsworth, S. V., Rutty, G. N., Int J Legal Med, 2018, 132, pp.229–236

[6] Hainsworth, S. V., Delaney R. J. & Rutty, G. N., Int J Legal Med, 2008, 122, pp. 281–291

[7] Nolan, G., Hainsworth, S. V., Rutty, G. N., J Forensic Sci, 2013, 58, 2, pp. 372-381

[8] Payne, T., O'Rourke, S., Malbon, C., CAST Publication Number 012/17, UK Home Office 2017

https://doi.org/10.52202/078352-0055

AUTHOR INDEX